GEOLOGY OF
SCOTLAND

GEOLOGY OF
SCOTLAND

THIRD EDITION

Edited by

G. Y. CRAIG

Department of Geology and Geophysics
University of Edinburgh

1991

Published by

The Geological Society

London

THE GEOLOGICAL SOCIETY

The Society was founded in 1807 as the Geological Society of London and is thus the oldest geological society in the world. It received its Royal Charter in 1825 for the purpose of 'investigating the mineral structure of the Earth'. The Society is Britain's national learned society for geology with a Fellowship exceeding 6000. It has countrywide coverage and approximately one quarter of its membership resides overseas. The Society is responsible for promoting all aspects of the geological sciences and will also embrace professional matters on the completion of the reunification with the Institution of Geologists. The Society has its own publishing house to produce its international journals, books and maps, and is the European distributor for materials published by the American Association of Petroleum Geologists.

Fellowship is open to those holding a recognized honours degree in geology or a cognate subject and who have at least two years relevant postgraduate experience, or have not less than six years relevant experience in geology or a cognate subject. A fellow who has not less than five years relevant postgraduate experience in the practice of Geology may apply for validation and subject to approval will be able to use the designatory letters C. Geol (Chartered Geologist). Further information about the Society is available from the Membership Manager, Geological Society, Burlington House, Piccadilly, London W1V 0JU.

Published by The Geological Society from:
The Geological Society Publishing House
Unit 7
Brassmill Enterprise Centre
Brassmill Lane
Bath
Avon BA1 3JN
UK
(*Orders*: Tel. 0225 445046)

First published 1965
Second edition 1983
Third edition 1991

Distributors
USA
  AAPG Bookstore
  PO Box 979
  Tulsa
  Oklahoma 74101–0979
  USA
(*Orders*: Tel: (918)584–2555)

Australia
  Australian Mineral Foundation
  63 Conyngham St
  Glenside
  South Australia 5065
  Australia
(*Orders*: Tel: (08)379–0444)

British Library Cataloguing in Publication Data
A catalogue record for this book is available from the British Library

ISBN 0–903317–63–x (hardback)
ISBN 0–903317–64–8 (paperback)

Printed in Great Britain at the Alden Press, Oxford

CONTENTS

CONTRIBUTING AUTHORS

R. Beveridge	formerly National Coal Board, Edinburgh
G. S. Boulton	Department of Geology and Geophysics, University of Edinburgh
P. E. Brown	Department of Geography and Geology, University of St Andrews
S. Brown	The Petroleum Science and Technology Institute, Edinburgh
H. Emeleus	Department of Geological Sciences, University of Durham
E. H. Francis	Department of Earth Sciences, University of Leeds
M. J. Gallagher	British Geological Survey, Edinburgh
A. Hallam	Department of Geological Sciences, University of Birmingham
A. L. Harris	Department of Geological Sciences, University of Liverpool
M. R. W. Johnson	Department of Geology and Geophysics, University of Edinburgh
J. P. B. Lovell	BP London
J. Merritt	British Geological Survey, Edinburgh
W. Mykura*	British Geological Survey, Edinburgh
G. J. H. Oliver	Department of Geology and Geography, University of St Andrews
J. D. Peacock	British Geological Survey, Edinburgh
R. G. Park	Department of Geology, University of Keele
A. D. Stewart	Paoluccio, 05020 Porchiano del Monte, Italy
D. Sutherland	2 London Street, Edinburgh
E. K. Walton	Department of Geography and Geology, University of St Andrews

* deceased

PREFACE to THIRD EDITION

The first edition of this book was published in 1965. It was largely based on the results of geological field-work in Scotland over the last hundred years. But long before the second edition appeared eighteen years later I and my contributors had had to adapt to the unexpected discovery of North Sea oil and the unifying theory of plate tectonics. In this third edition, seven years further on, I am no longer surprised at the rapid evolution in our understanding of Scottish geology, brought about by the increasing number of research workers and the evermore sophisticated laboratory techniques.

The text has been fully revised and largely rewritten by a happy blend of University and Survey geologists. Four of the present team of nineteen contributed to the first edition and eight to the second. Following the sad death of Janet Watson, Graham Park undertook the considerable task of rewriting the Lewisian. Wally Mykura was tragically killed in a road accident shortly after completing his revision of the Old Red Sandstone and I am much indebted to Matt Armstrong for checking and updating that chapter. Overall there are four completely new chapters (Lewisian, Torridonian, Quaternary and Economic Geology) and over 200 new or modified figures, in a book now a third longer than its predecessor.

Scotland has a great variety of geological environments, repeated orogenic, metamorphic and igneous events, a long geological history and considerable geological wealth. Not surprisingly it continues to be a mecca for geologists and students from all over the world.

G. Y. Craig

University of Edinburgh
February 1991

Acknowledgements

The authors gratefully acknowledge help from colleagues and permission from the listed organisations to reproduce certain figures and tables.

      Chapter 1 – Professor B. J. Bluck, Dr D. Barr, Dr Paula Haselock, Dr R. E. Holdsworth, Dr N. G. Lindsay, Dr Graeme Rodgers and Dr A. M. Roberts.
      Chapter 2 – Dr R. F. Cheeney who found suitable photographs.
      Chapter 4 – Dr A. W. Baird, Dr D. Hutton.
      Chapter 5 – Dr R. Anderton, Professor D. Flinn (especially for repeated help with Shetland), Dr J. Treagus.

Chapter 6 – Dr A. Owen.

Chapter 8 – Barry Fulton who drafted the figures.

Chapter 14 – B. R. Bell, B. G. J. Upton, C. Blair and K. Gittins.

Chapter 16 – Oil and Gas, Construction, and Industrial Minerals and Metalliferous Minerals contributions are made with the permission of the Director, British Geological Survey and the UK Department of Energy.

The editor is particularly indebted to Angela McQuillin for her shrewd drafting of most of the new text-figures and to Colin Will for his careful work with the index. He cannot let this occasion pass without congratulating Dr John C. Crowell for his sterling work in remaining at his post on the Moine–Lewisian contact (p. 90) for the last 25 years! All contributors owe a considerable debt of gratitude to Dr Douglas Grant of Scottish Academic Press and to the Officers and Council of the Geological Society.

Organisations

American Geophysical Union	4.11, 4.12, 4.21
J. Blackie	4.5
Blackwell Scientific Publications	10.34
British Geological Survey	2.1, 2.4, 2.5, 2.6, 2.17, 2.19, 2.20, 4.17, 8.36, 9.2, 9.4, 9.7, 9.9, 9.10, 9.11, 9.17, 10.4, 10.5, 10.8, 10.9, 10.11, 10.14, 10.21, 11.7, 14.2, 14.8, 14.15, 14.16, 14.24a
Geologists' Association	14.14, 14.22
Geological Society	1.2, 2.8, 2.9, 2.11, 2.12, 2.14, 2.15, 4.6, 4.8, 4.10, 4.15, 4.22, 4.23, 5.1a, 5.3, 5.9a, 5.12, 5.16, 6.14, 7.2, 7.7, 10.28c, 14.7a, b
Geological Society of America	4.1, 4.2, 4.4, 4.14, 4.16
Geological Society of Glasgow	10.35, 14.25
Kluver	10.28a, b, 10.32
NERC	14.21
Oxford University Press (*J. Petrology*)	14.6
Palaeontographical Society	Table 9.3
Pergamon Press (*J. Structural Geology*)	4.18, 4.19, 4.20
Royal Society of Edinburgh	6.10, 6.24, 6.25, 7.14, 7.15, Table 9.2, 11.13, 11.18, 11.19
University of Chicago Press (*J. Geology*)	14.29, 14.30
J. Wylie and Sons	14.27
Yorkshire Geological Society	10.36

BGS photographs are reproduced by permission of the Director of the British Geological Survey and are NERC copyright reserved.

Every reasonable effort has been made to identify and acknowledge the copyright owners of material reproduced in this book. If further acknowledgement is required, this will be done at the earliest opportunity.

map of Scotland

Sedimentary rocks

Mesozoic

Permo–Trias

Carboniferous
(including small undifferentiated
volcanic areas)

Devonian (O.R.S.)
(including undifferentiated
volcanic areas)

Lower Palaeozoic (undivided)

Wenlock

Llandovery

Ordovician

Cambro–Ordovician

Torridonian

Igneous rocks

Tertiary extrusions

Carboniferous volcanics

Devonian volcanics

Major Tertiary intrusions

Caledonian granitoids

Ophiolite complex
(B. Ballantrae S Shetland)

Major gabbro intrusions

Metamorphic rocks

Foliated granitoid

Southern Highland Gp. ⎤
Argyll Gp.   ⎬ Dalradian
Appin Gp.   ⎦

Moine of the Shetlands &
Grampian Group

Migmatitic and granitic
complexes Sutherland

Locheil Gp.

Glenfinnan Gp.   ⎬ Moine

Morar Gp.

Areas of Moine with
abundant Lewisian inliers

Lewisian

——— Fault

——▲— Thrusts

Insets show the main structural units, faults and

THE GROWTH AND STRUCTURE OF SCOTLAND

A. L. Harris

Since the late Professor Neville George wrote the equivalent chapter in the first edition of this book, our understanding of geological processes has advanced immeasurably. We know much more about the rates of geological activity, the longevity of major structural lineaments and their influence on processes, and the depths and temperatures at which igneous rocks melt, crystallise and fractionate. Isotopic age-dating has greatly refined the sequence of events within the geological time scale. The nature of the deep crust of Scotland has been revealed by seismic refraction experiments and palaeomagnetic methods have detected the changing palaeo-latitudes in Palaeozoic and younger rocks. Our appreciation of metamorphic processes and the evolution of sedimentary basins has been much helped by work on heat flow, and the recent exploration of the continental shelves has considerably increased our understanding of the Scottish Mesozoic and Tertiary.

A very large part of Scotland is underlain by Caledonian and older rocks and it is, therefore, appropriate to set the Scottish area in the context of the Caledonide/Appalachian orogen (Fig. 1.2). From this map it will be noted that Scotland and the remainder of the British Isles do not occupy merely a rather small sector of the former Iapetus Ocean but lie at a key location at or near the apex of the inverted Y which is now regarded as the shape of the orogen. This position explains why the geology of N Scotland can be related in a general way to the geology of east Greenland and to the NW margin of the Appalachian belt in N America. It also shows why, through the interposition of the Tornquist Sea forming one of the branches of the inverted Y, little direct resemblance between the British Isles and Scandinavia is forthcoming.

Nevertheless, in terms of the plate-tectonics hypothesis the geology of Scotland might simply be regarded as a study of the nature and evolution of a rather small, albeit critical, part of the north-western margin of the Caledonian Iapetus Ocean and its subsequent history, because the Scottish Border is generally thought to coincide with the suture which supposedly marks the site of Iapetus closure. However, one of the reasons why the geology of Scotland has been seminal in influencing geological studies throughout the world is the extraordinary diversity of its geological environments and the very large time span covered (Figs. 1.1, 1.3).

Many orogenic zones, notably the North American Cordillera, have been recognised as comprising a collage of far-travelled 'terranes' of diverse nature and early history. This means that the major units of Scottish geology separated by such major structures as the Great Glen Fault, the Highland Border Fault and the Southern Uplands Fault (inset, Fig. 1.1) should be scrutinised for evidence that their contemporary histories are sufficiently disparate to justify 'terrane' status. It must be recognised, however, that in North America major strike–slip movements of hundreds or thousands of kilometres have contributed to the assembling of the terrane collage; if this were the case in Scotland the limited along-strike (NE–SW) extent of Scotland may well prevent us from reaching firm conclusions about the original status of the different blocks of Scottish crust.

The diversity referred to above, which may be partly due to terrane assembling, is summarized below.

The basement of at least the northern part of the country comprises Archaean-to-middle Proterozoic gneisses and intrusions which are extensively exposed in the Hebridean craton where they are partly covered by late Precambrian-to-Ordovician sediments. The classic Moine thrust zone separates the Hebridean craton from the orthotectonic zone of the Caledonides which, comprising the Moine and Dalradian metamorphosed sediments and volcanics, probably accumulated in a largely ensialic setting at the south-eastern edge of the Laurentian–Greenland continent from c. 1,200 Ma to c. 600 Ma ago.

To the north-west of the Great Glen Fault the rocks of the Moine Supergroup, floored wholly or in part by Lewisian basement, probably suffered orogenesis rather less than 1,000 Ma ago and were further deformed and recrystallised during thrust-related orogenesis, between c. 470 and 400 Ma. To the south-east of the Great Glen

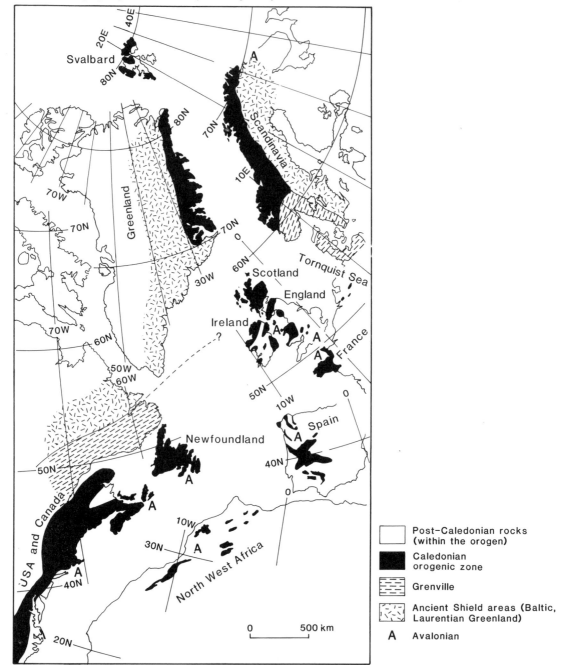

Fig. 1.2. Distribution of Caledonian rocks in the present N. Atlantic area (restoration after Bullard).

the metamorphic rocks comprising the late Precambrian Dalradian Supergroup and possibly pre-Dalradian rocks of Moine aspect may be floored by basement of diverse nature, perhaps rather different from and younger than the Lewisian. The Grampian orogeny which affected the Dalradian rocks was formerly regarded as a wholly Cambro–Ordovician event but is now believed to have reached its peak in the late Precambrian (c. 590 Ma), although later Caledonian orogenesis also occurred during the Ordovician and Silurian. Orogenesis of Moine and Dalradian rocks alike included metamorphism of moderate pressure and migmatisation, in many places, in the middle-upper amphibolite facies. Early granites, probably in part of migmatitic origin, punctuated the deformation episodes and enable them to be isotopically dated.

As the Ordovician–Silurian Iapetus Ocean closed it accumulated at its northern edge, lower Palaeozoic cherts, black shales and turbidites from Arenig to Wenlock in age, which now comprise the Southern Uplands. These constitute the paratectonic Caledonides of Scotland which may be partly underlain by Lower Palaeozoic oceanic lithosphere, although recent models for the evolution of the Southern Uplands involve the underthrusting of English continental crust.

Emplacement of the Silurian to Lower Devonian Newer granites with their swarms of minor intrusions lent buoyancy to the crust especially in the othotectonic zone and the orogen was uplifted in the lower Devonian, or possibly somewhat earlier, and shed Old Red Sandstone molasse deposits, locally accompanied by the extrusion of calc–alkaline volcanics, into vigorous fault bounded-and-controlled extensional basins. Important and long-lived faults such as the Great Glen, Highland Border and Southern Uplands faults were active at this time. Complex, structurally controlled, marine sedimentation and associated mildly alkaline basic volcanism occurred during the Carboniferous with cyclothems and lavas, tuffs and intrusions now exposed in classic sections in the Midland Valley. Deposition and preservation of the New Red Sandstone deposits were also structurally controlled, mainly in the Midland Valley and Southern Uplands.

Onshore Mesozoic sediments are largely confined to the margins of the present land and their deposition was closely related to that of the offshore basins with the Scottish land area forming a significant source of detritus.

The last important activity in the evolution of the solid geology of the country involved the early Tertiary development of major volcanic centres with classic intrusive relationships and lava stratigraphy along the western seaboard and in the Inner Hebrides. Here crustal thinning during the formation of extensional basins may have given rise to temporary high heat flow and the upwelling of copious basic magma which, in turn, by melting adjacent continental crust, created large quantities of acid magma.

Differential movements on faults still active in the Tertiary have to some extent controlled the form of the coast especially in the north and west of the country. General uplift related to Alpine orogenesis stimulated subaerial erosion which led to the sculpting of Tertiary erosion surfaces. Everywhere, the landforms show the erosional and depositional effects of several Pleistocene glaciations. Isostatic rebound following the melting of the last ice has combined with marine erosion and deposition to produce the raised beaches which are a feature of many parts of the Scottish coastline, but which seem to be absent from Shetland and the Outer Hebrides. The last major advance of the ice at about 10,000 B.P. was a corrie-and-valley glaciation which produced many of the glacial landforms of the Highlands including the major landslips on northern and eastern facing slopes which probably accompanied the down-wasting and retreat of the valley glaciers. This brief but intense glacial episode was followed by a general amelioration of the climate which is recorded by extensive spreads of upland peat.

In preparing this introductory chapter I am conscious that the material and interpretation are very largely derived from other workers and that they are seldom properly acknowledged. To have made proper acknowledgement would have involved repeated punctuation of the story with reference, but where I am particularly indebted to workers for their help in specific aspects I have acknowledged them by name in the running text. Much of my work has involved the study, in proof, of the other contributions to this volume and I refer the reader to these sections for a more complete and thorough documentation of events in the growth of Scotland. My task has been to provide a framework for the more detailed chapters which follow and to highlight the main features of the story. In addition Memoir 9 of the Geological Society of London has proved a valuable source of information, and the radiometric dates quoted in this chapter all relate to the dates which were recalculated by G. C. Brown for that Memoir.

The formation of the crystalline craton

Hebridean craton

Scotland lies at the south-eastern margin of the slab of ancient continental crust which makes up most of Greenland and Laurentian Shield (Fig. 1.2) and which incorporates the Amitsoq gneiss – the oldest (~3,800 Ma) terrestrial rocks yet recorded. In particular, work in Greenland and Canada has shown that its geological history has involved the periodic creation of new continental crust and the reworking and recrystallisation of already formed crust. The Lewisian rocks of Scotland record the latest stages in this process of crustal evolution and accretion, during the late Archaean and early-mid Proterozoic. Addition of new material, post 2,900 Ma, is probably limited to the Scourie dykes and Laxfordian vein complexes, and to the Loch Maree Series sediments and volcanics which were probably post-Badcallian (Badcallian = early Scourian) and which were deformed in the late Scourian (Fig. 1.3). The Lewisian complex during its activity spanning the 2·9 Ga–1·1 Ga period is a microcosm of the continental crust in many parts of the Earth and contains features initiated as the Badcallian gave way to Inverian activity which may reflect a general change of regime and tectonic process at the Archaean–Proterozoic boundary.

Age	HEBRIDEAN CRATON	CALEDONIDES NW of GREAT GLEN	CALEDONIDES SE of GREAT GLEN

Age

HEBRIDEAN CRATON

- 0 Ma — Skye and Rhum with acid and basic volcanics
- 100 Ma
  Skye
- 200 Ma — Skye and Raasay; Stornoway Beds of Lewis
- 300 Ma — ? Permo-Carboniferous
- 400 Ma — Renewed extensional movements; Ross-of-Mull granite; ← Moine thrust →; Break of unknown duration
- 500 Ma — Quartzites and carbonates (possible major hiatus); unconformity
- 600 Ma — Tilting of Torridonian (date uncertain)
- 700 Ma
- 800 Ma — Torridon Group unconformity (possibly older); Tilting of Stoer Group
- 900 Ma — Stoer Group unconformity (possibly 100 Ma older)
- 1000 Ma

Torridonian

Uplift
? Early movements on Outer Isles Thrust
Late
Laxfordian A
Early
Scourie dykes 10-20 km
Invarian Loch Maree Series
? unconformity

- 2000 Ma

Badcallian orogenesis A–G 35–50 km
Scourian parent rocks

- 3000 Ma

CALEDONIDES NW of GREAT GLEN

? Late movements on Gt Glen ?
Ardnamurchan Mull with acid and basic volcanics
Discontinuous scattered outcrops (espec. Mull & Morven)
Brora and Easter Ross (also part of Mull sequence)
Mull
Scattered New Red Sst outcrops – western seaboard and Inner Hebrides

Old Red Sandstone (e.g. Caithness)
Break of unknown duration
Newer granites uplift
Borolan Syenite
Strontian granite (late orogenic) upright Gr
Dessary syenite
Ductile thrusting A–Gr (e.g. Sgurr Beag Slide)
? Loy and Scaddle gabbros
Carn Chuinneag granite

Strath Halladale granite

Knoydartian/Morarian "Orogenesis"

Gr ——— A
Ardgour Orthogneiss
Moine: Morar, Glenfinnan Locheil groups
(Possibly <200 Ma younger)
Uplift Glenelg

Possible equivalance

Major displacement apparent on LISPB section (Age of displacement uncertain) ?

Badcallian orogenesis A–G 35–50 km
Scourian parent rocks

CALEDONIDES SE of GREAT GLEN

Arran
? Elgin
Moray Firth and Arran
Arran
Kintyre
Highland Boundary Fault

Old Red Sandstone; granites breach surface continuing uplift; Newer granites; ? Highland Border downbend; Glen Kyllachy gra. (late orogenic); Uplift and waning metaphism Strichen gra.
? Juxtaposi of Metamor Caledonide Midland Va

? Rising metamorphic grade — Crustal thickening

Ben Vuirich granite
*Grampian orogenesis *(Tay Nappe) Gr-A
S Highland Group
Argyll Group
Appin Group — Dalradian deposition
Grampian Group
? unconformity
A Orogenesis
? unconformity ? C Highland Division (may not exist as a separate unit)

NATURE AND AGE OF BASEMENT LARGELY UNKNOWN

Plutonic basement of Islay and Colonsay

NO DIRECT EVIDENCE
Possibly older basement

* Possibly <200 Ma older *

Major displacement apparent on LISPB section

According to R. G. Park and the late Janet Watson the existence of rocks as old as the Lewisian complex at the margin of a widening ocean (the present Atlantic) is a remarkable feature which reflects the abnormal buoyancy of the Lewisian and which may originate in the Scourian itself. They regarded this buoyancy as possibly accounting for the persistency of the erosion surface marking the base of the Torridonian, the Mesozoic and present sea level.

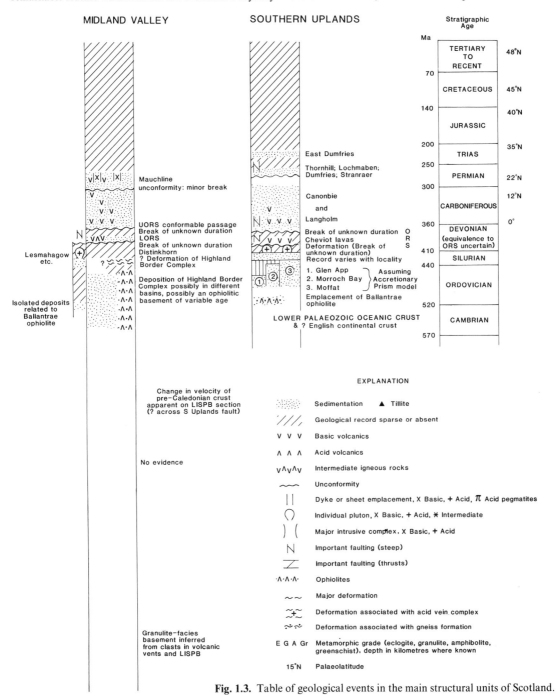

Fig. 1.3. Table of geological events in the main structural units of Scotland.

Isotopic evidence suggests that most material involved in Badcallian orogenesis ($\sim$2,700 Ma) was derived from the mantle no more than 200 Ma earlier. These rocks, the oldest in Scotland, have tonalitic bulk chemical compositions which suggest that prior to metamorphism they were calc–alkaline volcanics and/or the subvolcanic intrusive complexes which fed such volcanics. The volcanic origin of these rocks is indicated by close association with metasediments which have undergone the same structural and metamorphic history. Estimates of their metamorphism and deformation at 15 kbar and >1,000°C during Scourian (Badcallian) orogenesis may be spurious in that they possibly reflect temperatures of trondhjemitic intrusion, and 11 kbar and 950°C may reflect more accurately PT conditions at the peak of Badcallian metamorphism. Temperatures seem to have fallen to <600°C before the emplacement of the Scourie dykes (2,400 Ma). The very high temperatures at the peak of Badcallian metamorphism may be the result of a high thermal regime during the Archaean.

Deformation and partial melting during metamorphism have made the original relationships of these rocks quite obscure. They carry high-grade, locally granulite-facies, mineral assemblages and are depleted in such heat-producing elements as U, Th, Rb and K, a depletion which they have in common with highly metamorphosed Archaean complexes elsewhere. The nature of these rocks after they had undergone early Scourian deformation and metamorphism can be determined in localised *massifs* in the Outer Hebrides and on the Scottish Mainland, and an important zone about 65 km across between Scourie and Gruinard Bay (inset, Fig. 1.1) has largely survived later modification. Here the late Archaean gneisses and granulites are largely unchanged since 2,700 Ma, as is witnessed by the occurrence within them of abundant 2,400–2,200 Ma Scourie dykes – a swarm of tholeiitic dykes of WNW trend which retain sharp, quenched, cross-cutting margins against the Scourian gneisses. There is some evidence of contemporary ductile shearing of the country rocks during dyke emplacement and M. J. O'Hara has indicated that metamorphism of the Scourie dykes is consistent with intrusion at depth of 10–20 km. D. J. Fettes has suggested that the depth probably varied across the complex with emplacement being at greater depths in the Outer Hebrides than on the mainland. Nevertheless the field relationships indicate substantial crustal extension and emplacement at a much higher level in the crust than that at which Scourian metamorphism occurred. Hence, the Scourian gneisses and granulites which had been carried to great depths, possibly as much as 50 km, and transformed after their initial formation in the late Archaean were then uplifted and eroded prior to 2,200 Ma. Some of the uplift may have been accomplished by steep late-Scourian (Inverian) and early Laxfordian ductile shear zones. R. G. Park and J. Tarney have recently pointed out that the shear zones established at the close of the Archaean (Inverian) 2,600 Ma continued

as sites of tectonic, metamorphic and igneous activity for 1·5 Ga.

Areas on the Hebridean craton consisting of gneisses carrying unmodified Scourie dykes are essentially zones or augen where Laxfordian ($\sim$1,800 Ma) tectonic strain was low. They correlate in a general way with similar more extensive areas in Greenland and the Laurentian Shield which stabilised at *c.* 2,400 Ma. Elsewhere in the Hebridean craton the Lewisian rocks were severely, though heterogeneously reworked and metamorphosed in ductile shear zones during the Laxfordian; such areas (inset, Fig. 1.1) between Laxford Bridge and the north coast, and south-east of Gruinard Bay generally correlate with substantial areas of south and central Greenland which became stable at $\sim$1,800 Ma. The areas of Laxfordian reworking are marked by amphibolite–facies metamorphism and by the injection in many places of large amounts of tonalitic and granodioritic material some of which may be mantle-derived. Strain analysis of the Laxfordian shear zones indicates a lateral slip of the order of 18 km on individual shear zones. The cumulative effect of such displacements if maintained across the whole of the Laxfordian zone north of Laxford Bridge would produce at least 200 kilometres of lateral displacement and tens of kilometres of vertical displacement. By this means rocks derived from markedly different crustal sites have been juxtaposed.

Cratonic cover rocks

The Lewisian rocks of the Hebridean craton (inset, Fig. 1.1) had been uplifted and the top $\sim$45 km of Scourian rocks eroded to about their present level by about 1,100 Ma ago. Two groups of Torridonian rocks were subsequently deposited on the craton – viz. the Stoer and Torridon groups. The former has been dated at 995$\pm$17 Ma although, on palaeomagnetic grounds, J. D. A. Piper regards the groups as being at least 100 Ma older. The unconformity which separates the two groups may represent a substantial time gap. This unconformity and time gap is consistent with the change in palaeolatitude identified between the groups: the $\sim$2 km of red beds which comprise the Stoer Group are thought to have been deposited at a latitude of between 15° and 33°; the $\sim$7 km of the Torridon Group, consisting of red beds with minor grey shales were deposited at a latitude of 28°–50° C. Downie has indicated that the nannofossils from grey shales in the Torridon Group are upper Riphean. Pebbles in this group were derived from a westerly source and have yielded isotopic ages which indicate a Laxfordian source. On the craton, erosion is stripping off the Torridonian cover and large areas of the present landscape comprising low rounded hills (Fig. 1.4) and shallow hollows containing peaty lochs are probably an exhumed Torridonian terrain. The Torridonian rocks are unconformably overlain and are overstepped by Cambro–Ordovician quartzites and carbonates some 400 m thick.

Metamorphic Caledonides

Thrust zone

The rocks of the Hebridean craton are limited to the east by the NNE-trending, E-dipping Moine Thrust zone which consists of a series of thrusts of which the Moine thrust (*sensu stricto*) (Fig. 1.1) is considered by D. Elliott and M. R. W. Johnson to be the oldest. It is the highest, and crops out furthest to the east. The lower and successively younger thrusts crop out successively westwards each carrying a slice of rock or nappe which is named after the thrust on which it has ridden. Thus in Assynt (inset, Fig. 1.1) where a culmination or bulge of the highest thrust is produced above a major duplex, a sequence of nappes each successively younger and lower can be distinguished – the Ben More (Fig. 1.5), Glencoul and Sole nappes. Similarly at Eriboll on the north coast of Scotland (inset, Fig. 1.1) there is a westward sequence of Upper Arnaboll, Arnaboll and Sole nappes. Although the thrust sequence generally applies throughout the thrust zone, detailed work by M. P. Coward and his coworkers has revealed 'out of sequence' thrusting and renewed, but extensional, displacements on early thrust planes.

The general tectonic transport direction is WNW – approximately 290°–300°. This is parallel to the elongation lineation in the more deformed rocks. In southern Assynt, the Borolan igneous complex postdates early thrust movements but pre-dates the late movement. It has yielded an age of ~ 430 Ma thus dating some of the displacement in the Moine thrust zone. At levels seen at the present day in the Moine thrust zones are fault rocks the nature of which depends on the depth and strain rate at which they formed. Textures, deformation style and intensity vary across the thrust zone from sharp fault breaks in the lower thrusts in the west to thick mylonite zones in the east.

Work to link the thrusting history in the Moine thrust zone to the now widely recognised ductile thrusts within the Metamorphic Caledonides, notably by R. E. Holdsworth,

Fig. 1.4. Exhumed Torridonian topography exposing Lewisian rocks in Assynt. Monadnock (Quinag) consists of gently dipping Torridonian sandstone with peaks of Cambrian quartzite. *Scotsman* photograph.

A. L. HARRIS

has shown that the Moine thrust zone is the westernmost, lowest and youngest of a series of thrusts on the Scottish mainland which are well displayed along the north coast as the Ben Hope, Naver and Swordly ductile thrusts. The Outer Hebrides thrust and possibly the Flannan thrust are possibly still lower and younger thrusts which transect the Hebridean craton. Southwards, the Naver and Swordly thrusts merge to become the Sgurr Beag ductile thrust (slide) which has been suggested by Simon Kelley, working in Fannich, to be ~25 Ma older than the Moine thrust. In western Inverness-shire the Sgurr Beag thrust which has

been traced southwards as far as Ardnamurchan appears to be above the Knoydart thrust. Similar ductile thrusts have been recorded by Derek Flinn on Mainland and on Yell in Shetland. Recognition of the widespread development in the Northern Highlands of the ductile thrusts which reworked already metamorphosed and crystalline Moine rocks has suggested to many workers that the whole Northern Highlands might be regarded as a duplex structure on a crustal scale, possibly floored by a detachment at the base of the continental crust. It would seem therefore that the Moine thrust zone itself might have a similar

Fig. 1.5. The Ben More nappe and thrust. The topography reflects the structure in a contrast between deformed Torridonian and Cambrian rocks forming the hills of the centre and right of the photograph, and the low ground of Cambrian rocks to the left. The approximate line of the thrust runs at the foot of the hills at the left and runs back into the distance towards the top left. The small lochans occupy corries in the glacially scalloped hills. Photograph by J. K. St. Joseph. Crown copyright.

ductile expression in depth and that its spectacular present expression is a function of its intersection at a particular structural and stratigraphic level with the present erosion surface.

Basement

There seems little doubt that material similar to the Lewisian gneisses of the Hebridean craton extends south-eastwards at depths varying from 0–16 km as far as the Great Glen Fault (Fig. 1.6). Rocks of both Scourian and Laxfordian affinity occur sporadically at the surface among Moine rocks north-west of the Great Glen while Lewisian rocks are exposed immediately to the north-west of the fault in lower Glen Urquhart and in the Rosemarkie inlier. These Lewisian rocks are unmistakable lithologically, except where they are acid in composition and have been severely reworked by Caledonian and/or pre-Caledonian deformation.

South-east of the Great Glen, evidence for the distribution and the nature of the basement rocks is less direct in that basement rocks are exposed only in the Rhinns of Islay and as a small inlier in northern Colonsay. Indirect evidence for the nature of the basement comes from the geophysical LISPB (Fig. 1.7) and other experiments, from petrographic, chemical and isotopic studies of Caledonian granites which are believed to have sampled the lower parts of the crust during their formation and emplacement and from xenolith populations found in late Palaeozoic, deeply derived vent agglomerates in the Midland Valley and adjacent areas.

Roddy Muir and other workers at Aberystwyth regard the basement rocks of Islay and Colonsay as different from and younger than Lewisian, while the LISPB geophysical experiment which crossed Scotland about 200 km north-east of Islay suggested no significant change of seismic velocity of the lower crustal rocks across the Great Glen although their upper surface was stepped down by a few kilometres to the south-east at the Great Glen. According to the LISPB results basement rocks of >6.4 km sec^{-1} continue to the Southern Uplands fault. Jeremy Hall and others have shown that there is no significant seismic break

Fig. 1.6. The Great Glen Fault. (B.G.S. photograph.)

at the Southern Uplands fault in that basement, seis-mologically indistinguishable from that beneath the Midland Valley, continues southwards with continental-type basement having velocities > 6 km sec^{-1} underlying the Southern Uplands at shallow depths.

Petrographic and geochemical studies have led W. E. Stephens and A. N. Halliday to divide the Caledonian granites of Scotland to the south-east of the Great Glen into three suites – the Argyll, Cairngorm and South of Scotland suites (Fig. 1.8). In the sense that the well-defined boundaries between these suites must partly reflect changes in source-rock composition they may well have a bearing on the distribution of basement of different types. The boundary between the Argyll and South of Scotland suites is close to the mid-Grampian line (Fig. 1.8) which was drawn by Halliday on the basis of limited studies of Nd. These studies allied to those of Pb in Caledonian granite zircons point to a change in crust across the mid-Grampian line. Xenolith population studies in the Carboniferous vents suggested to B. G. U. Upton and R. H. Hunter that the Dalradian of the Southern Highlands and the Lower Palaeozoic of the Midland Valley and the Southern Uplands overlie a metamorphic basement of granulite-facies rocks varying in composition from ultramafic, locally garnetiferous, granulites through anorthosites to quartzo–feldspathic granulites. They concluded that the Highland Boundary and Southern Uplands faults do not separate strongly, widely contrasting lithosphere segments and thus confirmed the continuity across the Southern Uplands fault hinted at by Jeremy Hall's geophysical results. It is, nevertheless, worth noting the marked break in the LISPB profile, approximately coinciding with the Highland Boun-

dary fault which gives rise to a basement high at depth below the more superficial Midland Valley graben (Fig. 1.7).

Continental crustal rocks with which the crust of Scotland was originally continuous to the south are unknown, if indeed they existed, having presumably been substantially displaced by oblique opening and/or closing of the Caledonian Iapetus Ocean. In England, Wales and southern Ireland the Caledonian sedimentary and volcanic sequences contemporary with the lower Palaeozoic of Scotland were deposited unconformably on late Precambrian calc-alkaline volcanics and on generally low-grade metamorphic rocks. These are probably not continuously underlain by dense continental crust although the Rosslaire gneisses of SE Ireland and possibly some of the gneisses of Anglesey and the Lleyn peninsula may indicate that such crust is not entirely absent. Nevertheless, it is arguable that a fundamental change in the deep structure of Britain takes place approximately at the Scottish Border.

Metasedimentary cover rocks

In the Northern Highlands the Lewisian basement is overlain by the Moine Supergroup comprising late Proterozoic, largely metasedimentary rocks within which the Morar (oldest), Glenfinnan and Locheil (youngest) groups have been distinguished. This scheme replaces the tectonic/stratigraphic divisions originally proposed by G. S. Johnstone and colleagues on the Geological Survey (Figs. 1.1, 1.3; Fig. 4.1). The groups, established in western Inverness-shire and traced into Ross-shire have now been recognised by R. E. Holdsworth in the northern coastal

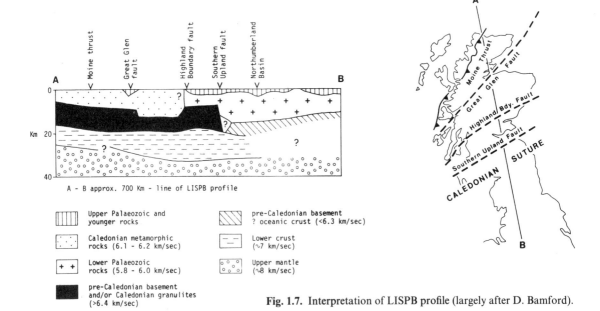

Fig. 1.7. Interpretation of LISPB profile (largely after D. Bamford).

areas from the Moine thrust eastwards to Bettyhill. Conglomerate has been recorded at the base of the Morar Group at Glenelg and near Talmine where it rests unconformably on Lewisian gneisses. Until recently the relative age of the rocks comprising the original Morar and Glenfinnan divisions was unknown because they were believed to be separated everywhere by the Sgurr Beag ductile thrust (Fig. 1.1; Figs. 4.1, 4.14). The Moine rocks on the Ross-of-Mull, however, show a continuous upward stratigraphic succession from Morar Group sediments into Glenfinnan. A. M. Roberts has shown unequivocally in the Loch Quoich area that psammites assigned to the Locheil Group are younger than the Glenfinnan Group gneisses and schists and with R. Strachan and others has provided a lithostratigraphy for all the Moine rocks above the Sgurr Beag thrust.

Allochthonous Lewisian rocks, to a greater or lesser extent modified by Caledonian and Precambrian orogenesis, occur within the Moine outcrop of the Northern Highlands, in fold cores and along the trace of the Sgurr Beag and other ductile thrusts but have not been recorded as thrust slices to the south of Loch Hourn. The Sgurr Beag thrust itself, lacking a Lewisian strip, is believed to continue southwards as far as Ardnamurchan. Parautochthonous Lewisian slices occur as isoclinal fold cores within the Morar Group of Glenelg, Morar and A'Mhoine itself.

Rocks similar to the Northern Highland Moine have been recognised in straths Spey and Nairn and designated as Central Highland Division by M. A. J. Piasecki and others. These rocks have yielded radiometric ages which suggest affinities to the Northern Highland Moine and an unconformity greatly modified by ductile displacement zones has been invoked to separate them from the younger and regionally overlying Grampian Group. Work by Paula Haselock and N. G. Lindsay has, however, suggested that the apparent contrasts between Grampian Group and Central Highland Division may be gradational and marked by neither unconformity nor tectonic discontinuity. Their work suggests continuity of tectonic style, fabrics and structural facing throughout the Grampian Group and Central Highland Division.

Dalradian rocks are unknown to the north of the Great Glen fault although some Dalradian units may be laterally equivalent to the Torridonian of the foreland in NW Scotland. W. S. Pitcher and the present author divided the Dalradian into the Appin, Argyll and Southern Highland groups. An additional older Grampian group has been included by the present author with the Dalradian and comprises the thick sequence of monotonous psammitic rocks which stratigraphically underlie the Appin Group. These are commonly called Moine, a practice which caused confusion in view of the correlation thus implied with the Moine rocks of the Northern Highlands which were believed for various reasons to be much older. If the suggestions of Haselock and Lindsay mentioned above are accepted, all the metamorphic rocks of the Central Highlands can be assigned to a single lithostratigraphic succession having a common structural and metamorphic history. A large (upper) part of this succession comprises the Appin, Argyll and Southern Highland groups which all workers would include with the Dalradian.

Deposition of this many kilometre thick succession was marked by a change from a stable shallow marine and tidal mud flat environment of the Grampian and Appin groups to one of rapid facies variation and instability with deposition dominated by turbidites. Deposition of tillites – locally named the Portaskaig Tillite on Islay and Schiehallion Boulder Bed in Perthshire – marked the end of deposition of the Appin Group and also the end of widely prevailing stable conditions. The top of the Tayvallich Subgroup is marked by the outbreak of widespread basic vulcanicity although central Perthshire (Tayside) had earlier experienced basic volcanic activity in the Crinan Subgroup. The close resemblance of the Erins Quartzite formation of the Crinan Subgroup to the Cambrian Eriboll Quartzite of the Hebridean craton and the fact that the Tayvallich Limestone is commonly turbiditic, and possibly, therefore, derived from a Durness Limestone source formerly led to speculation about correlation between foreland and basin deposition.

Charles Downie now regards the acritarch flora that was found in the Tayvallich Subgroup and which was formerly believed to be diagnostic of the Lower Cambrian to have a much greater and older stratigraphic range. This, together with the new isotopic age of 590 ± 2 Ma on the Ben Vuirich granite, suggests strongly that the deposition of the Dalradian succession was entirely Precambrian and that the *Pagetia*-bearing Leny Limestone of the Callander area occurs in deposits of the Highland Border complex rather than within the Dalradian. Severing links between the Dalradian and the Durness Sequence and the Leny Limestone also implies that the Dalradian sequence has no established links with the North American faunal province and as discussed below introduces possibilities of regional tectonic significance.

The facies variation which accompanied increasing instability in the Dalradian basin is put down to the movement of fault-bounded blocks within the basement presumably during lithosphere stretching. Thus rapid accumulations of coarse clastic sediment may be the lateral equivalent of thin sequences which were starved of abundant detritus. A. D. Stewart has shown that fault control of facies is also a feature of the Torridonian groups of the foreland.

Major tectonic discontinuities within the Central Highlands such as the Fort William and Iltay Boundary slides are marked by younger rocks resting on older, a feature which has led R. Anderton and N. J. Soper to suggest that, whatever their sense of displacement during orogenesis, these slides may mark the site of former extensional faults which were active during the formation of the Dalradian basin.

Nature and timing of orogenesis in the Metamorphic Caledonides

The new radiometric dating of the Dalradian rocks using evidence from the volcanics (c. 596 Ma) Tayvallich Sub-group and the synorogenic Ben Vuirich granite strongly suggests that all of the metamorphic sedimentary and extrusive rocks to the north of the Highland Boundary are Precambrian in age. Much of the deformation and metamorphism which affected these rocks were also Precambrian although the rocks on both sides of the Great Glen also suffered Lower Palaeozoic orogenic activity.

NW of the Great Glen the Moine Supergroup carries evidence of at least three major episodes of deformation. Of these the first two probably produced large-scale isoclinal folds which, it is thought likely, had regionally flat-lying axial planes. Some of these isoclinal folds are cored by Lewisian slices, notably in the Morar and Glenelg districts. The azimuth of facing of these structures is unknown, a problem exacerbated by the fact that their hinges are curvilinear on all scales apparently through angles of up to 180°. Thus the apparent eastward-facing direction of the Lewisian-cored isoclinal folds of Morar and Glenelg may be a local feature, not applicable to the whole orogen. A small body of fabric evidence suggests that the extension direction about which the isoclines are probably curvilinear is approximately N–S. It is worth noting that the Moine rocks after these episodes were still regionally right-way-up, because the vast majority of the later upright folds which overprinted the complex are upward-facing. Accompanying the early deformation there was emplacement of locally abundant basic dykes and sheets, now rendered largely concordant with foliation by tectonic strain, and major bodies of granitic magma, now the West Highland adamellitic orthogneiss. Coeval with this orthogneiss was regional metamorphism of Barrovian type locally rising to middle-to-upper amphibolite facies with the widespread development of pelitic and psammitic paragneisses. The orthogneiss has yielded a radiometric age of 1030 ± 45 Ma and paragneisses have been dated by M. Brook and others as 1,004 Ma, a date argued to be that of metamorphism, but believed by others to be that of sedimentary diagenesis. Together these dates constitute the evidence for the timing of early orogenesis in the Moine at 1,100–1,000 Ma on which the correlation with the Grenvillian is based. Nevertheless there is compelling evidence that Grenvillian events preceded deposition of the Moine in the Glenelg area: I. S. Saunders and others have obtained Sm:Nd ages of just over 1,000 Ma in eclogites within the eastern Glenelg Lewisian inlier which is overlain by comparatively low-grade Moine of the Morar Group. Precambrian activity is further suggested by dates of ∼750 Ma obtained on pegmatite bodies such as those of Knoydart, N Morar and Carngorm – one such body which is now deformed has been shown by

Derek Powell to postdate the fabrics of at least one episode of deformation. The Strath Halladale granite (649 ± 30 Ma) cuts deformed and migmatised Moine-like rocks in the far NE of the Highlands. It is on the pegmatite ages that the proposal for Knoydartian/Morarian orogenesis is based, although coeval deformation and metamorphism of this age has not been demonstrated in the N Highlands.

The thrust-related deformation involving ductile and brittle thrusting and major upright folds mentioned earlier in this chapter reworked the crystalline Moines, their Precambrian fabrics, metamorphic mineral assemblages and intrusions. This deformation to varying degrees re-distributed and reset the metamorphic zones while much of the evidence for this activity preserved at the present erosion surface would suggest continued Barrovian metamorphism and no major interruption of the thermal episode responsible for the Precambrian metamorphism; i.e. the rocks we now see appear not to have undergone significant uplift from c. 1,000–800 Ma until the time of the thrust-related orogenesis during the Ordovician and Silurian.

The thrusting must postdate the emplacement of the Carn Chuinneag granite (555 ± 10 Ma) while folding related to the episode postdates the emplacement of the Glen Dessary syenite (456 ± 5 Ma); evidence for continuing crustal shortening until ∼435 Ma derives from deformation of the Strontian granite (Fig. 1.8) reported by Donny Hutton, the deformation of Caledonian sheets of micro-diorite described by Chris Talbot and the presence of open folds and weak fabrics in the regional pegmatites (c. 440 Ma). The isotopic evidence for Caledonian orogenesis NW of the Great Glen seems strong, but the timing of Precambrian orogenesis is on much less sure ground and intuitively the time interval between the Precambrian and Caledonian episodes seems inappropriately long.

The steep inclination of the Sgurr Beag slide at its present crop is the result of folding by later Caledonian structures and of passive rotation during Caledonian crustal shorten-ing. Major structures such as the Morar anticline and sub-vertical, strongly curvilinear reclined folds characteristic of much of the Glenfinnan Division* are essentially the result of Caledonian folding. At the other major boundary within the Moine – the Loch Quoich line – the steeply inclined Glenfinnan Division gives way eastwards to the regionally flat-lying Locheil Division psammites. In all probability this line which has variously been regarded as an uncon-formity, nappe root zone and a zone of severe tectonic strain, is simply the eastern limit of strong Caledonian deformation in the Northern Highlands. Thus, it is proposed that the Locheil rocks have essentially the same horizontal disposition, albeit locally intensely deformed and moderately metamorphosed, that they had at the close of the Precambrian episode. The open asymmetric synform across which one division gives way to the other

---

* Division is used here in the context of the tectono-stratigraphic divisions of the Moine first introduced by G. S. Johnstone and his Geological survey colleagues.

marks the line – the Loch Quoich line – along which the Caledonian structures fail towards the east (Fig. 1.1).

To the SE of the Great Glen lies a major nappe complex, which includes the Tay Nappe, and which is characterised by polyphase deformation, and by Barrovian metamorphism with PT gradients varying between 15° and 30°C per km. The timing of the deformation and metamorphism is indicated for central Perthshire by the 590 ± 2 Ma date on the Ben Vuirich granite which was emplaced later than the local D_2, and earlier than the local D_3, fabrics. There is evidence that deformation during D_2 contributed substantially to the emplacement of the nappe. Both D_1 and D_2 must have been Precambrian. In Perthshire, both the Tay and Atholl nappes of the Precambrian Grampian orogeny face south-south-east, but in Argyll a steep zone separates similar SSE-facing structures from apparently primary structures which face north-west. The full significance of this steep zone is not entirely understood and its continuation north-eastwards from Argyllshire is uncertain, although it has been thought likely to lie in the Strath Ossian steep zone (inset, Fig. 1.1) described by P. R. Thomas. Other steep zones within the Tay Nappe are probably late stage Caledonian secondary structures.

That orogenesis to the SE of the Great Glen continued into the Lower Palaeozoic is borne out by the relationships of the late-orogenic Aberdeenshire gabbros to adjacent Dalradian rocks which carry the distinctive andalusite–cordierite mineral assemblages of Buchan metamorphism which was, in part at least, induced by the gabbros themselves. Work at the SURRC (East Kilbride) by G. Rodgers and T. Dempster seems to confirm the emplacement of the Insch gabbro at 489 ± 17 Ma. Continuing crustal shortening in the Central Highlands is demonstrated by the deformation of the gabbros themselves and of the Glen Kyllachy granite (443 Ma) described by M. Piasecki. The Highland Border downbend may be a manifestation of major displacements within the Highland Border zone which B. J. Bluck has cogently argued lies at a major terrane boundary.

Work by M. A. J. Piasecki and O. van Breemen, in particular, detected elements in their Central Highland Division which are Precambrian, largely based on dates obtained on small pegmatite bodies found in shear zones and believed by these workers to be coeval with their formation. This orogenic episode has been distinguished from later Grampian and Caledonian episodes. However, the late Precambrian age implied for at least early deformation and metamorphism in the Tay Nappe may also serve to date the structures and fabrics identified in the underlying Atholl Nappe first described by P. R. Thomas and now re-examined and traced by N. G. Lindsay from Glen Garry through Strathspey and into Strathnairn where they match closely those of Paula Haselock working in Strathnairn and the Corrieyairack. These workers suggest that a consistent pattern of structural style, sequence and structural facing, presumably late Precambrian in age, can be traced from the upper limb of the Atholl Nappe in Tayside to the Great Glen.

The above brief review of the evidence for the timing of Precambrian-to-Lower Palaeozoic orogenesis in the Highlands is intended as a framework within which the status of the Great Glen fault which transects the area as a possible terrane boundary can be examined. Are there sufficient contrasts in the coeval geological features of the crustal segments on either side of the Great Glen to justify a proposal that these were acquired far from one another prior to their juxtaposition? The detailed columns comprising Fig. 1.3 should be consulted to assess the history of the two segments from c. 1,200 Ma to 400 Ma.

Given the contrasts in timing that emerge from a comparison of the two columns and given the probable contrasts in the basement rocks on either side, at first sight it seems probable that the Great Glen constitutes a terrane boundary, especially given the displacement of the pre-Caledonian basement detected by the LISPB experiment. This displacement appears comparable to that shown by the same experiment at the Highland Border Fault where the geological evidence for a terrane boundary is stronger. The problem with accepting many of the apparent contrasts in the timing of events expressed in the Moine and Dalradian lies in the fact that almost all are based on isotopic ages in rocks which have been both deformed and metamorphosed and which have recently been shown by the Ben Vuirich evidence to be liable to substantial revision.

Bearing this in mind the case for a terrane boundary along the Great Glen must be regarded as non-proven. Given the continuing uncertainty as to the reliability of evidence for the $\sim 1,000$ Ma events in the N Highlands and the probability that Grampian orogenesis is late Precambrian, ~ 80 Ma older than the earlier evidence from Ben Vuirich had suggested, there is the intriguing possibility that the apparently sharp distinction between the age of early orogenesis on opposite sides of the Great Glen could become much less significant. If this were to happen it cannot be ruled out that the Moine and Dalradian Supergroups comprise a continuous sequence perhaps with Loch Eil and Grampian groups being equivalent, as suggested several years ago for somewhat different reasons by M. A. J. Piasecki. Further speculation might lead to the reassessment of the Albynian marbles and pelites, geochemically investigated by N. M. S. Rock, as possible correlatives of Appin Group Dalradian.

Removal of the constraints linking the Dalradian sequence to the North American Province and Laurentia via the Leny Limestone fauna and given the possibilities of major strike–slip displacements, the Dalradian terrane at least might be considered to have potential links with Pan-Africa whose rocks underwent widespread orogenesis c. 600 Ma and which are thought to be a feature of elements on the other side of Iapetus Ocean, such as for example, the Mona Complex of Anglesey. Previous tentative correlation of Grampian orogenesis with Finnmarkian orogenesis in

Scandinavia and with Penobscotian orogenesis of late Cambrian–early Ordovician age in the Appalachians for which a growing body of evidence is accumulating now appears to be ruled out.

Ophiolites and flysch

Ophiolites occur in three separate areas of Scotland. Much of Unst in Shetland is underlain by ultramafic rocks which have been interpreted by Derek Flinn as part of an ophiolite complex, but which are difficult to fit into the context of the Caledonian rocks of the Scottish mainland. The classic sections of Ballantrae display many of the features of an ophiolite complex and it is possible that these rocks represent Iapetus oceanic crust obducted from the south.

Remains of ophiolitic rocks have been recorded at many localities along the Highland Border – the Highland Border Series or Complex. They include at various localities ultramafic rocks, spilites, jasper, black shales and cherts. The recent convincing interpretation by B. J. Bluck and others of the Highland Border as a major terrane boundary has suggested these as the remnants of one or more basins, partly floored by oceanic crust, smeared out, together with their fill, as the Highland terrane was emplaced from the south-west, finally to 'dock' in Middle Old Red Sandstone times. Painstakingly acquired palaeontological evidence from many localities in the Highland Border Series, largely by G. Curry, has demonstrated that the Arenig age attributed to the Series was a great over-simplification and that a discontinuous sequence(s) can be established from the Upper Cambrian to the Upper Ordovician. The age of the rocks comprising the Series may now have to be extended downwards to uppermost Lower Cambrian in that the Leny Limestone and Shales together with adjacent clastic sediments must be included with the Highland Border Series, having been excluded from the Dalradian by the new Ben Vuirich evidence. Throughout much or all of this period, the Dalradian rocks would have been rising and would inevitably have been shedding detritus into these basins had they occupied their present relative position. Ordovician coarse clastic detritus of Dalradian provenance is notably absent from the Highland Border complex sequences. The subject of the role of the Highland Border as a terrane boundary will be considered further when relevant evidence derived from the Southern Uplands Lower Palaeozoic sequences and the Midland Valley Lower Palaeozoic and Old Red Sandstone sequences has been discussed.

Following the LISPB experiment the Lower Palaeozoic sediments of the Southern Uplands were thought to be underlain by Caledonian oceanic crust, but the geophysical work of Hall referred to earlier suggested that denser continental crust continues across the Southern Uplands fault to the south. N. J. Soper has suggested that some of the crust below the Lower Palaeozoic rocks is English crust including rocks continuous with those of the Lake District. The Ordovician and Silurian rocks of the Southern Uplands are remarkable in that, while the oldest rocks (Arenig) occur to the north-west and the youngest (Wenlock) lie to the south-east, sedimentation structures throughout the area indicate younging toward the north-west. This apparent anomaly can be explained by considerable dip–slip movements on strike faults.

Dip–slip displacements of apparently great magnitude have been interpreted as consistent with the development of the Southern Uplands as an accretionary wedge at a destructive plate margin. J. K. Leggett and others have identified as many as ten separate tracts of rocks each of which is characterised by a distinctive stratigraphy marked by fine-grained sediments such as cherts and/or black shales, floored by basalt in the NW part of the Southern Uplands. The fine-grained sediments in each tract are overlain by turbidites, the lowest of which become generally younger from NW to SE. In terms of the accretionary wedge model it is envisaged that as the Ordovician–Silurian oceanic lithosphere was subducted towards the north-west, progressively younger oceanic lithosphere carrying its distally accumulated thin cover of chert and/or graptolitic shales was brought within the depositional area of turbidites with local slumped rudites. Periodically the turbidites with their underlying pelagic layer, having undergone subduction and low-grade metamorphism, rose by buoyancy. Each segment, thus formed accreted to the continental margin to face away from the ocean, and eventually away from the next, but younger, segment to be emplaced.

The presence of bedded cherts and black shales of Arenig age resting on oceanic basalts in the oldest sediments of the accretionary wedge would suggest that the Iapetus Ocean was wide in the Arenig – possibly betwen 600 km and 2,000 km wide. The presence of Caradocian greywackes in the same segments shows that closure was well advanced by mid-Ordovician.

Since the initial proposal of the accretionary wedge model by W. S. McKerrow and J. K. Leggett, it has been realised that at least one of the faults bounding the various segments of the Southern Uplands has been the site of major sinistral lateral-slip movements, the Orlock Bridge fault described by T. B. Anderson and G. H. Oliver presenting clear evidence of this. Detailed provenance and palaeocurrent studies, moreover, have shown that the turbidites making up the clastic sequences within the wedge have at least two distinct sources.

It might be argued in support of the simple accretionary wedge model that the differing provenance of the clastic successions could be explained by strike–slip during their accumulation so that they were derived from source rocks of varying type *en passant*; subsequent strike–slip between segments could have heightened the contrasts between segments. However, contrasts exist *within* segments and there is clear indication throughout much of the accumulation of the Southern Uplands sediments that two coeval sources existed of clastic sediment which invaded basinal areas in which the condensed sequences of Moffat Shales facies were accumulating. One source lay to the south and yielded fresh andesite debris weakly diluted by quartzitic continental material; another lay to the north and contributed almost exclusively continental crustal material. P. Stone and his Geological Survey colleagues have argued strongly that provenance and palaeocurrents point to the Southern Uplands as the site of a former back-arc basin, the andesitic debris being derived from a now underthrust island arc which lay to the south of the basin. The northward-directed underthrusting of this basin by the extensively eroded arc and its continental host occurred as Iapetus Ocean closed to the south of the area. Such underthrusting is envisaged by Stone and others as causing, within the Southern Upland belt, the development of a southward rising and propagating thrust stack from which late Llandovery and Wenlock debris comprising recycled greywacke detritus was eroded to pour southwards. This eroded debris filled a foreland basin ahead of the advancing thrust stack and became the Hawick and Riccarton groups.

A problem with the thrusting hypothesis is the extraordinary lateral persistence of the individual segments which might be thought to be more reconcilable with the accretionary, subduction-related wedge hypothesis than with the thrust hypothesis. Both hypotheses would predict successively later tectonic structures from NW–SE and both might involve an eventual closure-related final overprint of cleavage, folds and faults. However, comparison of fabrics indicating sense of displacement at the margins of each segment might be useful in distinguishing between the two main hypotheses which would predict opposite senses of dip–slip, late-stage displacement; ironically, where T. B. Anderson and G. H. Oliver have employed such fabric techniques in the case of the Orlock Bridge fault the fabrics indicate lateral-slip rather than dip–slip displacements.

Although closely related to the geological evolution of the Southern Uplands, the Girvan–Ballantrae district has had a different history. B. J. Bluck has suggested that the Arenig sediments of Ballantrae had accumulated in a basin at the north-west margin of the Iapetus Ocean and that the basin was invaded by an obducted ophiolite nappe which advanced into it from the south during the middle Arenig. Mass-flow deposits derived by the erosion of the rising and advancing ophiolite were subsequently deformed as the nappe overrode them. The obduction of the ophiolite may have marked the onset of closure of the Iapetus Ocean, although it has been interpreted by P. Stone among others as a remnant of a continental margin transform zone.

Granites

The abundant granites emplaced within the metamorphic rocks of the Scottish Highlands have been traditionally classified as 'Older' and 'Newer'. With the former have been included both early intrusive bodies and the migmatitic complexes which should be more realistically related to the peak of regional metamorphism. The Newer granites are abundantly developed in the Grampian Highlands but are sparse to the NW of the Great Glen fault while they also occur sparsely in the Southern Uplands (Fig. 1.8).

The early granitic plutons in the Highlands are broadly contemporary with the tectonic and metamorphic events which affected the country rocks into which they were emplaced. Thus the Ardgour granitic orthogneiss ($1,030 \pm 45$ Ma) displays the same intense tectonic fabrics and metabasic sheets as the surrounding now migmatitic country rocks. The similarity between the granitic gneiss and some of the migmatitic country rocks led to early speculation as to the metasomatic origin of the granite which has since been discredited by the work of D. Barr and L. M. Parson. By contrast, although the Carn Chuinneag (550 ± 10 Ma) and Ben Vuirich (590 ± 2 Ma) granites were emplaced after deformation had begun in their country rocks, they retain their originally sharply defined margins and indeed imposed thermal aureoles on the fairly low-grade schists in their envelopes. Large tectonic strains and contemporary high-grade metamorphism may obscure such relationships in the case of the Ardgour granitic orthogneiss. Trace-element geochemistry of the older granites is similar to that of the adjacent Moine or Dalradian, which supports an origin largely *via* migmatites.

Although a few of the Younger Granites in the N Highlands were apparently emplaced as late as 400 Ma (e.g. Fearn and Migdale) many of those showing entirely post-tectonic passive emplacement such as Ross-of-Mull, Cluanie and Helmsdale are somewhat older than the abundant *c.* 400 Ma granites of the Central Highlands. The stockworks of granitic and pegmatitic veins described below are somewhat older than Cluanie (425 ± 4 Ma) and like Strontian (435 ± 10 Ma) show signs of residual deformation late in the Caledonian regional thrust-related deformation referred to earlier as characteristic of the whole N Highland terrane. The late Caledonian granites of the southern part of the Southern Uplands at least are younger than those of the Central Highlands, and are apparently geochemically distinct in having S-type characteristics; those of Shetland apart from Graven (405 ± 14 Ma) are much younger.

A feature of the Highlands north of the Great Glen and of parts of the Central Highlands such as around the Foyers pluton is the occurrence of granitic (*s.l.*) material which occurs as stockworks of intrusive veins over areas measurable in tens of square km. The emplacement of these veins was normally preceded by basic and intermediate intrusions, commonly hornblende-bearing, and thus they show a similar sequence to that displayed by major Newer granite plutons. Indeed their presence in the Northern Highlands may be linked to the apparent absence over large areas of major Newer granite bodies – the vein complexes may be a roof stockwork above still buried plutons, or, conceivably in some cases, a root zone through which granitic magma passed on its way to segregate as a large body at a level above the present erosion surface. Thus, the contrast in abundance of Newer granites may well be a function of crustal level on either side of the Great Glen fault zone.

Closely associated in time with the Newer granites and the granitic (*s.l.*) vein complexes are the major swarms of microdiorite–felsic porphyrite minor intrusions reported

Fig. 1.8. Distribution of main Caledonian and pre-Caledonian granites in Scotland, showing isotopic ages (after Memoir 9, Geol. Soc. 1985).

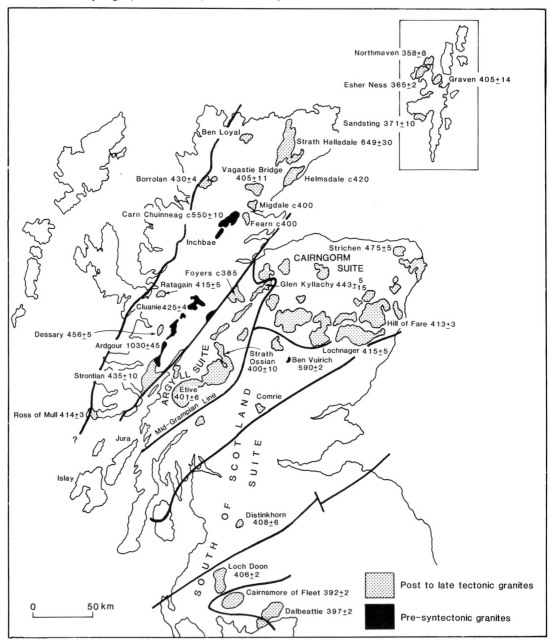

by D. I. Smith. These intrusions, which are of very wide distribution in the Northern Highlands and which also occur near the Great Glen fault in the Grampian Highlands, are commonly foliated and show a pattern of mineral paragenesis which may reflect regional thermal patterns in the Highlands during the Silurian as the regional metamorphism waned. An alternative view suggested by A. J. Highton, on geochemical grounds, is that these minor intrusions do not form a single swarm but constitute a series of overlapping and amalgamating intrusive complexes each reflecting a unique differentiation pattern but having similar mineral compositions. In terms of this model each intrusive complex, including the granitic vein stockworks could be related to a major Newer granite pluton either at or below the erosion surface. If this is the case the th mal patterns recorded by the minor intrusions simply reflect the overlapping thermal effects of major granitic bodies. Such thermal patterns will not only influence mineral parageneses in the country rocks and in the ; most contemporary but normally slightly older minor intrusions, but may well affect palaeomagnetic and isotopic results on Moine schists and older igneous intrusions over large areas.

Insofar as granites have sampled the lower crust and mantle it is interesting that, as mentioned earlier, W. E. Stephens and A. N. Halliday have distinguished the Argyll, Cairngorm and South of Scotland suites within the high-K, calc–alkaline granitoid province of Scotland to the SE of the Great Glen fault. The distribution of the suites is shown on Fig. 1.8; they have been distinguished on petrographic, major and trace-element and isotopic criteria, the details of which are described in Chapter 8. Dioritic types (including appinites) occur in varying abundance in all the suites and Stephens and Halliday have concluded that while the more acidic granitoids are dominantly the product of I-type crustal sources, the intermediate and basic members have a substantial mantle component. Stephens and Halliday have concluded that because of its profound influence on granitoid composition the variable nature of the deep crustal and mantle basement beneath the southern two-thirds of Scotland may control the position of boundaries drawn between the granitoid suites. The changes in the crust and mantle are thought to coincide. Such boundaries may originate at ancient terrane boundaries, while the boundary between the Argyll and South of Scotland suites coincides with Halliday's mid-Grampian line. Possibly this line which is projecting to pass through Jura and Islay (Fig. 1.1) may mark the boundary of the Islay/Colonsay terrane which A. Maltman and W. R. Fitches believe is characterised by the basement exposed in the Rhinns of Islay and which is plutonic, younger than and dissimilar to Lewisian. The South of Scotland suite, however, although lacking appinites to the south, appears to pay little heed to the presence of such major faults as the Highland Boundary and Southern Uplands, although the stratigraphic evidence of the Highland Border Old Red Sandstone is suggesting to B. J. Bluck and others sufficiently large Middle Old Red

Sandstone sinistral displacements at the Highland Border as to suggest a terrane boundary.

The major- and trace-element geochemistry and the isotope geochemistry of the younger granitoids has suggested to Stephens and Halliday that, with the exception of the S-type Criffel–Dalbeattie and Cairnsmore of Fleet plutons (Fig. 1.8), the suites comprising the whole plutonic province are I-type plutons in the sense of B. W. Chappell and A. J. R. White and that the Cairngorm suite is geochemically more evolved than the others. As I-type granites the magmas forming the plutons might be expected to have been generated during the subduction of oceanic lithosphere and to have been modified en route to their eventual mid-upper crustal level by contamination with ancient mantle and overlying lower crustal material. To the south of the Central Highlands, however, there is little evidence of contemporary subduction which could have given rise to these magmas. Even if the Southern Uplands were formed in accordance with the simple subduction-related accretionary prism model, the stratigraphic evidence shows that much of the subduction would have taken place in the mid-upper Ordovician, whereas the Central Highland plutons are largely mid-Silurian to early Devonian. Even allowing for substantial magma dwelling time in the lower crust, the time gap seems too long.

Possibly displacement on terrane boundaries can offer an explanation because two putative terrane boundaries occur within Scotland to the south of the Central Highlands – the Highland Boundary fault on which the evidence for major strike–slip displacement is strong, and the Southern Uplands fault. The evidence for subduction coeval with Central Highland plutonism, which could take the form of forearc/accretionary prism deposits, would in these circumstances have been removed by strike–slip. Pre-Upper Devonian events recorded in the Southern Uplands and in the Central Highlands could thus be quite unconnected and their relative timing, irrelevant.

Whatever the origin of the magmas it is likely that the emplacement of granitoid bodies, especially in the metamorphic crustal rocks of the Highlands caused general uplift and displacement across already existing faults; this late-orogenic uplift stimulated the molasse deposition of the Old Red Sandstone period.

The Highland Border complex – the fate of small basins

Given the new evidence adduced by B. J. Bluck and others for the history of the Highland Boundary zone, it would be naïve to suppose that a simple relationship can be adduced between the Highland area of Dalradian metamorphic rocks and granitic to dioritic plutons and the Devonian sediment and volcanic fill of the Midland Valley graben. It is appropriate at this stage, therefore, to look at the history of the Highland Border.

At many localities along the Highland Border, Arran,

Bute, Aberfoyle, Callander, Edzell and Stonehaven, there exist considerably deformed igneous and sedimentary sequences referred to collectively as the Highland Border complex, comprising serpentinites, amphibolite, pillow basalts, 'greenstones', cherts, limestones, black shales, turbiditic sandstones and rudites. At several of these localities the different lithologies are absent or sparse. It is believed that the complex was floored by an ophiolite, now represented by the sporadically preserved serpentinite and pillow basalts. This ophiolite must be older than the Ballantrae ophiolite (500–480 Ma) with which it has formerly been correlated.

Fossil evidence recorded in the Highland Border complex has formerly been used to indicate the maximum date of orogenesis for the adjacent Dalradian rocks in that it was believed by M. R. W. Johnson and A. L. Harris and others that the polyphase deformation which the Highland Border complex had suffered in the Edzell, Aberfoyle and Arran outcrops, in particular, was coeval with that of Grampian orogenesis to the north. The age of Grampian deformation, in the Dalradian block, now recalibrated to take account of the 590 ± 2 Ma age of the Ben Vuirich granite, and the palaeontological evidence in the Highland Border complex which indicates deposition from at least late Tremadoc to early Ashgill make this hypothesis untenable. Using the known relationships we can only conclude that the rocks comprising the Highland Border complex were strongly deformed, not necessarily for the first time, later than the early Ashgill. They were at some stage overthrust by the Dalradian from the north, so that their extent below the Central Highlands is uncertain and Lower Old Red Sandstone strata rest unconformably upon them, so that their extent to the SE is likewise uncertain.

The above analysis of the situation, for the essence of which this author is indebted to B. J. Bluck, did not take into account the significantly greater age of emplacement which in 1988/89 was attributed by Rodgers and others to the Ben Vuirich granite. This age implied that the Dalradian succession is wholly Precambrian and that the fossiliferous Leny Limestone of the Callander district cannot be part of the Dalradian, but was also deposited in the same basin or set of basins that now comprise the Highland Border complex. Its inclusion in the complex is, however, of some significance in that for many years workers, commencing with C. T. Clough early this century, were satisfied that the rocks in the quarry where J. Pringle in the late 1930s found the topmost Lower Cambrian *Pagetia* trilobites were continuous with the Leny Limestone and Shales of the Keltie Water some 3·5 km to the NE. In the Keltie Water the Leny Limestone and Shales separate the inverted Upper Leny Grits from the inverted Lower Leny Grits with apparently unbroken contacts. Both grit formations which are very similar to Southern Highland Group grits elsewhere in the Dalradian, carry upward- and downward-facing structures and thus have a history consistent with that of the Dalradian tract with which they have

formerly been included. If the continuity from quarry to stream is a reality, then there is a strong implication that an unknown amount of the supposed Dalradian slates and grits of the Callander area is also part of the Highland Border complex.

Bluck has argued convincingly that the Dalradian terrane would have been rising through the Ordovician and Silurian period and that such uplift would have involved the stripping by erosion of tens of kilometres of Dalradian metamorphic rocks, none of which is found in the Ordovician basin(s) which comprise(s) the Highland Border complex. We must conclude from this that the complex and the Dalradian were distant at this period and that the emplacement of the Dalradian block into its present position must have been partly responsible for the deformation and the attenuation of the basin(s). Lower Old Red Sandstone rocks, comprising volcanics, including the Lintrathen Porphyry ashflow, and sediments rest unconformably on already downward-facing Dalradian, for example in the Dunkeld area, and this suggests that early ductile displacements in the Highland Border zone had taken place to downbend the Tay Nappe. The same ashflow deposits also rest unconformably on the Highland Border complex in the north-east near Edzell. Thus the Dalradian and Highland Border complex were juxtaposed by the late Silurian suggesting the possibility that their juxtaposition and the formation of the Highland Border downbend in the Dalradian comprised the same event. If this is the case fabrics in the Highland Border complex should yield evidence of ductile transpression and the steeply plunging reclined folds recorded by Johnson and Harris at Edzell may be consistent with such deformation.

Further reactivation in this zone occurred during the Middle Old Red Sandstone hiatus of deposition in the Midland Valley and must have controlled the asymmetry of the Strathmore Syncline to the south of the Highland Border.

Late Caledonian Molasse and structure

Sedimentation and volcanicity

With the closure of the Iapetus Ocean and the establishment of the Rheic Ocean to the south, northern Britain was land until the start of the Carboniferous period. Thus the whole of the Devonian in Scotland reflects terrestrial conditions involving fluvial, aeolian and lacustrine deposition – the Old Red Sandstone. Such terrestrial sediments have been largely dated on fish and plant fossils but are regarded as broadly equivalent to the marine sediments which were deposited during the Devonian in SW England. The deposits and faunas of both terrestrial and marine Devonian suggest a warm arid climate, consistent with the palaeomagnetic data which show that Britain lay less than 30° south of the equator at that time.

In Scotland the Old Red Sandstone rocks closely reflect

the vigour of the environment of deposition as the area of the Highlands emerged as high rugged mountains accompanied by faulting which created active fault scarps and valleys. In places volcanic activity accompanied the faults as the last Newer granitoid plutons such as Glencoe breached the surface *via* ring faults which defined caldera. The timing of these events which gave rise to ignimbritic deposits both in Glencoe and within the Lorne Plateau volcanics provides a stratigraphic link between the latter stages of Caledonian igneous activity and the deposition of the Old Red Sandstone sediments which were initiated by the rising mountain zone. Differential movements on faults, probably along inherited lines, must have controlled the rise of blocks of the crust while the clasts in local sequences of Old Red Sandstone sediments such as those established in fanglomerates near the Highland Border reflect the uplift of Dalradian schists.

B. J. Bluck has established the existence during the Lower Old Red Sandstone of two major NE–SW trending valleys on the present site of the Midland Valley; bounding faults to these valleys are the Highland Border and Southern Uplands faults, neither of which at that time necessarily had the same hinterland as at present. Both faults being pre-Devonian in origin emphasise the inherited grain which controlled the Devonian processes. Local control of facies depended on the disposition and degree of maturity of alluvial fans along the fault-controlled escarpments. On a larger scale the influence of the Highland Border fault may have been exerted on the whole of the basin – the fluvial sandstones flanked by fault–scarp conglomerates overlap south-eastwards along the Highland Boundary fault towards the south-west.

Substantial Middle Old Red Sandstone displacements of the Highland Border zone must have taken place to the south of the rocks described by I. B. Paterson and M. Armstrong who have shown very large thickness and facies variations in Lower Old Red Sandstone sequences across the fault at the Highland edge which subsequently warped the sediments and volcanics into a monocline. It is a striking feature of the Lower Old Red Sandstone stratigraphy, which well illustrates the role of faulting at the Highland edge that the Lintrathen Porphyry – a dacitic ash flow deposit, and therefore a precise chronostratigraphic marker – lies almost horizontal to the north of the fault on a reddened Dalradian landsurface at Dunkeld, while its stratigraphical level occurs almost 4,000 m above the base of the sequence at Stonehaven to the south of the fault. Further, the Highland Border diorites such as those at Comrie and Callander which probably fed adjacent Lower Old Red Sandstone andesites, now vertical or overturned at Callander, thermally metamorphosed already cleaved and downward-facing Dalradian sediments.

Response to a tectonically controlled topography is not confined to the Midland Valley, and Lower Old Red Sandstone deposits influenced by structure occur widely but sporadically elsewhere in Scotland. Notable among these is the thick Shetland sequence consisting of *c.* 3 km of conglomerates and fluvial sands which are succeeded by calc–alkaline volcanics.

While high-energy conditions involving deposition of conglomerates and fluvial sands continued during the Middle Old Red Sandstone, deposits of this age are largely confined to the area of Brora, Easter Ross and the Great Glen. In the last locality sedimentation was probably controlled by the strike–slip fault systems as suggested by H. G. Reading. The late W. Mykura demonstrated the existence of > 2 km of Middle Old Red Sandstone in the NE part of the Great Glen zone, and has shown that periodic rejuvenation linked to faulting controlled sedimentation in a piedmont and alluvial flood plain environment. He has further shown that late in the Middle Old Red Sandstone period thrusting has disrupted the sediments and interleaved them with their crystalline basement. The end of the Lower Old Red Sandstone period in the Midland Valley was marked by faulting, folding and general uplift, with the result that erosion of much of the Lower Old Red Sandstone of this area took place during the Middle Old Red Sandstone times. By contrast the Middle Old Red Sandstone of Caithness and Orkney is marked by quiescent lacustrine deposits of which upward-shallowing deposition cycles are a feature. This may have been in response to the distant earth movements which were taking place in the Grampian Highlands and which created the youthful topography of the contemporary Brora, Great Glen and Easter Ross region. In Shetland the Walls Formation (< 9 km) flysch is also thought to be lacustrine, but here the thickness and nature of the sediments argue for sediment accumulation in a rapidly subsiding fault-bounded basin. The entirely different environment of deposition in Shetland from that elsewhere in Scotland at this time, is emphasised by the polyphase tectonic structures which at the end of the Middle Old Red Sandstone severely deformed the Walls formation and which were preceded by the emplacement of the 360 ± 11 Ma Sandwick plutonic complex.

Middle Devonian earth movements ensured that the Upper Old Red Sandstone is everywhere unconformable. The folding, faulting and subsequent erosion which were a feature of the Middle Old Red Sandstone period in the Midland Valley and adjacent upland areas produced spectacular overstepping relationships. Among the most notable of these are the overstepping of members of the Lower Old Red Sandstone to the southwest of Loch Lomond where the Upper Old Red Sandstone comes to rest on the Dalradian to the northwest of the Highland Border fault, and the overstepping of the Lower Old Red Sandstone on the flanks of the Sidlaw anticline in Tayside.

The contrast in the provenance of Lower and Upper Old Red Sandstone conglomerates is very marked at the Loch Lomond localities. Here the Lower Old Red Sandstone contains no clasts that can be readily related to the adjacent Dalradian while the Upper Old Red Sandstone contains abundant clasts of the Dalradian rocks on which it rests unconformably. It is perhaps significant that the Loch Lomond Upper Old Red Sandstone and the not-very-distant Dunkeld Lower Old Red Sandstone both contain clasts of Dalradian metasediments from approximately the same stratigraphic formation and of much the same metamorphic grade. This might argue that however much lateral displacement along the Highland Border might be supposed to have taken place in the Middle Old Red Sandstone period, that significant dip–slip displacement and down-cutting erosion of the Dalradian block did not intervene between the two deposits.

During deposition of the Upper Old Red Sandstone, fault control of sedimentation is again evident. In the Midland Valley this is largely the consequence of movement on the Highland Boundary fault, on this occasion not dip–slip as formerly, but sinistral slip; these movements probably coincided with sinistral slip on faults to the north of the Highland Border where the NE–SW trending Loch Tay and several related faults (inset 1, Fig. 1.1) were, or had been, undergoing sinistral wrench displacements. According to B. J. Bluck the sinistral displacements on the curving Highland Boundary fault plane in the Clyde area so stressed the crust to the south that it broke into a series of fault-bounded, tilted blocks. The development of each fault block saw the deposition on the downthrow side of upward-fining and maturing sequences, starting with alluvial-fan conglomerates and finishing with sandstones having caliche beds. This sequence exceeds 3 km in thickness.

At first sight the pattern of crustal behaviour, volcanicity and sedimentation in the Scottish Devonian appears to be one of decreasing vigour with time. Considered on the scale of Scotland this is indeed the case. However, the restricted Scottish area reveals a number of facies variants which, viewed on the scale of the whole emerging orogen, were contemporary. Further, whereas in much of Scotland the scene is one of powerful erosion, aided by high level brittle structures and by high energy deposition and transport it may be that this stage had been passed or had not been reached elsewhere during the declining Caledonian orogenic activity. A hint of this lies in Shetland where conditions of high heat flow conducive to ductile deformation persisted until the end of the Middle Old Red Sandstone.

The lateral-slip faults of the Highlands

The set of NE–SW trending sinistral faults (Fig. 1.1) which transect the Highland area were first described by W. Q. Kennedy who interpreted them as Hercynian structures produced by compression from the south during the upper Palaeozoic. Geological evidence indicates much of the displacement on the faults must indeed be upper Palaeozoic in age and amounts of sinistral displacement have been worked out in some cases, such as the 5–8 km on the Loch

Tay Fault. The extent to which these structures coincide with long-lived displacements in the deep crust has not been determined, although patterns of Dalradian sedimentation suggest that the movement of large blocks of basement had an important influence; moreover, there are indications, adjacent to the Loch Tay and Strathconon faults at least, that the faults are linked to the later stages of ductile deformation. By far the largest of the sinistral faults is that in the Great Glen (Fig. 1.6 and Fig. 1.1, inset). Kennedy's early work suggested a lateral-slip displacement of about 65 miles (104 km), but this was largely based on the supposed equivalence of the Strontian and Foyer granites (Fig. 1.8) although there was supporting evidence from the apparent displacement of metamorphic isograds across the fault. Geochemical studies have shown that the two granites are probably not fragments of an original whole and are of different age (Fig. 1.8). Subsequent correlations across the fault on metamorphic grounds by J. A. Winchester and on lithostratigraphic grounds by M. A. J. Piasecki seem to be inconclusive. Hence it must be admitted that the displacement across the Great Glen is not known although many lines of geological evidence constrain it to tens or at most a few hundred kms rather than the thousands of kms which have been suggested by palaeomagnetic workers. A sinistral rather than dextral displacement is likely on the grounds that the other faults of similar age and trend have this sense of movement although N. Holgate and J. M. Speight and J. G. Mitchell have advanced evidence of late stage (Tertiary) dextral slip. A detailed study by the writer, D. I. Smith, P. A. Rathbone, L. M. Parson, M. S. Stoker and A. J. Highton of all the metamorphic rocks adjacent to the fault in Scotland has failed to reveal any structures or fabrics which could irrefutably be related to deep-level ductile deformation along the fault zone although the Rosemarkie Inlier contains some evidence of such deformation. Hence we are left with the conclusion, remarkable in itself, that the fault zone as we see it has remained at much the same brittle level in the crust at least since the Middle Devonian.

The development of ensialic basins

By the start of the Carboniferous the palaeoequator probably lay across the Southern Uplands which separated the two main fault-bounded basins in which the Scottish Carboniferous was deposited – the Midland Valley and the Northumberland troughs. During the Carboniferous the Scottish area drifted northwards by some 12° of latitude. Scotland then lay towards the south-eastern margin of a major cratonic area comprising the present North America, Greenland and Northern Europe. This craton has been thought to be bounded to the present south-east by the proto-South Atlantic (Phoibic) and Rheic oceans which were apparently closing throughout the Carboniferous.

D. M. Mackenzie envisaged that basins caused by

lithosphere stretching and faulting would be characterised by a tensional stress regime, high heat flow and the development of narrow graben basins defined by faults. Strong control by faults over sedimentation is predicted. Volcanism might be the result of extreme crustal attenuation which by narrowing the isotherm intervals would of itself produce enhanced heatflow. Given that many of the faults were already features of the pre-Carboniferous lithosphere, Mackenzie's model fits well with much of the Scottish Carboniferous, although compression as well as tension seems to have operated from time to time.

The margins of the Midland Valley basin were topographically marked, as was the case for much of the Devonian, by the Highland Border and Southern Uplands faults (Fig. 1.1). The northern margin of the Northumberland Trough extends from the Solway Firth to Liddesdale, and being strongly fault-controlled may lie on a line inherited from the latest stages of subduction of the Iapetus Ocean at the south-eastern edge of the Southern Uplands.

Sedimentation and the volcanicity which occurred mainly in the early Carboniferous were to a great extent controlled by movements along faults which largely reflected the Caledonian NE–SW trend in the basement on which the Carboniferous was laid down. Major variations in thickness and facies occurred across faults extending upwards to the contemporary surface; in Ayrshire, in particular, there are marked changes in thickness across fault zones which indicate continuous syn-deposition displacements. The extraordinary relationship between fold structures and facies and thickness in the Namurian Westfield Basin is a further indication of tectonic influence over sedimentation. Periodic rejuvenation caused by fault-controlled uplift was recorded by C. E. Deegan at the northern margin of the Northumberland Trough where Dinantian sediments record the uplift and unroofing of the adjacent late-Caledonian Dalbeattie pluton (Fig. 1.8) and its lower Palaeozoic envelope.

To the fault control of sedimentation must be added control exerted by thick volcanic sequences, such as the Dinantian Clyde Plateau lavas, which separate areas of maximum subsidence and are themselves marked by thin sedimentary deposits or by the non-deposition of sediments. The maintenance of shallow-water conditions throughout the period of the Carboniferous is evidenced by the cyclic deposition, eustatically and tectonically influenced, of the coals, limestones, mudstones, siltstones and sandstones in lagoonal, deltaic and fluvial environments. That such sediments attained a thickness of ~ 3.5 km is sufficient evidence that movements on the contemporary faults kept pace in a general way with deposition. Such a relationship would be equally valid if the major cycles or mesotherms are caused by world-wide inundations providing widely recognised time-significant events, as suggested by W. H. C. Ramsbottom, or in the event of more localised tectonic controls of cyclicity.

Balance between sedimentation and tectonics in the Carboniferous may point to a further significant contrast between Devonian and Carboniferous activity. Whereas in the former, active fault-tectonics commonly stimulated the sedimentation process by creating vivid imbalance, during the Carboniferous the active faults as a general rule merely kept pace with sedimentation, tectonic displacement and deposition possibly regulating one another. Indeed the whole upper Palaeozoic history of Scotland may be characteristic of the evolution from youth to maturity of a rifted basin; the later stages of the Carboniferous, with the decrease in volcanicity, may well record the 'sagging' stage of basin evolution which Mackenzie thinks accompanied the relaxation of isotherms and the cooling of underlying asthenosphere causing general subsidence which largely ignores local faulting.

Permo–Carboniferous volcanicity in Scotland can be related to two cyclic episodes. Both of these Dinantian–Namurian and late-Namurian–Permian were marked by mildly undersaturated alkaline basalts while the centres to which the volcanic rocks are related were strongly controlled by tectonic structures, both folds and faults. Volcanicity in each cycle became more phreatic and the Namurian volcanic rocks of Scotland, for example, are almost entirely pyroclastic. R. Macdonald has suggested that the basaltic magmas, which during each episode became increasingly silica undersaturated, were produced by fusion of the mantle at progressively greater depth. Many workers including T. N. George and E. M. Anderson suggested variably compressive and tensional regimes for the Carboniferous of the Midland Valley and possibly the variable regimes proposed were related to activity far to the south where Hercynian orogenesis had sporadically occurred since the Devonian. The varying stresses to which the crust of the Scottish area was subjected during the Carboniferous is consistent with the hypothesis advanced by J. P. N. Badham and others of Hercynian strike–slip orogeny. Distant orogenesis of this type, capable of imposing differently directed stress at different times, may also explain the change in the orientation of the structures controlling the volcanicity – whereas the older Dinantian–Namurian followed a long-established Caledonian grain, the late Carboniferous volcanics were largely influenced by NW-trending structures. Furthermore, the Permo–Carboniferous dykes which indicate tension but which by no means necessarily opened normal to their margins seem to have been emplaced during a stress regime important well beyond the Scottish area. E. H. Francis has suggested that they may be continuous with dykes of comparable type but of north-westerly trend in southern Scandinavia.

Permo–Triassic desert basins

The onset of the Permian was marked by considerable uplift, not only on the site of the Hercynian orogenic zone, but in the adjacent cratonic areas. Scotland in common with most of the British Isles and Europe underwent this uplift such that the Highlands and much of the Midland Valley and Southern Uplands formed a land area albeit increasingly eroded and peneplained throughout the Permian. Uplift may have been at least partly responsible for the onset of widespread dessication which succeeded the generally humid Carboniferous period as the area moved gradually northwards; palaeolatitude of Scotland in the Rotliegender (Lower Permian) was probably between 22° and 26°N.

The uplifted Scottish area shed molasse deposits into adjacent basins. One of these coincided with the northern Irish Sea and the other lay in the North Sea to the north of the mid-North Sea High which was the eastward extension of what are now the Southern Uplands and Grampian Highlands. The bulk of the Scottish Permian seen at the present day was deposited adjacent to the 'Irish Sea' basin in deep fault-controlled troughs of general north-west trend where it was preserved during the Permian–Triassic erosion of the adjacent high land. The north-westerly trend of these troughs, notable among which are the Arran, Mauchline and Stranraer troughs, was inherited from the late Carboniferous when, as mentioned above, this trend seems to have been initiated. It is notable, moreover, that the north-west-trending basins are in marked contrast to the north-east-trending basins identified on the continental shelf in a zone running from Shetland to the Sea of the Hebrides which the late A. C. McLean and others have suggested was related to the contemporary continental margin.

Basic volcanicity which had been widespread in the Carboniferous persisted in SW Scotland with volcanics recorded in the oldest Permian on Arran and in the Mauchline and Thornhill basins (Fig. 1.1). Although continuous deposition from the Carboniferous into the Permian was formerly believed to have occurred in the Mauchline basin, this is now thought unlikely; W. Mykura has shown that the underlying Carboniferous is Westphalian in age, the implication being that much of the Stephanian is missing. The break is marked here, as elsewhere, by the deep reddening in the underlying rocks which is the mark of the extremely arid conditions under which the Permian erosion occurred.

The sediments which overlie the volcanics are characteristic continental deposits, consisting of breccias as in the case of the Sanquhar basin and much of the Arran basin and aeolian sands blowing from the east in the case of the Mauchline, Dumfries and Thornhill basins. Widespread amelioration of the climate was linked to the glacioeustatic rise in sea level at the start of the Upper Permian proposed by D. B. Smith. This in turn may be linked to deposition, in several of the basins, of water-lain sandstones. It is on the basis of this that Lower is distinguished from Upper Permian in Scotland. Only in Arran is there evidence for the Permian passing up into the Trias.

Northerly drift of the Scottish area continued during the Triassic such that by the end of the period Scotland probably lay between 35° and 40° N. By the beginning of the Trias much of the British Isles had been denuded to an almost featureless and arid peneplain which was subjected to some rejuvenation in Scotland as elsewhere. Because of the impoverished Permian and Triassic faunas of Scotland, the notable extinction of Palaeozoic forms and the widespread introduction of a Mesozoic fauna recorded in the contemporary marine deposits of Europe cannot be documented from the Scottish rocks. Nevertheless, the vertebrate fossils of the Lossiemouth area are representatives of the new Mesozoic reptilean fauna. Isolated deposits of New Red Sandstone occur, many of which are only tentatively assigned to the Triassic. The principal areas of deposition were the Moray Firth basin, with important on-shore deposits occurring in the Elgin–Lossiemouth area, and the western fringes of Scotland, notably in the south-west and in the Inner Hebrides. Permian deposition had taken place in SW Scotland and on Arran, but this basin which had been restricted in the Permian extended northwards during the Trias. The main onshore deposits of Triassic age in the Hebrides are locally derived breccias and sands, probably alluvial-fan and scree deposits, a facies marginal to silty flood plain or possibly lacustrine sediments. A similar marginal relationship is thought to have existed in the Moray Firth area, but here the marginal sediments consist dominantly of aeolian deposits, notably dune sands, which have yielded both very early and late Triassic reptile faunas. Here early and late Triassic aeolian rocks are apparently separated by waterlain sandstones carrying abundant well rounded pebbles, possibly marking the local passage of a river system draining into a Moray Firth lake.

On Arran there is evidence of a passage from the underlying late Permian and the boundary between the two systems is arbitrarily drawn within the Sherwood Sandstone. The Arran sections record a sequence which passes upwards from breccias and sandstones of marginal facies to mudstones and siltstones, of probably fluviatile origin, which can be correlated with part of the Penarth Group. Such a transgression may reflect the general extension of areas of New Red Sandstone deposition from the rather restricted Permian basins over, for example, the Hebridean area.

Mesozoic incursions and recessions

A large part of Scotland remained land during the Jurassic although most of the British Isles was inundated by a shallow sea which flooded the Inner Hebridean area and the Moray Firth basin. This event was part of a worldwide marine transgression. Mull may have experienced little or no break in deposition from the underlying Trias, but further north in the Hebridean area there was a distinct break although most areas had been inundated by the end of the Hettangian. The geological story of Scotland during Jurassic times must be thought of largely in terms of the evolution of the adjacent long-lived basins of marine deposition. Onshore outcrops are restricted to the margins of the Hebridean Sea and the Moray Firth. Inevitably, since the country lay at the limits of an epicontinental ocean subject to periodic eustatic changes these exposures commonly show evidence of discontinuous deposition of marginal deposits with non-sequences and even unconformities marking marine recessions. The other major factor which influenced Scottish Jurassic deposits was the rejuvenation of the land area which accompanied renewed activity on major faults such as the Great Glen, Camusunary and Helmsdale faults (inset, Fig. 1.1). This activity was probably contemporary with major doming and subsequent graben collapse which occurred in the North Sea during the Middle Jurassic. This tectonic activity has been linked to major basaltic volcanism which appears to have centred in the eastern part of the Moray Firth basin.

The rejuvenation stimulated the supply of terrigenous sandy material to fluviodeltaic and shallow marine marginal areas, coming not only from the Scottish mainland but from important offshore features such as the Halibut horst in the Moray Firth basin. An important indication of more direct fault control of sedimentation comes from the spectacular boulder beds of the Brora area. Here the Helmsdale Fault formed an active fault scarp at the foot of which boulders from adjacent shallow-water deposits were periodically transported by tsunamis to interdigitate with Kimmeridge clay during the Upper Jurassic. The Kimmeridge Clay which is so important as the source rock for the North Sea oil has a thin development in the west of Scotland on Skye.

The uppermost part of the Jurassic – the Portlandian – is not represented in Scotland as a consequence of a general retreat of the sea and the emergence as land of most of Britain. At the close of the Jurassic, Scotland lay about 40°N and had drifted a few degrees further north by the time that the Upper Cretaceous marine incursion took place in the Cenomanian. The interval between the deposition of the youngest Jurassic preserved and the formerly extensive, but now very restricted, Cretaceous sediments, was evidently marked by fault-controlled tilting and extensive erosion. Cretaceous rocks which are both clastic and chalky are thin but in various places lie on rocks of various Mesozoic ages. They owe their sporadic occurrence to further uplift, tilting and erosion in the late Cretaceous, such that the Palaeocene basalts rest unconformably on a great variety of geological formations from the Precambrian basement to the Cretaceous.

Major volcanoes

Although several hundreds of metres of Tertiary sediment occur offshore Scotland, the significant on-shore activity involved several major volcanic centres (Fig. 13.1). These – Arran, Mull, Ardnamurchan, Rhum and Skye – are strung out in a NNW-trending zone along the western seaboard. Other centres lie to the west and include St. Kilda, Blackstones and recently discovered submarine centres. Palaeontological and isotopic evidence suggest that the volcanoes were early Tertiary (Palaeocene–Eocene) emplaced in and erupting onto metamorphic basement and Torridonian and Mesozoic cover rocks. Each probably remained active for some 2–3 Ma within a general timespan from 55 to 61 Ma ago. Their occurrence is recorded by ash bands discovered in the sediments of the North Sea basin – in the Thanet, Woolwich and London Clay formations. Palaeomagnetic investigations, so far carried out, suggest that the lavas in the Scottish part of the Tertiary volcanic province show reversed remnant magnetisation. It has been suggested that this relates almost entirely to the period of anomaly 26 in the North Atlantic. This anomaly is somewhat older than that assigned to the East Greenland lavas which, although also showing reversed magnetisation, correspond to anomaly 24.

Different products of the igneous activity are displayed at the different centres depending on the level to which erosion has cut. Thus Skye and Mull offer a shallow section through the effusive products of the volcanoes; thick and extensive basaltic flows are traceable onto the adjacent mainland and are preserved far out to sea as, for example, Canna, Staffa and the Treshnish islands. The geometric perfection of the trap topography of Skye and the Wilderness of Burg, on Mull, together with the scarcity of pyroclastic rocks bear witness to the quiet outpouring, probably from fissures, of extremely fluid magmas. Lateritic bole, lacustrine sediments, plant beds and thin coals between flows indicate prolonged pauses in volcanic activity.

By contrast, the layered complex of Rhum and major ring dykes of Ardnamurchan are revealed by a deeper cut into the volcanic roots, and here the chambers in which the magma segregated are dissected. Major episodes of magmatic replenishment are recorded in the layering on Rhum. The large ring dykes form steeply inclined concentric structures, especially well displayed on Ardnamurchan and indicate large-scale subsidence in depth during volcanic activity. On Ardnamurchan and elsewhere the abundant cone sheets focus more shallowly inwards and downwards below the volcanic centres, indicating steeply directed crustal distension. Subsidence related to caldera formation during magmatism occurred within the area defined by the Loch Ba ring dyke on Mull. Concentric features, both dykes and sheets, define centres so precisely that an intricate history can be traced, with one centre giving way to another as the focus of activity shifted within individual major complexes.

Although tempting, a disarmingly simple history that relates the major intrusive centres to the lavas that are preserved is ruled out by the conclusion that lavas both predate and postdate intrusive centres. Thus the thermal effects of the intrusive centres on Skye overprint the patterns of zeolitisation recognised in the lava pile; con-glomerates interleaved with lavas on Skye, Rhum and Canna contain clasts, both acid and ultrabasic, which were derived from major intrusive centres.

The meticulous geological mapping and careful petrography that marked the Geological Survey work on Mull and Ardnamurchan, in particular, revealed the sequence of volcanic events and showed that magmas of composition ranging from acid to mildly undersaturated were contemporary with one another and were available throughout the volcano's history. It is not surprising that this work, carried out in such well exposed areas, stimulated international interest in the mechanisms of formation of volcanic structures and in the evolution of magmatic liquids. From the initial simple schemes of petrogenesis and magmatic descent proposed by J. E. Richey and by W. Q. Kennedy involving one or more parental magmas have come, by experiment and observation, more involved and realistic models of magmatic evolution. Thus, one may now envisage beneath the major volcanic centres differing degrees of partial melting at variable depths in the mantle and varying intervals and depths of magmatic segregation with fractionation taking place during the uprise of the resulting liquids. This activity, while capable of giving rise to a range of basaltic liquids, cannot of itself explain the presence of all the acid magmatism. But segregation in Lewisian basement of large amounts of liquid basalt, at several hundreds of degrees above the liquidus temperature of granite, could have produced by melting the copious quantities of acid material which form, for example, the Red Hills of Skye. It is believed that the origin of the granites by anatexis, rather than by the fractional crystal-isation of basic magmas increased in importance with time; many of the magma chambers at a high level in the crust must have been zoned. The rapid evacuation of such zoned magma bodies would give rise to the mixed acid/basic pyroclastic material that fills the Loch Ba ring dyke on Mull. The nature and size of the positive gravity anomalies which mark the volcanic centres suggest that ultrabasic material and not acid is predominant in the complexes. Acid plutons seem to be rather high-level bodies while acid extrusive rocks are sparse.

The factors which instigated this volcanic activity in separate volcanic centres in the west of Scotland have been a matter of conjecture. Workers such as J. R. Vann and G. P. L. Walker have suggested that the emplacement of major basic/ultrabasic complexes was preceded by the diapiric uprise of acid magma. Such magmas, it is thought, may have developed at the *loci* of high heat flow where intense dyke swarms intersect major NE–SW trending faults, some of which originated in the Caledonian and have been sporadically active since. There is, however, a lack of precision in the correlation between the known faults and the major centres and probably some other control should be sought.

It is important to recognise that, unlike the East Green-land centres, the centres of NW Scotland and Ireland do not lie at the margin of a major zone of seafloor spreading. Their position well within the continental margin of NW Europe does suggest that they may be linked to a frustrated attempt to form a constructive oceanic plate margin which if successful would have cast the Hebridean area adrift as a micro-continent.

Nevertheless, it is likely that the formation of the centres was the culmination of a sequence of events going back into the Mesozoic whereby thinned continental crust had become the site of shallow Mesozoic basins which, after the early Tertiary crustal warping affecting the whole of NW Europe, received the largest accumulations of early Ter-tiary volcanics. Further, with their thinned crust they were the site of high heat flow and of early Tertiary dyke swarms. The central complexes may simply mark the position within the extended crust where the maximum thinning had occurred.

2

THE LEWISIAN COMPLEX
R. G. Park

The north-western corner of Scotland has a crystalline basement overlain by undeformed or gently tilted sediments of various ages ranging from late Precambrian (Torridonian) to Mesozoic. The obvious contrast between basement and cover led to the early recognition of the Lewisian gneisses as a 'fundamental complex' on which younger sediments had been deposited. The first comprehensive account of the Lewisian rocks appears in the famous Geological Survey '1907 memoir' (Peach *et al.* 1907) in which the complex was clearly recognised to be made up of a wide range of different igneous rocks of various ages, together with some sediments, subjected to deformation and metamorphism. At this time therefore, the nature of the Lewisian outcrop as a 'complex', involving a series of events over a lengthy time span, was first clearly recognised.

This tract of country (see Fig. 2.1) is admirably summarised by J. Horne (Peach *et al.* 1907, pp. 1–10): 'along the western seaboard of the counties of Sutherland and Ross, the Lewisian or fundamental gneiss forms an interrupted belt stretching from Cape Wrath to Loch Torridon, and thence to the islands of Rona and Raasay. Throughout this belt of country bare rounded domes and ridges of rock, with intervening hollows, follow each other in endless succession, forming a singularly sterile tract, where the naked rock is but little concealed under superficial deposits, and where the surface is dotted over with innumerable lakes and tarns. Over wide areas the elevation of this undulating rocky plateau is comparatively uniform, save near the great escarpments of Torridon Sandstone and Cambrian quartzite, where the Lewisian Gneiss sometimes forms prominent peaks and lofty crags, as on Ben Stack (2364 ft.), near Loch Laxford in Sutherland, and on Ben Lair, near Loch Maree (2817 ft.) in Ross-shire.'

Peach *et al.* (1907) established a simple chronological sequence. An older group of rocks, mainly acid gneisses. which they referred to as the 'fundamental complex', was intruded by a younger group consisting of various basic and acid intrusions including the well-known Scourie dyke suite. Both sets of rocks were then affected by deformation, which caused severe modifications to the complex in certain areas, particularly the northern and southern regions of the mainland, but left the central mainland region comparatively unscathed. These movements did not affect the oldest of the overlying sedimentary formations, the Torridonian, and were termed the 'Pre-Torridonian movements'.

The chronological subdivision of the Lewisian, so clearly foreshadowed in the work of Peach and his Survey colleagues, was addressed again half a century later by Sutton & Watson (1951) in a classic paper based on their work on the Loch Torridon and Scourie areas. Sutton and Watson interpreted the chronology of the complex in terms of successive orogenic cycles: the older (corresponding to the fundamental complex of Peach *et al.* (1907) was termed the *Scourian* and the younger, the *Laxfordian*. The two cycles were separated by the intrusion of the Scourie dykes, which Sutton and Watson regarded as anorogenic. Thus the concept arose of Scourian *rocks*, formed during Scourian *time* and succeeded by Laxfordian rocks formed during Laxfordian time (see Table 2.1). The role of the Scourie dykes as stratigraphic markers in this chronology was absolutely critical; no other method of relative dating of rocks or events was available then. The time-equivalence of the various basic dykes attributed to the Scourie swarm was assumed; but there was no means of checking the validity of this assumption.

The first radiometric ages produced on a variety of Lewisian rocks by Giletti *et al.* (1961) provided a firm basis for Sutton and Watson's time scale. Rocks from the Scourian complex yielded late Archaean ages while in areas of Laxfordian reworking, granites, gneisses and Scourie dykes gave Proterozoic ages ranging from 1,610 to 1,160 Ma. Subsequent intensive geochronological studies have refined the Scourian time period to between *c*. 2,900 and *c*. 2,400 Ma, the age of the oldest dated Scourie dykes (see

Table 2.1). The main Laxfordian event is dated at *c.* 1,800 Ma, but younger events up to possibly *c.* 1,100 Ma may be represented by K–Ar reset ages. Thus on the basis of the radiometric ages, the history of the Lewisian complex spans an enormously long period of time, possibly as long as 1,900 Ma, and more than three times as long as the period from the Cambrian to the present.

It will be clear that the methods used to establish a stratigraphy based upon sedimentary successions cannot be applied to a crystalline basement complex such as the Lewisian. Here many of the processes responsible for the observed rocks and structures took place at deep crustal levels. Stratigraphy in the Lewisian therefore depends primarily on the recognition of sequences of events: of deformation, metamorphism and igneous intrusion, and by as accurate radiometric dating as possible of these events. The degree of precision so far attained is in no way comparable with the palaeontological precision of the Phanerozoic.

Distribution

The main outcrops of the Lewisian complex on land occur on the islands of the Outer Hebrides and on a coastal strip of the mainland from Cape Wrath in the N to Loch Torridon in the S, W of the Moine thrust zone (Fig. 2.2). To the SW of these mainland outcrops, Lewisian rocks occur on several islands of the Inner Hebrides: Rona, Raasay, Coll, Tiree and Iona. Geophysical and other evidence as to the nature of the continental shelf indicates that these surface outcrops form part of a broad region extending from the W and N of the Hebrides to the edge of the continental shelf. Lewisian rocks also form inliers within the younger Precambrian Moine complex of the Caledonian orogenic belt, E of the Moine thrust belt.

The offshore region is crossed by a number of deep seismic reflection profiles including both BIRPS and commercial lines. From these and from the geophysical evidence on the physical properties of Lewisian rocks (see

Fig. 2.1. Typical Lewisian terrain with relict hills of Torridonian, near Stoer. BGS photograph.

Table 2.1. Simplified Lewisian chronology

Ga (approx.)	*Events*
2·9	Formation of early Scourian sediments and mafic/ultramafic igneous rocks (oceanic crust?).
2·9–2·7	Emplacement of tonalitic/granodioritic plutonic complex; strong sub-horizontal shear-zone deformation (subduction?).
2·7	Badcallian high-grade metamorphism; granulite–facies at deeper levels, amphibolite–facies at higher.
2·5	Post-Badcallian biotite–pegmatites; initiation of Inverian shear zones, associated with uplift and isolation of Archaean blocks.
2·4	Emplacement of earlier Scourie dykes; continuing Inverian metamorphism ?and shear-zone movement.
2·0	Emplacement of later Scourie dykes; formation of Loch Maree Group supracrustal assemblage (crustal extension or transtension).
1·9	Early Laxfordian deformation and metamorphism (only proved in South Harris).
1·7	Main Laxfordian deformation and metamorphism (NW–SE movements on shear zones); emplacement of Laxfordian granites and pegmatites, and formation of migmatites.
1·5	Late Laxfordian retrogressive metamorphism and deformation (dextral transpression on steep shear zones).
1·4–1·1	Late Laxfordian and/or Grenville brittle deformation and retrogressive metamorphism (movements on crush belts).

See text for details. N.B. 1 Ga = 1,000 Ma = 10^9 years.

Hall 1987) it can be concluded that Lewisian or similar rocks probably form the basement to Scotland up to the line of the Great Glen fault at least, and possibly beneath the Midland Valley and Southern Uplands as well (Bamford *et al.* 1977). According to Hall, the properties of the upper crust of this region correspond to those of the Lewisian rocks of the Laxfordian belts, whereas those of the middle crustal layer correspond with those of the Scourian complex of the Central mainland region.

The disruption of the Lewisian outcrop is mainly attributable to the formation of fault-controlled extensional basins in Mesozoic and early Tertiary times, of which the most obvious examples are in the Minch separating the Outer Hebrides from the mainland (see Fig. 2.2). In this chapter, four main sub-divisions of the Lewisian complex will be descibed: the *mainland*, the *Outer Hebrides*, the *Inner Hebrides*, and the *Caledonian inliers*.

Stratigraphic subdivision

Lewisian stratigraphy is essentially based on the work of Peach *et al.* (1907) and Sutton & Watson (1951) on the mainland outcrops. The division of the Lewisian complex of the mainland into three separate regions – Northern, Central and Southern – originated with the recognition by Peach *et al.* that the Central region had largely escaped the effects of the post-Scourie dyke deformation and metamorphism now known as the Laxfordian.

The Central region extends from N of Scourie to S of Gruinard Bay, and is about 65 km across. It is composed of rocks of the Scourian complex relatively unmodified by younger Laxfordian effects. This complex is intruded by mafic and ultramafic dykes of the *Scourie dyke swarm* with a generally NW–SE to E–W trend. These dykes, although typically metamorphosed to some degree, are generally undeformed or only locally deformed where affected by the numerous narrow shear zones that traverse the region.

The Northern and Southern regions, on the other hand, represent Laxfordian belts or complexes where the original Scourian material has been modified during the Laxfordian period (Fig. 2.3). The *Northern region* extends from Loch Laxford to Cape Wrath on the N coast. It is separated from the Central region by a transition zone several km wide in which typical granulite–facies gneisses of the Scourian complex are progressively transformed to amphibolite–facies hornblende – and biotite gneisses in a zone of intense late Scourian and Laxfordian deformation.

The *Scourian region* extends from S of Gruinard Bay to Loch Torridon, and includes the islands of Rona and Raasay. In an 8-km-wide transition zone extending from Gruinard Bay to Fionn Loch, N of Loch Maree, late Scourian deformation and amphibolite–facies metamorphism have severely modified the Scourian gneisses but the Scourie dykes are little affected by Laxfordian deformation, which however becomes intense around Loch Maree and Gairloch.

The Scourie dykes of the Central region become increasingly deformed and metamorphosed as they are followed north-eastwards into the Northern region and south-westwards into the Southern region. It was thus logical to conclude that the Laxfordian belts were essentially zones of reworked Scourian crust. Apart from the Scourie dykes themselves, post-Scourian rocks appear to be relatively uncommon in the Laxfordian belts. Around Loch Laxford in the Northern region, and S of Loch Torridon, veins and sheets of granite and pegmatite occur and are locally abundant. Laxfordian deformation in these belts is marked by NW–SE folds and foliation and is in places very intense.

The only post-Scourian metasediments within the Laxfordian belts for which there is radiometric evidence, occur in the early Proterozoic Loch Maree Group. However the original extent of the Scourian complex, and the proportion of post-Scourian material within the Lewisian outcrop, has been the subject of considerable debate. Peach *et al.* (1907) followed by Sutton & Watson (1951) considered that the fundamental or Scourian complex (now known to be Archaean in age) extended throughout the whole of the mainland belt. They based their belief on the presence of amphibolite sheets similar to the Scourie dykes within both

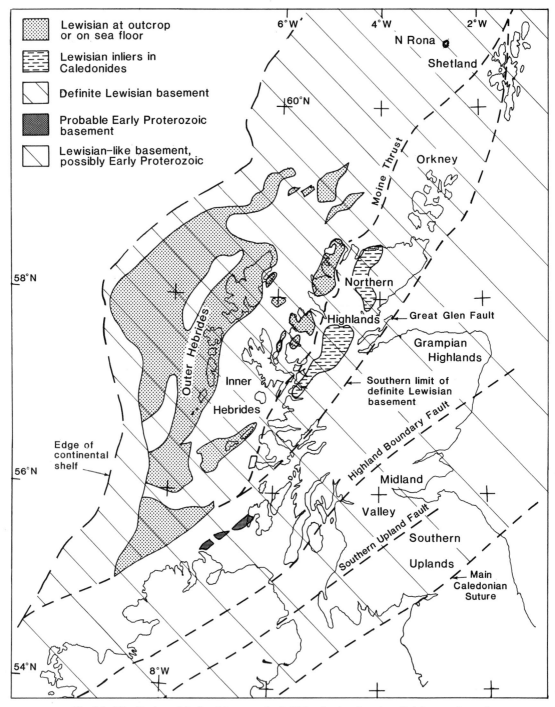

Fig. 2.2. Distribution of the Lewisian complex in N Scotland and on the adjoining continental shelf. After Dunning (1985).

the Northern and Southern regions. Others have suggested (e.g. see Bowes 1968, Holland & Lambert 1973) that much of the material of the Laxfordian complexes may have represented post-Scourian supracrustal sequences, and that the amphibolite sheets were not all of the same age as the Scourie dykes. The whole-rock Pb/Pb isotopic data of

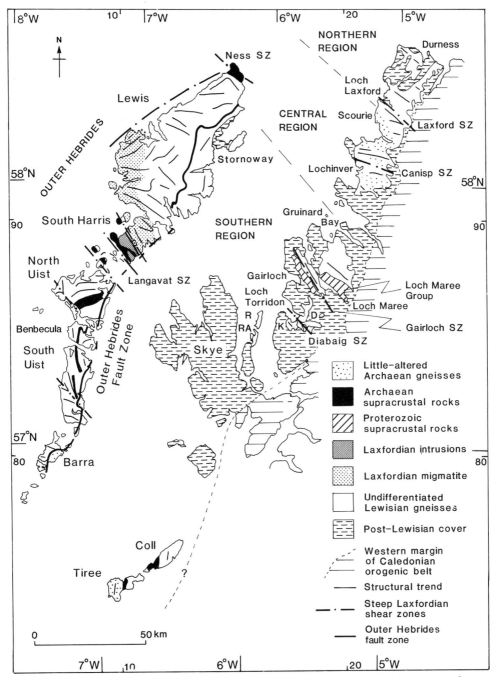

Fig. 2.3. Simplified map of the Lewisian complex showing main rock units and structures, and the regional subdivision. After Park & Tarney (1987) and Fettes & Mendum (1987).

Moorbath *et al.* (1969) and Whitehouse & Moorbath (1986) proved decisive in this debate, indicating that the crust of the whole of the mainland region, together with the Outer Hebrides, originated at about 2,900 Ma, and that the younger radiometric dates characteristic of the gneisses of the Laxfordian complexes were attributable to re-working. The Archaean age of the 'fundamental complex' of the mainland is therefore now firmly established.

The mainland outcrops

The Scourian complex

The nature of the Scourian complex can only be clearly determined within the Central region of the mainland which has largely escaped the effects of Laxfordian reworking. The complex is of Archaean age, and shows many similarities to other high-grade Archaean gneiss terrains throughout the world.

Rock types

Virtually the whole of the complex is made up of rocks with plutonic igneous affinities or appearance, and rocks interpreted as metasediments or other supracrustal remnants are rare. The presumed meta–igneous gneisses cover a broad spectrum. In the words of J. Horne (in Peach *et al.* (1907): 'the rocks that have affinities with plutonic igneous products have a wide petrographic range, and comprise ultrabasic, basic, and more acid materials . . . (including) pyroxenites, hornblendites together with pyroxenic, hornblendic and micaceous gneisses. Sometimes they appear in an amorphous form, like ordinary eruptive masses, sometimes with crude banding due to a rough parallel arrangement of the constituents, and yet again, over wide areas, with well-developed foliation. The term gneiss is not strictly applicable to many of the members of this series, owing partly to their massive character and partly to the absence in places of mineral banding.'

These rocks contain, as their chief mineral constituents, olivine, clinopyroxene, orthopyroxene, hornblende, biotite, plagioclase, microcline, orthoclase and quartz.

The *acid gneisses* in the unmodified parts of the Central region make up probably 75–80 per cent of the complex, and contain pyroxene and/or hornblende. Pyroxene–hornblende gneisses where hornblende aggregates apparently replace pyroxene are the most abundant type. Biotite and muscovite gneisses are apparently confined to areas of later modification. In the Laxfordian belts of the Northern and Southern regions, hornblende–, biotite– or biotite–muscovite gneisses are universal. In the Central region, the banding of these gneisses is characteristic and individual bands are typically several cm in breadth, broader than the bands in the Laxfordian gneisses. The banded structure (see Fig. 2.4) is due to variation in the proportions of darker and lighter material, and is accentuated by the alignment of

mineral aggregates. Most of the quartz occurs in parallel lenses or rods connected in a network which often stands proud of the weathered rock face. The quartz is typically opalescent.

These gneisses enclose frequent lenses and lumps entirely composed of hornblende or pyroxene. The gneissose banding may either bend around these masses or may be truncated by them. Many of the lenses themselves contain banding or foliation which may be cut by the margins of the host gneiss. Some basic lenses occur in strings suggesting boudinage.

The great majority of the acid gneisses are tonalitic to granodioritic in composition, but occasional sheets and veins of granitic gneiss characterised by abundant alkali feldspar are found in the Scourian complex of the Central region and, more commonly, in the Northern and Southern regions, where they are difficult to distinguish from Laxfordian granites unless cut by Scourie dykes (e.g. see Davies & Watson 1977). Pegmatites consisting of quartz and perthite, sometimes with graphic intergrowth, and with accessory biotite and magnetite, are widely distributed. These bodies cut the gneissose banding, and are associated with local retrogression of the granulite–facies assemblage. These pegmatites have been dated at 2,500-2,400 Ma (Giletti *et al.* 1961, Evans & Lambert 1974) and have been taken as marking the boundary between the Badcallian and Inverian events (see Table 2.1).

The more mafic enclaves within the acid gneisses are collectively known as the *early basic* bodies, to distinguish them from the post-Scourian basic intrusions. These masses vary in size from a few cm to about a km across, and are particularly common in the Scourie and Assynt areas. In the Central region, such bodies typically contain both clinopyroxene and orthopyroxene, and variable amounts of hornblende, in addition to plagioclase. In the strongly modified areas, partial or complete replacement of pyr-

Fig. 2.4. Banding in Scourian gneisses with intrafolial isoclinal folds and lensoid basic inclusions, Scourie. BGS photograph.

oxene by hornblende occurs, and in the Northern and Southern regions, pyroxene is normally absent. Garnet and some quartz are typically present also. The larger masses usually project from the surrounding gneisses in the form of small knolls or low ridges. They are often cut or veined by acid gneiss, and sometimes grade into agmatite (see Fig. 2.5) or, ultimately, to patches of acid gneiss enriched in small mafic blobs.

The ultramafic enclaves vary compositionally from monomineralic masses of hornblende or pyroxene, to large bodies of peridotitic or dunitic material, either homogeneous or banded, and with varying proportions of hornblende. The larger ultramafic bodies are often associated with mafic masses and are described by Bowes *et al.* (1964). A good example occurs at Loch Drumbeg, 1 km W of the village of Drumbeg, in the northern part of the Assynt district. This mass is about 700 m across, and occupies a shallow synform with acid gneiss in the core. It consists of lensoid bodies of banded metaperidotite and either banded or massive garnet–pyroclasite (garnet–plagioclase–pyroxene gneiss). The pyroclasite generally overlies the metaperidotite where both are seen in contact. The banding is concordant with that in the host gneisses, and consists of alternating layers 2–10 cm thick with varying proportions of olivine and pyroxene (Fig. 2.6). It would appear that an original igneous layering is accentuated and disrupted by tectonic banding similar to that affecting the acid gneisses. Similar ultramafic/mafic masses

near Scourie (Davies 1974) contain anorthosite layers and are intimately associated with pelitic metasediments. These mafic/ultramafic bodies appear to be generally older than the acid gneisses and to have been invaded by them; they may be plausibly interpreted as disrupted pieces of oceanic crust.

Metasedimentary rocks are rare in the Scourian complex of the mainland, but form prominent outcrops in the Inner and Outer Hebrides and in the Caledonian inliers. The only large outcrops of metasediments on the mainland are of the Loch Maree Group at Gairloch and Loch Maree, which are almost certainly of post-Scourian (Early Proterozoic) age (see pp. 41–44). A few narrow bands of metasediment are associated with the mafic/ultramafic layered complexes already described. These bands consist mainly of semi-pelitic gneisses but include minor calc–silicate rocks and possible arkoses (Okeke *et al.* 1983, Cartwright & Barnicoat 1987). The metasedimentary gneisses described by Cartwright and Barnicoat contain a corundum–kyanite–staurolite–plagioclase assemblage partly retrogressed to margarite–paragonite–sillimanite–chlorite and are interpreted as restites arising from partial melting. A narrow band of kyanite gneiss occurs within the basement gneisses at Fionn Loch, N of the contact with the Loch Maree Group metasediments. This band has yielded a Badcallian age (Bickerman *et al.* 1975) and is therefore not an outlier of the Loch Maree Group. Several bands of marble occur within the basement gneisses at Gairloch, again close to the

Fig. 2.5. Scourian agmatite, Heights of Kinlochewe. BGS photograph.

Fig. 2.6. Banding (modified igneous layering?) in Scourian early basic body, South Uist. BGS photograph.

Loch Maree Group contacts. These could also be Scourian in age but could equally be tectonically enclosed outliers of the Loch Maree Group; definite evidence is lacking.

Deformation and metamorphism

The extreme heterogeneity of the complex, coupled with the almost ubiquitous compositional banding, indicate a generally high level of deformation comparable with that achieved in the younger Laxfordian shear zones. Unfortunately the microscopic evidence of high strain has been largely obliterated by the high-grade metamorphic re-

crystallisation. The presence of platy shape fabrics, the regularity of the planar banding, and the fact that individual bands generally lens out laterally, are all familiar features of plutonic complexes deformed in zones of intense strain, such as the Precambrian shear zones of S Greenland (e.g. see Escher & Watt 1976). Tight to isoclinal intrafolial folds are common in the acid gneisses (see Fig. 2.4) although less so in the basic and ultrabasic varieties. These folds, often rootless, are usually of the order of cm across. The banding or foliation is typically sub-horizontal or gently inclined over large areas of the Central region (see Sheraton *et al.* 1973) although it is locally steepened in late-Scourian shear zones.

The granulite-facies metamorphism is characteristic of the Scourian complex and its effects can be seen throughout the Central region, although local retrogression to amphibolite facies is widespread. This metamorphic event is termed the *Badcallian*, and is dated at 2,700 Ma from Sm–Nd and zircon ages (Pidgeon & Bowes 1972, Lyon *et al.* 1973, Humphries & Cliff 1982). On the other hand, whole-rock Pb–Pb data for the gneisses over the whole Lewisian outcrop (Moorbath *et al.* 1969) and for the amphibolite-facies gneisses in particular (Whitehouse & Moorbath 1986) together with the Sm–Nd isotopic data of Hamilton *et al.* (1979) suggest that the separation of the igneous bodies making up the complex from their original mantle source occurred at *c.* 2,900 Ma. Thus it is likely that the formation of the Scourian complex, and inferentially of the Lewisian crust, took place over a period of *c.* 200 Ma from *c.* 2,900 to *c.* 2,700 Ma.

Pressure–temperature conditions deduced for the Badcallian event are amongst the highest recorded for granulite

Fig. 2.7. Temperature–time path for the Assynt block. After Sills & Rollinson (1987).

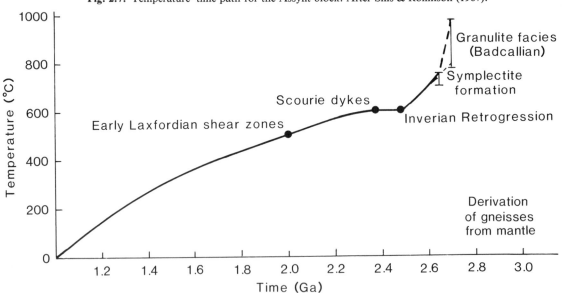

terrains (see Barnicoat 1987) with temperatures of over 1,000°C and pressures of up to 15 kb recorded from trondhjemite sheets. However the high temperatures may partly reflect magmatic conditions, and other estimates based on cores of crystals within mafic rocks yield temperatures of 950°C and 11 kb (Cartwright & Barnicoat 1987). These pressures imply an Archaean crustal thickness of 35–50 km. Crystal-rim compositions yield lower estimates (< 800°C and < 8 kb) indicating re-equilibration to lower grades of metamorphism by the end of the Badcallian (see Fig. 2.7).

Geochemically, the Scourian granulites are distinctive in showing extreme depletion in their contents of U, Th, Rb and K – the main radioactive heat-producing elements (Table 2.2). The usual model to explain this depletion – a common feature of granulite terrains – is lower-crustal melting, which extracts a component enriched in these elements. However there is no sign that the depletion is confined to the deeper parts of the Scourian complex; it is equally evident at Gruinard Bay, where the gneisses show no evidence of deep-seated origin. An alternative theory discussed by Tarney & Weaver (1987) is that these elements have been removed by CO_2-rich fluids. The cause of the very high pressures and temperatures of the granulite-facies metamorphism is thought to be primarily tectonic, related to the underplating model discussed below.

Origin of the Scourian complex

In common with many other Archaean high-grade terrains, the Scourian complex is composed dominantly of banded, or less commonly massive, acid gneisses of tonal-

Table 2.2. Representative analyses of Scourian igneous rocks

	1	2	3	4	5	6	7	8	9	10	11	12
SiO_2	61·22	67·1	68·00	68·72	69·4	71·00	70·87	50·53	51·4	37·66	42·09	47·0
TiO_2	0·54	0·34	0·36	0·32	0·4	0·34	0·29	0·72	1·5	0·26	0·37	1·20
Al_2O_3	15·64	15·48	15·29	15·07	14·7	15·00	15·23	11·61	13·0	5·93	7·34	18·29
Fe_2O_3	3·07	1·26	1·70	0·73	0·9	—	—	—	3·7	6·46	3·06	4·14
FeO	2·57	2·38	1·65	1·87	2·1	2·40*	1·77*	10·82*	11·1	5·88	7·12	6·40
MnO	0·08	0·05	0·04	0·04	0·05	0·03	0·03	0·21	0·25	0·12	0·31	0·14
MgO	3·36	1·44	1·20	1·19	1·6	0·69	0·59	8·64	5·3	30·53	24·44	54·77
CaO	5·57	4·81	3·51	2·29	3·1	3·40	2·19	10·88	9·8	2·76	10·73	8·46
Na_2O	4·42	4·62	4·20	4·27	4·4	4·60	4·20	2·40	1·8	0·45	0·76	3·83
K_2O	1·03	1·5	2·27	2·70	2·0	1·66	3·87	1·01	0·4	0·13	0·18	1·73
P_2O_5	0·18	—	0·13	0·11	0·1	0·07	0·07	0·05	0·12	0·04	0·03	0·49
Ba	757	809	787	797	795	1,013	1,909	207	48	20	50	1,100
Ce	48	—	—	—	65	51	75	19	—	—	—	—
Co	—	66	19	28	35	—	—	—	72	220	220	40
Cr	88	26	25	39	< 50	9	7	411	138	>3,000	3,000	25
Cu	—	25	15	31	35	—	—	—	14	—	—	65
Ga	—	15	10	12	15	21	20	16	0	0	0	24
La	20	43	50	53	55	30	43	10	—	—	—	—
Li	—	22	14	29	—	—	—	—	—	0	3	10
Nb	—	—	—	—	4	3	2	3	—	—	—	—
Ni	58	23	21	17	25	3	2	106	87	2,500	2,500	41
Pb	13	6	46	19	—	9	20	7	0	—	—	—
Rb	11	41	83	122	85	31	67	15	8	0	0	—
Sr	569	370	466	564	530	387	546	220	158	0	50	1,100
V	—	—	—	—	—	—	—	—	—	7	220	180
Y	9	11	7	8	—	1	3	17	32	—	—	—
Zn	—	35	71	48	45	26	28	106	101	—	—	—
Zr	202	185	176	171	135	190	198	48	191	45	45	100

* Total Fe as FeO.

Acid rocks: (1) tonalitic gneiss, Drumbeg, Central region (average of 254 analyses; (2) tonalitic gneiss, Uist (average of 5 analyses); (3) granodioritic gneiss, eastern Outer Hebrides (average of 118 analyses); (4) granodioritic gneiss, western Outer Hebrides (average of 31 analyses); (5) granodioritic gneiss, mainland Laxfordian belts (estimated average); (6) trondhjemitic gneiss, Gruinard Bay (average of 5 analyses); (7) granitic gneiss, Gruinard Bay (average of 3 analyses); *mafic/ultramafic rocks:* (8) amphibolite, Gruinard Bay (average of 5 analyses); (9) metadunite, Scourie; (10) metaperidotite, Scourie; (11) late Scourian microdiorite, Barra.

Sources: 1, Tarney *et al.* (1972); 2–4, 9, 12, Fettes *et al.* (in press); 5, Bowes (1972); 6–8, Rollinson & Fowler (1987); 10, 11, O'Hara (1961).

itic to trondhjemitic composition, with numerous mafic and ultramafic layers and enclaves. Metasedimentary gneisses are relatively uncommon. The relative proportions of these components vary in different parts of the mainland belt. The granulite-facies terrain of Scourie and Assynt in the Central region is characterised by a high proportion of intercalated ultramafic and mafic material, and the composition of the acid gneisses varies from mafic diorite to tonalite with only a small proportion of silicic trondhjemites (Sheraton *et al.* 1973). On the other hand, the southern part of the Central region around Gruinard Bay consists predominantly of amphibolite–facies trondhjemitic gneisses with numerous mafic enclaves. The Laxfordian belts of the Northern and Southern regions have even less mafic material, and few ultramafic enclaves, and the host gneisses are more silicic and richer in potash, with a significant proportion of granodioritic material (see Table 2.2). The geochemistry of the gneisses of the Northern region led Sheraton *et al.* (1973) to conclude that they had not reached granulite facies during the Badcallian event.

These petrological and geochemical differences may be largely explicable by differences in original crustal level; the Assynt gneisses experienced granulite–facies metamorphism at more than 11 kb pressure, corresponding to depths of at least 35 km, whereas the gneisses to the N and S were metamorphosed at much lower pressures and represent originally higher crustal levels. A traverse from Scourie to, say, Loch Maree may represent a cross-section through the Archaean crust from the lowest crustal level in the N to a mid-crustal level in the S.

It is now generally believed that the bulk of the gneisses are of plutonic igneous origin as originally suggested by Peach *et al.* (1907). Geochemical studies (Weaver & Tarney 1980) indicate that the gneisses are essentially bimodal, where the two components display different petrogenetic characteristics. The mafic components show a range of Fe/Mg ratios and their trace-element patterns are consistent with low-pressure tholeiitic crystal fractionation (Fig. 2.8). The common association of ultramafic/mafic bodies with metasedimentary layers suggests that this material

Fig. 2.8a. Comparison between incompatible-element patterns for amphibolite–facies and granulite–facies tonalites.

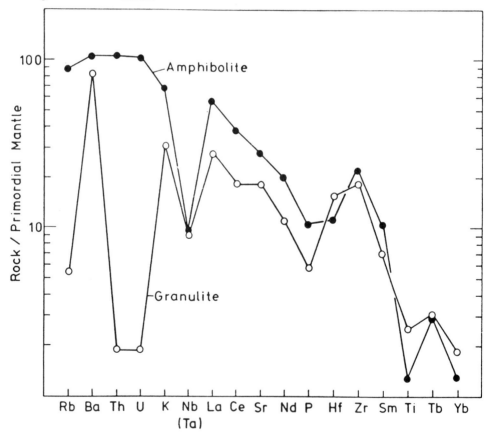

represents fragments of ocean-floor crust intercalated tectonically within the continental crust. The tonalitic to trondhjemitic gneisses on the other hand have rare-earth-element patterns consistent with partial melting of a mafic source under high-pressure hydrous conditions, in which hornblende and possibly garnet would be stable. Tarney & Weaver (1987) suggest that a subduction zone is the only environment where large volumes of mafic material could be melted in order to generate the tonalitic crustal material. They envisage a process of relatively shallow melting at a low-angle subduction zone (see Fig. 2.9a) where melts generated under hydrous conditions would yield relatively dense tonalitic magmas which would solidify at deep levels and progressively thicken the crust by underplating. Thus the mafic/metasediment (oceanic crust) association would first experience a high-grade metamorphic phase at the base of the continental crust before being uplifted by further underplating. The severe tectonic disruption of the deeper parts of the complex may be attributed to long periods of ductile-shear deformation affecting the base of the accret-

Fig. 2.8b. Rare-earth element patterns for mafic, intermediate, tonalitic and trondhjemitic granulites. A, B from Tarney & Weaver (1987).

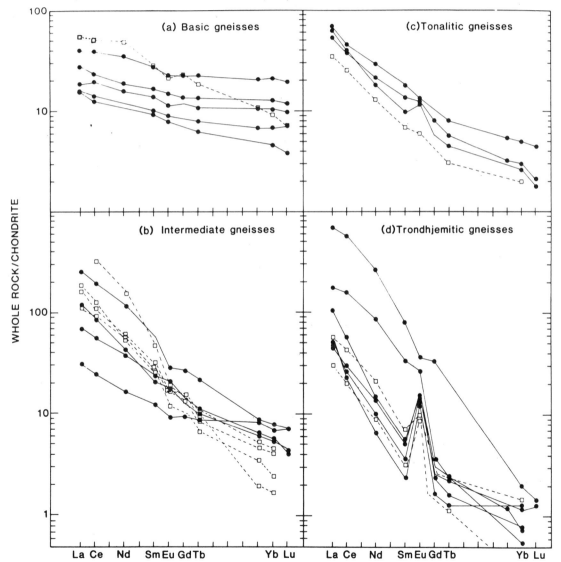

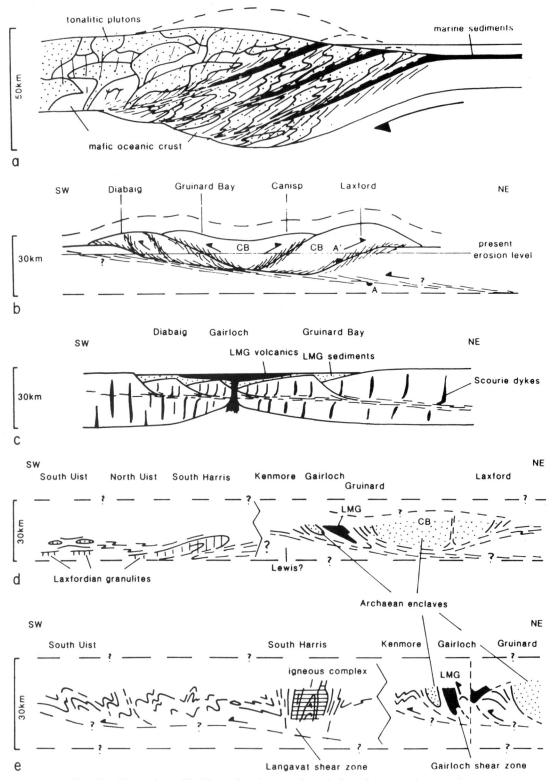

Fig. 2.9. Schematic profiles illustrating the tectonic evolution of the Lewisian complex. a, Badcallian; b, Inverian; c, emplacement of Scourie dykes and Loch Maree Group; d, Early Laxfordian; e, Late Laxfordian. After Park & Tarney (1987). CB, central block; LMG, Loch Maree Group. Note that in d, the transport direction is *into* the page.

ing crust as the underplating proceeded. This model would explain the otherwise puzzling concentration of mafic and sedimentary material in the deeper parts of the complex.

Late Scourian modifications

Peach *et al.* (1907) and Sutton & Watson (1951) considered that the fundamental chronological break within the Lewisian complex was represented by the intrusion of the Scourie dykes; but it subsequently became clear from the work of Tarney (1963), Park (1964), Evans & Tarney (1964) and Evans (1965) that a major tectono-metamorphic event took place after the Badcallian granulite–facies event of the Central region and before the Scourie dyke emplacement. This event was named the *Inverian* by Evans (1965) and has now been generally recognised throughout much of the Lewisian complex. The similarity of structural style and orientation, and of metamorphic facies, between the Inverian and the later Laxfordian has led to considerable confusion and debate. Structures can only be confidently

Fig. 2.10. Simplified map of the Northern region and the transitional zone in the N of the Central region, showing the principal structures and the boundaries of the major Inverian and Laxfordian shear zone. Dotted lines represent trend of banding; Scourie dykes in black; Laxfordian granite sheets hachured. Note generalised dip of the planar fabric and plunge of linear fabric in the Laxford shear zone. After Watson (1983).

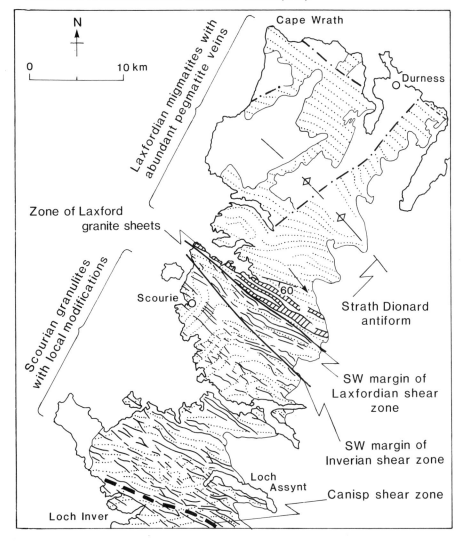

assigned to one or the other event where Scourie dykes can be seen to either cut or be affected by the structures in question.

Park (1970) suggested that the division between the early Scourian (Badcallian) and the Inverian events represented the major tectonic break within the Lewisian timespan, and this break is now recognised as the Archaean–Proterozoic boundary in Scotland (see Table 2.1). Nevertheless the simple subdivision into Scourian and Laxfordian is in some ways a more useful field classification, and is therefore still generally retained.

The Inverian can only be dated in relation to intrusive events before and after the metamorphism. Pre-Inverian pegmatites at Scourie yield a Rb–Sr age of *c.* 2,500 Ma (Giletti *et al.* 1961, Evans 1965, Evans & Lambert 1974) and the earliest dated Scourie dykes are *c.* 2,400 Ma in age

Fig. 2.11. Simplified map of the Southern region showing the principal structures. After Park *et al.* (1987). Inset map shows location; NR, Northern; CR, Central; SR, Southern regions.

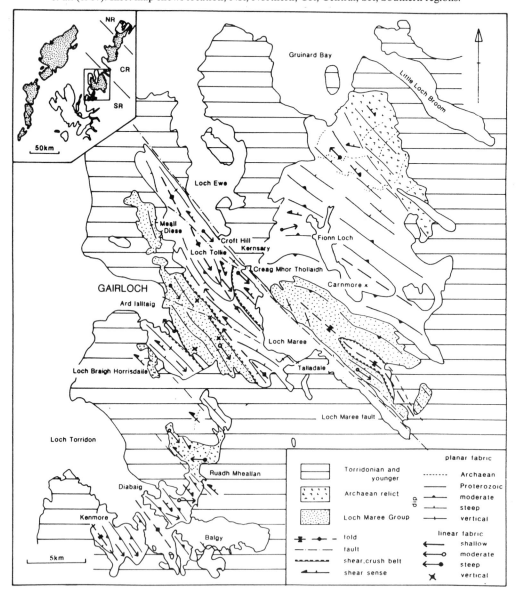

(Chapman 1979). According to Humphries & Cliff (1982) ultimate closure of the Sm–Nd isotopic system in the granulites at 2,490 Ma is associated with uplift, marking the commencement of Inverian tectonic activity. However, as Tarney (1963) recognised, the metamorphism affecting the shear zones into which the Scourie dykes were intruded persisted during the emplacement of at least the earlier members of the suite, so that the Inverian event overlaps the dyke emplacement.

Confusion is therefore inevitable between Inverian and Laxfordian events, since on the basis of the Sutton & Watson (1951) definition, all post-dyke events are Laxfordian. The likelihood is that deformation and metamorphism persisted intermittently at this crustal level from before the first dykes were emplaced at *c.* 2,400 Ma until the main Laxfordian event at *c.* 1,700 Ma. It is not yet possible to isolate the Inverian part of this activity by radiometric dating, since the systems did not become closed until 1,700 Ma. Until more dates become available, it is probably more useful to restrict the term 'Inverian' to cover events up to 2,400 Ma, and 'Laxfordian' for subsequent events, recognising the possibility that anomalies of nomenclature will arise where late Scourie dykes (*c.* 2,000 Ma, say) can be shown to cut early Laxfordian structures.

The Inverian structures of the mainland (see Fig. 2.9b) comprise three categories: (1) broad shear zones relatively unaffected by Laxfordian activity; (2) narrow shear zones cutting the Badcallian complex of the Central region; and (3) regions of inferred high Inverian strain intensely reworked during the Laxfordian. Major zones of type 1 have been recognised at the northern and southern margins of the Central region. The zone at the northern margin of the Central region (Fig. 2.10) is approximately 4 km wide and extends from near Scourie to Loch Laxford (see Beach *et al.* 1974, Davies 1978). It corresponds to the Claisfearn and Foindle zones of Sutton & Watson (1951). On its SW side, it cuts undeformed Badcallian structures and causes retrogression of the granulite–facies gneisses to amphibolite facies. On its NE side, it become intensely affected by Laxfordian deformation in the Laxford shear zone, and a zone of inferred Inverian high strain extends from there to Durness on the N coast, co-extensive with but obscured by, the Laxfordian belt of the Northern region.

The second type 1 zone occurs at the SW margin of the Central region between the Gruinard river and Fionn Loch, with a width of about 8 km (Fig. 2.11). This zone is a mirror image of the first, being overprinted and obscured by the major Laxfordian belt of the Southern region on its SW side (see Crane 1978). As in the Northern region, a zone of high Inverian strain appears to be generally co-extensive with the Laxfordian belt of the Southern region (see Park *et al.* 1987).

The third major Inverian shear zone is the 1–2-km-wide Canisp shear zone (Tarney 1963, Evans 1965, Attfield 1987) which cuts through the middle of the Central region in the Assynt district (see Fig. 2.3). This zone has also been subject to considerable Laxfordian activity.

These major zones run NW–SE and exhibit tight folding on NW–SE steep axial planes, with associated penetrative fabrics produced during amphibolite–facies retrogression. This metamorphism was accompanied by an influx of volatiles and by metasomatic activity (see Beach 1976, Beach & Tarney 1978).

The minor zones cutting the Badcallian gneisses of the Central region (i.e. type 2) are similar to the major zones but vary more widely in trend. They are usually marked by the development of monoclines with steep thinned limbs, in which new planar and linear fabrics are developed in association with tight minor folding of the pre-existing banding.

Thus, in Inverian times, the mainland Lewisian complex consisted of a central stable block, cut by many minor steep zones and by the larger Canisp zone, and bounded on both sides by major steep NW–SE zones. Coward & Park (1987) have suggested that these zones may connect at depth with a major flat-lying zone at a mid-crustal level, underlying the Central region (Fig. 2.9b). The sense of movement on these major belts indicates an overall reverse dip–slip (overthrust) movement with a small dextral component, indicative of a dextral transpressive tectonic regime.

The Scourie dyke swarm

The emplacement of the Scourie dykes was one of the most remarkable events in the evolution of the Lewisian complex. The dykes cover a minimum area of 120×250 km. Large volumes of mafic magma were emplaced, and in areas adjacent to the outcrop of the Loch Maree supracrustal belt at Gairloch, up to one-third of the surface area of the complex is occupied by them. They are typically steep, with a NW–SE to E–W trend, and for the most part appear to have been emplaced dilationally, implying considerable crustal extension. In the region S of the Gruinard river, the dykes show clear signs of intrusion under dextral shear (Fig. 2.12, Park *et al.* 1987) and this condition may well be general.

The dykes decrease in abundance northwards towards Durness, and are apparently absent in Coll and Tiree to the S. They are thickest and most numerous between Gruinard Bay and Torridon in the Southern region, where they are also significantly affected by pre-existing structure within the areas affected by the Inverian (see Park & Cresswell 1972, 1973).

The petrography, geochemistry and petrogenesis of the dykes have been investigated in detail. Tarney & Weaver (1987) define four petrological/geochemical types: bronzite–picrites, norites, olivine–gabbros, and quartz–dolerites. The quartz–dolerites are by far the most abundant and correspond to the main 'epidiorite' suite recognised by Peach *et al.* (1907). There is evidence in the Central region of emplacement at depth into hot country rock (O'Hara 1961, Tarney 1963).

Geochemically, the dykes show enrichment of light rare-earth and large-ion lithophile elements, but with prominent

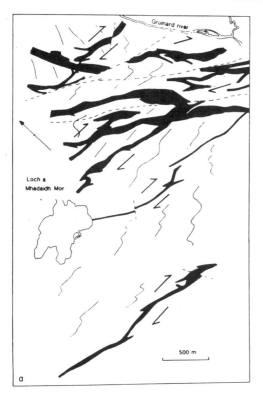

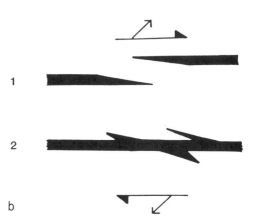

Fig. 2.12. a. Dyke outcrop pattern S of Gruinard River, in the northern part of the Southern region, showing evidence for emplacement during dextral shear. **b.** Criteria indicating dextral shear during emplacement: (1) en-echelon arrangement; (2) inclined apophyses. After Park *et al.* (1987).

Nb anomalies (Fig. 2.14, Weaver & Tarney 1981). These characteristics are typical of island-arc lavas, and of crustal material generally, and are ascribed by Weaver and Tarney to inheritance from the sub-continental lithosphere. The geochemical evidence indicates that the dyke magmas may have been derived from at least two distinct sources; one more Mg-rich and refractory, producing the picrites and

norites, and the other more Fe-rich and fertile, yielding the olivine-gabbros and quartz-dolerites. Tarney and Weaver suggest that the sub-continental lithosphere may have become enriched in crustal components during the generation of the Scourian crust in the Badcallian event at *c.* 2,900 Ma. They note that similar geochemical features characterise other Early Proterozoic dyke swarms and also Phanerozoic flood basalts.

The Rb–Sr date of *c.* 2,400 Ma obtained by Chapman (1979), which is a combined isochron from three typical quartz–dolerites from Scourie and Kylescu (13 km SE of Scourie), is generally considered to represent the date of intrusion of the main part of the swarm. However no emplacement ages have been published from the dykes in the very extensive areas of Laxfordian reworking, and therefore the assumption of contemporaneity of the main 'epidiorite' swarm remains to be tested. The older K–Ar and Rb–Sr dating of Evans & Tarney (1964) gave a range of ages interpreted as indicating a date of *c.* 2,200 Ma for the emplacement of the main swarm and *c.* 2,000 Ma for two younger alkali–basalt and tholeiite dykes. In addition, Humphries & Cliff (1982) obtained a date of 2,260 Ma on a thin dyke at Scourie from a Sm–Nd isochron. These data are consistent with the interpretation that the main Scourie dyke swarm was emplaced around 2,400 Ma ago, at depths of 10–20 km, during the Inverian metamorphism (see Dickinson & Watson 1976) and that certain members of the

Fig. 2.13. Undeformed Scourie dyke cutting granitic gneiss, near Shieldaig, Gairloch. R. G. Park photograph.

swarm were emplaced much later (*c.* 2,000 Ma?) into cooler crust in the Central region.

This interpretation is supported by recent U–Pb baddeleyite dating by Heaman & Tarney (1989).

In the Southern region there is evidence both of continuing tectonic activity within the Inverian shear zone during dyke emplacement, and of dykes cutting supracrustal rocks of the Loch Maree Group dated at *c.* 2,000 Ma (see below). This evidence supports the indications from the Assynt dating work that the dyke emplacement spanned a very long period of time (perhaps 400 Ma) and suggests that the earlier and later members of the swarm may be separated by tectonic activity and by the deposition of the Loch Maree Group.

The possibility that the Scourie dyke suite may consist of two or more chronologically separate swarms separated by major tectonic events is put forward by Bowes (1968).

However, much of the evidence on which this view was based could be explained in other ways (see Park 1970), and there is no undisputed evidence of deformed Scourie dykes cut by undeformed. Nevertheless, in view of the geochronology, it is likely that such tectonic activity did take place, and may ultimately be proved by more adequate dating.

The Loch Maree Group

Large tracts of metasediment occur within the Lewisian complex of Loch Maree and Gairloch (Southern region, see Fig. 2.11) as two narrow belts with a combined outcrop area of about 130 km^2 (Peach *et al.* 1907). The Loch Maree outcrop occupies a late Laxfordian synform, and the Gairloch outcrop forms a steep NW–SE belt within a wide shear zone. Both outcrops exhibit intense polyphase deformation. The metasediments were first described by

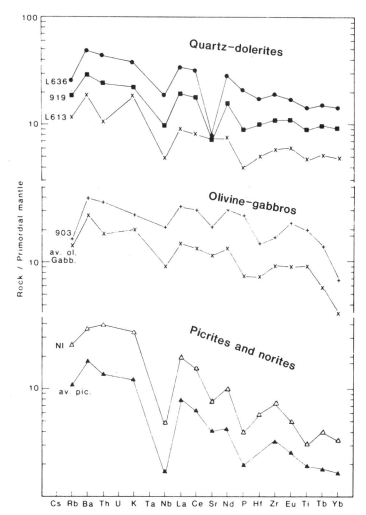

Fig. 2.14a. Trace element abundance patterns, normalised to primordial mantle. After Weaver & Tarney (1981).

Hicks (1880) and detailed descriptions are provided by Clough (*in* Peach *et al.* 1907) and Park (1964).

Peach *et al.* were unable to decide whether the sediments were older than the igneous rocks of the fundamental complex, or were unconformable upon them (see discussion in Peach & Horne 1930). They noted that the boundaries were tectonically modified and that gneisses appeared to be thrust over metasediments at Loch Maree. They also found no examples of Scourie dykes cutting the metasediments, although amphibolite sheets of Scourie dyke type have subsequently been found (see Johnson *et al.* 1987). Park (1964) investigated the Loch Maree supracrustal rocks at Gairloch and concluded from comparison of their structural and metamorphic history with that of the neighbouring Ialltaig gneisses, that the sediments were younger. Subsequently Park (1965) assigned the whole of the acid gneiss complex of the Gairloch district to a basement, to which the Loch Maree Group formed a supracrustal cover. This view was eventually confirmed by a *c.* 2,000 Ma Sm–

Nd model age based on the maximum provenance age of the clastic metasediments (O'Nions *et al.* 1983). Other published data represent either Laxfordian metamorphic ages or probable compromises between the provenance and metamorphic ages. The younger age limit of the supracrustal formation is usually taken to be the *c.* 1,800 Ma date of the Laxfordian metamorphism, although the probable age is *c.* 2,000 Ma, contemporaneous with the younger Scourie dykes.

The supracrustal assemblage consists of a thick sequence of amphibolites, of probable volcanic origin (Park 1966, Johnson *et al.* 1987), intercalated with semipelitic quartz–biotite–plagioclase schists and minor layers of marble, banded-iron formation and graphitic schist. The semipelitic schists are chemically equivalent to greywacke (Table 2.3). Floyd *et al.* (1989) interpret the chemical variation in light-ion lithophile- and rare-earth elements in terms of mixing between Lewisian basement with an acid to intermediate, continental-crust composition, and amphibolites with a

Fig. 2.14b. Chondrite-normalised rare-earth-element patterns, for the main types of Scourie dykes. After Weaver & Tarney (1981).

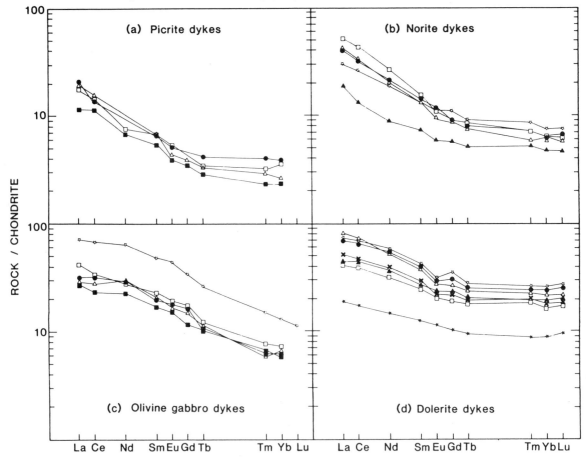

primitive tholeiitic composition, regarded as a probably local volcanogenic source. The amphibolites are geochemically divisible (Fig. 2.15) into a major group of mid-ocean-ridge (MORB) type, oceanic, tholeiitic basalts, and a minor group consisting of thin amphibolite sheets with light-rare-earth- and light-ion–lithophile-element enrichment trends similar to those of the Scourie dykes (Johnson et al. 1987).

Several extensive stratiform sulphide deposits occur within the metavolcanics; one of these contains significant amounts of Cu, Zn and Au, and has been prospected in detail (Jones et al. 1987).

The geochemical work on the supracrustal rocks suggests that the Loch Maree Group represents the fill of an early Proterozoic extensional rift basin (see Fig. 2.9c) in which a phase of early, relatively rapid, extension was marked by voluminous outpourings of primitive tholeiitic basalts, and a later phase by intrusive sills within the supracrustal pile, fed by an enriched magma source. The relationship between the early primitive tholeiites of the Loch Maree Group and

Table 2.3. Representative analyses of Lewisian metasedimentary rocks

	1	2	3	4	5	6	7	8	9
SiO₂	54·31	68·22	8·7	11·3	65·45	8·7	4·5	54·65	53·77
TiO₂	1·15	0·48	0·10	0·08	0·77	0·10	0·04	0·07	0·69
Al₂O₃	20·33	15·98	1·6	1·1	13·07	1·6	0·36	0·24	4·66
Fe₂O₃	0·10	0·18	—	—	—	—	—	22·77	0·75
FeO	9·12	3·45	0·75*	0·93*	5·83*	0·75*	1·6*	16·89	24·49
MnO	0·18	0·08	0·14	0·10	0·05	0·14	0·25	0·84	6·60
MgO	2·53	1·74	4·1	18·3	2·98	4·1	17·2	0·75	4·74
CaO	2·99	2·61	45·7	31·8	2·36	45·7	31·8	0·94	0·99
Na₂O	4·13	3·78	0·43	0·04	2·79	0·43	0·03	0·07	0·04
K₂O	2·61	2·48	0·45	0·33	2·73	0·45	0·05	—	—
P₂O₅	0·10	0·07	0·05	0·11	0·18	0·05	0·04	0·63	0·26
F	—	—	0·10	0·08	—	0·10	0·04	—	—
S	—	—	0·09	0·12	1,583	0·09	0·04	—	—
Cl	—	—	—	—	342	—	—	—	—
Ag	—	—	3	1	—	3	1	—	
Ba	891	1,993	95	153	646	95	898	20	29
Ce	—	—	13	17	65	13	9	—	—
Co	—	—	5	2	—	5	3	—	—
Cr	303	77	10	7	220	10	5	62	117
Cu	178	49	41	5	26	41	8	—	—
Ga	—	—	8	—	13	8	5	—	—
La	—	—	12	3	28	12	4	—	—
Mo	—	—	1	1	—	1	1	—	—
Nb	—	—	0	<1	12	0	<1	13	12
Ni	185	55	13	7	54	13	4	14	65
Pb	—	—	4	2	31	4	10	6	11
Rb	37	36	7	8	86	7	2	—	—
Sb	—	—	1	2	—	1	2	—	—
Sn	—	—	3	<1	—	3	<1	—	—
Sr	394	473	178	123	164	178	212	5	6
Th	15	4	<1	<1	—	<1	<1	—	—
U	—	—	1	2	—	1	<1	—	—
V	264	68	3	10	111	3	6	6	74
Y	31	20	5	—	28	5	2	6	27
Zn	157	49	7	17	97	7	14	—	
Zr	206	263	13	15	241	13	4	—	63

* Total Fe as FeO.

Scourian metasediments: (1) pelitic gneiss, Scourie (average of 2 analyses); (2) semipelitic gneiss, Scourie (average of 2 analyses); (3) calc. marble (average of 9 analyses); (4) dolomitic marble (average of 26 analyses). *Post-Scourian metasediments:* (5) semipelitic schist, Loch Maree group (average of 9 analyses); (6) calc. marble, Loch Maree group (average of 9 analyses); (7) dolomitic marble, Loch Maree group (average of 17 analyses); (8) banded iron formation (oxide facies), Loch Maree group (average of 4 analyses); (9) banded iron formation silicate facies (average of 4 analyses).

Sources: 1, 2, Okeke et al. (1983); 3, 4, 6, 7, Rock (1987); 5, Floyd et al. (1989); 8, 9, Al-Ameen (1979.

the adjacent very numerous and densely packed Scourie dykes within the basement is unclear. There are at least three possibilities.

(1) The Scourie-type dykes in the basement may be wholly unrelated and much earlier (*c.* 400 Ma earlier), in which case their massive concentration around Gairloch is co-incidental.

(2) Some (or possibly all) of the Scourie dykes of the Loch Maree–Gairloch district may be younger than those of the Central region and related to the crustal extension immediately preceding the deposition of the Loch Maree Group (i.e. *c.* 2,000 Ma).

(3) Some (or possibly all) of the Scourie dykes of this district may be younger than the Loch Maree Group (younger than *c.* 2,000 Ma) and correspond to the younger set that cut the Loch Maree Group. Until more geochronological work is done in the Southern region, it is not possible to distinguish between these three possibilities, although the second is currently favoured.

The Loch Maree Group shares the same Laxfordian deformational and metamorphic history as the basement

gneisses of the Southern region but has not been migmatised or veined by granite or pegmatite. The metamorphic assemblages are typical of middle to upper amphibolite facies, with the development of almandine, aluminous hornblende and oligoclase–andesine in the amphibolites, clinopyroxene in calc–silicates and grunerite–garnet in the iron-formation. Extensive retrogression to greenschist and lower facies is seen, with widespread development of chlorite, albite, epidote, muscovite and stilpnomelane.

Laxfordian modifications and younger events

The problem of naming those events post-dating the early Scourie dykes and preceding the later has already been discussed. In practice, structures and metamorphic effects affecting any of the Scourie dykes have been regarded as Laxfordian, following Sutton & Watson (1951). Laxfordian modifications on the mainland can be simply divided into an earlier set associated with amphibolite–facies metamorphism and emplacement of granites and pegmatites, and a later set accompanied by retrogressive alteration to greenschist facies or lower (see Table 2.1). In South Harris, a granulite–facies metamorphism is followed by an amphibolite–facies retrogression, but is not distinguishable in other areas, where the Laxfordian metamorphism apparently produced only amphibolite–facies assemblages. On the mainland, the earlier Laxfordian deformations produced fabrics in the Scourie dykes associated generally with amphibolite–facies recrystallisation of the original igneous assemblages. In many parts of the mainland, recrystallisation has occurred in the absence of deformation, producing the typical 'epidiorite' dyke textures of the Central region. It is possible that this static recrystallisation was general in the mainland, representing a continuation of the Inverian metamorphic event, and continued until the first Laxfordian deformation took place.

The theoretical older age limit to the earliest Laxfordian events is thus 2,400 Ma, the age of the earlier Scourie dykes; however, the main phase of widespread Laxfordian deformation post-dates the Loch Maree group, and is therefore post- *c.* 2,000 Ma in age (see also the *c.* 1,800 Ma age of the early metamorphism in South Harris, discussed below). A younger age limit is given by the granite/pegmatite suite (which is post-tectonic to the early structures) dated at *c.* 1,700 Ma (van Breemen *et al*. 1971, Lyon *et al.* 1973, Taylor *et al.* 1984). The end of the main (high-grade) Laxfordian metamorphism is well dated by a concentration of metamorphic ages, including Rb–Sr whole-rock, lead-isotope and zircon ages in the range 1,860–1,630 Ma, but concentrating around a date of about 1,700 Ma, corresponding to the granite and pegmatite emplacement (Lambert & Holland 1972, Lyon *et al.* 1977). K–Ar ages on hornblendes, biotites and muscovites fall in the range 1,750–1,560 Ma in the Northern region; however, the much larger data set of hornblende ages for the Southern region may be divided into an older group of ages around 1,700 Ma, correlated with the main Laxfordian metamorphic

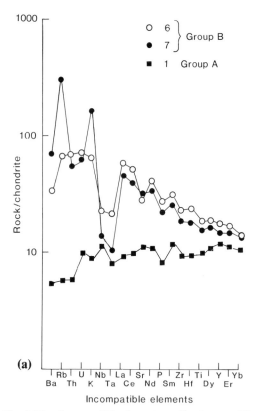

Fig. 2.15a. Incompatible-element profiles for amphibolites of the Loch Maree Group, normalised to chondrite values. After Johnson *et al.* (1987).

phase, and a younger group around 1,500 Ma, correlated with the later, retrogressive, phase (Moorbath & Park 1972). Biotites and muscovites give rather younger ages, around 1,400 Ma.

The earlier Laxfordian structures are heterogeneous. In the Central region, they are confined to narrow shear zones, generally of the order of m in width, with the exception of the Canisp shear zone where the belts of Laxfordian reactivation are much wider (see Attfield 1987). The main Laxfordian belts are situated in the Northern and Southern regions, as previously explained (see Fig. 2.3). In both regions, the first Laxfordian deformation is generally subconcordant with the Scourie dykes, progressing generally

from narrow marginal zones, eventually to encompass the whole width of the dyke, and spreading out into the host gneisses. The development of the first Laxfordian fabric is associated with variable, locally very large, strains, and is typically steep near the margins of the Laxfordian belts. In the Southern region, Park *et al.* (1987) show that this fabric, formed during the D1 phase of the Laxfordian deformation, is folded and becomes flat-lying between Carnmore and Gairloch, and S of Loch Torridon (Figs. 2.9d, 2.16). The trend of linear fabrics in the zones of high strain is consistently NW–SE to E–W, after allowing for the effects of subsequent refolding, implying that the movement direction during the early Laxfordian had a large strike-

Fig. 2.15b. Rare-earth-element profiles for amphibolites of the Loch Maree Group, normalised to chondrite values. Note marked differences between profiles for bodies 1–5 (group A) and bodies 6 and 7 (group B). After Johnson *et al.* (1987).

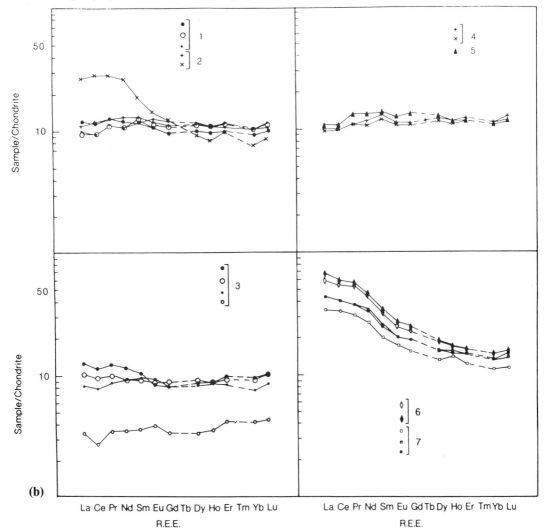

parallel component. The Northern region is similar; the fabrics in the Laxford shear zone are steeply inclined to the SW at Loch Laxford, but become flat-lying further N and are refolded by younger Laxfordian structures (Figs. 2.9d, 2.10).

The Laxford shear zone of Beach *et al.* (1973) and Davies (1978) is steeply inclined. It separates the very deep-seated Scourian complex to the S from the quite different, undepleted gneisses to the N, which represent much higher levels in the original Scourian complex (Davies & Watson 1977). The net movement on the shear zone therefore appears to involve a large component of thrusting to the

N. The principal component of movement in the early Laxfordian deformation however, as in the Southern region, is NW–SE, parallel to the strike, and there is also a significant normal-slip component, down to the SW (see Coward & Park 1987). The thrust-sense movements are thought to be largely Inverian in age. Rocks with similar physical properties to the granulites of the Central region are inferred at depths of <4 km beneath the Laxfordian complex of the Northern region (Hall 1987). These rocks represent the footwall to the gently inclined section of the Laxford shear zone.

In the zones of highest strain, original discordances

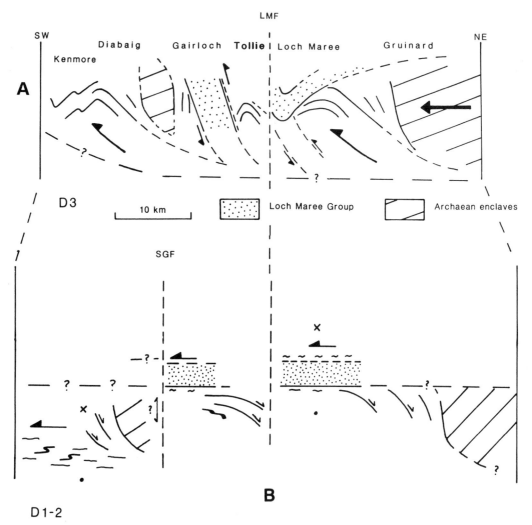

Fig. 2.16. Schematic structural profiles across the Southern region during D3 (A) and D1–2 (B) to show the interpretation of the shear-zone movements responsible for the respective structures. Note that in B, the movement direction is largely into the page (towards NW).

between dykes, gneisses and pegmatite veins become much reduced or even obliterated, and the dykes are markedly thinned. Unlike the Laxfordian of the Outer Hebrides, however, boudinage is uncommon. Thorough reworking of the Scourian complex produces a finely banded 'Laxfordian' gneiss containing concordant amphibolite sheets (produced from Scourie dykes), and pervaded by granitic migmatite of Laxfordian age (see Fig. 2.17). Thoroughly 'Laxfordianised' gneisses of this kind are confined to the Northern region, N of Loch Laxford, and to the southern-

Fig. 2.17. Typical finely banded gneiss, affected by intense Laxfordian deformation, with discordant granite sheets. Near Badcall, Loch Laxford. BGS photograph.

most parts of the Southern region round Kenmore, and on the islands of Rona and Raasay. In the Outer Hebrides, however, such gneisses are much more widespread. On the mainland, large bodies of granite are confined to the Loch Laxford area. Here several thick sub-concordant sheets of pink, gneissose granite occur within a zone 2–4 km wide. The granites contain both oligoclase and microcline, in addition to minor amounts of quartz, biotite, occasional hornblende, and haematite. N of the granite sheets, narrow bands of granite and pegmatite are abundant. The granites are typically foliated, and petrographically similar to the large sheets. The pegmatites are unfoliated, and contain albite or oligoclase, biotite, quartz and magnetite.

The main Laxfordian belts of the mainland, according to Coward & Park (1987), are linked in a major mid-crustal shear zone network which separates and encloses more stable crustal blocks, whose relative movement gives rise to the observed structures. The D1 and D2 deformations recognised in the mainland Laxfordian sequence probably represent earlier and later stages of progressive deformation involving transport towards the NW of higher-level

crustal blocks relative to lower, on a major sub-horizontal mid-crustal shear zone. The major shear zone is presently exposed in the N and S of the mainland, passing beneath the Central region, and is more widely represented in the Outer Hebrides (p. 50) where a lower Laxfordian crustal level appears to be represented (see Fig. 2.9d).

The later Laxfordian structures are prominent NW–SE major folds that dominate the outcrop pattern of the Laxfordian belts in both the Northern and Southern regions. In the N, the Strath Dionard antiform (Fig. 2.10; Dash 1969) and in the S, the Carnmore, Tollie and Torridon antiforms and the Letterewe synform (Fig. 2.11; Park *et al.* 1987) are of this generation. In the Southern region, Park *et al.* have linked these major folds with the appearance of a new, locally developed, planar fabric accompanied by retrogression to greenschist facies, with the development of albite–epidote–actinolite assemblages in amphibolite, and albite–epidote–biotite–chlorite–muscovite assemblages in acid gneisses. These retrogressive assemblages have yielded the *c.* 1,500 Ma K–Ar dates associated with the later Laxfordian metamorphism (Moorbath & Park 1972). In Gairloch, these structures are attributed to the Laxfordian D3 deformation, and are associated with the formation of the major Gairloch shear zone (Odling 1984, Park *et al.* 1987) which is about 6 km wide. The sense of movement on this shear zone is dextral, NE up, indicating an important kinematic change from the NW–SE transport direction of the earlier Laxfordian period to a dextral transpressional regime resulting in the refolding of the previously flat-lying Laxfordian fabrics.

The regional D3 Laxfordian folds were superseded by later, more localised structures of various styles and orientations, together with crush zones, often containing pseudotachylite. These later structures were assigned to the 'late phase' of the Laxfordian by Park (1964) in Gairloch, and have been recognised and described by subsequent workers in the Laxfordian belts (e.g. see Bhattacharjee 1968, Dash 1969, and Cresswell 1972). The folds are typically steeply plunging, and vary from cm to km in size. Several large km-scale folds in the Gairloch and Torridon areas have sinistral asymmetry, and Park *et al.* (1987) suggest that the late deformation occurred in response to sinistral strike–slip movements, possibly around 1,400 Ma, to correspond with the period of significant closure of K–Ar systems in biotites and muscovites.

Two younger K–Ar dates of 1,148 Ma and 1,169 Ma were obtained by Moorbath & Park (1972) from chloritised biotite in acid gneisses from Torridon. These ages are close to a biotite Rb–Sr age of 1,160 Ma reported by Giletti *et al.* (1961), and suggest the possibility that some of the later structures in the Lewisian complex (e.g. certain crush belts) may result from movements at around 1,100 Ma or later. Such movements are likely to relate to the Mid-Proterozoic Grenville–Sveconorwegian orogeny. This orogeny resulted from continental collision at about this time, and affects a broad belt which projects across Scotland not far S of the mainland Lewisian outcrop (see Fig. 2.22).

The Outer Hebrides

The Scourian complex

The islands of the Outer Hebrides collectively form the largest Lewisian outcrop in NW Scotland (see Fig. 2.3) and the island of Lewis gave its name to the complex. Those rocks received very little attention from the early geologists, and their chronology was more difficult to interpret because of the absence of large areas of unmodified Scourian. For these reasons, the main advances in understanding of the Lewisian complex came through studies on the mainland. The early descriptive work was carried out by Jehu & Craig (1923, 1925, 1926, 1927, 1934) and by Dearnley (1962, 1963). The latter extended to the Outer Hebrides the chronological subdivision established by Sutton & Watson (1951) on the mainland. Detailed mapping by the Geological Survey carried out in the 1970s was published at a scale of 1 : 100,000 in 1981 (see Fettes et al., in press) and a useful summary of the geology of the islands is given by Fettes & Mendum (1987).

Like the mainland Lewisian, the bulk of the Outer Hebrides outcrop is composed of rocks of the Scourian complex, although post-Scourian granites make up a significant proportion of the outcrop area. Amphibolite sheets correlated with the Scourie dyke swarm, on the other hand, are generally less numerous, smaller, and more deformed and disrupted than their mainland counterparts. Two groups of rocks have apparently no precise equivalent on the mainland. These are the South Harris igneous complex and the Corodale gneiss, both apparently of Early Proterozoic age.

The dominant rock types, as on the mainland, are tonalitic to granodioritic gneisses ('grey gneiss') containing a wide variety of ultramafic, mafic and acid bodies in the form both of inclusions and of later intrusions. The general character of these gneisses is more reminiscent of the Laxfordian belts of the mainland than of the Central region. The gneisses typically contain biotite and/or hornblende as their principal ferromagnesian constituent. Pyroxene-bearing varieties are confined to relatively small areas in Barra and South Uist (Coward 1973) where granulite–facies assemblages are preserved, showing varying degrees of retrogression to the amphibolite–facies assemblages typical elsewhere. Large areas of migmatite surround the Laxfordian granites of Lewis and South Harris, and are generally more common than on the mainland.

The geochemistry of the gneisses is similar to that of the mainland gneisses. Depletion of lighter elements resulting in high values of K/Rb, Ca/Sr, Ca/Y and Ba/Rb, and low values of K/Sr, K/Ba, Rb/Sr and Ba/Sr ratios are characteristic of the Uist gneisses (Fettes et al., in press). A systematic change is noted by Fettes et al. from SSE to NNW in certain element ratios: an increase in K/Ba and a decrease in K/Rb, Ca/Sr, Ca/Y and Ba/Rb (see Table 2.2). The authors believe that the pattern of variation in lithophile elements is

original and not due to Laxfordian metamorphism and migmatisation. This variation is analogous to that shown in the Scourian complex between the Central region of the mainland and the Laxfordian belts. The gneisses of E Uist are not as strongly depleted as those of the Scourie–Assynt area, but are intermediate in their chemistry between these and the Lewis gneisses.

The Outer Hebrides differs from the mainland in containing a significant proportion of supracrustal rocks within the Scourian complex. The distribution of these rocks is described by Coward et al. (1969) and the main outcrops are shown on the BGS maps. They comprise perhaps 5 per cent of the total outcrop area (see Fig. 2.3) and include quartzites, marbles, graphitic schists, kyanite- or sillimanite–garnet–pelites, quartzo–feldspathic gneisses, magnetite-rich gneisses, and fine-grained, banded amphibolites interpreted as metavolcanics. Many of these metasediments exhibit a distinctive rusty weathering due to the abundance of pyrite.

The largest and best preserved outcrops of metasediment occur in two belts at Langavat and Leverburgh in South Harris, flanking the South Harris igneous complex (Fig. 2.18). Elsewhere, they occur in the form of thin bands or lenses, a few m in width, but in zones traceable for several km. The banded basic rocks associated with the metasediments are also composed of isolated bodies usually tens of m in size and also traceable for large distances along strike. These bodies display banding on all scales from cm to m, and range in composition from ultramafic to anorthositic. They contain hornblende, garnet (clinopyroxene) and plagioclase, and their chemistry shows a strongly iron-enriched trend typical of modern tholeiites. The association is similar to that of the ultramafic/mafic bodies of the mainland, and presumably originated in the same way, as fragments of ocean-floor crust.

The arrangement of these supracrustal bands, together with the pattern of geochemical and gravity variation (see Fettes & Mendum 1987, fig. 1) indicates a strong NNE–SSW tectonic grain in the Outer Hebrides, apparently predating the Laxfordian deformation. Two possible explanations for this metamorphic and geochemical pattern are discussed by Fettes and Mendum. Either the variation is an original (Badcallian) feature reflecting differences in crustal level from SSE to NNW, or it is due to post-Badcallian (Inverian or Laxfordian) metamorphic/migmatitic effects. A strong argument in favour of the former interpretation (Moorbath et al. 1975) is that the gneisses of South Uist suffered extreme depletion like those of Barra, despite subsequent retrogression to amphibolite facies. It is believed therefore that Badcallian granulite–facies rocks were confined to South Uist and Barra at the present level of exposure, in the same way that the mainland equivalents were confined to the Central region, and that the bulk of the Outer Hebrides gneisses, like those of the Northern and Southern regions of the mainland, represent originally higher crustal levels of the Scourian complex. Fettes & Mendum (1987) suggest that much of the migmatisation seen in the west of the Outer Hebrides may be of Badcallian age, and represents the high-level consequences of fluid migration from the deeper levels undergoing granulite–facies metamorphism.

The granulite–facies metamorphism exhibited by the South Harris metasediments differs from the other occurrences of granulite–facies rocks in being of Laxfordian age, related possibly to the local effect of the igneous complex (Fettes & Mendum 1987).

The early deformation of the Scourian complex, like that of the mainland, is represented by the intense foliation accompanying the Badcallian metamorphism. Fettes *et al.* (in press) recognise an earlier foliation confined to the metasediments and predating the emplacement of the plutonic igneous precursors of the acid gneiss complex.

Late Scourian modifications

The period after the end of the Badcallian metamorphism and deformation, and before the intrusion of the Scourie

dykes, was marked by the emplacement of a suite of 'late Scourian intrusions' and by metamorphism and deformation of Inverian age. The late Scourian intrusions consist of two distinct widespread suites, the first consisting of diorites, monzodiorites and microdiorites (Table 2.2), and the second of potash-rich granites, monzonites and pegmatites. Both exhibit typical alkaline trends and contain plagioclase, K-feldspar, hornblende, biotite, clinopyroxene and quartz in varying amounts. These rocks are best seen in Barra, where the effects of Laxfordian reworking are minimal. Here, a number of diorite dykes, cm to m in width, are cut by members of the younger basic dyke suite (correlated with the Scourie dykes). The pegmatites have yielded an Rb–Sr whole-rock age of 2,610 Ma (Moorbath *et al.* 1975). Further N, on the island of Fuday, one of the dioritic dykes is folded prior to the injection of an undeformed younger basic dyke. This outcrop demonstrates the only clear geochronological evidence in the Outer Hebrides of the Inverian event.

Other structures attributed to the Inverian by Fettes &

Fig. 2.18. Simplified map of the South Harris igneous complex and the Langavat shear zone. After Graham (1980).

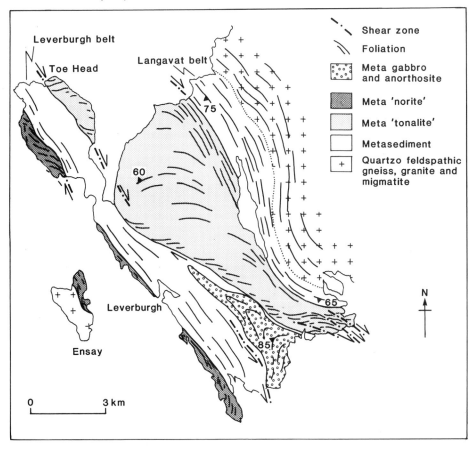

Mendum (1987) are regionally developed asymmetrical folds with steep NNE–SSW long limbs, and a number of sub-vertical NW–SE (and occasionally NE–SW) shear zones. Several of these shear zones are intruded by members of the younger basic dyke suite. The intensity of the Laxfordian deformation over most of the Outer Hebrides outcrop precludes any attempt to assess the regional extent or intensity of the Inverian event. However the absence of Inverian deformation in Barra suggests that, as on the mainland, areas of Badcallian granulite–facies rocks acted as resistant blocks during both Inverian and Laxfordian deformation.

The pattern of subsequent Proterozoic tectonic development was apparently established in the Inverian when undeformed Archaean blocks were separated by major steep shear zones. These influenced the emplacement of the younger basic dykes and acted as a focus for the Laxfordian deformation. Thus the Inverian marks a fundamental change from the regionally penetrative deformation of the Badcallian to localised deformation in major shear zones.

The younger basic (Scourie dyke) suite

Basic and ultrabasic intrusions believed to correlate with the Scourie dyke suite of the mainland are widespread throughout the Outer Hebrides, and are termed there the 'younger basics' by Fettes & Mendum (1987). They comprise three main groups: (1) ultrabasic bodies; (2) a group of noritic and picritic bodies termed the 'Cleitichean Beag dykes'; and (3) the abundant metadolerites of the main swarm.

The ultrabasic bodies comprise peridotites with subordinate dunites, and are generally massive. Their age relationship to the other two groups is uncertain. The Cleitichean Beag dykes contain two pyroxenes (olivine), hornblende, plagioclase and ore, and are concentrated in two E–W belts in Lewis and Harris. They are broadly similar, petrographically and chemically, to the norites and picrites of the mainland suite (see Tarney 1973) and have been dated at *c.* 2,400 Ma by Lambert *et al.* (1970) which correlates well with the date of the older mainland dykes (Chapman 1979). Members of this group are cut by metadolerites of group 3, but reliable intrusion dates of the latter are not available. The metadolerites typically have continental tholeiite chemistry with strong iron-enrichment trends similar to those of the mainland swarm (see Table 2.4).

The dykes exhibit a variety of textures and mineral assemblages interpreted as the result of emplacement and crystallisation under amphibolite–facies conditions at a deep crustal level (see Fettes *et al.*, in press). Original ophitic and sub-ophitic textures produced by the co-precipitation of clinopyroxene and plagioclase are modified by 'immediate' recrystallisation to granular metamorphic assemblages including garnet and orthopyroxene, which appear initially at crystal interfaces but result ultimately in complete recrystallisation to a granular texture. The in-crease in modal hornblende observed towards the margins of the dykes is interpreted as the effect of an increasing degree of equilibration with wall rock.

Ambient conditions at the time of intrusion are given by Fettes *et al.* (in press) as *c.* 700°C and 6–7 kb, equivalent to depths of *c.* 25 km (Fig. 2.7). These conditions differ significantly from those inferred at Scourie on the mainland, where O'Hara (1977) suggests a value of *c.* 450°C for the ambient temperature. Recrystallisation during the Laxfordian metamorphism resulted in equigranular hornblende–plagioclase assemblages. There is no indication of significant uplift of the Outer Hebrides between the end of the Badcallian event and the Laxfordian.

The South Harris igneous complex and the Corodale gneiss

The *South Harris igneous complex* (Fig. 2.18) is essentially a thick sheet of anorthosite and metagabbro, with associated bodies of tonalitic and pyroxene–granulitic gneiss. The complex was emplaced mainly within supracrustal gneisses. The granulite–facies metamorphic assemblages exhibited by the rocks of the complex suggested to some observers (e.g. see Dickinson & Watson 1976) a comparison with the Badcallian of the mainland, a comparison strengthened by the occurrence of narrow dykes of the younger basic suite within the complex. Dearnley (1963), on the other hand, regarded the complex as contemporaneous with the Scourie dykes, and this view has recently been confirmed by Sm–Nd geochronological dating by Cliff *et al.* (1983) who obtained an age of *c.* 2,200 Ma for the anorthosite (from whole-rock samples) but a much younger maximum age of 2,060 Ma for the tonalite. Mineral isochrons indicate a date of 1,860 Ma for the end of the granulite-facies metamorphism. Their work therefore supports Dearnley's interpretation of a Laxfordian age for the granulite–facies metamorphism, and suggests that deeper levels of the Laxfordian complex are preserved here compared with the mainland. The emplacement of the igneous complex falls within the broad age range (2,400–2,000 Ma) occupied by the Scourie dyke suite.

The *Corodale gneiss*, a large metagabbroic body metamorphosed in granulite facies, occupies almost the whole Lewisian outcrop in South Uist above the Outer Isles fault (see Coward 1972). It was tentatively correlated by Coward with the pyroxene-granulites of the South Harris complex on the grounds that it is cut by dykes of the younger basic suite and shows no signs of pre-dyke (Scourian) deformation.

Laxfordian modifications

The effects of Laxfordian deformation and metamorphism are much more widespread in the Outer Hebrides than on the mainland. Recrystallisation of the Scourie dykes to granular fabrics is general, and Laxfordian deformation is ubiquitous except in small areas in South Uist and Barra.

Fettes & Mendum (1987) recognise two metamorphic events, the early and late Laxfordian respectively, following Dearnley (1962). The 'early Laxfordian' metamorphism can be distinguished only in the South Harris igneous complex and in its host gneisses, where a high-pressure granulite–facies assemblage characterises a variety of meta-igneous and metasedimentary rocks (see Dearnley 1963). Cliff et al. (1983) report an Sm–Nd date of 1,870 Ma for the closure of that isotopic system marking the end of the high-grade metamorphic event.

The 'late Laxfordian' metamorphism is characterised by general amphibolite–facies assemblages present over the remainder of the Outer Hebrides outcrop, and is associated with regional migmatisation and emplacement of granites

and pegmatites. The extensive granite–migmatite complex of N Harris and S Lewis (see Fig. 2.00; Myers 1971) was formed during this event, and has been dated at c. 1,700 Ma (Van Breemen et al. 1971). Rb–Sr and K–Ar mineral ages show a similar range to those of the mainland, from about 1,700 to 1,500 Ma (Moorbath et al. 1975). The record of Laxfordian events in the Outer Hebrides is thus more complete than in the mainland, insofar as the main metamorphic event of the mainland can be divided in the Outer Hebrides into an earlier high-grade and later lower-grade component, the former rather precisely dated at 1,860 Ma, and marking the earliest geochronologically recorded movements on one of the major Laxfordian shear zones.

Table 2.4. Representative analyses of post-Scourian igneous rocks

	1	2	3	4	5	6	7	8	9	10	11
SiO_2	49·60	50·21	46·1	46·2	49·45	51·25	44·06	53·90	55·11	70·09	51·7
TiO_2	2·02	1·46	0·31	1·84	1·12	1·59	0·82	0	0·88	0·28	2·0
Al_2O_3	12·54	13·57	4·70	10·05	14·20	15·70	13·15	28·08	18·54	14·0	14·4
Fe_2O_3	—	3·60	3·92	4·2	2·39	1·47	2·67	0·31	1·63	1·0	5·4
FeO	16·76*	10·17	6·40	10·34	9·97	9·64	13·48	0·62	5·39	1·5	6·7
MnO	0·18	0·25	0·17	0·18	0·17	0·15	0·28	0	0·11	0·3	0·17
MgO	5·42	5·86	27·57	13·5	6·93	5·10	10·31	tr	4·33	0·52	4·2
CaO	9·38	9·63	4·76	8·5	9·78	7·84	13·31	11·64	7·91	1·42	6·5
Na_2O	2·68	2·65	0·83	2·9	2·43	3·25	1·02	3·88	4·04	3·7	3·2
K_2O	0·69	0·84	0·34	0·56	0·30	1·23	0·30	0·62	0·79	5·2	2·7
P_2O_5	0·23	0·13	0·05	0·13	0·10	0·22	—	0	1·13	0·09	1·6
S	1,068	—	487	2,930	—	—	—	—	—	—	—
Cl	—	—	463	293	—	—	—	—	—	—	—
Ba	317	199	153	203	75	328	35	—	300	1,191	1,719
Ce	38	—	—	—	19	36	—	—	—	—	—
Co	—	67	—	—	—	—	30	—	7	22	35
Cr	127	128	2,707	907	159	130	150	—	35	34	61
Cu	—	173	—	—	82	69	—	—	—	11	69
Ga	—	7	—	—	18	20	1	—	7	12	22
La	—	7	8	11	4	15	—	—	—	132	142
Li	—	15	—	—	—	—	< 30	—	5	20	49
Nb	—	—	4	7	5	6	—	—	—	—	—
Ni	82	78	2,003	753	96	65	40	—	6	3	39
Pb	8	120	18	9	0·5	2	—	—	—	17	0
Rb	15	27	12	12	7	35	—	—	—	288	92
Sr	206	222	108	273	149	435	45	—	600	334	1,341
Th	—	—	—	—	1	3	—	—	—	—	—
V	—	352	—	—	315	318	600	—	80	—	214
Y	26	22	6	13	24	30	—	—	—	10	35
Zn	—	121	86	96	103	112	—	—	—	45	140
Zr	135	125	46	118	73	126	4	—	16	357	310

* Total Fe as FeO.

Scourie dyke suite: (1) metadolerite, Assynt (average of 23 analyses); (2) 'younger basic' metadolerite, Outer Hebrides (average of 24 analyses); (3) picrite, Assynt (average of 3 analyses); (4) olivine gabbro, Assynt (average of 3 analyses). *Loch Maree Group volcanic rocks*: (5) type A amphibolite (average of 119 analyses); (6) type B amphibolite (average of 35 analyses). *South Harris igneous complex:* (7) metagabbro; (8) meta-anorthosite; (9) meta-tonalite. *Laxfordian igneous rocks:* (10) granite, Uig hills, Outer Hebrides (average of 59 analyses); quartz–microdiorite dyke, Benbecula.

Sources: 1, 3, 4, Tarney (1973); 2, 10, 11, Fettes & Mendum (1987); 5, 6, Johnson et al. (1987); 7–9, Dearnley (1963).

Six phases of deformation are recognised by Fettes & Mendum (1987). D1 is confined to planar fabrics in certain Scourie dykes. Fettes & Mendum relate these fabrics to the continuation of activity on the Inverian shear zones during or shortly after dyke emplacement. However these D1 fabrics correspond to the widely developed D1 structures on the mainland which are assigned to the Laxfordian but, as has been explained, may relate to deformation at any time during the very long interval between 2,400 Ma and *c.* 1,800 Ma.

The main, regionally developed, phases are D2 and D3. D2 produced open to tight folds with sub-horizontal axial planes, and variably developed axial-planar fabrics. In areas of high Laxfordian strain, these fabrics become more intense, the folds tighten, Scourian structures become rotated into parallelism, and the dykes become thinned and boudinaged (see Fig. 2.19). D3 is characterised by upright folds, locally with strongly attenuated limbs trending NW–SE, or locally NE–SW, in NW Lewis.

A tectonic summary of the Laxfordian structure of the Outer Hebrides by Coward *et al.* (1970) shows the arrangement of the major Laxfordian D3 structures. They point out that the antiforms correspond to regions of low Laxfordian strain, whereas the synforms correspond to regions of high Laxfordian strain. In more recent accounts (e.g. see Coward 1980) these high-strain regions are recognised to be major shear zones (see Fig. 2.4).

Coward & Park (1987) present a tectonic interpretation of the Laxfordian of the whole of the Lewisian complex, in which the Outer Hebrides D2 structure is regarded as the outcrop of a major D2 sub-horizontal shear zone which is represented on the mainland only in the N and S, descending beneath the Central region (see Fig. 2.9d). The orientation of elongation lineations in the Hebrides, and the sense of overthrusting of the South Harris granulites over the gneisses to the NE, suggest that the transport direction during the D2 event was towards the NW.

The D3 structures, like those of the mainland, represent a change to a dextral transpressional regime, with a strong component of NE–SW compression resulting in the folding

Fig. 2.19. Boudinage and folding of Scourie dykes intensely deformed in the Laxfordian, South Uist. BGS photograph.

of the previous flat-lying fabrics into major upright NW–SE D3 folds (see Fig. 2.9e).

The granite–migmatite complexes of N Harris and S Lewis consist of biotite–granite bodies (see Table 2.4) ranging from porphyritic varieties to leucogranites, aplites and pegmatitic veins. The emplacement of these rocks is accompanied by extensive metasomatic activity involving particularly the remobilisation of potash (Myers 1970, 1971). Studies of zircons from the larger granite sheets suggest that they are derived from juvenile sources, rather than from the Scourian gneisses (Van Breemen et al. 1971, Pidgeon & Aftalion 1978). The granite suite shows varying degrees of deformation associated with low-temperature retrogression. The deformation may be contemporaneous with emplacement (Fettes & Mendum 1987).

The youngest Laxfordian intrusions are thin micro-diorite dykes (Table 2.4), one of which cuts a pegmatite of the granite–migmatite suite described above and has yielded a K–Ar mica age of 1,409 Ma (Fettes et al., in press). No equivalents are known on the mainland.

The Outer Hebrides fault zone

The Outer Hebrides fault zone extends for over 200 km from Sandray, S of Barra, to Tolsta Head in N Lewis (Fig. 2.3) and is one of the most spectacular features of the geology of the Outer Hebrides. The zone trends NNE–SSW, and dips between 20° and 30° ESE below the Minch. The presence of pseudotachylite was noted by MacCulloch (1819) and Jehu & Craig (1923) who first mapped the course of the fault. It was termed the 'Hebridean thrust' by Kursten (1957) and has been referred to as a thrust, of probable Caledonian age, by most subsequent workers until seismic evidence for normal movement offshore was obtained (see Smythe et al. 1982).

The rocks and structures of the fault zone are described in detail by Sibson (1977) and White & Glasser (1987). The zone embraces structures ranging from crushed or brec-ciated gneiss through mylonite, ultramylonite and pseudo-tachylite, to phyllonite. This sequence is interpreted as an effect of increasing strain by White & Glasser (1987) and

Fig. 2.20. Pseudotachylite in crush breccia from the Outer Hebrides fault, Barra. BGS photograph.

is similar to that recorded from faults in other deeply exposed gneiss terrains, reflecting a strain-dependent increase in comminution and retrogression. The passage of fluids through the cataclastic rocks ultimately converts a quartzo–feldspathic gneiss to a quartz–mica–chlorite phyllonite. The increase in ductility which this transition represents is attributed by White and Glasser purely to the effect of the passage of increasing volumes of fluid.

The ultracataclasites form prominent crush zones (see Fig. 2.20) near the footwall of the fault zone in certain areas. The progressive increase in retrogression into the hangingwall suggests a normal rather than thrust-sense displacement. This is confirmed by the evidence from the accompanying minor structures such as fold asymmetry and the orientation of shear bands.

In the Uists, there are two clear phases of movement on the fault zone. The first is associated with the formation of the crush melange and produces ultracataclasite and pseudotachylite; the second appears to be responsible for the concentration of deformation into the more retrogressed cataclasites and phyllonites in the hangingwall.

The presence of the granulite–facies Corodale gneiss in the hangingwall was thought by the earlier workers to be evidence of deeper basement thrust over higher level cover, supporting the original assumption that the fault zone was a thrust. But the preservation, during the Laxfordian, of granulite–facies rocks in the footwall of the fault zone in the SE renders this argument invalid.

Radiometric dates from the fault zone are difficult to interpret. K–Ar dates range from 2,056 Ma to 442 Ma from the pseudotachylite, and 1,926 Ma to 1,000 Ma from the ultracataclasite. The phyllonites yield a more concentrated range of 471–394 Ma. The age scatter undoubtedly represents in part the protolith content of the zone, although the younger ages, particularly in the more recrystallised phyllonites, probably approach the real age of the fault, or at least of the latest movement on the fault, indicating a late Caledonian (early Devonian?) age.

Lailey et al. (1989) report that cataclastic structures within the fault zone at Scalpay, Harris, are cut by amphibolite sheets correlated with the Scourie dykes, and infer an Inverian age for the initiation of the fault zone. This evidence is difficult to reconcile with the view, discussed earlier, that the complex remained at a comparatively deep level during the Inverian, and through the period of Scourie dyke emplacement. Fettes & Mendum (1987) also suspect older movements and cite as evidence the presence of aplogranitic veins, displaying only slight deformation, cutting strongly sheared granite. They conclude that the initiation of the fault zone was 'closely connected' with the later stages of Laxfordian granitic activity (i.e. c. 1,700 Ma). It should also be remembered that the late fault rocks of the mainland are assigned a possible Grenville age by Park (1970). In summary, there is currently no universally accepted view as to either the age of initiation, or the number and dates of the main phases of movement on the fault zone, although the late Caledonian normal movement appears to be well established.

The Inner Hebrides and Caledonian inliers

Lewisian rocks occur in a number of the smaller islands of the Inner Hebrides: from N to S, Rona, Raasay, Coll, Tiree and Iona. Further S, rocks attributed to the Lewisian but of more doubtful status are found on Colonsay and Islay. The outcrops on Rona and Raasay are a southward continuation of the Laxfordian belt of the Southern mainland region, and have already been discussed.

The Lewisian of Coll, Tiree and Iona exhibits several distinctive features. Modern views on the Lewisian of Coll and Tiree are based on the work of Drury (1972, 1973) and Westbrook (1972). The following account is based primarily on Drury (1972). Both Coll and Tiree display a generally N–S arrangement of outcrops of acid and basic gneisses, together with prominent layers of metasediments, comprising quartzites, marbles, calc–silicate rocks and garnet–biotite–(kyanite) pelites. The rocks exhibit variably retrogressed granulite–facies assemblages correlated with those of the Badcallian on the mainland and on nearby Barra. However the Scourie dykes appear to be absent from the islands, making correlation with the mainland sequence difficult in the absence of radiometric dates.

Amphibolite–facies retrogression is characteristic of the eastern half of Tiree and of the whole of Coll. In the absence of Scourie dykes, it is difficult to divide the retrogressive effects into Inverian and Laxfordian elements. Drury (1972) recognises six phases of deformation, of which D1 and D2 are probably Badcallian. D2 produced the dominant N–S major folds. An interesting feature of the sequence here is the presence of a suite of Scourian basic dykes that not only cut the earliest foliation but are involved in the main Badcallian granulite–facies metamorphism and deformation. The D3 structures are associated with widespread retrogression to amphibolite facies, and comprise asymmetric folds producing local attenuation of the pre-existing banding and deformation of post-D2 pegmatites. These structures are similar to the Inverian structures both on the mainland and in the Outer Hebrides.

The D4 deformation produced widespread monoclines and shear zones, also associated with retrogression to amphibolite facies, attributed to the Laxfordian. They are followed by open folds and younger shear zones (D5) associated with lower-temperature retrogression, and finally by the development of crush belts with pseudotachylite (D6). The latter sequence (D4–D6) shows marked similarities to that of the Laxfordian further N, in those areas unaffected by the intense Laxfordian D2 deformation, for example between Loch Torridon and Gairloch.

A set of biotite–pegmatites emplaced between D4 and D5 (Drury 1972) may represent the local equivalent of the c.

1,700 Ma Laxfordian granite–migmatite suite abundantly developed farther N in Raasay and Rona.

The Caledonian inliers

Lewisian basement is exposed in numerous inliers within the Caledonides of the Northern Highlands (see Fig. 2.2). The inliers in the lower nappes of the Moine thrust zone are parautochthonous, and carry unconformable Torridonian or Cambrian cover. East of the Moine thrust, the Lewisian inliers are overlain by unconformable Moine cover. They occur mainly above the Moine thrust itself, around Glenelg, and above the structurally higher Sgurr Beag thrust. Modifications produced by Caledonian deformation have obscured many of the original Lewisian fatures, but general petrological and geochemical characteristics of the original rocks can still be recognised. Amphibolite bodies correlated with the Scourie dykes are usually identifiable. High Na/K ratios and low levels of incompatible elements are among the geochemeical features that distinguish the Lewisian material from the surrounding Moine metasediments (Winchester & Lambert 1970, Johnstone et al. 1979). An Rb–Sr date of c. 2,700 Ma obtained from the Scardroy inlier (Moorbath & Taylor 1975) confirms the Scourian age of the basement.

The Glenelg inlier shows some unusual features not found in the foreland outcrops. Typical Scourian grey gneisses are accompanied by bands of pelitic and semi-pelitic metasediment carrying kyanite and garnet, by marbles and calc–silicate rocks containing forsterite, diopside, tremolite and phlogopite, and by iron-rich metasediments (eulysites) containing the assemblage fayalite, hedenbergite, Fe-hypersthene, garnet and magnetite (Tilley 1936). The marble bands can be followed continuously for over 10 km. These metasediments are associated with migmatitic gneisses and are veined by pegmatites which are themselves cut by Scourie-type amphibolites, indicating that the metasediments are part of the Scourian complex, rather than equivalents of the Loch Maree group. Thin metasedimentary layers consisting of marble, calc–silicate gneiss, rusty-weathering pelite, and graphitic gneiss are also common in the inliers above the Sgurr Beag thrust, farther east.

In the eastern Glenelg inlier, lenticular masses and bands of eclogite are enclosed within the acid gneisses. The eclogites are composed essentially of omphacite and pyrope–almandine, and typically are partially amphibolitised. The status of these rocks is uncertain, and they do not appear to be present elsewhere in the Lewisian. Watson (1983) regarded the eclogite facies as Badcallian but the local development of higher pressure assemblages within a granulite–facies terrain is difficult to explain unless the eclogite assemblage records a deeper level metamorphic history within an ocean slab prior to its incorporation in the continental crust. Sanders et al. (1984) report a Sm–Nd date of c. 1,100 Ma from the eclogite, interpreted as indicating a Grenvillian age for the eclogite–facies metamorphism. This interpretation does not preclude a Lewisian age for the inlier as a whole, or an earlier Badcallian granulite–facies event.

Several general conclusions may be drawn from the Lewisian inliers. Typical Scourian grey gneiss appears to be ubiquitous in the Northern Highlands, and the Badcallian metamorphism appears everywhere to be at least in amphibolite facies. On the basis of structural evidence (e.g. see Ramsay 1958, 1963) and geochemistry (Harrison & Moorhouse 1976) the Laxfordian effects are weak compared with those of the Laxfordian belts of the foreland. In particular, granite–migmatite complexes such as those of Laxford and Harris have not been identified.

The intrusive Caledonian granites in both the Northern and Grampian Highlands yield evidence on the nature of the hidden basement. Isotopic compositions of zircons, and of lead in feldspars suggest derivation from both Early Proterozoic and Archaean crustal sources (Pidgeon & Aftalion 1978, Blaxland et al. 1979). Johnstone et al. (1979b) suggest that the 'Lewisian' basement of the Caledonides may contain late- or post-Laxfordian granite plutons. It has also been suggested (e.g. see Marcantonio et al. 1988, Menuge & Daly 1989) that the southern part of this hidden basement may correspond to the extension of the Ketilidian orogenic belt of S Greenland where a substantial proportion of new granitoid material was added to the crust during the Early Proterozoic (see below).

Regional context

The Lewisian basement of NW Britain is a comparatively small part of a very extensive region of Precambrian continental crust which includes the Laurentian shield of N America, Greenland and Scandinavia. The north-western part of the British Isles became detached from N America during the early Tertiary opening of the North Atlantic ocean, and is separated from Scandinavia both by the Caledonian orogenic belt and by extensional rifts and basins of Devonian to Mesozoic age.

The relative positions of the various Precambrian orogenic belts of the N Atlantic region at the end of the Grenville–Sveconorwegian orogeny (at c. 1,100 Ma) can be reconstructed approximately from palaeomagnetic evidence (see Stearn & Piper 1984) and by removing the effects of the Atlantic opening (Fig. 2.21). This reconstruction shows the Grenville 'front' crossing NW Britain immediately S of the Lewisian outcrop. Possible Grenvillian tectonic effects in the southern Lewisian have already been discussed. However a more relevant reconstruction is based on the palaeomagnetically determined relative positions of N America and Scandinavia prior to about 1,200 Ma, when a major change in orientation of Scandinavia with respect to N America took place (see Stearn & Piper 1984). This reconstruction (Fig. 2.21a) enables us to compare the relative positions of the Early Proterozoic belts of N America and Scandinavia. Fig. 2.22, based on a modified reconstruction by Winchester (1988), shows the possible arrangement of these belts. The Lewisian complex can be seen to form part of a continuous Early Proterozoic belt linking the Churchill province of the Canadian shield with the Nagssugtoqidian of Greenland, and the Svecokarelian of Scandinavia. An important distinction has long been recognised (e.g. see Sutton & Watson 1987) between the Ketilidian belt of S Greenland and the Nagssugtoqidian belt on the north side of the S Greenland Archaean craton

(see Escher & Watt 1976). The Nagssugtoqidian belt may be intracratonic rather than the product of continental collision with associated subduction of intervening oceanic lithosphere. The belt is typified by intense deformation, high-pressure metamorphism, and subdued igneous activity – in particular, by a lack of primitive calc–alkaline magmas that could be subduction-related. This belt bears many similarities to the Lewisian complex, and its equivalent can be found to the E in the Karelian belt of Finland

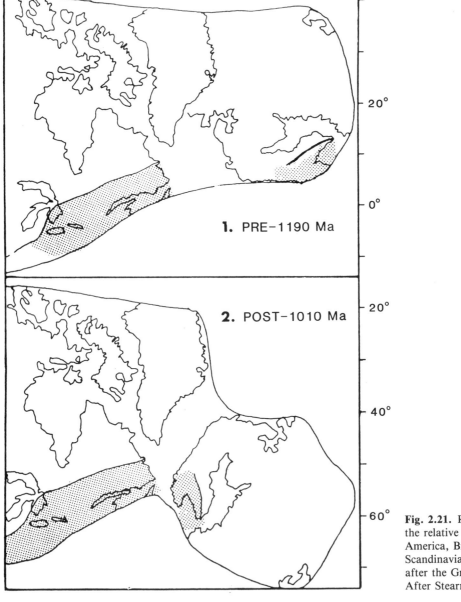

Fig. 2.21. Reconstruction of the relative positions of N America, Britain and Scandinavia before and after the Grenville orogeny. After Stearn & Piper (1984).

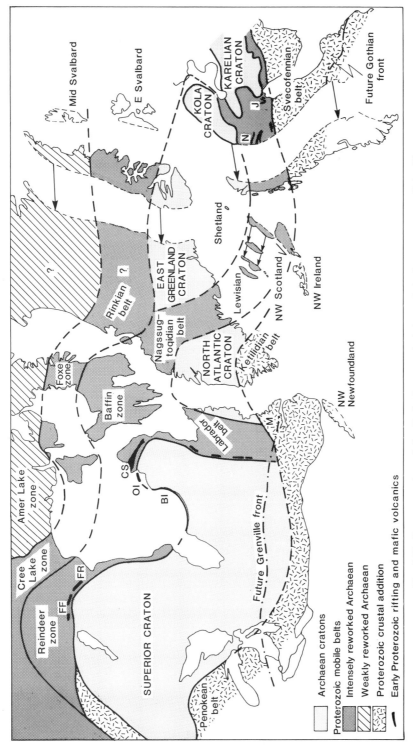

Fig. 2.22. Reconstruction of the Early Proterozoic belts and Archaean cratons of the N Atlantic region. After Winchester (1988).

and Sweden, and to the W in the Churchill province of the Canadian shield. Specific similarities have been pointed out between the Loch Maree Group in the Lewisian, and the Cape Smith and Labrador belts of the Churchill province (see Johnson *et al.* 1987, Floyd *et al.* 1989).

The Ketilidian belt, on the other hand, is generally interpreted as the site of the contemporary Early Proterozoic plate margin. It is characterised by high-temperature, low-pressure metamorphism and abundant magmatism, particularly of calc–alkaline affinities, which can be related to a volcanic arc of Andean type. The belt can be followed westwards into the Makkovik orogen of Labrador, which itself is interpreted as part of a formerly continuous belt linking with the Penokean of SE Canada and northern USA, but overprinted in the intervening ground by the Grenville belt (see Winchester 1988 for a summary of the relevant published work). To the E, the continuation of this belt is found in the Svecofennian belt of Sweden and Finland. All these belts display similar features to the Ketilidian.

The main part of the Lewisian outcrop has obvious similarities to the Nagssugtoqidian belt (see Myers 1987), but the continuation of the Ketilidian belt across Britain is indicated by the 'Lewisian' gneisses in Islay and Colonsay (Wilkinson *et al.* 1907). These gneisses, currently being re-investigated by Muir, Fitches & Maltman (pers. comm.) comprise a variably deformed bimodal igneous complex of

Laxfordian age. The isotopic evidence of Marcantonio *et al.* (1988) and of Muir *et al.* appears to preclude an Archaean crustal component. The rocks and the isotopic evidence from the Caledonian granites referred to earlier, suggest that the northern boundary of the Ketilidian terrain may extend from NW Ireland (Menuge & Daly, in press) across the Grampian Highlands. Much, or indeed all, of the 'Lewisianoid' crust of Scotland S of the Great Glen fault may thus be of Proterozoic rather than Archaean age. Perhaps the term 'Lewisian' should be restricted to the type area of the complex on the mainland and Outer Hebrides and to those neighbouring areas (including the Inner Hebrides and the Moine inliers) where the rocks and sequences are readily correlatable with the type area.

At the end of the Archaean, continental crust extended continuously over the whole of the N Atlantic region N of the Ketilidian belt, which can be regarded as the Early Proterozoic continental margin. The various Early Proterozoic intracontinental belts fragmented this early continent into several separate cratons: the Superior, N Atlantic (Godhaab), E Greenland and Kola cratons by means of displacements along belts such as the Nagssugtoqidian. By studying the Inverian and Laxfordian shear zone system (see Coward & Park 1987) it is possible to relate the deformation patterns to displacements on these zones, and ultimately to shed light on Early Proterozoic plate movements.

REFERENCES

ATTFIELD, P.   1987   The structural history of the Canisp Shear Zone. *In* Park, R. G. & Tarney, J. (Eds.) *Evolution of the Lewisian and comparable Precambrian high grade terrains.* Geol. Soc. Lond. Spec. Publ., **27**, 165–173.

AL-AMEEN, S. I.   1979   Mineralogy, petrology and geochemistry of the banded iron formation, Gairloch, N.W. Scotland. Unpublished M.Sc thesis, Univ. Keele.

BAMFORD, D., NUNN, K.,   1978   LISPB-IV. Crustal structure of northern Britain. *Geophys. J.* PRODEHL, C. &   *Roy. Astr. Soc.*, **54**, 43–60. JACOBS, B.

BARNICOAT, A. C.   1987   The causes of the high-grade metamorphism of the Scourie complex, NW Scotland. *In* Park, R. G. & Tarney, J. (Eds.) *Evolution of the Lewisian and comparable Precambrian high-grade terrains.* Geol. Soc. Lond. Spec. Publ., **27**, 73–79.

BHATTACHARJEE, C. C.   1968   The structural history of the Lewisian rocks north-west of Loch Tollie, Ross-shire, Scotland. *Scott. J. Geol.*, **4**, 235–264.

BEACH, A.   1976   The interrelationships of fluid transport, deformation, geochemistry and heat flow in early Proterozoic shear zones in the Lewisian complex. *Phil. Trans. R. Soc. London.* **A280**, 569–604.

BEACH, A., COWARD, M. P. &   1973   An interpretation of the structural evolution of the Laxford GRAHAM, R. H.   front. *Scott. J. Geol.*, **9**, 297–308.

BEACH, A. & TARNEY, J.   1978   Major and trace element patterns established during retrogressive metamorphism of granulite-facies gneisses, NW Scotland. *Precambrian Res.*, **7**, 325–348.

BICKERMAN, M., BOWES, D. R. & VAN BREEMEN, O. 1975 Rb–Sr whole rock isotopic studies of Lewisian metasediments and gneisses in the Loch Maree region, Ross-shire. *J. geol. Soc. Lond.*, **131**, 237–254.

BLAXLAND, A. B., AFTALION, M. & VAN BREEMEN, O. 1979 Pb isotopic composition of feldspars from Scottish Caledonian granites and the nature of the underlying crust. *Scott. J. Geol.*, **15**, 139–154.

BOWES, D. R. 1968 An orogenic interpretation of the Lewisian of Scotland. *XXIII International Geological Congress*, **4**, 225–236.

1972 Geochemistry of Precambrian crystalline basement rocks, Northwest Highlands of Scotland. *24th Int. geol. Congr.*, **1**, 97–103.

BOWES, D. R., WRIGHT, A. E. & PARK, R. G. 1964 Layered intrusive rocks in the Lewisian of the north-west Highlands of Scotland. *Quart. J. geol. Soc. Lond.*, **120**, 153–184.

CARTWRIGHT, I. & BARNICOAT, A. C. 1987 Petrology of Scourian supracrustal rocks and orthogneisses from Stoer, NW Scotland: implications for the geological evolution of the Lewisian complex. *In* Park, R. G. & Tarney, J. (Eds.) *Evolution of the Lewisian and comparable Precambrian high-grade terrains.* Geol. Soc. Lond. Spec. Publ., **27**, 93–107.

CHAPMAN, H. J. 1979 2390 Myr Rb–Sr whole-rock age for the Scourie dykes of north-west Scotland. *Nature, Lond.*, **277**, 642–643.

CLIFF, R. A., GRAY, C. M. & HUHMA, H. 1983 A Sm–Nd isotopic study of the South Harris Igneous Complex, the Outer Hebrides. *Contrib. Mineral. Petrol.*, **82**, 91–98.

COWARD, M. P. 1972 The eastern gneisses of South Uist. *Scott. J. Geol.*, **8**, 1–12.

1973 Heterogeneous deformation in the development of the Laxfordian complex of South Uist, Outer Hebrides. *J. geol. Soc. Lond.*, **129**, 139–160.

1984 Major shear zones in the Precambrian crust; examples from NW Scotland and southern Africa and their significance. *In* Kröner, A. & Greiling, S. R. (Eds.) *Precambrian tectonics illustrated*, Stuttgart, 207–235.

COWARD, M. P., FRANCIS, P. W., GRAHAM, R. H. & WATSON, J. 1970 Large-scale Laxfordian structures of the Outer Hebrides in relation to those of the Scottish mainland. *Tectonophysics*, **10**, 425–435.

COWARD, M. P., FRANCIS, P. W., GRAHAM, R. H., MYERS, J. S. & WATSON, J. 1969 Remnants of an early metasedimentary assemblage in the Lewisian Complex of the Outer Hebrides. *Proc. Geol. Assoc.*, **80**, 387–408.

COWARD, M. P. & PARK, R. G. 1987 The role of mid-crustal shear zones in the Early Proterozoic evolution of the Lewisian. *In* Park, R. G. & Tarney, J. (Eds.) *Evolution of the Lewisian and comparable Precambrian high grade terrains.* Geol. Soc. Lond. Spec. Publ., **27**, 127–138.

CRANE, A. 1978 Correlation of metamorphic fabrics and the age of Lewisian metasediments near Loch Maree. *Scott. J. Geol.*, **14**, 225–246.

CRESSWELL, D. 1972 The structural development of the Lewisian rocks on the north shore of Loch Torridon, Ross-shire. *Scott. J. Geol.*, **8**, 293–308.

DASH, B. 1969 Structure of the Lewisian rocks between Strath Dionard and Rhiconich, Sutherland, Scotland. *Scott. J. Geol.*, **5**, 347–374.

DAVIES, F. B. 1974 A layered basic complex in the Lewisian, south of Loch Laxford, Sutherland. *J. geol. Soc. Lond.*, **130**, 270–284.

1976 Early Scourian structures in the Scourie–Laxford region and their bearing on the evolution of the Laxford Front. *J. geol. Soc. Lond.*, **118**, 143–176.

1978 Progressive simple shear deformation of the Laxford shear zone, Sutherland. *Proc. Geol. Ass.*, **89**, 177–196.

DAVIES, F. B. & WATSON, J.   1977   Early basic bodies in the type Laxfordian complex, NW
                                       Scotland, and their bearing on its origin. *J. geol. Soc. Lond.*, **133**,
                                       123–132.

DEARNLEY, R.   1962   An outline of the Lewisian complex of the Outer Hebrides in
                                       relation to that of the Scottish mainland. *Quart. J. geol. Soc.
                                       Lond.*, **118**, 143–176.

                                1963   The Lewisian complex of South Harris. *Quart. J. geol. Soc.
                                       Lond.*, **119**, 243–307.

DICKINSON, B. B. &   1976   Variations in crustal level and geothermal gradient during the
   WATSON, J.   evolution of the Lewisian complex of northwest Scotland.
                                       Precamb. Res., **3**, 363–374.

DRURY, S. A.   1972   The tectonic evolution of a Lewisian complex on Coll, Inner
                                       Hebrides. *Scott. J. Geol.*, **8**, 309–333.

                                1973   The geochemistry of Precambrian granulite facies rocks from the
                                       Lewisian complex of Tiree, Inner Hebrides. *Chemical Geology*,
                                       11, 167–188.

DUNNING, F. W. (Ed.)   1985   *Geological structure of Great Britain, Ireland and surrounding
                                       seas.* Mapchart, Geol. Soc. Lond.

ESCHER, A. &   1976   *Geology of Greenland.* Grønlands Geol. Unders.
   WATT, W. S. (Eds.)

EVANS, C. R.   1965   Geochronology of the Lewisian basement near Lochinver,
                                       Sutherland. *Nature, Lond.*, **204**, 638–641.

EVANS, C. R. &   1974   The Lewisian of Lochinver, Sutherland; the type area for the
   LAMBERT, R. ST. J.   Inverian metamorphism. *J. geol. Soc. Lond.*, **130**, 125–150.

EVANS, C. R. & TARNEY, J.   1964   Isotopic ages of Assynt dykes. *Nature, Lond.*, **204**, 638–641.

FETTES, D. J. &   1987   The evolution of the Lewisian complex in the Outer Hebrides. *In*
   MENDUM, J. R.   Park, R. G. & Tarney, J. (Eds.) *Evolution of the Lewisian and
                                       comparable Precambrian high grade terrains.* Geol. Soc. Lond.
                                       Spec. Publ., **27**, 27–44.

FETTES, D. J.,   In press *The geology of the Outer Hebrides.* Mem. Geol. Surv. UK.
   MENDUM, J. R.,
   SMITH, D. I. &
   WATSON, J.

FLOYD, P. A.,   1989   Geochemistry and tectonic setting of Lewisian clastic
   WINCHESTER, J. A. &   metasediments from the early Proterozoic Loch Maree Group of
   PARK, R. G.   Gairloch, NW Scotland. *Precambrian Res.*, in press.

GILETTI, B.,   1961   A geochronological study of the metamorphic complexes of the
   MOORBATH, S. &   Scottish Highlands. *Quart. J. geol. Soc. Lond.*, **117**, 233–264.
   LAMBERT, R. ST. J.

GRAHAM, R. H.   1980   The role of shear belts in the structural evolution of the South
                                       Harris igneous complex. *J. struct. Geol.*, **2**, 29–37.

HALL, J.   1987   Physical properties of Lewisian rocks: implications for deep
                                       crustal structure. *In* Park, R. G. & Tarney, J. (Eds.) *Evolution of
                                       the Lewisian and comparable Precambrian high grade terrains.*
                                       Geol. Soc. Lond. Spec. Publ., **27**, 185–192.

HAMILTON, P. J.,   1979   Sm–Nd systematics of Lewisian gneisses: implications for the
   EVENSEN, N. M.,   origin of granulites, *Nature, London*, **277**, 25–28.
   O'NIONS, R. K. &
   TARNEY, J.

HEAMAN, L. M. &   1989   U–Pb baddeleyite ages for the Scourie dyke swarm, Scotland:
   TARNEY, J.   evidence for two distinct intrusion events. *Nature*, **340**, 705–708.

HARRISON, V. E. &   1976   A possible early Scourian supracrustal assemblage within the
   MOORHOUSE, S. J.   Moine. *J. geol. Soc. Lond.*, **132**, 461–466.

HOLLAND, J. G. &   1973   Comparative major element geochemistry of the Lewisian of the
   LAMBERT, R. ST. J.   mainland of Scotland. *In* Park R. G. & Tarney, J. (Eds.). *The
                                       early Precambrian of Scotland and related rocks of Greenland.*
                                       Univ. Keele, 51–62.

HUMPHRIES, F. J. &   1982   Sm–Nd dating and cooling history of Scourian granulites,
   CLIFF, R. A.   Sutherland. *Nature, London*, **295**, 515–517.

JEHU, T. J. & CRAIG, R. M. 1923– Geology of the Outer Hebrides, Parts I–V. *Trans. R. Soc. Edinb.*
                            34   **53**, 419–441 (1923); **53**, 615–641 (1925); **54**, 46–89 (1926); **55**,
                                   457–488 (1927); **57**, 839–874 (1934).

JOHNSON, Y., PARK, R. G. & 1987   Geochemistry, petrogenesis and tectonic significance of the
   WINCHESTER, J.   Early Proterozoic Loch Maree amphibolites. *In* Pharoah, T. C.,
                                   Beckinsale, R. D. & Rickard, D. T. *Geochemistry and Min-
                                   eralization of Proterozoic volcanic suites.* Geol Soc. Lond. Spec.
                                   Publ., **33**, 255–269.

JOHNSTONE, G. S.,   1979a Regional geochemistry of the Northern Highlands of Scotland.
   PLANT, J. & WATSON, J.   *In* Harris, A. L. & Leake, B. L. (Eds.) *The British Caledonides:
                                   reviewed.* Geol. Soc. Lond. Spec. Pub., **8**, 117–128.

                            1979b Caledonian granites in relation to regional geochemistry in
                                   northern Scotland. *In* Harris, A. L. & Leake, B. E. *The British
                                   Caledonides: reviewed.* Geol. Soc. Lond., 663–667.

JONES, E. M., RICE, C. M. & 1987   Lower Proterozoic stratiform sulphide deposits in Loch Maree
   TWEEDIE, J. R.   Group, Gairloch, northwest Scotland. *Trans. Instn. Min. Metall.
                                   B*, **96**, 128–140.

KURSTEN, M.   1957   The metamorphic and tectonic history of parts of the Outer
                                   Hebrides. *Trans. Edinb. geol. Soc.* **17**, 1–31.

LAILEY, M., STEIN, A. M. & 1989   The Outer Hebrides fault, Scotland: a major Proterozoic struc-
   RESTON, T. J.   ture in NW Britain. *J. geol. Soc. Lond.*, **146**, 253–260.

LAMBERT, R. ST. J. &   1972   A geochronological study of the Lewisian from Loch Laxford to
   HOLLAND, J. G.   Durness, Sutherland, NW Scotland. *Scott. J. Geol.*, **128**, 3–19.

LAMBERT, R. ST. J.,   1970   An apparent age for a member of the Scourie dyke suite in Lewis,
   MYERS, J. S. & WATSON, J.   Outer Hebrides. *Scott. J. Geol.*, **6**, 214–220.

LYON, T. B. D.,   1973   Geochronological investigation of the quartzofeldspathic rocks
   PIDGEON, R. T.,   of the Lewisian of Rona, Inner Hebrides. *J. geol. Soc. London*,
   BOWES, D. R. &   **129**, 389–402.
   HOPGOOD, A. R.

MACCULLOCH, J.   1819   *A description of the Western Isles of Scotland, including the Isle of
                                   Man* (3 volumes). Hurst, Robinson & Co., London.

MARCANTONIO, F.,   1988   A 1800-million-year-old Proterozoic gneiss terrane in Islay with
   DICKIN, A. P.,   implications for the crustal structure evolution of Britain.
   McNUTT, R. H. &   *Nature, Lond.*, **335**, 62–64.
   HEAMAN, L. M.

MENUGE, J. F. &   In press Proterozoic evolution of the Erris complex, NW Mayo, Ireland:
   DALY, J. S.   neodymium isotope evidence. *In* Gower, C. F., Rivers, T. &
                                   Ryan, B. (Eds.) *Mid-Proterozoic geology of the southern margin
                                   of proto Laurentia-Baltica.* Geol. Ass. Canada Spec. Paper.

MOORBATH, S. &   1972   The Lewisian chronology of the southern region of the Scottish
   PARK, R. G.   Mainland. *Scott. J. Geol.*, **8**, 51–74.

MOORBATH, S. &   1974   Lewisian age for the Scardroy mass. *Nature, Lond.*, **250**, 41–43.
   TAYLOR, P. N.

MOORBATH, S., WELKE, H. &   1969   The significance of lead isotope studies in ancient high grade
   GALE, N. H.   metamorphic basement complexes, as exemplified by the Lewis-
                                   ian rocks of NW Scotland. *Earth planet. Sci. Lett.*, **6**, 245–256.

MOORBATH, S.,   1975   Isotopic evidence for the age and origin of the 'grey gneiss'
   POWELL, J. L. &   complex of the southern Outer Hebrides, north-west Scotland. *J.
   TAYLOR, P. N.   geol. Soc. Lond.*, **131**, 213–222.

MYERS, J. S.   1970   Gneiss types and their significance in the repeatedly deformed
                                   and metamorphosed Lewisian complex of Western Harris,
                                   Outer Hebrides. *Scott. J. Geol.*, **6**, 186–199.

                            1971   The Late Laxfordian granite–migmatite complex of western
                                   Harris, Outer Hebrides. *Scott. J. Geol.*, **7**, 254–284.

MYERS, J. S.   1987   The East Greenland Nagssugtoqidian mobile belt compared
                                            with the Lewisian complex. *In* Park, R. G. & Tarney, J. (Eds.)
                                            *Evolution of the Lewisian and comparable Precambrian high grade
                                            terrains*. Geol. Soc. Lond. Spec. Publ., **27**, 235–246.

ODLING, N. E.   1984   Strain analysis and strain path modelling in the Loch Tollie
                                            gneisses, Gairloch, NW Scotland, *J. Struct. Geol.*, **6**, 543–562.

O'HARA, M. J.   1961   Zoned ultrabasic and basic gneiss masses in the early Lewisian
                                            metamorphic complex at Scourie, Sutherland. *J. Petrol., Oxford*,
                                            2, 248–276.

                                     1977   Thermal history of excavation of Archaean gneisses from the
                                            base of the continental crust. *J. geol. Soc. Lond*, **134**, 185–200.

OKEKE, P. O.,   1983   A geochemical study of Lewisian metasedimentary granulites
  BORLEY, G. D. &   and gneisses in the Scourie–Laxford area of north-west
  WATSON, J.   Scotland. *Mineral. Mag.*, **47**, 1–9.

O'NIONS, R. K.,   1983   A Nd isotope investigation of sediments related to crustal
  HAMILTON, P. J. &   development in the British Isles. *Earth planet. Sci. Lett.*, **63**, 229–
  HOOKER, P. J.   240.

PARK, R. G.   1964   The structural history of the Lewisian rocks of Gairloch, Wester
                                            Ross. *Quart. J. geol. Soc. Lond.*, **120**, 397–434.

                                     1965   Early metamorphic complex of the Lewisian north-east of
                                            Gairloch, Ross-shire, Scotland. *Nature, Lond.*, **207**, 66–68.

                                     1966   Nature and origin of Lewisian basic rocks at Gairloch, Ross-
                                            shire. *Scott. J. Geol.*, **2**, 179–199.

                                     1970   Observations on Lewisian chronology. *Scott. J. Geol.*, **6**, 379–
                                            399.

PARK, R. G., CRANE, A. &   1987   Early Proterozoic structure and kinematic evolution of the
  NIAMATULLAH, M.   southern mainland Lewisian. *In* Park, R. G. & Tarney, J. (Eds.)
                                            *Evolution of the Lewisian and comparable Precambrian high grade
                                            terrains*. Geol. Soc. Lond. Spec. Publ., **27**, 139–151.

PARK, R. G. &   1972   Basic dykes in the early Precambrian (Lewisian) of NW
  CRESSWELL, D.   Scotland: their structural relations, conditions of emplacement
                                            and orogenic significance. *24th Int. Geol. Congr., Montreal*, **1**,
                                            238–245.

                                     1973   The dykes of the Laxfordian belts. *In* Park, R. G. & Tarney, J.
                                            (Eds.) *The Early Precambrian of Scotland and related rocks of
                                            Greenland*. Univ. Keele, 119–130.

PEACH, B. N., HORNE, H.,   1907   The geological structure of the north-west Highlands of
  GUNN, W., CLOUGH, C. T.,   Scotland. *Mem. Geol. Surv. UK.*
  HINXMAN, L. W. &
  TEALL, J. J. H.

PEACH, B. N. & HORNE, J.   1930   *Chapters on the geology of Scotland*. Oxford.

PIDGEON, R. T. &   1972   Zircon U–Pb ages of granulites from the central region of the
  BOWES, D. R.   Lewisian of north-western Scotland. *Geol. Mag.*, **109**, 247–258.

PIDGEON, R. T. &   1978   Cogenetic and inherited zircon U–Pb systems in granites:
  AFTALION, M. A.   Palaeozoic granites of Scotland and England. *Geol. Journ. Spec.
                                            Issue*, **10**, 183–220.

RAMSAY, J. G.   1958   Moine–Lewisian relations at Glenelg, Inverness-shire. *Quart. J.
                                            geol. Soc. Lond.*, **113**, 487–520.

                                     1963   Structure and metamorphism of the Moine and Lewisian rocks
                                            in the north-western Caledonides. *In* Johnson, M. R. W. &
                                            Stewart, F. H. (Eds.). *The British Caledonides*. Oliver and Boyd.

ROCK, N. M. S.   1987   The geochemistry of Lewisian marbles. *In* Park, R. G. & Tarney,
                                            J. (Eds.) *Evolution of the Lewisian and comparable Precambrian
                                            high grade terrains*. Geol. Soc. Lond. Spec. Publ., **27**, 109–126.

ROLLINSON, H. R. &   1987   The magmatic evolution of the mainland Lewisian complex. *In*
  FOWLER, M. B.   Park R. G. & Tarney, J. (Eds.) *Evolution of the Lewisian and*
                                  comparable Precambrian high grade terrains. Geol. Soc. Lond.
                                  Spec. Publ., **27**, 81–92.

SHERATON, J. W., TARNEY, J.,   1973   The structural history of the Assynt district. *In* Park, R. G. &
  WHEATLEY, T. H. &   Tarney, J. (Eds.) *The early Precambrian of Scotland and related*
  WRIGHT, A. E.   *rocks of Greenland.* Univ. Keele, 31–44.

SANDERS, I. S.,   1984   A Grenville Sm–Nd age for the Glenelg eclogite in north-west
  VAN CALSTEREN, P. W. C. &   Scotland. *Nature, Lond.*, **312**, 439–440.
  HAWKESWORTH, C. J.

SIBSON, R. H.   1977   Fault rocks and fault mechanisms. *J. geol. Soc. Lond.*, 191–214.

SILLS, J. D. &   1987   Metamorphic evolution of the mainland Lewisian complex. *In*
  ROLLINSON, H. R.   Park, R. G. & Tarney, J. (Eds.) *Evolution of the Lewisian and*
                                  comparable Precambrian high grade terrains. Geol. Soc. Lond.
                                  Spec. Publ., **27**, 81–92.

SMYTHE, D. K.,   1982   Deep structure of the Scottish Caledonides revealed by the
  DOBINSON, A.,   MOIST reflection profile. *Nature, Lond.*, **299**, 338–340.
  McQUILLIN, R.,
  BREWER, J. A.,
  MATTHEWS, D. H.,
  BLUNDELL, D. J. &
  KELK, B.

STEARN, J. E. F. &   1984   Palaeomagnetism of the Sveconorwegian mobile belt of the
  PIPER, J. D. A.   Fennoscandian shield. *Precambrian Res.*, **23**, 201–246.

SUTTON, J. &   1951   The pre-Torridonian metamorphic history of the Loch Torridon
  WATSON, J.   and Scourie areas in the North-west Highlands and its bearing
                                  on the chronological classification of the Lewisian. *Quart. J.*
                                  geol. Soc. Lond., **106**, 241–307.

                        1962   Further observations on the margin of the Laxfordian complex
                                of the Lewisian near Loch Laxford, Sutherland. *Trans. R. Soc.*
                                Edinb., **65**, 89–106.

                        1987   The Lewisian complex: questions for the future. *In* Park, R. G. &
                                Tarney, J. *Evolution of the Lewisian and comparable Precambrian*
                                high grade terrains. Geol. Soc. Lond. Spec. Publ., **27**, 7–11.

TARNEY, J.   1963   Assynt dykes and their metamorphism. *Nature, Lond.*, **199**, 672–
                                674.

                        1973   The Scourie dyke suite and the nature of the Inverian event in
                                Assynt. *In* Park, R. G. & Tarney, J. (Eds.) *The early Precambrian*
                                of Scotland and related rocks of Greenland. Univ. of Keele, 105–
                                118.

TARNEY, J. &   1987a Geochemistry of the Scourian complex: petrogenesis and tec-
  WEAVER, B. L.   tonic models. *In* Park, R. G. & Tarney, J. (Eds.) *Evolution of the*
                                Lewisian and comparable Precambrian high grade terrains. Geol.
                                Soc. Lond. Spec. Publ., **27**, 45–56.

                        1987b Mineralogy, petrology and geochemistry of the Scourie dykes:
                                petrogenesis and crystallisation processes in dykes intruded at
                                depth. *In* Park, R. G. & Tarney, J. (Eds.) *Evolution of the*
                                Lewisian and comparable Precambrian high grade terrains. Geol.
                                Soc. Lond. Spec. Publ., **27**, 217–233.

TAYLOR, P. N.,   1984   Isotopic assessment of relative contributions from crust and
  JONES, N. W. &   mantle sources to the magma genesis of Precambrian granitoid
  MOORBATH, S.   rocks. *Phil. Trans. R. Soc. Lond.*, **A310**, 605–625.

TILLEY, C. E.   1936   Eulysites and related rocks types from Loch Duich, Ross-shire.
                                Mineral. Mag., **24**, 331–342.

VAN BREEMEN, O.,   1971   The age of the granite injection-complex of Harris, Outer
  AFTALION, M. A. &   Hebrides. *Scott. J. Geol.*, **5**, 269–285.
  PIDGEON, R. T.

WATSON, J.   1983   Lewisian. *In* Craig, G. Y. (Ed.) *Geology of Scotland.* Scottish
                                         Academic Press, Edinburgh.

WEAVER, B. L. & TARNEY, J.   1980   Rare-earth geochemistry of Lewisian granulite-facies gneisses,
                                         northwest Scotland: implications for the petrogenesis of the
                                         Archaean lower continental crust. *Earth planet. Sci. Lett.,* **51,**
                                         279–296.

                                  1981   The Scourie dyke suite: petrogenesis and geochemical nature of
                                         the Proterozoic sub-continental mantle. *Contrib. Mineral.
                                         Petrol.,* **78,** 175–188.

WESTBROOK, G. K.   1972   Structure and metamorphism of the Lewisian of east Tiree, Inner
                                         Hebrides. *Scott. J. Geol.,* **8,** 13–30.

WHITE, S. H. & GLASSER, J.   1987   The Outer Hebrides fault zone: evidence for normal movements.
                                         In Park, R. G. & Tarney, J. (Eds.) *Evolution of the Lewisian and
                                         comparable Precambrian high grade terrains.* Geol. Soc. Lond.
                                         Spec. Publ., **27,** 175–183.

WHITEHOUSE, M. J. &   1986   Pb–Pb systematics of Lewisian gneisses – implications for crustal
    MOORBATH, S.   differentiation. *Nature, London,* **319,** 488–489.

WILKINSON, S. B.   1907   *The geology of Islay, including Oronsay and portions of Colonsay
                                         and Jura.* Mem. Geol. Surv. Scotland.

WINCHESTER, J. A.   1988   Later Proterozoic environments and tectonic evolution in the
                                         northern Atlantic lands. *In* Winchester, J. A. (Ed.) *Later
                                         Proterozoic stratigraphy of the Northern Atlantic Regions.*
                                         Blackie, Glasgow and London, 253–270.

WINCHESTER, J. A. &   1970   Geochemical distinctions between the Lewisian of Cassley,
    LAMBERT, R. ST. J.   Durcha and Loch Shin, Sutherland and the surrounding Moin-
                                         ian. *Proc. Geol. Ass.,* **81,** 275–302.

3

TORRIDONIAN

A. D. Stewart

The red sandstone mountains of north-west Scotland (Fig. 3.1) are remnants of deposits which once filled late Proterozoic rifts on the eastern margin of Laurentia. These rifts may have been connected with a prolonged phase of crustal extension which preceded the opening of the Iapetus ocean. Closure of Iapetus in the Palaeozoic seems to have transformed some of the old normal faults into thrusts, such as the Ben More thrust and the Outer Isles thrust.

By their very nature these sandstones are unlikely to have precise lithostratigraphic correlatives in other parts of Scotland. In particular, the Torridon Group and the Moines, though often equated (Geikie 1895; Kennedy 1951), had different source areas. This is shown by their diverse detrital zircon suites (MacKie 1923), and their very different tourmaline contents – reflected in the boron content of stream sediment (Plant 1984). Time correlation with part of the Moine is conceivable, but if the Moines originated far from their present position and later accreted onto the edge of Laurentia then such correlation, even if true, might not be very significant.

The outcrop of the sandstones, mainly based on Geological Survey mapping, is shown in Fig. 3.2. The sub-crop, however, is considerably more extensive. Westwards as far as the Minch fault seismic data show that 'Torridon-ian' underlies the Triassic (Smythe *et al.* 1972; Chesher *et al.* 1983). The Tertiary rocks of Canna, west of Rhum, include sub-angular blocks of red sandstone and metamorphics from the underlying basement. Tertiary agglomerate on Eigg also contains blocks of red sandstone from the basement (Harker 1908). To the south-west of Rhum 'Torridonian' rocks form the sea floor for 125 km, reaching as far south as the latitude of Colonsay (McQuillin & Binns 1973; Evans *et al.* 1982, Fig. 1). This explains the occurrence of 'red Torridonian sandstone' and fossiliferous Durness Limestone in the Triassic conglomerates of Mull (Rast *et al.* 1968). The 500 m thick sequence of breccias and sandstone unconformably overlying Lewisian basement in Iona (Stewart 1962) may also belong to the 'Torridonian' but definite proof is lacking.

Stratigraphically the 'Torridonian' can be divided into two parts (Fig. 3.3). Red beds up to 2 km thick, out-cropping at Stoer and along the coast to the south (Stoer Group), have been shown to be much older than the rest, from which they are separated by a clear angular uncon-formity and a 90° change in direction of magnetization. The beds above (Torridon Group), up to 5 km thick, are responsible for the spectacular scenery of north-west Scotland, and especially the mountains around Torridon

Fig. 3.1. The mountains of north-west Scotland drawn by Dr John MacCulloch (1819). The sketch shows the unconformable relationship of the strata to the gneisses beneath, which MacCulloch was the first to recognise. Quinag is on the left, Suilven in the middle and Cùl Mòr on the right, with the observer looking eastward from the sea.

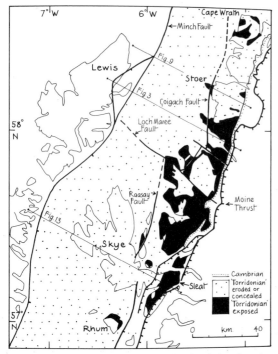

Fig. 3.2. Sketch map of north-west Scotland showing the present and former extent of the 'Torridonian', together with the more important pre-Palaeozoic faults which controlled sedimentation.

which Nicol (1866) took as the type area for his 'Torridon Sandstone'. Clastic sediments 3 km thick in the Kishorn nappe (Sleat Group), conformably beneath the Torridon Group, are not seen in contact with the Stoer Group but are nevertheless believed to be younger. They may have accumulated in a sub-graben.

The age of the three Groups is bracketed by a metamorphic event at 1100 Ma in the basement gneisses which they unconformably overlie (Moorbath *et al.* 1967), and the Lower Cambrian fossils in beds which they unconformably underlie (Cowie & McNamara 1978). Algal remains in grey shales of the Stoer Group are consistent with a Middle Riphean age according to Downie (pers. comm.), while those in the grey shales of the Torridon Group are Upper Riphean. These attributions fit in quite well with whole-rock Rb-Sr isochron ages obtained by Moorbath (1969), respectively, from siltstones in the Stoer Group (968 ± 24 Ma) and the Torridon Group (777 ± 24 Ma). However, these isochron ages date diagenesis and may, therefore, be as much as 100 Ma too young, as suggested by Smith *et al.* (1983).

Stoer Group

The Group was named from the peninsula of Stoer (Stewart 1969) because the rocks there are sedimentologically representative, structurally simple, and superbly exposed. The main features of the succession at Stoer are shown in

Fig. 3.3. Cross-section of the North Minch basin showing the relationship between the Stoer and Torridon Groups and the overlying Triassic & Jurassic. There is no definite evidence that the Stoer Group exists west of the Coigach Fault. All faults shown are believed to be Proterozoic in origin. The Minch Fault was re-activated during the Mesozoic. The Coigach fault was re-activated as a thrust during the Palaeozoic and identifies with the easterly dipping reflector which on the MOIST profile reaches the surface near shot point 2000 (Blundell *et al.* 1985, Fig. 4).

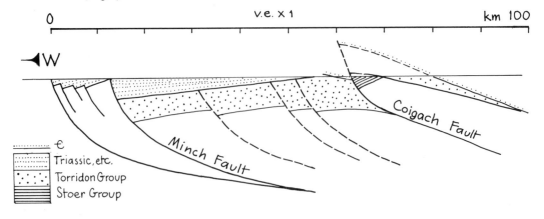

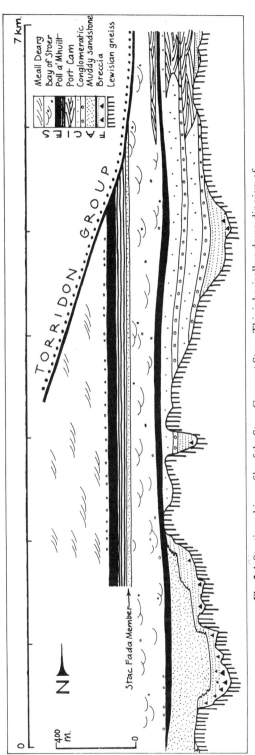

Fig. 3.4. Stratigraphic profile of the Stoer Group at Stoer. This is basically a down-dip view of the strata exposed on the peninsula. The Stoer Group is truncated unconformably by the Torridon Group.

Fig. 3.4. The key stratigraphic element is the Stac Fada Member – a unique volcanic sandstone which identifies outcrops of the Group along the coast south of Stoer, as far as Poolewe. The seven red-bed facies which build the Stoer succession are described briefly below.

The *breccia facies* immediately overlies a hilly landscape of Lewisian gneiss which at Stoer has over 300 m of relief. The clasts come exclusively from the Lewisian and it is significant that the proportion of basic (and even ultrabasic) rock types in the breccia is much the same as in the basement nearby. The degree of rounding suggests transport distances of generally less than a kilometre, except near the unconformity where clasts have obviously moved only a few metres. The breccia in contact with the Lewisian is usually massive, with clasts up to about half a metre in size. Stratigraphically upwards a crude stratification appears and the breccias pass into pebbly red sandstone, and sometimes red shale.

There can be no doubt that the facies represents a series of fanglomerates. Maximum fan radius seems to have been about 300 m, the upward fining resulting from upstream retreat of the fan heads.

The *muddy sandstone facies* (including the Stac Fada Member) consists of reddish-brown rocks, texturally greywackes with about 40 per cent matrix. They are always lateral equivalents of the finer, distal part of the breccia facies, which geochemically they closely resemble. Sedimentologically, however, there is no resemblance at all. The lowest 130 m of the muddy sandstone facies at Clachtoll are completely devoid of bedding or lamination. This massive development is succeeded upwards by a bedded subfacies in which the beds are about half a metre thick, defined by desiccated sheets of red siltstone or carbonate. The muddy sandstone also shows desiccation patterns (Fig. 3.5) suggesting that the structureless nature of the facies is due to repeated wetting and drying of a sediment originally rich in smectitic clay.

The ponded water mudflats of Hardie *et al.* (1978) would be a suitable setting for the facies, the sediments recording periodic flushes of weathered material from a source area with abundant basic rocks. Percolating ground-water rich in Mg and Ca may also have contributed to smectite production. Very similar massive reddish-brown siltstones are developed stratigraphically close to lacustrine shales in the Lower Jurassic East Berlin Formation of Connecticut. They are probably those described by Demicco & Kordesch (1986) as 'disrupted mudstones' and attributed by them to repeated wetting and drying of lake-marginal clay-rich sediment. In Connecticut the facies contains dinosaur footprints.

The *conglomeratic facies* consists of upward-fining sequences of coarse, trough cross-bedded red sandstone, and occasionally multi-storey conglomerate. Such sequences are typically tens of metres thick. The bases are erosional (Fig. 3.6) and the tops marked by a metre or so of red siltstone (rarely exposed). The conglomerates were

Fig. 3.5. Desiccation patterns in the upper part of the muddy sandstone facies at Clachtoll [NC 037272]. The ruler is 20 cm long.

derived from local basement, but unlike the breccia facies only acid Lewisian detritus is present. This, together with the rotund shape of the pebbles, suggests 5–10 km of transport.

The facies was probably deposited in shallow braided channels. The fining-upward sequences closely resemble those of the Donjek River (Miall 1977, Fig. 12). However, one of the mapped conglomerates occupies the full width of the palaeovalley (see Fig. 3.4), so that conglomerate deposition must have been essentially synchronous over the whole of the flood plain. There is no evidence of lateral accretion in the conglomerate units, so that an origin by episodic source rejuvenation is preferred to avulsion.

The *Port Cam facies* is striking because of its cross-bedding (Fig. 3.7). Set thickness is generally a few decimetres, but sometimes reaches as much as 10 m. The thinner sets persist laterally for tens of metres and the thicker ones for much further. The cross-beds are only a few millimetres thick but can be followed for many metres as they asymptotically approach the base of the set. The cross-beds originally dipped eastwards at angles usually less than 20 degrees and only rarely more than 25 degrees. The grains forming the rock are well sorted, subangular in shape, and average 0·2 mm in diameter. The maximum is about 2 mm.

In contrast to these well-laminated sandstones there are also decimetre or metre-thick intercalations of relatively massive sandstone with irregular, erosional, bases. Occasionally these sandstones incorporate gneiss fragments, and even lumps of the Port Cam facies, which must have been already partially lithified. The intercalations are quite common where the Port Cam facies is in contact with the breccia and conglomeratic facies (see Fig. 3.4). The base of the Bay of Stoer facies (described below) is also highly erosive where it overlies the Port Cam facies.

The cross-bedding described above is identical to that formed by migrating barchan dunes, while the massive sands evidently record periodic invasions of the dune field by torrential flood water.

The *Poll a'Mhuilt facies* basically consists of thinly-bedded red siltstone and fine sandstone. Wave ripples and desiccation cracks are characteristic. These fine-grained sediments form several intercalations, each only a few metres thick, within the Bay of Stoer facies (see below). Though thin, the intercalations can be traced right across the peninsula for about 6 km, with only slight change in thickness. Deposition of the facies was preceded in every case by a decimetre-thick bed of muddy sandstone. The most spectacular example of this association is afforded by

Fig. 3.6. Erosive base of a multi-storey conglomerate unit (conglomeratic facies) at Stoer [NC 047329]. The underlying beds belong to the Port Cam facies. The ruler is 20 cm long.

Fig. 3.7. Cross bedding in the Port Cam facies at Stoer [NC 048328]. The ruler is 20 cm long.

the Stac Fada Member and the overlying sequence of red siltstone and sandstone. As mentioned earlier, the Stac Fada Member belongs to the muddy sandstone facies, differing only from the rest of the facies in containing about 30 per cent of devitrified volcanic glass. Petrographic details have been published by Lawson (1972). A new analysis of the glass shows that it was olivine normative and probably undersaturated – a typical feature of rift volcanics.

The silty sediments following the Stac Fada Member, which belong to the Poll a'Mhuilt facies, are about 100 m thick. In addition to the usual red beds there are also limestones (Upfold 1984), laminated black shales containing poorly preserved organic-walled microfossils (Cloud & Germs 1971), and abundant gypsum pseudomorphs (Stewart & Parker 1979).

The interpretation of the Poll a'Mhuilt facies by Stewart & Parker (1979), based on sedimentology and boron-in-illite data, is that it formed in temporary lakes. The lake associated with the Stac Fada Member must have covered hundreds of square kilometres and had a maximum depth of at least 100 m and perhaps twice as much. For the first half of its life it must have been perennial and stratified.

The *Bay of Stoer facies* simply consists of trough cross-bedded sandstones. Soft-sediment contortions and overturned cross-bedded are common (Fig. 3.8). Well-rounded, centimetre-sized pebbles of gneiss and orthoquartzite, in roughly equal proportions, are sporadically present throughout the facies.

The *Meall Dearg facies* is built entirely of sandstones petrographically indistinguishable from those in the Bay of Stoer facies. Both planar cross-bedding, and planar bedding with extensive wave-rippled surfaces, are equally common. Pebbles are absent except at the very base at Stoer.

Both the Bay of Stoer and Meall Dearg facies are

Fig. 3.8. Contorted bedding in the Bay of Stoer facies at Clachtoll [NC 035272]. The 20 cm ruler marks the top of a cusp.

believed to have been deposited by braided rivers, the latter perhaps deposited in wider channels, on gentler palaeoslopes than the former. A modern analogue of the Bay of Stoer sandstones might be the predominantly trough cross-bedded sands deposited by powerful floods in central Australia (Williams 1971). In contrast the Meall Dearg resembles the predominantly planar cross-bedded sands deposited in the transverse bars of the relatively sluggish Platte River (Smith 1970).

The three facies found laterally adjacent to basement hills, *viz.* the breccia, conglomeratic and Port Cam facies, find close modern analogues in areas such as South Yemen (Moseley 1971). There, gravel fans fringe basement hills of Precambrian gneiss 300–600 m high. The gravels interfinger with the fluvial deposits of an ephemeral river system, over which drift barchanoid dunes. South Yemen lies in latitude 15 degrees, like that deduced from the palaeomagnetism of the Stoer Group (Stewart & Irving 1974), and has a semi-arid climate.

Palaeocurrents within the Stoer Group at Stoer reverse through 180 degrees at the base of the Bay of Stoer facies

and again at the base of the Stac Fada Member (Stewart 1982), suggesting fault-controlled deposition in a rift valley (Fig. 3.9). The repetition of lacustrine interludes (Poll a'Mhuilt facies) within the fluvial Bay of Stoer facies can also be attributed to episodic fault-tilting and disruption of the fluvial drainage net.

All the sandstones in the Stoer Group are arkosic, with oligoclase the dominant feldspar, as would be expected if Scourian granulitic gneisses formed the source area. However, the geochemistry of the sandstones forming the breccia and muddy sandstone facies, which might have been expected to most closely resemble the parent gneisses, tells a different story. The average geochemistry of sandstone from these facies is compared with that for Scourian gneiss in Table 3.1. This shows that the sandstones are much richer in Mg and Rb. The model source shown in Table 3.1 explains the excess Mg and Rb, but the sediments lack both the Ni which the basic rocks would have provided, and also the light rare earth enrichment shown by the amphibolite gneisses. As there is no geological evidence for the former existence of such a model source near Stoer it

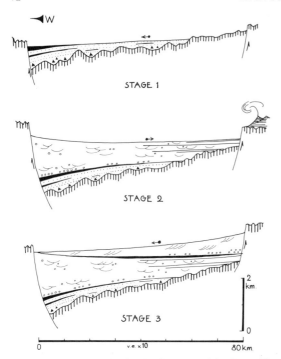

Fig. 3.9. Three stages in the development of the Stoer rift. Stage 1 shows locally derived sediments such as the breccia and conglomeratic facies. Stage 2 shows the Bay of Stoer and Stage 3 the Meall Dearg facies. Stoer is located roughly in the middle of the rift. Arrows show palaeocurrent directions.

seems more probable that the sediments are contaminated by tuff, rich in Mg and Rb, rather like that in the Stac Fada Member. The sandstones show only slight depletion in Ca and none at all in Na, suggesting poor drainage together with aridity. The aridity is in line with the palaeolatitude of 15° deduced from palaeomagnetism (Stewart & Irving 1974).

The sandstones forming the Bay of Stoer and Meall Dearg facies have K/Rb ratios like the others, so that the feldspars, at least, probably came from the same source. The Bay of Stoer and Meall Dearg facies, however, have a much higher ratio of silica to alumina (7·3 as against 4·7), that is, they are much more mature. This is due not just to the breakdown of original mafic minerals to clays and their loss from the depositional system, but to an additional source of silica, namely siliceous sediments. To produce the observed dilution, these siliceous sediments must have formed about a quarter of the catchment. They also provided the quartzite pebbles so characteristic of the Bay of Stoer facies. The petrography of these pebbles suggests that the parent rocks were fine-grained red beds, quite unlike any exposed in western Scotland today.

Table 3.1. Comparative geochemistry of Scourian Basement and adjacent Stoer Group sandstones

	A	B	C
SiO_2	61·5	60·3	58·1
TiO_2	0·6	0·4	0·7
Al_2O_3	15·5	13·9	13·0
T Fe_2O_3	6·2	6·0	7·5
MnO	0·1	0·1	0·1
MgO	3·5	7·1	9·0
CaO	5·9	4·6	3·3
Na_2O	4·0	3·8	4·0
K_2O	1·0	1·4	1·5
P_2O_5	0·2	0·1	0·1
Vol.	1·7	1·5	1·5
Total	100·2	99·2	98·8
Rb	9	41	41
K/Rb	922	277	304

Column A: average Scourian
Holland & Lambert (1975) Table I, anal. 6.

Column B: model source
30% average Scourian + 50% amphibolite gneiss (Sheraton *et al.* 1973, Table 4 C) + 20% ultrabasic rock (Sheraton *et al.* 1973, Table 3 I–J).

Column C: average sandstone
Based on representative analyses 81SO80 & 83SO69. Analyst Franz Street, Univ. Reading, Geol. Dept.

Sleat Group

The Group consists of 3500 m of coarse grey fluviatile sandstones, with some subordinate grey shales, best exposed between Loch na Dal and Kylerhea in the Sleat of Skye. Although the beds are confined to the Kishorn nappe their stratigraphic position is secured by a conformable relationship with the overlying Torridon Group.

The absence of any sequence like the Sleat Group outside the Kishorn nappe may mean that the sediments were deposited in an independent rift, the western edge being a listric normal fault which during Palaeozoic compression was transformed into the Kishorn–Suardal thrust.

The Sleat Group is nowhere seen in contact with the Stoer but, as mentioned above, it does conformably underlie the Torridon Group in eastern Skye. In view of the long hiatus between the Stoer and Torridon Groups it seems almost certain that the Sleat Group must be younger than the Stoer. No Rb–Sr isotopic dating of the Group has ever been attempted because of the lower greenschist facies metamorphism which affected the rocks during the

Palaeozoic and probably reset the isotopic system. This low-grade metamorphism also changed the colour of the rocks from red to grey. The colour change is partly due to growth of chlorite, but may also stem from the transformation of some detrital haematite into magnetite (Bailey 1955, pp. 97 & 134). Coward & Whalley (1979) report significant amounts of magnetite, as well as haematite, in both the Sleat Group and the Applecross Formation of the overlying Torridon Group.

Stratigraphic names, palaeocurrents and framework mineralogy of the sandstones are given in Fig. 3.10. Brief sedimentological details follow.

The *Rubh Guail Formation* consists almost entirely of coarse sandstone, coloured green by its content of chlorite and epidote. The sandstone is underlain by gneiss breccia close to the unconformity with the Lewisian north of Loch Alsh (Peach *et al.* 1907, p. 343), but unfortunately neither unconformity nor breccia is exposed in the type section.

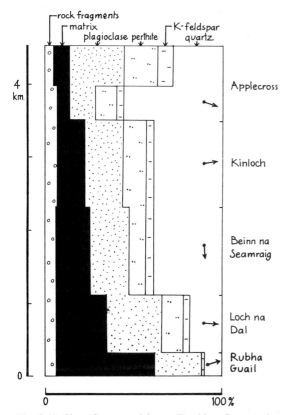

Fig. 3.10. Sleat Group and lower Torridon Group mineralogy and vector mean palaeocurrent directions in Skye. The mineralogy is based on 50 modal analyses by Byers (1972) and the palaeocurrents on 229 cross bedding directions measured by Sutton & Watson (1960, 1964). Constituent Formations of the Groups are named.

Trough cross-bedding is typical of the coarser beds, the palaeocurrents coming consistently from the west. Fine-grained, banded sediments, which become more abundant towards the top of the Formation, contain wave ripples and desiccation cracks (Sutton & Watson 1960; Stewart 1962, Fig. 11). They are followed by laminated dark grey siltstones and sandstones of the Loch na Dal Formation.

The tendency for the Formation to fine upwards into grey shales suggests that we are looking at a large alluvial fan, building out from the flank of a basement hill or fault scarp into a lake.

The lower 200 m of the *Loch na Dal Formation* is composed of laminated, dark grey siltstones, often phosphatic and frequently punctuated by coarse, or very coarse, sandstone laminae. This unusual juxtaposition of fine and very coarse grain sizes has been noted by several workers (Clough, in Peach *et al.* 1907, p. 354; Sutton & Watson 1964) and is identical to that seen in the Diabaig Formation at Camas a' Chlarsair on the south side of Upper Loch Torridon. The upper part of the Loch na Dal Formation is dominated by trough cross-bedded sandstones, still showing palaeocurrents from the west.

The shales probably mark the maximum expansion of a lacustrine or shallow marine phase, terminated by outward building deltas. The interbanded coarse and fine sediments at Upper Loch Torridon, mentioned above, result from fan toes reaching out only 100 m from the side of a palaeovalley into a lake. Perhaps the Loch na Dal Formation was as close as this to its source.

The *Beinn na Seamraig* and *Kinloch Formations* can conveniently be considered together for they are much alike. A substantial proportion of both is made of strongly contorted, cross-bedded sandstones like those in Torridon Group above. Ripple lamination forms metre-thick sequences, especially in the Kinloch Formation. Less commonly there are grey, shaly intercalations resembling those found in the lower part of the Loch na Dal Formation. In the upper part of the Kinloch these shales form the upper parts of cycles roughly 10 m thick (Stewart 1966a). Palaeocurrents measured by Sutton & Watson (1964) show directions in the Kinloch from the west but in the Beinn na Seamraig they come from the north.

These two Formations are thought to be braided river deposits, like the Applecross Formation of the overlying Torridon Group. In Beinn na Seamraig times the channels were apparently constrained to follow the rift margin.

Despite the similarity between the upper formations of the Sleat Group and the overlying Torridon Group there are two significant differences. Firstly, the pebble suite in the Sleat contains none of the metasedimentary pebbles which are so common in the Torridon Group of Skye and elsewhere. The vast majority of Sleat pebbles are porphyry, of rhyolitic or rhyodacitic composition. The remainder are acid gneiss, typically quartz – plagioclase ± microcline ± biotite. This suggests different source areas for the Groups. The second important difference lies in the composition of the plagioclase, which is variably calcic in the Sleat but always albite in the Torridon.

The Sleat Group clearly derives from an upper crustal source, for the sediments have an average K/Rb ratio of 285. The Group as a whole shows a marked upward increase in the proportion of quartz at the expense of plagioclase (Fig. 3.10). Moreover, plagioclase composition becomes progressively less calcic up into the lower part of the Applecross Formation, probably due to more effective weathering in the source area. Source area rejuvenation in Applecross times, responsible for the sudden increase in feldspar content half-way through the Formation, was not, however, accompanied by the reappearance of calcic plagioclase. The most immature sediments in the Group are found, as might be expected, at the base (Rubha Guail Formation). This Formation differs from the rest in having Fe, Ni, Ti, Ca and Mg enriched two-fold, the result of its proximity to unusually basic source rocks.

The Tarskavaig Moines, though petrographically and geochemically similar to the Sleat Group, have K/Rb ≃ 500, indicating a completely different source, probably granulitic. This rather undermines the lithostratigraphic correlation formerly advanced (Cheeney, *in* discussion of Sutton & Watson 1964; Stewart 1982), and suggests that they may have been deposited in a different trough.

Torridon Group

The Group rests on an old land surface which has a relief of 600 m around Loch Maree, declining to almost nothing in the Cape Wrath area (Geikie 1888, pp. 400–401; Stewart 1972). An example of the relief on the unconformity, exhumed in geologically recent time, is shown in Fig. 3.11. There is now no trace of the weathering which generated this ancient topography; the rotten gneiss beneath the Torridon Group near Cape Wrath attributed by Williams (1968) to Precambrian weathering probably formed in the Cenozoic (*cf.* Hall 1985). Cenozoic weathering also affects

Fig. 3.11. Precambrian topography exhumed from beneath the gently dipping beds of the Torridon Group on the north side of Quinag; a pen and ink sketch by Sir Archibald Geikie (1906).

the Torridon Group itself at some localities (Stewart, *in* Barber *et al.* 1978, pp. 35 & 80).

The unconformity generally cuts Lewisian gneiss, but near the mainland coast it truncates the westward dipping beds of the Stoer Group. There are good exposures at Stoer (Williams 1966; Stewart *in* Barber *et al.* 1978), Achiltibuie (Stewart *in* Barber *et al.* 1978), Stattic Point (Lawson 1976), Bac an Leth-Choin (Stewart 1966b) and Rubha Reidh (Lawson 1976). There is no doubt, however, that the key locality is Enard Bay (Gracie & Stewart 1967). The superb coastal section here shows the Stoer Group, including the unique Stac Fada Member, overlain by the two lowest formations of the Torridon Group, one of them containing its diagnostic suite of exotic pebbles. Both here, as well as at Achiltibuie and Rubha Reidh, the direction of magnetization of the beds changes abruptly across the unconformity (Stewart & Irving 1974; Smith *et al.* 1983).

The Torridon Group can be divided into the Diabaig, Applecross, Aultbea and Cailleach Head Formations as shown in the restored stratigraphic profiles, Figs. 3.12 & 3.13. From these profiles it will be noticed that the Diabaig is confined to the lower half of the palaeovalleys at the base of the Group. The lack of physical continuity with the type area means that this Formation generally has only facies status. Brief descriptions of the Formations follow.

The *Diabaig Formation* is excellently exposed around Loch Torridon, and especially in the eponymous township (Peach *et al.* 1907, p. 324; Stewart *in* Barber *et al.* 1978, pp. 78–81). The sedimentology of the Diabaig facies in Raasay has been described by Selley (1965a, 1965b). There are four component subfacies.

Red breccias mantle the gneiss landscape and choke the lower parts of the palaeovalleys (Fig. 3.14). They are quite similar to those at the base of the Stoer Group except that the clasts are more angular. Clasts are of sandstone where the facies overlies the Stoer Group, otherwise they are made of local gneiss. Transport distances never exceed 3 km eastward from the source rock (e.g. Peach *et al.* 1907, p. 315) and are usually negligible. The breccias pass upwards, and also laterally away from the palaeovalley walls, into tabular red sandstone.

The tabular sandstones, usually a few decimetres thick and separated by films of red silt, often show trough and planar cross-bedding, horizontal lamination and extensive wave-rippled surfaces (Fig. 3.15). Shallow channels are locally quite common. However, the sandstones forming many of the beds are well sorted and internally featureless at first glance. Stratigraphically upward and away from the palaeovalley walls this facies interfingers with grey shales.

The grey shales comprise millimetre-thick graded units, possibly seasonal and usually desiccated, together with fine sandstone bands, millimetres to centimetres thick, showing wave-ripples (Fig. 3.16). These fine sandstones fill the desiccation cracks. Phosphatic

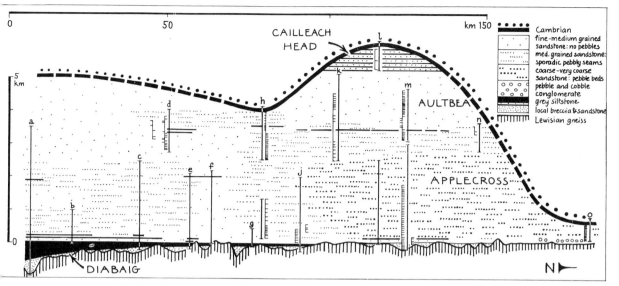

Fig. 3.12. Longitudinal profile of the Torridon Group between Rhum and Cape Wrath, perpendicular to the palaeocurrent direction. Key sections are *a*, Rhum; *b*, Soay; *c*, Scalpay; *d*, Toscaig; *e*, Raasay; *f*, Shieldaig to Applecross; *g*, Diabaig; *h*, Torridon, west and east of the Fasag Fault; *j*, Gairloch; *k*, Aultbea; *l*, Cailleach Head and Scoraig to Dundonnell; *m*, Summer Isles and Achiltibuie to Strath Kanaird; *n*, Rhubh Stoer; *o*, Cape Wrath. All available palaeomagnetic reversal data are also plotted, SE positive to the right, NW negative to the left.

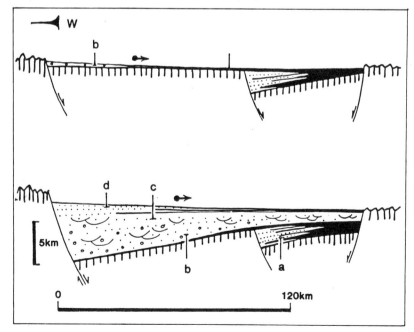

Fig. 3.13. Transverse profile of the Sleat and Torridon Groups in the latitude of Skye, restored to show their condition prior to Palaeozoic thrusting. Unit *a* belongs to the Sleat Group. Units *b–d* are the Applecross, Aultbea and Cailleach Head Formations of the Torridon Group. Arrows show palaeocurrent directions.

Fig. 3.14. Roughly stratified basal breccia of the Diabaig Formation on Beinn Dearg Bheag [NH 020825]. The photograph shows a bedding plane, about 30 m laterally from a Lewisian hill slope. Some of the loose pebbles strewn across the bedding plane have been liberated from the breccia by recent weathering.

laminae and pods are common. Cryptarchs are abundant in the shale and are particularly well-preserved in the phosphate (Naumova & Pavlovski 1961; Peat & Diver 1982). Grey shale sequences can be over 100 m thick. Grey sandstone beds appear in the upper part of the facies, increasing in frequency and thickness towards the top.

The grey sandstones contain about 15 per cent of clayey matrix and are petrographically, therefore, subgrey-wackes. They are usually several decimetres thick. The beds are typically massive, with sharp bottoms. Ripple-drift lamination due to currents flowing from the west is common near the tops of the beds (Fig. 3.17). It is significant that the shales interbedded with the grey sandstones are still desiccated.

The simplest interpretation of the Diabaig facies is that the breccias and tabular sandstones are fan deposits which accumulated in the palaeovalleys, with the grey shales recording ephemeral lakes in the valley bottoms. The arrival of Applecross rivers (see below) before the end of Diabaig times is evidenced by turbidites (the grey sand-

stones) in the shales. Ultimately these rivers completely filled the lakes and buried the remaining Lewisian hill tops. Boron-in-illite studies suggest that the shales are non-marine (Stewart & Parker 1979), as does the lack of primary carbonate and complete absence of evaporites. However, temporary marine influences cannot be entirely excluded. The significance of the southward thickening shown by the Formation as a whole, and the grey shale in particular (Fig. 3.12), is not clear. It could simply result from increasing palaeorelief. There seems to be no evidence for trunk streams during Diabaig times, except in the sector Inver-polly Forest–Cam Loch where gneiss-cobble conglomerate fills some palaeovalleys (Stewart 1972, Fig. 8).

The *Applecross Formation* consists of red sandstones in which trough cross-bedding is slightly more abundant than the planar type (but *cf.* Selley 1969, Table 1 & Fig. 2). Average width-to-depth ratio for the troughs is 10. Some troughs are 30 m wide, but most are only 1–2 m. The sandstones usually contain pebbles, including highly distinctive types such as porphyry and jasper. Pebble abundance provides a ready index to the Applecross subfacies recognised at Cape Wrath by Williams (1969b), and also used in Fig. 3.12. Red siltstone beds are virtually absent from all but the lowest 100 m of the Formation. One or two grey siltstone beds mark the top of the Formation and appear to be of regional significance for they correlate with a sequence of rapid palaeomagnetic reversals (see Fig. 3.12). The Formation north of the Loch Maree fault is remarkable geochemically in showing a monotonic upward decline in the ratio Na_2O/K_2O from unity at the base to almost zero at the top, due to the progressive disappearance of plagioclase.

About half the beds show soft sediment contortions which usually take the form of open synclines 0·5–2 m wide, linked by sharp cusps. The cusps frequently have structureless cores, suggesting fluidisation by upward moving pore water (type B pillars of Lowe 1975). Isolated cusps also occur, along with complex recumbent folds and overturned cross-bedding. Most cusps tend to lie perpendicular to the palaeocurrent direction, slightly overturned in the down-current (or down-slope) direction. The great majority of the beds, though not all, have been mobilised only once, for the contortions are usually truncated by the base of the next bed.

The origin of these contortions, which are far more abundant here than in any comparable clastic sequence, is still obscure despite a substantial amount of research (Selley & Shearman 1962; Selley et al. 1963; Stewart 1963; Selley 1969). Earthquake shocks are certainly capable of creating structures like these, but it would require an incredible degree of seismic activity to ensure that the deposition of every second bed coincided with an earthquake. More likely, perhaps, is the idea of liquefaction resulting from a flood-related process.

The *Aultbea Formation* consists of red sandstones which differ from the Applecross in being generally finer. Average grain size is slightly less than 0·5 mm and pebbles are generally absent except for a lens south of Applecross village. Virtually all the beds are contorted. A few grey shale units, individually no more than a metre or two thick, are notable for their sphaeromorphic acritarchs and filamentous sheaths (Zhang Zhongying et al. 1981; Zhang Zhongying 1982), perhaps the remains of a lake flora.

The interpretation of the Applecross Formation in terms of braided river deposition is well established from the studies of Selley (1965a, 1969) and Williams (1969b). Similar reasoning can be applied to the Aultbea. According to Miall (1977, Table V) Applecross sedimentation is Platte type, *i.e.* deposition was by linguoid and transverse sand

Fig. 3.15. Tabular sandstones of the Diabaig Formation showing rippled surfaces, from Balgy Bay [NG 852547].

Fig. 3.16. Desiccated grey shales of the Diabaig Formation on Diabaig shore [NG 796601].

bars in very shallow river channels. However, the lowest 500 m of the Applecross south of the Loch Maree fault shows Donjek type fining-upward cycles with red siltstone tops and erosive bases. These are well exposed on the northern coast of Loch Gairloch, immediately west of Big Sand fishing station.

North of the Loch Maree fault the lowest 500 m of the Applecross is built of fining-upward alluvial fan cycles, of the order of 100 m thick, which Williams (1969b) showed had their apices along the Minch fault, roughly 40–50 km from the present outcrop.

This difference in Applecross stratigraphy north and south of the Loch Maree fault suggests that it was active in Torridon Group times. It is significant that the line formed by the intersection of the base of the Torridon Group and the base of the Stoer Group is dextrally displaced by the fault at least 17 km, only 5 km of which is post Cambrian (Peach *et al.* 1907, pp. 192 & 548).

The *Cailleach Head Beds* are only exposed on the cliffs of ,Cailleach Head. The base of the Formation, concealed by the sea at this point, may be seen in the north-eastern part of Gruinard Island. The Formation consists of cyclothems, averaging 22 m in thickness (Fig. 3.18). Each begins with laminated dark-grey shales which pass up into tabular red

sandstones, internally containing planar cross-bedding and often covered with wave ripples. This tabular facies gives way above to trough cross-bedded sandstones, red or green in colour and often very micaceous. Deep desiccation cracks commonly occur near the top of the grey shale, but just as in the Diabaig Formation there are no evaporites or carbonates. Teall (*in* Peach *et al.* 1907, p. 287) described and figured microfossils from phosphatic laminae and pods in the shales – the first Precambrian fossils described in Britain.

The lack of evaporites and carbonates from the cyclothems suggests that they represent repeated delta advance into fresh-water lakes. The thickness of the tabular facies suggest maximum water depths around 5–6 m.

Provenance of the Sleat & Torridon Groups

The Diabaig Formation was clearly derived from gneisses which can still be seen outcropping close by, but the rest of the sediments come from areas now inaccessible, the nature of which can only be inferred from the geochemistry and mineralogy of the sediments themselves. For this purpose 300 whole rock chemical analyses and 220 modal analyses are now available, sampling the whole of the two Groups.

A comparison of the geochemistry of Laxfordian gneisses around Diabaig with the average composition of the adjacent Diabaig Formation is shown in Table 3.2. It is clear from this that 80 per cent of the calcium and 25 per cent of the sodium present in the source rocks have disappeared from the system. Virtually all the mafic minerals in the gneiss, together with about 40 per cent of the plagioclase, were destroyed either in the weathering profile or by interstratal dissolution from precursors of the present Diabaig sediments. At the same time the magnetite originally present in the gneisses must have been transformed to the haematite now found in the sediments. The Si and Al released as a result of plagioclase breakdown were removed as clay and redeposited downstream in the shales, along with chlorite and iron minerals from the mafics. The Ca and Na were probably removed in solution in a hydraulically open system. Loss of Ca and Na from the Formation

during the diagenetic transformation of smectite to illite can be shown to have been relatively slight.

This loss of soluble components during Diabaig times is in strong contrast to their conservation in the lowest sediments of the Stoer Group, suggesting a climate in Diabaig times much wetter than that during the early part of the Stoer Group.

The source areas for the arkoses of the rest of the Torridon Group and the Sleat Group have to be compatible with the following facts:

Gross mineralogy of the sand-sized material and schist pebbles; quartz, orthoclase, microcline, perthite, albite (in the Torridon Group), oligoclase (in the Sleat Group).

Pebble suite; this comprises muscovite schist, porphyritic rhyolite and dacite, chert and jasper (sometimes oolitic and possibly containing greenalite), banded iron formation, and tourmalinized quartzite (Williams 1969a;

Fig. 3.17. Grey sandstones near the top of the Diabaig Formation on Diabaig shore [NG 79276027].

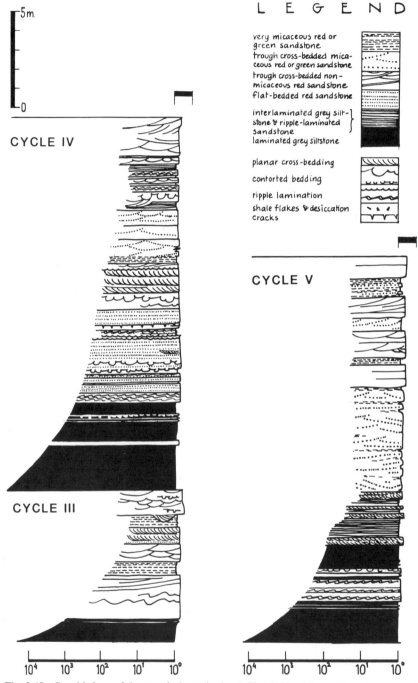

Fig. 3.18. Graphic logs of three cyclothems in the Cailleach Head Formation. Fifteen such cyclothems are perfectly exposed on the cliffs and beach south of the Head. The base of cyclothem III is at NG 98499848 and the top of cyclothem V at NG 98469835. The horizontal scale gives the lateral persistency of the beds, defined as lateral extent divided by maximum thickness. The grain-size bar represents 0–4 ϕ units.

Table 3.2. Comparative geochemistry of Laxfordian Basement and the adjacent Diabaig Formation

	A	B
SiO_2	68·8	70·7
TiO_2	0·4	0·5
Al_2O_3	15·4	13·5
T Fe_2O_3	3·7	4·6
MnO	0·1	0·0
MgO	1·8	3·0
CaO	3·4	0·7
Na_2O	4·3	3·3
K_2O	2·3	3·1
P_2O_5	—	0·0
Vol.	—	0·9
Total	100·2	100·3

Column A: Laxfordian basement
Based on 90% biotite gneiss + 10% metadolerite (Holland & Lambert 1973, Tables 4 G & 6 K).

Column B: average Diabaig sediment
Based on 29 analyses by Rodd (1983).

Anderton 1980). In the Sleat Group, however, metasedimentary pebbles are completely lacking.

Laxfordian ages for the volcanic, muscovite schist and microcline pebbles (Moorbath *et al.* 1967).

Scourian ages for some of the tourmaline-bearing quartzite pebbles (Allen *et al.* 1974).

K/Rb ratio for the sediments of about 300, similar to that in the upper crust.

Bearing in mind that the source area for most of the Torridon Group was over the present position of the Outer Hebrides it is not surprising to find that the Laxfordian granite gneiss and pegmatites of Harris and Lewis could have provided geochemically and mineralogically suitable sand-grade material. Details of these rocks have been published by Jehu & Craig (1927, 1934), Dearnley (1963), Myers (1971) and Van Breeman *et al.* (1971). The granite gneisses contain about 35 per cent of albite–oligoclase. The absence of oligoclase (and Ca) from the Applecross and Aultbea Formations is partly due to albitization, but also to its destruction during weathering in the mountainous source regions. The Ca released was removed from the system in solution even more effectively than from the Diabaig, doubtless the result of the higher rainfall in the mountains. The formidable amounts of clay which must have been produced have also vanished.

The upward decline in albite through the Applecross, mentioned earlier, is not matched by any significant reduction in grain-size and probably results from gradual reduction of source area relief after rapid initial uplift. This would mean that as time progressed there would be more effective weathering of the source rocks. First the oligoclase would go, then the albite, in accord with Goldich's stability series. A rather similar trend of increasing maturity with time is seen in the Sleat Group (Fig. 3.10). However, the source area for the Sleat Group, as already explained, seems to have been somewhat different in composition, with more calcic plagioclase than the gneisses in the Outer Hebrides today.

The source of the 'exotic' volcanic and metasedimentary pebbles is less easily specified. According to Moorbath *et al.* (1967) they represent Laxfordian supracrustals. This explains the Rb–Sr ages, the well-preserved igneous textures in the volcanic pebbles, and the fine grain-size of the greenalite-bearing chert pebbles which suggests $T \simeq 200°C$, and $P < 2$ kb (Klein 1983, pp. 434–440 & Figs. 11–16). On this basis the igneous pebbles could be giving unmodified eruptive ages. The Laxfordian mineral assemblage of the Outer Hebrides today is in the amphibolite facies, corresponding to $P \simeq 5$ kb. So the total uplift of the Outer Hebrides block since Sleat group times would need to be about 10 km. Considering that 4 km of uplift occurred during the Permo–Triassic alone (Steel & Wilson 1975) this is perfectly feasible. But the hypothesis fails to account for the Scourian ages for tourmaline in quartzite pebbles. Nor does it explain why the pebble suite is so persistent through over 3 km of Applecross and Aultbea sediment.

An alternative hypothesis is that the metasediments and volcanics are *Scourian* supracrustals. The association of porphyritic rhyolite and dacite with cherts, banded iron formation and quartzites is like that found in Archaean greenstone belts and certain early Proterozoic basins. Any basic volcanics or greywackes originally present would not be expected to survive as pebbles, though Black and Welsh (1961) record greywacke pebbles in the Applecross of Rhum. The hypothesis requires the Rb–Sr ages to have been reset during Laxfordian metamorphism by temperatures not much more than the 200°C mentioned above, while at least some of the $^{40}Ar/^{39}Ar$ tourmaline ages remained unaltered. If these Scourian sediments and volcanics were deeply infolded, or thrust into the basement as a result of Laxfordian deformation then they would be able to contribute clasts to the sediments for a long time. The early Proterozoic sediments around Gairloch (Williams *et al.* 1985) and in the Outer Hebrides may represent deeper infolds of similar supracrustals.

The proposition that the Sleat and Torridon Groups fill ancient rifts (Fig. 3.13) satisfies all the provenance data given above, and also the relationship to Palaeozoic thrusting. This relationship involves thrusts lying perpendicular to the palaeocurrents in the Torridon Group, and thrust location just where rift-bounding faults would be expected. If the thrusts are reactivated normal faults then perhaps the eastern boundary of the Torridon Group was not a major fault (*cf.* Stewart 1982, Fig. 4 with Fig. 3.13 herein).

REFERENCES

ALLEN, P., SUTTON, J., & WATSON, J. V. — 1974 — Torridonian tourmaline-quartz pebbles and the Precambrian crust northwest of Britain. *J. geol. Soc. London*, **130**, 85–91.

ANDERTON, R. — 1980 — Distinctive pebbles as indicators of Dalradian provenance. *Scott. J. Geol.*, **16**, 143–152.

BAILEY, E. B. — 1955 — Moine tectonics and metamorphism in Skye. *Trans. geol. Soc. Edinburgh*, **16**, 93–166.

BARBER, A. J., BEACH, A., PARK, R. G., TARNEY, J., & STEWART, A. D. — 1978 — The Lewisian and Torridonian rocks of north-west Scotland. *Geol. Assoc. Guide*, **21**, 1–99.

BLACK, G. P. & WELSH, W. — 1961 — The Torridonian succession of the Isle of Rhum. *Geol. Mag.*, **98**, 265–276.

BLUNDELL, D. J., HURICH, C. A. & SMITHSON, S. B. — 1985 — A model for the MOIST seismic reflection profile, N Scotland. *J. geol. Soc. London*, **142**, 254–258.

BYERS, P. N. — 1972 — *Correlation and provenance of the Precambrian Moine and Torridonian rocks of Morar, Raasay, Rhum, and Skye, northwest Scotland.* Univ. Reading, Ph.D. thesis, 2 vols.

COWARD, M. P. & WHALLEY, J. S. — 1979 — Texture and fabric studies across the Kishorn Nappe, near Kyle of Lochalsh, western Scotland. *J. struct. Geol.*, **1**, 259–273.

COWIE, J. & McNAMARA, K. J. — 1978 — Olenellus (Trilobita) from the Lower Cambrian strata of north-west Scotland. *Palaeontology, London*, **21**, 615–634.

CHESHER, J. A., SMYTHE, D. K. & BISHOP, P. — 1983 — The geology of the Minches, Inner Sound and Sound of Raasay. *Rep. Inst. geol. Sci. London*, **83/6**, 1–29.

CLOUD, P. & GERMS, A. — 1971 — New Pre-Paleozoic nannofossils from the Stoer Formation (Torridonian), Northwest Scotland. *Bull. geol. Soc. Am.*, **82**, 3469–3474.

DEARNLEY, R. — 1963 — The Lewisian complex of South Harris with some observations on the metamorphosed basic intrusions of the outer Hebrides, Scotland. *Q. J. geol. Soc. London*, **119**, 243–312.

DEMICCO, R. V. & KORDESCH, E. G. — 1986 — Facies sequences of a semi-arid closed basin: the Lower Jurassic East Berlin Formation of the Hartford Basin, New England, USA. *Sedimentology*, **33**, 107–118.

EVANS, D., CHESHER, J. A., DEEGAN, C. E. & FANNIN, N. G. T. — 1982 — The offshore geology of Scotland in relation to the IGS shallow drilling program, 1970–1978. *Rep. Inst. geol. Sci. London*, **81/12**, 1–36.

GEIKIE, A. — 1888 — Report on the recent work of the Geological Survey in the north-west Highlands of Scotland based on the field notes and maps of Messrs. B. N. Peach, J. Horne, W. Gunn, C. T. Clough, L. Hinxman and H. M. Cadell. *Q. J. geol. Soc. London*, **44**, 378–439.

—— 1895 — *Rep. geol. Surv. Mus. London* for 1894.

—— 1906 — The history of the geography of Scotland. *Scott. geogr. Mag.*, **22**, 117–134.

GRACIE, A. J. & STEWART, A. D. — 1967 — Torridonian sediments at Enard Bay, Ross-shire. *Scott. J. Geol.*, **3**, 181–194.

HALL, A. M. — 1985 — Cenozoic weathering covers in Buchan, Scotland and their significance. *Nature, London*, **315**, 392–395.

HARDIE, L. A., SMOOT, J. P. & EUGSTER, H. P. — 1978 — Saline lakes and their deposits: a sedimentological approach. *Spec. Publ. Int. Assoc. Sedimentol.*, **2**, 7–41.

HARKER, A. — 1908 — The geology of the small isles of Inverness-shire. *Mem. geol. Surv. Scotland*, 1–210.

HOLLAND, J. G. & LAMBERT, R. ST. J. — 1973 Comparative major element geochemistry of the Lewisian of the mainland of Scotland. *In* Park, R. G. & Tarney, J. (Eds.), *The Early Precambrian of Scotland and related rocks of Greenland*, Dept. Geology, Univ. of Keele, Newcastle, 51–62.

1975 The chemistry and origin of the Lewisian gneisses of the Scottish mainland: the Scourie and Inver assemblages and sub-crustal accretion. *Precambrian Res.*, **2**, 161–188.

JEHU, T. J. & CRAIG, R. M. — 1927 Geology of the Outer Hebrides. Part IV – South Harris. *Trans. R. Soc. Edinburgh*, **55**, 457–488.

1934 Geology of the Outer Hebrides. Part V – North Harris and Lewis. *Trans. R. Soc. Edinburgh*, **57**, 839–874.

KENNEDY, W. Q. — 1951 Sedimentary differentiation as a factor in the Moine–Torridonian correlation. *Geol. Mag.*, **88**, 257–266.

KLEIN, C. — 1983 Diagenesis and metamorphism of Precambrian banded iron-formations. *In* Trendall, A. F. & Morris, R. C. (Eds.) *Iron-formation; facts and problems.* Elsevier, Amsterdam, 417–469.

LAWSON, D. E. — 1965 Lithofacies and correlation within the lower Torridonian. *Nature, London*, **207**, 706–708.

1972 Torridonian volcanic sediments. *Scott. J. Geol.*, **8**, 345–362.

1976 Sandstone-boulder conglomerates and a Torridonian cliffed shoreline between Gairloch and Stoer, northwest Scotland. *Scott. J. Geol.*, **12**, 67–88.

LOWE, D. R. — 1975 Water escape structures in coarse-grained sediment. *Sedimentology*, **22**, 157–204.

MACCULLOCH, J. — 1819 *A description of the western islands of Scotland, including the Isle of Man, comprising an account of their geological structure, with remarks on their agriculture, scenery, and antiques.* 3 vols., London.

MACKIE, W. — 1923 The source of purple zircons in the sedimentary rocks of Scotland. *Trans. geol. Soc. Edinburgh*, **11**, 200–213.

McQUILLIN, R. & BINNS, P. E. — 1973 Geological structure in the Sea of the Hebrides. *Nature (phys. Sci.)*, **241**, 2–4.

MIALL, A. D. — 1977 A review of the braided-river depositional environment. *Earth Sci. Rev.*, **13**, 1–62.

MOORBATH, S. — 1969 Evidence for the age of deposition of the Torridonian sediments of north-west Scotland. *Scott. J. Geol.*, **5**, 154–170.

MOORBATH, S., STEWART, A. D., LAWSON, D. E. & WILLIAMS, G. E. — 1967 Geochronological studies on the Torridonian sediments of north-west Scotland. *Scott. J. Geol.*, **3**, 389–412.

MOSELEY, F. — 1971 A reconnaissance of the Wadi Beihan, South Yemen. *Proc. geol. Assoc. London*, **82**, 61–69.

MYERS, J. S. — 1971 The late Laxfordian granite–migmatite complex of western Harris, Outer Hebrides. *Scott. J. Geol.*, **7**, 234–284.

NAUMOVA, S. N. & PAVLOVSKY, E. V. — 1961 The discovery of plant remains (spores) in the Torridonian shales of Scotland. *Dokl. Acad. Sci. USSR*, **141**, 181–182.

NICOL, J. — 1866 *The geology and scenery of the north of Scotland: being two lectures given at the Philosophical Institution, Edinburgh, with notes and an appendix.* Edinburgh, 1–96.

PEACH, B. N., HORNE, J., GUNN, W., CLOUGH, C. T., HINXMAN, L. W. & TEALL, J. J. H. — 1907 *The geological structure of the north-west highlands of Scotland.* Mem. geol. Surv. G.B., 1–668.

PEAT, C. J. — 1984 Comments on some of Britain's oldest microfossils. *J. Micropalaeontol.*, **3**, 65–71.

PEAT, C. J. & DIVER, W. — 1982 First signs of life on Earth. *New Scientist*, **95**, 776–780.

PLANT, J. A.   1984   Regional geochemical maps of the United Kingdom. *NERC News J.*, **3(4)**, 1 & 5–7.

RAST, N., DIGGENS, J. N. &   1968   Triassic rocks of the Isle of Mull; their sedimentation, facies,
RAST, D. E.   structure and relationship to the Great Glen Fault and the Mull caldera. *Proc. geol. Soc. London*, **1645**, 299–304.

RODD, J. A.   1983   *The sedimentology and geochemistry of the type Diabaig Formation in the Upper Proterozoic Torridon Group of Scotland.* Univ. Reading, Ph.D. thesis, 1–596.

SELLEY, R. C.   1965a Diagnostic characters of fluviatile sediments of the Torridonian Formation (Precambrian) of northwest Scotland. *J. sediment. Petrol.*, **35**, 366–380.

                                      1965b The Torridonian succession on the islands of Fladday, Raasay, and Scalpay, Inverness-shire. *Geol. Mag.*, **102**, 361–369.

                                      1969   Torridonian alluvium and quicksands. *Scott. J. Geol.*, **5**, 328–346.

SELLEY, R. C. &   1962   Experimental production of sedimentary structures in quick-
SHEARMAN, D. J.   sands. *Proc. geol. Soc. London*, **1599**, 101–102.

SELLEY, R. C.,   1963   Some underwater disturbances in the Torridonian of Skye and
SHEARMAN, D. J.,   Rhum. *Geol. Mag.*, **100**, 224–243.
SUTTON, J. & WATSON, J.

SHERATON, J. W.,   1973   The geochemistry of the Scourian gneisses of the Assynt district.
SKINNER, A. C. &   *In* Park, R. G. & Tarney, J. (Eds.) *The Early Precambrian of*
TARNEY, J.   *Scotland and related rocks of Greenland.* Geol. Dept., Univ. of Keele, Newcastle, 13–30.

SMITH, N. D.   1970   The braided stream depositional environment; comparison of the Platte River with some Silurian clastic rocks, north-central Appalachians. *Bull. Am. Assoc. Petrol. Geol.*, **81**, 2993–3014.

SMITH, R. L.,   1983   Palaeomagnetic studies of the Torridonian sediments, NW
STEARN, J. E. F. &   Scotland. *Scott. J. Geol.*, **19**, 29–45.
PIPER, J. D. A.

SMYTHE, D. K.,   1972   Deep sedimentary basin below northern Skye and the Little
SOWERBUTTS, W. T. C.,   Minch. *Nature (phys. Sci.)*, **236**, 87–89.
BACON, M. &
McQUILLIN, R.

STEEL, R. J. &   1975   Sedimentation and tectonism (?Permo-Triassic) on the margin
WILSON, A. C.   of the North Minch Basin, Lewis. *J. geol. Soc. London*, **131**, 183–202.

STEWART, A. D.   1962   On the Torridonian sediments of Colonsay and their relationship to the main outcrop in north-west Scotland. *Liverpool Manchester geol. J.*, **3**, 121–156.

                                      1963   On certain slump structures in the Torridonian sandstones of Applecross. *Geol. Mag.*, **100**, 205–218.

                                      1966a On the correlation of the Torridonian between Rhum and Skye. *Geol. Mag.*, **103**, 432–439.

                                      1966b An unconformity in the Torridonian. *Geol. Mag.*, **103**, 462–465.

                                      1969   Torridonian rocks of Scotland reviewed. *Mem. Am. Assoc. Petrol. Geol.*, **12**, 595–608.

                                      1972   Precambrian landscapes in northwest Scotland. *Geol. J.*, **8**, 111–124.

                                      1975   'Torridonian' rocks of western Scotland. *Spec. Rep. geol. Soc. London*, **6**, 43–51.

                                      1982   Late Proterozoic rifting in NW Scotland: the genesis of the 'Torridonian'. *J. geol. Soc. London*, **139**, 413–420.

STEWART, A. D. &   1974   Palaeomagnetism of Precambrian sedimentary rocks from NW
IRVING, E.   Scotland and the apparent polar wandering path of Laurentia. *Geophys. J. R. astron. Soc.*, **37**, 51–72.

STEWART, A. D. & PARKER, A. — 1979 Palaeosalinity and environmental interpretation of red beds from the late Precambrian ('Torridonian') of Scotland. *Sediment. Geol.*, **22**, 229–241.

SUTTON, J. & WATSON, J. — 1960 Sedimentary structures in the Epidotic Grits of Skye. *Geol. Mag.*, **97**, 106–122.

1964 Some aspects of Torridonian stratigraphy in Skye. *Proc. Geol. Assoc. London*, **75**, 251–289.

UPFOLD, R. L. — 1984 Tufted microbial (cyanobacterial) mats from the Proterozoic Stoer Group, Scotland. *Geol. Mag.*, **121**, 351–55.

VAN BREEMEN, O., AFTALION, M. & PIDGEON, R. T. — 1971 The age of the granite injection complex of Harris, Outer Hebrides. *Scott. J. Geol.*, **7**, 139–152.

WILLIAMS, G. E. — 1966 Palaeogeography of the Torridonian Applecross Group. *Nature, London*, **209**, 1303–1306.

1968 Torridonian weathering, and its bearing on Torridonian palaeo-climate and source. *Scott. J. Geol.*, **4**, 164–184.

1969a Petrography and origin of pebbles from Torridonian strata (late Precambrian), Northwest Scotland. *Mem. Am. Assoc. Petrol. Geol.*, **12**, 609–629.

1969b Characteristics and origin of a Precambrian pediment. *J. Geol. Chicago*, **77**, 183–207.

1971 Flood deposits of the sand-bed ephemeral streams of central Australia. *Sedimentology*, **17**, 1–40.

WILLIAMS, P. J., TOMKINSON, M. J. & CATTELL, A. C. — 1985 Petrology and deformation of metamorphosed volcanic-exhalative sediments in the Gairloch Schist Belt, NW Scotland. *Mineralium Deposita*, **20**, 302–308.

ZHANG ZHONGYING — 1982 Upper Proterozoic microfossils from the Summer Isles, NW Scotland. *Palaeontology, London*, **25**, 443–460.

ZHANG ZHONGYING, DIVER, W. L. & GRANT, P. R. — 1981 Microfossils from the Aultbea Formation, Torridon Group, Tanera Beg, Summer Isles. *Scott. J. Geol.*, **17**, 149–154.

4

MOINE

A. L. Harris and M. R. W. Johnson

After about 100 years of controversy it is now generally accepted that the Moine shows polyorogenic evolution spanning some 600 Ma. The acceptance of polymetamorphism vindicates H. H. Read and J. Horne, who dissented from B. N. Peach and E. B. Bailey who favoured an entirely Caledonian orogenic history for the Moine. However, many problems remain; in particular the style and tectonic significance of the earliest, late Mid-Proterozoic, orogenic phase, is not clear and the role of Proterozoic *vis-à-vis* Caledonian metamorphism is controversial.

Moine of the Northern Highlands

Stratigraphy

The Moine rocks of the north Highlands of Scotland are predominantly metasedimentary and comprise monotonous psammitic, pelitic and semipelitic lithostratigraphic units, or units which are characteristically striped and banded with psammitic, pelitic or semipelitic layering on all scales. Units identified on the basis of their gross lithostratigraphic nature can be further distinguished by the presence or absence of minor lithologies such as calc–silicate lenses or bands, heavy-mineral laminae and amphibolite bodies which are commonly lensoid. Using such lithostratigraphic criteria and the way-up evidence of sedimentary structures preserved in many of the formations, Johnstone *et al.* (1969, table 1) distinguished several formations within a structural–stratigraphic framework which they erected for the Moine rocks of western Inverness-shire (1969, figs. 1 and 2).

Inverness-shire

The structural–stratigraphic model for the rocks of western Inverness-shire introduced by Johnstone *et al.* (1969) and developed by Tanner *et al.* (1970) and Johnstone

(1975) placed earlier work on comparatively small areas into a larger context. Three divisions were recognised, the Morar, Glenfinnan and Loch Eil (Figs. 4.1, 4.2). The early Morar work (Richey *et al.* 1936, 1937; Richey & Kennedy 1939) and that around Loch Hourn (Ramsay & Spring 1962) lies almost entirely within the Morar Division; that in Lochailort (Powell 1964) lies across the Morar/Glenfinnan boundary, while in Ardgour Dalziel (1966) was concerned with the Glenfinnan/Loch Eil boundary. The high grade, severely deformed sub-Moine of Richey & Kennedy (1939) occurs in the core of the Morar Anticline, where it is spatially closely associated with acid and mafic Lewisian gneisses, and was thought to be unconformably overlain by the Morar cover succession of lower grade, more weakly deformed rocks. Subsequently Kennedy (1954) recognised that the lithostratigraphy of the sub-Moine is the same as that of the Morar succession but believed, it transpires incorrectly, that they occurred in different nappes, separated by a thrust or slide.

The status and interpretation of Lewisian inliers interleaved within the Moine as slices of late Archaean basement were crucial in Moine stratigraphic interpretation. Major inliers within the Morar Division were regarded (Johnstone *et al.* 1969) as parautochthonous basement, forming isoclinal fold cores, generally adjacent to the oldest Moine rocks of the Morar Division (Fig. 4.3). Moine stratigraphic sequences in this context are symmetrically disposed on either side of inliers, as is the case in Morar and Glenelg. Other inliers, notably those emplaced as slices in major ductile thrust zones, are characterised by marked asymmetry of Moine stratigraphic sequence on either side, as is exemplified on a regional scale by the inliers along the Sgurr Beag Slide (ductile thrust) which separates the Morar and Glenfinnan lithostratigraphies. The structural–stratigraphic model proposed by Johnstone *et al.* (1969), Tanner *et al.* (1970) and Johnstone (1975) for the area mainly to the south of the Strathconon fault has subsequently been applied in modified form to the whole of the mainland Moine including the north coast region (see Moorhouse & Moorhouse 1983, Holdsworth 1987). This, and the Moine rocks of Shetland Flinn *et al.* (1972) and Flinn (1987) will be described in more detail later.

In 1970 the contacts between divisions were thought to be everywhere tectonic but most Moine geologists believed that whereas stratigraphic relationships could be established within divisions, the original stratigraphic relationships between divisions had been obscured, possibly irretrievably. The Sgurr Beag Slide (ductile thrust) was

Fig. 4.1. Western Highlands showing the distribution of divisions of groups of the Moine, Lewisian inliers and major Caledonian structures.
A, Assynt; B, Bettyhill; C, Carn Chuinneag; F, Fannich; G, Glenfinnan; K, Kintail; Kn, Knoydart; KT, Kyle of Tongue; M, Morar; Mo, Monar; Q, Quoich; Rm, Ross of Mull; S.B.T., Sgurr Beag thrust (Slide); X—X', Line of section in Fig. 4.14

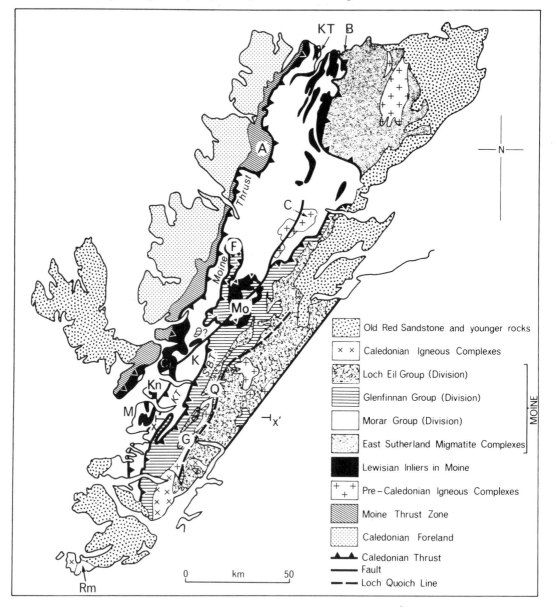

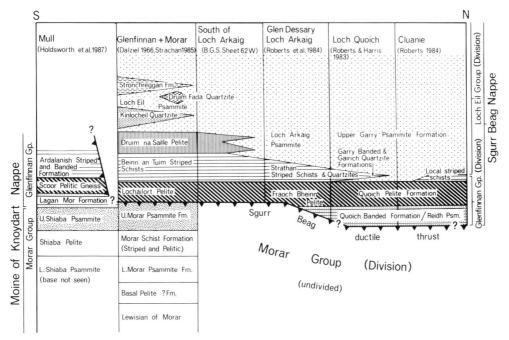

Fig. 4.2. Stratigraphic successions and correlations in the Sgurr Beag and Moine of Knoydart nappes (adapted from Roberts *et al.* 1987).

believed to have many tens of kilometres displacement; it was regarded as a structure of such magnitude that stratigraphy on either side could not be matched or related (Fleuty *in* Johnstone 1975), although Powell (1964) working at Lochailort had noted consistent younging of the Upper Morar Psammite (Morar Division) towards the Lochailort Pelite (Glenfinnan Division), a relationship subsequently found to hold true for much of the south-western Moine region. The slide and its associated Lewisian basement inliers was interpreted by Tanner *et al.* (1970) as marking the site of a former basement high separating basins within each of which the Morar and Glenfinnan divisions respectively accumulated. The Loch Quoich line (Clifford 1957) forming the Glenfinnan–Loch Eil contact was a structure of uncertain status. Hence, the apparent younging of Loch Eil Division rocks away from the Glenfinnan could not be relied upon as having original stratigraphic significance, although Dalziel (1966) recognised that the Garvan Psammite (= Loch Eil Psammite) formed a stratigraphical and structural outlier within rocks which were later to be assigned to the Glenfinnan Division.

In the 1980s several investigations into the structure and stratigraphy of the Moine of the south-west N Highlands have done much to elucidate Moine stratigraphy, notably Stoker (1983), Roberts *et al.* (1984, 1987), Strachan (1985) and Holdsworth *et al.* (1987). The stratigraphy of the rocks above the Sgurr Beag ductile thrust in Inverness-shire was

set out by Roberts *et al.* (1987, fig. 4) (Fig. 4.2), the former Loch Eil and Glenfinnan structural–stratigraphic *divisions* being assigned the status of lithostratigraphic *groups*. A four-fold sequence is recognised in the Glenfinnan Group (older) with formations having only local status and having rapid facies variations both lateral and vertical into Loch Eil Group lithologies. Thus the Upper Garry Psammite Formation of the Loch Eil Group at Cluanie appears to have a lateral passage into typical striped and pelitic formations at Glenfinnan (Fig. 4.2). The essentially monotonous Loch Eil Group psammites have been subdivided by Strachan (1985) (Fig. 4.2).

On the Scottish mainland the rocks of the Morar Division are everywhere separated from the Glenfinnan Group by the Sgurr Beag ductile thrust, but in the Ross of Mull this thrust is missing at the Morar–Glenfinnan boundary which here is a normal stratigraphic passage, with good sedimentary way-up evidence. On the basis of this evidence, Holdsworth *et al.* (1987) assigned the Morar Division also to lithostratigraphic *group* status.

In the light of this new work, the Moine of the Northern Highlands may now be regarded as a lithostratigraphic *Supergroup* comprising three groups and numerous *formations* (Fig. 4.2).

Metabasic rocks are abundant locally in the Glenfinnan and Loch Eil groups but are generally absent from the Morar Group of Inverness-shire. Amphibolites of two types have been distinguished in the Glenfinnan Group. Para-amphibolites occur in the core of rare thick calc–silicate bands or lenticles and grade outwards into normal hornblende–garnet–plagioclase ± diopside-bearing calc–silicates. Ortho-amphibolites in the Glenfinnan Group are much more abundant than para-, especially in southern Inverness-shire and on Mull. They commonly occur as coarsely garnetiferous mafic rocks which vary from a centimetre to several metres in thickness. While they are normally concordant with foliation-banding in the country rocks, many bodies are locally cross-cutting, although original discordances between ortho-amphibolite sheets and modified bedding planes in metasediments are commonly obliterated by large tectonic strains. It is likely that they were originally semi-concordant sheets which locally changed horizon, cutting up-section.

Ortho-amphibolite sheets in both Loch Eil and Glenfinnan groups of western Inverness-shire and the Ross of Mull apparently suffered most if not all of the polyphase deformation and metamorphism of the Moine country rocks, although the early-orogenic West Highland granitic orthogneiss bodies (Barr *et al.* 1985) are cut by non-concordant ortho-amphibolite sheets. Amphibolite sheets have been recorded from Morar Group rocks of Ross-shire where Winchester (1976) showed them to be chemically distinct from those of the Glenfinnan Group rocks of

Ross-shire. Morar Group amphibolites are not known in Inverness-shire.

Sutherland

Holdsworth (1987) has provided a synthesis of the Moine rocks of the north coastal Moine rocks of Scotland. Figure 4.4 shows a section along the north coast between the Moine thrust zone in the west and the East Sutherland migmatite complex in the east. Several units are distinguished. From west to east these are in ascending structural order:

1. *The Moine* nappe comprising (a) in the west the *A'Mhoine Psammites* overlain across a thrust to the east by (b) the *Altnaharra Psammites* (≡ Morar Group).
   (a) arkosic commonly gritty psammites with local conglomerates especially towards the base, abundant heavy mineral laminae and rare calc–silicates.
   (b) banded psammites, gritty and arkosic at base; thin semipelitic and pelitic units and locally calc–silicate bearing.
2. The *Naver* nappe comprising the *Bettyhill Assemblage* (Moorhouse & Moorhouse 1983).

   This is a unit of mixed migmatised banded, locally highly feldspathic psammites, semipelites and pelites with sparse calc–silicates. It is similar to the high-grade

Fig. 4.3a. Moine (right)–Lewisian (left) contact, Loch Hourn near Glenelg. (Photograph by M. R. W. Johnson.)

Morar Group rocks of the Knoydart nappe (Barr *et al.* 1986), and is limited to the east by the Swordly thrust.

3. The *Swordly* nappe comprising (a) in the west the *Kirtomy Assemblage* which is overlain in the east by (b) the *Portsherra Assemblage* (Moorhouse & Moorhouse 1983).

    (a) highly migmatised striped psammitic/semipelitic with major pelitic units and common calc–silicates, the whole resembling Glenfinnan Group rocks.

    (b) migmatised micaceous psammites with subordinate semipelites (? = Loch Eil Group).

These Caledonian nappes involved the interleaving of Moine and Lewisian, both as thrust slices and as fold cores. It is generally recognised (Evans & White 1984, Barr *et al.* 1986, Holdsworth 1987) that the Moine rocks involved had already suffered deformation and amphibolite–facies metamorphism during Late Proterozoic orogenesis. Strachan & Holdsworth (1988) have subsequently shown that the Moine and Lewisian of central and southern

Sutherland have a similar history and structural style to those of the north coast.

Units of highly differentiated, tholeiitic ortho-amphibolites are widely recognised within both Moine and Lewisian rocks of the Sutherland region (Moorhouse & Moorhouse 1979, Winchester & Floyd 1984). The largest and regionally most significant group is the garnetiferous Ben Hope 'sill' suite (Moorhouse & Moorhouse 1979) which crops out in the Kyle of Tongue region. Holdsworth (1987) has revealed that the igneous precursors were emplaced as dilational sills, sub-parallel to bedding/banding, prior to all phases of deformation and metamorphism affecting the Moine rocks; cross-cutting relationships are locally preserved in regions of low strains. The largest amphibolites occur adjacent to the stratigraphic Moine–Lewisian boundary (e.g. the Ben Hope 'sill' itself), possibly because such a planar surface would have constituted a prominent anisotropy between basement gneisses and Moine sediments.

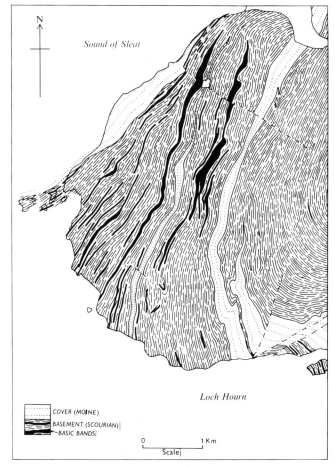

COVER (MOINE)
BASEMENT (SCOURIAN)
BASIC BANDS
0   1 Km
Scale

Fig. 4.3b. Interbanding of basement (Lewisian) and cover (Moine) in the Glenelg area. After Ramsay, in Johnson & Stewart (1963).

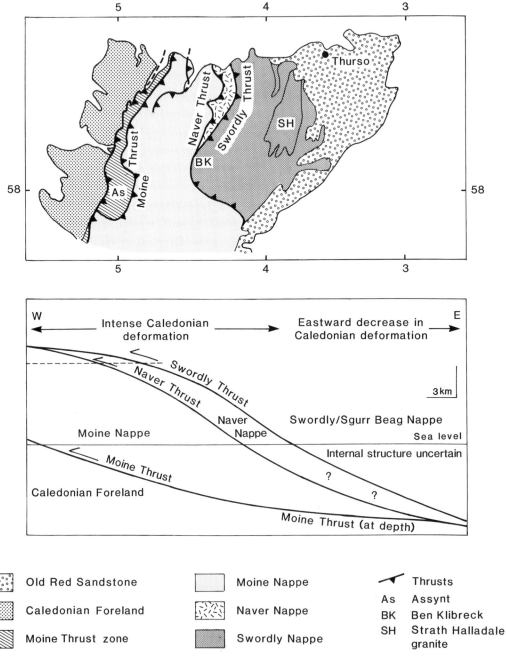

Fig. 4.4. Geological sketch map and section of N Scotland to show the main structural units. Mainly after Barr *et al.* (1986, figs. 1B and 3A).

Table 4.1. Simplified tectonic and metamorphic history of the N Highland Moines

	Igneous Activity		Deformation	Metamorphism
	Post-tectonic granites (414–400 Ma)		Open monoclines and kinkfolds	Cooling history ? 440–400 Ma General retrogression
	Borrolan syenite (430 ± 4 Ma)	MOINE THRUST c. 430 Ma	Isoclinal minor folds and coaxial stretching lineation plunging ESE	?
			Mylonite belts; Loch Alsh– Sgonnan Mor folds	
Caledonian Orogeny	Pegmatites (~440 Ma) microdiorites Strontian Granite 435 ± 10		Open-to-tight upright folds on all scales axial planes N/S – NE–SW	
	Glen Dessary syenite (456 ± 5 Ma)			High grade regional
			Interleaving of Lewisian basement along Sgurr Beag and other thrust sheets (localised coeval folding). Formation of blastomylonites in Sgurr Beag, Knoydart, Naver and Swordley thrust zones; stretching lineation plunges ESE	metamorphism in central N Highlands; low grade in West
Morarian/Knoydartian	Carn Chuinneag granite (550 ± 10 Ma) Strath Halladale granite (649 ± 30 Ma) Local pegmatites (c. 760 Ma)		None recorded	Temperature of rocks at present erosion level probably remained high
? Grenville Orogenesis			2. Tight-isoclinal major and minor folds; axial planar crenulation schistosity N–S extension. Folds curvilinear through < 180°.	Regional high-grade metamorphism and migmatisation
	Mafic dykes and sheets		Both episodes probably involve interfolding of Moine cover and Lewisian basement	
	Ardgour Granitic Orthogneiss (1,028 ± 43 Ma)		1. Isoclinal major and minor folds	

(leftmost bracket spanning Morarian/Knoydartian and ? Grenville Orogenesis: **PRECAMBRIAN**)

A later and distinct suite of metadolerites, the Loch a'Mhoid suite (Moorhouse & Moorhouse 1979), is also present in Sutherland. Its chemistry is mildly alkaline, within-plate type, and intrusion was syn-tectonic with respect to the Caledonian deformation.

Shetland

Two occurrences of Moine-like rocks at low–mid amphibolite facies have been recorded on Shetland. These crop out on either side of the Walls Boundary Fault (Fig. 4.5) and may have been separated by as much as 200 km before faulting occurred (Flinn 1987).

To the west of the fault in the northern part of the Mainland, thrust slices of psammitic metasediments of Moine aspect are interleaved with slices of basement gneiss and are in marked contrast to the Yell Sound Division which underlies most of Yell and part of Lunna Ness and which also forms a N–S strip some two kilometres wide which runs down the eastern side of the Walls Boundary

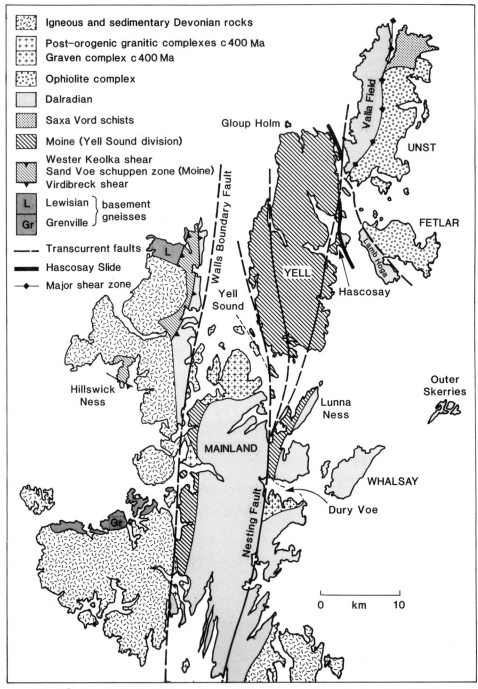

Fig. 4.5. Moine in Shetland (for location of Shetland see frontispiece). After Flinn (1987).

Fault on Mainland (Fig. 4.5). The Yell Sound Division carries Lewisian inliers and is also dominated by psammitic schists, with lesser quantities of quartzite and pelite.

Orogenic evolution

Variation in metamorphic grade of the Moine, as shown in Figure 4.6, increases from epidote amphibolite facies in the west to mid-amphibolite facies in the east in a NNE–SSW-trending zone, declining to lower amphibolite facies towards the Great Glen. Adjacent to the Moine thrust in the N Highlands there is a belt of greenschist–facies metamorphism indicated by the presence of garnet + biotite in pelitic schists and zoisite + albite or oligoclase in calc–silicates (Fig. 4.6). Farther east, rare kyanite appears in pelites and calc–silicates show hornblende + plagioclase assemblages. Within the Glenfinnan Group outcropping in the N Highland steep belt (Leedal 1952) sillimanite-bearing gneisses associated with hornblende ± pyroxene-bytownite-bearing calc–silicates indicate peak metamorphic grade in the middle amphibolite facies (Barr 1985). The Loch Eil Group rocks of the N Highland flat belt to the east of the Loch Quoich Line, while partly within the sillimanite zone, are for the most part of only lower amphibolite facies.

The simplicity of metamorphic zonation in the Moine was first demonstrated by Kennedy (1948) who relied mainly on the mineral assemblages in calc–silicates for the investigation of regional distribution of metamorphic grade. Tanner (1976) expanded Kennedy's scheme to give the following sequence: albite + zoisite + calcite + biotite→ oligoclase + zoisite + calcite + biotite →andesine + zoisite + biotite + hornblende→bytownite/anorthite + hornblende/ bytownite/anorthite + pyroxene. Although Winchester's (1974) metamorphic map, based on calc–silicates, is more complicated than Kennedy's it confirms the overall pattern, as does the map of Fettes *et al.* (1985), which is based on the anorthite content of plagioclase. The broad metamorphic character of the Moine is Barrovian (Harte 1988): it shows a simplicity which is belied however by two factors:

(a) if the isotopic ages used to calibrate orogenic activity are correct the metamorphic zonation is the result of at least two regional metamorphisms – late Proterozoic (~1000 Ma) and Caledonian (~460 Ma). The evidence rests heavily on the supposition that the radiometrically dated West Highland granitic gneiss was emplaced coevally with regional metamorphism and that fabrics associated with this metamorphism have been overprinted by fabrics of Caledonian age (see p. 100). Although there are still great difficulties, at least on a regional scale, in distinguishing the thermal effects of the two metamorphisms it is likely that the metamorphic zonation referred to above is largely of Proterozoic age.

(b) the attitude and distribution of the zones strongly reflect Caledonian events especially ductile thrusting and

related major folding, as well as local thermal 'highs' associated with Caledonian vein complexes (p. 281, Chapter 8). For example the fall-off in metamorphic grade passing eastward from the main outcrop of Glenfinnan Group into the Loch Eil Group is probably a reflection of the folding of isograds. Caledonian folds have folded the isograds in such a way that high-grade rocks comparable to those of the N Highland Steep Belt lie beneath the regionally flat-lying Loch Eil Group. That the latter has a high-grade 'infrastructure' is shown by the gneisses in the Achnacarry dome near the Great Glen.

It might be argued that the west-to-east increase in metamorphic grade in the western part of the N Highlands is merely a consequence of thrust stacking and fore-shortening. Certainly metamorphic grade is uniformly higher above the Sgurr Beag thrust which is probably the major Caledonian ductile thrust. What is significant, however, is that the difference in metamorphic grade between the footwall and hangingwall of this thrust varies from place to place and may be slight, suggesting that the thrust has disrupted a pre-existing west-to-east increase in metamorphic grade.

This account has promoted the importance of the older metamorphism at the expense of the Caledonian event. Yet the latter must have profoundly affected the Moine, not only in the form of vein complexes and pegmatite suites but also by widespread recrystallisation. This is beyond doubt where a time-marker permits clear recognition of Caledonian events. One such locality is in the aureole of the Carn Chuinneag granite which is radiometrically dated at 550 ± 10 Ma. Pre-hornfels garnet suggests that Precambrian metamorphic grade was low in that region. The contact-minerals were replaced by kyanite-bearing assemblages presumably during Caledonian regional metamorphism. Powell *et al.* (1982) viewed the Caledonian metamorphism in the West Highlands as increasing in grade from low greenschist facies in the west to middle amphibolite facies in the east, the metamorphism being roughly synchronous with displacement on the Sgurr Beag thrust. Elsewhere it would seem that the peak of Caledonian metamorphism was lower grade: for example, in Glen Dessary the Caledonian metamorphism appears to be of lower amphibolite facies, and contrived to retrogress high grade gneisses. Likewise in the northernmost Highlands the Caledonian metamorphism occurs as a weak overprint on high-grade rocks (Fettes *et al.* 1985).

In conclusion, although the Caledonian thermal event in the West Highlands is given a minimum age of 467 ± 20 Ma (Brewer 1979, Powell 1981) the regional thermal gradients existed as late as 450–420 Ma (Smith 1979) – the presumed time of intrusion of a suite of microdiorites (see Chapter 8). Late thermal events in the Moine continued at least to the time of major uplift along the Moine thrust dated as roughly 430 Ma by Johnson *et al.* (1985), a date which corresponds to 'closure' temperatures in muscovite and biotite.

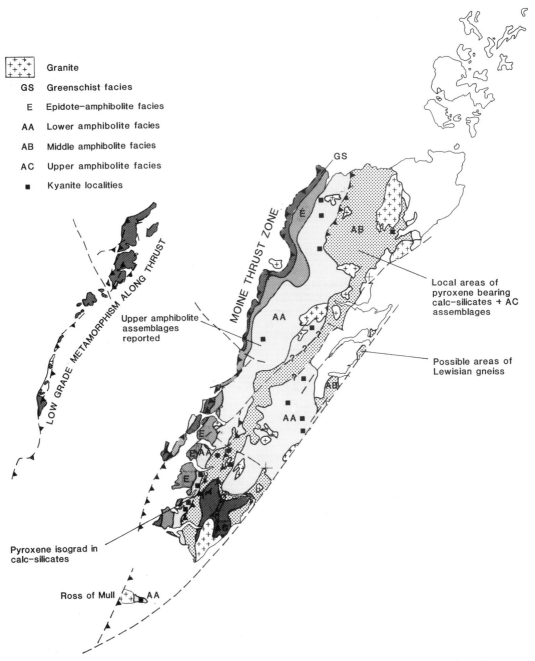

Fig. 4.6. Metamorphic zones in the Moine of the Western Highlands (adapted from Fettes *et al.* 1985).

Proterozoic orogenesis

Perhaps the most significant advance in Moine studies in recent years has been the recognition of pre-Caledonian orogenic events. The Precambrian orogeny is dated at ~1000 Ma and has been referred to variously as the Ardgourian and the Grenville orogeny, the latter term linking it with the important orogenic belt in eastern North America. The main evidence comes from radiometric dating of the West Highland granitic gneiss (Fig. 4.7A *and* F), viz. a whole rock Rb/Sr age of 1028±43 Ma (Brook *et al.* 1976, Brewer *et al.* 1979) and a U/Pb zircon age of 1028± 43 Ma (Aftalion & van Breemen 1980). Without a clear understanding of petrogenesis and structural setting of the granite gneiss these dates are equivocal. Fortunately, the granitic gneiss has been reinvestigated by Barr *et al.* (1985) who favour a magmatic origin, with geochemical affinity to an S-type granite (see also Barr 1985). They discarded the metasomatic or migmatitic origins proposed in the past (cf. Dalziel 1966). The proposed magmatic nature is consistent with the discordancy of contact against stratigraphy on a *regional* scale (at outcrop scale there is concordancy between the gneissic foliation and the Moine foliation). A syn–metamorphic age of intrusion is consistent with the lack of aureole around the granitic gneiss. The gneiss shows two foliations, a strong S_1 and an S_2 foliation; coeval with the latter, local melting gave rise to pegmatitic segregations (Fig. 4.8). Within the granitic gneiss xenoliths of metasediment and hornblende schist carry a D_1 foliation (Fig. 4.8).

Further support for the ~1000 Ma tectono–thermal event comes from whole-rock Rb/Sr dating (Brewer *et al.* 1979) of Morar Group pelitic schists situated outside the broad belt of two-mica oligoclase–gneisses with which the West Highland granitic gneiss is commonly associated. It is inferred that the gneiss-producing metamorphism developed during the 1000 Ma event and now is seen in a belt of high-grade rocks traceable along the regional strike as far north as Sutherland into the large area of migmatites made classic by the work of H. H. Read (1931). The Sutherland migmatites are intruded by the little-deformed Strath Halladale granite radiometrically dated at 649±30 Ma (M. Brook *in* Pankhurst 1982). If correct, this date must provide a younger age limit for the sillimanite-growth, migmatisation and intense deformation seen in the Sutherland migmatites.

Thus two broad features are apparent concerning the parts of the Grenville orogen in northern Scotland. First of all, the high-grade gneissose belt is consistently present along the regional (Caledonian) strike. Secondly, metamorphic grade decreases towards the west because in the western part of the Morar Group outcrop medium-grade metamorphism prevails, expressed by garnet growth (Macqueen & Powell 1977). This pattern is well shown in the area around the Carn Chuinneag granite which was intruded at 550±10 Ma into schistose, garnetiferous Moine rocks (Long & Lambert 1963, Pidgeon & Johnson 1973, Wilson & Shepherd 1979). East of the gneissose belt come medium-grade, Loch Eil Group rocks.

Before leaving the thermal aspects of the Proterozoic event in the Moine, reference should be made to the pegmatite suite dated at 740–780 Ma and found at scattered localities across the Moine outcrop (Long & Lambert 1963, van Breemen *et al.* 1974, Barr 1985). Except possibly the Carn Gorm pegmatite, all the pegmatites occur in the Morar Group and have been taken to indicate a major tectono–thermal event ('Knoydartian' or 'Morarian') (van Breemen *et al.* 1974, Piasecki *et al.* 1981). Powell *et al.* (1983) questioned this because there are no obvious structures related to the episode of pegmatitisation. Soper & Anderton (1984) suggested another possibility, namely that the Morarian event was associated with lithospheric stretching, perhaps marking the initiation of the Dalradian basin.

The pattern or tectonic style of the Grenville orogen in northern Scotland is unclear, perhaps due in the main to extensive disruption by folding and thrusting of Caledonian age. It is evident, however, that the Grenville deformation was both widespread and intense, being responsible for D_1 and D_2 fabrics. The most impressive demonstration of this lies in the western part of the Moine outcrop between Morar and Glenelg and Arnisdale where the Grenville deformation involved strong deformation of the Lewisian basement as well as the Moine cover (Figs. 4.1, 4.9).

The history of the recognition of the earliest structures dates back to 1910 when Clough recognised that the Lewisian at Glenelg forms a number of anticlinal fold-cores with the younger Moines occupying synclines. Ramsay (1958) modified Clough's interpretation (*in* Peach *et al.* 1910) of the structural relations of the Moine and Lewisian rocks. He regarded the Lewisian rocks as occupying the cores of nappes, which have a south-easterly vergence. The Lewisian belt that extends from Loch Carron to Loch Hourn is, according to Ramsay, a root-zone and the nappe-movement is to the south-east (see also Simoney 1973).

The Lewisian rocks occurring to the east of the Moine thrust were the subject of a major debate – the so-called 'Lewisian inlier controversy'. H. H. Read and others disputed their Lewisian origin and instead viewed them as integral parts of the Moine succession. But other geologists, notably W. Q. Kennedy, never accepted Read's position, and the controversy ended only when Ramsay (1958) rediscovered and amplified Clough's (1910) work in the Glenelg 'inlier' and was able to separate the Lewisian basement from the Moine cover, viz. (a) the preservation of an angular discordance between the Lewisian and Moine – <12° (Fig. 4.3b); the preservation of a metamorphic discordance, the Lewisian showing evidence of a granulite-facies event not present in the adjacent Moine and (c) the existence of dykes which cut the Lewisian foliation but never penetrate the Moine.

Elsewhere in the Northern Highlands these criteria are

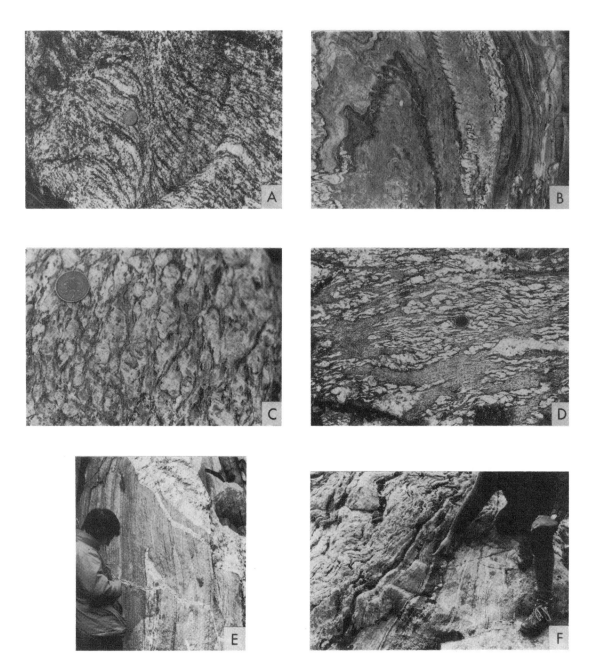

Fig. 4.7.

A. Coarse Ardgour/Quoich granitic orthogneiss. Quoich Dam Spillway, Inverness. (Photograph by A. L. Harris.)

B. The overprinting by upright folds of earlier isoclinal folds in the Glenfinnan Group of Monar. (Photograph by A. L. Harris.)

C. Deformed megacrystic granite from Carn Chuinneag. (Photograph of a boulder on Rosemarkie foreshore by A. L. Harris.)

D. Coarse pelite to semipelitic gneiss from near Brin Rock, Upper Strathnairn (? Central Highland Division). (Photograph by D. Barr.)

E. Upright tight folds formed during the deformation of early (*c.* 750 Ma) pegmatitic veins are cut by late, almost undeformed pegmatites (*c.* 450 Ma). Loch Eilt, Inverness. (Photograph by A. L. Harris.)

F. Western contact of Ardgour granitic orthogneiss (on right) against Glenfinnan Group metasediments. (Photograph by A. L. Harris.)

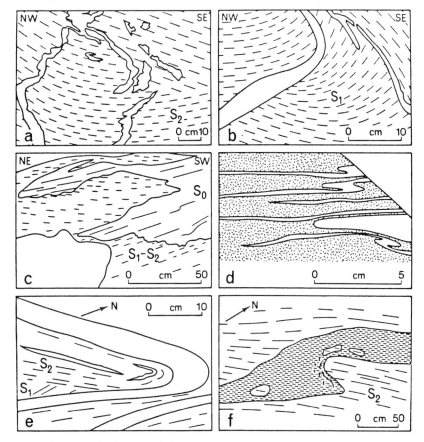

Fig. 4.8. Structures in the West Highland granitic gneiss.
a. Early pegmatitic lits cut by the second foliation.
b. Early pegmatitic lit cutting first foliation.
c. Psammite xenolith: banding in psammite (S₀) is truncated by granitic gneiss. Composite foliation (S₁–S₂) terminates at boundary with xenolith.
d. Intrafolial F₂ folds in pegmatitic lits.
e. F₂ fold of pegmatitic lits: S₂ is axial plane foliation which has largely obliterated S₁.
f. Amphibolite sheet with xenoliths of granitic gneiss.
   After Barr *et al.* (1985).

usually blurred or lost due to high-grade metamorphism and/or intense deformation involving both Moine and Lewisian rocks. The Lewisian inlier controversy arose as a result of the neglect of the Glenelg inlier and the consequence was a long delay in understanding the fundamental structure of the Moine.

The first folds at Glenelg are usually sub-isoclinal, the Moines being tightly squeezed into long-limbed synclines, the shape of which implies enormous differential movements between, and within, the Lewisian and Moine rocks (Fig. 4.9). Over large areas the original angular discordance between the Lewisian and the Moines has been obliterated. At Glenelg, Clough noted a Moine basal

conglomerate and a discordance between the basement and cover rocks, whereby different members of the Lewisian complex come to be adjacent to the Moine basal semipelitic schist. Within the two rock groups angular discordances of various kinds (e.g. angle between foreset beds and the truncation plane in cross-bedded Moine psammites, and the angle between dykes and the foliation of the gneisses in the Lewisian) are partly or completely eliminated. The mechanism is one of regional ductile deformation of both basement and cover rock. Confirmation of the flat-lying recumbent style of the Grenville belt is found to the east of the Loch Quoich line where, according to Roberts & Harris (1983), the weak Caledonian overprint allows the recogni-

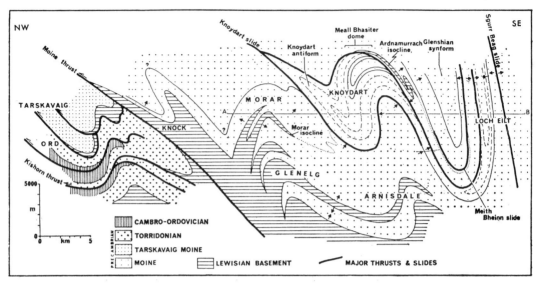

Fig. 4.9. Diagrammatic section across the NW Caledonian front: Skye to Loch Eilt showing tectonic setting of Lewisian inliers of Morar, Glenelg and Arnisdale. After Powell (1974).

tion of major D_2 (Proterozoic) recumbent folds. The vergence of these is unknown and the only indication here of Proterozoic strain is that given by Holdsworth & Roberts (1984) who recorded a N–S stretching fabric which they assign to D_2.

An assembly of the rather meagre evidence reveals a paradox – the Grenville belt appears to be an asymmetrical orogen on the evidence of basement/cover deformation yet the asymmetry or vergence is towards the high-grade metamorphic zone (? internal zone). In this sense the belt cannot be compared to the Alps or Himalayas or any classic collision orogens. The Caledonian thrusts have carried the Grenville belt onto a foreland which shows no trace of the Grenville event. From this it can be inferred that a western front to the Grenville belt lies beneath the Moine thrust, perhaps coinciding with the eastern limit of foreland basement detected by Watson & Dunning (1979) from aeromagnetic and gravity maps of northern Scotland. But the observed Grenville structures provide no guidance on whether to look eastwards or westwards for the front in the allochthon.

Caledonian orogenesis

The long-delayed recognition that Lewisian basement was actively involved in the orogenic deformation of the Moine has led to a revolution in the understanding of the fundamental structure of the NW Highlands. One example of this is the recognition of basement-cored fold nappes which have been assigned to the Grenville orogeny; another is the discovery that the Moine outcrop is traversed by major thrusts or slides which carry basement wedges. This

in turn has led to the modern view of the Moine in terms of a thrust-related model based on external foreland thrust belts like the Canadian Rockies. The first step in this was taken by Elliott & Johnson (1978, 1980) and Barton (1978) with reference to the Moine thrust belt. Subsequent work not only along the Moine thrust but also in the internal Moine has greatly extended the approach (Barr *et al.* 1986, Butler 1986).

The structure of the north-west Highlands is regarded as a stack of at least three major thrust nappes forming a thrust system which evolved from east to west and in time-ranges from syn- to post-Caledonian metamorphism (Fig. 4.10). The foreland-propagating model is strongly supported by evidence of folded thrusts, e.g. Dundonnell (Elliott & Johnson 1980) and in south Assynt (Elliott & Johnson 1980, Butler & Coward 1984). The simple duplex model of Elliott & Johnson has been revised due to the discovery of back-thrusting and 'pop-up' structures and demonstrations that lower and later thrusts cut up through higher ones. In addition, evidence for differential motion has been recorded within thrust sheets. For example, in north Assynt oblique folds and non-plane strain in the Cambrian Pipe Rock testify to differential sinistral slip which perhaps amounted to 3–5 km (Coward & Kim 1981).

During the final episode of thrusting the whole stack of sheets was transported towards WNW on the Moine thrust onto a foreland composed of Lewisian, Torridonian and Cambro–Ordovician shelf sediments. According to Butler & Coward (1984) the width of the Cambrian shelf was at least 54 km. In terms of deep structure the model has been viewed as 'thick-skinned' by Soper & Barber (1982) who suggested that the Moine thrust cuts steeply down into the mantle, but 'thin-skinned' by others (Barr *et al.* 1986, Butler 1986). On the whole it is the 'thin-skinned' model which fits best the known geology.

At present the geophysical evidence does not offer any clear-cut solutions to the problems of the deep structure of northern Scotland (Fig. 4.11). The LISPB profile indicates horizontal seismic layering and does seem to favour a flat or thin-skinned model for the Moine thrust. However, electrical conductivity gives no support to this. Seismic reflection profiles (MOIST) on off-shore lines indicate a Moho at ~28 km but no clear indication of the trajectory of the Moine thrust. Butler & Coward (1984) use MOIST data to suggest imbricated structures in the Lewisian basement beneath the Moine thrust as well as a possible linkage with the Outer Isles thrust (Fig. 4.12).

In the following account of the slides within the Moine the outcrop is divided into two regions: (a) a Southern Region from the Fannich Mountains as far south as the island of Mull and (b) a Northern Region: Fannich to the north coast.

Southern Region

Here the main structural units of the Moine are most clearly established. In structural order they are:

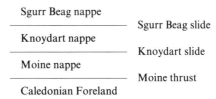

The dual terminology for the tectonic surfaces dividing these units follows Hutton's (1981) scheme and recognises the fact that whereas the Sgurr Beag slide and Knoydart slide are entirely ductile shear zones (or 'ductile thrusts'), the Moine thrust is both a mylonite zone and a clean-cut fracture surface. Evidence that the Sgurr Beag slide antedates the Moine thrust was given by Kelley & Powell (1985) who demonstrate that the fabrics associated with the Sgurr

Fig. 4.10. Schematic cross-section through the southern Moine (adapted from Barr *et al.* 1986). 600 – Temperature estimate for pre-Caledonian metamorphism. For modification to the western part of the section see Holdsworth (1987).

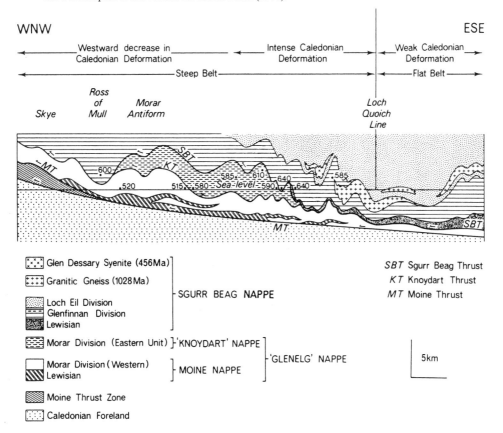

Beag slide are overprinted by those belonging to the Moine thrust episode. A regional suite of pegmatites that post-dates the slide was deformed prior to and during the movement on the Moine thrust. Also major and minor D₃ folds post-dated the Sgurr Beag Slide but are pre- or syn-movement on the Moine thrust.

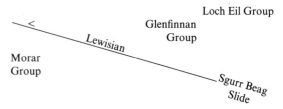

1. *Sgurr Beag slide*

This is a mid-crustal ductile shear zone with a strike length of at least 180 km. It is likely that displacement on it was in excess of 50 km which is not substantially less than the displacement envisaged for the Moine thrust. The significance of the slide is most readily seen in northerly parts of the region where Lewisian basement appears above the slide and a structural succession is thus:

Lewisian forms the basal unit of the Sgurr Beag nappe and reaches a thickness of several hundred metres. Towards the south the Lewisian is much thinner than this and eventually disappears south of Kinlochhourn so that the Glenfinnan Group rocks are directly in contact with the Morar Group. In such places Rathbone & Harris (1979) have shown that the slide can still be recognised by the

Fig. 4.11. Geophysical data for NW Scotland.
Location map (*a*) for 'MOIST' seismic reflection profile (*b*), for 'LISPB' seismic refraction profile (*c*), and electrical conductivity profile (*d*).
FT, Flannan 'thrust'; GGF, Great Glen Fault; MT, Moine thrust; OIT, Outer Isles thrust; 1–5 in (*d*) – increasing conductivity. After Butler & Coward (1984).

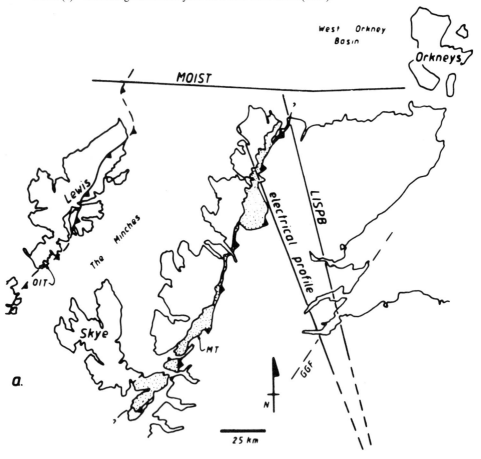

increase in strain near its outcrop. In the Ross of Mull, however, the slide appears to be absent and the Glenfinnan Group is in sedimentary contact with the Morar Group (Holdsworth *et al.* 1987). The slide is unlikely to have terminated, however, and may well have 'ramped' between Ardnamurchan and Mull, leaving the Morar/Glenfinnan boundary intact *within*, probably, the Knoydart nappe.

Deformation and metamorphism in the Sgurr Beag nappe

The prevalence of steep dip in the Sgurr Beag nappe is an indication that the Sgurr Beag slide has been subjected to folding after the cessation of active displacement (Fig. 4.13). Barr *et al.* (1986) consider that the Sgurr Beag and Knoydart slides have been folded by the movements on a detachment surface lying beneath the NW Highlands. This detachment is equated with the Moine thrust (Fig. 4.10). It is clear that the slide is folded by at least one set of major folds but there is some disagreement concerning the precise relationship between the slide and the two sets of major folds (D_3 and D_2) which have been recognised in the region. Roberts & Harris (1983) and Kelley & Powell (1985)

consider that D_3 post-dates the slide in the Quoich–Fannich–Arkaig area. A steep belt, some 12 km across strike, brings up the high-grade Glenfinnan Group rocks (Fig. 4.14). The eastern limit of this steep belt has been known for many years as the Loch Quoich Line. It has been variously interpreted as a slide, an unconformity and a root zone and as a detachment surface between a mobile infrastructure and a more rigid superstructure. Roberts & Harris suggested that it defines an eastern limit of Caledonian reworking of the Grenville belt (Fig. 4.14). To the east of the Loch Quoich Line the Loch Eil Group is flat-lying, little affected by Caledonian deformation (D_3). This event can be given a maximum age of 456 ± 5 Ma, the age of intrusion of the Glen Dessary syenite which is strongly deformed by the D_3 (Roberts *et al.* 1984). This is consistent with the indirect age of 467 ± 20 Ma given for the Sgurr Beag slide (Powell *et al.* 1981).

A rather different view is presented by Baird (1984) for the Glenfinnan area. D_3 in his structural sequence are developed as semi-recumbent folds associated with the motion on the Sgurr Beag slide. They underwent intense

Fig. 4.11. See caption on facing page.

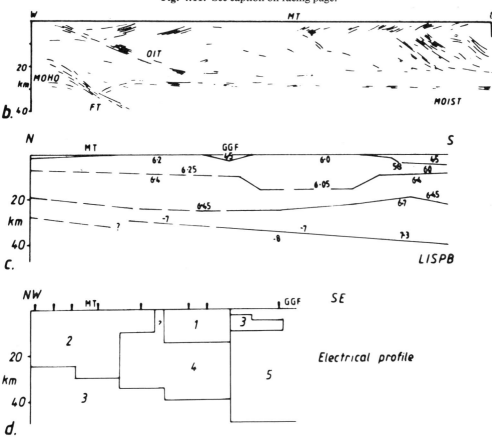

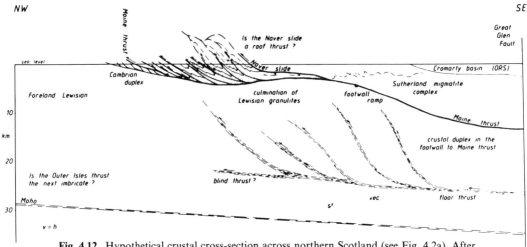

Fig. 4.12. Hypothetical crustal cross-section across northern Scotland (see Fig. 4.2a). After Butler & Coward (1984).

axial rotation and sheath-fold formation during translation on the slide. Subsequently D_4 major folds rotated the D_3 folds into their present steep axial plane and axial orientation and also folded the Sgurr Beag slide. In other words the steep belt referred to above is assigned to D_4 by Baird but D_3 by Roberts & Harris (1983). Roberts et al. (1985) contested Baird's hypothesis on the grounds that it predicts fold interference patterns on a regional scale which are not found and that there are systematic minor fold vergences related to only one set of late folds.

Caledonian strains in the Moine of the West Highlands are documented by various intrusive igneous rocks notably the microdiorites which are unstrained and unmetamorphosed in the west but are often strongly deformed and metamorphosed to greenschist (albite + actinolite + green mica) or amphibolite facies (calcic–plagioclase + green hornblende ± biotite) in the east (Smith 1979 and Chapter 8). Talbot (1983) recognised four generations of microdiorites which were intruded along dilated shear, or in the case of last generation, tensile fractures during several regional penetrative deformations, indeed spanning the Morarian and Caledonian orogenies. The microdiorites behaved as incompetent sheets which became involved in a sequence of co-axial irrotational plane strains, causing rotations of the sheets and the development of foliation and lineation. It seems very probable that the microdiorites are wholly Caledonian in age. In addition more recent work does not support Talbot's view that the Loch Eil 'Division' rests unconformably on previously deformed and metamorphosed Morar and Glenfinnan 'divisions'.

Dr A. W. Baird comments as follows:

'I have not observed individual microdiorites cross-cutting other members of the suite though they frequently cut amphibolites and are themselves cut by camptonites

and dolerites. Microdiorite sheets were introduced after D_3 deformation (Baird's D_3) into relatively flatlying easterly and westerly dipping fractures. Metamorphism of the microdiorites post-dates the Sgurr Beag slide because the metamorphic facies boundary drawn by Smith (1979) for the microdiorites is not displaced by the slide. Subsequently horizontally applied WNW–ESE D_4 maximum compression rotated and deformed the sheets so that they have their highest dips in the areas of most intense D_4 strain. D_4 deformation has passively rotated the relatively flat lying D_3 isoclines and sheath folds into upright and reclined orientations to produce the Moine "steep belt".' (N.B. D_4 of Dr Baird is equivalent to the D_3 of Roberts et al. 1984.)

Late Caledonian strain has also been recorded in the Strontian granite intruded at 435 ± 10 Ma. Dr D. Hutton states:

'Early syn-magmatic deformation fabrics are preserved in the northern parts of the Strontian tonalite and granodiorite. The foliation has the shape of a south-plunging synform: flat in the middle and dipping inwardly around the margins where it is concordant with the country rock–granite contact. A relatively narrow zone of concordant deformation is recorded in the immediately adjacent Moine and this deforms members of the microdiorite dyke suite. An approximately north–south stretching lineation and xenolith alignment is best seen in the central flat area of the granite. This, together with syn-granite sense of shear indicators in the granite and the Moine contact zone, suggests top-to-the-south (extensional) overshear. This is inconsistent with syn-granite Caledonian folding of the pluton on a north–south axis. However, this and other data are consistent with emplacement of the magma into a pre-existing

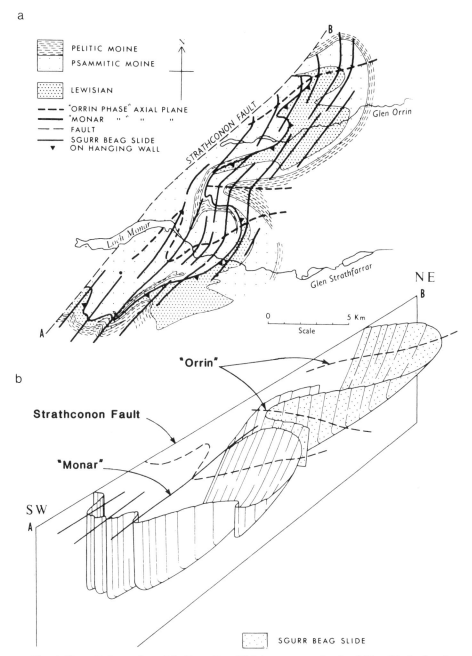

Fig. 4.13. a. Deformation of the Sgurr Beag Slide by two sets of major folds – 'Orrin phase' and 'Monar phase' (Tobisch *et al*. 1970) at Monar (see Fig. 4.1 for location). **b.** Block diagram of Monar area. After Ramsay, in Johnson & Stewart (1963).

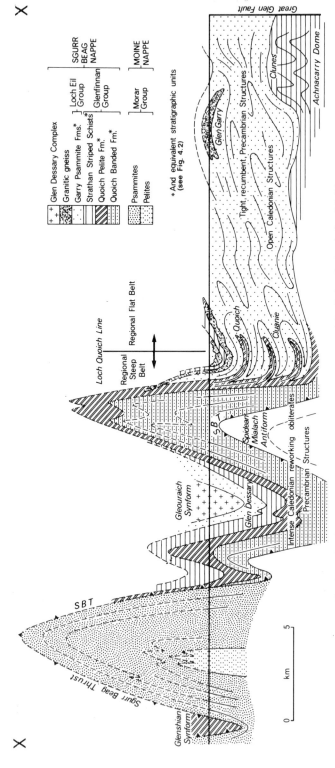

Fig. 4.14. Section X–X' (see Fig. 4.1) showing the steep and flat belts, the Loch Quoich Line and folding of the Sgurr Beag thrust (SBT). After Roberts *et al.* (1984).

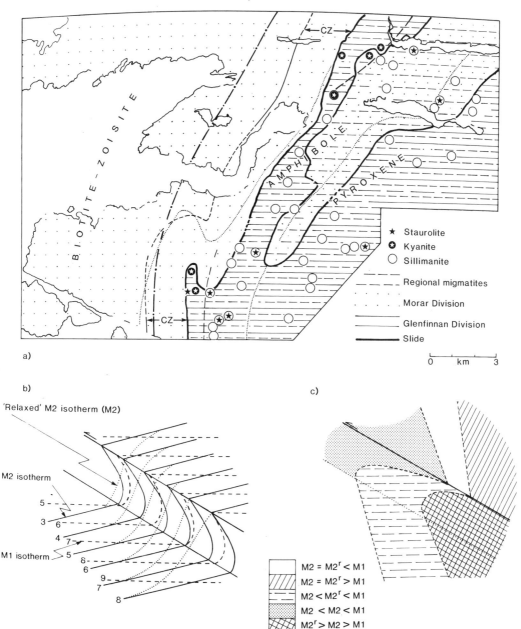

Fig. 4.15. Metamorphism and the Sgurr Beag Slide.

(a) Morar to Glenfinnan area: Dotted lines: limits of calc–silicate mineral assemblages. Dot–dash line: western limit of amphibolite facies assemblages in microdiorites. Dashed line: western limit of sillimanite in pelitic rocks. CZ – clinozosite zone assemblages determined from calc–silicate.

(b) Hypothetical metamorphic patterns arising from displacement of pre(M_1)- and syn(M_2)-shearing isotherms across an asymmetric shear zone, after displacement and thermal relaxation. Numbers show relative temperature values.

(c) Metamorphic fields resulting from model b. $M2^r$ – relaxed M_2 isotherms. After Powell *et al*. (1981).

Caledonian synform (which lies immediately east of the Loch Quoich Line). Emplacement space could have been created by the removal through southward extensional shear of a Moine hanging wall block in the core of the synform (Hutton, 1988a). Following this the biotite granite was emplaced further to the south by a similar mechanism (Hutton, 1988b). This generated a zone of extensional sheeting along the northern margins of the biotite granite and dextral shearing on its western side. After emplacement the biotite granite was marginally affected by this deformation.'

Metamorphism in the Sgurr Beag nappe

The absence of cataclastic fabrics along the Sgurr Beag slide shows that displacement took place during regional metamorphism which can be dated at about 465 Ma. The An content of plagioclase feldspar and the mineral assemblages in calc–silicate rocks provide sensitive indicators of metamorphic grade used by Powell et al. (1981) to show that grade changes abruptly across the slide. Apparently it put hotter rocks (~600°C) onto cooler (500°C). Powell's thermal modelling of the Sgurr Beag shows a pattern of thermal relaxation of geotherms and from this he deduces that the Sgurr Beag nappe was thick (Fig. 4.15).

The Sgurr Beag nappe contains extensive pegmatite suites which are also found in the Morar Division to the west of the Sgurr Beag slide (Barr 1985). If the ages (450–440 Ma given by van Breemen et al. 1974), are not cooling ages then these pegmatites post-date the Sgurr Beag slide.

2. Knoydart slide

The Knoydart slide carried migmatitic Knoydart Pelite over lower grade Morar Group of the Moine nappe. It may have a smaller displacement than the overlying Sgurr Beag slide and can be traced from Inverie in Knoydart to Ardnamurchan, a distance of about 50 km, and possibly extended as far as the Great Glen near the Ross of Mull where the metamorphic grade of the Moine rocks is consistent with grade in the Knoydart nappe rather than that in the Moine nappe.

Northern Region

It has been realised that the Moine, north of Fannich and occupying most of Sutherland, differs in structural style and even age, from the southern region. Although structural distinctions are still valid it is now accepted that both regions contain a Proterozoic metamorphic complex which has been disrupted by thrusting/sliding during the Caledonian orogeny. What is unusual in the north is the pile of gently dipping sheets of interbanded Moine and Lewisian, apparently forming a belt of either tight folding (Barr et al. 1986, Strachan & Holdsworth 1988) or imbricate faults in the hanging wall of the Moine thrust (Butler & Coward 1984).

Two major slides are found in Sutherland: the Naver slide and the Swordly slide, which appear to converge to the south and may continue as the Sgurr Beag slide (Fig. 4.4).

Since the work of Coles Phillips (1937) it has been known that in Sutherland a prominent grain-shape and mineral lineation describes a broad arc-like pattern, trending roughly WNW near to the Moine thrust, but nearly NW–SE in central Sutherland. For some workers this is a pre-Caledonian fabric, others have questioned its unity. Barr et al. (1986) supported the latter view and suggested that the pattern of lineation in Sutherland may indicate a change in transport direction, which was NW–SE on the early Sgurr Beag, Naver and Swordly slides but nearly WNW on the later Moine thrust. Kelley & Powell (1985) made a similar point in the Fannich area.

Metamorphism

Caledonian metamorphism in the northern region increases from chlorite grade in the west, eastwards to amphibolite facies in central Sutherland. Near the Naver slide Barr et al. (1986) referred to the over-printing of amphibolite facies assemblages, which are assigned to the Precambrian orogeny, by amphibolite facies assemblages of Caledonian age (Fig. 4.16a). Nearer to the Moine thrust, while the Precambrian metamorphism appears to be of lower grade, e.g. garnet, the Caledonian metamorphism is even lower (chlorite) grade (Fig. 4.16b). Butler & Coward (1984) suggested that the metamorphism of Morar Division rocks beneath the Naver–Swordly thrust sheets may have been induced as a result of the tectonic loading by these sheets.

Moine thrust

The Moine thrust is now seen as the most westerly and youngest of a system of Caledonian thrusts (Fig. 4.17). It serves therefore as the north-west front of the Caledonides. It carried the polyorogenic Moine rocks, with their previous slides, over the Foreland; it can be seen as a detachment surface for the entire complex. As well as an on-land strike length of at least 200 km, it seems possible that it extends north-eastwards towards Shetland and south-westwards beyond the island of Mull, a distance of over 500 km. This length is consistent with the rather tentative estimates for displacement in the Moine thrust – over 70 km (Elliott & Johnson 1980). A date of 430–425 Ma for active motion on the Moine thrust is fairly well constrained by radiometric datings on the Assynt igneous suite (van Breemen et al. 1979, Halliday et al. 1986) and by cooling ages from the mylonites (Johnson et al. 1985). In terms of the regional structural history, the Moine thrust is thought to be syn- or post-D_3 (Kelley & Powell 1986).

Although in several places the Moine thrust is a simple structure, with mylonitic or low-grade Moine schists resting more or less directly on the Foreland, elsewhere there is a complex array of thrust sheets separating the Moine rocks from the Foreland (Fig. 4.18). In such places the thrust belt is thickened causing an upwarp or culmination in the Moine thrust. Assynt is the classic example of a culmination and other, mostly smaller ones, are spaced along the length of the Moine thrust. Substantial displacements have been estimated for some of the sub-Moine thrust sheets, e.g. 20–25 km for the Glencoul sheet of the Assynt region. The sub-Moine sheets carry Lewisian, Torridonian and Cambro–Ordovician rocks, which are not usually intensely deformed and show only slight metamorphism. The lowest thrust or sole, is unlikely to be a major slip surface.

As well as thrusts, the footwall to the Moine thrust shows major folding in places, for example the Sgonnan Mor syncline developed in the Ben More thrust sheet in the Assynt region and, even more impressively, the Loch Alsh syncline of the Loch Carron to Skye sector. The Loch Alsh syncline has an upper limb measuring some 10 km across strike, consisting of inverted Lewisian and Torridonian rocks. Both folds can be interpreted as footwall synclines, that is they formed during motion on the immediately overlying Moine thrust.

A final point concerns the evidence for extensional tectonics in the Moine thrust belt as shown by normal faulting notably in the southern part of the Assynt region. In addition, surge zones (Fig. 4.19) exhibiting displacements of 1–2 km to WNW have been found to affect the already emplaced Glencoul sheet of north Assynt (Coward 1982, 1983).

(a)

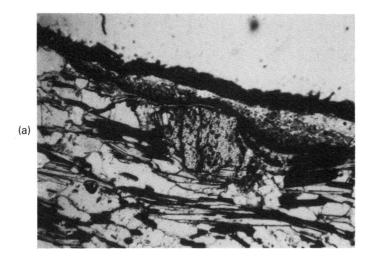

(b)

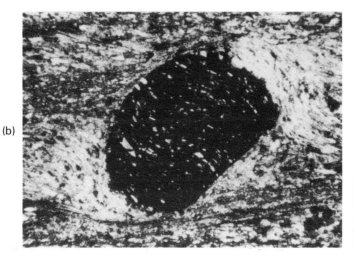

Fig. 4.16. Early (Precambrian) garnets in Moine. (a) Morar Division; (b) mylonite from near to the Moine thrust. In (a) and (b) the matrix fabric is of Caledonian age. From Barr *et al.* (1986).

In south Assynt late extensional slip on the Moine thrust caused a truncation of the underlying Ben More thrust (Fig. 4.19). In seeking an explanation for the extensional regime Butler & Coward (1984) pointed out that broad open warps within the Moine outcrop, earlier highlighted by Elliott & Johnson, are clearly later than the thrusts and slides and could be located above culminations formed by imbricated Lewisian basement beneath the Moine thrust. One culmi- nation roughly coincides with a negative Bouguer anomaly (Fig. 4.21a). Such culminations would cause uplift and hence gravity-spreading towards the west. Thus the exten- sional features in the Moine thrust belt may reflect thrust- induced uplift (Fig. 4.21b).

A number of studies of the microstructure of the mylonites associated with the Moine have made advances on the classic work of Christie (1963). Weathers *et al.* (1979)

Fig. 4.17. View looking eastwards along Loch Glendhu (left), and Loch Glencoul (right) showing the Foreland (F), Glencoul thrust sheet (G), Moine nappe (M) in the northern part of the Assynt region of Sutherland. Most of the Foreland and Glencoul sheet is composed of Lewisian gneiss. B.G.S. photograph. Crown copyright reserved.

report progressive increase in finite strain and degree of recrystallisation in mylonites as the Moine thrust is approached from below. They also claimed that palaeo-stress was independent of distance from the Moine thrust and therefore strain rate increased towards the thrust. This is because there is a constancy of dislocation densities through the mylonites. Ord & Christie (1984), however, disputed this and suggested that the densities reflect late annealing recovery after the main deformation. In the mylonites there is a correlation between shape fabric symmetry (S > L, L − S or L > S tectonite) and crystallo-graphic preferred orientation which typically shows a cross-girdle c-axis fabric, symmetrical to foliation and lineation and probably reflecting plane strain deformation.

Further in certain locations fabric symmetry varies according to distance from the Moine thrust, with non-co-axial asymmetric c- and a-axis fabrics near to the thrust. This fabric pattern is consistent with thrusting to WNW (Law *et al.* 1986). However, farther away from the thrust co-axial fabrics are found and these are taken to be associated with extensional flow and enormous vertical shortening (86 per cent) occurring during or after nappe emplacement (Butler 1984, Law *et al.* 1986).

Moine of the Central Highlands

The metamorphic rocks of the Central Highlands not unequivocally assigned to the Dalradian Supergroup (Fig. 4.22) are not yet properly understood in terms of their lithostratigraphy, their structural geometry or the timing of

their deformation and metamorphism. Much of the eastern part of the area underlain by these rocks was mapped by the Geological Survey in the late 19th century and early 20th century and has not subsequently been reassessed in the light of recent stratigraphic models or structural tech-niques, although detailed stratigraphic and structural information is available locally within the area (e.g. Piasecki 1975, 1980). Neither Sheet 63 nor Sheet 73 (Foyers) has been published on 1:10,000 or 1:50,000 scale by the Geological Survey and most small-scale published maps of the area of these sheets are based on the reconnaissance work of Anderson (1956) and on work in progress by Dr P. Haselock and by the Geological Survey. No regional synthesis based on detailed knowledge of the whole area is yet available, although Thomas (1979) has advanced a structural model for its western part.

Stratigraphy

Harris & Pitcher (1975, figs. 12 and 13) interpreted major, dominantly pelitic bodies in the Central Highlands (A–G on Fig. 4.22) as part of the Appin Group Dalradian, in the main as Lochaber Transition Subgroup, although Johnstone (1975, pp. 36 and 38) touched on the possibility that some of these are Moine. Johnstone (op. cit.) and Harris & Pitcher (op. cit.) included the largely psammitic Central Highland Granulites with the Moine, no distinc-tion being drawn by Johnstone (1975, fig. 6) between these Moine rocks of the Grampian Highlands and that of the Loch Eil Division north-west of the Great Glen fault. The stratigraphy envisaged by Johnstone (1975) and Harris &

Fig. 4.18. Cross-sections through the central part of the Assynt region (Fig. 4.1) to show folding of the Ben More thrust and imbricate thrust faults and folded thrusts in its footwall. After Coward (1984).

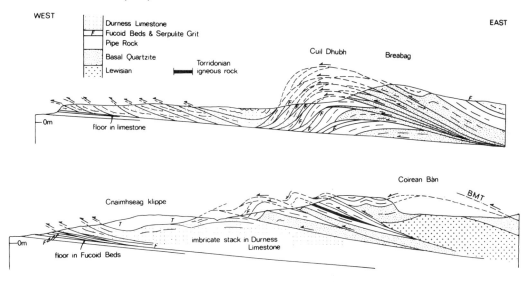

Pitcher (1975) is consistent with the well-documented stratigraphies shown in the columns included in Figure 4.22, the Moine being variously given lithostratigraphic names of only local significance. Most of the workers, on whose results these columns are based, refer the Moine rocks of their areas to the Grampian Division of Piasecki (1980), preferring to retain the term Moine for the major lithostratigraphic unit within which they occur. Because they are stratigraphically continuous with Appin Group Dalradian, however, they all lie within the Grampian Group of the Dalradian as defined by Harris et al. (1978).

Within the Appin Group Dalradian, Harris & Pitcher (1975) included the bodies of dominantly pelitic rock with minor marble, calc–silicate and/or amphibolite which crop out at Loch Laggan, Ord Ban, Grantown-on-Spey and

Kincraig (A, B, C, D respectively on Fig. 4.22). The body at Loch Laggan was interpreted by Anderson (1956) and Smith (1968) as Appin Group Dalradian rocks lying in the core of a tight upright syncline within an envelope of older Eilde Flags, but Piasecki & van Breemen (1983, pp. 123–125) report work by Temperley which interprets these rocks as Moine (see also Piasecki & Temperley 1988). Work by Piasecki (1980) on the Kincraig and Ord Ban localities has brought out a stratigraphy and structure of far greater complexity than that shown on published Geological Survey Sheet 74 (mid-Strathspey). Using the evidence of rhythmically banded metasediments as a way-up indicator, Piasecki (1980, pl. 2) concluded that the marble, calc–silicate and amphibolite bodies with associated quartzite are stratigraphically older than

Fig. 4.19. Extensional faults in the Assynt region (Fig. 4.1). (a) Extensional faults together with contractional and strike–slip faults, define a surge zone. (b) Imbricate extensional faults with floor fault, near Elphin. After Coward (1982).

(a)

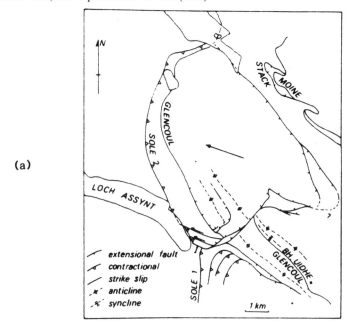

(b)

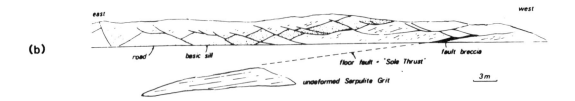

adjacent psammite and semipelitic rocks of Eilde Flags aspect. If this conclusion is correct, correlations based on general lithostratigraphic similarities between the main Appin group outcrop and bodies A–D (Fig. 4.22) are an over-simplification in that lithologies characteristic of Appin group must occur stratigraphically interleaved with Eilde Flags, near the base of the Grampian Division as defined by Piasecki (1980), perhaps many hundreds of metres below the true Appin Group.

To the dubious status of these pelitic units (A–D) with subsidiary calcareous and mafic rocks must be added the uncertainty as to the stratigraphic affinity of other major pelitic bodies (e.g. E, F and G on Fig. 4.22). Certainly the very coarse, sillimanite-grade migmatitic pelites with garnetiferous amphibolites of the Upper Strathnairn dis-

trict (1 on Fig. 4.22) bear little resemblance to the Appin Group of Glen Roy and the continuity of the latter with the rocks of Upper Strathnairn and the Upper Findhorn (Fig. 4.22) must be open to considerable doubt.

These lithostratigraphic uncertainties together with recognition of the Grampian slide, a zone of ductile displacement, has led Piasecki & van Breemen (1979) and Piasecki (1980) to propose a new lithostratigraphic/structural framework for the Moine rocks of the Central Highlands (Fig. 4.22). This framework recognises two divisions – the Grampian and Central Highland divisions which have a tectonically modified cover–basement relationship; the Grampian cover rocks (younger) have been brought to lie above the Central Highland basement rocks (older) by the ductile Grampian slide which thus

Fig. 4.20. Sequence (a–e) of thrusting in the northern part of the Moine thrust zone. After Coward (1980).

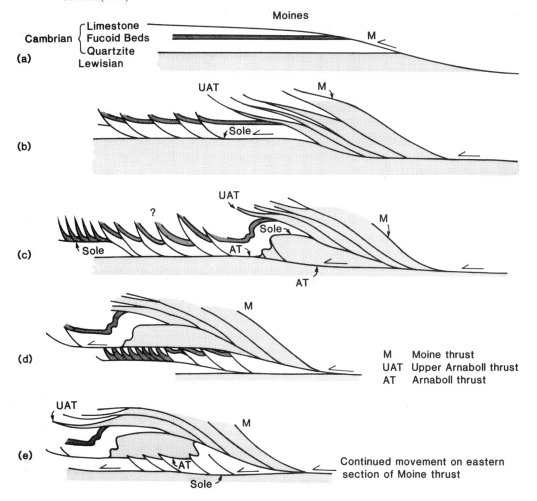

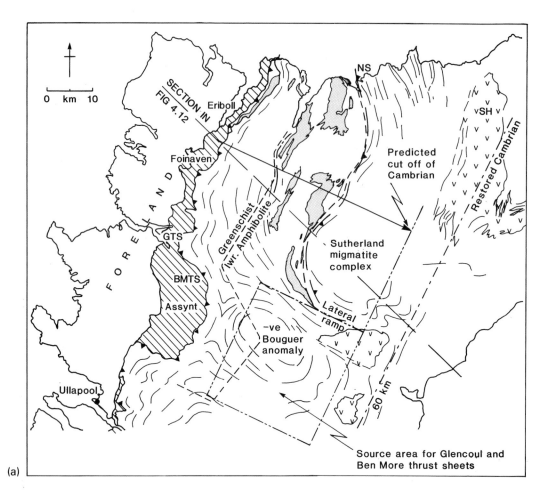

(a)

(b)

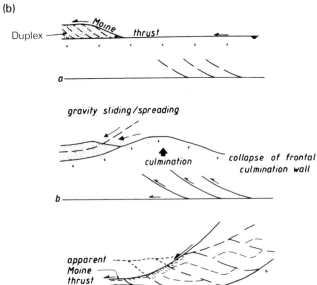

Duplex

gravity sliding / spreading

collapse of frontal
culmination wall

culmination

apparent
Moine
thrust

Fig. 4.21. Interpretation of crustal structure in northern part of Moine thrust sheet based on geophysical data and restored sections.

(a) Foliation trends define culminations or bulges in the footwall of the Moine thrust. The approximate width (~54 km/) of the Cambrian shelf is indicated by the line showing cut off of Cambrian in the footwall of the Moine thrust. In a restored section the thrust belt Cambrian would lie SE of the line marked 'Restored Cambrian'. GTS, Glencoul thrust sheet; BMTS, Ben More thrust sheet; NS, Never slide; SH, Strath Halladale granite. Lewisian 'inliers' are stippled.

(b) Culminations give rise to gravity spreading and an extensional regime. After Butler & Coward (1986).

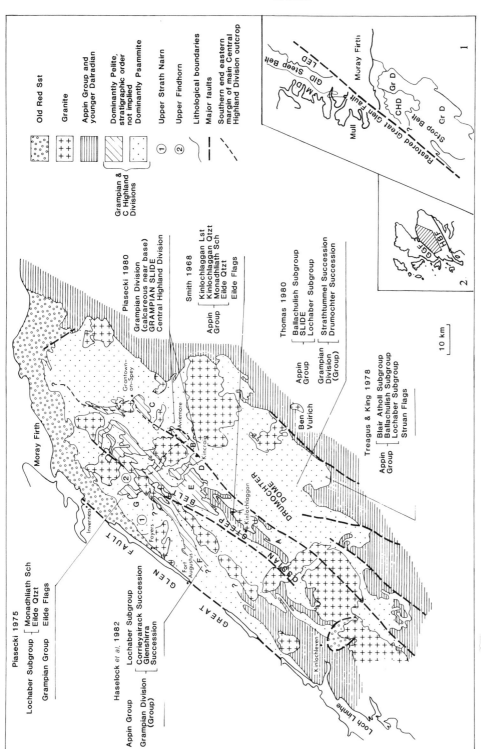

Fig. 4.22. Map to show the Caledonian and pre-Caledonian geology of the Central Highlands (lines mainly after Geological Survey 1:625,000 Sheet 1). A–G refer to rock units mentioned in the text as occurring at specific localities. Rocks lying to the north and west of the trace of the Grampian Slide form the main outcrop of the Central Highland Division. 1, Upper Strathnairn; 2, Upper Findhorn. References to published work appear in the reference list.

Inset 1 (largely after van Breemen & Piasecki (1983); igneous rocks omitted) to show the disposition of the main Moine divisions following an hypothetical 160 km dextral restoration of the Great Glen Fault (but see Discussion in text). MD, Morar Division; GID, Glenfinnan Division; LED, Loch Eil Division; CHD, Central Highland Division; GrD, Grampian Division.

Inset 2 to show location of area.

disrupts a putative unconformity. Both divisions have similar rather monotonous psammitic, semipelitic and pelitic lithologies although the calcareous and metabasic rocks are apparently confined to the Grampian Group. In addition to the structural criteria which will be outlined below, the main distinction between the two divisions is that of metamorphic grade – the Central Highland Division rocks are distinctively gneissose and migmatitic with local coarse quartzites, whereas the Grampian Division is not on the whole gneissic, although the division is reported as of amphibolite facies (Piasecki & van Breemen 1983, table 1) reaching kyanite/sillimanite grade near its base (Piasecki 1980, p. 51). Work so far published shows the Central Highland Division to occupy several hundred square kilometres of Inverness-shire, east and south of Inverness and three other small inliers at Kincraig, Laggan and Ord Ban (Piasecki & van Breemen 1983, fig. 3, Piasecki & Temperley 1988) (Fig. 4.22). The remaining rocks of the northern Central Highlands are shown by Piasecki & van Breemen (op. cit.) as Grampian Division.

Some at least of the Grampian Division falls within the definition proposed by Harris et al. (1978) for the Grampian Group Dalradian. These are rocks of unknown

base which are stratigraphically continuous with Appin Group Dalradian (e.g. Eilde Flags of the Kinlochleven area: Hickman 1975; Struan (or Strowan Flags) of the Schiehallion–Rannoch district: Treagus & King 1978, Thomas 1980; Glen Shirra and Corrieyairack successions of the southern Monadhliaths: Haselock et al. 1982 (F on Fig. 4.22). No greater sedimentological change was involved at the Grampian/Appin Group boundary than at any other group boundary within the Dalradian and in the southern and western parts of the Central Highlands at least the Grampian Group, as defined, first suffered orogenesis in the late Precambrian with the Appin and younger Dalradian. The continued use of Moine to describe the late Precambrian Grampian Group metasediments which suffered only late Precambrian–Ordovician orogenesis is to invite confusion among the international community of geologists who will read elsewhere in this chapter the consensus view that rocks called Moine, after their type locality in the N Highlands, suffered c. 1100–1000 Ma orogenesis and were subsequently reworked by thrusting and folding in the mid-Ordovician-to-Silurian. The N Highland Moine protolith, although possibly comparable with that of the Central Highland Division is, on the basis

Fig. 4.23. Structural block diagram of major structures in the West Central Highlands. After Thomas (1979).

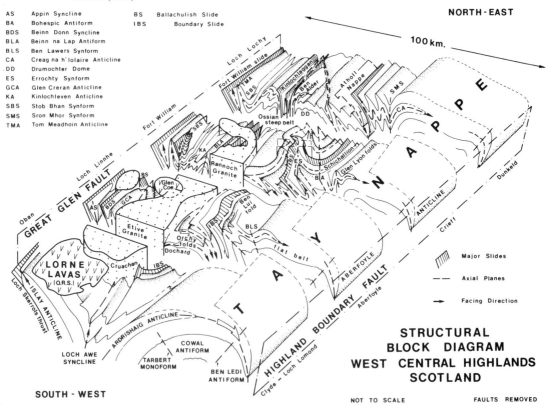

AS   Appin Syncline
BA   Bohespic Antiform
BDS   Beinn Donn Syncline
BLA   Beinn na Lap Antiform
BLS   Ben Lawers Synform
CA   Creag na h'Iolaire Anticline
DD   Drumochter Dome
ES   Errochty Synform
GCA   Glen Creran Anticline
KA   Kinlochleven Anticline
SBS   Stob Bhan Synform
SMS   Sron Mhor Synform
TMA   Tom Meadhoin Anticline

BS   Ballachulish Slide
IBS   Boundary Slide

NORTH-EAST

100 km.

SOUTH-WEST

STRUCTURAL BLOCK DIAGRAM WEST CENTRAL HIGHLANDS SCOTLAND

Major Slides
Axial Planes
Facing Direction

NOT TO SCALE   FAULTS REMOVED

of published isotopic ages, part of an entirely different sedimentary/tectonic cycle from the Grampian Group rocks which pass up without a break into the late Precambrian–Dalradian.

Structure of the 'Moine' of the Central Highlands

A regional structural synthesis is available only for the western part of the area shown on Figure 4.22 (Thomas 1979, figs. 1–5). Localised deformation sequences are available for several districts (McIntyre 1951; Whitten 1959; Smith 1968; Piasecki 1975, 1980; Thomas 1980; Haselock *et al.* 1982; van Breemen & Piasecki 1983). East of the area synthesised by Thomas (op. cit.) few major structures have been identified which can be rigorously related in terms of geometry, facing and orientation to sets of minor structures. These include the Kinlochlaggan–Monadhliath syncline and related structures in Upper Strathspey and the Upper Findhorn (Anderson 1956, plate 4; Smith 1968, fig. 1; Piasecki 1975, fig. 1), the Corrieyairack Syncline (Piasecki 1975, fig. 1; Haselock *et al.* 1982, fig. 1) and the Grampian slide (Piasecki 1979 and later papers).

The Geal Charn–Ossian Steep Belt (Thomas 1979, fig. 5) (Figs. 4.22, 4.23) forms a zone of steeply inclined folds one of which is the Kinlochlaggan Syncline. On either side of this zone the sense of overturning and facing of major primary folds diverges (Fig. 4.23). To the south-east major structures pass over the Drumochter Dome to form the Atholl nappe which closes and faces down to the south-east below the major slide which in Central Perthshire marks the base of the Appin Group Dalradian. To the north-west of the steep zone primary folds face to the north-west – these were traced in Dalradian rocks to the west and south-west of Loch Laggan by Thomas (1979, figs. 3–5) (Fig. 4.23), and may also include the early almost isoclinal folds that pass around the Corrieyairack Syncline in the southern Monadhliath Mountains (Dr P. Haselock, pers. comm. 1986). The trace of the Geal Charn–Ossian Steep Belt has been extended north-eastwards from Loch Laggan by van Breemen & Piasecki (1983, fig. 1) where it is represented as the Grampian Steep Belt. The stratigraphy and the north-west overturned and facing primary structures of the rocks of the Upper Findhorn as interpreted by Piasecki (1975, fig. 1 cross-sections) would have been consistent with the role of the Geal Charn–Ossian–Grampian Steep Belt as a zone of structural facing divergence. However, the subsequent assignment of these rocks to the Central Highland Division (van Breemen & Piasecki 1983) makes this simple interpretation improbable.

The G1–G3 deformation sequence restricted to the Central Highland Division and the D1–D5 sequence suffered by the Central Highland and Grampian Divisions have been distinguished by Piasecki & van Breemen (1979, 1983) and Piasecki (1980) on grounds of style and orientation and of contemporary metamorphic grade. The Grampian Slide Zone which separates the two divisions is regarded (Piasecki 1980, table 2) as a D2–D3 event in the Grampian Division (cover) sequence and entirely later than the G1–G3 events of the Central Highland Division (basement). It is described in detail by Piasecki (1980, p. 51) who characterises it as a zone up to 200 m thick within which slices of basement occur in tectonised cover rocks and vice versa. In this zone the normal foliation of the basement gneisses and the cover schists develops into a distinctive and intense thin platy fabric.

Work in progress in Upper Strathnairn (1 on Fig. 4.22) by Dr P. Haselock and Dr N. G. Lindsay suggests that the distinction between the Grampian Division and the Central Highland Division should be viewed with some caution. Here weakly deformed cross-bedded and coarsely pebbly psammites occur below apparently younger, extremely coarse-grained pelitic gneisses which are clearly of mid-upper amphibolite facies. Such psammites can be traced sporadically to the south of the Great Glen as far as Fort Augustus and to Glen Shirra (F on Fig. 4.22); in the former locality they are scarcely of garnet grade and very weakly deformed (Parson 1979). Pebbly cross-bedded psammites likewise can be traced into migmatitic feldspathic psammites via heightened tectonic strain and recrystallisation. Elsewhere, P. Haselock (pers. comm. 1987) has demonstrated coarse migmatitic pelitic rocks in Glen Shirra (F on Fig. 4.22) which overlie coarse pebbly psammites and which pass up gradually over a hundred metres or so into quite fine-grained pelitic mica schists in a manner reminiscent of the exposures of Central Highland Division and Grampian Division at Kincraig. Such evidence suggests that distinctions between Central Highland and Grampian Divisions may, in some cases at least, be an artefact of apparent contrasts in metamorphic grade created by variation in protolith and intensity of deformation.

Age of structures

Muscovite-bearing pegmatites which are thought to be confined to the Grampian Slide Zone were emplaced into active planes of D2 sliding (Piasecki 1980, p. 56) and were further sheared by D3, implying that the sliding episode was a D2–D3 event broadly coeval with pegmatite emplacement (Piasecki & van Breemen 1979, 1983). The age of muscovite samples from pegmatites in the Grampian Slide Zone is given by Piasecki (1980, table 3) as ranging from 573 ± 13 Ma to 718 ± 19 Ma (Rb–Sr muscovite), although a pegmatite in the zone subsequently yielded ages of 736 ± 8 to 784 ± 8 Ma (Piasecki & van Breemen 1983, table 2). These dates are interpreted by Piasecki & van Breemen (op. cit.) as indicating a *c.* 750 Ma episode of pegmatite emplacement broadly contemporary with the onset of the sliding that brought the Central Highland and Grampian divisions together. Younger pegmatite ages are thought to indicate subsequent resetting by continuing or renewed deformation and thermal activity.

Using the pegmatites to calibrate the age of deformation in the Central Highlands, Piasecki & van Breemen (1979, 1983, table 1) concluded that the G1–G3 events are pre-750 Ma, that the early events in the Grampian Division are 750 Ma, that D3 and D4 in the Grampian Division relate to the Grampian Orogeny (Lambert & McKerrow 1976) and that D5 was Caledonian. The youngest deformation in the Central Highlands is constrained by the 443^{+5}_{-15} Ma age for the emplacement of the late-orogenic Glen Kyllachy Granite of the Upper Findhorn (van Breemen & Piasecki 1983). In areas such as Rannoch–Schiehallion the age of structures in the Grampian Division rocks is related to the Grampian Orogeny by analogy with contemporary structures in the Dalradian rocks which contain the syn-orogenic 590±2 Ma Ben Vuirich granite.

Relationship to Northern Highland Moine

Pegmatites of the suite yielding muscovite Rb–Sr ages of c. 750 Ma were noted (Piasecki & van Breemen 1983, fig. 8) as coeval with Morarian/Knoydartian pegmatites from different parts of the N Highlands. This has led to the Grampian Slide being interpreted as a largely Morarian structure. It is argued further (Piasecki & van Breemen 1983) that the rocks of the Central Highland Division should be correlated with the Glenfinnan Division rocks of the N Highlands on grounds of general lithological similarity, the presence at Ord Ban of a body of granitic gneiss similar to the 1030 ± 45 Ma Ardgour Gneiss and because they record a prolonged pre-750 Ma orogenic history which is based on pegmatite ages from shear zones. The Grampian Division rocks, regarded as a cover sequence on the Central Highland Division, were correlated with the Loch Eil Division rocks because the latter were regarded by them as unconformable on the Glenfinnan Division (Piasecki et al. 1981, Piasecki & van Breemen 1983). Correlation of Grampian Division with the Morar Division was advanced for a variety of reasons (Piasecki & van Breemen 1979, pp. 142–143) but princi-

pally because both contain deformed pegmatites of comparable age. Correlation of the Central Highland Division with the Glenfinnan Division suggested original continuity of their outcrop to Piasecki & van Breemen (1983, fig. 1) (inset 1 Fig. 4.22), implying a sinistral displacement of c. 160 km on the Great Glen Fault. Supporting evidence was offered by a similar apparent continuity (Piasecki & van Breemen 1983) of the Geal Charn–Ossian–Grampian Steep Belt with the N Highland Steep Belt (Leedal 1952).

Subsequent work on the N Highland Moine rocks has led the present authors to believe that regardless of relationships identified in the Central Highlands, none of these correlations across the Great Glen Fault is firmly established, although there are tempting similarities between the Central Highland Division pelitic gneisses and those of the Glenfinnan Division. Structural and stratigraphic work in the N Highlands, allied to isotopic work and summarised earlier in this chapter (Brook et al. 1976, Powell et al. 1983, Roberts & Harris 1983, Barr et al. 1985, Barr et al. 1986, Holdsworth et al. 1987, Roberts et al. 1987) suggest strongly that by contrast with the model advanced for the Central Highlands by Piasecki & van Breemen (1983) for the Central Highlands, the N Highland Moine formed a continuous stratigraphic sequence all of which underwent orogenesis early in the late Proterozoic, was locally intruded by 760–740 Ma pegmatites and eventually was reworked by a Caledonian foreland-propagating thrust and fold system in the Ordovician–Silurian. The Ossian Steep Belt (Thomas 1979) in the Central Highlands is regarded as a root-zone of major folds formed during late Precambrian orogenesis and appears to have a 'mushroom' geometry. By contrast the N Highland Steep Belt is an asymmetric zone comprising steep upright folds which postdated the late Ordovician Glen Dessary Syenite (456 ± 5 Ma) (Roberts et al. 1984).

Editorial Footnote: An excursion guide to the Moine geology of the Scottish Highlands (eds. Allison, I., May, F. & Strachan, R. A.) has been published (1988) by Scottish Academic Press, Edinburgh. 270 pp.

REFERENCES

AFTALION, M. &   1980   U–Pb zircon, monazite and Rb–Sr whole-rock systematics of
    VAN BREEMEN, O.   granitic gneiss and psammitic to semipelitic host gneiss from
                              Glenfinnan, northwestern Scotland. *Contrib. Mineral. Petrol.*,
                              72, 87–98.

ANDERSON, J. G. C.   1956   The Moinian and Dalradian rocks between Glen Roy and the
                              Monadhliath Mountains *Trans. R. Soc. Edinburgh*, **63**, 15–36.

BAIRD, A. W.   1982   The Sgurr Beag Slide within Moine rocks at Loch Eilt, Inverness-
                              shire. *J. geol. Soc. London*, **139**, 647–654.

                    1985   A structural and metamorphic study of Moine rocks between
                              Loch Eil and Loch Eilt, Inverness-shire. *Unpub. Ph.D. thesis,
                              Univ. of London*.

BAIRD, A. W.   1985   Discussion of Roberts, Smith & Harris. *J. geol. Soc. London*, **142**, 713.

BAMFORD, D., NUNN, K.,   1977   LISPB. – III Upper crustal structure of northern Britain. *J. geol.*
PRODEHL, C., &   *Soc. London*, **133**, 481–488.
JACOB, B.

BARR, D.   1985   Migmatites in the Moine. *In* Ashworth, J. R. (Ed.) *Migmatites*. Blackie, Glasgow and London, 226–264.

BARR, D., ROBERTS, A. M.,   1985   Structural setting and geochronological significance of the West
HIGHTON, A. J.,   Highland Granitic Gneiss, a deformed early granite within
PARSON, L. M., &   Proterozoic, Moine rocks of NW Scotland. *J. geol. Soc. London*,
HARRIS, A. L.   **142**, 663–675.

BARR, D.,   1986   Caledonian ductile thrusting in a Precambrian metamorphic
HOLDSWORTH, R. E., &   complex: The Moine of northwestern Scotland. *Bull. geol. Soc.*
ROBERTS, A. M.   *Am.*, **97**, 754–764.

BARTON, C. M.   1978   An Appalachian view of the Moine thrust. *Scott. J. Geol.*, **14**, 247–257.

BREWER, J. &   1986   Deep structure of the Caledonian orogen, NW Scotland: results
SMYTHE, D.   of the BIRPS Winch profile. *Tectonics*, **5**, 171–194.

BREWER, M. S., BROOK, M., &   1979   Dating of the tectono-metamorphic history of the southwestern
POWELL, D.   Moine, Scotland. *In* Harris, A. L., Holland, C. H., & Leake, B. E. (Eds.) *The Caledonides of the British Isles Reviewed*. Spec. Publ. geol. Soc. London, **8**, 129–137.

BROOK, M., BREWER, M. S., &   1976   Grenville age for rocks in the Moine of north-western Scotland.
POWELL, D.   *Nature, London*, **260**, 515–517.

BROOK, M., POWELL, D., &   1977   Grenville events in Moine rocks of the northern Highlands,
BREWER, M. S.   Scotland. *J. geol. Soc. London*, **133**, 489–496.

BUTLER, R. W. H.   1984   Structural evolution of the Moine thrust belt between Loch More and Glen Dubh, Scotland. *Scott. J. Geol.*, **20**, 161–179.

                                      1986   Structural evolution in the Moine of NW Scotland: A Caledonian linked thrust system? *Geol. Mag.*, **123**, 1–11.

BUTLER, R. W. H. &   1984   Geological constraints, structural evolution and deep geology of
COWARD, M. P.   the northwest Scottish Caledonides. *Tectonics*, **3**, 347–365.

CHRISTIE, J. M.   1960   Mylonitic rocks of the Moine thrust-zone in the Assynt region, North-west Scotland. *Trans. Edinb. geol. Soc.*, **18**, 79–93.

                                      1963   The Moine thrust zone in the Assynt region northwest Scotland. *Univ. Calif. Pubs. in Geol. Sci.*, **40**, 345–440.

CLIFFORD, T. N.   1957   The stratigraphy and structure of part of the Kintail district of southern Ross-shire; its relation to the Northern Highlands. *Q. J. geol. Soc. London*, **113**, 57–92.

COWARD, M. P.   1982   Surge zones in the Moine thrust zone of NW Scotland. *J. Structur. geol.*, **4**, 247–256.

                                      1983   The thrust and shear zones of the Moine thrust zone and NW Scottish Caledonides. *J. geol. Soc. London*, **140**, 795–811.

                                      1984   The strain and textural history of thin-skinned tectonic zones: examples from the Assynt region of the Moine thrust zone, NW Scotland. *J. structur. geol*, **6**, 89–99.

COWARD, M. P. & KIM, J. H.   1981   Strain within thrust sheets. *In* McClay, K. R. & Price, N. J. (Eds.) Thrust and Nappe Tectonics. *Spec. Publ. geol. Soc. London*, **9**, 275–292.

DALZIEL, I. W. D.   1966   A structural study of the granitic gneiss of western Ardgour, Argyll & Inverness-shire. *Scott. J. Geol.*, **2**, 125–152.

ELLIOTT, D. &   1978   Discussion on structures found in thrust belts. *J. geol. Soc.*
JOHNSON, M. R. W.   *London*, **135**, 259–260.

                                      1980   The structural evolution of the northern part of the Moine thrust belt. *Trans. R. Edinburgh*, **71**, 69–96.

EVANS, D. J. &
  WHITE, S. H.
1984 Microstructural and fabric studies from the rocks of the Moine nappe, Eriboll, NW Scotland. *J. structur. geol.*, **6**, 369–390.

FETTES, D. J., LONG, C. B.,
  MAX, M. D., &
  YARDLEY, B. W. D.
1985 Grade and Time of Metamorphism in the Caledonide orogen of Britain and Ireland. *Geol. Soc. London Mem.*, **9**.

FLINN, D., MAY, F.,
  ROBERTS, J. L., &
  TREAGUS, J. E.
1972 A revision of the stratigraphic succession of the East Mainland of Shetland. *Scott. J. Geol.*, **8**, 335–343.

FLINN, D.
1988 The Moine rocks of Shetland. *In* Winchester, J. A. (Ed.) *Later Proterozoic stratigraphy of the N. Atlantic regions.* J. Blackie, Glasgow, 74–85.

FLINN, D., FRANK, P. L.,
  BROOK, M., &
  PRINGLE, I. R.
1979 Basement-cover relations in Shetland. *In* Harris, A. L., Holland, C. H., & Leake, B. E. (Eds.) The Caledonides of the British Isles – reviewed. *Spec. Publ. geol. Soc. London*, 109–115.

HALLIDAY, A. N.,
  AFTALION, M.,
  PARSON, I.,
  DICKIN, A. P., &
  JOHNSON, M. R. W.
1987 Syn-orogenic alkaline magmatism and its relationship to the Moine thrust zone and the thermal state of the lithosphere in North-west Scotland. *J. geol. Soc. London*, **144**, 611–617.

HARRIS, A. L.,
  BALDWIN, C. T.,
  BRADBURY, H. J.,
  JOHNSON, H. D., &
  SMITH, R. A.
1978 Ensialic basin sedimentation: the Dalradian Supergroup. *In* Bowes, D. R. & Leake, B. E. (Eds.) Crustal evolution in northwestern Britain and adjacent regions. *Geol. J. Spec. Issue*, **10**, 115–138.

HARRIS, A. L. &
  PITCHER, W. S.
1975 The Dalradian Supergroup. *In* Harris, A. L. *et al.* (Eds.) A correlation of Precambrian rocks in the British Isles. *Spec. Rep. geol. Soc. London*, **6**, 52–75.

HARTE, B.
1988 Lower Palaeozoic metamorphism in the Moine–Dalradian belt of the British Isles. *In* Harris, A. L. & Fettes, D. J. (Eds.) The Caledonian–Appalachian Orogen. *Spec. Publ. geol. Soc., London*, 38.

HASELOCK, P. J.,
  WINCHESTER, J. A., &
  WHITTLES, K. H.
1982 The stratigraphy and structure of the southern Monadhliath Mountains between Loch Killin and upper Glen Roy. *Scott. J. Geol.*, **18**, 275–290.

HICKMAN, A. H.
1975 The stratigraphy of late Precambrian metasediments between Glen Roy and Lismore. *Scott. J. Geol.*, **11**, 117–142.

HOLDSWORTH, R. E.,
  HARRIS, A. L., &
  ROBERTS, A. M.
1987 The stratigraphy, structure and regional significance of the Moine rocks of Mull, Argyllshire, W Scotland. *Geol. J.*, **22**, 83–107.

HOLDSWORTH, R. E.
1987 Basement-cover relationships, reworking and Caledonian ductile thrust tectonics of the Northern Moine, NW Scotland. Ph.D. thesis (unpubl.) Univ. of Leeds.

HOLDSWORTH, R. E. &
  ROBERTS, A. M.
1984 Early curvilinear fold structures and strain in the Moine of the Glen Garry region, Inverness-shire. *J. geol. Soc. London*, **141**, 327–338.

HUTTON, D. H. W.
1988a Igneous emplacement in a shear-zone termination: The biotite granite at Strontian, Scotland. *Bull. geol. Soc. Am.*, **100**, 1392–1399.
1988b Granite emplacement mechanisms and tectonic controls: inferences from deformation studies. *Trans. R. Soc. Edinburgh,* **79**, 245–255.

JOHNSON, M. R. W.,
  KELLEY, S. P.,
  OLIVER, G. J. H., &
  WINTER, D. A.
1985 Thermal effects and timing of thrusting in the Moine thrust zone. *J. geol. Soc. London*, **142**, 863–873.

JOHNSTONE, G. S., 1969 The Moinian Assemblage of Scotland. *In* Kay, M. (Ed.) North
    SMITH, D. I., &   Atlantic geology and continental drift. *Am. Assoc. Petrol. Geol.*,
    HARRIS, A. L.   **12**, 159–180.

JOHNSTONE, G. S.   1975 The Moine Succession. *In* Harris, A. L. *et al.* (Eds.) A
                            correlation of Precambrian rocks in the British Isles. *Spec. Rep.*
                            geol. Soc. London, **6**, 30–42.

KELLEY, S. P. & POWELL, D. 1985 Relationships between marginal thrusting and movement on
                            major internal shear zones in the northern Highland Caledon-
                            ides, Scotland. *J. structur. Geol.,* **7**, 161–174.

KENNEDY, W. Q.   1954 The tectonics of the Morar Anticline and the problem of the
                            northwest Caledonian front. *Q. J. geol. Soc. London,* **110**, 357–
                            390.

LAMBERT, R. ST. J. 1969 Isotopic studies relating to the Pre-Cambrian history of the
                            Moinian of Scotland. *Proc. geol. Soc. London,* **1652**, 243–245.

LAMBERT, R. ST. J. & 1976 The Grampian Orogeny. *Scott. J. Geol.,* **12**, 272–292.
    MCKERROW, W. S.

LAW, R. D., KNIPE, R. J., & 1984 Strain path partitioning within thrust sheets: microstructural
    DAYAN, H.   and petrofabric evidence from the Moine thrust zone at Loch
                            Eriboll, northwest Scotland. *J. structur. Geol.,* **6**, 477–497.

LAW, R. D., CASEY, M., & 1986 Kinematic and tectonic significance of microstructures and
    KNIPE, R. J.   crystallographic fabrics within quartz mylonites from the Assynt
                            and Eriboll regions of the Moine thrust zone, NW Scotland.
                            Trans. R. Soc. Edinburgh, **77**, 99–125.

LEEDAL, G. P.   1952 The Cluanie igneous intrusion, Inverness-shire and Ross-shire.
                            Q. J. geol. Soc. London, **108**, 35–63.

LONG, L. E. &   1963 Rb–Sr isotope ages from the Moine Series. *In* Johnson, M. R. W.
    LAMBERT, R. ST. J.   & Stewart, F. H. (Eds.) *The British Caledonides.* Oliver & Boyd,
                            Edinburgh.

MACQUEEN, J. A. &   1977 Relationships between deformation and garnet growth in Moine
    POWELL, D.   (Precambrian) rocks of western Scotland. *Bull. geol. Soc. Am.,*
                            88, 235–240.

MCINTYRE, D. B.   1951 The tectonics of the area between Grantown and Tomintoul
                            (mid-Strathspey). *Q. J. geol. Soc. London,* **107**, 1–22.

MOORHOUSE, S. J. & 1979 The Moine amphibolite suites of central and northern Suth-
    MOORHOUSE, V. E.   erland, Scotland. *Mineralog. Mag.,* **43**, 211–225.

MOORHOUSE, V. E. & 1983 The geology and geochemistry of the Strathy Complex of north-
    MOORHOUSE, S. J.   east Sutherland, Scotland. *Mineralog. Mag.,* **47**, 127–137.

ORD, A. & CHRISTIE, J. M. 1984 Flow stresses from microstructures in mylonitic quartzites of the
                            Moine thrust zone, Assynt area, Scotland. *J. structur. Geol.,* **6**,
                            639–654.

PANKHURST, R. J.   1982 Geochronological tables for British igneous rocks in Sutherland.
                            In Sutherland, D. S. (Ed.) *Igneous rocks of the British Isles.* John
                            Wiley & Sons, Chichester, England, 575–581.

PARSON, L. M.   1979 The state of strain adjacent to the Great Glen fault. *In* Harris,
                            A. L., Holland, C. H., & Leake, B. L. (Eds.) The Caledonides
                            of the British Isles – reviewed. *Spec. Publ. geol. Soc. London,*
                            8, 287–289.

PEACH, B. N., HORNE, J., 1907 The geological structure of the northwest Highlands of Scotland.
    GUNN, W., CLOUGH, C. T.,   *Mem. geol. Surv. U.K.*
    HINXMAN, L. W., &
    TEALL, J. J. H.

PEACH, B. N., HORNE, J. *et al.* 1910 The geology of Glenelg, Lochalsh & south-east part of Skye.
                            Mem. geol. Surv. Scotland.

PHILLIPS, F. C.   1937 A fabric study of some Moine schists and associated rocks. *Q. J.*
                            geol. Soc. London, **93**, 581–620.

PIASECKI, M. A. J. 1975 Tectonic and metamorphic history of the Upper Findhorn,
                            Inverness-shire, Scotland. *Scott. J. Geol.,* **11**, 87–115.

PIASECKI, M. A. J. 1980 New light on the Moine rocks of the Central Highlands of Scotland. *J. Geol. Soc. London*, **137**, 41–59.

PIASECKI, M. A. J. & 1979 The 'Central Highland Granulites': cover-basement tectonics in
VAN BREEMEN, O. the Moine. *In* Harris, A. L. *et al.* (Eds.) The Caledonian of the British Isles – reviewed. *Spec. Publ. geol. Soc. London*, **8**, 139–144.

1979 Field and isotopic evidence for a *c.* 750 Ma tectonothermal event in Moine rocks in the Central Highland region of the Scottish Caledonides. *Trans. R. Soc. Edinburgh*, **73** (for 1982), 119–134.

1979 A Morarian age for the 'younger Moines' of central and western Scotland. *Nature*, **278**, 734–736.

PIASECKI, M. A. J., 1981 The late Precambrian geology of Scotland, England and Wales.
VAN BREEMEN, O., & *In* Kerr, J. W. & Fergusson, A. J. (Eds.) Geology of the North
WRIGHT, A. E. Atlantic Borderlands. *Can. Soc. Petrol. geol. Memoir*, **7**, 57–94.

PIASECKI, M. A. J. & 1988 The northern sector of the Central Highlands. *In* Allison, I.,
TEMPERLEY, S. May, F. & Strachan, R. A. (Eds.) *An excursion guide to the Moine geology of the Scottish Highlands.* Scottish Academic Press, Edinburgh, 51–79.

PIDGEON, R. T. & 1973 Isotopic evidence for 'early' events in the Moines. *In* Pidgeon,
JOHNSON, M. R. W. R. T., MacIntyre, R. M., Sheppard, S. M. F., & van Breemen, O. (Eds.) *Geochronology Isotope geology of Scotland.* Scottish Univ. Res. & Reactor Centre.

POWELL, D. 1964 The stratigraphical succession of the Moine schists around Lochailort (Inverness-shire) and its regional significance. *Proc. Geol. Assoc. London*, **75**, 223–250.

1974 Stratigraphy and structure of the western Moine and the problem of Moine orogenesis. *J. geol. Soc. London*, **130**, 575–593.

POWELL, D., BROOK, M., & 1983 Structural dating of a Precambrian pegmatite in Moine rocks of
BAIRD, A. W. northern Scotland and its bearing on the status of the 'Morarian orogeny'. *J. geol. Soc. London*, **140**, 813–823.

POWELL, D., 1981 The metamorphic environment of the Sgurr Beag Slide: a major
CHARNLEY, N. R., & crustal displacement zone in Proterozoic, Moine rocks of
JORDAN, P. J. Scotland. *J. geol. Soc. London*, **138**, 661–673.

POWELL, D. & 1976 Relations between garnet shape, rotational inclusion fabrics and
MACQUEEN, J. A. strain in some Moine metamorphic rocks of Skye, Scotland. *Tectonophysics*, **35**, 391–402.

RAMSAY, J. G. 1958 Moine–Lewisian relations at Glenelg, Inverness-shire. *Q. J. geol. Soc. London*, **113**, 487–523.

RAMSAY, J. D. & 1962 Moine stratigraphy in the Western Highlands of Scotland. *Proc.*
SPRING, J. *Geol. Assoc. London*, **73**, 295–326.

RAMSAY, J. G. 1963 Structure and metamorphism of the Moine and Lewisian rocks of the North-West Caledonides. *In* Johnson, M. R. W. & Stewart, F. H. (Eds.) *The British Caledonides.* Oliver & Boyd, Edinburgh & London.

RATHBONE, P. A. & 1979 Basement-cover relationships at Lewisian inliers. *In* Harris,
HARRIS, A. L. A. L., Holland, C. H., & Leake, B. E. (Eds.) The Caledonides of the British Isles – reviewed. *Spec. Publ. geol. Soc. London*, 101–107.

READ, H. H. 1931 The geology of Central Sutherland. *Mem. Brit. Geol. Surv.*

RICHEY, J. E., 1936 The West Highland district. *Summ. Prog. Geol. Surv. Gt. Br.*,
SIMPSON, J., & 70–80.
KENNEDY, W. Q.

RICHEY, J. E., 1937 The West Highland district. *Summ. Prog. Geol. Surv. Gt. Br.*,
EAGLES, V. A. & 74–76.
KENNEDY, W. Q.

RICHEY, J. E. & KENNEDY, W. Q. 1939 The Moine and sub-Moine series of Morar, Inverness-shire. *Bull. Geol. Surv. Gt. Br.*, **2**, 26–45.

ROBERTS, A. M. & HARRIS, A. L. 1983 The Loch Quoich Line – a limit of early Palaeozoic crustal reworking in the Moine of the Northern Highlands of Scotland. *J. geol. Soc. London*, **140**, 883–892.

ROBERTS, A. M., SMITH, D. I. & HARRIS, A. L. 1984 The structural setting and tectonic significance of the Glen Dessary Syenite, Inverness-shire. *J. geol. Soc. London*, **141**, 1033–1042.

ROBERTS, A. M., STRACHAN, R. A., HARRIS, A. L., BARR, D. & HOLDSWORTH, R. E. 1987 The Sgurr Beag nappe: a reassessment of the stratigraphy and structure of the northern Highland Moine. *Bull. geol. Soc. Am.*, **98**, 497–506.

SIMONY, P. S. 1973 Lewisian sheets within the Moines around 'The Saddle' of Northwest Scotland. *J. geol. Soc. London*, **129**, 191–201.

SMITH, D. I. 1979 Caledonian minor intrusions of the N. Highlands of Scotland. *In* Harris, A. L., Holland, C. H., & Leake, B. E. (Eds.) The Caledonides of the British Isles – reviewed. *Spec. Publ. geol. Soc. London*, **8**, 683–697.

SMITH, T. E. 1968 Tectonics in Upper Strathspey, Inverness-shire. *Scott. J. Geol.*, **4**, 68–84.

SOPER, N. J. & BARBER, A. J. 1982 A model for the deep structure of the Moine thrust zone. *J. geol. Soc. London*, **139**, 127–138.

STOKER, M. S. 1983 The stratigraphy and structure of the Moine rocks of eastern Ardgour. *Scott. J. Geol.*, **19**, 369–385.

STRACHAN, R. A. 1985 The stratigraphy and structure of the Moine rocks of the Loch Eil area, West Inverness-shire. *Scott. J. Geol.*, **21**, 9–22.

STRACHAN, R. A. & HOLDSWORTH, R. E. 1988 Basement-cover relationships and structure within Moine rocks of central and southeast Sutherland. *J. geol. Soc. London*, **145**, 23–36.

TALBOT, C. J. 1983 Microdiorite sheet intrusions as incompetent time- and strain-markers in the Moine assemblage NW of the Great Glen fault, Scotland. *Trans. R. Soc. Edinburgh*, **74**, 137–152.

TANNER, P. W. G. 1976 Progressive regional metamorphism of thin calcareous bands from the Moinian rocks of N.W. Scotland. *J. Petrol.*, **17**, 100–134.

TANNER, P. W. G., JOHNSTONE, G. S., SMITH, D. I., & HARRIS, A. L. 1970 Moinian stratigraphy and the problem of the Central Ross-shire inliers. *Bull. Geol. Soc. Am.*, **81**, 299–306.

THOMAS, P. R. 1979 New evidence for a Central Highland Root Zone. *In* Harris, A. L. *et al.* (Eds.) The Caledonides of the British Isles – reviewed. *Spec. Publ. geol. Soc. London*, **8**, 205–211.

1980 The stratigraphy and structure of the Moine rocks N of the Schiehallion complex, Scotland. *J. Geol. Soc. London*, **137**, 469–482.

TREAGUS, J. E. & KING, G. 1978 A complete Lower Dalradian succession in the Schiehallion district, Central Perthshire. *Scott. J. Geol.*, **14**, 157–166.

VAN BREEMEN, O., PIDGEON, R., & JOHNSON, M. R. W. 1974 Pre-Cambrian and Palaeozoic pegmatites in the Moines of northern Scotland. *J. geol. Soc. London*, **130**, 493–507.

VAN BREEMEN, O., HALLIDAY, A. H., JOHNSON, M. R. W., & BOWES, D. R. 1978 Crustal additions in late Pre-Cambrian times. *In* Bowes, D. R. & Leake, B. E. (Eds.) *Crustal evolution in north-western Britain and adjacent regions.* Seel House Press, Liverpool, 81–106.

WINCHESTER, J. A. & FLOYD, P. 1984 Geochemistry of the Ben Hope sill suite, Northern Scotland, UK. *Chem. Geol.*, **43**, 49–71.

5

DALRADIAN

M. R. W. Johnson

The Dalradian rocks are mostly metamorphosed marine sediments of late Precambrian and Lower Palaeozoic age and they crop out on the Scottish mainland from the Banffshire coast through the central Highlands into the south-west Highlands (Fig. 5.1). They are also present on the large islands of Islay and Jura and in the Shetland Isles.

Knowledge of the stratigraphy, tectonics and metamorphism of the Dalradian rocks has been accumulating since the late 19th century. The history of the controversies on the tectonics and stratigraphy (e.g. the conflicting views of Sir Edward Bailey and J. F. N. Green and others) and on the metamorphism (e.g. Barrow, Bailey, Clough and Read) is well documented in earlier accounts. The Dalradian, long established as a classic region for the study of orogenic deformation and metamorphism has emerged as a result of modern work as a fine example of late Proterozoic–Cambrian rifting and passive margin development lasting well over 100 Ma.

The Dalradian Supergroup

Cambrian	Southern Highland Group ('Upper Dalradian')	Greywackes, shales and volcanics
	LOCH TAY LIMESTONE	
?	Argyll Group ('Middle Dalradian')	Well-sorted quartzites, pelites and limestones. Turbidite facies in places.
Vendian	**PORT ASKAIG TILLITE**	
?	Appin Group ('Lower Dalradian')	Limestones, pelites, quartzites and laminated psammite-pelite
Riphean	Grampian Group	Generally passes up without a break into Appin Group but in tectonic contact in places.
	? ? ?	

Stratigraphy

For many years it was thought that there were two or more distinct successions, the Ballappel and Iltay successions (see Read 1948, pp. 14–16 for a historical review of research on Dalradian stratigraphy). Following Harris & Pitcher (1975) the Dalradian is recognised as a supergroup which includes four groups. The generally accepted succession is seen in the following table.

The scheme resembles that first proposed by Knill (1963) and Rast (1963), who recognised Upper, Middle and Lower parts in the Dalradian, but it also includes, in the Appin Group, the former Ballappel succession (Rast & Litherland 1970, Harris & Pitcher 1975). The addition of a fourth group, the Grampian Group (formerly 'Central Highland Granulites' or 'Younger Moines'), is a logical step (but see Anderton 1986) in view of the sedimentary transition between it and the Appin Group. The work of Piasecki

(1980) and Piasecki & van Breemen (1979a, b) has revealed stratigraphical complexities in the central Highlands involving not only a basement ('Central Highlands Division') to the Grampian Group, but also a possible unconformity within the latter (see Chapter 4). This does not invalidate the scheme shown above, but it makes the point that the Grampian Group, as described, does not have a well-defined base (see also Harris et al. 1978). Along the Highland Border the Dalradian Supergroup is in tectonic contact with the Highland Border Complex which is of Lower Ordovician age (see Chapter 6).

Following the important work of Vogt (1930), Tanton (1930) and Bailey (1934) sedimentary structures have been widely employed in working out sedimentary sequences and thereby the structural geometry. More recent work has brought to the Dalradian the insights of modern sedimentology with a rapid increase in the understanding of depositional environments, notably in the less metamorphosed parts of the Dalradian.

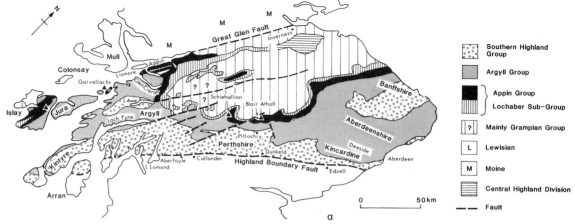

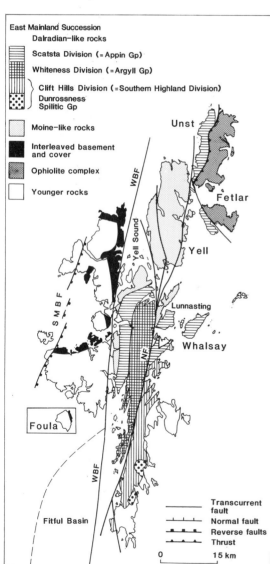

Fig. 5.1. (a) Map showing the distribution of the Dalradian Supergroup and its subdivisions. Modified after Harris & Pitcher (1975), and (b) in Shetland, after Flinn (1985). SMBF, St Magnus Bay Fault; WBF, Walls Boundary Fault; NF, Nesting Fault.

Mention should be made of the Colonsay Group found on the islands of Colonsay and Oronsay and in the Rhinns of Islay, and the Bowmore Group of central Islay. The tectonic setting for these groups is near to the western margin of the Scottish Caledonides and south of the Great Glen fault. The Colonsay Group consists of 5 km of possibly deltaic and turbidite sandstones, shales and minor limestones, the deposits of a basin which subsided on Lewisian crust. The Bowmore Group contains monotonous grey-brown feldspathic sandstones which probably formed as near-coastal marine platform deposits. These groups are stratigraphical enigmas – often in the past referred to the Torridonian in recent years correlations have been made with parts of the Dalradian Supergroup, e.g. Stewart's (1975) suggestion that the Colonsay Group could be equivalant to part of the Appin Group (but see Bentley 1988).

In Shetland, about 300 km north of the Dalradian outcrops in Banffshire, Flinn (1985) has shown that the eastern part of the East Mainland Succession can be correlated with the Dalradian Supergroup. Thus the Scatsta and Whiteness divisions contain limestones, feldspathic sandstones, quartzites and shales and are assigned to the Appin and Argyll groups, while the Clift Hills Division which contains spilites, turbidite quartzites and siltstones is correlated with the Southern Highland Group. The Dunrossness Spilite Group with pillow lavas, ultramafics and a komatiitic serpentine breccia, appears to be younger than the Clift Hills Division.

The age of the Dalradian is given by the sparse fossils and by radiometric analysis. The time-span involved must be nearly 200 Ma. No precise data for the start of Dalradian sedimentation are available: the Varanger tillite (= Port Askaig Tillite) in Norway has been dated c. 650 ± 23 Ma and a mere 2 km of Appin Group separate this horizon from the Grampian Group. The Appin Group has yielded stromatolites and oncoliths (Spencer & Spencer 1972) but these are not stratigraphically diagnostic. In the Argyll Group Vendian acritarchs have been obtained from the **Bonahaven Dolomite** and long-range Vendian/Cambrian forms from the Easdale Slates (Downie et al. 1971). The Islay Quartzite contains possible worm burrows of Vendian age. Stromatolites from the Islay Quartzite Dolomitic Beds (Hackman & Knill 1962) are thought by Downie (1975) to resemble the Vendian form *Aldania*; other stromatolites from this horizon are likened to *Jurusania*, an Upper Riphean–Vendian genus (Downie et al. loc. cit., Spencer & Spencer 1972).

The uppermost part of the succession is Cambrian as evidenced by the following: the Leny Limestone (Southern Highland Group) near Callander contains late Lower Cambrian trilobites (Pringle 1940); acritarchs of Lower Cambrian age from the Tayvallich Limestone (Downie et al. 1971); Lower Ordovician microfossils from the Highland Border Series (Curry et al. 1985); *Protospongia hicksi* (Rushton & Phillips 1973) a middle Cambrian sponge from

the Dalradian of Clare Island, Western Ireland. Somewhat controversial chitinozoans from the Macduff Slates (Southern Highland Group) in Banffshire are even put as young as Ordovician by Downie et al. (1971). Downie (1975) places the base of the Cambrian in the Argyll Group, perhaps as low as the Bonahaven Dolomite.

Applications of plate tectonics theory have given rise to differing views on the nature of the basement to the Dalradian Supergroup and on the position of the huge pile of these sediments in relation to the northern continental margin of the Iapetus ocean. The view that the Dalradian was set on a continental rise, and partly floored by oceanic crust (Dewey 1969, Lambert & McKerrow 1976) is not consistent with the LISPB profile (Bamford et al. 1977), which indicates substantial continental basement beneath the Dalradian, or with the evidence for continental basement to the Midland Valley (Upton et al. 1976). In the plate model proposed by Phillips et al. (1976) a trench is placed to the south of the Southern Uplands, along the 'Solway Line'. According to these writers the Dalradian represents an ensialic basin (see Chapter 1) which was divided from the ocean by a horst of continental crust (see also Smith 1976). This model has important implications for the ensuing comments on the depositional history of the Dalradian Supergroup (Harris et al. 1978).

Geophysical evidence from the WINCH and LISPB seismic reflection profiles (see Hall et al. 1984, Hall 1985), from magnetic anomalies (Westbrook & Borradaile 1978) and from regional gravity analysis (Hipkin & Hussain 1983) shows that the continental crust in the Dalradian area varies from 25 to 35 km in thickness with significant variation in its character. In the SW Highlands 'Lewisian-like' basement to the Dalradian Supergroup occurs at a depth of 10–15 km. To the north-east 'Lewisian-like' crust is covered by upper crustal acid plutons and metasediments which are the Moine and Dalradian. Separating these two differing types of basement is the Cruachan Line (Graham 1986) a NW–SE-trending lineament, marked by a strong gravity gradient. This lineament is one of several now recognised in the Dalradian outcrop (see later) and according to Fettes et al. (1986) they have exerted a pervasive tectonic control on the development of the Dalradian belt, controlling at different times the sedimentation, deformation, metamorphism and post-orogenic plutonism.

The Dalradian evolution is viewed as a rifted margin involving attenuation of continental crust (Anderton 1980, 1982), rifting commencing in the later Proterozoic prior to the formation of Iapetus ocean (Fig. 5.2a). The traditional view is that depositional basins in the Dalradian are elongated in a 'Caledonoid' direction so that along-strike facies changes are far less significant than those occurring across the belt. This view has been challenged by the recognition of important transverse structures such as the Cruachan Lineament and some syndepositional faults. Graham (1986) envisages a dextral transcurrent regime which prevailed for a considerable proportion of the

a

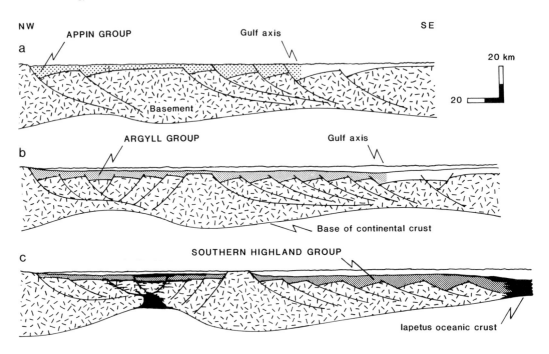

b

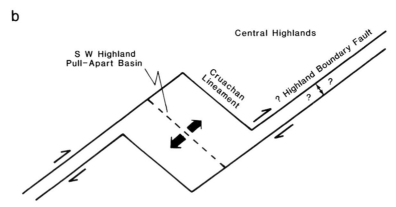

Fig. 5.2.

a. Schematic cross-section of the Dalradian in the SW Highlands to show the tectonic setting in (a) late Appin Group times, (b) late Argyll Group times and (c) Southern Highland Group times. In (c) Basaltic intrusions and volcanics are shown in solid ornament. After Anderton (1985.)

b. Cartoon to show the transcurrent pull-apart model for the evolution of the early Dalradian, from Graham (1986).

history of the Dalradian belt (Fig. 5.2b). This regime resulted in the opening of pull-apart basins floored by quasi-oceanic crust. The Cruachan Lineament is thought to mark the north-eastern margin of one such basin.

Depositional history

Modern sedimentological investigations of the Dalradian are founded on the pioneering papers by Knill (1963) and Sutton & Watson (1955). Knill identified sedimentary environments and constructed lithofacies maps in parts of the south-west Highlands, while Sutton and Watson recognised the profound change in conditions that took place near the boundary between Argyll and Southern Highland Groups, marked by the incoming of greywacke facies which reflects a possible uplift in a land-mass bordering the Dalradian basin. Although recent progress has been mainly in low-grade parts of the Dalradian it is evident that a combination of plate modelling and sedimentology is now yielding a much clearer picture of the late Precambrian to early Palaeozoic history of this part of Scotland. The following account is largely based on work in the SW Highlands, but it is reasonable to suggest that environments mentioned are characteristic of the Dalradian elsewhere. This is not to imply contemporaneity for these environments.

Little is known about the sedimentology of the Grampian Group but the transitional, and possibly diachronous contact between this and the Appin Group is well established. Hickman (1975) infers a shallow-water, probably marine, environment for the Grampian Group (Eilde Flags) of the Glen Roy–Lismore area, citing cross-stratification, ripple marks, grading and erosional features in immature quartz and feldspathic sands. The presence of mud cracks and rain pits suggests intermittent shoreline or terrestrial conditions. Winchester & Glover (1988) provide a more extensive review.

That shallow-water marine conditions were established early on is shown by the cross-bedded, often ripple-marked quartzites with polymodal palaeocurrent directions (Glen Coe and Binnein Quartzites, Table 5.1). The sandstone-shale-cream limestone ensemble in the Leven schists, heralding the carbonate deposition marked by the Ballachulish Limestone, reinforces this interpretation. In contrast the succeeding pyritiferous (Ballachulish) black slates indicate stagnant conditions, but this was only a pause before a return to open marine, oxidised, conditions shown by the cross-stratified Appin Quartzite Formation. Hickman (1975) has proposed that a deltaic facies existed in the south-west and has shown that there is a systematic thinning and fining, with accompanying increase in maturity towards the north-east. These facies changes are recorded between Lismore and Glen Roy, a distance of about 40 km. The uppermost rocks of the Appin Group show a return to a 'black slate' facies. The Cuil Bay Slate

and Lismore Limestone formations are thought by Hickman (1975) to be shallow marine or lagoonal deposits.

It is worth commenting here that the subsidence of the Dalradian basin and the facies changes recorded in the Appin Group and the lower part of the Argyll Group, may have been echoed in the Caledonian Foreland by the late Precambrian upwarps.

The Argyll Group was ushered in by a glacial episode and the deposition of the widespread Port Askaig Tillite (Fig. 5.3), which has been traced from western Ireland as far as Banffshire (Spencer & Pitcher 1968) but not into Shetland. Spencer (1971) has shown that this deposit, which attains a thickness of 750 m in the Islay–Garvellach area, suggests continental or marginal marine conditions. Some 47 beds of tillite (0·5–65 m thick) have been identified, separated by siltstones, dolomites, conglomerates and sandstones. A variety of glacial deposits has been recognised, including massive tillites, reworked marine tillites and, possibly, outwash sands. There are also varved beds containing dropstones from floating ice. In the lower part boulders are predominantly dolomite, while in the upper part granite clasts suggest the arrival of extra-basinal material. Radiometric dating of the granites gives ages of c. 1,000 Ma suggesting a Grenvillian source area.

Spencer (1971) argued that the tillites were deposited from grounded ice-sheets which advanced into a shallow sea. Eyles & Eyles (1983) however suggest a glaciomarine origin for at least part of the Port Askaig Tillite. Anderton (1980, 1982) has pointed out that the evidence for a north-westerly direction of ice movements, presumably away from a large landmass, is incompatible with the existence of Iapetus ocean at this time. Other glacial episodes in the Dalradian are suggested by the Kinlochlaggan Boulder Bed (Lochaber Sub-group) (Treagus 1969, 1981) and by the dropstone horizon in the Macduff slates exposed on the Banffshire coast (Sutton & Watson 1954, Hambrey & Waddams 1981).

The emergent conditions indicated by the Port Askaig Tillite are echoed by the Vendian tillites around the North Atlantic (Føyn 1937, Reading & Walker 1966) and elsewhere. Likewise the transition upwards from the Vendian tillite into the tidally influenced shoreline environment shown by the Bonahaven Dolomites is widely recognised in other areas (Reading & Walker 1966, Edwards 1972). The Port Askaig Tillite and its equivalents suggest that the late Precambrian glaciation was unusual in that low latitudes were affected: this is shown by the warm-water dolomites associated with the tillite and by palaeomagnetic data.

The Jura Quartzite (Table 5.1) marks a striking change in conditions with the sudden influx of cross-bedded sands some 5 km in total thickness but thinning rapidly north-eastwards and south-westwards. Tidal shelf conditions in an open sea seem to have prevailed at this time (Anderton 1976, 1977) and NE currents were flowing parallel to the shoreline of a landmass lying to the north-west (Fig. 5.4a). The sea, while connected to the ocean and subjected to strong tidal currents and storm processes, was a partly enclosed gulf, divided from the deep ocean by a ridge to the south-east. Comparisons with modern tidal seas indicate that deeper water existed to the north-east. Anderton (1976) records liquefaction structures in the Jura Quartzite and attributes these to earthquake activity, suggesting that contemporaneous faulting has controlled sediment thickness. Harris & Pitcher (1975) note that the Jura Quartzite, and indeed some other Dalradian quartzites have a lens shape suggestive of delta-form built out, according to them, from a south-eastern landmass. However, this interpretation is incompatible with the palaeocurrent patterns and

Table 5.1. Stratigraphic successions in the Dalradian

	Groups	Subgroups	Idealised Sequence	Garvellachs Islay-Jura Tayvallich	Kintyre-Arran Easdale-Appin	Loch Awe, Dalmally Loch Rannoch
Ordovician	Southern Highland				Highland Border Series ? Leny Limestone Lochranza Slates Green Beds	Loch Avich Lavas <300
						Loch Avich Grit <1100
				Kells Grit Tayvallich volcanics	Glen Sluan Schist	Tayvallich volcanics 2000
Cambrian	Argyll	Tayvallich		Tayvallich Limestone (with local volcanics)	Tayvallich Limestone	Kilchrenan Conglomerate Tayvallich Slate & Limestone
		Crinan		Crinan Grits	Stonefield Schist Crinan Grits	Crinan Grits 600-3000
		Easdale		Laphroaig Quartzite	Craignish (=Ardrishaig) Phyllite	Shira Limestone 5-300
				Port Ellen Phyllite Scarba Conglomerate	Easdale Slate 500	Ardrishaig Phyllite >400 Degnish Limestone >20 Easdale Slate Pebbly quartzite
Vendian		Islay		Islay (=Jura) Quartzite 5000 Bonahaven Dolomite 295 Port Askaig Tillite 750	Islay Quartzite Dolomites 300 Tillite	
Riphean	Appin	Blair Atholl		Islay Limestone Mullach Dubh Phyllite Ballygrant Limestone Baharradail Phyllite	Lismore Limestone 300 Cuil Bay Slate	
		Ballachulish		Cnoc Donn Phyllite Cnoc Donn Quartzite Cnoc Donn Slate Kintra Limestone	Appin Phyllite and Limestone 400 Appin Quartzite 300 Ballachulish Slate 200 Ballachulish Limestone 200	
		Lochaber (transition)		Kintra Phyllite Maol an Fhithich Quartzite	Leven Schist 500 Glencoe Quartzite Binnein Schist Binnein Quartzite Eilde Schist Eilde Quartzite	Leven Schist Glencoe Quartzite Passage to
	Grampian				Eilde Flags	Eilde Flags
			? Gneissose Basement			

(vertical note in Garvellachs/Kintyre columns: Mull of Oa Phyllite)

Key

V V	Volcanics	⊥⊤⊥	Limestone/Dolomite	∴∴	Quartzite
‒‒‒	Shales	Δ Δ	Conglomerate	░░	Sandstone
°•°•	Grits	▼ ▼	Tillite	500	Thickness in m

Anderton (1979) suggests the lens shape results from syndepositional faulting.

An important change in conditions followed the Jura Quartzite, with the incoming of deeper-water sediments indicating an unstable tectonic environment with syndepositional faulting (although Wright 1988 argued that instability started in Appin Group times). Anderton (1986) describes 'Caledonoid' listric extension faults with throws of several km and 'non-Caledonoid' – i.e. transverse, transfer faults which are marked by lateral facies changes (Fig. 5.4b). Mass flow and turbidite deposits which fine northwards, comprise the Scarba conglomerate. Blocks of

Callander, Loch Tay Kinlochlaggan, Schiehallion	Glen Esk	Banffshire	Shetland
hert & Shale (Highland order Series) pper Leny Grit eny Limestone <100 ower Leny Grit <1300 berfoyle Slate en Ledi Grit reen Beds itlochry Schist	Black shales, cherts, spilitic volcanics, limestone, grits } Highland Border Series Glen Lethnot Grit Glen Effock Pelite <1500	ALLOCHTHON Macduff Slate 2200 Boyndie Bay 'Group' Upper Whitehills 'Group'	Dunrossness Spilitic Group Dunrossness Phyllite ?
och Tay imestone	Tarfside Limestone	Lower Whitehills 'Group' Boyne Limestone 300	Dales Voe Grit
en Lui Schist		Cowhythe Gneiss-Basement Thrust	Clift Hills Phyllite 3000
ron Bheag Schist en Lawers Schist en Eagach Schist airn Mairg Quartzite		Portsoy Group	Asta Spilitic 'Group' Laxfirth Limestone* Micaceous psammites of Wadbista Ness 'Group'
illicrankie Schist chiehallion Quartzite olomites chiehallion Boulder Bed		Durnhill Quartzite Tillite 100	Girlsta Limestone East Burra Pelite 6000
ale Limestone anded Group ark Limestone ark Schist		Sandend Group 1000?	Whiteness Limestone Colla Firth 'Group' (micaceous psammites)
inlochlaggan Limestone ocal Boulder Bed		Cairnfield Actinolite Flags =Garron Point 'Group' Findlater Flags	Weisdale Limestone
inlochlaggan Quartzite inlochlaggan Boulder ed onadhliahth Schist ilde Quartzite		West Sands Mica Schist Cullen Quartzite <1100 =Ben Aigan Quartzite	Scatsta Quartzite 'Group' 4000 Scatsta Pelite 'Group'
truan Flags	"Moine"		"Moine"-Yell Sound Division

a

ROCK SUCCESSION
and INTERPRETATION

SPECULATION on the causes
of movements of sea-level
with respect to the surface
of the sediments

INTERBED (e.g. cross-bedded sandstones)
    Deposited in shallow (? marine)
    water. Icebergs usually absent.

Beneath sea-level

Eustatic rise of sea-level
greater than isostatic
rise of land

Pebble band or conglomerate. Non-sequence or erosion surface.
Marine transgression produces erosion surface with lag conglomerate and beach gravel.

Sandstone wedges.
Permafrost conditions.

Above sea-level

MIXTITE with a little internal bedding.
    Till deposited by the melting of a
    grounded ice sheet.

Eustatic fall of sea-level
greater than isostatic
depression of land

Sharp, usually conformable contact. Underlying strata rarely folded.
Surface over which a grounded (?) ice sheet advanced

INTERBED (e.g. cross-bedded sandstones)
    Deposited in shallow (? marine) water. Icebergs
    usually absent.

Beneath sea-level

b

Fig. 5.3.
a. Port Askaig Tillite. British Geological Survey photograph. Crown copyright reserved.
b. Glacial advance–retreat cycle. From Spencer (1971).

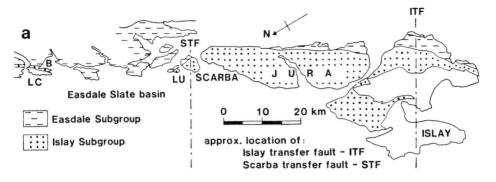

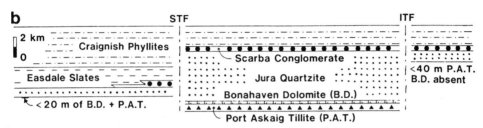

Fig. 5.4.
a. Islay and Easdale Subgroups of Argyll Group showing the approximate position of transfer faults. B, Benderloch; LC, Loch Creran; LU, Lunga.
b. Schematic NE–SW longitudinal section of the Islay and Easdale subgroups in the area depicted in (**a**). After Anderton (1985).

lithified quartzite, limestone and shale are found in slide, slump and debris flow deposits (Knill 1963, Anderton 1977) which overlie a slump scar. Graded beds, convolute lamination, partial Bouma sequences and channels are present in the turbidites.

The Easdale Slates were deposited in quieter conditions. The deepening phase shown by the Easdale Slates has been correlated with the base of the Cambrian (Harris & Pitcher 1975), in which case it is linked to the important marine transgression shown by the Cambrian rocks of NW Scotland, which, allowing for substantial sinistral displacement on the Great Glen Fault, and much smaller displacement on the sub-Moine thrusts, were not far off the strike of the SW Dalradian.

In Perthshire members of the Easdale Subgroup show different facies from the above, e.g. the contemporaneous vulcanicity in the Killiecrankie schists or the Farragon Beds, and the graded, turbiditic Cairn Mairg Quartzite. This subgroup contains most of the syngenetic mineral deposits of the Dalradian, e.g. the baryte-celesian deposit found near Aberfeldy.

The upper part of the Easdale Subgroup indicates regression, well shown by the Craignish Phyllites which according to Anderton (1975) contain laminated siltstones, sandstones and limestones, mostly deposited in tidal flats

and sub-tidal environments. Gypsum pseudomorphs indicate periodic emergence. These beds together with the Shira Limestone all represent very shallow-water environments.

Deepening of the sea seems to have occurred during the deposition of the Crinan Grits which are thick-bedded turbidites (Knill 1963). Harris et al. (1978) suggest that this is a proximal submarine fan-type deposit, derived from the north-west according to Knill (1959, 1963) and containing clasts from a granitic basement. There may be a basinward transition into distal deposits represented by the Ben Lui schists. These authors regard this as an important facies change and prefer this horizon to the Easdale Slates as the Precambrian–Cambrian boundary.

The Tayvallich Subgroup consists of both autochthonous limestone and limestones redeposited from turbidity currents, together with black pelites (Fig. 5.5). The limestones are interlayered with basic volcanic deposits which include pillow lavas, hyaloclastites and airfall tuffs (Knill 1963, Gower 1977). Graham's (1976) work on the sills and dykes intruded beneath and contemporaneously with, the Tayvallich lavas, suggests a tholeiitic affinity and together these rocks suggest mantle melting and magma migration upwards along fractures. The tectonic regime is tensional, and Wright (1976) has related the tholeiites of the upper

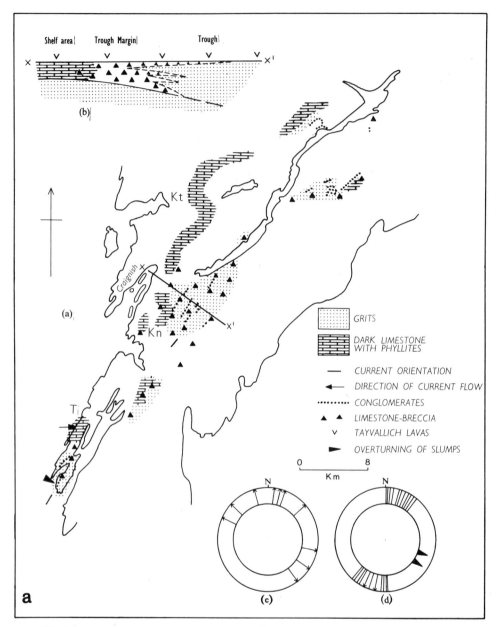

Fig. 5.5.
a. Tayvallich Subgroup.
   (a) Distribution of rock types. (b) NW–SE section, X–X on (a), through the Kilmartin area showing facies changes. (c) Palaeocurrent rose diagrams of small-scale cross-lamination, Kilmartin. (d) Palaeocurrent rose diagram of flow structures at the base of grits, and overturning of slumps, Kilmartin. Kt, Kilmelfort; Kn, Kilmartin; T, Tayvallich. From Knill (1963).
b. Tectonic cross-section of the Tayvallich–Loch Awe area during Tayvallich Subgroup times. From Anderton (1985).

NW   SE

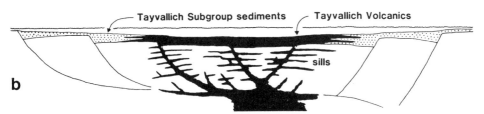

Fig. 5.5b. See caption on facing page.

part of the Dalradian to the opening of a marginal basin with an island arc placed to the south-east of it. In contrast Graham (1986) and Fettes *et al.* (1986) invoke a transtensional regime. The NW–SE Cruachan Lineament separates an area in which mafic igneous rocks are common (comprising <30 per cent of the Dalradian succession) from one to the north-east of it in which such rocks are rare. The Cruachan Lineament marks the north-eastern margin of a transverse basin which opened as a result of dextral transcurrent motion across the Dalradian belt. Dykes on Jura can be restored to a NW trend and they are thought to be an expression of transverse rifting (Graham & Borradaile 1984).

The sedimentology of the Southern Highland Group is poorly known. The general turbidite affinities of the greywackes were pointed out by Sutton & Watson (1955) and Harris *et al.* (1978) described a thick sequence (<2 km) of coarse turbidites with lenticular and tabular units up to 15 m thick, from the Highland Border region. The structure, small-scale sequences and sandstone geometries are compared with lower slope and inner zones of modern deep water submarine fans.

This deep-sea fan sedimentation is accompanied by basic volcanic extrusives, some showing pillow lavas (Borradaile 1973). The basic volcanic activity ('Green Beds') first 'tremors' of which are seen in the Argyll Group, may indicate increased tectonic instability in the basin.

Throughout most of the time of deposition of the Appin and Argyll Groups it seems likely that a landmass, north-west of the present outcrop, supplied mature sediment to the subsiding Dalradian basin (Anderton 1986). The sedimentary character of the Southern Highland Group signals an important change in this palaeogeography. Unlike most of the lower clastic rocks those of the Southern Highland Group are immature and contain abundant fresh metamorphic, granitic clasts and sizeable potassium feldspar fragments (Sutton & Watson 1955, Harris 1972, Borradaile 1973). There is also the blue quartz noted by Sutton and Watson.

This evidence has been interpreted as implying the creation, or uplift of a new southern continental source area, now forming the basement to the Midland Valley, which separated the Dalradian basin from the Iapetus ocean. The possibility that an uplifted northern landmass could serve as a source is unlikely because of the existence of

mature Cambro–Ordovician shelf sediments flanking the northern land. This southern land, postulated by Phillips *et al.* (1976) is, as mentioned previously, consistent with the deep seismic profile but as Yardley *et al.* (1982) point out, it seems too small to have served as a source for the Southern Highland Group. These authors suggest that much of this landmass was removed from proximity to the Dalradian basin by the strike–slip motions which they believe played an important role in the subduction processes following the deposition of the Southern Highland Group. The pattern of subsiding sialic blocks operating over some 200 Ma during deposition of the Dalradian, implies a tensional regime, which may also have controlled late upwarps in the Caledonian Foreland. Harris *et al.* (1978) have compared this tectonic framework to the graben-controlled sedimentation during the Mesozoic and Tertiary in the North Sea area, the point being that these are basins that evolved adjacent to an ocean.

Structure

Within a relatively short period in late Cambrian and early Ordovician times, the Dalradian Supergroup underwent polyphase deformation and regional metamorphism. Lambert & McKerrow (1976) have named this event the Grampian orogeny. The time-span of the Grampian orogeny is fairly well constrained by the radiometric data. The Ben Vuirich granite emplaced between D_2 and D_3 (Bradbury *et al.* 1976, Bradbury 1979) and close to the age of peak metamorphism, has given a U–Pb zircon age of 514 ± 6 Ma (Pankhurst & Pidgeon 1976). The large basic intrusions of Aberdeenshire which are thought to have been intruded shortly after the main regional metamorphism, i.e. post D_2, pre D_3, give a Rb–Sr whole-rock age of 489 ± 17 Ma (Pankhurst 1970). The Dunfallandy Hill granite which is considered to be syn-D_3 has given a Rb–Sr whole-rock age of 481 ± 15 Ma (Pankhurst & Pidgeon 1976). Lastly, several small post-metamorphic muscovite-granites in NE Scotland give Rb–Sr whole-rock ages in the range of $470 - 440$ Ma (Pankhurst 1970, 1974). In addition there are dates from the regional metamorphic rocks referred to later.

Our basic knowledge of the major structures in the Dalradian is largely due to the work of Bailey, Clough and

Shackleton (see Fig. 5.6). Clough (1897) in Cowal & Bailey (1910) in the Loch Leven area, realised the importance of large-scale recumbent folding. Later the work of Vogt (1930), Tanton (1930) and Bailey (1934) in the Loch Leven area introduced into Dalradian geology the technique of using sedimentary 'way-up' criteria to deduce the structural arrangement of deformed strata. Bailey (1934) proposed the use of the general terms antiform and synform with purely geometrical connotations, reserving the terms anticline and syncline for folds where the stratigraphical succession is known. Later work, e.g. Bailey & McCallien (1937) and Read (1936, 1955) employed this technique in other parts of the Dalradian. In 1958 Shackleton introduced the valuable concept of 'facing' of structures. Shackleton demonstrated a large downward-closing (therefore downward-facing) recumbent anticline (i.e. a synform with the oldest rocks in the core) in the Highland Border region.

Bailey's synthesis for the Dalradian tract (1922, 1934, 1940) involved two main structural units – the Iltay Nappe and the Ballappel Foundation – separated by an important dislocation plane, the Iltay Boundary slide which was traced from the Central Highlands to Loch Linnhe and into Islay. The Ballappel Foundation was thought to be divided from the underlying Moines by the Fort William slide and was disrupted by the Ballachulish slide.

In recent years the detailed structural histories of the Dalradian have been studied by several workers. The recognition of a single, unified, stratigraphical succession for the Dalradian has had implications for structural interpretation in particular in regard to the status of the slides mentioned above. The investigations tend to confirm earlier views that the recumbent folds and slides are early structures that have been refolded by later deformations. At least three or four sets of folds have been recognised (see Fig. 5.6). In this account the structural sequence is divided into Early Grampian and Late Grampian deformations each of these being polyphasal. This is partly for the sake of conciseness but also because this classification recognises the affinities of style shown by early structures on the one hand and by the later structures on the other. The Early Grampian deformations are responsible for recumbent folds and slides ('Nappe Phase' and 'Late Nappe Phase' of Bradbury (1985)) and resulted in the overall disposition of the stratigraphical units. The Late Grampian deformations resulted in mostly smaller scale folds with more upright axial surfaces.

Structural history

Early Grampian: D_1 and D_2

Modern geobarometric studies mentioned later make it clear that during orogenesis the Dalradian underwent substantial (< 400 per cent) tectonic thickening mostly by folding and sliding. There is little doubt that the early folds in the Dalradian include huge recumbent folds, the largest of which, the Aberfoyle anticline or Tay Nappe is comparable in size to the Pennine nappes (Fig. 5.7). The axes of these folds run in a general NE–SW or ENE–WSW direction. Another recumbent fold, the Ben Lui fold, can be traced for a distance of over 100 km from Dalmally to Braemar (Bailey 1922, 1925; Read 1955) and Roberts & Treagus (1977a, b) include it among a number of recumbent folds formed by the D_2 deformation (but see Thomas 1979).

Over large areas the Dalradian rocks lie on the lower limb of the Tay Nappe in a stratigraphical sequence proved by sedimentary way-up to be inverted, with the Loch Tay Limestone overlain by, for example the Ben Lui Schists. The largest bulk-strains occurred in the Early Grampian deformations and although the Tay Nappe is usually attributed to gravity collapse it is likely that plate motions were the ultimate cause (cf. Bradbury *et al.* 1979).

Shackleton (1958) and Harris & Fettes (1972) have studied the closure of the Tay Nappe in the Highland Border region (see Fig. 5.6). Here the anticline, originally sideways facing and recumbent, now as a result of later folding, closes downwards. The structure is therefore a synform with the older rocks in its core.

In the Loch Leven area (Bailey 1934, Roberts & Treagus 1977) major recumbent D_1 anticlines face, unexpectedly north-westwards. The Islay anticline (Bailey 1917) is another north-west-facing fold which, in part at least, formed during the early deformation. These folds imply transport in the opposite sense to that suggested by the much larger Tay Nappe. Most workers depict some form of steep belt between the divergent folds resembling the asymmetric mushroom of Sturt (1961) but opinions differ on the age and significance of these features. Thus, in the south-west Highlands the upright D_1 Loch Awe syncline separates early folds with opposed facing directions (Roberts & Treagus 1977). The SE-facing Ardrishaig anticline to the south of the Loch Awe syncline is thought to become increasingly recumbent as traced south-eastwards along its axial plane, and thereby becomes the Tay Nappe. Increase in tightness and recumbency along the axial planes of anticlines to the north is believed to account for the north-westwards-facing recumbent folds (Fig. 5.7). Thomas (1979) and Figure 4.23 depict a similar profile across the central part of the Dalradian tract with the steep belt ('Ossian steep belt') coinciding with a D_1/D_2 collision zone between two continental blocks. However Treagus (1987) views the change of facing in the Central Highlands as involving an age difference, the SE-facing Tay Nappe being earlier than the NW-facing (D_2) folds.

Revision of the stratigraphy has led to reassessment of the significance of Bailey's slides. The Boundary Slide could be merely a local detachment more or less along the contact between the Appin and Grampian Groups. However, something more significant is postulated by Roberts & Treagus (1977) who have correlated the Boundary Slide

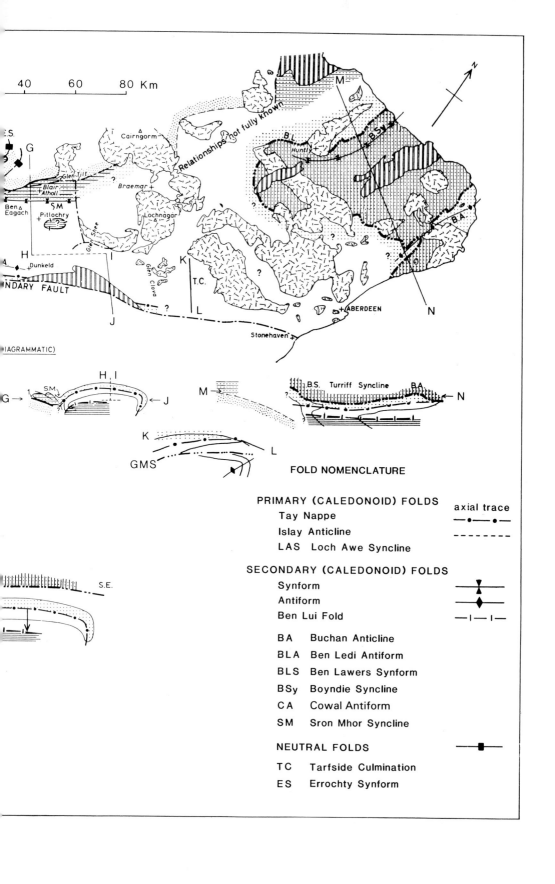

40   60   80 Km

M

Relationships not fully known

Cairngorm

E.S.
G

B.L.
Huntly

B.Sy

Glen Tilt
Blair
Atholl
Braemar

Lochnagar

B.A.

Ben
Eagach
SM
Pitlochry

H

Glen Shee

A
Dunkeld

Glen Clova

K

NDARY FAULT

T.C.

?

O

ABERDEEN

N

J

L

Stonehaven

(DIAGRAMMATIC)

H, I

SM

M

B.S. Turriff Syncline

B.A.

G →

← J

?

N

K

L

GMS

S.E.

FOLD NOMENCLATURE

PRIMARY (CALEDONOID) FOLDS

axial trace

Tay Nappe   —•—•—

Islay Anticline   — — — — —

LAS   Loch Awe Syncline

SECONDARY (CALEDONOID) FOLDS

Synform

Antiform

Ben Lui Fold   — | — | —

BA   Buchan Anticline

BLA   Ben Ledi Antiform

BLS   Ben Lawers Synform

BSy   Boyndie Syncline

CA   Cowal Antiform

SM   Sron Mhor Syncline

NEUTRAL FOLDS   ——■——

TC   Tarfside Culmination

ES   Errochty Synform

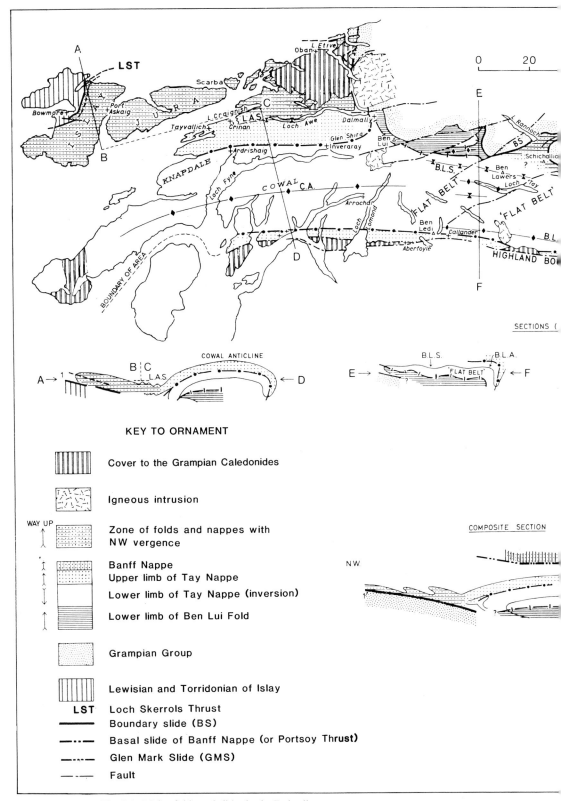

Fig. 5.6. Major folds and slides in the Dalradian.

and the Ballachulish Slide, giving a combined length of about 60 km. It is most probable that the slides belong to the Early Grampian deformations but not necessarily solely to D_1 or D_2. Thus while in the Central Highlands the Boundary Slide appears as a shear zone associated with the D_2 Ben Lui fold, elsewhere it appears to be a D_1 structure (see Fig. 5.7a). Soper & Anderton (1984) have suggested that these slides originated as syndepositional extension faults.

In north-east Scotland, Read (1923, 1955) presented evidence for a major slide, the Boyne lag, which has modified the upper normal limb of a great recumbent fold, probably the Tay Nappe (Fig. 5.7c). Sturt *et al.* (1977) accept this picture essentially but add an important new element. In their view the gneisses of NE Scotland (the Cowhythe, Ellon, Inzie Head gneisses) are not integral parts of the Dalradian succession, as all previous workers, including Read, have supposed. Instead they believe that these gneisses are basement (?Cadomian) that has been carried on the Portsoy thrust, which lies at a slightly lower structural level than the Boyne lag. Above the gneisses the Portsoy sheet carries a right-way-up sedimentary sequence with the Boyne Limestone at the base and the Macduff Slates on top (see Table 5.1). Although limited now in areal extent such a thrust sheet may be of immense significance. However, Ashcroft *et al.* (1984) dispute this and regard the Banffshire sequence as being autochthonous. Two other studies tend to support this view. Thus Treagus & Roberts (1981) consider that the monoclinal Boyndie Syncline (Fig. 5.6, Sutton & Watson 1956, Johnson 1962) is the dominant primary fold in NE Scotland while Harte *et al.* (1984) suggest that the Tay Nappe decreases in amplitude to the north-east and is not present in Banffshire.

To the south-west (T.C. on Fig. 5.6) Harte (1979) has proposed that the Tay Nappe is underlain by his Tarfside Nappe, which carries a right-way-up sequence. Thus in north-east Scotland the recent work has indicated a major reassessment of the tectonic style of the Dalradian.

The Early Grampian deformations are characterised by numerous linear structures (mullions, rods, mineral alignment and pebble-elongation lineation) mostly belonging to D_2. Small-scale folds (F_2) are generally common; they are usually tight, rarely isoclinal, structures which re-fold the limbs and axial planes of D_1 folds and the related slaty cleavage or schistosity. D_2 crenulation cleavages distort the schistosity related to the earliest folds (Fig. 5.8) which in places has been completely transposed during formation of a new schistosity parallel to the D_2 axial planes.

The 'D_2' structures may reflect the later stages of nappe emplacement. Near to the Highland Border, Harris *et al.* (1976) describe 'D_2' shear zones in which an earlier pressure-solution cleavage is deformed (Fig. 5.8c) and they suggest that these zones are a response to intense strain in the deeper levels of the Tay Nappe. According to them the main translation of this nappe, as such, took place in 'D_2', following autochthonous folding.

Voll (1960) drew attention to the evidence for stretching (grain elongation) and constriction in a NW–SE direction. The stretching may be associated with several fold episodes, which it ties together into a continuous sequence of events. Folding started on axes at high angles to the stretching direction and subsequently rotated towards it (Esher & Waterson 1974, Roberts & Sanderson 1974).

The interpretation of the major structure in the Dalradian still presents problems but recent work has escaped somewhat from the dominance once given to the Tay Nappe. Several workers (e.g. Coward 1982, Bradbury 1985) have expressed the view that SE motion of Tay Nappe was succeeded by NW motion on structures like the Portsoy Thrust and the Boundary Slide. An even more radical reassessment is presented by Dewey & Shackleton (1984) who relate the NW motion to large-scale ophiolite obduction which virtually eclipses the Tay Nappe in its significance (Fig. 5.9b).

Late Grampian deformations: D_3 and D_4

Major folds, measuring several km in wavelength, with steep axial planes striking between NE–SE and E–W, are common in the Dalradian (Fig. 5.6) e.g. in Perthshire the Ben Lawers syncline, the monocline [Highland Border Downbend] that has caused the downbending of the nose of the Tay Nappe near the Highland Border and the Cowal antiform in Argyll. The downbend ('D_4') is an important element in the Highland Border lineament: it appears to be situated above a major step (?fault) in the basement and is associated with uplift of the Dalradian Supergroup in the period 460–440 Ma.

Minor Late Grampian structures are plentiful, and consist of asymmetrical folds (Fig. 5.8d, 5.13) which in competent beds give the impression of having a strong concentric, buckling component. In incompetent beds crenulation cleavage is found to be parallel to the axial planes of the secondary folds. Axial-plane schistosities related to the primary deformation become distorted and transposed during the later movement.

Finally there are small-scale, open folds with NNE–SSW-trending axial planes. In this set are included small conjugate fold systems, 'kink-zones' or joint-drag structures (e.g. Knill 1961) which obviously belong to much less plastic phases of movement than the previous structures.

Reference was made earlier to the proposal that lineaments reflecting basement control, have affected the depositional evolution of the Dalradian belt. In addition Fettes *et al.* (1986) suggest that lineaments have played an important role during the ensuing orogenic deformation. This is shown by variations in structural profile, e.g. affecting the Tay Nappe across the Loch Lomond Lineament (Mendum & Fettes 1984) or in the supposed lack of continuity of the steep belt 'mushroom' structure across the

Fig. 5.7.

a. Diagrammatic cross-section to show the major structure of the Dalradian in the SW Highlands.

b. Cross-sections in the Central Highlands from north of Schiehallion to Aberfoyle. Note the re-folded D_1 folds and the NW vergence of D_2 around Schiehallion.

c. Restoration of D_1/D_2 structures shown in **b** to pre-D_3 attitudes. Note the location and sense of slip in the D_2 shear zones. G, Grampian Group; L, Ben Lawers schist; B, Blair Athol subgroup; P, Pitlochry schist.

a. From Roberts & Treagus (1977) modified by Treagus. **b** and **c**. From Treagus (1987).

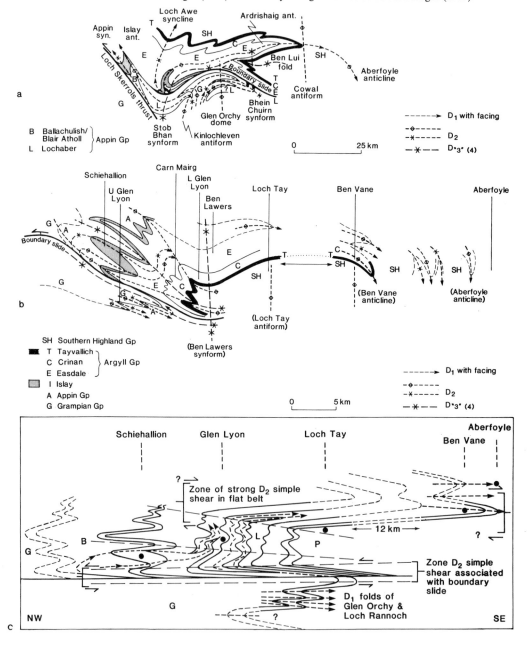

Cruachan Lineament. Another possible example of linea-ment control mentioned by Fettes *et al.* is the supposed north-eastwards diminution of the Tay Nappe towards

Banffshire which Ashcroft *et al.* (1984) suggest has been effectively 'de-coupled' from the rest of the Dalradian by a variably oriented pattern of shear zones. The most promi-

Fig. 5.8. Minor structures on the Dalradian.
  a. D_1 with strong axial plane schistosity.
  b. D_2 folds with crenulation cleavage.
  c. D_2 shear-zone fold which has deformed D_1 foliation.
  d. D_4 fold which has deformed earlier foliations.
    a *and* **b.** From Rast (1963).
    c and **d.** From Harris *et al.* (1976).

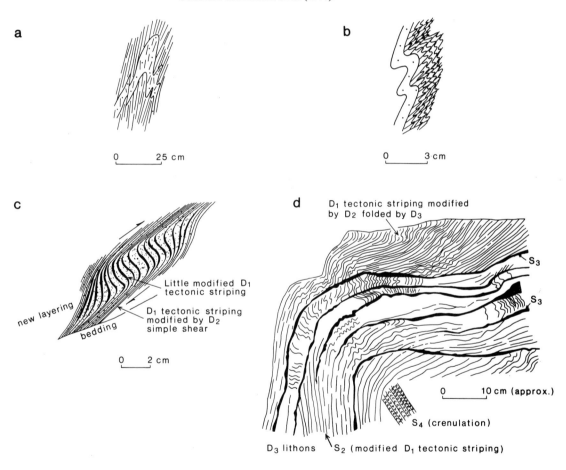

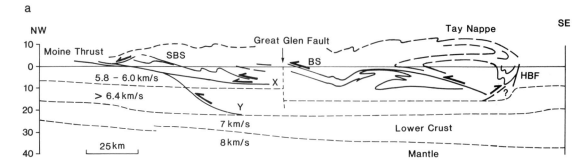

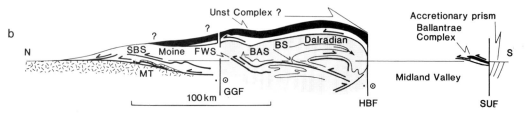

Fig. 5.9. Diagrammatic sections to show possible relationships of major structures in the NW and Grampian Highlands and the deep structure of these regions.
  a. Section drawn by Coward (1983).
  b. Section drawn by Dewey & Shackleton (1984).
  BAS, Ballachulish Slide; X–Y: alternative trajectories of the Moine thrust; BS, Boundary Slide; FWS, Fort William Slide; HBF, Highland Boundary Fault; MT, Moine Thrust; SBS, Sgurr Beag Slide; SUF, Southern Upland Fault.

nent of these is the Portsoy–Duchray shear zone which may be similar to the Cruachan Lineament except that it strikes roughly NNE–SSW.

Major structural features in Shetland

This account concentrates on East Mainland and north-east Shetland, in particular the small islands of Unst, Fetlar and Yell where detailed investigations have been made by Phillips (1927), Read (1933, 1934, 1937) and Flinn (1952, 1956, 1958, 1985). In eastern Shetland, Flinn *et al.* (1967) recognise 14 km of metasediments (phyllites, schists, gneisses, migmatites) and meta-igneous rocks termed the East Mainland Succession, part of which can be correlated with the Dalradian of the mainland of Scotland. In addition the Yell Sound Division is thought to be Moine (see Chapter 4). During orogenesis the rocks have suffered high-grade metamorphism including migmatisation dated at 526 Ma in the Dalradian.

The NNE–SSW general strike of foliation in the Shetland Isles is also the strike of the thrust planes that have been described on Unst and Fetlar. In these islands there are two nappes which have been emplaced over the East Mainland Succession (Flinn *et al.* 1979, see also Fig. 5.10).

The nappes which attain thicknesses of several km are slices of an ophiolite complex which can be compared to other obducted ophiolite complexes such as those in Newfoundland. Dewey & Shackleton (1984) press home this analogy by suggesting that the Unst ophiolite is but a part of a huge 10–15 km thick obducted ophiolite that originally covered most of the Scottish Highlands. The lower nappe exhibits a layered sequence with olivine-rich Lherzolite–Harzburgite at the base covered by a dunite layer, which is rhythmically banded at the junction, and a gabbro layer capping the sequence, which is also rhythmically banded in the lower part. In the upper part it is heavily intruded by fine-grained basic rocks including NE-trending vertical dykes and occasional trondhjemites. At the base of the gabbro and the dyke complex there are discontinuous irregular layers of wehrlite–clinopyroxenite. The upper nappe is composed entirely of olivine-rich lherzolite–harzburgite. The ophiolitic rocks are not tectonised but have suffered low-grade metamorphism including serpentinisation prior to thrusting. The emplacement of the lower nappe took place before that of the upper nappe and both nappes must have moved at least several km to reach their present positions. Nappe transport occurred after the high-grade metamorphism of the East Mainland succession which is

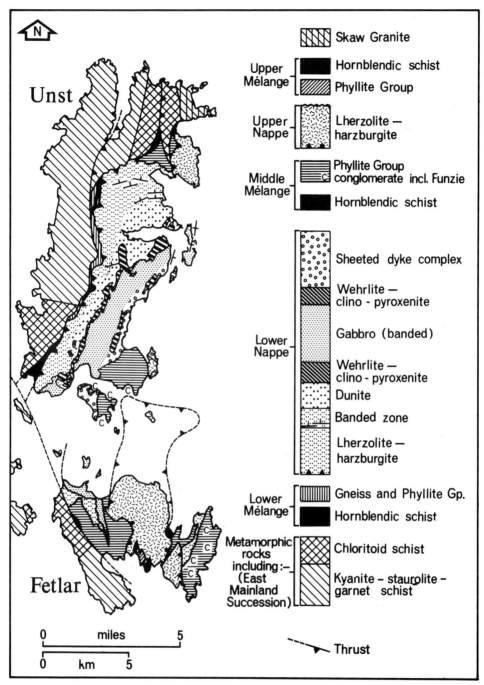

Fig. 5.10. The ophiolite complex of north-east Shetlands in thrust contact with the East Mainland succession (Scatsta Division). From Flinn (1985).

retrograded at the thrust planes (Fig. 5.15). In mainland Scotland, however, the nappes ante-date the climax of regional metamorphism.

Mélange zones occur between the nappes, below the lower nappe and above the upper nappe (Fig. 5.10). These zones contain blocks of serpentinised peridotite and gabbro derived from the nappes, as well as slices of migmatitic rocks sheared off from the substratum over which the nappes have travelled, together with slices of hornblende schists which locally contain garnet and diopside. The mélange zones also contain meta-volcanic rocks and low-grade metasediments, e.g. the pelites, semi-pelites and gritty sandstones of the Phyllite Group (Flinn *et al.* 1979) and conglomerates, some of which are dominantly composed of gabbro pebbles, and one, the Funzie conglomerate (Fig. 5.11), of quartzite pebbles. The conglomerates are poly-mictic and include pebbles of a spilite, quartz–albite–porphyry and of albite–granite (trondhjemite). These sediments are believed to be derived at least in part from the erosion of the nappes. The Phyllite Group found in the mélange zone intervening between the nappes has been deposited on the eroded upper surface of the lower nappe. Subsequently the upper nappe was thrust over these sediments.

The direction and amount of the displacement along the thrust planes of Unst and Fetlar is unknown, but an ESE–WNW transport has been suggested. The actual thrust planes lack directional structures and no root zone has been recognised in Shetland. The minor structures in the melanges provide very little information on the direction and sense of movement of the nappes, there being no systematic fold vergence. During nappe movement the rocks within these zones were folded and lineated on NNE–SSW axes, and, as shown by pebble elongation in the conglomerates of the melange zones (Fig. 5.11), they were elongated parallel to the fold axes as a result of constriction in the plane normal to the fold axes. This deformation in the melanges was not directly caused by movements on the thrust planes but by the flow of relatively incompetent rocks towards the direction of easiest relief, the flow being due to pressure from the relatively rigid masses of the nappes.

Fig. 5.11. Funzie conglomerate, Fetlar. British Geological Survey photograph. Crown copyright reserved.

Tear faults

The Dalradian rocks are cut by several conspicuous tear faults which trend NNE–SSW to NE–SW, that is roughly parallel to the Great Glen Fault but oblique to the Highland Boundary Fault. These faults are younger than the main folding and regional metamorphism in the southern Highlands. Although stratigraphical control is imprecise it is inferred that the faults were active in the Devonian at about the same time as large movements on the Great Glen Fault and the Highland Boundary Fault

(see Chapter 1). Some post-orogenic intrusions (see Chapter 8), e.g. the Moor of Rannoch granite and the Garabal Hill complex, are displaced by a few km along tear faults. In these cases the sinistral slip, which is assumed to have acted along most of the tear faults in the southern Highlands can be demonstrated clearly. Other tear faults (e.g. the Killin Fault) seem to antedate post-orogenic intrusions. Johnson & Frost (1977) have interpreted the fault pattern in the Grampian Highlands as a Riedel shear system associated with major sinistral slip on the Great Glen Fault.

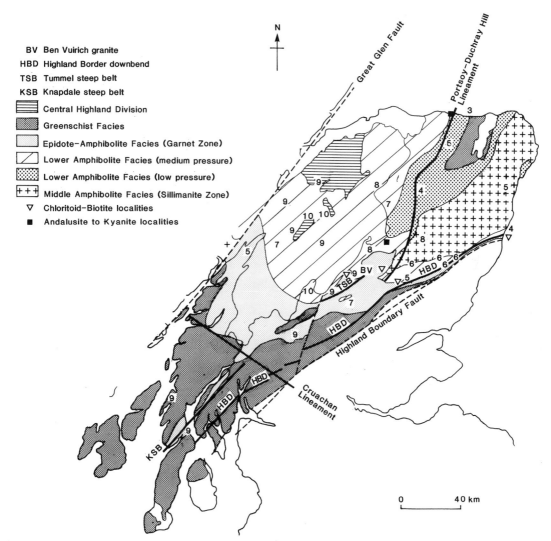

Fig. 5.12. Distribution of metamorphic facies in the Southern Highlands. Numbers indicate approximate pressures in Kilobars during peak metamorphic conditions. From Harte (1988).

Regional metamorphism

The Dalradian outcrop is classic for the study of regional metamorphism, the process whereby rocks of varying composition are transformed into slates, schists and gneisses, Barrow (1893) showed that in pelitic rocks of part of the SE Highlands differing metamorphic mineral assemblages reflect different conditions of metamorphism. He mapped lines that mark the incoming of critical index minerals, that is minerals that are presumed to be sensitive indicators of changes in physical conditions. Barrow's (1912) map of metamorphic zones shows the distribution of index minerals in pelitic rocks, in order of increasing metamorphic grade. Tilley (1925) modified the zonal scheme, substituting the chlorite zone for the first two of Barrow's zones and Elles & Tilley (1930) continued the zonal boundaries into the central and south-west Highlands and Islay and Jura. The grade of regional metamorphism in the Dalradian was found to increase rapidly as the rocks are traced northwards from the Highland Border, and in places a mere 7 km separates the chlorite and sillimanite zones.

The reality of the zonal maps (Fig. 5.12) based on mineral assemblages in pelitic rocks was confirmed by the study of the distribution of minerals in other rocks. Zonal schemes recognised by Wiseman (1934) in the epidiorites, and by Phillips (1930) in the Green Beds, roughly coincide with the pelite zones. The distribution of other metamorphic minerals has been studied, e.g. chloritoid (Barrow 1898) which occurs in a 35 km long zone in the SE Highlands (Williamson 1953), and in Shetland (Flinn 1967), stilpnomelane (Wiseman 1934, Mather & Atherton 1965) and paragonite (McNamara 1965, Baltatzis & Wood 1977). Both Barrow and Tilley equated the mapped zonal boundaries with thermal surfaces or 'isograds', but the disposition of the isograds is still controversial (Atherton 1977). The zonal pattern (Fig. 5.12) mostly reflects the variation in temperature in the Dalradian at the peak of regional metamorphism (Grampian orogeny). Low temperature, greenschist facies prevails in the south-west and high-temperature, mid-amphibolite facies, occurs in the north-east part of the Dalradian where local migmatites are associated with sillimanite-bearing rocks.

In addition to the Barrovian zonal scheme outlined above, Read (1923, 1952) has described the Buchan scheme which exists over an area of some 3,500 sq km in north-east Scotland. Read (1952) viewed the Buchan as a progressive, regional metamorphism with chlorite–sericite–slates in the lowest grade, cordierite–andalusite–biotite–phyllites and andalusite–cordierite–staurolite–schists in the middle-grade and cordierite–sillimanite–gneiss representing the highest grade. The Buchan metamorphism reflects the paramount control of temperature, pressure being lower than in the case of the Barrovian zones.

Careful petrochemical studies have attempted to refine the metamorphic zonal schemes. The development of the various index minerals has been investigated, e.g. biotite (Mather 1970), garnet (Atherton 1964), staurolite (Harte 1975), kyanite (Chinner 1965), sillimanite (Chinner 1965, 1966; Ashworth 1975a); chloritoid (Williamson 1953, Chinner 1967). Other studies of this kind are mentioned by Atherton (1977). This work and the ensuing construction of phase diagrams and petrogenetic grids give impressive testimony to the significance of Barrow's discovery (Fig. 5.12).

Over the years various opinions have been expressed on the relations of deformation and metamorphism in the Dalradian. Rast (1958b), Sturt & Harris (1961), Johnson (1962, 1963) and Harte & Johnson (1969) have employed minor structures as 'time markers' to study the periods of crystallisation of metamorphic minerals (Fig. 5.13). The progress of metamorphism can be studied using the structural time-scale, and the technique is probably reliable at least in a moderate-sized area. Table 5.2 shows the time-scale of metamorphic events in the central and north-eastern parts of the Dalradian outcrop.

The time-scale suggests that the metamorphic zonal assemblages of both Barrovian and Buchan type were imprinted on the rocks during the climax of metamorphism (M_3) that followed the F_2 fold episode (Fig. 5.14), that is after the nappe-forming movements. Although presented as episodic mineral growth it is likely that the history reflects a single continuous rise and fall in the metamorphism (Harte & Johnson 1969). In the NE Scotland it is possible that the intrusion of the huge Newer Gabbros intervened in time between F_2 and the static M_3 metamorphism (Stewart & Johnson 1961, Fettes 1970). Yardley et al. (1982) have interpreted the association of low-pressure assemblages with basic intrusions in Connemara and Buchan as indicating a late Cambrian/early Ordovician arc volcanic centre.

The seemingly coeval Barrovian (M_3) and Buchan mineral assemblages were formed at different structural levels in the Dalradian, with the Barrovian metamorphism occupying the deeper levels in the orogen. This depth arrangement of the two Al_2SiO_5 polymorphs, kyanite (Barrovian) and andalusite (Buchan), supports experimental data (Miyashiro 1961, Richardson, Gilbert & Bell 1969) which suggest that kyanite is the high, and andalusite the low pressure polymorph. Hudson (1985) gives the following pressure/depth estimates for the NE Dalradian:

Andalusite zone – 2 Kbar: 6 km
Staurolite zone – 2–3·5 Kbar: 10 km
Kyanite zone – 4·5 Kbar: 17 km

These figures are consistent with the 4–4·5 Kbar (15–18 km) estimated from a study of the aureoles of the Newer Gabbros (Droop & Charnley 1985).

A new tectono–thermal model for NE Scotland is given by Baker (1987) who observed that after unfolding the thermal and baric syncline, centred on the Boyndie–Turriff Syncline, high thermal gradient Buchan type rocks overlaid

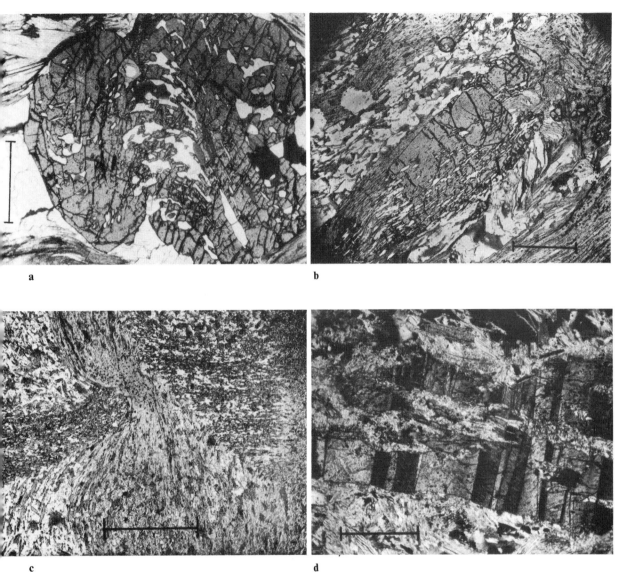

Fig. 5.13.

a. Garnet crystal with helicitic folds (D_2) defined by elongated quartz grains. Sillimanite felts in the matrix post-date garnet growth. Glen Clova, Angus.

b. Staurolite–garnet–schist showing helicitic (D_2) fabric in the porphyroblast and D_3 microfolds in the matrix. Portsoy, Banffshire.

c. Andalusite–schist, showing the preservation of an early (S_1) schistosity in the porphyroblasts. In the matrix a new schistosity, D_3, has developed. Rosehearty, Aberdeenshire.

d. Kyanite crystal deformed by late kink bands. Glen Clova, Angus.

higher pressure but lower thermal gradient rocks to the west. A tectonic model was suggested which envisaged the occurrence of the Buchan metamorphism before or during NW directed thrusting of the Buchan area relative to areas to the west. The tectonic burial consequent upon the emplacement of the Buchan rocks resulted in the transformation of andalusite to kyanite and the enhancement of the metamorphic grade in the footwall. This means that peak Barrovian metamorphism, at least in some areas, postdated Buchan metamorphism. Later deformation resulted in differential uplift and erosion of the thrust pile,

resulting in folding of isotherms, development of the Boyndie/Turriff syncline and steepening of the Portsoy slide zone. As Baker points out this conclusion conflicts with correlations of structural and metamorphic episodes published by earlier workers. While supporting the concept of an allochthonous Buchan area, Baker does not accept the Sturt & Ramsay (1977) view that the Cowhythe and Ellon gneisses are derived from basement.

Chinner (1966) showed how a P/T diagram for the metamorphic climax can be deduced from the stability fields of index minerals, particularly aluminium silicate

Table 5.2. Ages of metamorphic and related events in the Dalradian rocks of Scotland and the Shetland Isles

Age (Ma)	Scottish Dalradian eastern and south–central Highlands	Shetland Isles	Age (Ma)
		> 860 Ma metam. in Walls Metamorphic series	
550			550
—		↑ Peak regional	—
		? metamorphism and	
—		↓ migmatisation	—
—	D_1 and D_2 major nappe structures		—
	Ben Vuirich Granite*		
—	D_3 folding, Peak metam. (Grampian age) in Barrovian		—
500	and Buchan terranes and Newer Gabbro intrusions.		500
—			—
—	(Granitic intrusions)		—
—			—
—			—
—	K/Ar mica cooling ages and uplift with D_4	↑	—
450	structures and Highland	? Ophiolite nappe	450
	Border Downbend	↓ emplacement	
—	formation (460–440 Ma)		—
—			—
—	Low pressure contact metam. about		—
—	granitoid intrusions	K/Ar mica cooling ages; contact	—
400		metam. about granitoid	400
—		intrusions	—

D_1 – deformation episode.
metam. – metamorphism.
Dashed vertical line indicates a major range of K/Ar ages from micas and interpreted as cooling ages following peak metamorphism.
(Modified from Harte 1988.)

* A reinvestigation of U–Pb zircon age of the Ben Vuirich granite by G. Rogers (in press) indicates that this granite was intruded at about 590 Ma. If correct this means (a) the Dalradian Supergroup is entirely Precambrian and (b) the Grampian orogeny occurred mainly in Precambrian times.

polymorphs. Curves on the diagram describe the P–T conditions for the dehydration reactions that produced the alumino-silicates, the alumino-silicate inversions, the 'minimum melting' of granite and the muscovite–quartz breakdown reaction. A two-stage metamorphic history is postulated, firstly, the kyanite andalusite metamorphic event with a geothermal gradient of 30°C/Kbar and secondly, an isobaric rise in temperature to permit the growth of sillimanite and potash–feldspar along with partial melting. The geometry of the isobars and isotherms shown on Chinner's diagram has been modified by Porteous (1973) and Harte & Hudson (1979).

A highly significant feature in Chinner's model is the 'lateness' of sillimanite in relation to the kyanite–andalusite event. This has been confirmed by micro-textural investigations by Harte & Johnson (1969), and Fettes (1970). Ashworth (1975a, 1976) has argued that the sillimanite zone, and the migmatites, in north-east Scotland are causally related to the intrusion of the Younger Gabbros. Fettes (1970) brought the various events together under the umbrella of the single M_3 metamorphism in suggesting that the gabbros were intruded into hot country-rock, so that it is not possible to separate 'regional' and 'contact' metamorphisms in the vicinity of the gabbros. Harte & Hudson

(1979) indicated that sillimanite growth pre-dated the intrusion of the Newer Gabbros and the consensus view is opposed to Ashworth's conclusion.

Following the publication of Chinner's important paper there have been several attempts to illustrate pressure variations across the Dalradian outcrop. For example Hudson (in Fettes (1976)) envisages that pressure increases in the Buchan zones as the boundary with the Barrovian zones is approached, the evidence being the incoming of a staurolite–garnet assemblage near this boundary. Fettes *et al.* (1976) suggest that there was a decrease in pressure from SW to NE in the Dalradian detected by the varying unit cell dimensions (b_0 values) of potassic white mica. Harte (1975) and Harte & Hudson (1979) constructed petrogenetic grids for the north-east part of the Dalradian, including the Buchan area discussed by Chinner. Included on the grids is the Chloritoid zone which Harte places within a new facies series, the Stonehavian, in which rocks contain andalusite and sillimanite, but no kyanite, together with chloritoid–garnet–biotite assemblages.

The elaborate refinements of metamorphic facies are yielding results which have important implications for the tectonics of the Dalradian as well as for the understanding of the metamorphism. For example it is now possible to give estimates of the pressure and temperature conditions which prevailed during the metamorphic climax. Pressures in north-east Scotland varied from 3·5 Kbar in the Buchan area to 6·5 Kbar in the Barrovian area, with temperature in the range 300–700°C. Kerrich *et al.* (1977) used oxygen isotopes to estimate temperatures for the Barrovian zones, from 300°C for chlorite grade up to nearly 600°C for the kyanite zone. Richardson & Powell (1976) present results for the Dalradian near the Great Glen derived from studies of co-existing minerals in pelites and carbonates. These workers suggest a temperature of 535°C (for the garnet zone) and a pressure of 5 Kbar. In the central part of the Dalradian, Wells & Richardson (1979) have studied co-existing kyanite, garnet, biotite, plagioclase and quartz and conclude that temperatures of 550–620°C were reached at pressures of 9–12 Kbar, corresponding to depths of burial of 35–40 km (see also Baker & Droop 1983). Contrasting these results with those of Richardson & Powell (loc. cit.) they speculate that either the erosion rates during metamorphism were greater in the central Dalradian or that the tectonic cover there was much larger than in the more westerly parts. Baker (1985) records 6–7 Kbar from Glen Clova and Glen Muick and 9–10 Kbar from the Schiehallion area. Depths of burial of this order of magnitude indicate an original crustal thickness of 70 km in the Grampian Highlands. These results led to speculation on the eroded nappe-cover, e.g. Dewey & Shackleton (1984), the extension of the structures that are barely preserved in north-east Scotland (see p. 140) and on the ultimate cause of the regional metamorphism in the Dalradian.

The causal relationship, first postulated by Barrow between the zonary metamorphic mineral assemblages and

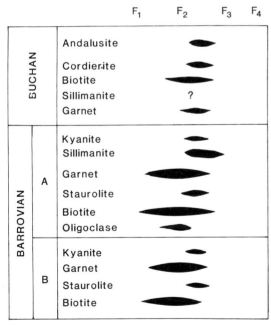

Fig. 5.14. Periods of crystallisation of metamorphic minerals in Buchan and Barrovian zones. F_1–F_4, fold episodes; A, High-grade Barrovian; B, Low-grade Barrovian.

the migmatitic 'Older Granites' – the zones of regional metamorphism forming a thermal aureole around 'hot-spots' represented by migmatites – cannot be accepted. There is force in the criticism implied in the verse, quoted by Read in his book *The Granite Controversy* 'when it (Older granite) is absent it alters them (the schists) most'. Kennedy (1948) interpreted the zonal map that he prepared for

the whole Highland region as showing an antiformal arrangement of the zones, or thermal surfaces, which cut across the recumbent folds. In the core of this gigantic thermal antiform the Moine and Dalradian sediments were thought to have been soaked in trondhjemitic magma (cf. Read 1927, 1937), derived from the melting of the sialic mountain 'root'. This migmatitic core forms a thermal

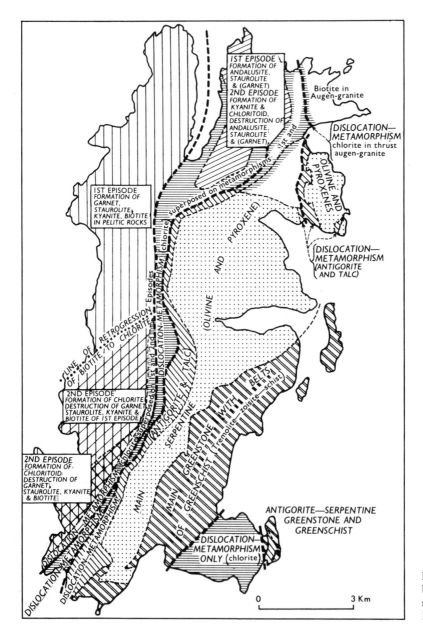

Fig. 5.15. Relationship between metamorphism and structure in Unst, Shetland. From Read (1934).

focus for the zonary structure. Accordingly the high-grade rocks are found in the core of the thermal antiform, in the central parts of the Moines and Dalradians; and the low-grade rocks are found on the limbs, on the margins to the Highlands.

In recent assessments (Fettes 1983, Harte 1986) this rather simple depth-controlled model appears to break down. For example, there is a lack of correlation between pressure and temperature, witness the discovery that pressures were higher in certain greenschist areas than in some high-grade zones. This points to the existence of local controls for the metamorphism which perhaps reflect variations in the basement to the Dalradian Supergroup (Fettes *et al.* 1986). There is also the suggestion that the

which is in accord with Harte's (1975) estimates in the Barrovian area. It is unlikely that the debate over causes will end there. Already, doubts have been expressed as to whether a depth-controlled metamorphism is consistent with evidence of high-level abrupt changes of temperature (Atherton 1977), and Harte & Hudson (1979) postulate widespread, deep-seated magma intrusion as an explanation of the regional metamorphic gradients and the migmatites.

After the relatively short metamorphic climax in the Dalradian geothermal gradients fell fairly rapidly from roughly $100^{\circ}C/km$ to perhaps something like $30–35^{\circ}C/km$. There followed a prolonged period of isostatic recovery and thermal adjustment. The waning heat-flow is shown by the

Table 5.3. Structural and thermal history of the Dalradian

Time (Ma)	Uplift events	Structural events	Metamorphic events	Igneous events
520–490	Local Uplift	Early ductile folding and tilting (Dalradian $D_1–D_3$)	Metamorphism (Dalradian)	Gabbros 'Older Granites'
490–460	<----------------------------------- Inactivity ----------------------------------->			
460–440	Regional Uplift (Cooling of Dalradian)	Late folding (Dalradian D_4)	Metamorphism (Moine)	Granites
440–410	<----------------------------------- Inactivity ----------------------------------->			
410–390	(Cooling of Dalradian + Moine	Major faulting	?Metamorphism? (Lower crust + mantle)	Granites

metamorphic zones are inverted locally (Chinner 1978, Watkins 1985).

In attributing Barrovian- and Buchan-type metamorphisms to the introduction of magmas or hot fluids into high-levels of the crust there is, of course, a danger of confusing cause and effect. Long ago Read (1927, p. 348) stated that 'injection' granites and sillimanite-grade metamorphism are localised effects of the deep burial in the mountain region. Furthermore the 'injection complexes' and migmatites of the Dalradian and Moine tracts may not reflect large-scale transference of material. Local partial melting of sedimentary rocks, perhaps induced by differential heat transport and the rise of volatiles, etc. may be important factors in the process of migmatisation (Chinner 1966, Ashworth 1975b).

A rival hypothesis is that there was little or no accession of heat from below the crust during the regional metamorphism of the Dalradian; instead the heat required for the Buchan and Barrovian metamorphism was generated within the thickened crust. Richardson & Powell (1976) have explored this hypothesis using the estimated tectonic thickness of the Dalradian, the likely basement (?Lewisian, plus Grenville?), the time available for metamorphism, and the likely depth of burial at the time of metamorphism. The steady-state geothermal gradient required is $100^{\circ}C/km$

retrogressive metamorphic effects, e.g. growth of albite porphyroblasts (Watkins 1983), associated with the late deformation (D_4) (Graham *et al.* 1983).

The cooling and uplift model developed by Harper (1967) and Dewey & Pankhurst (1970) envisages a long cooling history, until about 440 Ma, when the minerals in the deeper higher grade parts of the Dalradian eventually became 'closed systems'. Delayed cooling in the Shetland is suggested by the mineral ages of about 420 Ma (Miller & Flinn 1966).

Dempster (1985) has been able to document with a remarkable precision the cooling/uplift history of the Dalradian. From a study of mineral ages he sets out the events shown in Table 5.3.

The uplift is shown to be local and discontinuous rather than regional and steady state (Fig. 5.16). This concept has been developed by Harte (1988) who suggests that the whole metamorphic evolution of the Dalradian has been influenced by domains or provinces which are bounded by lineaments such as those referred to earlier.

Activity on the lineaments has caused variations in uplift rates and timing of metamorphism across the Dalradian tract, the differences in thermal conductivity and heat production between one domain and the next reflecting activity of the lineaments before the onset of orogeny.

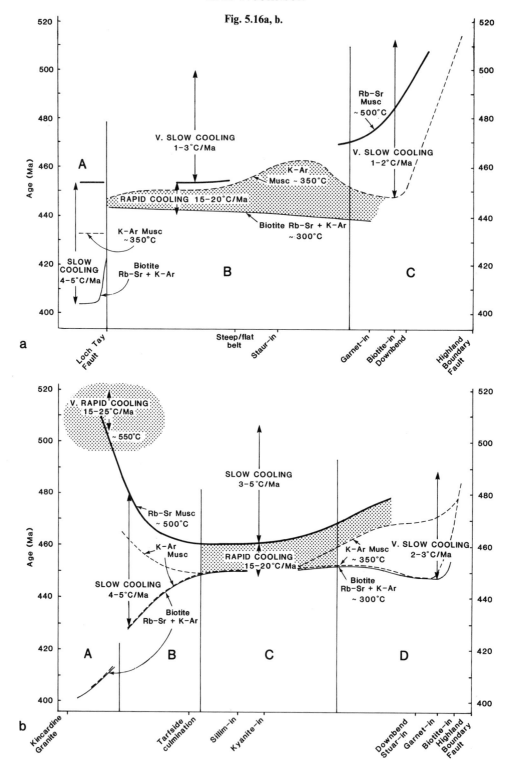

Fig. 5.16a, b.

Fig. 5.16c, d.

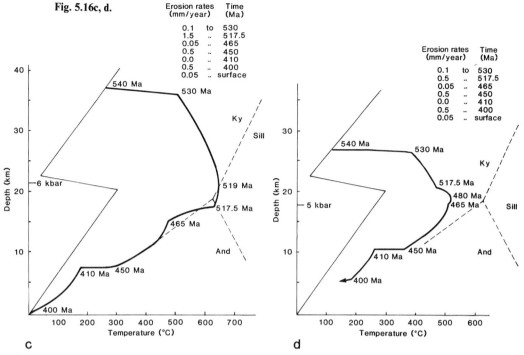

c   d

Fig. 5.16. Mineral age data from **a**. Perthshire and **b**. Angus, showing the approximate cooling rates and assumed closure temperatures. A–D refer to different sections in these areas. Periods of rapid cooling stippled. **c.** and **d.** show possible P–T paths simulating metamorphism and later cooling in the sillimanite zone, **c.**, and the staurolite zone, **d**. Light solid line shows the initial thermal profile. Dotted lines show aluminosilicate triple point for kyanite (Ky), Andalusite (And) and sillimanite (sill). From Dempster (1986).

REFERENCES

ANDERTON, R.

1971   Dalradian palaeocurrents from the Jura Quartzite. *Scott. J. Geol.*, **7**, 175–178.

1975   Tidal Flat and shallow marine sediments from the Craignish Phyllites, Middle Dalradian, Argyll, Scotland. *Geol. Mag.*, **112**, 337–340.

1976   Tidal-shelf sedimentation: an example from the Scottish Dalradian. *Sedimentology*, **23**, 429–458.

1977   A field guide to the Dalradian rocks of Jura, Argyll. *Scott. J. Geol.*, **13**, 135–142.

1979   Slopes, submarine fans and syn-depositional faults: sedimentology of parts of the Middle and Upper Dalradian in the S.W. Highlands of Scotland. *In* Harris, A. L., Holland, C. H., & Leake, B. E. (Eds.) The British Caledonides – reviewed. *Spec. Publ. geol. Soc. London*, **8**, 483–488.

1980a   Distinctive pebbles as indicators of Dalradian provenance. *Scott. J. Geol.*, **16**, 143–152.

1980b   Did Iapetus start to open during the Cambrian? *Nature, London*, **286**, 706–708.

ANDERTON, R. 1982 Dalradian deposition and the late Precambrian–Cambrian history of the N. Atlantic region: a review of the early evolution of the Iapetus Ocean. *J. geol. Soc. London*, **139**, 421–431.

1985 Sedimentation and tectonics in the Scottish Dalradian. *Scott. J. Geol.*, **21**, 407–436.

ASHCROFT, W. A., KNELLER, B. C., LESLIE, A. G., & MUNRO, M. 1984 Major shear zones and autochthonous Dalradian in the northeast Scottish Caledonides. *Nature*, **310**, 760–762.

ASHWORTH, J. R. 1975a The sillimanite zones of the Huntly–Portsoy area in the northeast Dalradian, Scotland. *Geol. Mag.*, **112**, 113–136.

1975b Staurolite at anomalously high grade. *Contrib. Mineral Petrol.*, **53**, 281–291.

1976 Petrogenesis of migmatites in the Huntly–Portsoy area, northeast Scotland. *Mineralog. Mag.*, **40**, 661–682.

ATHERTON, M. P. 1964 The garnet isograd in pelitic rocks and its relationship to metamorphic facies. *Am. Mineralog.*, **49**, 1331–1349.

1977 The metamorphism of the Dalradian rocks of Scotland. *Scott. J. Geol.*, **13**, 331–370.

BAILEY, E. B. 1910 Recumbent folds in the schists of the Scottish Highlands. *Q. Jl. geol. Soc. London*, **66**, 586–620.

1917 The Islay anticline (Inner Hebrides). *Q. Jl. geol. Soc. London*, **72** (for 1916), 132–164.

1922 The structure of the south-west Highlands of Scotland. *Q. Jl. geol. Soc. London*, **78**, 82–127.

1925 Perthshire Tectonics: Loch Tummel, Blair Atholl and Glen Shee. *Trans. R. Soc. Edinburgh*, **53**, 671–698.

1934 West Highland Tectonics: Loch Leven to Glen Roy. *Q. Jl. geol. Soc. London*, **90**, 462–523.

1940 Discussion of paper by A. Allison: Loch Awe succession and tectonics: Kilmartin–Tayvallich–Danna. *Q. Jl. geol. Soc. London*, **96**, 423–449.

BAILEY, E. B. & McCALLIEN, W. J. 1937 Perthshire tectonics: Schiehallion to Glen Lyon. *Trans. R. Soc. Edinburgh*, **54**, 79–118.

BAKER, A. J. 1985 Pressures and temperatures of metamorphism in the eastern Dalradian. *J. geol. Soc. London*, **142**, 137–148.

1987 Models for the tectonothermal evolution of the eastern Dalradian of Scotland. *J. metamorphic Geol.*, **5**, 101–118.

BAKER, A. J. & DROOP, G. T. R. 1983 Grampian metamorphic conditions deduced from mafic granulites and sillimanite-K-feldspar gneisses in the Dalradian of Glen Muick, Scotland. *J. geol. Soc. London*, **140**, 489–497.

BALTATZIS, E. & WOOD, B. J. 1977 The occurrence of paragonite in chloritoid schists from Stonehaven/Scotland. *Mineralog. Mag.*, **41**, 211–216.

BAMFORD, D., NUNN, K. PRODEHL, C., & JACOB, B. 1977 LISPB–III. Upper crustal structure of northern Britain. *J. geol. Soc. London*, **133**, 481–488.

BARROW, G. 1893 On an intrusion of muscovite biotite gneiss in the southeast Highlands of Scotland and its accompanying metamorphism. *Q. Jl. geol. Soc. London*, **49**, 330–358.

1898 Chloritoid schists from Kincardineshire. *Q. Jl. geol. Soc. London*, **54**, 149–155.

1912 On the geology of lower Deeside and the southern Highland Border. *Proc. Geol. Assoc. London*, **23**, 268–273.

BENTLEY, M. 1988 The Colonsay Group. *In* Winchester, J. A. (Ed.) *Later Proterozoic stratigraphy of the Northern Atlantic regions*. Blackie, Glasgow & London, 119–129.

BORRADAILE, G. J.   1973   Dalradian structure and stratigraphy of the northern Loch Awe district, Argyllshire. *Trans. R. Soc. Edinburgh*, **69**, 1–21.

1976   Thermal anisotropy – a factor contributing to the distribution of Caledonian metamorphic zones in the southwest Scottish Highlands. *Geologie Mijnb.*, **42**, 121–142.

1979   Pretectonic reconstruction of the Islay anticline: implications for the depositional history of the Dalradian rocks in the S.W. Highlands. *In* Harris, A. L., Leake, B. E., & Holland, C. H. (Eds.) The Caledonides of the British Isles reviewed. *Geol. soc. London, Spec. Publ.*, **8**, 229–238.

BORRADAILE, G. J. &   1973   Finite strain estimates from the Dalradian Dolomitic Forma-
JOHNSON, H.   tion, Islay, Argyll, Scotland. *Tectonophysics*, **18**, 249–259.

BRADBURY, H. J.   1979   Migmatisation, deformation and porphyroblast growth in the Dalradian of Tayside, Scotland. *In* Harris, A. L., Holland, C. H., & Leake, B. E. (Eds.) The Caledonides of the British Isles – reviewed. *Spec. Publ. Geol. Soc. London*, **8**, 351–356.

BRADBURY, H. J.   1985   The Caledonian metamorphic core: an Alpine model. *J. geol. Soc. London*, **142**, 129–136.

BRADBURY, H. J.,   1979   Geometry and emplacement of nappes in the Central Scottish
HARRIS, A. L., &   Highlands. *In* Harris, A. L., Holland, C. H., & Leake, B. E.
SMITH, R. A.   (Eds.) The Caledonides of the British Isles – reviewed. *Spec. Publ. geol. Soc. London.*

BRADBURY, H. J.,   1976   'Older' granites as time-markers in Dalradian evolution. *Jl. geol.*
SMITH, R. A., &   *Soc. London*, **132**, 677–684.
HARRIS, A. L.

VAN BREEMEN, O.,   1979   Age of the Glen Dessary syenite, Inverness-shire: Diachronous
AFTALION, M.,   Palaeozoic metamorphisms across the Great Glen. *Scott. J.*
PANKHURST, R. J., &   *geol.*, **15**, 49–62.
RICHARDSON, S. W.

VAN BREEMEN, O. &   1972   A radiometric age for pegmatite cutting the Belhelvie mafic
BOYD, R.   intrusion, Aberdeenshire. *Scott. J. Geol.*, **8**, 115–120.

VAN BREEMEN, O.,   1978   Crustal additions in late Precambrian times. *In* Bowes, D. R. &
HALLIDAY, A. N.,   Leake, B. E. (Eds.) Crustal evolution in northwestern Britain
JOHNSON, M. R. W., &   and adjacent regions. *Geol. Jour.* Special issue No. 10, Seel
BOWES, D. R.   House Press.

CHINNER, G. A.   1960   Pelitic gneisses with varying ferrous/ferric ratios from Glen Clova, Angus, Scotland. *J. Petrol.*, **1**, 178–217.

1961   The origin of sillimanite in Glen Clova, Angus. *J. Petrol.*, **2**, 312–323.

1962   Regional metamorphic sequences in the light of the Al_2SiO_5 polymorphs. *Proc. geol. Soc. London*, **1594**, 15–16.

1965   The Kyanite isograd in Glen Clova, Angus, Scotland. *Mineralog. Mag.*, **34**, 132–143.

1966   The distribution of pressure and temperature during Dalradian metamorphism. *Q. Jl. geol. Soc. London*, **122**, 159–186.

1967   Chloritoid and the isochemical character of Barrow's zones. *J. Petrol.*, **8**, 268–282.

1978   Metamorphic zones and fault displacements in the Scottish Highlands. *Geol. Mag.*, **115**, 37–45.

1980   Kyanite isograds of Grampian metamorphism. *J. geol. Soc. London*, **137**, 35–39.

CHINNER, G. A. &   1979   The Grampide andalusite–Kyanite isograd. *Scott. J. Geol.*, **15**,
HESELTINE, F. J.   117–127.
CLOUGH, C. T.   1897   The geology of Cowal. *Mem. geol. Surv. Scotld.*
COWARD, M. P.   1983   The thrust and shear zones of the Moine thrust zone and the NW Scottish Caledonides. *J. geol. Soc. London*, **140**, 795–811.

CURRY, G. B., BLUCK, B. J.,   1985   Age, evolution and tectonic history of the Highland Border
  BURTON, C. J.,   Complex, Scotland. *Trans. R. Soc. Edinburgh*, **75**, 113–133.
  INGHAM, J. K.,
  SIVETER, D. J., &
  WILLIAMS, A.

DEMPSTER, T. J.   1985   Uplift patterns and orogenic evolution in the Scottish Dalradian.
                                       J. geol. Soc. London, **142**, 111–128.

DEWEY, J. F.   1969   Evolution of the Appalachian/Caledonian orogen. *Nature
                                       London*, **222**, 124–129.

DEWEY, J. F. &   1970   The evolution of the Scottish Caledonides in relation to their
  PANKHURST, R. J.   isotopic age pattern. *Trans. R. Soc. Edinburgh*, **68**, 361–389.

DEWEY, J. F. &   1984   A model for the evolution of the Grampian tract in the early
  SHACKLETON, R. J.   Caledonides and Appalachians. *Nature*, **312**, 115–120.

DOWNIE, C.   1975   Precambrian of the British Isles: Palaeontology. *In* Harris, A. L.
                                       et al. (Eds.) A correlation of Precambrian rocks in the British
                                       Isles. *Spec. Rep. geol. Soc. London*, **6**, 113–115.

DOWNIE, C.,   1971   A palynological investigation of the Dalradian rocks of
  LISTER, T. R.,   Scotland. *Rep. Inst. geol. Sci.*, No. 71/9, 30 pp.
  HARRIS, A. L. &
  FETTES, D. J.

DROOP, G. T. R. &   1985   Comparative geobarometry of pelitic hornfelses associated with
  CHARNLEY, N. R.   the Newer Gabbros: a preliminary study. *J. geol. Soc. London*,
                                       142, 53–62.

DUNNING, F. W.   1972   Dating events in the Metamorphic Caledonides: impression of
                                       the symposium held in Edinburgh, September 1971. *Scott. J.
                                       Geol.*, **8**, 179–192.

EDWARDS, M. B.   1972   Glacial, interglacial and postglacial sedimentation in a Late
                                       Precambrian shelf environment. Finnmark, North Norway.
                                       Univ. of Oxford D.Phil. Thesis.

ELLES, G. L. &   1930   Metamorphism in relation to structure in the Scottish High-
  TILLEY, C. E.   lands. *Trans. R. Soc. Edinburgh*, **56**, 621–646.

ESCHER, A. &   1974   Stretching fabrics, folds and crustal shortening. *Tectonophysics*,
  WATTERSON, J.   **22**, 223–231.

EYLES, C. H. & EYLES, N.   1984   Glaciomarine model for upper Precambrian diamictites of the
                                       Port Askaig Formation, Scotland. *Geology*, **11**, 692–696.

FETTES, D. J.   1970   The structural and metamorphic state of the Dalradian rocks
                                       and their bearing on the age of emplacement of the basic sheet.
                                       Scott. J. Geol., **6**, 108–118.

                               1983   Metamorphism in the British Caledonides. *In* Schenk, P. E. (Ed.)
                                       Regional trends in the geology of the Appalachian–Caledonian–
                                       Hercynian–Mauritanide Orogen. NATO ASI series, C, *Math-
                                       ematical and Physical Sciences*, **116**, 205–219. Reidel Publ. Co.,
                                       Holland.

FETTES, D. J.,   1986   Lineaments and basement domains: an alternative view of
  GRAHAM, C. M.,   Dalradian evolution. *J. geol. Soc. London*, **143**, 453–464.
  HARTE, B. &
  PLANT, J. A.

FETTES, D. J.,   1976   The basal spacing of potassic white micas and facies series
  GRAHAM, C. M.,   variation across the Caledonides. *Scott. J. Geol.*, **12**, 227–236.
  SASSI, F. P., &
  SCHOLARI, A

FLINN, D.   1952   A tectonic analysis of the Muness Phyllite block of Unst and
                                       Uyea, Shetland. *Geol. Mag.*, **89**, 263–272.

                               1954   On the time relations between regional metamorphism and
                                       permeation in Delting, Shetland. *Q. Jl. geol. Soc. London*, **110**,
                                       177–201.

FLINN, D.   1956   On the deformation of the Funzie Conglomerate, Fetlar,
                                          Shetland. *J. Geol.*, **64**, 480–505.

                                   1958   On the nappe structure of North-east Shetland. *Q. Jl. geol. Soc.
                                          London*, **114**, 107–136.

                                   1967   The metamorphic rocks of the southern part of the Mainland of
                                          Shetland. *Geol. J.*, **5**, 251–290.

                                   1985   The Caledonides of Shetland. *In* Gee, D. G. & Sturt, B. A. (Eds.)
                                          The Caledonide orogen–Scandinavia and related areas. John
                                          Wiley & Sons.

FLINN, D., FRANK, P. L.,   1979   Basement-cover relations in Shetland. *In* Harris, A. L., Holland,
   BROOK, M. &   C. H., Leake, B. E. (Eds.) The Caledonides of the British Isles –
   PRINGLE, I. R.   reviewed. *Spec. Publ. geol. Soc. London*, **8**, 109–115.

FLINN, D., MAY, F.,   1972   A revision of the stratigraphic succession of the east Mainland of
   ROBERTS, J. L. &   Shetland. *Scott. J. Geol.*, **8**, 335–343.
   TREAGUS, J. E.

FLINN, D. & MOFFAT, D. T.   1985   A peridotitic Komatiite from the Dalradian of Shetland. *Geol.
                                          J.*, **20**, 287–292.

FOYN, S.   1937   The Eocambrian Series of the Tana district, Northern Norway.
                                          Norsk geol. Tidsskr, **2**, 65–164.

GOWER, P. J.   1977   Dalradian rocks of the west coast of the Tayvallich Peninsula.
                                          Scott. J. Geol., **13**, 125–134.

GRAHAM, C. M.   1976   Petrochemistry and tectonic significance of Dalradian
                                          metabasaltic rocks of the S.W. Scottish Highlands. *J. geol. Soc.
                                          London*, **132**, 61–84.

                                   1986   The role of the Cruachan Lineament during Dalradian evolu-
                                          tion. *Scott. J. geol.*, **22**, 257–270.

GRAHAM, C. M. &   1984   The petrology and structure of Dalradian metabasaltic dykes of
   BORRADAILE, G. J.   Jura: implications for early Dalradian evolution. *Scott. J. Geol.*,
                                          20, 257–276.

GRAHAM, C. M. &   1981   Cambrian and late Precambrian basaltic igneous activity in the
   BORRADAILE, G. J.   Scottish Dalradian: a review. *Geol. Mag.*, **118**, 27–37.

GRAHAM, C. M.,   1983   Genesis and mobility of the $H_2O–CO_2$ fluid phase during
   GREIG, K. M.,   regional greenschist and epidote amphibolite facies metamorph-
   SHEPPARD, S. M. F. &   ism: a petrological and stable isotope study in the Scottish
   TURI, B.   Dalradian. *J. geol. Soc. London*, **140**, 577–599.

HACKMAN, B. D. &   1962   Calcareous algae from the Dalradian of Islay. *Palaeontology*, **5**,
   KNILL, J. L.   268–271.

HALL, J.   1985   Geophysical constraints on crustal structure in the Dalradian
                                          region of Scotland. *J. Geol. Soc. London*, **142**, 149–155.

HALL, J., BREWER, J. A.,   1984   Crustal structure across the Caledonides from the 'WINCH'
   MATTHEWS, D. H. &   seismic reflection profile: influences on the evolution of the
   WARNER, M. R.   Midland Valley of Scotland. *Trans. R. Soc. Edinburgh*, **75**, 97–
                                          109.

HALLIDAY, A. N.,   1989   The depositional age of the Dalradian Supergroup: U–Pb and
   GRAHAM, C. M.,   Sm–Nd isotopic studies of the Tayvallich Volcanics, Scotland.
   AFTALION, M. &   *J. Geol. Soc. London*, **146**, 3–6.
   DYMOKE, P.

HAMBREY, M. J. &   1981   Glacigenic boulder-bearing deposits in the Upper Dalradian
   WADDAMS, P.   Macduff Slates, North Eastern Scotland. *In* Hambrey, M. J. &
                                          Harland, W. B. (Eds.) *Earth's pre-Pleistocene Glacial Record*,
                                          571–575. Cambridge University Press, Cambridge.

HARLAND, W. B.   1972   The Ordovician Ice Age. *Geol. Mag.*, **109**, 451–456.

HARLAND, W. B. &   1972   The Arctic Caledonides and earlier Oceans. *Geol. Mag.*, **109**,
   GAYER, R. A.   289–314.

HARPER, C. T.   1967   The geological interpretation of potassium–argon ages of
                                          metamorphic rocks from the Scottish Highlands. *Scott. J. Geol.*,
                                          3, 46–66.

HARRIS, A. L.   1972 The Dalradian rocks at Dunkeld, Perthshire. *Bull. geol. Surv. Gt. Br.*, **38**, 1–10.

HARRIS, A. L.,   1978 Ensialic basin sedimentation: the Dalradian Supergroup. *In*
    BALDWIN, C. T.,   Bowes & Leake (Eds.) Crustal evolution in north-western
    BRADBURY, H. J.,   Britain and adjacent regions. Seel House Press, Liverpool.
    JOHNSON, H. D. &
    SMITH, R. A.

HARRIS, A. L.,   1976 The evolution and transport of the Tay nappe. *Scott. J. Geol.*, **12**,
    BRADBURY, H. J. &   103–113.
    McGONIGAL, M. H.

HARRIS, A. L. &   1972 Stratigraphy and structure of the Upper Dalradian rocks at the
    FETTES, D. J.   Highland Border. *Scott. J. geol.*, **8**, 253–264.

HARRIS, A. L. &   1975 The Dalradian Supergroup. *In* Harris, A. L. *et al.* (Eds.) A
    PITCHER, W. S.   correlation of Precambrian rocks in the British Isles. *Spec. Rep. geol. Soc. London*, **6**, 52–75.

HARTE, B.   1975 Determination of a pelite petrogenetic grid for the eastern
                                   Scottish Dalradian. *Yb. Carnegie Instn. Wash.*, **74**, 438–446.

                             1979 The Tarfside succession and its bearing on the structure and
                                   stratigraphy of the Eastern Scottish Dalradian. *In* Harris, A. L.,
                                   Holland, C. H., & Leake, B. E. (Eds.) The Caledonides of the
                                   British Isles – reviewed. *Spec. Publ. geol. Soc. London*, **8**, 221–
                                   228.

                             1988 Lower Palaeozoic metamorphism in the Moine–Dalradian belt
                                   of the British Isles. *In* Harris, A. L. & Fettes, D. J. (Eds.) The
                                   Caledonian–Appalachian Orogen. *Spec. Publ. geol. Soc. London.*

HARTE, B., BOOTH, J. E.,   1984 Aspects of the post depositional evolution of the Dalradian &
    DEMPSTER, T. J.,   Highland Boundary rocks in the Southern Highlands of
    FETTES, D. J.,   Scotland. *Trans. R. Soc. Edinburgh*, **75**, 151–163.
    MENDRUM, J. R., &
    WATTS, D.

HARTE, B. &   1979 Pelite facies series and the temperatures and pressures of
    HUDSON, N. F. C.   Dalradian metamorphism in eastern Scotland. *In* Harris, A. L.,
                                   Holland, C. H., & Leake, B. E. (Eds.) The Caledonides of the
                                   British Isles – reviewed. *Spec. Publ. geol. Soc. London.*

HARTE, B. &   1969 Metamorphic history of Dalradian rocks in Glens Clova, Esk
    JOHNSON, M. R. W.   and Lethnot, Angus, Scotland. *Scott. J. Geol.*, **5**, 54–80.

HENDERSON, W. G. &   1982 The Highland Border rocks and their relation to marginal basin
    ROBERTSON, A. H. F.   development in the Scottish Caledonides. *J. geol. Soc. London*,
                                   139, 435–449.

HICKMAN, A. H.   1975 The stratigraphy of late Precambrian metasediments between
                                   Glen Roy and Lismore. *Scott. J. Geol.*, **11**, 117–142.

                             1978 Recumbent folds between Glen Roy and Lismore. *Scott. J. Geol.*, **14**, 191–212.

HIPKIN, R. G. &   1983 Regional gravity analysis: North Britain. *Rep. Inst. Geol. Sci.*
    HUSSAIN, A.   *London*, 82.

HUDSON, N. F. C.   1985 Conditions of Dalradian metamorphism in the Buchan area, NE
                                   Scotland. *J. geol. Soc. London*, **142**, 63–76.

JOHNSON, M. R. W.   1962 Relations of movement and metamorphism in the Dalradians of
                                   Banffshire. *Trans. geol. Soc. Edinburgh*, **19**, 29–64.

                             1963 Some time relations of movement and metamorphism in the
                                   Scottish Highlands. *Geologie Mijnb.*, **42**, 121–142.

JOHNSON, M. R. W. &   1977 Fault and Lineament patterns in the Southern Highlands of
    FROST, R. T. C.   Scotland. *Geologie Mijnb.*, **56**, 287–294.

JOHNSON, M. R. W. &   1967 Dalradian–?Arenig relations in parts of the Highland Border,
    HARRIS, A. L.   Scotland and their significance in the chronology of the Caledon-
                                   ian orogeny. *Scott. J. Geol.*, **3**, 1–16.

KENNEDY, W. Q. — 1948 On the significance of thermal structure in the Scottish Highlands. *Geol. Mag.*, **85**, 229–234.

KERRICH, R., BECKINSALE, R. D., & DURHAM, J. J. — 1977 The transition between deformation regimes dominated by intercrystalline diffusion and intracrystalline creep evaluated by oxygen isotope thermometry. *Tectonophysics*, **38**, 241–257.

KLEIN, G. DE V. — 1970 Tidal origin of a Precambrian quartzite – the Lower Fine-Grained Quartzite (Middle Dalradian) of Islay, Scotland. *J. sediment. Petrol*, **40**, 973–985.

KNELLER, B. C. — 1985 Dalradian basin evolution and metamorphism (abstr.). *J. geol. Soc. London*, **142**, 4.

KNILL, J. L. — 1959 Palaeocurrents and sedimentary facies of the Dalradian metasediments of the Craignish–Kilmelfort district. *Proc. Geol. Assoc. London*, **70**, 273–284.

1961 Joint-drags in Mid-Argyllshire. *Proc. Geol. Assoc. London*, **72**, 13–19.

1963 A sedimentary history of the Dalradian Series. *In* Johnson, M. R. W. & Stewart, F. H. (Eds.) *The British Caledonides*. Oliver & Boyd, London and Edinburgh.

LAMBERT, R. ST.J. & MCKERROW, W. S. — 1976 The Grampian orogeny. *Scott. J. Geol.*, **12**, 271–292.

MCNAMARA, M. J. — 1965 The lower greenschist facies in the Scottish Highlands. *Geol. För. Stockh. Förh.*, **87**, 347–389.

MATHER, J. D. — 1970 The biotite isograd and lower greenschist facies in the Dalradian rocks of Scotland. *J. Petrol.*, **11**, 253–275.

MATHER, J. D. & ATHERTON, M. P. — 1965 Stilpnomelane from the Dalradian. *Nature*, **207**, 971–972.

MENDUM, J. R. & FETTES, D. J. — 1984 The Tay Nappe and associated folding in the Ben Ledi–Loch Lomond area. *Scott. J. Geol.*, **21**, 41–56.

MILLER, J. A. & FLINN, D. — 1966 A survey of the age relations of Shetland rocks. *Geol. J.*, **5**, 95–116.

MIYASHIRO, A. — 1961 Evolution of metamorphic belts. *J. Petrol.*, **2**, 277–311.

PANKHURST, R. J. — 1970 The geochronology of the basic igneous complexes. *Scott. J. Geol.*, **6**, 83–107.

1974 Rb–Sr whole rock chronology of Caledonian events in Northeast Scotland. *Bull. Geol. Soc. Am.*, **85**, 345–350.

PANKHURST, R. J. & PIDGEON, R. T. — 1976 Inherited isotope systems and the source region prehistory of early Caledonian granites in the Dalradian Series of Scotland. *Earth Planet. Sci. Lett.*, **31**, 55–68.

PHILLIPS, F. C. — 1927 The serpentines and associated rocks and minerals of the Shetland Islands. *Q. Jl. geol. Soc. London*, **83**, 622–652.

1930 Some mineralogical changes induced by progressive metamorphism in the Green Bed group of the Scottish Dalradian. *Mineralog. Mag.*, **22**, 239–256.

PHILLIPS, W. E. A., STILLMAN, C. J., & MURPHY, T. — 1976 A Caledonian plate tectonic model. *J. geol. Soc. London*, **32**, 579–609.

PIASECKI, M. A. J. — 1980 New light on the Moine rocks of the Central Highlands of Scotland. *J. geol. Soc. London*, **137**, 41–59.

PIASECKI, M. A. J. & VAN BREEMEN, O. — 1979a A Morarian age for the 'Younger Moines' of central and western Scotland. *Nature*, **278**, 734–736.

PIASECKI, M. A. J. & VAN BREEMEN, O. — 1979b The 'Central Highland Granulites' – cover–basement tectonics in the Moine. *In* Harris, A. L., Holland, C. H., & Leake, B. E. (Eds.) The Caledonides of the British Isles – reviewed. *Spec. Publ. geol. Soc. London*.

PORTEOUS, W. G. — 1973 Metamorphic index minerals in the eastern Dalradian. *Scott. J. Geol.*, **9**, 29–43.

PRINGLE, J.   1940 The discovery of Cambrian trilobites in the Highland Border
                             rocks near Callander, Perthshire. *Advanc. Sci. London*, **1**, 252.

RAST, N.   1958a Tectonics of the Schiehallion Complex. *Q. Jl. geol. Soc. London*,
                             114, 25–46.

                       1958b Metamorphic history of the Schiehallion complex Perthshire.
                             Trans. R. Soc. Edinburgh, **64**, 413–431.

                       1963 Structure and metamorphism of the Dalradian rocks of
                             Scotland. *In* Johnson, M. R. W. & Stewart, F. H. (Eds.) The
                             British Caledonides. Oliver & Boyd, Edinburgh and London.

RAST, N. &   1970 The correlation of the Ballachulish and Perthshire (Iltay)
   LITHERLAND, M.   Dalradian successions. *Geol. Mag.*, **107**, 259–272.

READ, H. H.   1923 The geology of Banff, Huntly, Turriff. *Mem. geol. Surv. Scotld.*

                       1927 The igneous and metamorphic history of Cromar, Deeside,
                             Aberdeenshire. *Trans. R. Soc. Edinburgh*, **55**, 317–352.

                       1933 On the quartz–kyanite rocks in Unst, Shetland Islands, and their
                             bearing on metamorphic differentiation. *Mineralog. Mag.*, **23**,
                             317–328.

                       1934 The metamorphic geology of Unst in the Shetland Islands. *Q. Jl.
                             geol. Soc. London*, **90**, 637–688.

                       1936 The stratigraphical order of the Dalradian rocks of the Banff-
                             shire coast. *Geol. Mag.*, **73**, 468–473.

                       1937 Metamorphic correlation in the polymetamorphic rocks of the
                             Valla Field block, Unsts. Shetland Islands. *Trans. R. Soc.
                             Edinburgh*, **59**, 195–221.

                       1948 Brit. reg. geol. – The Grampian Highlands (revised by A. G.
                             Macgregor). *Geol. Surv. Museum.*

                       1952 Metamorphism and migmatization in the Ythan Valley, Aber-
                             deenshire. *Trans. geol. Soc. Edinburgh*, **15**, 265–279.

                       1955 The Banff Nappe. *Proc. geol. Assoc. London*, **66**, 1–29.

READING, H. G. &   1966 Sedimentation of Eocambrian tillites and associated sediments
   WALKER, R. G.   in Finnmark, northern Norway. *Palaeoggeogr. Palaeoclimatol.
                             Palaeoecol.*, **2**, 177–212.

RICHARDSON, S. W.,   1969 Experimental   determination   of   kyanite–andalusite   and
   GILBERT, M. C. &   andalusite–sillimanite equilibria: the aluminium silicate triple
   BELL, P. M.   point. *Am. J. Sci.*, **267**, 259–272.

RICHARDSON, S. W. &   1976 Thermal causes of the Dalradian metamorphism in the central
   POWELL, R.   Highlands of Scotland. *Scott. J. Geol.*, **12**, 237–268.

ROBERTS, J. L. &   1974 Oblique fold axes in the Dalradian rocks of the South-west
   SANDERSON, D. J.   Highlands. *Scott. J. Geol.*, **9**, 281–296.

ROBERTS, J. L. &   1964 A re-interpretation of the Ben Lui Fold. *Geol. Mag.*, **101**, 512–
   TREAGUS, J. E.   516.

                       1977a The Dalradian rocks of the South-west Highlands – introduc-
                             tion. *Scott. J. geol.*, **13**, 87–99.

                       1977b Polyphase generation of nappe structures in the Dalradian rocks
                             of the southwest Highlands of Scotland. *Scott. J. Geol.*, **13**, 237–
                             254.

                       1979 Stratigraphical and structural correlation between the Dalradian
                             rocks of the S.W. and Central Highlands of Scotland. *In* Harris,
                             A. L., Holland, C. H., & Leake, B. E. (Eds.) The Caledonides of
                             the British Isles – reviewed. *Spec. geol. Soc. London.*

RUSHTON, A. &   1973 A *Protospongia* from the Dalradian of Clare Island, Co. Mayo,
   PHILLIPS, W. E. A.   Ireland. *Palaeontology*, **16**, 231–237.

SHACKLETON, R. M.   1958 Downward-facing structures of the Highland Border. *Q. Jl. geol.
                             Soc. London*, **113**, 361–393.

SKEVINGTON, D.   1971 Palaeontological evidence on the age of the Dalradian deforma-
                             tion and metamorphism in Ireland and Scotland. *Scott. J. Geol.*,
                             7, 285–288.

SMITH, A. G.	1976	Plate tectonics and orogeny: a review. *In* Briden, J. C. (Ed.) Ancient Plate Margins. *Tectonophysics*, **33**, 215–285.
SMITH, R. A. & HARRIS, A. L.	1976	The Ballachulish rocks of the Blair Atholl district. *Scott. J. Geol.*, **12**, 153–157.
SOPER, N. J. & ANDERTON, R.	1984	Did the Dalradian slides originate as extensional faults? *Nature*, **307**, 357–360.
SPENCER, A. M.	1971	Late Precambrian glaciation in Scotland. *Mem. geol. Soc. London*, **6**, 100p.
SPENCER, A. M. & SPENCER, M. O.	1972	The Late Precambrian–Lower Cambrian Bonahaven Dolomite of Islay and its stromatolites. *Scott. J. Geol.*, **8**, 269–282.
SPENCER, M. O. & PITCHER, W. S.	1968	Occurrence of the Portaskaig Tillite in north-east Scotland. *Proc. geol. Soc. London*, **1650**, 195–198.
STEWART, A. D.	1975	'Torridonian' rocks of Western Scotland. *In* Harris, A. L., Shackleton, R. M., Watson, J. V., Downie, C., Harland, W. B., & Moorbath, S. (Eds.) Precambrian. *Spec. Report Geol. Soc. London*, **6**.
STEWART, F. H. & JOHNSON, M. R. W.	1961	The structural problem of the younger gabbros of North-east Scotland. *Trans. geol. Soc. Edinburgh*, **18**, 104–112.
STURT, B. A.	1961	The geological structure of the area south of Loch Tummel. *J. geol. Soc. London*, **117**, 131–156.
STURT, B. A. & HARRIS, A. L.	1961	The metamorphic history of the Loch Tummel area, Central Perthshire. *Liverpool Manchester geol. J.*, **2**, 689–711.
STURT, B. A., RAMSAY, D. M., PRINGLE, I. R., & TEGGIN, D. E.	1977	Precambrian gneisses in the Dalradian sequence of NE Scotland. *J. geol. Soc. London*, **134**, 41–44.
SUTTON, J. & WATSON, J.	1954	Ice-borne boulders in the Macduff Group. *Geol. Mag.*, **91**, 391–398.
	1955	The deposition of the Upper Dalradian rocks of the Banffshire coast. *Proc. Geol. Assoc. London*, **66**, 101–133.
	1956	The Boyndie Bay syncline of the Dalradian of the Banffshire coast. *Q. Jl. geol. Soc. London*, **112** (for 1955), 103–128.
TANTON, T. L.	1930	Determination of age-relations in folded rocks. *Geol. Mag.*, **67**, 73–76.
THOMAS, P. R.	1979	New evidence for a Central Highland Root Zone. *In* Harris, A. L., Holland, C. H., & Leake, B. E. (Eds.) The Caledonides of the British Isles – reviewed. *Spec. Publ. geol. Soc. London.*
	1980	The stratigraphy and structure of the Moine rocks N of the Schiehallion Complex, Scotland. *J. geol. Soc. London*, **137**, 469–482.
TILLEY, C. E.	1925	A preliminary survey of metamorphic zones in the southern Highlands of Scotland. *Q. Jl. geol. Soc. London*, **81**, 100–112.
TREAGUS, J. E.	1969	The Kinlochlaggan Boulder Bed. *Proc. geol. Soc. London*, **1654**, 55–60.
	1974	A structural cross-section of the Moine and Dalradian rocks of the Kinlochleven area, Scotland. *J. geol. Soc. London*, **130**, 525–544.
	1981	The Lower Dalradian Kinlochlaggan Boulder Bed, central Scotland. *In* Hambrey, M. J. & Harland, W. B. (Eds.) *Earth's pre-Pleistocene glacial record.* Cambridge University Press, Cambridge, 637–639.
	1987	The structural evolution of the Dalradian of the Central Highlands of Scotland. *Trans. Roy. Soc. Edinburgh*, **78**, 1–15.
TREAGUS, J. E. & KING, G.	1978	A complete Lower Dalradian succession in the Schiehallion district, Central Perthshire. *Scott. J. Geol.*, **14**, 157–166.
TREAGUS, J. E. & ROBERTS, J. L.	1981	The Boyndie Syncline: a D_1 structure in the Dalradian of Scotland *Geol. J.*, **16**, 125–135.

UPTON, B. G. J., ASPEN, P., GRAHAM, A. M. & CHAPMAN, N. A. 1976 Pre-Palaeozoic basement of the Scottish Midland Valley. *Nature London*, **260**, 517–518.

VOGT, T. 1930 On chronological order of deposition of Highland schists. *Geol. Mag.*, **67**, 68–73.

VOLL, G. 1960 New work on petrofabrics. *Liverpool Manchester geol. J.*, **2**, 503–567.

WATKINS, K. P. 1983 Petrogenesis of Dalradian albite porphyroblast schists. *J. geol. Soc. London*, **140**, 601–618.

1985 Geothermometry and geobarometry of inverted metamorphic zones in the W. Central Scottish Dalradian. *J. geol. Soc. London*, **142**, 157–165.

WELLS, P. R. A. & RICHARDSON, S. W. 1979 Thermal evolution of metamorphic rocks in the central Highlands of Scotland. *In* Harris, A. L., Holland, C. H., & Leake, B. E. (Eds.) The Caledonides of the British Isles – reviewed. *Spec. Publ. geol. Soc. London*.

WESTBROOK, G. K. & BORRODAILE, G. J. 1978 The geological significance of the magnetic anomalies in the region of Islay. *Scott. J. Geol.*, **14**, 213–224.

WILLIAMSON, D. H. 1953 Petrology of chloritoid and staurolite rocks north of Stonehaven, Kincardineshire. *Geol. Mag.*, **90**, 353–361.

WINCHESTER, J. A. & GLOVER, B. W. 1988 The Grampian Group, Scotland. *In* Winchester, J. A. (Ed.) *Later Proterozoic stratigraphy of the Northern Atlantic regions*. Blackie, Glasgow & London, 146–161.

WISEMAN, J. D. H. 1934 The central and south-west Highland epidorites. *Q. Jl. geol. Soc. London*, **90**, 355–417.

WRIGHT, A. E. 1976 Alternating subduction direction and the evolution of the Atlantic Caledonides. *Nature, London*, **264**, 156–160.

1988 The Appin Group. *In* Winchester, J. A. (Ed.) *Later Proterozoic stratigraphy of the Northern Atlantic regions*. Blackie, Glasgow & London, 146–161.

YARDLEY, B. W. D., VINE, F. J., & BALDWIN, C. T. 1982 The plate tectonic setting of NW Britain and Ireland in late Cambrian and early Ordovician times. *J. geol. Soc. London*, **139**, 455–463.

6

LOWER PALAEOZOIC – STRATIGRAPHY
E. K. Walton and G. J. H. Oliver

Lower Palaeozoic rocks in the Dalradian Supergroup have been treated in Chapter 5; here we are concerned with those successions which have suffered rather less metamorphism. Such sequences are found in the north-west Highlands along the Highland Boundary Fault zone, in small inliers in the Midland Valley and over much of the Southern Uplands. This geographical distribution provides a convenient mode for treating lithologies and their development.

In the north-west Highlands, Cambrian and lower Ordovician rocks occur with no marked physical break between the systems. South of the Dalradian tract, the Highland Border Complex, of mostly Ordovician rocks, is found in a series of small discontinuous, badly faulted exposures associated with the Highland Boundary Fault. With the possible exception of the Girvan area no Cambrian rocks are exposed in the south of Scotland; there, the early Ordovician is separated from the later Ordovician by unconformity in the Girvan area and perhaps a pause or slow pelagic sedimentation in the Southern Uplands. Silurian rocks cover the southern part of the Uplands as well as forming all the inliers of the Midland Valley.

This chapter gives an account of the regional stratigraphy. Structure is considered in Chapter 7 and a synthesis presented of palaeotectonics and palaeogeography.

North-west Highlands

Cambrian and lower Ordovician strata occupy a narrow band up to 20 km across. The band runs in a direction slightly west of south for about 200 km from Durness on the north coast to the Isle of Skye. The rocks are bounded to the west by Torridonian and Lewisian rocks on which they lie with strong unconformity. To the east they are involved in the nappes associated with the Moine Thrust and limited by that Thrust.

The succession divides simply into a lower, arenaceous and argillaceous suite and an upper, carbonate sequence (Fig. 6.1). Little variation in thickness has been detected and 1,500 m has been given as a probable maximum for the whole sequence (Gobbett and Wilson 1960, Swett 1969). The succession is well seen at only a few places. In the north, an almost complete sequence can be obtained around Durness and Loch Eriboll; to the south, although broken up by faulting, parts of the sequence are well exposed in the Assynt area and in the neighbourhood of Ullapool and Kinlochewe.

The main stratigraphical divisions were established by Peach and Horne (1907) whose Arenaceous Series and Middle Series comprised beds up to the Serpulite (Salterella) Grit and these were topped by the Calcareous Series. Swett (1969) attempted to formalise the nomenclature to conform with modern practice. He regarded the 'Arenaceous Series' as a formation – the Eriboll Sandstone with two members, the False-bedded Quartzite and the Pipe Rock. The 'Middle Series' was named the An t-Sron Formation, again with two members, the Fucoid Beds and the Salterella Grit. The Calcareous Series also became a formation with the seven Peach and Horne divisions as seven members. Cowie (1974) proposed that the carbonate sequence was better termed the Durness Group made up of the seven divisions as formations while the Salterella Grit, Fucoid Beds, Pipe Rock and False-bedded Quartzite are better regarded as formations. If the last named are accepted as formations, they should be amalgamated as the Eriboll Group.

Eriboll Group

The *False-bedded Quartzite* derives its name from its characteristic ubiquitous sets of cross-strata. The formation lies unconformably on Lewisian and Torridonian rocks, the former mostly in the south and the latter predominantly in the northern part of the outcrop (Fig. 6.2). In a number of places the basal conglomerate transgresses over the Torridonian on to the Lewisian. Peach and Horne (1907) commented on the uniform nature of the Cambrian plane of unconformity contrasting it with the strongly undulating sub-Torridonian surface. Their suggestion that the sub-Cambrian plane is one of marine erosion has been generally accepted. They also pointed to the intense pre-Cambrian weathering which reduced the Lewisian rocks below the unconformity to a peculiar soft aggregate which they called agalmatolite. The silicates of the gneiss have been altered to a mass of white mica with some iron oxides and epidote. The exceptional nature of the rock is ascribed to the strongly alkaline conditions of weathering consequent on the absence of land plants and the humic acids which control weathering of minerals today (Russell and Allison 1985).

The basal bed is a pebble conglomerate usually less than 1 m in thickness. Well-rounded pebbles of brown, red or pink clasts of quartz, quartzite, felsite and feldspars are set in a pale-cream or greenish matrix. The overlying beds are quartz arenites with a few sub-arkoses and very rare arkoses. Sets of cross-strata average 30 cm in thickness; they are mostly planar with less than 10 per cent appearing

as trough-cross sets. Some herringbone patterns are present and palaeocurrent directions tend to be bipolar. These and other features are consistent with formation in a shallow tidal, inter-tidal or sub-tidal environment (Swett *et al.* 1971). The *Pipe-Rock* is another mature sandstone, medium or fine in grain and well bedded below, becoming thinner bedded above. As the name suggests its appearance is dominated by cylindroidal structures which run through the rocks at right angles to the bedding for up to 1 m (Fig. 6.3). The simple pipes of *Skolithus* are parallel-sided with a diameter between 3 and 15 mm. *Monocraterion* forms the so-called trumpet pipes which have a funnel-like shape flaring up to 4 cm or so across. Both types of pipe may have been produced in inter-tidal sediments by individuals of the same, worm-like species (Hallam and Swett 1966). The different forms may have been determined by rates of sedimentation and adjustment by the organisms to changing conditions. *Salterella* and *Serpulites* have also been recorded. The *Fucoid Beds* are usually less than 20 m thick. In the field they are readily distinguished from the resistant splintery quartzites of the underlying beds in that they weather more easily and take on a dark buff colour. They

Fig. 6.1. Cambro–Ordovician succession in the NW Highlands (based on Swett 1969, Palmer *et al.* 1980 and Wright 1985).

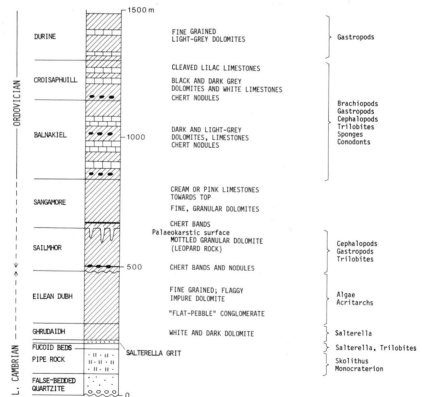

are predominantly dolomitic siltstones with an exceptionally high proportion of fine-grained potash feldspar. Shale interbeds are common; there are some sandstones and even a pisolitic ironstone. Olenellid trilobites have been found at a number of horizons throughout the sequence (Cowie 1974). In addition to the trilobites *Olenellus, Olenelloides, Paedeumias* and *Wanneria*, the fauna includes *Salterella* together with brachiopods, molluscs and fragments of echinoderms. *Skolithus* also occurs but the beds are characterised particularly by *Planolites*. The latter was earlier misidentified as fucoid markings – hence the name of the beds. The formation has a surprising content of potash which averages 8 per cent and may reach 12 per cent. The potash is held in authigenic sanidine and its discovery led to the suggestion of derivation from volcanic sources (Bowie *et al.* 1966). Swett (1966) pointed to a lack of volcanic sources in both Lewisian and Torridonian rocks. He preferred the source to be illite originally deposited in the overlying carbonates. During dolomitisation (see p. 165) potash was released from the illite to be redeposited as feldspar in the Fucoid Beds. A novel proposal relates the high potash content to the peculiar weathering conditions of the period. The production of agalmatolite in the Lewisian rocks released large amounts of potash, alumina and silica to the Cambrian seas. Sanidine was precipitated in a lagoonal environment and excess silica went to form the overgrowths so common on the quartz of the sandstones (Russell and Allison 1985).

The *Salterella Grit* was originally called the Serpulite Grit because of a misidentification of the small gastropod. Here, the revised name is used. The sandstone beds are mature, orthoquartzitic, coarse or medium-grained with cross lamination and separated by thin shales. More siliceous below they tend to be dolomitic above and show honeycomb weathering to a buff colour. *Olenellus* has been recorded from basal shales; *Salterella* is present in the dolomitic sandstones and *Skolithus* also occurs.

If the Fucoid Beds record a lagoonal phase superimposed on the tidal sandstones below, the Salterella Grit would perhaps represent a transgressing barrier. But the sand of the barrier did not last long in so far as the Grit is relatively thin and was soon replaced by carbonates.

Fig. 6.2. Looking north from 1·2 km north of Archfary, Mountains of Cambrian quartzite unconformably overlying Lewisian Gneiss. BGS photograph.

Fig. 6.3a. Thin-bedded 'Pipe rock' (tubes of *Skolithus*) in section. Scale in feet. Photograph by K. Swett.

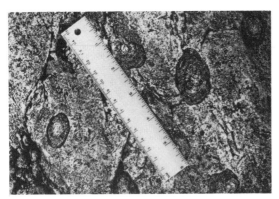

Fig. 6.3b. 'Trumpet pipes' (Monocraterion) on bedding plane. Upper Quartzite, Loch Eriboll. Photograph by K. Swett.

Durness Group

Seven formations have been recognised as making up the Durness Group in the type area at Durness (Fig. 6.1). The three lowest formations have been identified in Skye but the overlying carbonates cannot be correlated with the type section with any certainty. Elsewhere along the outcrop, the sequence has been considerably reduced by faulting and the formations are not easily discernible.

The *Ghrudaidh Formation* consists of leaden-coloured structureless coarse-grained dolomites along with some burrow-mottled dolomites. Less common are laminated (algal) dolomites and there are minor amounts of grainstones. The latter comprise some oolites and some bioclastic laminae of *Salterella* and echinoderm fragments. Stylolites are common. As well as *Salterella*, *Planolites* occurs.

Eilean Dubh rocks show a rather greater variety of sediments and structures. Stromatolites are abundant and of different forms (Fig. 6.4). Flaggy, rippled and cross-laminated horizons occur as well as flake breccias which are often found in erosional hollows; load-structures are occasionally conspicuous. Oolitic horizons occur and well-rounded millet-seed quartz grains are dispersed through dolomitic layers. Although not common a number of small-scale cycles have been recognised (Wright 1985). Each cycle begins with an emergent karstic surface which is often stylolitised; then a laminated dolomite is overlain by massive dolomites, a domal stromatolitic layer and finally a fenestral dolomite. Evaporitic minerals are absent but the fenestrae may be taken to represent their pseudomorphs. This variety of sedimentary features indicates a shoaling of the area from the sub-tidal burrowed dolomites of the Ghrudaidh Fm. to the tidal-flat sediments of the Eilean Dubh Fm. A biota, extracted from a silicified dolomite, consisted mostly of microscopic forms. The fossils include trilobites, hyolithids, sponge spicules and archaeocyathids. The algae present compare with *Corbularia conglutinata*; *Renalcis* and acritarchs are also present.

The overlying *Sailmhor Formation* displays massive dolomites boldly mottled in dark grey or black with white and graphically described as the 'Leopard Rock'. The mottling derives from patches of fine-grained (dark) and coarse-grained (light) dolomite. Stromatolitic masses as well as stromatactoid structures are present. Cherts form layers as well as conspicuous nodules (Fig. 6.5). Cycles 2–10 m thick represent repeated shoaling events (Wright 1985). Like the Eilean Dubh succession, emergent surfaces form the boundaries of the cycles. Such surfaces are overlain discontinuously by thin lenses of grainstone or conglomerate. Above, a burrow-mottled dolomite occurs, followed by a stromatactoid dolomite and pale-grey laminated dolomites.

The top of the formation is distinctive in having breccia-filled fissures up to 1 m wide and extending to depths of 150 m (Palmer *et al.* 1980). Wright (1985) estimated that depths of this order could be developed by meteoric water with the emergent surface only 4 m above sea level. The boundary is evidently an important hiatus and it was taken to represent both Middle and Upper Cambrian (Palmer *et al.* 1980). The fauna reported by Peach and Horne (1907) was regarded as Ordovician but equivocal. Wright (1985) collected cephalopods and trilobites which confirm the Ordovician age. The hiatus is, therefore, intra-Ordovician rather than Cambrian.

The *Sangomore Formation* records a change in that limestones occur as well as dolomites. Pink and cream grainstones with early cement have avoided dolomitisation. Lenticular and bedded cherts are an important component of this formation. Megaquartz in geodes and nodules is rather less important volumetrically but has an environmental significance in that its textures suggest that it has replaced evaporites. Similarly bedded breccias have formed by collapse after dissolution of evaporites. The environment of formation compares with that of the Trucial Coast, sabkha-type sedimentation taking place with banks and islands interspersed on a carbonate platform. A sparse fauna of gastropods, cephalopods and conodonts indicates an Ordovician (Canadian) age (Palmer *et al.* 1980, Wright 1985).

The *Balnakiel* beds are dominantly thinly bedded dolomites although with appreciable limestone. Some bioherms of gastropods and cephalopods are found and thinly laminated rocks have alternating laminae of limestone and dolomite. The limestone laminae are bioclastic with trilobite and echinoderm fragments. Stromatolites occur especially in the lower part and periods of emergence are indicated by dissolution surfaces. Higher in the succession stromatolites are rare and more burrows are found.

The *Croisaphuill Formation* is typically mottled in which the burrows of *Spongeliomorpha*-type have been replaced by sucrosic dolomite and stand out against a limestone

Fig. 6.4. Pillow-shaped stromatolite. Eilean Dubh beds, Balnakiel Bay. Photograph by K. Swett.

Fig. 6.5. Chert nodules in Sailmhor Beds, Balnakiel Bay. Photograph by K. Swett.

matrix of micrite or microspar. The abundant gastropod faunas of this and the Balnakiel Formation are of late Canadian age. A conodont fauna from the Croisaphuill Fm. was inferred to be middle and upper Canadian age (Bergstrom and Orchard 1985).

The *Durine Formation* consists of pale-grey, burrow-mottled dolomites below with laminated dolomites above. Limestone and chert are present with some evaporitic textures. Macrofossils are uncommon and only *Hormotoma* has been reported. Conodonts suggest a middle to late Arenig age with a possible range into the Llanvirn (Bergstrom and Orchard 1985).

Diagenesis

The lithologies of the Durness carbonates have been determined to a large degree by diagenetic changes; these have produced the predominant dolomites, the dolomitic limestones, the chert nodules and bands. In so doing they have determined the appearance of the rocks. In dolomitic limestones, separation of dolomite from paler calcareous patches has produced a breccia-like appearance with raised dolomite and calcite hollows. In rocks like the Leopard Rock different dolomite crystallinity gives dark and light areas. Dissolution of original evaporitic minerals has led to collapse breccias and siliceous nodules.

The diagenetic sequence is complicated. One interpretation sees a series of at least five phases. Recrystallisation of original calcium carbonate (1) was followed by (2) dolomitisation; (3) silicification; (4) calcitisation and finally (5) a second dolomitisation (Swett 1969). The changes in the carbonates were linked with the breakdown of illite originally deposited with the carbonates. The illite also yielded potassium to form the sanidine in the Fucoid Beds although as noted above Allison and Russell (1985) have recently countered this.

An alternative interpretation of the diagenesis sees repetition of phases during the accumulation of the carbonates. Some limestones were subject to early calcite cementation and these have been preserved. Penecontemporaneous dolomitisation took place in the supra-tidal sabkha sediments which now appear as thinly laminated beds often below an emergent surface affected by dissolution. On emergence a schizohaline mixing zone spread through the sediments producing first silicification and then dolomitisation. An input of meteoric water is indicated by oxygen-isotopes of the dolomite. Episodes of chert and dolomite formation therefore occurred *pari-passu* with sedimentation. Silicification was also determined by the availability of silica as sponge spicules. These were more abundant in the Ordovician; so is the chert. Stylolites were a late development. In addition dolomite in 'saddle' crystals has replaced earlier sediment and cuts across original features (Wright 1985). These late effects are better regarded as metamorphism reaching the anchizone grade. The colour alteration of acritarchs and conodonts together with the crystallinity of

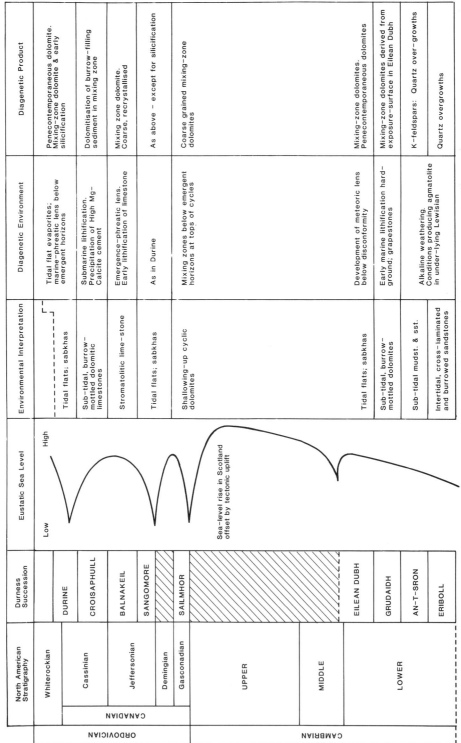

Fig. 6.6. Durness sequence, environments and diagenesis in relation to eustatic changes (based on Wright 1985, Allison & Russell 1985).

illites in the Durness beds indicate temperatures of about 300°C below the Moine and associated nappes (Johnson *et al.* 1985). Burial beneath the Moine succession must have been responsible for considerable recrystallisation of the dolomites and quartzites as well as perhaps the growth of the feldspar in the Fucoid Beds.

Correlation

The close affinity of the Durness and North American rocks was recognised long ago (Salter 1859). The olenellid fauna of the Fucoid Beds is an association characteristic of the North American, Pacific Province and suggests a position in the middle part of the Lower Cambrian (Cowie 1974). *Salterella* is found in beds up to and including the Eilean Dubh Formation and this would point to Lower Cambrian. Silicified fossils from Eilean Dubh carbonates can be correlated with the Lower Cambrian Forteau Formation of Newfoundland (Brasier 1977).

Balnakiel and Croisaphuill faunas correlate with those of the Beekmantown Group of the Appalachians of early Ordovician age (cf. Sando 1957). Species are common to Beekmantown, East Greenland, the St. Lawrence region and the Durness rocks and Poulsen (1951) took them to be Arenigian. The cephalopods collected by Peach and Horne (1907) probably belong to the *Cassinoceras wortheni* zone (Flower, reported by Wright 1985). Sailmhor Fm. fossils are meagre and not too helpful although they have been taken to be Ordovician (Poulsen 1951, Cowie 1974, Palmer *et al.* 1980). Wright (1985) collected poorly preserved cephalopods and ?trilobites suggestive of an early Ordovician (lower Canadian) age. Evidence to date points to an absence of both Middle and Upper Cambrian sediments. Cowie saw similarities amongst the Sailmhor, the Cass Fjord Formation of East Greenland and the L. Oslabreen Limestone of Spitzbergen. The last two probably correlate with the Tremadoc Series of the Anglo–Welsh succession. Brachiopods extracted from the Ben Suardal Fm. in the Durness Group of Skye are Arenig and have strong affinities with the American faunal province (Curry and Williams 1984). Conodonts from the Croisaphuill and Durine Fms. at Durness compare closely with North American Mid-Continent faunas (Bergstrom and Orchard 1985). There seems little doubt that the Durine beds extend into the Llanvirn.

Faunal similarities in Cambrian rocks of the Boreal area are paralleled by lithological comparisons. The differing lithologies have been interpreted above in terms of variation in sea level and periods of emergence. These in turn can be seen as the combined result of eustatic changes and tectonism. The Ghrudaidh and Eilean Dubh beds record a cycle of deepening and shoaling controlled by eustatic rise punctuated by a fall (Fig. 6.6). The subsequent eustatic rise was countered in middle and late Cambrian times by tectonic uplift preventing sedimentation in Scotland, East Greenland and Spitzbergen. The Sailmhor Fm. reflects the early Canadian eustatic rise and fall with sedimentary effects producing a cyclicity in the beds. The prominent Sailmhor karstic surface then suggests a eustatic rise augmented locally perhaps by the effects of the Grampian orogeny. Finally the Sangomore–Durine Fms. record another eustatic rise and fall (Wright 1985).

Highland Border

Along the southern boundary of the Highlands a distinctive suite of rocks is distinguishable from the main Dalradian succession. The rocks of the Highland Border Complex (HBC) are varied in lithology and are found in discontinuous fault-bound wedges and lenses. They are best exposed at Stonehaven on the north-east coast, in the North and South Esk valleys in Angus, from Callander south through Aberfoyle to Loch Lomond, and on Bute and Arran. Age and structural relations have been subject to long debate but recent work has produced some resolution (Gunn 1900, Barrow 1901, Pringle 1941, Anderson 1947, Johnson and Harris 1967, Harris 1969, Curry *et al.* 1984, Harte *et al.* 1984, Robertson and Henderson 1984). Re-examination of structural features led Harte *et al.* (1984) to conclude that the deformation history of the HBC was different from that of the Dalradian. The HBC faunas confirm distinctive ages for the two groups. Four rock assemblages have been recognised (Fig. 6.7). Although these assemblages are based on the rocks in the Aberfoyle district the lithologies can be correlated along the entire belt (Curry *et al.* 1984, Ingham *et al.* 1985).

Rock Assemblage 1 consists of serpentinite and associated rocks. Commonly the serpentinite is represented only by carbonate rock and quartz or jasper. Where carbonation has not occurred original ultrabasic textures may be preserved in lizardite, or antigorite may have replaced lizardite irrespective of the original texture. Original olivine may be represented now by a mesh texture and pyroxene by bastite. The relative proportions of the two textures suggests that the original ultramafic rocks were predominantly harzburgites (Henderson and Robertson 1982). Isotopic proportions of D/H and $^{18}O/^{16}O$ suggest that serpentinisation took place during a phase of uplift and accession of meteoric waters. The alteration to antigorite, on the other hand was accomplished by metamorphic fluids (Ikin and Harmon 1983).

The overlying rocks of Assemblage 2 are unconformable and mid Arenig in age. This gives a minimum age for the serpentinite. The correlation with the serpentinite of the Ballantrae Igneous Complex (put forward long ago by Nicol (1850) and generally accepted) may mean that obduction of the HBC serpentinite took place at the same time. The obducted mass may have included gabbro and lavas now represented by epidiorites and hornblende and tremolite schists associated with the serpentinites. The schists have been affected by metamorphic fluids in the same way as the serpentinite and on the Isle of Bute for example have been regarded as original basalts metamorphosed at the base of the obducted mass (Ikin 1983).

Rock Assemblage 2 At Aberfoyle and Loch Lomond serpentinite is followed unconformably by conglomerate or breccia with serpentinite clasts. Some of the serpentinite clasts were carbonated before incorporation into the sediment. The conglomerate at Aberfoyle has a carbonate matrix and lies below the dolomitic Dounans Limestone. A varied silicified fauna separated from this limestone is mid Arenig in age. Trilobites are the most abundant element and include *Ischyrotoma* as the most common form (Curry *et al.* 1982). A brachiopod fauna has also been extracted (Ingham *et al.* 1985).

Rock Assemblage 3 is a distinctive group made up of pillow lavas, tuffs and breccias, black shales and cherts as well as quartz–wacke turbidites. A fauna of brachiopods and chitinozoans from Stonehaven prescribe a Llanvirn to ?Llandeilo age. A sequence in Arran of well-exposed spilitic pillow lavas and black shales is the same assemblage. Inarticulate brachiopods were described from here (Anderson and Pringle 1944) but the fauna has not been subject to modern assessment. Microfossils from the Aberfoyle district point to a range of upper Arenig to Llandeilo.

Although spilitised and metamorphosed, some lavas from Arran and other localities at Aberfoyle, River North Esk and Stonehaven have geochemical characteristics which suggest origin at a mid-ocean ridge; others are alkaline in character, perhaps ocean-island and yet others are more like arc tholeiites (Henderson and Robertson 1982, Ikin 1983). The associated sediments, shales, cherts and turbiditic sandstones complete an oceanic association.

Rock Assemblage 4 is composed of sandstones and conglomerates with occasional carbonates. In the Aberfoyle district the Archroy Sst. Fm. is unconformable on the underlying assemblage and a basal breccia is developed. An assemblage of chitinozoans from siltstones in this area shows two types. One comprises darkened derived fossils of Llanvirn–?Llandeilo age and the second is contemporary with the sediments and Caradoc in age. Evidently metamorphism (post-Llanvirn, pre-Caradoc) affected assemblages 1 to 3 and these were reworked above the unconformity into assemblage 4. In the North Esk valley buff-coloured sandstones of this last assemblage include the Margie Limestone. Two chitinozoans from this horizon, *Pogonochitina* cf. *spinifera intermedia* and *Desmochitina juglandiformis*, are of Caradoc–Ashgill in age. The siliciclastic rocks suggest a quartzofeldspathic source area. Although their trace elements are comparable with those of the Dalradian, the textural maturity of the Margie rocks makes derivation from that group unlikely when it is recognised that at this time the Dalradian rocks were experiencing rapid uplift (Henderson and Robertson 1982, Robertson and Henderson 1984, Harte *et al.* 1984) in an area possibly remote from the Highland Border basin.

South of Scotland

Early Ordovician

Lower Ordovician rocks are known in two areas; the first lies near Girvan in south Ayrshire, and the second near Abington in Lanarkshire.

Girvan area

This relatively small area of around 80 km² has played and continues to play a role in interpretations of the Caledonian orogeny in Scotland quite disproportionate to its size. Its importance lies in its variety of rock types and the insights it gives, albeit tantalisingly incomplete, into the early Ordovician history of the region (Cameron *et al.* 1986, Stone and Smellie 1986).

The rocks comprising the Ballantrae Igneous Complex (BIC) form an ophiolitic assemblage with the Steinmann Trinity of serpentinite, pillow lavas and cherts. Associated with the lavas and cherts are graptolitic shales and mudstones, conglomerates including olistostromes, and sandstones. In simplified terms the rock distribution shows two

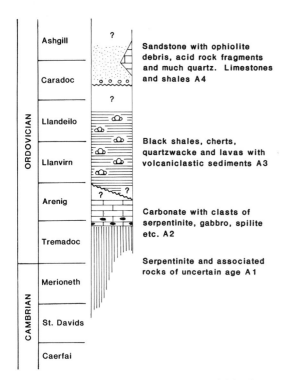

Fig. 6.7. Rock assemblages in the Highland Border Complex (based on Bluck 1985).

north-east–south-west-trending belts of serpentinite separating three bands of lavas (Fig. 8.12).

The *Serpentinite* was derived largely from peridotite with either orthopyroxene now bastite, and originally a harzburgite or with two pyroxenes, i.e. lherzolite. Very often serpentinisation has reached 95 per cent of the peridotite. In the northern belt the original ultramafic rock was lherzolite along with some websterite while in the southern belt harzburgite occurs as well as dunite, wehrlite and pyroxenite. Olivine and pyroxene-rich layers probably represent cumulates from the gabbros described below. At the base of both serpentinite masses but especially well developed on the southern margin of the northern belt, is a metamorphic zone or sole up to 100 m thick. The metamorphic grade of the sole decreased downwards from the serpentinite, from garnet metapyroxenite through amphibolite to greenschist. This arrangement is taken to be the result of the obduction of a hot ultramafic slab over a cool rocks. Pyroxenites in serpentinite probably formed at depths of 60 km. A metapyroxenite of the sole gives a cooling age of 505–+11 Ma reflecting early accretion on to the rising obducting mass. As obduction proceeded, accretion of basaltic crust occurred at higher levels and dynamothermal metamorphism produced the amphibolites of the sole. A date of 476+–14 Ma (lower Llanvirn) for an amphibolite therefore gives a relatively short obduction period of some 30 Ma (Jelinek *et al.* 1984).

Foliated mafic intrusions

Associated with the ultramafics, often as tectonised lenses within the serpentinite, is a series of gabbroic and doleritic intrusions. That these are pre-obduction is shown by their foliation and metamorphism. The dykes referred to as beerbachites and the gabbros have been altered to an assemblage of clinopyroxene and titanium-rich hornblende. The mineral chemistry of this assemblage indicates a high temperature (900–1,000°C) consistent with formation at an oceanic ridge and being held there for some time. This pre-obduction phase is succeeded by a low temperature metamorphism during which actinolite and chlorite were developed presumably as a result of obduction or subsequent burial metamorphism (Jelinek *et al.* 1980, Oliver *et al.* 1984).

The gabbro with dykes may represent layer 2 of an ocean crust although in the BIC dykes make up less than 30 per cent in all cases except one. At Millenderdale dykes form up to 60 per cent of a gabbro–dolerite mass.

Post-obduction intrusions

Unfoliated gabbros and dolerites cut both the serpentinite and its metamorphic sole. The dolerites are typically clinopyroxene- and plagioclase-bearing although the latter is commonly albitised. Alteration of the dykes to rodingite is common especially at their margins. Intrusion, although post-obduction, evidently took place while the peridotite slab was cooling and during serpentinisation. The latter

provided Ca-rich fluids which produced prehnite then hydrogarnet and finally rodingite.

At the north-west margin of the northern serpentinite belt gabbro grades through diorite to trondjemite. The last, made up of albite and biotite or amphibole, contains accessory zircons from which an age of 482–+5 Ma was obtained (Aftalion *et al.* 1984).

Volcanics

Volcanic rocks in the BIC comprise lavas, agglomerates and tuffs. The lavas are tholeiitic basalts mostly spilitised and often pillowed. Aphyric lavas may show a flow texture or are sub-variolitic. Porphyritic lavas have plagioclase phenocrysts up to 10 mm in length. A block of lava in an agglomerate which Balsillie (1932) termed 'pre-spilite' has labradorite and pyroxene phenocrysts but albitisation of the feldspars in the majority of the basalts has been accompanied by alteration of the mafics to chlorite. Continued burial under younger volcanics plus the cover of Ordovician sediments has produced pumpellyite-actinolite, prehnite–pumpellyite or zeolite metamorphism.

Most of the tuffs and agglomerates are water-laid and presumably deep water, with two exceptions. At Pinbain in the northern belt shallow-water conditions led to the development of hyalotuff deltas and in another tuff, accretionary lapilli point to sub-aerial conditions (Bluck 1982, Smellie 1984).

A range of ages has been derived from lavas at different localities. 501+–12 Ma years was found for a lava from Mains Hill in the southern belt north of the R. Stinchar; to the north in the central belt, a lava at Games Loup gave 476+–14 Ma. South of the Stinchar Fault, the Downan Point lavas are distinct from the BIC in giving a Llandeilo age of 468 Ma (Thirlwall and Bluck 1984).

The *Sediments* interbedded with the volcanics are cherts and conglomerates with minor amounts of claystones bearing sparse graptolites. Cherts are thin and evenly bedded, radiolarian and red, green or grey to black. Conglomerates have been described from a number of localities as typically massive, poorly sorted and often with a highly sheared lutite matrix. Clasts vary in size up to several metres and are distributed irregularly through the beds. Shale fragments have often been deformed within the sediments adding to the features suggesting mass flow and an origin as olistostromes or melanges. Not all the clastic horizons are thick conglomerates; some sandstones and pebbly sandstones interbeddded with cherts are only a few cm thick.

The range of ages indicated by the radiometric dates is paralleled by the graptolites. Lower and Middle Arenig forms have been found in a number of fault-bound sections but some sequences have possible Llanvirn graptolites (Stone and Strachan 1981, Stone and Rushton 1983, Rushton *et al.* 1986). Conodonts from cherts associated with Arenig graptolitic shales belong to the *D. extensus* zone (Lindstrom 1971).

The clasts in the sediments include ophiolitic debris at least in part from the earlier phases of the complex. Thus there are fragments of serpentinite, gabbro, dolerite, pyroxenite, spilite, amphibolite, epidote schist, chert and black shale. There are, however, blocks exotic to the BIC. Greywackes and limestones point to a sedimentary complex older than anything now exposed in the south of Scotland. A very early subduction episode is indicated by a glaucophane-bearing olistolith and perhaps from the same Precambrian or Cambrian active margin, another olistolith of eclogite has an age of 576 + -32 Ma.

Geochemical investigations have attempted to define the tectonic settings of various rocks of the BIC. Within-plate hot-spot basalts were diagnosed by Wilkinson and Cann (1974) as well as low-K tholeiites of island-arc type. Lewis and Bloxam (1977) thought that the geochemistry pointed to island arc or ridge settings and preferred the former because of the supposed great thickness of the lavas. Although structural relationships are still rather obscure it is now recognised that the apparently very thick sequence of volcanics is most probably due to faulting (Stone 1984). The known variation in the lavas even within individual pillows probably vitiates any precise diagnosis of tectonic settings; dykes may be more useful and beerbachites have been taken to indicate a ridge environment of formation (Jelinek *et al.* 1980). Post-obduction dykes, on the other hand, are mostly 'within-plate'; a subordinate group resemble modern island-arc basalts (Holub *et al.* 1984).

Raven Gill

At the head of a small stream – Raven Gill – near Abington about 50 m of strata are exposed in a small inlier. Their relationship to the surrounding greywacke sequences in the enclosing Abington block (see Fig. 6.16) is obscure. To the south basalts of the inlier are faulted against greywackes. The basalts have albite laths (sometimes showing flow texture and partially replaced by pumpellyite) set in a matrix of chlorite, prehnite and iron oxides. North of the basalts, conformable cherts and mudstones occur with thin dolerite sills. The dolerites are aphyric with pyroxene, as salite and augite, resembling the more abundant Bail Hill lavas (see p. 181). Vesicles are abundant suggesting shallow intrusion perhaps in to soft sediment (Hepworth 1981); their mineralogy, carbonate, chlorite and pumpellyite points to subsequent metamorphism. The cherts form packets (up to 2 m) of thin, light-grey bands (4–6 cm). Recrystallisation has been patchy and radiolaria are picked out in a darker-grey colour. The brownish mudstones are very fossiliferous with inarticulate brachiopods such as *Acrotreta nicholsoni* most common. Graptolites occur but are indeterminate. Conodonts suggest a lower Arenig age (Lamont and Lindstrom 1957). The overlying Kirkton beds of red and green, sometimes cherty, mudstones may be *N. gracilis* or pre-*N. gracilis* in age.

Later Ordovician

Stinchar Valley and north

The varied sediments of the area between the R. Stinchar and the Craighead Inlier just north of Girvan have excited interest ever since Moore's discovery of fossils in the Stinchar Limestone in 1849. Amongst many stratigraphical and palaeontological workers, Lapworth (1882) and Williams (1962) provided notable advances. Recent sedimentological and geochemical work includes Hubert (1966), Ince (1984) and Longman *et al.* (1979).

Later Ordovician rocks transgress northwards over the Ballantrae Igneous Complex (Fig. 6.8). The lowermost formation of the Barr Group, the *Kirkland Conglomerate*, is exposed in an anticlinal core in the Stinchar Valley. Massive purple conglomerates contain clasts of the BIC with basalts predominating. A lower sequence comprising clast-supported, locally imbricated, fluvial sediments is followed by marine, fan-delta pebble and cobble conglomerate with abundant matrix and scattered boulders (Ince 1984).

The varied fossiliferous sediments above the conglomerate record a delta-retreat phase culminating in abandonment and the formation of the *Stinchar Limestone*. In the type area of this important horizon, impure limestones pass up into hard compact limestones. The lower rubbly beds are calcilutites with greenish siltstone and mudstone surrounds. The fauna resembles that in the beds below; limy bands have *Girvanella* and other algae encrusting shells while the silt laminae consist mostly of serpentine and quartz. The higher compact limestones are varied with calcilutites, bioclastic calcarenites and oolites. The commonest fossils are *Girvanella problematica*, *Saccaminopsis carteri* and brachiopods. Trilobites are also an important faunal element indicating inshore conditions with the illaenid–cheirurid community (Tripp 1979). Varying conditions over the sea floor are implied by lateral variations in thickness and lithology (see Fig. 7.14).

The *Superstes Mudstones* are grey or greenish-blue, brown-weathering, hard, nodular siltstones and mudstones. Laminae alternate in graded units and the coarse silt grains are of serpentinite, quartz, feldspar and magnetite along with pieces of *Girvanella*. Fossils are scarce but important because the graptolites indicate a Llandeilo or early Caradoc age. Trilobites are also found and these suggest an outer-shelf environment (Tripp 1976, Ince 1984).

The *Benan Conglomerate* is, in part, laterally equivalent to the mudstones although it is difficult to define precisely from field exposures. The full sequence cannot be seen in any one area; boundaries are usually gradational and similar conglomerates developed within the Stinchar Limestone and *Confinis* Flags horizons cause some confusion. The Benan Conglomerate is also diachronous and shows large changes in thickness due partly to overstep on to a variable BIC basement and partly to channelling. Boulder, cobble and pebble beds, pebbly sandstones, sandstones with occasional intercalations of siltstone and mudstone reach a maximum thickness of 640 m. Pebble to boulder conglomerates form massive beds. They are usually disorganised in texture and matrix-supported. The mega-clasts are set in a sandy matrix of grains of spilite, serpentinite and quartz. Overall the colour of the beds is a dull greenish grey to black contrasting with the purple of the Kirkland Conglomerate. Higher beds are thinner and finer in grain with some imbrication and cross-stratification. The lower part records growth of submarine fan by mass-flow processes. Upwards, the finer grain and sedimentary structures suggest shallow-waters becoming fluviatile (Ince 1984). Fossils are rare. In addition to BIC clasts there are pebbles of Stinchar Limestone and a calcarenite boulder has yielded a Tremadocian fauna (Rushton and Tripp 1979). Dating of granitic clasts produced ages of 560 and 470 Ma (Longman *et al.* 1979).

The *Balclatchie Formation* forms the base of the Ardmillan Group. It is mainly a dark, bluish mudstone with some sandy lenses. In addition to an abundant shelly fauna some graptolites occur and in places they are particularly well preserved (Bulman 1944–47). Graptolites and conodonts prescribe an early Caradocian age (Bergstrom 1971). Brachiopods and trilobites are indigenous deep-water forms from an outer-shelf or slope environment (Tripp 1980a, Harper and Owen 1986). In coarser beds ophiolitic clasts along with granite, microgranite, quartz and shells are set in a matrix which may be calcareous and occasionally fossiliferous. Fractured, entombed trilobites along with brachiopods, ostracods and other shells record catastrophic events forming some conglomerates. On the other hand, the Kilranny Conglomerate, a well-known Balclatchie member exposed on the coast has an organised texture with some imbrication suggesting shallow marine if not fluvial conditions.

The *Ardwell Formation* is dominated by flags. They are mainly sandy or silty, greenish-grey mudstones with thin sandy bands and occasional black graptolitic partings; some calcareous concretions occur. The dominant units are of mudstone and sandstone forming varve-like rhythms up to 5 cm in thickness. Grading, however, is rare in the sandstones which are mostly ripple-laminated with some ripple-drift and occasional convolute lamination. There are numerous examples of intraformational, flat-pebble breccias. These vary from a few cm to, exceptionally, 1·6 m in thickness and are usually of very limited lateral extent. Clasts are of surrounding sediments. Evidently the breccias formed contemporaneously either in response to earthquake shocks and associated tsunamis (Henderson 1935) or, more likely, to local current shearing and erosion (Hubert 1966). Rather larger folds than those associated with the breccias have been regarded as part of a distinct

Fig. 6.8. Geological sketch-map of the Girvan area (after Williams 1962). Stippled ornament within the Lower Ardmillan Group represents conglomeratic horizons.

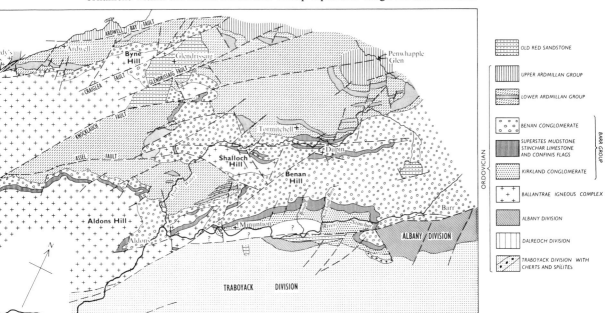

tectonic phase ('Ardwell phase' of Williams 1959) but they show some features suggesting soft-sediment slippage down a basinward slope (Ingham 1978). Flagstones pass laterally into a number of lenticular conglomerates especially towards the north-east and some of these con- glomerates have a varied shelly fauna. The coarse-grained beds which form the top of the sequence inland (the Cascade conglomerates and grits) are equivalent to grapto- litic shales on the coast which were previously taken to be the base of the succeeding Whitehouse Formation (Ingham 1978).

The *Whitehouse Formation* comprises three divisions, the Limestone, Sandstone and Shale members; although it should be noted that Harper (1984) regards the Whitehouse succession as a Group with six Formations. The limestone and sandstone members are flysch-like in character being dominated by graded units. The limestone beds are about 50 cm or less in thickness, the sandstones being thinner, generally less than 10 cm. Both members have grey turbiditic material distinguishable from the greenish muds of the 'background' sedimentation. Most of the pelite units are grey, turbiditic rather than hemipelagic in origin. The usual divisions of turbidites are present, the modal sequence in both members being CDE (Hubert 1966). In the limestone flysch the sequence ABCDE is almost equally common. Even more remarkable is the replacement of the

A and B divisions by a unit made up of dune cross- lamination with amplitudes up to 20 cm. The coarser units of the limestone flysch tend to be sandy or pebbly pelletiferous biosparudites or calcite-cemented fossili- ferous, volcano-clastic, feldspathic, pebbly sandstones. Apart from shell fragments the clasts are ophiolitic with some metamorphic components. Arenaceous laminae are sandy biopelmicrites and the finer beds are silty, micaceous pelmicrites. The sandstone flysch has at least one strongly lenticular conglomerate with large-scale cross-lamination apparently built out towards the south-east. Similar lenses are common in the upper Whitehouse, Shale Member. Red mudstones in this member contain a peculiar fauna of large-eyed or blind trilobites as well as a restricted fauna belonging to the relatively deep-water *Foliomena* commun- ity (Harper 1979). The presence of graptolites as well as trilobites and brachiopods allow the placing of the upper Whitehouse beds in the *D. complanatus* graptolite zone and equating this with the Pusgillian stage while the lower beds are late Caradoc (Ingham 1978, Harper 1984, Williams 1987).

The *Shalloch Formation* (or Barren Flags) is mostly fine- grained, laminated, ungraded, greenish sandstones with frequent ripple and/or convolute lamination. The inter- bedded pelites change from lower portions of greenish-grey to dark-grey above. Sole markings and ripple marks

Fig. 6.9. Stratigraphic location of middle Ordovician Girvan rocks based on conodont and graptolite zonations (Ingham 1978).

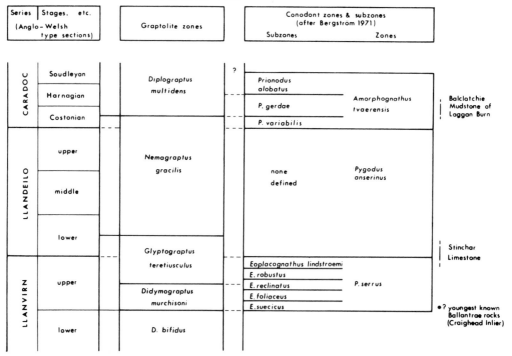

indicate flow directed predominantly towards the south-west. Three thin biosparitic, graded limestones are found in the lower, middle and upper portions of the sequence. Graptolites of the *D. complanatus* zone have been found near the base of the 330 m thick succession and from the *D. anceps* zone higher up.

The uppermost Ordovician beds are represented in the Craighead inlier, a few kilometres north-east of Girvan. Here the *Drummock Formation* (400 m) has mudstones above a basal conglomerate. The mid-Ashgill fauna of trilobites and brachiopods was, at least in the lower part of the sequence, resedimented in large mud-flows. Although the evidence is less clear, mass flowage may also have played a part in the accumulation of the upper beds. Amongst these the Starfish Bed is noteworthy. Named from the presence of *Tetraster wyville-thomsoni*, it has a prolific fauna of echinoderms, brachiopods, corals, trilobites and gastropods. The Ordovician is completed by the Hirnantian High Mains Sandstone, a shallow-water deposit with a rich brachiopod fauna (Harper 1981, Owen 1986). The succession at Craighead is also important in its lower members. The *Craighead Limestone* which lies unconformably on Ballantrae volcanics and cherts is very similar to the Stinchar Limestone but its fauna shows that it is contemporaneous with the Ardwell Fm. It therefore represents, like the Stinchar Limestone, an inshore equivalent of basinal shales and sandstones. Even so, a deeper-water phase – the *Sericoidea Mudstones* – is inter-polated in the Craighead Limestone (Tripp 1980b).

The faunas of the Girvan sequences allow the recognition of the various stages of the Caradocian and the Ashgillian in the Ardmillan and upper Barr beds. The lower beds of the Barr Group extend into the Llandeilo; if Bergstrom's (1971) conclusions from the conodont faunas are correct they may stretch down into the Llanvirn (Fig. 6.9).

The provinciality of the Girvan faunas is marked, a strong link with southern Appalachian faunas (Williams 1962, Tripp 1980a) contrasting with remoteness to Welsh faunas. By late Ordovician some convergence of the two British faunas occurred although some elements, e.g. some species of *Cryptolithus* and *Flexicalymene*, present in the Girvan area are not found in the Ashgillian faunas of England and Wales.

Stinchar Valley to Glen App

The successions to the south of the R. Stinchar provide a contrast to the varied facies to the north. Thick sequences are made up of greywackes, shales and mudstones with occasional conglomerates and basalt–chert horizons. Similar lithologies, faulting and lack of faunal control mean that stratigraphical relations are still in doubt (Fig. 6.10).

The *Glen App Conglomerate* and its lateral equivalent across Loch Ryan, the *Corsewall Conglomerate*, are made up of rudites with disorganised texture and clasts up to boulder size. Interbedded with the conglomerate horizons

are massive greywackes, some flaggy greywackes and silty and shaly partings with a sparse *N. gracilis* fauna. The clast composition is similar to the Benan (Walton 1956, Kelling 1962).

Other greywacke sequences (*Dalreoch, Darley, Changue*) are quartzo–feldspathic and lack the hornblende and pyroxene grains present in the Corsewall and Glen App rocks. The Darley Fm. has spilitic and chert horizons at its base. Somewhat similar horizons on the coast at Currarie have been interpreted as olistostromic (Leggett *et al.* 1984).

The *Traboyack* rocks are distinctive, striped, red and green beds with red muddy laminae and greenish grey silty and sandy bands. Occasional interbedded lenticular con-glomerates are reminiscent of the Kirkland lithology.

The *Albany* beds have Benan-like conglomerates and mudstones. Their fauna is equivalent to the upper Stinchar Limestone and the 1,000 m succession may equate with the Stinchar Lst, the *Superstes* Mudstones and the Benan Conglomerate (Williams 1962, Tripp 1965). Also in the Tappins Complex are carbonate lenses which resemble the Auchensoul Lst. They would be shallow water in origin; Albany rocks may be outer shelf; slope deposits may be represented by the Changue Fm. with the Darley Fm. as base-of-slope, ocean-trench sediments. Glen App–Corsewall are inner-fan, proximal, mass-flow deposits.

Southern Uplands

The northern boundary of the Southern Uplands proper is taken as the Southern Uplands Fault (*sensu lato*) of which the Glen App Fault forms the south-westerly sector. The Southern Uplands Fault (*sensu stricto*) occupies the central sector while in the north-east the line is formed by the Lammermuir Fault. In the Southern Uplands three divi-sions have long been recognised (Peach and Horne 1899). The Northern Belt consists entirely of Ordovician rocks; the Central Belt has Silurian (Llandovery) as well as Ordovician sequences while the Southern Belt has only Wenlock beds. The Ordovician of the Central Belt is in the form of a condensed sequence of fine-grained beds which formed the basis of Lapworth's classic graptolite zonation (Lapworth 1878). It is convenient to describe these first and then discuss the development of the arenaceous sequences in the Northern Belt.

Moffat sequence (Ordovician section)

The Ordovician portion of the Moffat 'Series' of Lap-worth (1878) comprises two divisions, the *Glenkiln* and *Hartfell Shales* (Figs. 6.11, 6.12). In the Moffat Valley, the Glenkiln beds (6 m) are orange-weathering mudstones, a few, fine-grained sandy ribs, grey cherts, tuffs and black pyritous graptolitic shales of the *N. gracilis* and *C. peltifer* zones.

The Hartfell Shales (30 m) have a very fossiliferous lower division but the upper part has fewer graptolite bands. In

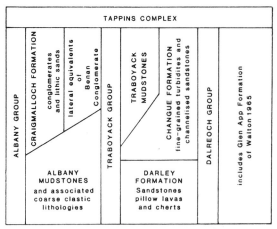

Fig. 6.10. Stratigraphy of the rocks between the Stinchar Valley and Glen App (from Ince 1984).

the lower beds the *Climacograptus wilsoni* zone has hard cherty ribs separating dark, flaky, graptolitic shales. The *Dicranograptus clingani* zone is made up of distinctive rusty-weathering beds which are often siliceous and markedly platy. Individual bedding planes tend to be crowded with a few or even single species. In upward succession there are planes with *Climacograptus caudatus, Dicellograptus caduceus,* and *Climacograptus styloideus.* Similar horizons of mass entombment are found in the succeeding zone with the zone fossil, *Pleurograptus linearis* prolific at one horizon; another plane has hosts of *Orthograptus truncatus pauperatus;* another, *Leptograptus capillaris* and yet another, *Dicellograptus pumilis* and *D. morrisi.* The beds are black, pyritous mudstones with occasional soft, pale-coloured (meta-bentonitic) claystones. The Barren Mudstones above are grey and greenish beds which have been taken to represent a regressive episode due to the Ashgillian glaciation (McKerrow 1979). They have yielded a sparse fauna of blind trilobites (Williams 1980). Two thin bands with *Dicellograptus complanatus* occur near the base of the upper Hartfell shale and define the zone rather poorly. Above, thin bands in barren grey mudstones denote the *Dicellograptus anceps* zone (Williams 1982a, b). Sparse graptolitic bands towards the top of the Hartfell Shales denote the *extraordinarius* zone and black shales above form the base of the overlying Birkhill Shales (see p. 183). The most spectacular lateral variation in the Hartfell shales occurs at Ettrickbridgend where 60 m of greywackes are interpolated into the succession. Current structures from the north-west suggest an exceptional channel deposit penetrating across basin-plain muds (Casey 1983).

Fig. 6.11. Lapworth's (1878) map of Dob's Linn, Moffat Water. Scale 1:1,500 approximately.

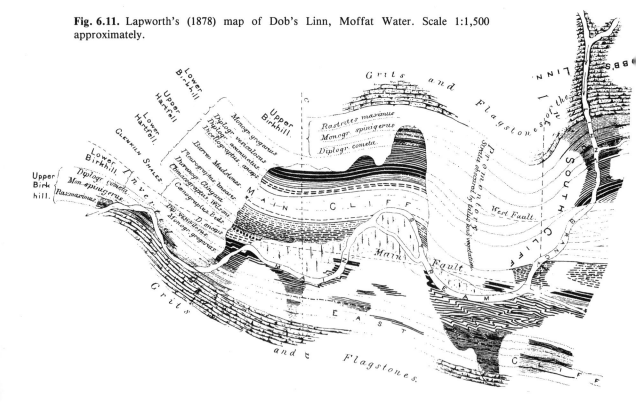

Table. 6.1. Correlation of formations in the Ordovician of the Northern Belt of the Southern Uplands.

SW. Rhinns of Galloway		Wigtonshire	W. Nithsdale	Abington	Gala / NE.
Portayew Fm		Boreland Fm.	Shinnel Fm.	?	Heriot Fm.
-----------------------		---------------------	---------------------	-------------------	---------------------
Portpatrick Fm.	Basic Clast Division	Glenwhan Fm.	Scar Fm.	Glencaple Fm.	Falahill Fm.
	Acid Clast Division			Elvan Fm.	
-----------------------		---------------------	---------------------	-------------------	---------------------
Galdenoch Fm.		Cairnerzean Fm.	Blackcraig Fm.	-	
Kirkcolm Fm.	U. Barren Division	Loch Ryan Fm. — Upper Division	Afton Fm. — Upper Division	Abington Fm.	
	Metaclast Division	Lower Division	Lower Division	-------------------	
	L. Barren Division			Crawfordjohn Fm.	
-----------------------		---------------------	---------------------	-------------------	
Corsewall Fm.	Conglomeratic Div.	Glen App Fm.	March Burn Fm.	Mill Burn Fm.	
	Flaggy Div.				

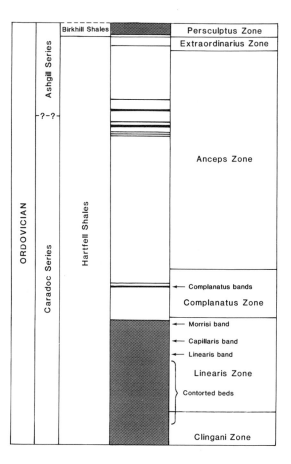

Fig. 6.12. Vertical section (22 m) through the Hartfell Shales in Dob's Linn showing the distribution of fossiliferous black mudstones (black) and barren grey mudstones (uncoloured) (based on Toghill 1970; Williams 1982a, b, 1983, 1987).

Northern Belt

The particular nature of the Southern Uplands requires a different account of the stratigraphy than is usual. The distribution and delineation of rock sequences is dominated by strike faults. These divide the area into tracts trending north-east–south-west and up to about 5 km across strike. In each tract the sequence usually begins with a basal sequence of graptolitic shales, often with cherts and sometimes with lavas and/or tuffs. Greywackes and shales follow. In addition the tracts are often distinguished by the composition of the greywackes; this allows the recognition of formations although the upper limits of these are usually not definable because of faulting.

The blocks or tracts which form the Northern Belt were first delineated by Leggett et al. (1979) mainly on the basis of work inland. The sequences are much better displayed on the south-west coast, south of the Stinchar Valley and in the Rhinns of Galloway (Kelling 1962). This area will therefore be used as a typical section. Using greywacke composition and the major fault lines correlative sequences can be traced strikewise towards the north-east (Table 6.2, Figs. 6.13, 6.14).

Table 6.2. Summary of lithofacies attributes of Ordovician stratigraphic units (non-pelagic) on the Rhinns of Galloway (based on Kelling *et al.* 1987).

Tract	Local stratigraphic units	Dominant Grain-size range	Dominant Bouma sequence	Bed-geometry	Cyclic character	General transport direction	Petrographic character	Interpreted depositional morphology/processes
1	Corsewall Group — Conglomeratic 'Division'	C-GSd		S. Lent.	TNU cycles common (5-20 m)	To ESE-SW sector	Fe-Mg rich, igneous clasts	Channel-fills, terraces; interchannel remnants
	Flaggy 'Division'	FSd, St.	Tbc/Tcde	Tab.	Acyclic	To S-SW sector	Fe-Mg rich, igneous clasts	Interchannel to open basin plain
2	Kirkcolm Group	(a)C-GSd;	Ta/Tab	Lent.	TNU cycles common (18-30 m)	To SE	Intrabasinal cobbles; sand-matrix siliceous	Channelized proximal mid-fan
		(b)F-MSd	Tab/Tbcd	Tab.	TKU cycles common (5-25 m)	To SW or NE	Siliceous, meta-clast rich	Lobes (distal mid-fan)
		(c)St, Md; minor FSd	Tcde/Tde	S. Tab.	Acyclic	To SW		Inter-channel
	Galdenoch Group	M-Cd.	Tab/Tb	Tab.	TKU cycles common (3-8 m)	To? SW-WNW sector	Fe-Mg rich and siliceous alternations	Stacked small lobes
3a	Port-patrick Group — Basic-Clast 'Division'	(a)M-GSd	Ta/Tab/Tabc(d)	Lent.-Tab.	Mainly acyclic	To NE; some to NW	Fe-Mg rich; volcanilithic	Stacked proximal lobe and shallow channel fills
		(b)St-FSd;	Tcd/Tcde Tab/Tabc/ Tbc/Tc	Tab.-Lent.		To NE	Fe-Mg rich; volcanilithic	Inter-channels and channel-fills
		(c)F-CSd;	Ta/Tab	Tab. Tab.-Lent.		?	Siliceous units	Open sheet-flow
	Acid-clast 'Division'	(a)C-GSd.		Tab.		To NE; some to SW-NW	Fe-Mg rich; more siliceous than Basic-Clast	Shallow channel-fills and proximal lobes
		(b)M-CSd	Tab/Tabc/ Tbcd/Tcd	Tab.				Lobes and inter-channel sectors
3b	'Portayew Rocks'	(a)C-GSd	Ta/Tab/	Tab.-Lent.	Mainly acyclic	? to N-NW sector	Siliceous moderate-low Fe-Mg content	Proximal lobes and ? broad channels
		(b)Md-CSd	Tabc/Tbcd/ Tde	Tab.				Distal lobes and open sheet flows

Md, Mud; St, silt; FSd, fine sand; MSd, medium sand; CSd, coarse sand; GSd, granular sand; pkts, packets; Lent., lenticular bedding; Tab., tabular bedding; S., strongly; Irreg., irregular; TNU, thinning up; TKU, thickening up; Fe-Mg, ferromagnesian grains (unaltered pyroxenes, amphiboles, etc.).

Tract 1 (late Llandeilo–early Caradoc)

Strictly the Corsewall, Glen App and Tappins rocks lie outside the Southern Uplands but north-eastwards and south of the Southern Uplands Fault (*s.s.*) greywackes have compositions which suggest correlation with the former sequences. The *March Burn Fm.* of Nithsdale includes basic clasts reminiscent of Corsewall–Glen App and includes volcanics and cherts resembling those of the Tappins complex (Floyd 1982, Stone *et al.* 1987). Greywacke and conglomeratic sequences with such compositions resemble the Basic Clast Petrofacies (BCP; Morris 1987), although the BCP is defined by the Portpatrick Group described below. Lithological features of this, and other Northern Belt tracts are summarised in Table 6.2.

Tract 2 (late Caradoc)

In the Rhinns of Galloway, the *Kirkcolm Fm.* lies between the Glen App and Killantringan Faults. The formation (up to 2,000 m thick) is composed of greywackes each up to 2 m separated by pale-blue silty bands and occasionally thicker packets of flaggy blue-grey siltstone and dark blue-grey mudstone. These packets, 10 m, occasionally 30 m thick, have seams of graptolitic shale of *N. gracilis* age. In the lowermost part of the sequence two impersistent horizons of spilitic agglomerate and chert occur. The greywackes are generally coarse to medium grain, occasionally pebbly. They show typical turbidite features; grading is generally good with amalgamation common in the thicker beds; sole markings are abundant and varied. Packets of greywacke beds are arranged in thinning and fining upwards cycles up to 30 m thick. These represent channel deposits of a mid-fan position while occasional thickening-up cycles (up to 25 m) are taken to have formed as distal mid-fan lobes (Table 6.2).

Modal composition allows a threefold classification into Upper and Lower Barren and 'Metaclast' divisions although this cannot be sustained along strike from the Rhinns. The middle division forms the type sequence for the Metaclast Petrofacies (MCP) while the Barren divisions form a subfacies of that petrofacies (Morris 1987). The colour of the greywackes and their vitreous lustre reflect the high content of quartz and siliceous fragments. Some basic lava fragments are present but pyroxene and hornblende are absent or very rare. Quartzite and phyllite clasts are common especially in the middle (MCP) division along with schists (garnet–muscovite; garnet–talc, graphite, andalusite), gneiss and hornfels (Kelling 1962).

Paleocurrent directions from the Kirkcolm Fm. and correlatives along strike to the north-east vary. The majority are from the north-east and many from the north-west but recent observations have given some indications of south-easterly currents (Kelling 1962, Welsh 1964, Kelling *et al.* 1987, Stone *et al.* 1987).

Similar southerly currents are reported from the overlying Galdenoch Formation in the Rhinns and in the Barrhill area. The dominant currents from the south in this

Fig. 6.13. The Southern Uplands divided into tracts by major faults and showing the distribution of Ordovician and Silurian Formations (based on Leggett *et al.* 1979 and others).

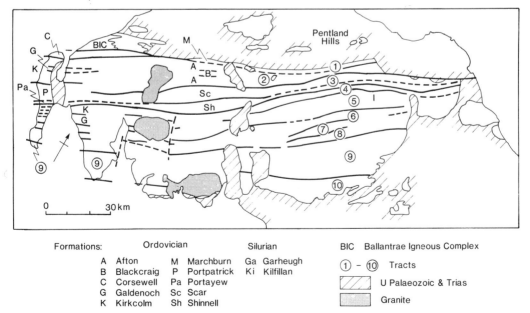

Formations:

	Ordovician			Silurian	
A	Afton	M	Marchburn	Ga	Garheugh
B	Blackcraig	P	Portpatrick	Ki	Kilfillan
C	Corsewell	Pa	Portayew		
G	Galdenoch	Sc	Scar		
K	Kirkcolm	Sh	Shinnell		

BIC Ballantrae Igneous Complex

① – ⑩ Tracts

▨ U Palaeozoic & Trias

▦ Granite

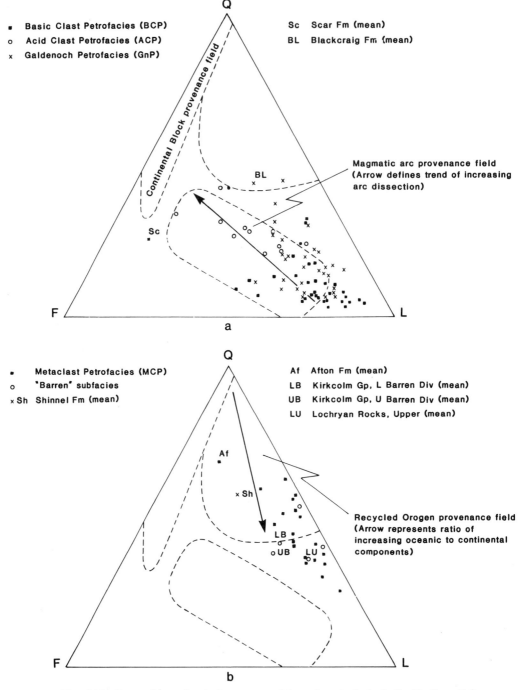

Fig. 6.14. Composition of rocks from some of the main petrofacies in the Northern Belt in relation to provenance fields of Dickinson & Suczek 1979). Q, total quartz grains; F, total feldspar; L, total unstable lithic fragments. (a) Basic clast and associated pertofacies. (b) Metaclast and associated facies (from Morris 1987).

formation are accompanied by a change in composition. Hornblende and pyroxene grains as well as volcanic fragments of andesite with phenocrysts of the two ferromagnesian minerals become abundant. The andesitic and mineral fragments along with variable quartz and felsic fragments led Morris (1987) to separate this sequence as the Galdenoch Petrofacies (GnP).

The Galdenoch greywackes tend to be arranged in packets. Each packet is about 8 m thick and made up of four or five medium-grained beds separated by siltstone and topped by a thick (up to 2 m) band of coarse-graded greywacke. These thickening-up cycles suggest progradations of sandstone lobes on small fans.

Greywacke sequences are generally lacking in shelly fossils but a few metres of mudstones and conglomerates within the dominantly greywacke sequence in the Biggar area, at Kilbucho and Wallace's Cast, contain a diverse shelly fauna. Over 20 species of brachiopods and 12 of trilobites are known and include species in common with the mid-Caradoc, Ardwell and Balclatchie groups at Girvan and the Bardahessiagh Fm. in Pomeroy, Northern Ireland. The shelly material was probably carried from shallower water settings within large submarine debris flows. Wallace's Cast lies some 12 km to the south-west of Kilbucho along the regional strike. Similar but less well exposed fossiliferous beds occur at geographically intermediate localities and also at Duntercleuch and Snar near Wanlockhead a further 16 km to the south-west of Wallace's Cast. The debris flows were evidently quite extensive (E. Clarkson and A. Owen pers. comm. 1988).

Tract 3 (mid-Caradoc–?basal Llandovery)

The arenites of Tract 3a form the *Portpatrick Fm.* in the Rhinns, and belong to the *P. linearis* zone. Bedding thicknesses and sedimentary features are similar to those of the Kirkcolm except that sole markings are rather fewer and cycles are lacking. The Portpatrick rocks have a darker blue-grey colour and a less vitreous lustre, a reflection of their more basic character. They resemble the Galdenoch rocks in having high proportions of hornblende and pyroxene (up to 15 per cent) and andesitic fragments as well as turbid feldspars; but in addition, blueschist fragments occur. Basic clasts are more abundant in the upper beds so that two divisions (Basic- and Acid clast) were recognised in the Rhinns. These form the type BCP and Acid Clast Petrofacies (ACP). The differences between the divisions are not always sustained across country and Morris recognised that one might grade into the other. More quartzose beds also occur interbedded with the basic-clast greywackes. Palaeocurrents are mostly 'axial', i.e. from the south-west although there is some indication of flows from the south-east.

The *Portayew* rocks of Tract 3b in the Rhinns are not so well documented (but see Table 6.3) as their correlatives of the *Shinnel Fm.*, lying inland between the Fardingmullach

Fault and the Silurian boundary, another major fault, the Orlock Bridge Fault (Floyd 1982, see also p. 196).

Greywackes are thin-bedded with abundant ripple cross-lamination and quartzose in composition. Interbedded with the greywackes in the Nithsdale area are conglomerates with clasts sometimes exclusively of penecontemporaneous shales and greywackes but otherwise enclosing fine-grained acid igneous fragments.

Volcanic rocks and cherts

Volcanics and cherts tend to occur at the base of greywacke sequences with black shales although the field evidence is not always clear. There are exceptions for example in the Marchburn Fm. and two notable occurrences are at Bail Hill and Wrae.

The complex at Bail Hill near Sanquhar is the largest in the Northern Belt (Fig. 6.15). At its thickest it is 1·8 km and probably had a diameter reaching 3 km although it is now cut by the fault which brings in the Carboniferous of the Sanquhar Coalfield. Early lavas of mugearite and hawaiite (referred to since Peach and Horne as andesites) were erupted on to *N. gracilis* shales but pyroclastic rocks predominate. Clasts include gabbro, diorite and spilite along with idiomorphic crystals of hornblende and pyroxene. Outwith the main mass there are a number of bands of tuff similar to the Bail Hill rocks. These are interbedded with sediments and were clearly derived from the Bail Hill centre. The present orientation and setting of the complex suggest that it represents a sea-mount enclosed in sediments and sheared from its roots by faulting (McMurtry 1980, Hepworth *et al.* 1982).

Trachytic and non-trachytic keratophyres are found in the Tweed Valley near Peebles (Eckford and Ritchie 1931). The lavas, pyroclastic rocks and fossiliferous limestones at Wrae Hill lie within beds of the *clingani* zone. This indicates an episode later than Bail Hill but the position is complicated because their conodont fauna is Llanvirn. The exposures have an historical interest in that Sir James Hall discovered shells in the limestone at Wrae in 1772 thus proving the contention of his friend James Hutton that the 'schistus' of the district had been formed under normal marine conditions (Hutton 1795). Both sediments and volcanics are interpreted as an olistostrome which brought a Llanvirnian sequence from a sea-mount down into deepwaters in mid-Caradoc times (Leggett 1980a). Geochemically the lavas are peralkaline in nature, a composition consistent with an origin as an oceanic sea-mount (Thirlwall 1981), although Stone *et al.* (1987) claim that the chemistry can be linked with formation in a back-arc area.

Two types of chert have been distinguished, the Hartree and Abington (Leggett 1978). The former include true cherts with siliceous micro-vein systems and those beds without microveins which are perhaps better referred to as siliceous mudstones. The cherts (*s.s.*) are reddish-brown and appear structureless except when weathering or etching

Table 6.3. Summary of lithofacies attributes of early Silurian stratigraphic units (non-pelagic) on the Rhinns of Galloway (based on Kelling *et al.* 1987).

Tract	Local stratigraphic unit	Dominant grain-size range	Dominant Bouma sequence	Bed-geometry	Cyclic character	General transport direction	Petrographic character	Interpreted depositional morphology/processes
4a	Money Head Formation	(a) C-GSd	Ta/Tab/Tb	Lent.	Acyclic	To SW-SSW sector	Siliceous with low Fe-Mg content	Stacked channels
		(b) C-GSd	Tab/Tabc/Tbcd	Tab.	Acyclic			Channel-fills
		(c) F-CSd	Tab/Tabc	Tab-Lent.	Common TKU cycles (5-7m) into (a) or (b)			Channel-margin/inter-channel deposits
4b	Float Bay Formation	(a) M-CSd	Tab/Tabc/Tbcd	Tab.	Acyclic	To SW	Siliceous/lithic; very low Fe-Mg content	Lobes/minor channel-fills
		(b) F-MSd	Tbcd/Tcd					Open sheet-flow deposits
		(c) Md-FSd	Tde					Interlobe/? basin-plain sheet-flows
5a	Stinking Bight Beds	(a) M-CSd	Tab/Tabc	Lent-Tab.	Acyclic	To NE or SW ?	Siliceous and lithic	Channel and lobe deposits
		(b) F-MSd	Tbc/Tbcd	Tab.				Channel-margin/inter-channel deposits
5b	Grennan Point Formation	M-CSd	Tab/Tabc/Tabc/Tcd		Rare TKU cycles (5-5m)	To SW	Siliceous and lithic	Lobes/minor channels
6a	Mull of Logan Formation — The Chair Member	(a) MSd	Tab/bc	Tab.	Mainly acyclic	?		Lobe deposits
		(b) Md-St	Tcde/Tde			?		Open sheet-flows
	Duniehinnie Member	(a) Intraclast gravels, blocks to 10 x 15 m; matrix muddy M-CSd		Irreg.	Overall coarsening-up from facies (b) to (a)	To NE (?)	Siliceous/lithic low-moderate Fe-Mg content in matrix and some clasts	Large debris flow 'lobes' slump/slides
		(b) Intraclast gravels, clasts to 10 x 40 cm; matrix C-GSd		Irreg.				High density sheet/flows; slumps/slides
	Daw Point Member	(a) C-GSd	Tab	Lent.	TKU cycles	?		Channel-fills
		(b) F-MSd	Tbcd/Tcd	Tab.	Acyclic	?	Siliceous and lithic	Inter-channel/channel-fills
		(c) C-GSd	Ta/Tab	Lent.		?		Channel-fills
	Cairnie Finnart Member	F-CSd; St	Tabc/Tbc/Tcd	Tab.	Mainly acyclic, rare cycles (1-5 m)	To SE		Distal suprafan lobes
6b	Port Logan Formation	(a) Csd	Tab/Tabc	Lent.	Acyclic	To SSW	Siliceous and lithic	Channel-fills
		(b) St-MSd	Tabcd/Tbcd	Lent.		?		Inter-channel and channel-fills
		(c) Md-FSd	Tcde/Tde	Tab.				Outer-fan open sheet-flows

Abbreviations as in Table 6.2

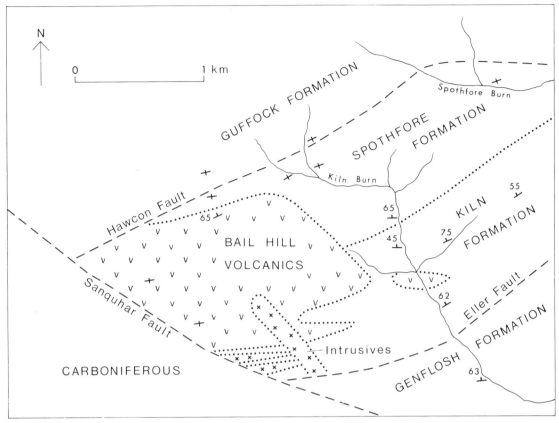

Fig. 6.15. Geological sketch map of the Bail Hill area, near Sanquhar, showing volcanic complex and associated sedimentary formations. The rocks are vertical or steeply dipping and the outcrop is interpreted as a section through a volcanic sea-mount surrounded and covered by trench sediments (after McMurtry 1980).

reveals bands rich and poor in radiolaria. The radiolaria are spheres up to 400 microns across and are often crowded at the base of the layers. The layers averaging 1–2 cm in thickness may therefore be turbidites. Ungraded, radiolarian-rich layers may be due to reworking. The siliceous mudstones probably represent hemipelagic sedimentation; they are sometimes interbedded with grey-wackes and shales. The Hartree cherts owe their colouring to iron oxides and some have up to 10 per cent Fe_2O_3. Iron-rich cherts and mudstones may be interstitial to or interbedded with pillow-basalts.

Abington cherts are grey and interbedded with khaki-coloured siliceous mudstones and soft, grey, clay partings. Up to 64 m have been recorded in one sequence. The chert tends to be structureless with few radiolaria but with some sericite and calcite.

In summary, the Northern Belt is made up predominantly of sequences of greywackes with finer-grained inter-beds. Some interbedded conglomerates like some of the greywackes were built out from the north although many currents are axial to the trend of the basin; exceptionally some currents from the south-east are found. All the arenaceous sequences thin to the south-east and, except for the channel sands at Ettrickbridgend, are absent in the Central Belt. Clasts indicate a mixed provenance of acid and basic igneous as well as metamorphic rocks. Variable supply of andesitic and metamorphic debris allows the recognition of formations most of which can be traced strikewise along the length of the Southern Uplands. Conformable volcanics often with cherts are of various ages. At Raven Gill the rocks are of Arenig age and undoubtedly basal to the greywacke sequence. Other volcanics of *gracilis* age are usually taken to be basal to the arenaceous rocks but this is not always the case as in the Marchburn Fm. The Bail Hill Complex was a sea-mount developed on shales rather than forming basal oceanic crust. The Tweeddale volcanics are olistostromic in *clingani* sediments.

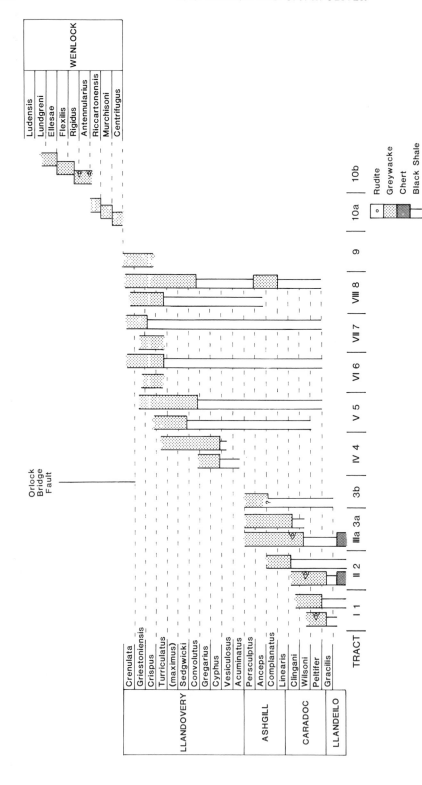

Fig. 6.16. Time–stratigraphic diagram showing positions of coarse clastic sediments, hemipelagic and pelagic shales, cherts (based on Leggett *et al.* 1979, Casey 1983, Kemp 1986, Kelling *et al.* 1987; and Barnes *et al.* 1987). Tract numbers – Roman – Rhinns of Galloway; Arabic – Central section through Moffat.

Tracts: 1, Coulter/Noblehouse; 2, Abington; 3, Tweeddale; 4, Talla; 5, Hartfell; 6, Dob's Linn; 7, Craigmichan; 8, Ettrickbridgend; 9, Hawick; 10, Southern Belt.

I, Corswall; II, Kirkcolm; IIIa, Portpatrick; IIIb, Portayew; IV, Money Head; V, Ardwell Point; VI, Mull of Logan; VII, Port Logan; VIII, Clanyard.

Age relations

The graptolites of the black shales in the imbricate zones of the major faults and the sparse occurrences interbedded with the greywackes point to a delay in the onset of arenaceous debris from northerly to southerly blocks (Fig. 6.16). The compositional differences may then be ascribed to successive source rocks becoming available through time by erosion, contemporaneous volcanicity and possibly strike-slip movement (see pp. 205ff. for further discussion). Thus Floyd (1982) was able to suggest a younger age for the Scar Fm. than the Afton. Stratigraphic control is not entirely satisfactory in so far as the faunas are scarce. Along strike from the Afton and Scar Fms. the correlative Abington and Elvan Fms. have been regarded as coeval. This perhaps points to geographically separate sources supplying contemporaneous overlapping fans (Hepworth 1981).

While the rise in the onset of sandy sedimentation is most marked from north-west to south-east, the base of the Kirkcolm MCP rises progressively from the south-west in Ireland as does the Portpatrick BCP from the *gracilis* zone or earlier in County Longford to the *clingani* zone in West Nithsdale (Kelling *et al.* 1987).

Silurian

Silurian rocks form most of the Central Belt and all the Southern Belt. They are represented at Girvan and occupy a number of other inliers in other parts of the Midland Valley. Sedimentation was similar to that of the later Ordovician but greywackes spread further south-eastwards in the early part of the period and eventually covered the graptolitic shales of the Moffat area. Thick arenite successions are also developed in the Southern Belt. Filling of the trough and emergence may have occurred in late Silurian times. Sedimentation took place in the area of the Midland Valley throughout the period; early turbiditic sequences gave way to marginal and terrestrial sediments later.

Central Belt

The Central Belt lies between the Ordovician rocks to the north and the Wenlock rocks to the south. Both boundaries are faults (Fig. 6.13). Conformable Ordovician–Silurian relations are seen only in the inliers of Moffat Shales. The Ordovician–Silurian boundary has been traditionally taken at the incoming of black Birkhill shales. The lowermost Birkhill Shales belonging to the *persculptus* zone however, have now been assigned to the Ordovician although this is the subject of continuing debate (Williams 1983, Berry 1987, Lesperance *et al.* 1987).

The inliers of Moffat Shales are found in the middle of the Central Belt and can be traced almost from coast to coast although they are best seen inland. The northern part of the belt is occupied by greywacke sequences which are largely lateral equivalents of the Birkhill Shales, the lower Silurian part of the Moffat Shales. Arenites dominate in the Hawick Group which forms the southern part of the belt but paucity of fossils makes correlation uncertain.

Birkhill Shales

The Birkhill Shales follow conformably on the Hartfell Shales (Figs. 6.11, 6.17). Three lithologies are displayed, black pyritous mudstones, barren grey mudstones and soft, pale-grey claystones. The lower part is black, graptolitic and spans the upper *acuminatus, vesiculosus* and lower *cyphus* zones. This graptolite-rich portion passes up into grey and mostly barren beds with only occasional graptolite bands, a change similar to that seen in the Hartfell Shales. The claystones are meta-bentonites now made up largely of illite and quartz (Weir 1973, Stephens *et al.* 1975, Watson 1976, Batchelor and Weir 1988).

Tracts

The structural theme found in the Ordovician region is continued in the Silurian but with some variations. In the part of the Central Belt with the Moffat Shales, decollement along the shales frequently leads to a profusion of faults. Greywacke composition again allows the recognition of formations but the mineralogy is rather more variable along strike than in the Ordovician.

In the Peebles–Moffat area five tracts have been recognised between the Silurian–Ordovician boundary and the Hawick Rocks (Fig. 6.13, Legget *et al.* 1979). This area has the advantage of extensive exposures of the Moffat Shales. The Rhinns and Wigtown areas provide more continuous exposures of the greywacke sequences.

Talla Tract, 4 (early Llandovery)

The Talla Block is characterised by basic-clast greywackes (the 'Pyroxenous Group' of Peeblesshire, Walton 1955 and the Pyroxenous Petrofacies of Morris 1987); with similar components to the BCP but lesser amounts of augite and hornblende and no blueschist fragments. Often coarse-grained and occasionally conglomeratic, the rocks form thinning and fining-upward cycles of a mid-fan channelled association. The sandstone sequences together with those of the three subsequent blocks form the Gala Group of Lapworth and Wilson (1871), correlative at least in part with the Queensberry Grits of Dumfriesshire (Peach and Horne 1899).

Hartfell Tract, 5 (mid Llandovery)

Pyroxenes are absent in the greywackes of this block which is defined on its southern boundary by the imbricated Moffat Shales at Hartfell. Quartz content of the arenites lies between that of the pyroxenous greywackes and the siliceous greywackes of the succeeding block. They have been distinguished as the 'Intermediate Group' (Walton 1955) and form the Intermediate Petrofacies (IP) of Morris (1987). Intermediate and acid volcanic fragments are common along with low-grade metasediments and some granitic fragments. Some conglomeratic units are present south-west of the type locality near Peebles; sandstone/shale ratios are high and sediment environment was similar to the Talla Tract greywackes; palaeoflow was axial (Casey 1983).

Dob's Linn Tract, 6 (late Llandovery)

The classic locality at Dob's Linn (Lapworth 1878) has Birkhill Shales succeeded by greywackes characterised by a high silica content and in the heavy mineral assemblage, abundant garnet. They form the 'Garnetiferous Group'

Fig. 6.17. Vertical section through the Birkhill Shales showing the black graptolitic mudstones and increase in barren grey mudstones upwards. Claystones mostly metabentonites. Locations of sections indicated on sketch map Fig. 6.12, except North Cliff which is situated on the left bank of the Linn Branch (Toghill 1968, Williams 1987).

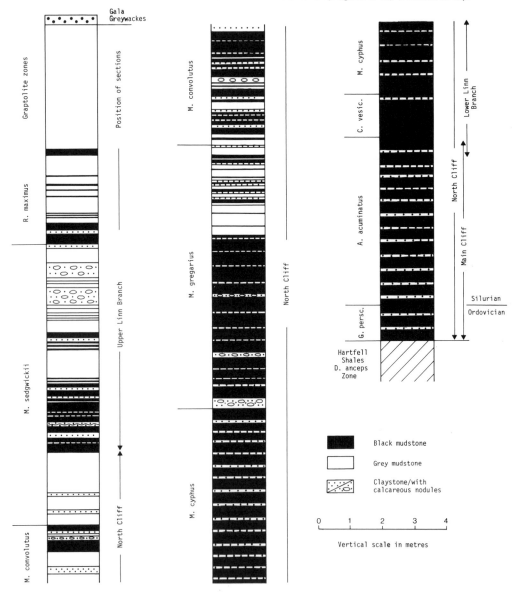

(Walton 1955) of the Tweed Valley and include the Buckholm Grits (Lapworth and Wilson 1871, Kassi 1984). *Crispus* and *turriculatus* zone graptolites occur in inter-bedded greenish shales along with some reddish mudstones and a considerable variety of trace fossils (Peach and Horne 1899). The traces are of the *Nereites* subfacies and domin-ated by *Dictyodora* (Benton 1982).

The greywackes are often coarse grained and massively bedded with fining-upwards units representing, again, a channelled mid-fan locus of deposition. Palaeoflow was predominantly axial towards the south-west.

Morris regards the garnetiferous greywackes as a sub-facies of the IP and the separation of a distinctive garneti-ferous set of greywackes may not be valid according to Stone *et al.* (1987) who regard greywackes of Intermediate type as occupying all the blocks from Hartfell to the Hawick Rocks. As well as the Hartfell and Dob's Linn this would include the Craigmichan and Ettrickbridgend Tracts (7 and 8).

Craigmichan Tract, 7 (late Llandovery)

The Birkhill Shales survive as late as the *turriculatus* zone in this tract. The arenites tend to be coarse grained and resemble those of the previous tracts.

Ettrickbridgend Tract, 8 (mid-Llandovery)

Although this is south-east of tract 7 the Birkhill Shales reach only to the *gregarius* zone with a late intercalation of shales in *maximus* times. In this the tract is an exception to the delay of onset of sandy sedimentation south-eastwards just as it was in the Ordovician.

The greywackes are quartzose resembling the IP. They tend to be fine grained and thin as the Abbotsford Flags (Lapworth and Wilson 1871). Channelling is absent and they were formed as distal mid-fan lobes by currents from the north-west and the north-east (Casey 1983).

Fig. 6.18. Modal composition of greywackes from the Central Belt of the Southern Uplands (Peebles–Hawick area). The formations from the different tracts show contrasting compositions with a trend towards maturity from Northern to Southern tracts (after Casey 1983). Q, total quartz; Qp, polycrystalline quartz; Qm, mono-crystalline quartz; F, total feldspar; L, total unstable lithoclasts; Lv, volcanic clasts; Lm, metamorphic clasts; Ls, sedimentary clasts; Lt, L + Qp.

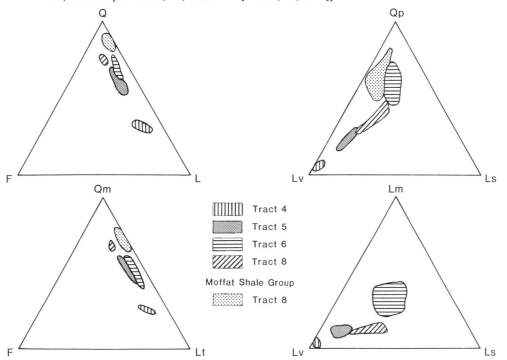

Hawick Tract, 9 (?Llandovery)

Hawick arenites tend to be rather finer grained than those to the north and have a rather more restricted variety of clasts, basic igneous types for example being scarce. More conspicuously, carbonate cement is characteristically high. Another distinguishing feature of the sequence is the appearance of laminae and beds of reddish mudstone in addition to the ubiquitous, unfossiliferous greenish mudstones and shales. Very often clastic white micas in the sandstones are stained red with a veneer of hematite.

In the Rhinns of Galloway and the neighbouring peninsula of Wigtown compositional differences amongst greywacke sequences are not so marked (Barnes *et al.* 1987). Pyroxenous greywackes for example are rather sporadically developed. Nevertheless the Kilfillan Fm. of Wigtown (Gordon 1962) and some of the Money Head rocks compare with the Pyroxenous greywackes of Peeblesshire. In addition the excellent exposures allow the recognition of more fault blocks than inland. Eight tracts have been listed of which the last two are Hawick Rocks (Barnes *et al.* 1987). The Hawick rocks comprise the Kirkmaiden and Carghidown Fms. While the latter contains red mudstones this lithology is absent from the Kirkmaiden sequence. North of the Hawick Group correlation has been made with the Peebles–Moffat tracts (Fig. 7.8).

Age relations

Graptolites are largely restricted to basal shales, the Birkhill and their equivalents. They enable the arenaceous deposits to be dated, helped by sporadic occurrences of thin graptolitic seams within the greywacke sequences for example in the Garheugh rocks south of Glenluce and in the Ettrickbridgend Block.

The overall trend of sands spreading with time southeastwards, demonstrated graphically by Leggett *et al.* (1979), has been substantiated by later work, particularly in the south-west and across the Ards peninsula in Northern Ireland (Barnes *et al.* 1987) (Fig. 6.16). As in the Ordovician, Ettrickbridgend is anomalous in having sands possibly as early as the *gregarius* zone. This anomaly is not present in the Rhinns where sands in correlative blocks are delayed till the *turriculatus* zone.

The modal composition of the greywackes in the different tracts also shows a variation from north-west to south-east with a general increase in quartzose content (Fig. 6.18).

Fig. 6.19. Sole markings on the base of a Silurian greywacke. Mainly prod and bounce casts produced by hard objects (? shells and shale fragments) impinging on the mud floor. Specimen about 20 cm across.

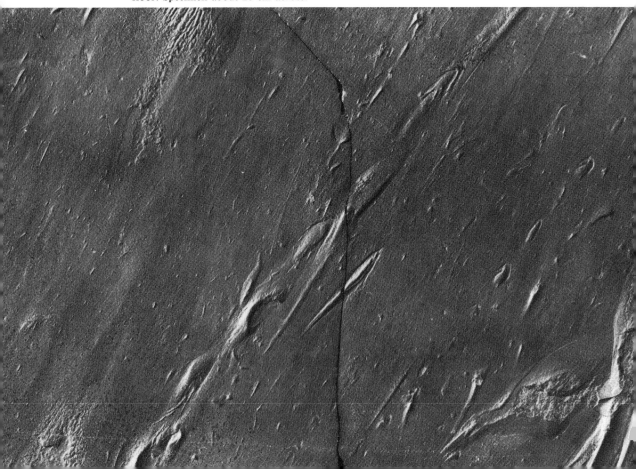

The *Linkum* and *Coldingham Beds* on the Berwickshire coast are anomalous in lithology, structure and metamorphism (Shiells and Dearman 1963, Casey 1983 and see p. 201); and acritarchs suggests an age difference from other beds of the Central Belt in that they appear to be early Wenlock (Molyneux 1987).

Southern Belt

Wenlock rocks of the Southern Belt lie conformably on Hawick Rocks only at Burrows Head at the tip of the Wigtown peninsula; elsewhere the northern boundary is faulted. Relations to the south are covered by the waters of the Solway Firth and by unconformable Carboniferous rocks. Leggett *et al.* (1979) took the Southern Belt to be essentially one but in the Kirkcudbright Bay area several blocks have been delineated (Kemp 1986, Fig. 6.19).

The arenites are similar to those of the Hawick Rocks in that they are relatively mature in composition and have a high content of carbonate. Red mudstones are, however, absent, as are the red micas of the Hawick Rocks. Distinctively the sequences are characterised by many graptolite horizons. These vary from mm-thick laminae and streaks to beds of 1 m. They are dark-grey when fresh and weather to brownish-khaki colour. Their texture is unique in the Southern Uplands in having a fibrous 'felted' appearance, ascribed to tiny lenses of silty grains surrounded by films of clay and opaque material (Warren 1963). But the lithology is identical to that in the Brathay Flags of the English Lake District.

An extensive sequence in the Hawick–Riccarton area comprises some 4,000 m of Wenlock beds in four divisions, with graptolites ranging from the *C. murchisoni* to the *C. lundgreni* zones (Warren 1964). Similar zones were recognised in the Langholm area although the four divisions were not sustained by the British Geological Survey. Observations from Langholm and Kirkcudbright have given rise to a twofold division into the Ross Fm. below and the Raeberry Castle Fm. above. The Ross Fm. is made up of a monotonous succession of turbidites. These are usually medium grained, of moderate thickness and often show well-developed sole markings (Fig. 6.20). The Fm. is thicker in Langholm (where the palaeocurrents are generally from the north-west) than in the Kirkcudbright area where palaeoflow is axial from the north-east. The acyclic nature of the successions and the predominance of sands suggest a slope-apron system rather than a fan (Kemp 1987),

The Raeberry Castle Fm. by contrast has much greater

Fig. 6.20. Time–stratigraphic diagram of the Southern Belt near Kirkcudbright (after Kemp 1987).

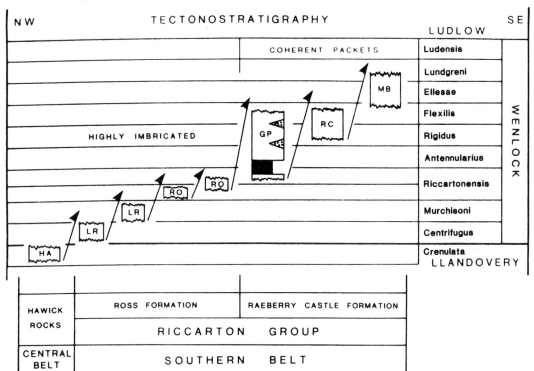

lithological variety. As well as classic turbidites there is a striking development of channelised conglomerates with thin overbank deposits and one massive slumped horizon topped by large sand volcanoes. Currents in the conglomerates tend to be from the north-west but, in keeping with a proximal fan environment there is considerable variation especially in the finer-grained beds (Scott 1967, Walton 1968, Lovell 1974, Kemp 1987).

Midland Valley

Silurian beds occur in several inliers in the Midland Valley (Fig. 6.21). Girvan and Craighead have Ordovician as well as Silurian rocks. Lesmahagow, Hagshaw Hills, Carmichael and Eastfield and the Pentland Hills have only Silurian. Though small, the inliers have a significance quite disproportionate to their size. Shelly and graptolitic faunas allow a linking of the two zonal schemes. Some of the inliers provide rich, often exquisitely preserved faunas of early fish and arthropods. Sedimentologically they provide a record of transition from early marine turbidite sequences to later terrestrial successions and they give evidence that the region was separated from the Southern Uplands during Silurian times. Correlation has proved difficult. Apart from Girvan and Craighead the successions have been placed in different positions in the Old Red Sandstone and the Silurian (compare for example Geikie 1873 with Robertson 1985). There seems little doubt now that the lower rocks of the inliers are Llandovery but the upper limit is still uncertain (Fig. 6.22).

Girvan and Craighead

The Silurian–Ordovician contact in these two inliers is one of overlap and overstep. At Craighead the basal conglomerate lies on the Hirnantian, High Mains Sandstone; and 12 km to the south-west on the shore at Girvan the basal rocks, younger than those at Craighead, rest on the Shalloch Fm. (Fig. 6.23) with slight angular unconformity. Inland higher parts of the same formation appear below the unconformity. The deposits thicken northwards – the reverse of the Ordovician – apparently in contradiction to the direction of overlap. This paradox is resolved if sedimentation is regarded as having taken place in a faulted trough (Bluck 1983).

Of the coarse-grained lithologies, the sedimentary features of the Mulloch Hill Formation suggest shallow-water conditions while those of the Craigskelly and Quartzite conglomerates are probably proximal subsea-fan in origin. Apart from the Mulloch Hill Sandstone and the Straiton Grits the 'grits' and 'flags' are turbidites of varying grain size and bed thickness but with typical grading and sole markings.

The fine-grained formations are mostly made up of greyish green mudstones and shales with occasional graptolitic bands. Some of the last are black and pyritous with graptolites preserved in the round. Silty interbeds are common and a few graded sandy bands are not uncommon. Red and purple shales and mudstones appear first in the *Maxwelltown Mudstones* (*R. Maximus* subzone) and are repeated in the *Penkill Shales* (*turriculatus*) and the *Lachlan Fm.* (*griestoniensis*). The red beds of the topmost *Straiton*

Fig. 6.21. Sketch map showing the Lower Palaeozoic inliers in the Southern part of the Midland Valley.

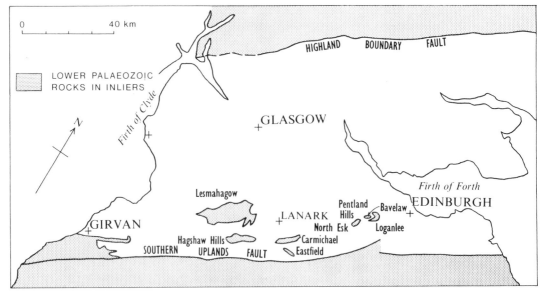

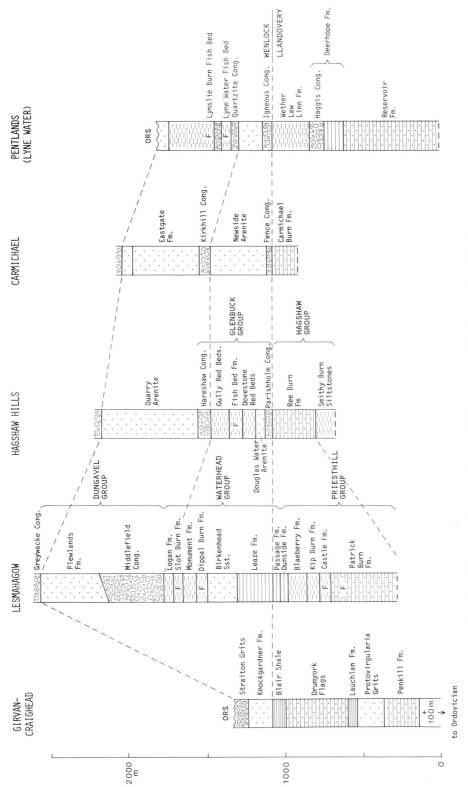

Fig. 6.22. Successions in Midland Valley inliers compared with Girvan–Craighead Sequence.

Fm. are of Old Red Sandstone facies with coarse sandstones and conglomerates.

Communities of brachiopods from various depth zones have been recognised and the Llandovery beds show three phases of deepening culminating in graptolite faunas. First, from the Lady Burn Conglomerate to the Glenwells Shales; second from the Glenwells Conglomerate to the Glenshalloch Shales; and the third from the Lower Camregan Grits to the Maxwellton Mudstones. A shelly fauna in the *Knockgardner Fm.* indicates shoal conditions and is Wenlock, an age confirmed by an acritarch assemblage (Dorning 1982). In the Straiton Grits a sparse fauna of bivalves and ostracods suggests continued regression with brackish or fresh-water conditions (Cocks and Toghill 1973).

Other inliers

The turbidite sequences forming the lower parts in each inlier tend to have two faunal assemblages, one associated with the arenite and another with the lutite interbeds. The first is resedimented and for example in the *Patrick Burn Fm.* (Lesmahagow) the shelly fauna has *Leptaena, Orthoceras, Encrinurus, Beyrichia*, bivalves, bryozoans and crinoids. The interbedded siltstones have a fauna which includes *Thelodus scoticus, Jaymoytius kerwoodi, Aniktozoon loganense, Ceratiocaris papilio, Slimonia acuminata, Platyschisma helicites* and *Palaeophonus caledonicus*. A little higher in the succession at Lesmahagow, the *Kip Burn Fm.* (Jennings 1961) contains abundant Ceratiocarids and Eurypterids.

In the Pentland Hills the *Reservoir Fm.* (Tipper 1976) is made up of fine-grained and thin turbidites with sparse graptolites and in its upper part bears two notable bands – the Eurypterid and the Starfish Beds. Higher in the succession, the *Weather Law Linn Fm.* has abundant shelly faunas. These have been divided into communities (Tipper 1975) and on a rather more rigorous statistical basis, associations (Robertson 1985). The *Isorthis mackenziei –* pelmatozoan association is followed upwards by a varied marine assemblage, the *Skenidioides lewisi–Cyrtia*

Fig. 6.23. Silurian successions at Girvan and Craighead. Inset: Sketch map showing Silurian outcrops in Girvan area (after Cocks & Toghill 1973).

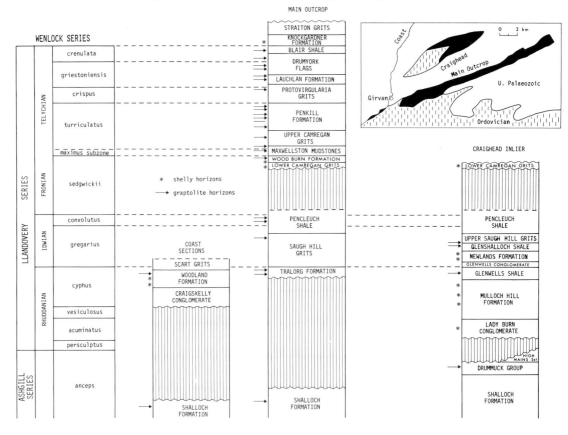

exporrecta association which is terminated by a white clay (pyroclastic) horizon. Two other associations follow in succession, the *Eoplectodonta penkillensis* and the *Liospira ?simulans–Synek(?) sp.*, Robertson relates the associations to energy level and substrate in the first instance. Overall there is a shoaling through time leading to a reduced fauna and eventually in the succeeding *Henshaw Fm.*, emergence with red beds and desiccation cracks.

In all the inliers, a regression, if not emergence, is indicated. McGiven (1968) claimed that emergence did occur and that it is recorded in the abrupt change to conglomeratic sedimentation. The conglomerates are the Igneous (Pentlands), Fence (Carmichael) and Parishholm (Hagshaws). No conglomerate is found at the equivalent horizon in Lesmahagow. Isopachs of the conglomerates show a thinning towards the north-west over about 8 km, from 50 m to 0 just short of Lesmahagow. Bedding and grain-size features are typical of both submarine and alluvial fans. McGiven preferred the latter because of the underlying break in sedimentation and the indications of shoaling in sediments and faunas; in particular a petrographic feature, the coating of particles with fine-grained material (?illite), which has only been reported from alluvial fans. As well as the abrupt change marked by the appearance of the conglomerate, there is some palaeontological evidence for a hiatus. Thin sections of calcareous pebbles from the Parishholm Conglomerate showed stromatoporoids and bryozoans of Wenlock age. The conglomerate may be late Wenlock and the hiatus span much of that period (Rolfe and Fritz 1966).

Igneous clast predominate. Spilites, keratophyres, andesites and quartz porphyries often make up 70–80 per cent of the clasts and are accompanied by chert, coarse-grained igneous fragments, sedimentary quartz and quartzite.

Between the lower igneous conglomerates and the quartzite conglomerates (*Middlefield, Hareshaw, Kirkhill and Quartzite*) the arenites have large-scale cross-lamination and the finer beds, mud cracks. Fish beds are located in finely laminated, olive-green sequences with sometimes as many as 20 organic-rich and -poor laminae per cm. The sandstones seem to be fluvial deposits in a flood plain while the finer beds are lacustrine even playa in origin. The enclosed fish faunas differ from the truly marine assemblages in the lower formations. Some of the common forms are *Birkenia elegans, Lasanius problematicus, L. armatus, Lanarkia horrida, L. spinosa, Ateleaspis tesselata.* Also found are arthropods, *Glauconome* and plant fragments.

The quartzite conglomerates show similar alluvial fan features to the igneous conglomerates with the beds thinning outwards from the Tinto area. Clasts are similar to those before but the proportion of quartzite and quartz is much higher, often reaching over 50 per cent.

Igneous clasts have been taken to indicate a source in the Ordovician conglomerates of the Southern Uplands and the quartzose debris in the upper conglomerates may have come from the same source. But calculations based on clast size and rate of decrease in thickness of the conglomerates point to fans of a size requiring a source well south of the present Southern Uplands. Furthermore the compositions of the inlier conglomerates differ in a number of the respects from the Ordovician; in particular greywackes fragments of compositions corresponding to those of the Northern Belt are difficult to find even in the Greywacke Conglomerate (Bluck 1983).

Metamorphism

Metamorphic effects have been inferred from mineral assemblages, illite crystallinity (Hb rel.) and lateral spacing (bo), graptolite reflectance together with conodont and acritarch colour alteration (Oliver and Leggett 1980, Oliver *et al.* 1984, Kemp *et al.* 1985).

In the Southern Uplands basalts and basic-clast greywackes show the development of prehnite and pumpellyite. These minerals are found in veins and in vesicles. Pumpellyite may also be found replacing albite and as euhedral green and brown crystals occurring in the matrix of some greywackes. Sphene and chlorite are common alteration products with the prehnite and pumpellyite along with quartz, sericite (phengite), calcite and hematite. The veins with prehnite–pumpellyite cut S1 cleavage but show microfolding; metamorphism is therefore syntectonic. Zeolites (analcite and thomsonite) have been found only in the Bail Hill volcanics; associated sediments at the same locality have kaolinite.

Graptolite reflectance shows a tendency to increase northwards across the Southern Uplands. This effect of burial metamorphism may have been augmented in places by the presence underground of granitic plutons (Watson 1976).

Illite crystallinity measurements along a number of traverses confirm the reflectance results in showing anchizone metamorphism. Generally no systematic trend can be detected either between or within blocks except in the extreme south-east near Kirkcudbright. There the Hb rel. shows an increase in grade towards the north-west. This is accompanied by an increasingly well-developed cleavage in the same direction. The same blocks along strike at Langholm show no change in crystallinity (Kemp *et al.* 1985). It is concluded, first, that burial metamorphism continued to take place after rotation of the blocks rather than during sedimentation. Second, relative slippage between blocks has been negligible since metamorphism so that similar erosion levels have been reached in adjacent blocks. In contrast the Wenlock blocks in Kirkcudbright-shire have suffered relative movement. The southernmost block which shows simple diagenetic effects and no cleavage is at a high tectonic level and blocks to the north are at successively lower levels.

The Coldingham and Linkum beds on the east coast are exceptional in showing low grades. Illite characteristics and

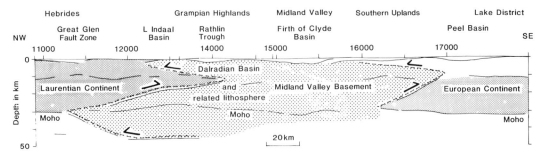

Fig. 6.24. Main structural features as derived from seismic experiments (based on Hall *et al.* 1984).

acritarch colouration are consistent in suggesting a grade transitional between zeolite and prehnite–pumpellyite facies (Oliver *et al.* 1984, fig. 5).

Metamorphism of the Ballantrae Igneous Complex has already been described (pp. 168–169). The cover sequences in the Girvan area of later Ordovician and Silurian rocks were affected only by burial metamorphism.

The Northern and Central belts generally experienced PT conditions around 350°C and 2·5–4 kb or depths of 9–14 km. Coldingham–Linkum indicators suggest temperatures less than 200°C and often not more than 150°C and similar conditions obtained in the Midland Valley inliers.

The colour alteration index of conodonts in the Abington block suggest relatively high temperatures approaching

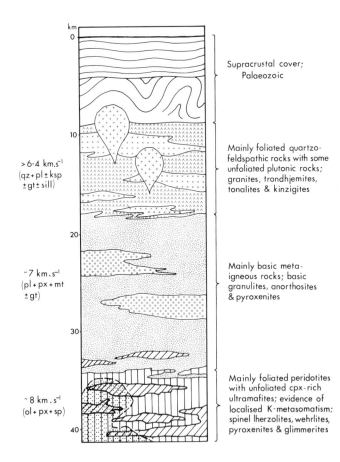

Fig. 6.25. Lithologies forming the upper lithosphere in the Midland Valley of Scotland based on volcanic xenoliths (from Upton *et al.* 1984). The xenoliths from the Southern Uplands, although few, suggest similar lithologies.

the transition to greenschist facies. If the metamorphic grade measured from surface exposures is dependent on tectonic level, the observed prehnite–pumpellyite facies in most blocks might be expected to change in depth to greenschist facies. In the Abington block the boundary must be close to the surface. Oliver (*in* Oliver *et al.* 1984) has suggested that the shallow velocity change identified in the Southern Uplands at depths down to 5 km records the change to foliated rocks of greenschist facies and not crystalline basement as interpreted by Hall *et al.* (1983).

Basement

A knowledge of the material underlying the greywacke successions is crucial to the understanding of the development of the south of Scotland. Evidence bearing on this question is geophysical, geochemical and petrological.

Earlier models based on a variety of techniques have been refined by Hall *et al.* (1984) and Beamish and Smythe (1986). Seismic experiments indicate that the Moho is not clearly marked under the Southern Uplands compared with its definition to the north and the south at about 30 km depth. Above the transition to the mantle the structure appears to be dominated by inclined surfaces. One boundary dips northwards under the Southern Uplands at 15–25 to a depth of 28 km (Fig. 6.24). Above this the crust has similar seismic properties to that under the Midland Valley. Below the surface the crustal segment appears to continue

southwards under the Lake District. A less-well-defined surface dips southwards from the Southern Uplands at depths of only a few kms. This has been taken to form the base of the Lower Palaeozoic successions (Hall *et al.* 1983) although an alternative interpretation sees this as a transition from prehnite–pumpellyite to greenschist facies (Oliver and McKerrow 1985).

The continuity of the crust from the Southern Uplands northwards is supported by geochemical data. Major and trace-element content of Caledonian granitoid plutons in the Southern Highlands, the Midland Valley and the Southern Uplands are similar to one another but distinguished from the plutons of the Lake District and southern Ireland (Stephens and Halliday 1984).

Xenoliths in vents from the Southern Uplands and the Midland Valley are similar petrologically. They show considerable variety which, taken together with the geophysical evidence, produces a plausible model of the crust (Fig. 6.25; Upton *et al.* 1984, Hunter *et al.* 1984). The Palaeozoic cover is pictured as lying above a quartzo-feldspathic Upper Crust. No amphibolite-facies xenoliths have been reported. The Lower Crust comprises basic granulites, anorthosites and pyroxenites. The mantle has foliated peridotites and unfoliated ultramafic rocks of various types, lherzolites, wehrlites, pyroxenites and glimmerites. These high-grade xenoliths may have originated in Lake District crust buried deeply below the Southern Uplands (Oliver and MacKerrow 1985).

References for Chapter 6 are incorporated with those of Chapter 7 (see p. 218).

7

LOWER PALAEOZOIC – STRUCTURE AND PALAEOGEOGRAPHY

E. K. Walton and G. J. H. Oliver

Structure

The vast majority of the beds in the Southern Uplands strike north-east–south-west and are highly inclined, even overturned. With the exception of a zone in Silurian rocks, younging is directed predominantly towards the north-west. But overall, Ordovician rocks are found to the north-west of the Silurian. This paradox is the result of strike faults which divide the area into blocks or tracts which consistently throw down to the south-east (Fig. 6.13).

Faulting

Faulting is discussed before folding because it plays the major role in determining rock distribution.

Southern Uplands Fault

The Southern Uplands Fault (SUF), arguably the most important in that it forms the boundary with the Midland Valley, has up to now not been subjected to modern structural analysis. In general, downthrow appears to be predominantly northwards with the Upper Palaeozoic rocks in the Midland Valley lying against Ordovician to the south. But there are anomalies. In the south-west portion (the Glen App Fault) Ordovician rocks have been thrown down to the south and the Carboniferous rocks of Loch Ryan have been similarly affected. Between the Glen App Fault and the central portion (Southern Uplands Fault *s.s.*) the Carboniferous rocks at Cumnock step over the line of the fault without any marked displacement but at Sanquhar they have been thrown down to the south. The north-eastern segment, the Lammermuir Fault, appears to have little effect on the Carboniferous. In many places along the length of the SUF (*s.l.*) slickensides are predominantly horizontal and small-scale fractures are lateral in their effect. A Lower Palaeozoic history similar to the Orlock

Bridge Fault is possible although many features of the Southern Uplands Fault indicate movement at higher brittle levels in the crust. Large dip-slip movements took place about Middle Old Red Sandstone times (George 1960). Movement affecting Carboniferous sequences seems to have been variable both in amount and direction of vertical throw.

The Glen App portion of the Fault forms a zone about 300 m wide on the coast of Loch Ryan. In this zone two sets of strike-slip shear planes occur, an earlier sinistral cut by a later dextral set. The latter set is parallel to the spaced fracture cleavage in a felsite dyke within the fault zone. To the south, just outside the fault zone, another dyke is uncleaved whereas the surrounding greywackes and shales display the main Southern Uplands S1 cleavage. Taking the dykes to be co-eval with the 400 Ma granites of the Southern Uplands and elsewhere gives the following chronology (Anderson and Oliver unpublished):

1. Regional S1 cleavage in the Northern Belt, pre-400 Ma, possibly end-Ordovician (since cleaved black shales with *Dicellograptus* occur as detritus in Silurian conglomerates at Pinstane Hill).
2. Sinistral shear planes formed in the fault zone later than S1 and earlier than the dykes.
3. Intrusion of the dykes at about 400 Ma.
4. Dextral strike-slip movement off-setting the sinistral planes and foliating the dyke in the fault-zone.

Stinchar Valley Fault

Some indication of the importance of this fault, which broadly forms the boundary between the Ballantrae Igneous Complex and the Girvan succession from the greywacke sequences to the south, can be gained by comparing estimates of palaeotemperatures on opposite sides. From reflectance data, it can be calculated that the

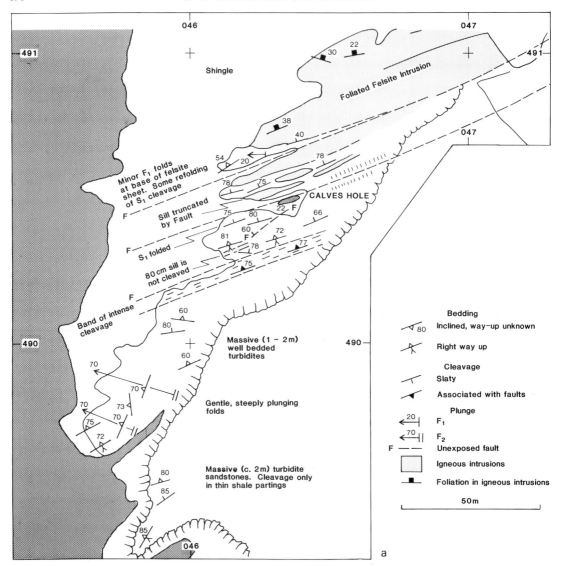

Fig. 7.1. Structure of major strike faults in maps and vertical sections. (a) Orlock Bridge–Kingledores Fault at Cairngarroch, west coast, Rhinns of Galloway. (b) Leadhills Fault near Abington. (c) Ettrick Valley Fault at Craigmichan Scaurs. (a) After Anderson & Oliver (1986); (b) after Hepworth *et al.* (1982); (c) after Fyfe & Weir (1976).

total upward throw on the south side is of the order of 4·5 km (Kemp *et al.* 1985). Similar considerations suggest a throw of some 5·8 km on the south side of the Glen App Fault.

Orlock Bridge Fault

The boundary between the Silurian and Ordovician is the Kingledores Fault near Edinburgh and the Cairngarroch Fault in the Rhinns of Galloway. It is best exposed in Northern Ireland as the Orlock Bridge Fault (Anderson and Oliver 1986).

When fully developed the fault zone is characterised by
1. numerous shear zones slicing arenites at low angles to the bedding,
2. phyllonitic texture,
3. quartz segregations parallel to the foliation and emphasising the schistosity,

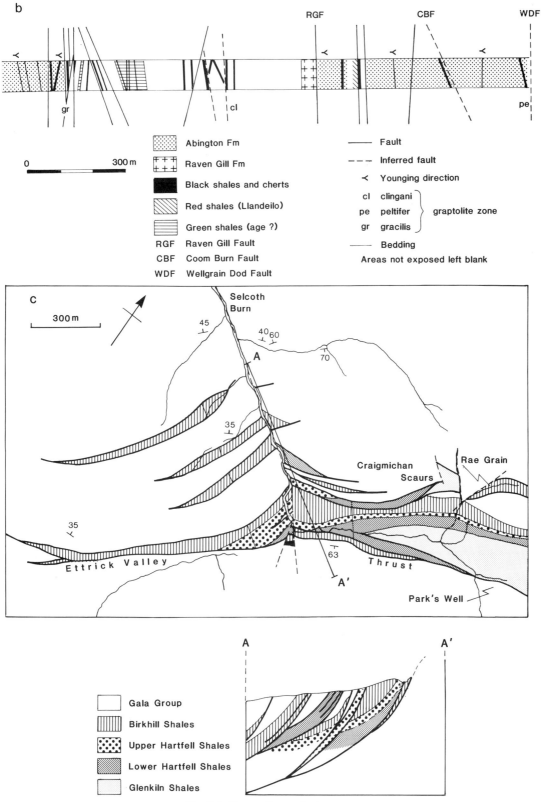

b

RGF   CBF   WDF

gr

cl

pe

	Abington Fm
	Raven Gill Fm
	Black shales and cherts
	Red shales (Llandeilo)
	Green shales (age ?)

RGF   Raven Gill Fault
CBF   Coom Burn Fault
WDF   Wellgrain Dod Fault

—— Fault
- - - Inferred fault
< Younging direction
cl clingani
pe peltifer   } graptolite zone
gr gracilis
—— Bedding
Areas not exposed left blank

0   300 m

c

300 m

Selcoth
Burn

45

40 60

A

70

35

Craigmichan
Scaurs

Rae Grain

35

35

Ettrick Valley

63

Thrust

A'

Park's Well

A   A'

	Gala Group
	Birkhill Shales
	Upper Hartfell Shales
	Lower Hartfell Shales
	Glenkiln Shales

Fig. 7.1. See caption on facing page.

4. folding of the fault fabric into minor folds with sinistral vergence and

5. C–S mylonites in the zone of greatest deformation, showing sinistral strike-slip movement.

These features imply significant sinistral movement but the metamorphic grade on either side of the Fault precludes a large vertical component of displacement. Across the Orlock Bridge Fault there is a rise of six graptolite zones to the base of the first greywackes, suggesting that perhaps four tracts have been cut out by thrusting. Greywacke successions cannot be correlated across the Fault along the whole of its 400 km trace. The Fault may therefore be more significant than any of the other tract-bounding faults. Furthermore it is cut by lamprophyre and felsite dykes thought to be associated with the 400 Ma granitic activity in the Southern Uplands. It may be that either two accretionary prisms were amalgamated around early Devonian/late Silurian times or one large prism was dissected by the Fault (Anderson and Oliver 1986, but see Floyd *et al.* 1987 for counter-opinions).

Other tract-bounding faults

The Orlock Bridge Fault is also exceptional in the Northern and Central Belts in not having graptolitic shales marking the line of outcrop. Other major tract-bounding faults, for example the Leadhills Fault and those in the central, Moffat area, are all characterised by a zone of imbricated slices of Moffat Shales. In the Northern Belt the shales are accompanied by cherts and sometimes basalts; in the Central Belt the basalts are absent. The section at Craigmichan Scaurs across the Ettrick Valley Fault (Fig. 7.1), provides a typical section showing slices of shales and greywackes bounded by faults with a listric geometry (Fyfe and Weir 1976). The thrust dips northwards at 70° and forms the base of a schuppen zone about a km across. Time relations to folding are not clear. In the south-west the Ettrick Valley Fault appears to be post-folding (Cook 1976) while in the Moffat area thrusting and isoclinal folding are synchronous (McKerrow *et al.* 1977). The Leadhills Fault in the type area also has a very wide imbricate zone some 1,700 m across (Fig. 7.1). Here the basal fault is vertical and associated faults have a variety of inclinations. At the coast it appears as the Killantringan Fault with a relatively low inclination to the north. Other tract-bounding faults in the Rhinns tend to be sub-vertical (Kelling 1961, Hepworth *et al.* 1982, McCurrie 1987).

Tracts in the Southern Belt are more difficult to pick out because they lack the imbricate zones of graptolitic shales. But careful mapping and fossil collecting allows the recognition of repeated zones in the Wenlock along with their boundary faults (Fig. 6.20, Kemp and White 1985, Kemp 1986).

The inclination of the major tract-bounding faults is sometimes low, but more often it is high-angle. On current interpretations (see p. 205) inclinations are variable because most originated as low-angle thrusts or slides and have since suffered rotation. Folding is often most intense in the imbricate zones of the major faults. Associated features of the zones are belts of intense disruption and brecciation. These belts have been ascribed to deformation before lithification and cleavage formation or alternatively to later deformation during continued thrust movements after folds have locked (Knipe and Needham 1986, McCurry 1987). The faulting is relatively brittle compared with the mylonitisation found in the Orlock Bridge Fault.

Minor strike faults

Minor strike or sub-strike faults occur in most sizeable exposures in the Southern Uplands. The high-angle faults may show normal or reverse throws but in view of the probable rotation of the blocks these designations may not be significant. The minor faults described by Knipe and Needham (1986) illustrate both their nature and possible origin (Fig. 7.2).

A distinct set of low-angle faults occur and form conjugate thrusts. They dip north-west and south-east and, cutting both folds and high-angle faults, are probably late in the deformation history.

Wrench faults

Wrench faults often form obvious topographic features because they tend to have wide zones of brecciation. The faults form a conjugate set with sinistrals, generally the more abundant, oriented approximately north–south and dextrals, west-north-west–east-south-east. Sinistral faults in the Langholm area have been called shatter belts. One, the Cue Sike, with a horizontal displacement of 1 km has disrupted strata up to 500 m on either side of the fault (Lumsden *et al.* 1967). The wrench system displaces major strike faults and many folds.

Iapetus Suture

North American faunas in the Girvan succession and European faunas in the Lake District have long been used to propose a wide Ordovician Iapetus Ocean although this view has recently been challenged (Wilson 1966, Williams 1972, but see Schallreuter and Siveter 1985, Freeman *et al.* 1988). Ultimate closure took place along the line of the Solway Firth forming the so-called Iapetus Suture (Phillips *et al.* 1976). The suture has been identified on deep seismic-reflection profiles and in magnetotelluric surveys (Brewer *et al.* 1983, Beamish and Smythe 1986). The surface expression of the suture is buried under the Carboniferous of the Northumberland Trough but in eastern Ireland it is identified as a 100 km wide zone occupied by an anastomosing system of late Caledonian sinistral faults. Precursors to this fault system may have originated in late

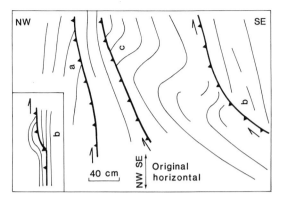

Fig. 7.2. Minor faults and associated folds. Faults interpreted originating as minor thrusts during accretion which were later rotated, as illustrated in the inset. (a) Part of a hanging-wall ramp; (b) a footwall ramp; (c) a folded footwall ramp (after Knipe & Needham 1986).

Ordovician during early phases of Iapetus closure (Murphy 1987).

Folding

A number of fold phases have been differentiated in the past on style and orientation but most phases are now generally regarded as episodes in a continuum of one main deformation along with minor subsequent folding (Kelling 1961, Rust 1965, Anderson and Cameron 1979, Knipe and Needham 1986, Knipe *et al.* 1988).

The most common folds (F1) seen in the field are syncline–anticline pairs which temporarily reverse the younging direction in steeply inclined beds (Figs. 7.3a, 7.4). The folds are intermediate in size (wavelengths about 5–50 m) and are often isolated although sometimes aggregated. Where the folds are concentrated the fold envelope may approach the horizontal but in many areas the sparsely distributed folds produce an inclined fold envelope.

Folds are asymmetrical in style with anticlines having steeply dipping northern limbs and shallow southern limbs; isoclines are relatively scarce. Vergence is towards the

Fig. 7.3. Styles of main-phase folds in the Southern Uplands: (a) impersistent syncline–anticline pair; (b) upright symmetrical anticline bifurcating; (c) anticline with variable axial surface; (d) steeply plunging fold affected by subsequent cleavage.

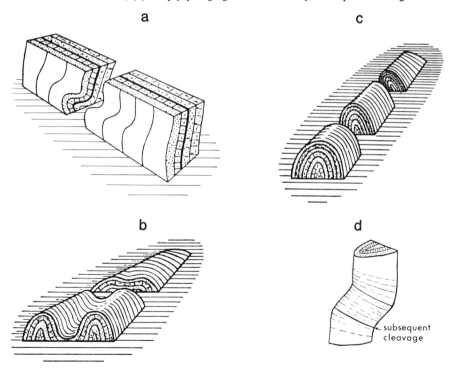

Fig. 7.4. Main-phase asymmetrical syncline showing hinge thickening and main cleavage. Hawick Rocks, Whithorn area. (Photograph by B. R. Rust.)

south-east. In the Hawick Rocks interlimb angles vary from 20° to 70° and folds tend to be upright with axial surfaces dipping generally between 60° and 90° to the north-west or south-east (Stringer and Treagus 1980). Hinges vary from angular to rounded mainly depending on lithology. Sandstones often show orthogonal thickening around hinges. Thickening in the hinges may be by a factor of three in lutites and two in arenites; locally Rust found a tenfold thickening in mudstones. Stringer and Treagus regarded the folds as flattened concentric and the strain as essentially oblate. Stretching both parallel and normal to the strike has been described by Anderson and Cameron (1979) from Northern Ireland. There has been no sys-

tematic investigation of this in the Southern Uplands but similar mineral lineations can be seen. In the Ettrick area the wavelength of folds is related to the thickness of the greywacke beds; folds involving greywackes are mostly concentric in style whereas siltstones and mudstones tend to form tight isoclines (Webb 1983).

Folds are often non-cylindroidal. Syncline–anticline pairs in steeply dipping sequences are often impersistent vertically as well as along the strike (Fig. 7.3a). Laterally folds may disappear or become multiple (Fig. 7.3b); the axial surface may change markedly in dip (Fig. 7.3c). Fold hinges are oriented generally north-east or south-west with minor variations due to impersistence of folds and more strikingly by rotation in the vicinity of actual or incipient wrench faults. Plunges are generally shallow and inclined towards both north-east and south-west so that many folds are periclinal. Adjacent fold pairs and even the complementary hinges of a fold pair may plunge gently in opposing directions.

For the past few decades, regional descriptions have emphasised the predominance of beds younging to the north-west and associated south-east-verging folds. This is still broadly true for most of the region but lately, a zone, a few km across strike has been recognised in Silurian blocks in the Central Belt. In this zone, younging is mostly to the south-east and north-west-verging folds are associated with northerly directed thrusts (McCurry and Anderson 1989; and see p. 205).

Steeply plunging folds

Folds with vertical or near-vertical fold-hinges have been observed in a number of regions but particularly in the south-west in Hawick and Wenlock rocks. Some of the folds are overturned and downward-facing. They may be concentrated in strikewise zones such as that at Brighouse Bay (Craig and Walton 1959) near Kirkcudbright and at Burrows Head (Stringer and Treagus 1980). Similar concentrations have been reported from the Rhinns of Galloway and Wigtown (Barnes *et al.* 1987). But other folds may be sporadic in their distribution (Rust 1965, see also Dearman *et al.* 1962).

The folds are similar in style to those with shallow plunges and this led Stringer and Treagus (1980) to reject Rust's (1965) contention that the folding was a separate, later phase. They regarded the arrangement as being due to disruption and rotation of early folds during thrusting along the major faults; some strike-slip movements may also be involved. More recently plunge variation in amount and orientation has been ascribed to differential movement and lateral ramping during accretion of soft-sediment slippage (Knipe and Needham 1986). Soft sediment deformation caused intense and complicated folding at Coldingham Bay (Fig. 7.5, Casey 1983).

Cleavage

S1 cleavage is most strongly developed in the Southern Belt; it is often very difficult to detect in the Northern Belt. In the south-west it is very marked and closely spaced in the lutites where it reflects a fabric formed by the combined

Fig. 7.5. Low-level aerial photograph of intense soft-sediment folding at Coldingham Bay. (Photograph by J. Allen.) Old buildings give scale.

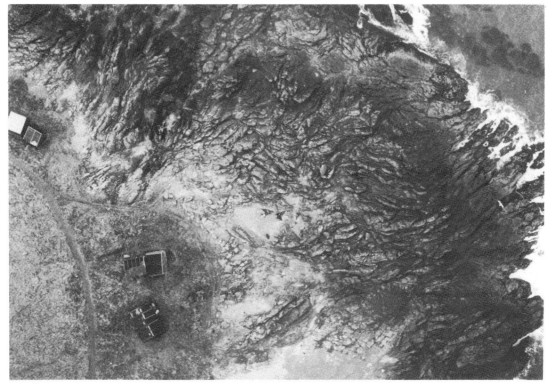

effects of pressure solution, mineral orientation and crenulation (Stringer and Treagus 1980). It is often refracted through graded arenites. In the fold profile the cleavage is divergent in the sandstones and convergent in the lutites. But the most interesting feature of the cleavage is that it is not coincident with axial surfaces. Cleavage has a strike which is displaced clockwise to the axial surface and the hinge (Fig. 7.6). This deviation of cleavage can be expressed as the angle between the bisector of the cleavage fan and the axial surface. The angle is usually smaller than 10° in mudstones and up to 25° in arenites. This arrangement could arise by folding and cleavage development in beds already inclined at an angle to the maximum direction of shortening (Stringer and Treagus 1980) or by the cleavage developing on folds initiated with axes already slightly oblique to maximum shortening but within the XZ-plane in a transpressive regime (Knipe and Needham 1986). Sinistral transpression was first effective in the Llandovery-age tracts in Ireland corresponding to a position in the middle of the Central Belt of the Southern Uplands. Sinistral movements continued episodically during the deformation of early pre- and post-S1 vein sets and such movements are recorded in the Orlock Bridge and Southern Uplands Faults as noted above. The final expression of the transpression was in the widespread wrench faulting which continued beyond the intrusion of the dykes around 400 Ma (Anderson 1987).

Subsequent phases of folding

Late phases of folding which cause minor flexuring of early folds and crenulation of S1 bedding and cleavage are commonly seen only in blocks of Hawick and Wenlock rocks. Folds are generally small (less than 10 m across) and may cause the formation of box or downward-facing folds depending on their position relative to F1 folds. Hinges plunge at low angles to the north-east and south-west; axial surfaces dip at low angles to north-west and south-east. Like those in Northern Ireland F2 folds are south-verging (Anderson and Cameron 1979, Knipe and Needham 1986). F3 folds have northerly vergence and appear to correspond to Knipe and Needham's second group of late folds which

tend to have shallower axial surfaces dipping south-east. They are associated with similarly inclined, northerly directed late thrusts and have tension gashes filled with quartz.

The final widespread phase is kink-bands. These small angular folds with steeply plunging axes are bounded by shear surfaces corresponding to sinistral or dextral wrenches (Fig. 7.7, Anderson 1968). The folds are particularly associated with thin-bedded flags. Although both sinistral and dextral kink-bands occur the latter are much more abundant.

General structure

The inclination of the axial surfaces of the main fold-phase had a profound effect on early interpretations of the general structure of the region (Lapworth 1889). Observations suggested that axial surfaces had a fan and an inverted fan arrangement around two 'lines' running north-east–south-west, respectively the Leadhills line in the north and the Hawick line in the south. The general structure was therefore an anticlinorium with the Leadhills line as the axis and a synclinorium with limbs on either side of the Hawick line.

Recent work has partly confirmed the orientation of axial surfaces (Fig. 7.8). In the Kirkcudbright area, for example, axial surfaces may dip as low as 40° to the south-east. But the recognition of the dominance of major faults has negated the anti-synclinorium model. The precise significance of the axial-surface dip is not clear; in general it would seem to be determined by the amount of rotation suffered by individual blocks.

The structure in the Ordovician section of the Rhinns of Galloway is dominated by belts made up almost exclusively of steeply dipping or overturned beds with minor anticlinal–synclinal pairs, as at Corsewall and the northern part of the Portpatrick block. These steep belts are separated by zones where the fold envelope is flat as in the Kirkcolm–Galdenoch block and the southern part of the Portpatrick block (Fig. 7.8). Elsewhere in the Northern Belt flat belts have not been recognised so clearly, perhaps because of the poorer exposure inland. Various inclinations have been

Fig. 7.6. Sketch illustrating the relationship of the main cleavage to the main folds (Hawick Rocks, Stringer & Treagus (1980)). Arrows and figures refer to the direction and amount of plunge of fold hinges. Cleavage in dashed lines, more closely spaced in mudstones.

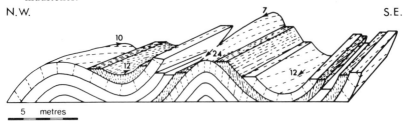

Fig. 7.7. Subsequent phases of folding: (a) Crenulation cleavage. Note that transverse ripples are also present on the main bedding surface. Hawick Rocks, Kirkcudbright. (b) Fold with NW dipping (left in photograph) axial surface developed on gently (SE) dipping southern limb of main-phase anticline. Hawick Rocks, Kirkcudbright. (c) Dextral kink-band in Hawick rocks, Kirkcudbright.

envisaged for fold envelopes in the Central and Southern Belts (Springer and Treagus 1981, 1983; Craig and Walton 1959, 1983).

Deformation synthesis

The structures of the main phase of the Southern Uplands may be interpreted as portraying a continuum of deformation in a developing accretionary prism (Fig. 7.9, Knipe and Needham 1986). Although it is often difficult to

separate structural episodes, at least in principle it is possible to pick out downslope gravity-driven structures. These are essentially slumps in unconsolidated sediments. As sedimentation and subduction proceeded, lithification was followed by low-angle thrusts, ramping and the formation of folds mostly with axes at right angles to the thrust direction but some with varying orientations dependent on lateral ramping. The main imbricate zones and tracts were delineated in this way but underthrusting was also accompanied by much minor faulting in the same

Fig. 7.8. Cross-sections, approximately NW–SE, through the Southern Uplands illustrating the main structural features (based on Casey 1983, Barnes *et al.* 1987). (a) Rhinns of Galloway–Newton Stewart (a1), Kirkcudbright (a2) areas. (b) West Nithsdale–Moffat–Hawick (b1) areas. (c) Berwickshire coast.

Tracts numbered as in Fig. 6.13.

Faults in Rhinns of Galloway: GA, Glen App; KT, Killantringen; OB, Orlock Bridge; SP, Salt Pans Bay; D, Drumbredan Bay; PL, Portlogan; CL, Clanyard Bay; NK, Knick of Kindram; TB, Tarbet.

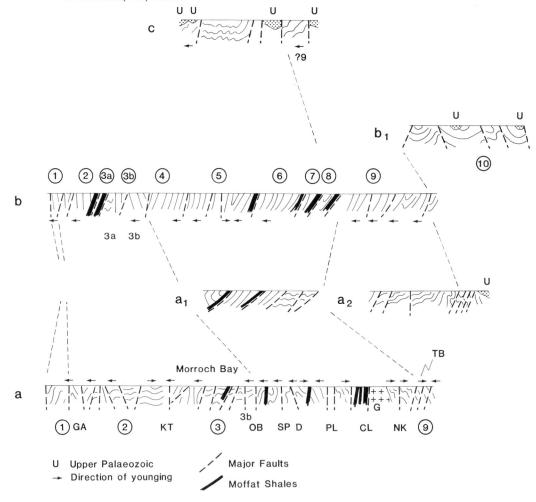

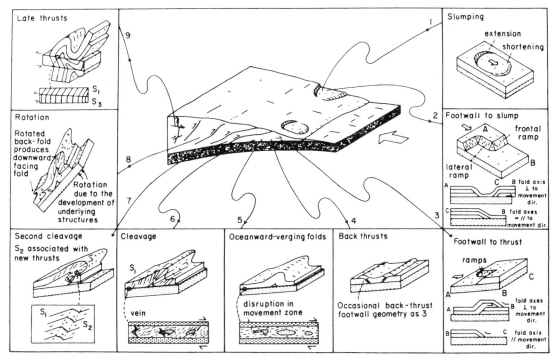

Fig. 7.9. Deformation features of the Southern Uplands interpreted in terms of a developing accretionary wedge (from Knipe & Needham 1986).

style. Subsequent rotation as tracts were accreted produced the dominant north-younging beds and south-east-verging folds.

The zone in the Silurian with south-younging beds and north-verging folds represents an interlude of landward obduction with northerly directed thrusts. The change in style is ascribed to increased rates of sedimentation in mid-Llandovery times and high pore-pressures associated with the Moffat Shales. Analogy is drawn with the Washington–Oregon margin which has suffered landward thrusting of accretion sediments after a phase of rapid sedimentation (McCurry and Anderson 1989).

The geometry of the accretionary prism model is dominated by underthrusting down to the north-west. On the other hand, the same geometry with overthrusting to the south-east would result from a continuum of deformation in a thrust-stack system developed from a prograding fan (Stone *et al.* 1987) or a successor basin (Murphy and Hutton 1986). These alternative models are discussed below.

Midland Valley

Structures in the Midland Valley Inliers present some contrasting styles. Girvan and the Pentland Hills bear comparison with the Southern Uplands in having mainly steeply dipping northward-younging beds. Folds are asym-

metrical with steep northern limbs and shallow southern limbs on the anticlines and axes tend to run north-east although in the Girvan area a prominent set of folds in the Ardwell beds have axes diverted to a more northerly direction (Williams 1962). Additionally the folds in Girvan have been linked with thrusting which is directed north-westwards (Fig. 7.10). A strong unconformity exists between the Silurian beds and Old Red Sandstone sequences in both Girvan and the Pentlands. By contrast folding in the Craighead, Lesmahagow and the Hagshaw Hills is relatively gentle. These inliers are anticlinal with relatively low dips in the first two; the Hagshaw anticline is markedly asymmetrical with a subvertical southern limb.

Post-Lower Palaeozoic faulting has affected Old Red Sandstone and Carboniferous rocks in the inliers. The faulting is normal in throw and, striking north-easterly, may represent reactivation along Caledonian fractures. Examples of these faults some of which exerted active control on Carboniferous sedimentation are the Kerse Loch and Inchgotrick faults (see Chapter 10).

Palaeogeography and Palaeotectonics

Early Ordovician

The limited exposures of early Ordovician rocks and their structural complexity mean that a satisfactory model has

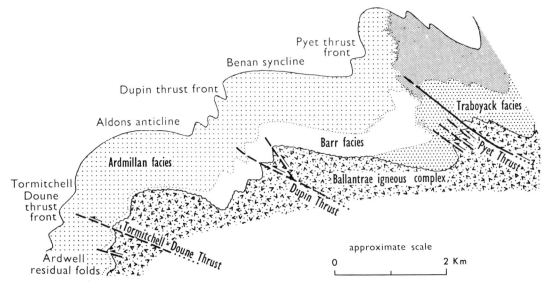

Fig. 7.10. Folding in the Girvan area. Idealised section inferred relationship between the principal folds and thrusts during the early stages of deformation (after Williams 1959).

still to be developed. There is agreement that the rocks of the Ballantrae Igneous Complex are of oceanic affinity but the geochemistry of lavas and intrusives offers alternative models. Accumulation essentially at an oceanic ridge would account for the closeness of all the ages of the various rock groups (Lambert and McKerrow 1976). On the other hand apparent lava thicknesses, intensity of metamorphism and juxtaposition of many varied rock masses, some with intra-plate geochemistry, point to the assembling by collision of a number of oceanic islands (Barrett *et al.* 1982). Any model involving open oceanic conditions may be too simple to account for the complexity of the Ballantrae area. Moreover some lavas and tuffs appear to have accumulated in shallow waters; and the geochemistry of some lavas indicates an arc origin. More eclectic models are necessary. Marginal basins and volcanic arcs would provide thick volcanic and volcaniclastic sequences with varied geo-chemical signatures in close proximity, local shallow water situations, relatively rapid development and a narrow spread of ages (Fig. 7.11).

North of Ballantrae, across the area of the present Midland Valley but at an unknown distance and position in Arenig times, lavas, cherts and black shales accumulated in a basin, possibly similar to that in the south. These now form Rock Assemblage 2 of the Highland Border Complex (see Fig. 6.7). The underlying carbonated serpentinite may compare in age and formation with the serpentinites of Ballantrae.

To the south of Ballantrae, again at an unknown distance and position, the cherts and mudstones of Raven Gill accumulated on what might be oceanic crust although

exposures are pitifully small. Other cherts and basalts in the Northern Belt representing early Ordovician ocean floor, developed through the Llanvirn and in some cases perhaps into the Llandeilo.

Later Ordovician and Silurian

From late Llandeilo onwards through the Silurian the geography and tectonics although clearer than in the early Ordovician is still subject to considerable debate.

Source rocks

The types of clasts present in conglomerates and sand-stones of Girvan and the Northern Belt denote an spilitic complex with intrusives comparable at least in part with the Ballantrae Igneous Complex (BIC) together with a plutonic and metamorphic basement (Fig. 7.12). Spilitic and acid igneous fragments are present throughout the successions. Metamorphic and andesitic fragments tend to be more restricted stratigraphically. As noted in Chapter 6, the Metaclast, Acid clast and Basic clast Petrofacies form the basis for stratigraphic divisions. The supply of andesitic debris from lavas or more probably pyroclasts was inter-mittent, a result of episodic volcanicity and/or varying sources due to contemporaneous strike-slip movement. Some of the basic clasts with pyroxene and hornblende may have been from intra-basinal sources like the Bail Hill volcanics (McMurtry 1980) but most of this detritus has the geochemical features of calc–alkaline arc volcanism (Kassi 1984).

Judging from the amount of spilitic clasts, the Ballantrae Igneous Complex must represent a tiny fragment of a much more extensive terrane. In addition to the basic rocks there must also have been in this source extensive outcrops of acid lavas and minor intrusives. None of these are currently exposed. They may at least in part represent the super-structure of the complexes which provided the granite clasts.

Some of the granitic fragments compare with the trond-jemitic granite in the Ballantrae Igneous Complex, but most do not. Some are microcline-bearing, others are either foliated or unfoliated. The unfoliated clasts from the conglomerates in the Girvan succession range in age from 595 to 450 Ma (Bluck 1983) and they bear both biotite and hornblende. Granitic clasts have also been dated from three localities in the Southern Uplands, viz. Corsewall Point (Tract 1), Glen Afton (Tract 2) and Pinstane Hill (Tract 4). Groups of clasts with similar mineralogy and chemistry

were assumed, rather questionably, to represent individual plutons; isochrons were constructed from the groups. Postulated ages from Northern Belt clasts (Tracts 1 and 2) range from 1,265 to 475 Ma while Tract 4 clasts gave 458 ± 26 Ma. It is suggested that no comparable granites with the older ages occur in Britain but there are some in Newfoundland. If the latter were the source of the Southern Upland debris then some 1,500 km strike-slip displacement took place between Caradoc and Lower Devonian times (Elders 1987).

The most common and widespread metamorphic clasts are quartzites but at certain horizons like the Kirkcolm Metaclast Division various schist and gneissose fragments from a regionally metamorphosed terrane are found. These have been compared with Dalradian and Grenvillian metamorphics but it now appears that the Dalradian terrane was not emplaced in its present position until Devonian times (Bluck 1984) and Halliday *et al.* (1985)

Fig. 7.11. Alternative models for the Ballantrae Igneous Complex, involving obduction (A) of volcanic islands a and b; (B) of oceanic crust; and (C) of a small marginal basin (from Bluck 1985 based on Barrett *et al.* 1982; Lambert & McKerrow 1976; Bluck *et al.* 1980; Bluck 1981).

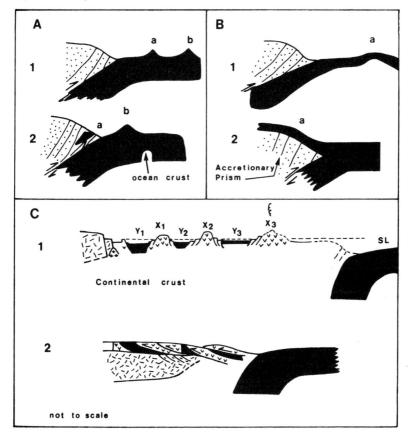

found no evidence of Grenville parentage in the granites in Scotland. This leaves the possibility that the clasts were derived from a regional–metamorphic complex associated with the foliated-granite source-rocks. The oldest unfoliated granite clasts are around 595 Ma, so that the associated complex would be older than any Dalradian metamorphics (Bluck 1983).

Early Silurian sees a recurrence of the basic-clast petrofacies, with fragments of pyroxene and hornblende-bearing andesite. This is very obvious in the north-east for example in the 'Pyroxenous Group' of the Peebles area but rather less clear in the south-west where the Basic-clast and Acid-clast petrofacies (BCP and ACP) tend to be inter-bedded. Metamorphic contributions are especially marked in the Peebles 'Garnetiferous Group' but again this distinctive group loses its identity towards the south-west where it merges with ACP of the 'Intermediate Group'. Hawick and Wenlock rocks have few basic volcanics. Rudites from the Wenlockian Gipsy Point member are highly quartzose with vein quartz and quartzitic clasts, along with granitic fragments including some large microperthitic grains. Especially distinctive are the abundant limestone clasts of Silurian age some enclosing fragments of shells and colonial corals. Inshore, shallow-water deposits evidently supplied some detritus along with recycled clasts from earlier trench sediments. The earlier sediments north of the Hawick boundary may have some recycled material but the fresh-ness of ferromagnesian grains in the basic-clast beds and the abundance of garnets in other formations points to first-cycle sources.

In the Midland Valley area the conglomerates show variations in source type. Fine-grained basic-igneous clasts are dominant in the lower beds which include the Igneous Conglomerate. Acid volcanic clasts from the basal Craigs-kelly Conglomerate have geochemical features which sug-gest derivation from island-arc volcanics (Loeschke 1985). In the Igneous and Quartzite conglomerates in the Hag-shaw Hills inlier, on the other hand acid volcanic and plutonic clasts showed some affinities with an intra-plate setting as well as others which were derived from an island arc or active continental-margin. Both conglomerates were derived from the direction of the Southern Uplands but the calc–alkaline rocks as well as other granite, gneiss and schist fragments are unknown in that region. The per-

Fig. 7.12. Composition of clasts from conglomerates in the Girvan area. Sediments successively overstepped the Ballantrae Igneous complex over a series of faults. Younger conglomerates show increasing granite–diorite clasts, which tend to vary in age in response to continuing magmatic activity and erosion (after Bluck 1983).

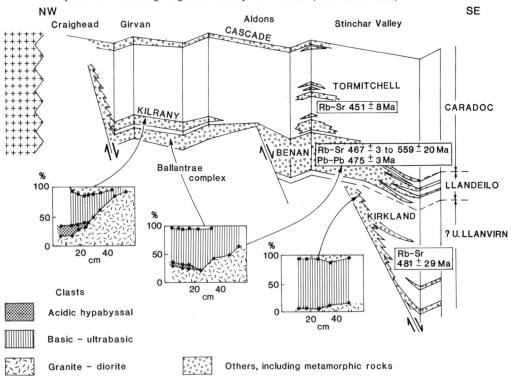

alkaline rhyolitic clasts compare with those of the Tweed-dale Lavas (p. 179) but the latter have an insignificant outcrop area. Moreover the clasts appear to be first- rather than polycyclic (Heinz and Loeschke 1988).

The uppermost Greywacke Conglomerate, as its name implies is mostly of clasts of that arenite. The greywackes suggest derivation from a Southern-Uplands-type parent and the same source might have supplied both igneous and quartzose components of the lower conglomerates since they are common in the Ordovician successions. But some caution is necessary in accepting this solution in view of the nature of the acid clasts discussed above and in so far as the greywacke clasts cannot be matched precisely with any from the Southern Uplands (Bluck 1983).

Sediment dispersal

The majority of current structures in the Ordovician of Girvan and the Northern Belt show longitudinal flow either from the NE or SW with the former predominating (Table 6.3). In addition the evidence from sole-markings indicates north-westerly currents. This, taken together with fans in the Northern Belt radiating from the same direction suggests dominant sources in that direction. Longitudinal flow, NE or SW, would be the result of the deflection of bottom-hugging, mostly turbidity currents, along the length of the trough; some of these flows may have been of distant derivation (Fig. 7.12). Observations by Kelling (1961) and Welsh (1964) suggesting currents from the SE were, until recently, regarded as anomalous and insignifi-cant. Investigations around Loch Ryan and inland near Barrhill, however, have revealed an association of basic clast petrofacies of the Galdenoch Fm. with SE currents (Stone *et al.* 1987). In addition, in the uppermost part of the Kirkcolm Fm. there is an interdigitation of quartzose, Kirkcolm greywackes with basic-clast beds, the former showing currents from the north and the latter from the south. Some workers infer (p. 211), therefore, that two sources were affecting sedimentation during the late Ordovician, a quartz-rich metamorphic–plutonic region to the NW and an active volcanic terrane to the SE.

In the Southern Uplands, Silurian dispersal mechanisms and directions continued as in the Ordovician with flow from north-west, south-west and especially from the north-east. Even in the extreme south on the shores of the Solway, the Gipsy Point conglomerates show flow from the north-west. Exceptionally, a significant flow from the south-east has been demonstrated recently in the Upper Llandovery tracts in the Rhinns of Galloway (McCurry 1987).

Sand-rich fans represented generally by channel and mid-fan lobe sectors were formed mainly by turbidity currents. Occasional debris flows are found locally as in the Duniehinnie Member in the Rhinns of Galloway (Fig. 7.13).

In the Midland Valley inliers the lower turbidite beds have E and ENE currents in the Pentlands while in the Lesmahagow and Hagshaw areas the flows came from the south. Later current directions tend to be from the south-easterly quadrant. The latter are associated with con-glomerates and other sediments interpreted as terrestrial in an environment involving sheet floods, stream flow and lagoonal or lacustrine conditions. The conglomerates formed alluvial fans spread out from point sources which, judging from size and shape of clasts, were situated at distances beyond the present Southern Uplands (McGiven 1968, Bluck 1983).

Palaeogeography and Palaeotectonics

The geographical setting for the Girvan sediments in the Ordovician is perhaps the least doubtful of all the sequences. Against a step-faulted margin, the fan deltas of the Kirkland and Benan Conglomerates alternated with abandonment phases during which the finer sediments and carbonates accumulated (Williams 1963, Ince 1984). The Stinchar Limestones is one extremely shallow, inshore sediment and as transgression proceeded northwards similar conditions produced the Craighead Limestone.

Interpretations of the Southern Uplands are much more controversial. McKerrow *et al.* (1977) published the innovative accretionary prism model which for the first time integrated palaeontological and structural evidence with investigations of modern active subduction zones. The model has stood virtually unchallenged for almost a decade. As further field and laboratory work has continued in Ireland as well as Scotland, more apparently contradic-tory evidence has come to light. In the following sections the different models which have been proposed will be outlined and their relative merits assessed.

Accretionary prism

One interpretation sees the Girvan sequences forming against the southern margin of the Laurentian continent. To the south-east the sandy sequences now forming the greywacke successions of the Northern Belt accumulated on oceanic crust in a trench. Progressive subduction northwards produced a growing accretionary prism (McKerrow *et al.* 1977). Detritus was spread mainly by bottom-hugging turbidity currents from fans along the north-west margin and diverted longitudinally along the trough by an outer rise on which sedimentation was restricted to the graptolitic muds of the Moffat Shales. The Barren Mudstones of the Hartfell Shales are exceptional in recording a period of aerobic conditions possibly linked with the regression due to the Ashgillian glaciation. Another exception is found at Ettrickbridgend where a channel, comparable perhaps to those of the present-day which extend onto abyssal plains (Casey 1983).

Fan complexes are situated at intervals across the area. The Corsewall–Glen App fan is accompanied to the north-east by other fans (Fig. 7.13). Using the widths of channels

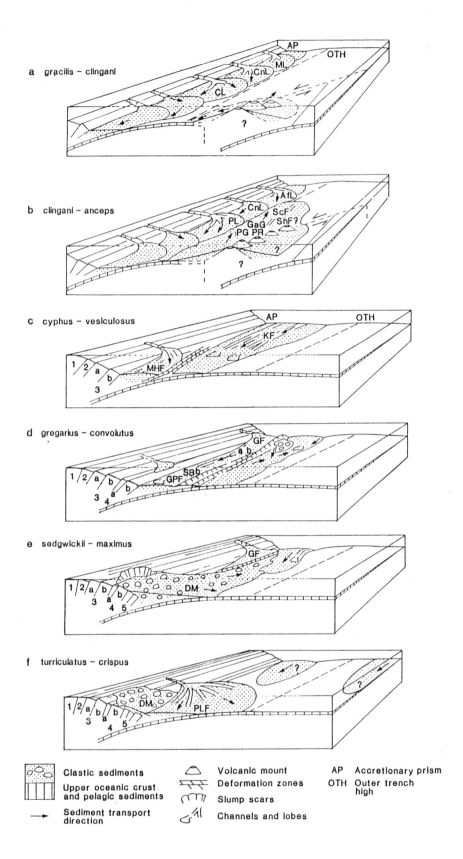

a grailis – clinganl

AP
OTH
CnL
ML
CL
?

b clinganl – anceps

AfL
CnL
ScF
PL
GaG
ShF?
PG PR
?
?

c cyphus – vesiculosus

AP
OTH
KF
1
2
a
b
3
MHF

d gregarius – convolutus

GF
a b
SBb
1 2 a
b
GPF
b
3 4
a

e sedgwickil – maximus

GF
1 2 a b
b
a
DM
3
4 5

f turriculatus – crispus

?
1 2 a b b
a
DM
3
PLF
a
4 5
?

Clastic sediments   ⌂ Volcanic mount   AP Accretionary prism
Upper oceanic crust   Deformation zones   OTH Outer trench
and pelagic sediments   high
→ Sediment transport   Slump scars
   direction   Channels and lobes

observed in the Corsewall, Afton, Carsphairn and Shinnel-Scar rocks, Kelling *et al.* (1987) have estimated a fan radii of between 150 and 250 km. Other notable features are olistostromic masses of ophiolitic material near Glen App in Tract 1 and remnants of sea-mounts at Bail Hill and Wrae Hill.

By Silurian times the Ordovician prism had become land, so separating the Midland Valley area of sedimentation from that of the Southern Uplands. Recycled debris was then shed to the north and south. The prism continued to grow and closure of the Iapetus Ocean did not take place until the Ludlovian when the first Southern-Uplands-like debris reached the Lake District (Barrett *et al.* 1982).

Objections to this reconstruction on the grounds of volcaniclastic debris associated with flows from the southeast are countered by reference to examples in the Nankai Trough off Japan where longitudinal currents are diverted landwards by local topography. Even if the south-easterly currents do indicate a source to the south, this might have been made up of oceanic islands. Moreover, the accretionary prism model is also modified in the light of the pervasive prehnite–pumpellyite metamorphism. The PT conditions responsible for the metamorphism are better explained by underplating rather than simple frontal accretion (Leggett 1987).

Accretionary prism with tectonically displaced arc

The ferromagnesian minerals of the basic clast greywackes in both the Ordovician and the Silurian from *clingani* to *cyphus* zones are of arc rather than ocean-island origin. Accepting this and that the volcanic debris was derived from the south led to the proposal that a portion of a volcanic arc (the Gowna–Portpatrick arc) was faulted into position and formed an active source for a relatively short time (Fig. 7.13a, b). Strike-slip movement progressed from the south-west so that the northward supply of clasts began in Scotland later than in Ireland. Continued movement and/or cessation of volcanicity and/or erosion through to a plutonic/metamorphic basement ended the supply of basic clasts (Kelling *et al.* 1987).

Midland Valley Landmass

Consideration of the Highland Border Complex requires an alternative view of the Laurentian margin to that envisaged in the simple accretionary-prism model. Rock Assemblage 3 indicates oceanic conditions through to Llandeilo times. The environment is best regarded as a marginal basin north of a volcanic arc (Curry *et al.* 1984). After deformation and low-grade metamorphism

Fig. 7.13. Palaeogeography and palaeotectonic setting of Southern Uplands and Ireland in later Ordovician times (after Kelling *et al.* 1987). (a) gracilis–clingani. CL, Corsewall Group 'lobe'; CnL, Carsphairn 'lobe'; ML, Marchburn Fm. 'lobe'. (b) clingani–anceps. Afl, Afton 'lobe' (Blackcraig Fm.); CnL, Carsphairn 'lobe'; GaG, Galdenoch Group; PG, Portpatrick Fm.; PL, Portslogan 'lobe' (Kirkcolm Fm.); Pr, Portayew Fm.; ScF, Scar Fm.; ShF, Shinnel Fm. (c) cyphus–vesiculosus. KF, Kilfillan Fm.; MH, Money Head Fm. (d) gregarius–convolutus. GF, Garheugh Fm.; GPF, Grennan Pt. Fm. (e) sedgwickii–maximus. DM, Duniehinnie Member; GF, Garheugh Fm. (f) turriculatus–crispus. PLF, Port Logan Fm. 1, 2, a, b, etc. mark earlier accreted tracts.

Assemblage 4 was deposited unconformably, possibly in shallower-water conditions. The mixture of acid clasts with some spilites in the conglomerates and sandstones bears some similarities to those in contemporary sediments in southern Scotland and both sequences may have been derived from a Midland Valley landmass (Bluck 1984).

The Midland Valley Landmass would then form the faulted margin, against which the Girvan sediments accumulated (Figs. 7.11, 7.14). Basin development and

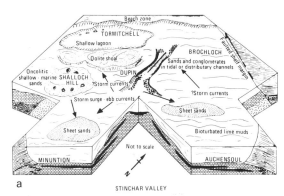

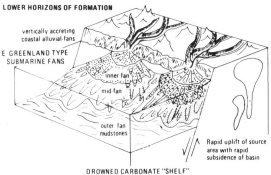

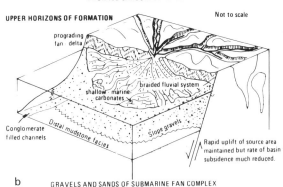

Fig. 7.14. Palaeographic reconstructions from the Girvan area (from Ince 1984). (a) Stinchar Limestone – facies types and relationships. (b) Benan conglomerate.

sedimentation went on *pari passu* with the emplacement of the later (460 Ma) granites in the sourcelands. There is some indication that the granite clasts become younger from the Kirkland Conglomerate (? U. Llanvirn) through the Benan to the Kilrany (Caradoc) Conglomerate. This presumably reflects progressive emplacement and unroofing of the granites (Longman *et al.* 1982). Source rocks as distant as the Aberdeenshire granites (Yardley *et al.* 1982) seem unlikely. Clast-size alone suggests minimal transport and the granites are hornblende-bearing rather than having two micas like those in Aberdeen. These considerations along with the evidence from the Highland Border Complex point to the Midland Valley Landmass as the source. The clast petrography prescribes it as a continental-margin arc rather than an ocean-island arc (Bluck 1983).

In this model the Girvan succession accumulated in a fore-arc basin. If the Girvan rocks are proximal fore-arc basin deposits, and the Northern Belt greywackes are maintained as trench sediments deformed in an accretionary prism, has faulting closed a gap of perhaps more than 60 km(?) (Bluck 1985). Some features suggest perhaps not. Parts of the Tappins Group in Tract 1 can be correlated across the Stinchar Valley Fault with, for example, Benan horizons, without striking changes in lithology apart from fining in grain-size. The size of the boulders in the Glen App, Finnarts and Corsewall conglomerates compares with the Girvan beds. If the latter were proximal, the former were similarly close to source. Lithological correlation of the Corsewall Fm. with the Marchburn Fm. south of the Southern Uplands Fault may mean that the gap, if one existed, may have been closed along that fault.

Whatever the distance between the Ordovician location of the Girvan area and the Northern Belt, sedimentation in Tracts 1 and 2 was dominated by south-easterly progradation of fans such as the Corsewall fan in Llandeilo–early Caradoc times. In time these were replaced by sediment, much of which flowed axially along the trench southwestwards from fans like the Afton (Fig. 7.13).

On this interpretation Silurian conditions were controlled by the emergence of land separating the main Southern Uplands from a basin of sedimentation south of the Midland Valley landmass. The sequences in the present inliers accumulated in this basin, perhaps in an inter-arc situation; there is evidence that lowermost Devonian rocks against the Highland Border were derived from a volcanic terrain which divided the Midland Valley area into two basins. Closure of Iapetus was accomplished by late Silurian, and compression thrust the Southern Uplands northwards against the fore-arc rocks at Girvan. This gains some support from seismic evidence of a shallow, southward dipping thrust as envisaged for the Southern Uplands (Fig. 6.24, Hall *et al.* 1984); on the other hand the seismic velocities have also been interpreted as indicating not a thrust but a change from a prehnite–pumpellyite metamorphic grade to greenschist facies (Oliver and McKerrow 1984).

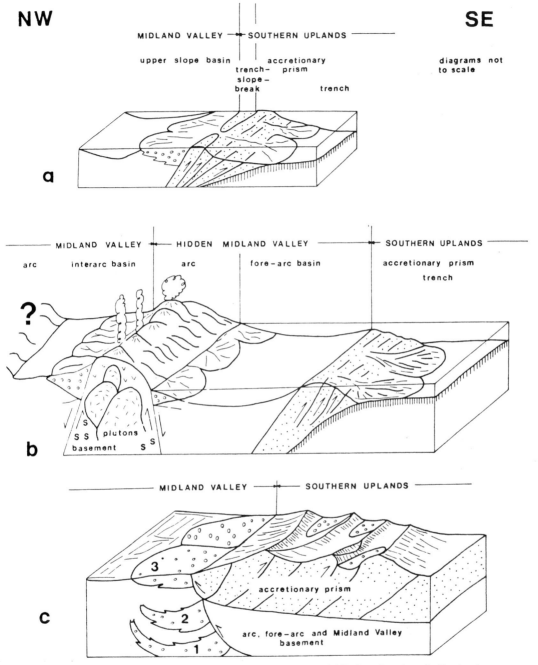

Fig. 7.15. Alternative interpretation of Ordovician and Silurian situations in Scotland (from Bluck 1983). (a) Accretionary prism model with sedimentation against Laurentian continent (Leggett 1980). (b) In Silurian times sedimentation in Midland Valley area as an upper slope basin against an emergent trench/slope break. (c) Midland Valley Landmass bordering Highland Border Basin in early Ordovician. Silurian in Midland Valley as inter-arc basin and accretionary prism at unknown distance to the South.

a

Late Ordovician

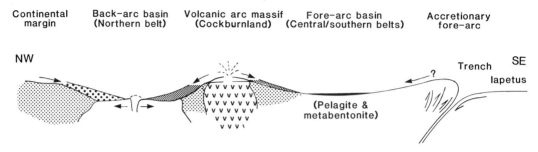

End Ordovician – Wenlock

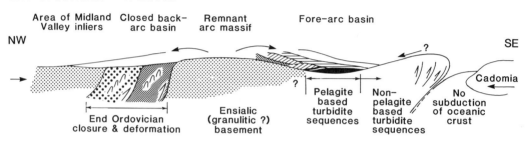

Late Silurian

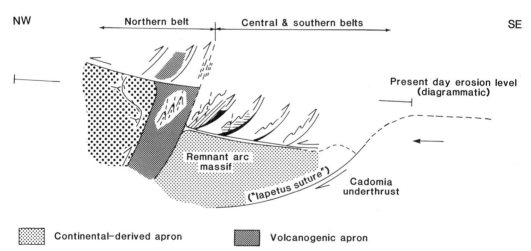

▨ Continental–derived apron   ▨ Volcanogenic apron

Fig. 7.16a. Models of the Southern Uplands sequences involving a back-arc basin with arc between Northern and Central Belts. Island arc in a position between Northern and Central Belts; deformation and closure of back-arc basin at end-Ordovician–Silurian sedimentation in fore-arc basin; closure with thrusting of Silurian sequence over remnant arc (after Morris 1987).

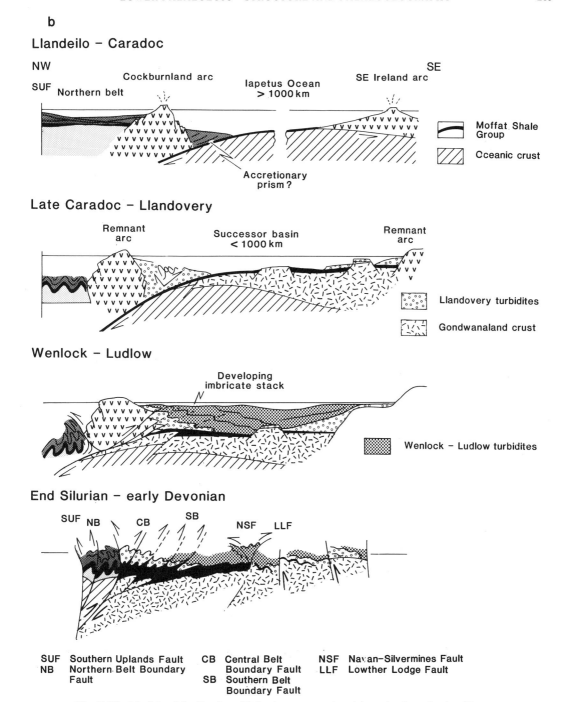

b

Llandeilo – Caradoc

NW

SUF Northern belt

Cockburnland arc

Iapetus Ocean
> 1000 km

SE Ireland arc

SE

Accretionary
prism?

Moffat Shale
Group

Oceanic crust

Late Caradoc – Llandovery

Remnant
arc

Successor basin
< 1000 km

Remnant
arc

Llandovery turbidites

Gondwanaland crust

Wenlock – Ludlow

Developing
imbricate stack

Wenlock – Ludlow turbidites

End Silurian – early Devonian

SUF NB   CB   SB   NSF LLF

SUF	Southern Uplands Fault	CB	Central Belt	NSF	Navan–Silvermines Fault
NB	Northern Belt Boundary Fault		Boundary Fault	LLF	Lowther Lodge Fault
		SB	Southern Belt Boundary Fault		

Fig. 7.16b. Models of the Southern Uplands sequences involving a back-arc basin with arc between Northern and Central Belts. As in (a) with closure of Iapetus at end of Ordovician; Silurian sedimentation in a 'successor' basin; ultimate deformation by end-Silurian transpressive movements (after Hutton & Murphy 1987).

LLANDEILO – EARLY LLANDOVERY *(N. gracilis – C. cyphus)*

Basin eventually filled by overlap from NW
although fans interdigitate at least during
N. gracilis **and** *P. linearis* **zones**

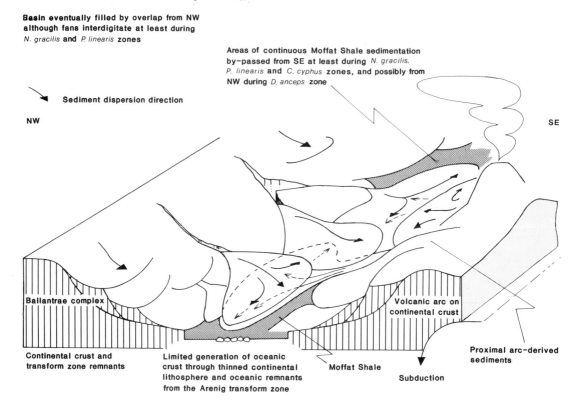

Areas of continuous Moffat Shale sedimentation
by-passed from SE at least during *N. gracilis,*
P. linearis and *C. cyphus* zones, and possibly from
NW during *D. anceps* zone

Sediment dispersion direction

NW   SE

Ballantrae complex

Continental crust and
transform zone remnants

Limited generation of oceanic
crust through thinned continental
lithosphere and oceanic remnants
from the Arenig transform zone

Moffat Shale

Volcanic arc on
continental crust

Subduction

Proximal arc-derived
sediments

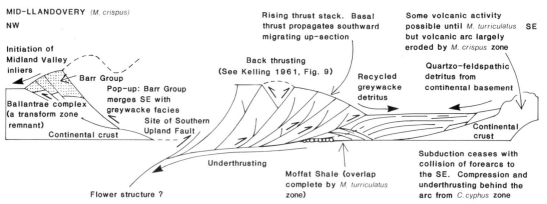

MID–LLANDOVERY *(M. crispus)*
NW

Initiation of
Midland Valley
inliers

Barr Group

Ballantrae complex
(a transform zone
remnant)

Continental crust

Pop-up: Barr Group
merges SE with
greywacke facies

Site of Southern
Upland Fault

Flower structure ?

Underthrusting

Rising thrust stack. Basal
thrust propagates southward
migrating up-section

Back thrusting
(See Kelling 1961, Fig. 9)

Recycled
greywacke
detritus

Moffat Shale (overlap
complete by *M. turriculatus*
zone)

Some volcanic activity
possible until *M. turriculatus* SE
but volcanic arc largely
eroded by *M. crispus* zone

Quartzo-feldspathic
detritus from
continental basement

Continental
crust

Subduction ceases with
collision of forearcs to
the SE. Compression and
underthrusting behind the
arc from *C. cyphus* zone

Fig. 7.17. Model with arc located within the Central Belt. Volcanic arc on continental
crust supplied sediment into back-arc basin until late Llandovery when continued
thrusting carried deformed sediments over the remnant arc. Hawick and Wenlock
sediments recycled from earlier sequences. Thrusting essentially southwards first over the
arc remnants and ultimately over under-thrust Cadomia (Lake District) continental crust
(after Stone *et al.* 1987).

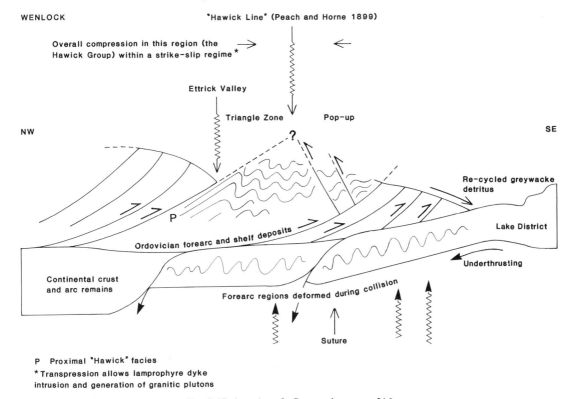

WENLOCK

"Hawick Line" (Peach and Horne 1899)

Overall compression in this region (the →
Hawick Group) within a strike–slip regime * ←

Ettrick Valley

Triangle Zone   Pop-up

NW   SE

?

Re-cycled greywacke
detritus

Lake District

P

Ordovician forearc and shelf deposits →

Underthrusting

Continental crust
and arc remains

Forearc regions deformed during collision

Suture

P   Proximal "Hawick" facies
* Transpression allows lamprophyre dyke
intrusion and generation of granitic plutons

Fig. 7.17. (*continued*). See caption on p. 216.

Southern Uplands as a back-arc/marginal basin

Further interpretations arise from emphasis of a southerly source for the basic clasts and of differences between the Northern and the Central and Southern Belts. This type of model sees the Northern Belt originally as a back-arc basin and prescribes that basic-clast and acid-clast petrofacies be derived from a volcanic arc (Fig. 7.15). The arc would have a continental basement to provide the acid-clast petrofacies, i.e. a continental-margin arc. The location of this arc would be either at the margin of the Northern Belt (Murphy and Hutton 1986, Morris 1987) or in the Central Belt at the northern boundary of the Hawick Rocks (Stone *et al.* 1987). The former position requires two separate locations for the Moffat shales to develop, north and south of the arc. At the end of the Ordovician the northern basin is envisaged as closing with the deformation and emergence of the sediments. On the Morris model, Silurian sedimentation continues in a fore-arc basin before closure at the end of the period. On the Murphy–Hutton model, closure occurred with deformation of the Ordovician prism and sedimentation continued in a successor basin between the deformed Ordovician to the north and the Cadomian continental mass to the south. At the

end of the Silurian transpression produced imbrication in the sedimentary pile.

Locating the arc within the Central Belt requires the shale to have accumulated in one basin with the basic clast petrofacies by-passing the area of shale accumulation (Fig. 7.15c). Volcanicity either continued to *turriculatus* time or at least volcanogenic material was available in the source until then. Erosion having removed all basic clast debris could, from then on, only deliver acid clasts from the plutonic–metamorphic basement. Hawick Rocks accumulated in a foreland basin, carbonate debris delivered from an inshore area accompanying recycled debris from the deformed sediments in the north. Continued thrusting from the north carried the deformed sediments over the remnant arc towards the southern Cadomian land mass.

Both versions of the model apparently ignore the fact that northerly derived detritus in Corsewall and equivalent formations are rich in basic clasts. Moreover there is a curious lack of thick volcaniclastic sediments which might be expected at the base of a marginal basin; instead the basal sequences are distal black shales and cherts (Leggett 1987). On the other hand, the abundance of dykes throughout the region has been taken to favour a back-arc model (Barnes *et al.* 1986).

Thrust–duplex models

The interpretations of the Southern Uplands as a back-arc basin also appeal to deformation of a thrust–duplex style. The structure and metamorphism in the Valley and Ridge Province in the Appalachians suggest a possible comparison. The thrusting and the range of temperatures (400–150°C) are similar (Harris and Milici 1977, Weaver 1984). But there are features which suggest that the Province may not be a valid analogue for the Southern Uplands:

1. The same stratigraphic sequences are repeated in thrust slices in the Appalachians where thrusting was imposed on a pre-existing succession and movement post-dated metamorphism. In the Southern Uplands differing tracts have distinctive petrographic sequences and metamorphism occurred *pari passu* with sedimentation and deformation; metamorphic fragments from earlier tracts are found in younger sediments.

2. The Province has a Lower Palaeozoic sequence < 15 km thick and this accounted for the burial metamorphism. Comparable sequences in Scotland would have had to reach 15 km in thickness and the Ordovician rocks at this depth metamorphosed before uplift and erosion in Silurian times. Fragments of dicellograptid-bearing shales are reported from the Silurian rocks of Craigenputtock Hill and Pinstane Hill (Peach and Horne 1899). Such a thickness of Ordovician rocks in the Southern Uplands is unlikely. Moreover there is very little variation in the metamorphism in individual tracts. Instead, isograds are sub-horizontal – an orientation more easily explained by burial during accretion.

3. In the Alps and the Appalachians thrusting of already metamorphosed sequences up ramps has left higher grades at the base of each thrust sheet. The reverse is found in the Southern Uplands where, in Tract 10 for example the illite crystallinity and graptolite reflectance increase towards the top of each thrust sheet (Kemp *et al.* 1985).

Basement to Central and Southern Belts

The Northern Belt contains slivers of ocean-floor basalt and the Bail Hill sea-mount; its basement was ensimatic. The Silurian floor is more problematic. On the Murphy–Hutton model it would be ensialic to the successor basin.

Seismic and magnetotelluric evidence suggests a NW-dipping feature in the Southern Uplands which is interpreted as the Iapetus Suture; it would also mark the top of the Lake District portion of the Cadomian plate (Beamish and Smythe 1986). During the Silurian, the Central and Southern Belts may have had an ensialic basement with thin continental crust being subducted. Such a system has been invoked for the Carpathian Flysch basin (Hesse 1982, Pescatore and Slaczka 1984). The proximity of the Lake District volcanic arc in Silurian times could provide the debris for greywackes particularly for the southerly derived sediments in the Rhinns of Galloway.

In summary it seems likely that:

1. Oblique collision during the closure of Iapetus caused major transcurrent sinistral movements as well as overthrusting. The Southern Uplands and Highland Border sequences are exotic terranes.

2. Sediment in the Highland Border and the Southern Uplands basins was derived from the same landmass situated in the region of the present Midland Valley.

3. The Highlands with Dalradian rocks played little or no part as Lower Palaeozoic source rocks in the Scottish region prior to the Lower Devonian; sources now in North America might be more relevant.

On the other hand, major questions have emerged:

1. When did the main phase of oceanic closure take place and how narrow was it in Silurian times?

2. Did Ordovician sedimentation take place in a back-arc setting and if so what was the location of the arc and where is it now?

3. Did the Silurian sequences accumulate in a fore-arc situation perhaps on the thin leading edge of continental basement or did they form in a foreland basin?

4. Given that the Southern Uplands appears to be a composite exotic terrane, does it lie as an allochthonous thin or thick skin on Cadomian or Laurentian continental crust? If so, was thrusting predominantly to the north or the south?

About a hundred years ago Lapworth, Peach and Horne presented remarkable insights into Scottish geology. Now a second century of investigations is opening with the emergence of challenging questions and the prospect of new revelations into the Lower Palaeozoic history of Scotland.

REFERENCES (Chapters 6 and 7)

AFTALION, M.,   1984   Age constraints on basement of the Midland Valley of Scotland.
    BREEMEN, O. VAN &   *Trans. R. Soc. Edinburgh: Earth Sciences*, **75**, 53–64.
    BOWES, D. R.

ANDERSON, J. G. C.   1947   The geology of the Highland Border: Stonehaven to Arran.
                            Trans. R. Soc. Edinburgh, **61**, 479–515.

ANDERSON, J. G. C. & PRINGLE, J. 1944 The Arenig rocks of Arran and their relationship to the Dalradian Series. *Geol. Mag.*, **81**, 81–87.

ANDERSON, T. B. 1968 The Geometry of a Natural Orthorhombic System of Kink Bands. *In*: Baer, A. J. & Norris, D. K. (Eds.) *Proc. Conf. on Research in Tectonics. Geol. Surv. Canada Paper*, **68–52**, 200–228.

1987 The onset and timing of Caledonian sinistral shear in County Down. *J. geol. Soc. London*, **144**, 817–825.

ANDERSON, T. B. & CAMERON, T. D. J. 1979 A structural profile across Co. Down. *In*: Harris, A. L., Holland, C. H. & Leake, B. E. (Eds.) *The Caledonides in the British Isles reviewed. Geol. Soc. London Sp. Publ.*, **8**, 263–267.

ANDERSON, T. B. & OLIVER, G. J. H. 1986 The Orlock Bridge Fault: a major late Caledonian sinistral fault in the Southern Uplands terrane, British Isles. *Trans. R. Soc. Edinburgh: Earth Sciences*, **77**, 203–222.

BALSILLIE, D. 1932 The Ballantrae Igneous Complex. *Geol. Mag.*, **69**, 107–131.

BARNES, R. P., ANDERSON, T. B. & McCURRY, J. A. 1987 Along-strike variation in the stratigraphical and structural profile of the Southern Uplands Central Belt in Galloway and Down. *J. geol. Soc. London*, **144**, 807–816.

BARNES, R. P., ROCK, N. M. S. & GASKARTH, J. W. 1986 Late Caledonian dyke swarms in Southern Scotland: new field, petrological and geochemical data for the Wigtown peninsula, Galloway. *Geol. J.*, **21**, 101–205.

BARRETT, T. J., JENKYNS, H. C., LEGGETT, J. K. & ROBERTSON, A. H. F. 1982 Comment and reply on 'Age and origin of Ballantrae ophiolite and its significance to the Caledonian orogeny and the Ordovician time-scale'. *Geology*, **9**, 331–333.

BARROW, G. 1901 On the occurrences of Silurian (?) Rocks in Forfarshire and Kincardineshire along the eastern Border of the Highlands. *Quart. J. geol. Soc. London.*, **57**, 328–345.

BATCHELOR, R. A. & WEIR, J. A. 1988 Metabentonite geochemistry; magmatic cycles and graptolite extinctions at Dob's Linn, Southern Scotland. *Trans. R. Soc. Edinburgh: Earth Sciences*, **79**, 19–41.

BEAMISH, D. R. & SMYTHE, D. K. 1986 Geophysical images of the deep crust; the Iapetus suture. *J. geol. Soc. London*, **143**, 489–497.

BENTON, M. J. 1982 Trace fossils from Lower Palaeozoic ocean-floor sediments of the Southern Uplands of Scotland. *Trans. R. Soc. Edinburgh: Earth Sciences*, **73**, 67–87.

BERGSTRÖM, S. M. 1971 Conodont biostratigraphy of the Middle and Upper Ordovician of Europe and Eastern North America. *In*: Sweet, W. C. & Bergstrom, S. M. (Eds.) *Symposium on Conodont Biostratigraphy. Geol. Soc. Amer. Mem.*, **127**, 83–157.

BERGSTRÖM, S. M. & ORCHARD, M. J. 1985 Conodonts of the Cambrian and Ordovician Systems from the British Isles. *In*: Higgins, A. C. & Austin, R. L. (Eds.) *A stratigraphical index of conodonts*. Ellis, Harwood, Chichester, 32–67.

BERRY, W. B. N. 1987 The Ordovician–Silurian boundary; new data, new concerns. *Lethaia*, **20**, 209–216.

BLUCK, B. J. 1982 Hyalotuff delta deposits in the Ballantrae ophiolite of SW Scotland: evidence for crustal position of the lava sequence. *Trans. R. Soc. Edinburgh: Earth Sciences*, **72**, 217–228.

1983 Role of the Midland Valley of Scotland in the Caledonian orogeny. *Trans. R. Soc. Edinburgh: Earth Sciences*, **74**, 119–136.

1984 Pre-Carboniferous history of the Midland Valley of Scotland. *Trans. R. Soc. Edinburgh: Earth Sciences*, **75**, 275–295.

1985 The Scottish paratectonic Caledonides. *Scott. J. Geol.*, **21**, 437–464.

BOWIE, S. H. U.,   1966   Potassium-rich sediments in the Cambrian of Northwest
    DAWSON, J.,   Scotland. *Trans. Instn. Min. Metall.* (*Sect. B. Applied Earth Sci.*)
    GALLAGHER, M. J.,   **75**, B109.
    OSTLE, D.,
    LAMBERT, R. ST. J. &
    LAWSON, R. I.

BRASIER, M. D.   1977   An early Cambrian chert biota and its implications. *Nature*, **268**,
                                   719–720.

BREWER, J. A.,   1983   BIRPS deep seismic reflection studies of the British Caledonides.
    MATHEWS, D. H.,   *Nature*, **305**, 206–210.
    WARNER, M. R., HALL, J.,
    SMYTHE, D. K. &
    WHITTINGTON, R. J.

BULMAN, O. M. D.   1944– A monograph of the Caradoc (Balclatchie) graptolites from
                            1947   limestones in Laggan Burn, Ayrshire. *Palaeont. Soc. London.*

CAMERON, T. D. J.,   1986   Geology of the country around Girvan. Explanation for the
    STONE, P. &   1:50,000 Geological Sheet 7 (Scotland). British Geological
    SMELLIE, J. L.   Survey.

CASEY, D. M.   1983   Geological Studies in the Central Belt of the Eastern Southern
                                   Uplands of Scotland. Ph.D. Thesis, Univ. of Oxford (unpubl.).

COCKS, L. R. M. &   1973   The biostratigraphy of the Silurian rocks of the Girvan district,
    TOGHILL, P.   Scotland. *J. geol. Soc. London*, **129**, 209–243.

COOK, D. R.   1976   The geology of the Cairnsmore of Fleet and its environs,
                                   southwest Scotland. Ph.D. Thesis, Univ. of St Andrews
                                   (unpubl.).

COWIE, J. W.   1974   The Cambrian of Spitsbergen and Scotland. *In*: Holland, C. H.
                                   (Ed.) *Cambrian of the British Isles, Norden and Spitsbergen.*
                                   Wiley, London, 123–155.

CRAIG, G. Y. & WALTON, E. K. 1983 Asymmetrical folding in the Hawick Rocks of the Galloway
                                   area, Southern Uplands: comment. *Scott. J. Geol.*, **191**, 103–106.
                            1959   Sequence and structure in the Silurian rocks of Kirkcudbright-
                                   shire. *Geol. Mag.*, **96**, 209–220.

CURRY, G. B.,   1982   The significance of a reliable Ordovician age for some Highland
    INGHAM, J. K., BLUCK, B. J.   Border rocks in Central Scotland. *J. geol. Soc. London*, **139**, 451–
    & WILLIAMS, A.   454.

CURRY, G. B., BLUCK, B. J., 1984   Age, evolution and tectonic history of the Highland Border
    BURTON, C. J., INGHAM, J. K.,   Complex, Scotland. *Trans. R. Soc. Edinburgh: Earth Sciences*,
    SIVETER, D. J. &   **75**, 113–133.
    WILLIAMS, A.

CURRY, G. B. & WILLIAMS, A. 1984   Lower Ordovician brachiopods from the Ben Suardal Limestone
                                   Formation (Durness Group) of Skye, western Scotland. *Trans.
                                   R. Soc. Edinburgh: Earth Sciences*, **75**, 301–310.

DEARMAN, W. R.,   1962   Refolded folds in the Silurian rocks of Eyemouth, Berwickshire.
    SHIELLS, K. A. G. &   *Proc. Yorks. geol. Soc.*, **33**, 273–285.
    LARWOOD, G. P.

DORNING, K. J.   1982   Early Wenlock acritarchs from the Knockgardner and Straiton
                                   Grit Formations of Knockgardner, Ayrshire. *Scott. J. Geol.*,
                                   18, 267–273.

ECKFORD, R. J. A. &   1931   The lavas of Tweeddale and their position in the Caradocian
    RITCHIE, M.   sequence. *Summ. Progr. Geol. Surv. G.B. for 1930*, 46–57.

ELDERS, C. F.   1987   The provenance of granite boulders in conglomerates of the
                                   Northern and Central Belts of the Southern Uplands of
                                   Scotland. *J. geol. Soc. London*, **144**, 853–863.

FLOYD, J. D.   1982   Stratigraphy of a flysch succession: the Ordovician of W
                                   Nithsdale, SW Scotland. *Trans. R. Soc. Edinburgh: Earth
                                   Sciences*, **73**, 1–9.

FLOYD, J. D., STONE, P., BARNES, R. P. & LINTERN, B. C. — 1987 — Constraints on the significance of the Orlock Bridge Fault within the Scottish Southern Uplands – a discussion of 'The Orlock Bridge Fault: a major late Caledonian sinistral fault in the Southern Uplands terrane, British Isles'. *Trans. R. Soc. Edinburgh: Earth Sciences*, **78**, 219–221.

FREEMAN, B., KLEMPERER, S. L. & HOBBS, R. W. — 1988 — The deep structure of Northern England and the Iapetus suture zone from BIRPS deep seismic reflection profiles. *J. geol. Soc. London*, **145**, 727–740.

FYFE, T. B. & WEIR, J. A. — 1976 — The Ettrick Valley Thrust and the upper limit of the Moffat Shales in Craigmichan Scaurs (Dumfries and Galloway Region; Annandale & Eskdale District). *Scott. J. Geol.*, **12**, 93–102.

GEIKIE, A. — 1873 — Explanation of Sheet 23, Lanarkshire: Central District. *Mem. Geol. Surv.*, Scotland.

GEORGE, T. N. — 1960 — The stratigraphical evolution of the Midland Valley. *Trans. geol. Soc. Glasgow*, **24**, 32–107.

GOBBETT, D. J. & WILSON, C. B. — 1960 — The Oslobreen Series, Upper Hekla Hoek of Ny Friesland, Spitsbergen. *Geol. Mag.*, **97**, 441–460.

GORDON, A. J. — 1962 — *The Lower Palaeozoic Rocks around Glenluce, Wigtownshire*. Ph.D. Thesis, Univ. of Edinburgh (unpubl.).

GUNN, W. — 1900 — Island of Arran. Probable Arenig rocks of the Highland Border. *Summ. Prog. for 1899. Geol. Surv. U.K.* 67–71.

HALL, J., BREWER, J. A., MATTHEWS, D. H. & WARNER, M. R. — 1984 — Crustal Structure across the Caledonides from the 'WINCH' seismic reflection profile: influences on the evolution of the Midland Valley of Scotland. *Trans. R. Soc. Edinburgh: Earth Sciences*, **75**, 97–109.

HALL, J., POWELL, D. W., WARNER, M. R., EL-ISA, Z. H. M., ADESANYA, O. & BLUCK, B. J. — 1983 — Seismological evidence for shallow crystalline basement in the Southern Uplands of Scotland. *Nature*, **305**, 418–420.

HALLAM, A. & SWETT, K. — 1966 — Trace Fossils from the Lower Cambrian Pipe Rock in the northwest Highlands. *Scott. J. Geol.*, **2**, 101–105.

HALLIDAY, A. N., STEPHENS, W. E., HUNTER, R. H., MENZIES, M. A., DICKIN, A. P. & HAMILTON, P. S. — 1985 — Isotopic and chemical constraints on the building of the deep Scottish lithosphere. *Scott. J. Geol.*, **21**, 465–491.

HARMON, R. S., HALLIDAY, A. N., CLAYBURN, J. A. P. & STEPHENS, W. E. — 1984 — Chemical and Isotopic systematics of the Caledonian intrusions of Scotland and Northern England: a guide to magma source region and magma-crust interaction. *Phil. Trans. R. Soc. Lond.*, A. **310**, 709–742.

HARPER, D. A. T. — 1979 — The environmental significance of some faunal changes in the Upper Ardmillan succession (upper Ordovician), Girvan, Scotland. *In*: Harris, A. L., Holland, C. H. & Leake, B. E. (Eds.) *The Caledonides of the British Isles reviewed. Geol. Soc. London Sp. Publ.*, **8**, 439–445.

1981 — The stratigraphy and faunas of the Upper Ordovician High Mains Formation of the Girvan district. *Scott. J. Geol.*, **17**, 247–255.

1982 — The stratigraphy of the Drummuck Group (Ashgill), Girvan. *Geol. J.*, **17**, 251–277.

1984 — Brachiopods from the upper Ardmillan succession (Ordovician) of the Girvan district, Scotland. Part 1. *Monogr. Palaeontogr. Soc. London*, 1–78.

HARPER, D. A. T. &   1986   A shelly biofacies from the graptolitic mudstones of the Lower
OWEN, A. W.   Balclatchie Group (lower Caradoc) near Laggan, Girvan district. *Scott. J. Geol.*, **22**, 271–283.

HARRIS, A. L.   1969   The relationship of the Leny Limestone to the Dalradian. *Scott. J. Geol.*, **5**, 187–190.

HARRIS, L. D. &   1977   Characteristics of thin-skinned style of deformation in the
MILICI, R. C.   Southern Appalachians and potential hydrocarbon traps. *U.S. Geol. Surv. Prof. Paper*, **1018**, 40 pp.

HARTE, B., BOOTH, J. E.,   1984   Aspects of the post-depositional evolution of Dalradian and
DEMPSTER, T. J., FETTES, D. J.,   Highland Border Complex rocks in the Southern Highlands of
MENDUM, J. R. &   Scotland. *Trans. R. Soc. Edinburgh: Earth Sciences*, **75**, 151–163.
WATTS, D.

HEINZ, W. &   1988   Volcanic clasts in Silurian Conglomerates in the Midland Valley
LOESCHKE, J.   (Hagshaw Hills Inlier, Scotland) and their meaning for the Caledonian Orogeny. *Geol. Rdsch.*, **77**, 453–466.

HENDERSON, S. M. K.   1935   Ordovician Submarine Disturbances in the Girvan District. *Trans. R. Soc. Edinburgh*, **58**, 487–509.

HENDERSON, W. G. &   1982   The Highland Border rocks and their relation to marginal basin
ROBERTSON, A. H. F.   development in the Scottish Caledonides. *J. geol. Soc. London*, **139**, 433–450.

HEPWORTH, B. C.   1981   *Geology of the Ordovician rocks between Leadhills and Abington, Lanarkshire.* Thesis, Ph.D. Univ. of St Andrews (unpubl.).

HEPWORTH, B. C.,   1982   Sedimentology, volcanism, structure and metamorphism of the
OLIVER, G. J. H. &   northern margin of a Lower Palaeozoic accretionary complex;
McMURTRY, M. J.   Bail Hill–Abington area of the Southern Uplands of Scotland. *In*: Leggett, J. K. (Ed.) *Trench-Fore-arc Geology. Geol. Soc. London. Sp. Publ.*, **10**, 521–534.

HESSE, R.   1982   Cretaceous–Palaeogene flysch zone of the East Alps and Carpathians: identification and plate-tectonic significance of 'dormant' and 'active' deep-sea trenches in the Alpine–Carpathian Arc. *In*: Leggett, J. K. (Ed.) *Trench-fore-arc geology. Spec. Publ. geol. Soc. London*, **10**, 472–494.

HOLUB, F. V., KLÁPOVÁ, H.,   1984   Petrology and geochemistry of post-obduction dykes of the
BLUCK, B. J. & BOWES, D. R.   Ballantrae complex, SW Scotland. *Trans. R. Soc. Edinburgh: Earth Sciences*, **75**, 211–223.

HUBERT, J. F.   1966   Sedimentary history of Upper Ordovician Geosynclinal Rocks, Girvan, Scotland. *J. Sediment. Petrol.*, **36**, 677–699.

HUNTER, R. H.,   1984   Meta-igneous granulite and ultramafic xenoliths from basalts of
UPTON, B. G. H. &   the Midland Valley of Scotland: petrology and mineralogy of the
ASPEN, P.   lower crust and upper mantle. *Trans. R. Soc., Edinburgh: Earth Sciences*, **75**, 75–84.

HUTTON, J.   1795   *Theory of the Earth*, **1**. Creech, Edinburgh.

IKIN, N. P.   1983   Petrochemistry and tectonic significance of the Highland Border Suite mafic rocks. *J. geol. Soc., London*, **140**, 267–278.

IKIN, N. P. & HARMON, R. S.   1983   A stable isotope study of serpentinization and metamorphism in the Highland Border Suite, Scotland, U.K. *Geochim. Cosmochim. Acta*, **47**, 153–167.

INCE, D.   1984   Sedimentation and tectonism in the middle Ordovician of the Girvan district, SW Scotland. *Trans. R. Soc. Edinburgh: Earth Sciences*, **75**, 225–237.

INGHAM, J. K.   1978   Geology of a continental margin 2: middle and late Ordovician transgression, Girvan. *In*: Bowes, D. R. & Leake, B. H. (Eds.) *Crustal evolution in northwestern Britain and adjacent regions.* Seel House, Liverpool, 163–176.

INGHAM, J. K., CURRY, G. B. & 1985   Early Ordovician Dounans Limestone fauna, Highland Border
WILLIAMS, A.   Complex. *Trans. R. Soc. Edinburgh: Earth Sciences*, **76**, 481–513.

JELINEK, E., SOUČEK, J., BLUCK, B. J., BOWES, D. R. & TRELOAR, P. J. 1980 Nature and significance of beerbachites in the Ballantrae ophiolite, SW Scotland. *Trans. R. Soc. Edinburgh: Earth Sciences*, **71**, 159–179.

JELINEK, E., SOUČEK, J., RANDA, Z., JAKEŠ, P., BLUCK, B. J. & BOWES, D. R. 1984 Geochemistry of peridotites, gabbros and trondjemites of the Ballantrae complex, SW Scotland. *Trans. R. Soc. Edinburgh: Earth Sciences*, **75**, 193–209.

JENNINGS, J. S. 1961 *The geology of the eastern part of the Lesmahagow Inlier.* Ph.D. Thesis, Univ. of Edinburgh (unpubl.).

JOHNSON, M. R. W. & HARRIS, A. L. 1967 Dalradian–?Arenig relations in part of the Highland Border, Scotland and their significance in the chronology of the Caledonian orogeny. *Scott. J. Geol.*, **3**, 1–16.

JOHNSON, M. R. W., KELLEY, S. P., OLIVER, G. J. H. & WINTER, D. A. 1985 Thermal effects and timing of thrusting in the Moine Thrust zone. *J. geol. Soc., London*, **142**, 863–873.

KASSI, A. M. 1984 Lower Palaeozoic Geology of the Gala area, Borders Region, Scotland. Ph.D. Thesis, Univ. of St Andrews (unpubl.).

KELLING, G. 1961 The Stratigraphy and Structure of the Ordovician rocks of the Rhinns of Galloway. *Quart. J. geol. Soc. London*, **117**, 37–75.

1962 The Petrology and Sedimentation of Upper Ordovician Rocks in the Rhinns of Galloway, South-West Scotland. *Trans. R. Soc. Edinburgh*, **65**, 107–137.

KELLING, G., DAVIES, P. & HOLYROYD, J. 1987 Style, scale and significance of sand bodies in the Northern and Central Belts, southwest Southern Uplands. *J. geol. Soc. London*, **144**, 787–805.

KEMP, A. E. S. 1986 Tectonostratigraphy of the Southern Belt of the Southern Uplands. *Scott. J. Geol.*, **22**, 241–256.

1987 Evolution of Silurian Depositional Systems in the Southern Uplands, Scotland. *In*: Leggett, J. K. & Zuffa, G. G. (Eds.) *Marine Clastic Sedimentology.* Graham & Trotman, London, 124–155.

KEMP, A. E. S., OLIVER, G. J. H. & BALDWIN, J. R. 1985 Low-grade metamorphism and accretion tectonics: Southern Uplands terrain, Scotland. *Mineral. Mag.*, **49**, 335–344.

KEMP, A. E. S. & WHITE, D. E. 1985 Silurian trench sedimentation in the Southern Uplands, Scotland: implications of new age data. *Geol. Mag.*, **122**, 275–277.

KNIPE, R. J. & NEEDHAM, D. T. 1986 Deformation processes in accretionary wedges – examples from the SW margin of the Southern Uplands, Scotland. *In*: Coward, M. P. & Ries, A. C. (Eds.) *Collision Tectonics. Geol. Soc. London Sp. Publ.*, **9**, 51–65.

KNIPE, R. J., CHAMBERLAIN, M. I., PAGE, A. & NEEDHAM, D. T. 1988 Structural histories in the SW Southern Uplands, Scotland. *J. Geol. Soc. London*, **145**, 679–684.

LAMBERT, R. ST. J. & McKERROW, W. S. 1976 The Grampian Orogeny. *Scott. J. Geol.*, **12**, 271–292.

LAMONT, A. & LINDSTROM, M. 1957 Arenigian and Llandeilian cherts identified in the Southern Uplands by means of conodonts. *Trans. geol. Soc. Edinburgh*, **17**, 60–70.

LAPWORTH, C. 1878 The Moffat Series. *Quart. J. geol. Soc. London*, **34**, 240–346.

1882 The Girvan Succession. *Quart. J. geol. Soc. London*, **38**, 537–666.

1889 On the Ballantrae Rocks of the South of Scotland and their place in the Upland Sequence. *Geol. Mag.*, **26**, 20–24, 59–69.

LAPWORTH, C. & WILSON, J. 1871 On the Silurian Rocks of the Counties of Roxburgh and Selkirk. *Geol. Mag.*, **8**, 456–464.

LEGGETT, J. K.   1978   *Studies in the Ordovician Rocks of the Southern Uplands with particular reference to the Northern belt.* D.Phil. Thesis, Univ. of Oxford (unpubl.).

                                  1980   Palaeogeographic setting of the Wrae Limestone: an Ordovician submarine slide deposit in Tweeddale. *Scott. J. Geol.*, **16**, 91–104.

                                  1987   The Southern Uplands as an accretionary prism: the importance of analogues in reconstructing palaeogeography. *J. geol. Soc. London*, **144**, 737–752.

LEGGETT, J. K.,   1979   The Southern Uplands: A Lower Palaeozoic accretionary prism.
  MCKERROW, W. S. &   *J. geol. Soc. London*, **136**, 755–770.
  EALES, M. H.

LEGGETT, J. K.,   1984   Evolution of the Ballantrae complex: evidence from the
  BARRETT, T. J., OGAWA, Y. &   Downan–Curraric terrane and analogues from the SW Japan
  ROBERTSON, A. H. F.   active margin (abstract). *Trans. R. Soc. Edinburgh: Earth Sciences*, **75**, 298.

LESPÉRANCE, P. J.,   1987   The Ordovician-boundary stratotype; consequences of its
  BARNES, C. R.,   approval by the I.U.G.S. *Lethaia*, **20**, 217–222.
  BERRY, W. B. N.,
  BOUCOT, A. J. &
  MU EN-ZHI

LEWIS, A. D. & BLOXAM, T. W.   1977   Petrogenetic environments of the Girvan–Ballantrae lavas from rare-earth element distribution. *Scott. J. Geol.*, **13**, 211–222.

LINDSTROM, M.   1971   Lower Ordovician Conodonts of Europe. *In:* Sweet, W. C. & Bergstrom, S. M. (eds.) *Geol. Soc. Amer. Mem.*, **127**, 21–61.

LOESCHKE, J.   1985   Geochemistry of acid volcanic clasts in Silurian conglomerates of the Midland Valley of Scotland; implications on the Caledonian orogeny. *Geol. Rdsch.*, **74**, 537–546.

LONGMAN, C. D.,   1979   Ordovician conglomerates and the evolution of the Midland
  BLUCK, B. J. &   Valley. *Nature*, **280**, 578–581.
  VAN BREEMEN, O.

LONGMAN, C. D., BLUCK, B. J.,   1982   Ordovician conglomerates: constraints on the time-scale. *In*:
  VAN BREEMEN, O. &   Odin, G. S. (Ed.) *Numerical dating in stratigraphy*. Wiley, New
  AFTALION, M.   York, 807–809.

LOVELL, J. P. B.   1974   Sand volcanoes in the Silurian rocks of Kirkcudbrightshire. *Scott. J. Geol.*, **10**, 161–162.

LUMSDEN, G. I., TULLOCH, W.,   1967   The Geology of Neighbourhood of Langholm. *Mem. Geol. Surv.*
  HOWELLS, M. F. &   *Scotland.*
  DAVIES, A.

MCCURRY, J. A.   1987   The Geology of the Rhinns of Galloway (south of the Portayew area). British Geological Survey. Southern Uplands Project. *B.G.S. open file*, Edinburgh.

MCCURRY, J. A. &   1989   Landward vergence in the Lower Palaeozoic, Southern
  ANDERSON, T. B.   Uplands–Down–Longford Terrane, British Isles. *Geology,* **17**, 630–633.

MCGIVEN, A.   1968   *Sedimentation and provenance of post-Valentian conglomerates up to and including the basal conglomerate of the Lower Old Red Sandstone in the southern part of the Midland Valley of Scotland.* Ph.D. Thesis, Univ. of Glasgow (unpubl.).

MCKERROW, W. S.   1979   Ordovician and Silurian changes in sea level. *J. geol. Soc. London*, **136**, 137–145.

MCKERROW, W. S.,   1977   Imbricate thrust model of the Southern Uplands of Scotland.
  LEGGETT, J. K. &   *Nature*, **267**, 237–239.
  EALES, M. H.

MCMURTRY, M. J.   1980   *The Ordovician Rocks of the Bail Hill area, Sanquhar, South Scotland: Volcanism and Sedimentation in the Iapetus Ocean.* Ph.D. Thesis, Univ. of St Andrews (unpubl.).

MOLYNEUX, S. G.	1987	Possible early Wenlock acritarchs from the Linkum Beds of the Southern Uplands. *Scott. J. Geol.*, **23**, 301–313.
MOORE, J. C.	1849	On some fossiliferous beds in the Silurian rocks of Wigtownshire and Ayrshire. *Quart. J. geol. Soc. London*, **5**, 7–12.
MORRIS, J. H.	1987	The Northern Belt of the Longford–Down Inlier, Ireland and Southern Uplands, Scotland: an Ordovician back-arc basin. *J. geol. Soc. London*, **144**, 773–786.
MURPHY, F. C.	1987	Evidence for late Ordovician amalgamation of volcanogenic terranes in the Iapetus suture zone, eastern Ireland. *Trans. R. Soc. Edinburgh: Earth Sciences*, **78**, 153–167.
MURPHY, F. C. & HUTTON, D. H. W.	1986	Is the Southern Uplands of Scotland really an accretionary prism? *Geology*, **14**, 354–357.
NICOL, J.	1850	Observations on the Silurian Strata of the SE of Scotland. *Quart. J. geol. Soc. London*, **6**, 53.
OLIVER, G. J. H. & LEGGETT, J. K.	1980	Metamorphism in an accretionary prism; prehnite–pumpellyite facies metamorphism of the Southern Uplands of Scotland. *Trans. R. Soc. Edinburgh: Earth Sciences*, **71**, 235–246.
OLIVER, G. J. H. & McKERROW, W. S.	1984	Comment on 'Seismological evidence for shallow crystalline basement in Southern Uplands, Scotland' by Hall *et al. Nature*, **309**, 89.
OLIVER, G. H. J., SMELLIE, J. L., THOMAS, L. J., CASEY, D. M., KEMP, A. E. S., EVANS, L. J., BALDWIN, J. R. & HEPWORTH, B. C.	1984	Early Palaeozoic metamorphic history of the Midland Valley, Southern Uplands–Longford–Down massif and the Lake District, British Isles. *Trans. R. Soc. Edinburgh: Earth Sciences*, **75**, 245–258.
OWEN, A. W.	1986	The uppermost Ordovician (Hirnantian) trilobites of Girvan, SW Scotland with a review of coeval trilobite faunas. *Trans. R. Soc. Edinburgh: Earth Sciences*, **77**, 231–239.
PALMER, T. J., McKERROW, W. S. & COWIE, J. W.	1980	Sedimentological evidence for a stratigraphic break in the Durness Group. *Nature*, **287**, 720–722.
PEACH, B. N. & HORNE, J.	1899	*The Silurian Rocks of Britain. 1 Scotland, Mem. Geol. Surv. U.K.*
	1907	The Geological Structure of the Northwest Highlands of Scotland. *Mem. Geol. Surv. U.K.*
PESCATORE, T. & SLACZKA, A.	1984	Evolution models of two flysch basins: the Northern Carpathians and the Southern Apennines. *Tectonophysics*, **106**, 49–70.
PHILLIPS, W. E. A., STILLMAN, C. J. & MURPHY, T.	1976	A Caledonian plate–tectonic model. *J. geol. Soc. London*, **132**, 579–609.
POULSEN, C.	1951	The position of the East Greenland Cambro–Ordovician in the Palaeogeography of the North Atlantic region. *Medd. Dansk. Geol. Foren.*, **12**, 161–162.
PRINGLE, J.	1941	On the relationship of the Green conglomerate to the Margie Grits in the North Esk, near Edzell. *Trans. geol. Soc. Glasgow*, **20**, 136–140.
ROBERTSON, A. H. F. & HENDERSON, W. G.	1984	Geochemical evidence for the origins of igneous and sedimentary rocks of the Highland Border, Scotland. *Trans. R. Soc. Edinburgh: Earth Sciences*, **75**, 135–150.
ROBERTSON, G.	1985	*Palaeoenvironmental Interpretation of the Silurian Rocks of the Pentland Hills*. Ph.D. Thesis, Univ. of Edinburgh (unpubl.).
ROLFE, W. D. I. & FRITZ, M. A.	1966	Recent evidence for the age of Hagshaw Hills Silurian Inlier, Lanarkshire. *Scott. J. Geol.*, **2**, 159–164.
RUSHTON, A. W. A. & TRIPP, R. P.	1979	A fossiliferous lower Canadian (Tremadoc) boulder from the Benan Conglomerate of the Girvan district. *Scott. J. Geol.*, **15**, 321–327.

RUSHTON, A. W. A., STONE, P., 1986   An early Arenig age for the Pinbain sequence of the Ballantrae
   SMELLIE, J. L. &   complex. *Scott. J. Geol.*, **22**, 41–54.
   TUNNICLIFF, S. P.

RUSSELL, M. J. & ALLISON, I.   1985   Agmatolite and the maturity of sandstones of the Appin and
                              Argyll groups and Eriboll Sandstone. *Scott. J. Geol.*, **21**, 113–
                              122.

RUST, B. R.   1965   The stratigraphy and structure of the Whithorn area of Wig-
                              townshire, Scotland. *Scott. J. Geol.*, **2**, 159–164.

SALTER, J. W.   1859   Durness Limestone Fossils. *Quart. J. geol. Soc. London*, **15**, 374–
                              381.

SANDO, W. J.   1957   Beekmantown Group (Lower Ordovician) of Maryland. *Geol.*
                              Soc. Amer. Mem., **68**.

SCHALLREUTER, R. E. L. &   1985   Ostracodes across the Iapetus Ocean. *Palaeontology*, **28**, 577–
   SIVETER, D. J.   598.

SCOTT, K. M.   1967   Intra-bed palaeocurrent variations in a Silurian flysch sequence,
                              Kirkcudbrightshire, Southern Uplands of Scotland. *Scott. J.*
                              Geol., **3**, 268–281.

SHIELLS, K. A. G. &   1963   Tectonics of the Coldingham Bay area of Berwickshire in the
   DEARMAN, W. R.   Southern Uplands of Scotland. *Proc. Yorks geol. Soc.*, **34**, 209–
                            234.

SMELLIE, J. L.   1984   Accretionary lapilli and highly vesiculated pumice in the Ballan-
                              trae ophiolite complex: ash-fall products of sub-aerial eruptions.
                            Rep. Br. Geol. Surv., **16**, 36–40.

STEPHENS, W. E. &   1984   Geochemical contrasts between late Caledonian granitoid
   HALLIDAY, A. N.   plutons of northern, central and Southern Scotland. *Trans. R.*
                            Soc. Edinburgh: Earth Sciences, **75**, 259–273.

STEPHENS, W. E.,   1975   Element associations and distributions through a Lower
   WATSON, S. W.,   Palaeozoic graptolite shale sequence in the Southern Uplands of
   PHILIP, P. R. & WEIR, J. A.   Scotland. *Chemical Geology*, **16**, 269–294.

STONE, P.   1984   Constraints on genetic models for the Ballantrae complex, SW
                              Scotland. *Trans. R. Soc. Edinburgh: Earth Sciences*, **75**, 189–191.

STONE, P. &   1983   Graptolite faunas from the Ballantrae ophiolite complex and
   RUSHTON, A. W. A.   their structural implications. *Scott. J. Geol.*, **19**, 297–310.

STONE, P. & SMELLIE, J. L.   1986   Ballantrae. Description of the solid geology of parts of 1:2500
                              NK08, 09, 18 and 19 Classical areas of British Geology. British
                              Geological Survey.

STONE, P. & STRACHAN, I.   1981   A fossiliferous borehole section within the Ballantrae ophiolite.
                              Nature, **293**, 455–456.

STONE, P., FLOYD, J. D.,   1987   A sequential back-arc and foreland basin thrust duplex model
   BARNES, R. P. &   for the Southern Uplands of Scotland. *J. geol. Soc. London*, **144**,
   LINTERN, B. C.   753–764.

STRINGER, P. & TREAGUS, J. E.   1980   Non-planar S1 cleavage in the Hawick Rocks of the Galloway
                                area, Southern Uplands, Scotland. *J. Struct. Geol.*, **2**, 317–331.

                       1981   Asymmetrical folding in the Hawick Rocks of the Galloway
                              area, Southern Uplands. *Scott. J. Geol.*, **17**, 129–148.

                       1983   Asymmetrical folding in the Hawick Rocks of the Galloway
                              area, Southern Uplands: reply. *Scott. J. Geol.*, **19**, 107–112.

SWETT, K.   1966   Authigenic feldspars and cherts resulting from dolomitization of
                              illitic limestones: a hypothesis (abs.) *Geol. Soc. Amer. Prog. 1966*
                            Ann. Mtg. San Francisco, p. 216.

                       1969   Interpretation of depositional and diagenetic history of Cam-
                            brian–Ordovician Succession of northwest Scotland. *In*: Kay,
                            M. (Ed.) *North-Atlantic Geology and Continental drift. Mem.* **12**,
                            Amer. Assoc. Petrol. Geologists. 630–646.

SWETT, K., KLEIN, G. DE V. &   1971   A Cambrian Tidal Sand body – The Eriboll Sandstone of
   SMITH, D. E.   northwest Scotland: An ancient-recent analog. *J. Geol.*, **79**, 400–
                            415.

THIRLWALL, M. F.   1981 Peralkaline rhyolites from the Ordovician Tweeddale lavas, Peeblesshire. *Geol. J.*, **16**, 41–44.

THIRLWALL, M. F. &   1984 Sm–Nd isotope and chemical evidence that the Ballantrae
BLUCK, B. J.   'ophiolite', SW Scotland is polygenetic. *In*: Gass, I. G., Lippard, S. J. & Shelton, A. W. (eds.) *Ophiolites and the ocean lithosphere. Sp. Publ. Geol. Soc. London*, **13**, 215–230.

TIPPER, J. C.   1975 Lower Silurian Communities – three case histories. *Lethaia*, **8**, 287–299.

                                1976 The stratigraphy of the North Esk Inlier, Midlothian. *Scott. J. Geol.*, **12**, 15–22.

                                1965 Trilobites from the Albany division (Ordovician) of the Girvan district, Ayrshire. *Palaeontology*, **8**, 577–603.

                                1976 Trilobites from the basal Superstes mudstones (Ordovician) at Aldons Quarry, near Girvan, Ayrshire. *Trans. R. Soc. Edinburgh*, **69**, 369–423.

                                1979 Trilobites from the Ordovician Auchensoul and Stinchar Limestones of the Girvan district, Strathclyde. *Palaeontology*, **22**, 339–361.

                                1980a Trilobites from the Ordovician Balclatchie and Lower Ardwell groups of the Girvan district, Scotland. *Trans. R. Soc. Edinburgh: Earth Sciences*, **71**, 123–145.

                                1980b Trilobites from the Ordovician Ardwell Group of the Craighead Inlier, Girvan district, Scotland. *Trans. R. Soc. Edinburgh: Earth Sciences*, **71**, 147–157.

UPTON, B. G. J.,   1984 Xenoliths and their implications for the deep geology of the
ASPEN, P. &   Midland Valley of Scotland and adjacent regions. *Trans. R. Soc.
HUNTER, R. H.   Edinburgh: Earth Sciences*, **75**, 65–70.

WALTON, E. K.   1955 Silurian greywackes in Peeblesshire. *Proc. R. Soc. Edinburgh*, **B65**, 327–357.

                                1956 The Ordovician Conglomerates in South Ayrshire. *Trans. geol. Soc. Glasgow*, **22**, 133–156.

                                1968 Some rare sedimentary structures in the Silurian rocks of Kirkcudbrightshire. *Scott. J. Geol.*, **4**, 356–369.

WARREN, P. T.   1964 The stratigraphy and structure of the Silurian rocks southeast of Hawick, Roxburghshire. *Quart. J. geol. Soc. London*, **120**, 193–218.

                                1963 The Petrography, Sedimentation and Provenance of the Wenlock Rocks near Hawick, Roxburghshire. *Trans. geol. Soc. Edinburgh*, **19**, 225–255.

WATSON, S. W.   1976 *The sedimentary geochemistry of the Moffat Shales, a carbonaceous sequence in the Southern Uplands of Scotland.* Ph.D. Thesis, Univ. of St Andrews (unpubl.).

WEAVER, C. E.   1984 Shale–slate metamorphism in Southern Appalachians. Development in Geology, 10. Elsevier.

WEBB, B.   1983 Imbricate structure in the Ettrick area, Southern Uplands. *Scott. J. Geol.*, **19**, 387–400.

WEIR, J. A.   1973 The Lower Palaeozoic graptolite facies in Ireland and Scotland; review correlation and palaeogeography. *Sci. Prog. R. Dublin Soc.*, Ser. A, **4**, 439–460.

WELSH, W.   1964 The Ordovician Rocks of Northwest Wigtownshire. Ph.D. Thesis, Univ. of Edinburgh (unpubl.).

WILKINSON, J. M. &   1974 Trace elements and tectonic relationships of basaltic rocks in the
CANN, J. R.   Ballantrae Igneous Complex, Ayrshire. *Geol. Mag.*, **111**, 35–41.

WILLIAMS, A.   1959 A structural history of the Girvan district, SW Ayrshire. *Trans. R. Soc. Edinburgh*, **63**, 629–667.

                                1962 The Barr and Lower Ardmillan Series (Caradoc) of the Girvan district, South-West Ayrshire. *Mem. geol. Soc. London*, **3**.

WILLIAMS, A. 1972 Distribution of brachiopod assemblages in relation to Ordovician palaeogeography. *Sp. Papers, Palaeontology*, **12**, 241–269.

WILLIAMS, S. H. 1980 An excursion guide to Dob's Linn. *Proc. geol. Soc. Glasgow*, **121/122**, 13–18.

1982a Upper Ordovician graptolites from the top Lower Hartfell Shale Formation (*D. clingani* and *P. linearis* zones) near Moffat, southern Scotland. *Trans. R. Soc. Edinburgh: Earth Sciences*, **72**, 229–255.

1982b The Late Ordovician graptolite fauna of the Anceps Band at Dob's Linn, southern Scotland. *Geologica Palaeont.*, **16**, 29–56.

1983 The Ordovician–Silurian boundary graptolite fauna of Dob's Linn, southern Scotland. *Palaeontology*, **26**, 605–639.

1987 Upper Ordovician graptolites from the *D. complanatus* Zone of the Moffat and Girvan districts and their significance for correlation. *Scott. J. Geol.*, **23**, 65–92.

WILSON, J. T. 1966 Did the Atlantic close and then re-open? *Nature*, **211**, 676–681.

WRIGHT, S. C. 1985 The study of depositional environments and diagenesis in the Durness Group of north-west Scotland. Ph.D. Thesis, Univ. of Oxford (unpubl.).

YARDLEY, B. W. D., VINE, F. J. & BALDWIN, C. T. 1982 The plate tectonic setting of NW Britain and Ireland in late Cambrian and early Ordovician times. *J. Geol. Soc. London*, **139**, 455–463.

8

CALEDONIAN AND EARLIER MAGMATISM

P. E. Brown

Early metabasites in the Moines

Numerous metamorphosed minor basic intrusions occur in the Moine rocks of the Northern Highlands (Fig. 8.1). The broad regional distribution of these bodies was described by Smith (1979) who referred to them as the 'amphibolite suite'. In more recent accounts (Rock *et al.* 1985) the term metabasite has replaced amphibolite as a suite name to avoid confusion with amphibolite as a specific rock name.

In the southern Inverness-shire area of the Moine outcrop the metabasite bodies are particularly well developed in the Glenfinnan and Loch Eil Groups (see Chapter 4), but not in the Morar Group. The time span of intrusion of these metabasites is not known, but they appear to have undergone most, if not all, of the polyphase deformation seen in their hosts.

Barr *et al.* (1985) have described how the *c.* 1,000 Ma West Highland Granitic Gneiss, which outcrops in the Glenfinnan and Loch Eil groups, is itself cut by metabasite sheets which, since they carry the early S1 fabric, are also of Grenvillian age. Rock *et al.* (1985) also refer to conformation of a Grenvillian age by an isochron obtained through exhaustive sampling of a metagabbro cutting granite-gneiss in the River Doe.

The latter authors describe the western Inverness-shire metabasites as an extensive differentiated sill complex with minor dykes. Differentiation trends show strong Fe and Ti enrichment and depletion in Cr and Ni, indicative of a tholeiitic suite. There is evidence of strong metabasite–country rock reactions during igneous emplacement or regional metamorphism and even the 'immobile' element chemistry can be unreliable in tracing magmatic affinities unless this factor is allowed for. Minerals present are plagioclase, hornblende and various combinations of garnet, biotite, epidote and quartz.

Metabasites in limestones gained Ca, Mg, Sr and lost Fe, Ti, Rb, Y and Zr, totally suppressing garnet; metabasites in pelites lost Na, enhancing garnet relative to plagioclase; metabasites in psammites gained Rb and K, enhancing biotite.

In the Fannich district of northern Ross-shire two suites of metabasites have been identified by Winchester (1976). The Meall an t'Sithe suite comprises massive bodies widely distributed within the Meall an t'Sithe pelite which is regarded as an allochthonous portion of the Glenfinnan Group; these metabasites were shown by Winchester, on the basis of a detailed chemical study, to be tholeiitic, synmetamorphic intrusions. The Sgurr Mor suite occurs as small schistose bodies in the Sgurr Mor pelite of the Morar Group and possibly originated as lavas or pyroclastic material, in this case of alkali basalt type.

Winchester (1984) has also made a geochemical study of amphibolites in the Strathconon area of Ross-shire, where the individual bodies range from small pods only 2–3 metres thick up to large bodies exceeding 200 metres thick. Whether the host rocks should be assigned to the Glenfinnan Group or the Loch Eil Group is not clear. These metabasites also represent a variously fractionated tholeiitic suite with some evidence of *in situ* fractionation.

In central northern Sutherland metabasites described by Read (1931) have been further studied by Moorhouse and Moorhouse (1979). These authors recognised an early suite of schistose garnetiferous amphibolites representing intrusions of basaltic magma early in the deformation sequence and also a later 'Loch a' Mhoid' amphibolite suite. Other rare epidotic hornblende schists in pelites possibly originated as penecontemporaneous volcanic ash. The early schistose amphibolites typically occur both as small sheets and as larger bodies, of which the Ben Hope sill is an example. Originally dykes or sills, all these early intrusions are essentially concordant with the regional foliation and usually have a strong mineral lineation and schistosity. They are believed to have undergone all the phases of deformation seen in the Moine host rocks (Winchester & Floyd 1984). There is no indication of the original igneous mineralogy, the common assemblage being hornblende, plagioclase, quartz, ± garnet, epidote, biotite and accessories.

The later Loch a' Mhoid metadolerites occur as small intrusive bodies in which evidence of original igneous texture is often preserved. Moorhouse and Moorhouse (1979) concluded that they were intruded between fold phases assigned to the Precambrian and Caledonian respectively. Ultramafic lenses of schistose tremolite–chlorite–serpentine–carbonate–ore rock, with relic olivine, are believed to be cumulates from the Loch a' Mhoid suite magma.

The chemistry of the Sutherland metabasites suggests

Fig. 8.1. Localities of metabasite suites referred to in the text.

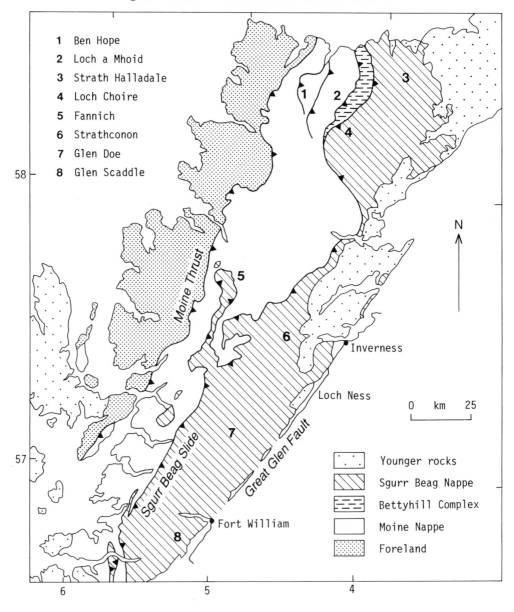

1 Ben Hope
2 Loch a Mhoid
3 Strath Halladale
4 Loch Choire
5 Fannich
6 Strathconon
7 Glen Doe
8 Glen Scaddle

Younger rocks
Sgurr Beag Nappe
Bettyhill Complex
Moine Nappe
Foreland

that the magma of the early suite was tholeiitic while that of the later Loch a' Mhoid amphibolites was mildly alkaline (Fig. 8.2).

The Ben Hope suite of Sutherland has been the subject of a further geochemical study by Winchester and Floyd (1984) who concluded that these metabasites represent a series of chemically discrete, but related, variously fractionated tholeiitic sills. Marked enhancement of K, Rb and Ba, with conversion of amphibole to biotite is found in meta-shears which occur intermittently throughout the Ben Hope amphibolite suite (Floyd & Winchester 1983).

Differing trace-element chemistry (Ti/Y) has been used by Winchester (1985) to match metabasite suites in the Caledonide nappes below and above the Sgurr Beag slide, from which it was deduced that westward movement of the overriding nappe exceeded 140 km.

A major basic intrusion of uncertain lineage and age is present in the Glen Scaddle area of Ardgour. This sheet, or sill, encloses xenoliths of amphibolite, quartzite, and gneiss and is folded to produce separate outcrops in Glen Saddle and nearby Glen Gour (Drever 1940). The intrusion was described by Drever as gabbro–diorite. In general it retains its original mineralogy and texture but in places is strongly foliated and regressed by post-intrusive movements. According to Stoker (1983) the intrusion post-dates Precambrian D1 folds but is deformed by later D2 and D3 structures. Drever (op. cit.) noted that the Glen Scaddle rocks have appinitic affinities and Bailey (1960) also noted similarities to appinitic rocks associated with the Newer Granites. Barr (1985) notes, however, that its structural setting is comparable to that of the 456 Ma Glen Dessary syenite (see below) and a

Fig. 8.2. Zr/P_2O_5–Nb/Y plot showing the tholeiitic affinities of the majority of the metabasite suites (after Rock *et al.* 1985).

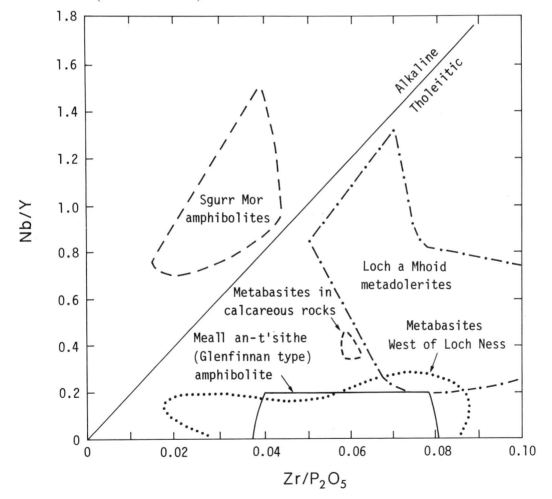

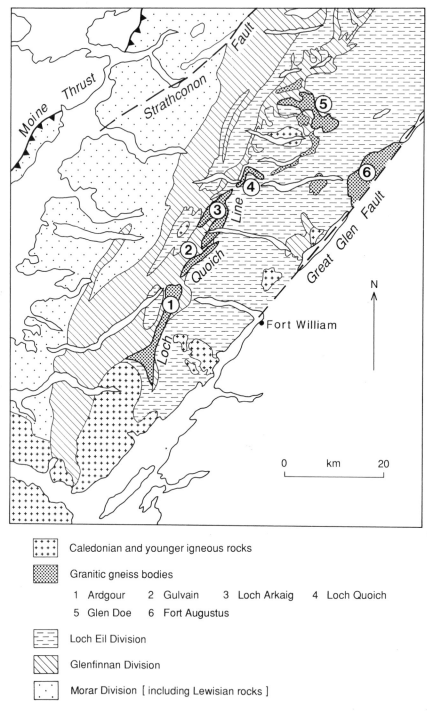

Fig. 8.3. West Highland Granite Gneiss localities.

Legend:

⊞ Caledonian and younger igneous rocks

▦ Granitic gneiss bodies

  1 Ardgour   2 Gulvain   3 Loch Arkaig   4 Loch Quoich
  5 Glen Doe   6 Fort Augustus

⊟ Loch Eil Division

▨ Glenfinnan Division

⋯ Morar Division [including Lewisian rocks]

later Caledonian Newer Granite association (cf. Ashworth & Chinner 1978) is not possible.

West Highland Granite Gneiss

The West Highland Granite Gneiss, sometimes referred to as the Ardgour granitic gneiss after its principal outcrop (Fig. 8.3), has a radiometric age of 1028 ± 43 Ma (Brooke et al. 1976, recalculated by Brewer et al. 1979) and has provided the most convincing evidence so far of Grenvillian events in the Highlands.

The meaning of the Grenvillian age for the regional Moine geology has, however, been controversial because of uncertainty over the nature of the granite gneiss which has been variously interpreted as metasomatic (Harry 1954; Dalziel 1963, 1966; Johnstone et al. 1969), magmatic (Mercy 1963) or a tectonically emplaced slice of pre-existing granitic basement (Harris in discussion of Winchester 1974).

Latterly, Barr et al. (1985) have concluded that the granite which formed the protolith of the gneiss was a magmatic intrusion emplaced either before or during the first phase of Moine deformation (D1). The gneiss has a restricted range of bulk composition with high SiO_2, low Na_2O/K_2O and Fe_2O_3/FeO, and is corundum normative. The essentially S-type chemistry and an initial $^{87}Sr/86_{Sr}$ ratio of 0.709 favour Moine metasediments as the immediate source of the granite protolith. The isotopic characteristics of zircons both in the gneiss and the nearby metasediments can be explained by a three-stage model involving initial homogenisation at c. 1,750 Ma followed by partial resetting at 1,030–1,100 Ma and c. 490 Ma. This suggests that the Moine sediments and therefore ultimately the zircons in the granite gneiss had a Laxfordian (c. 1,700 Ma) source (Aftalion & van Breemen 1980).

Outcrop-scale concordance of contacts between the granite gneiss and the Moine metasediments is a result of deformation and on a regional scale the gneiss is cross-cutting (Fig. 8.3). Evidence that the intrusion of granite magma was synorogenic (D1) is seen in the lack of a thermal aureole (cf. Carn Chuinneag). Following D1 deformation, pegmatitic segregations developed (Fig. 8.4) which were broadly contemporaneous with the climax of regional migmatisation (MP_1), followed by intense deformation (D_2) and localised MP_2 remelting.

Fig. 8.4. The Ardgour gneiss. Photo Scott Johnstone.

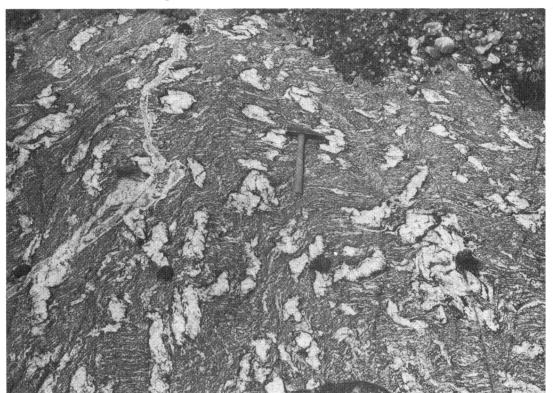

Moine migmatites

The main regional migmatite belt in the Moines trends northwards from the Sound of Mull, through Sunart, Morar, Moidart and Knoydart and presents a particularly difficult problem of interpretation because of the interaction of Precambrian and Caledonian tectono-thermal events. The development of these migmatites has been reviewed by Barr (1985) who described the earliest widespread migmatisation as associated with the peak of metamorphism in the kyanite and sillimanite zones, which gave rise in suitable lithologies to subsolidus, essentially trondhjemitic, leucosomes of Grenvillian (c. 1,000 Ma) age. Stevenson (1971) made a chemical study of pelites between Lochs Eilt and Ailort in Inverness-shire and by using the detailed stratigraphy deduced by Powell (1964, 1966) was able to compare the composition of the same pelite in the unmigmatised garnet zone and the migmatised sillimanite zone. It was found that the high and low grade pelites have a similar composition and, assuming that the two had the same pre-metamorphism composition, the migmatisation process was isochemical. This work has been further amplified by Barr (op. cit.).

In some areas these early migmatites underwent intense deformation before renewed local migmatisation, which at Loch Quoich involved partial melting of K-feldspar bearing rocks in structurally defined zones.

The migmatites encompass numerous pegmatites, and a combination of radiometric techniques has defined the age of these as c. 750 Ma (Long & Lambert 1963; van Breemen et al. 1974, 1978). The pegmatites are generally concordant, coarse-grained, c. 1 m thick, and are deformed by Caledonian structures. Their distribution has not been systematically mapped but they are widespread in the Morar and Glenfinnan Groups.

Lambert (1969) used the whole of the structural, fabric and isotopic evidence available from these pre-Caledonian pegmatites to define a Morarian tectonothermal event. On the other hand, Powell et al. (1983) argued that the pegmatites belong to an independent suite (c. 800 Ma) of non-orogenic origin, while Soper and Anderton (1984) suggested that the Morarian is an extensional event' related to lithospheric stretching prior to the opening of the Iapetus Ocean.

Caledonian pegmatites are also common in the eastern Morar Group and in the Glenfinnan and Loch Eil Groups. According to Barr (1985) they are typically cross-cutting, dilational, range from a few centimetres to a few metres thick, and carry two feldspars, one or two micas and rarely garnet. They have yielded ages in the range 430–445 Ma (van Breeman et al. 1974, 1978; Brewer et al. 1979) and were emplaced as Caledonian deformation waned in the last stages of D3 deformation. This episode of migmatisation probably resulted from melting due to tectonic thickening in the D3 deformation and it should be noted that there is a 40 to 50 My

difference between this metamorphism and migmatisation and comparable, earlier, events in the Dalradian rocks south-east of the Great Glen fault.

In central and eastern Sutherland the Moine rocks are extensively migmatised and the great 'injection' complexes of Loch Choire and Strath Halladale occupy some 1,500 km² (Read 1931). In Sutherland a major zone of ductile strain, probably the continuation of the Sgurr Beag slide, separates in the east and west respectively, rocks with Morar and Glenfinnan characteristics. Barr (1985) has demonstrated a Precambrian age for early migmatisation in the 'Glenfinnan Group' rocks and it is evident that the regional distribution of migmatisation results from the shortening of an E–W Precambrian metamorphic zonation by Caledonian slides, with tectonic emplacement of Precambrian migmatites on to lower-grade 'Morar Group' rocks after the migmatites had cooled down (cf. Soper & Brown 1971). Lit-par-lit leucosomes of trondhjemitic type were explained by Brown (1967, 1971) as subsolidus products of metasomatism, though as pointed out by Barr (op. cit.) more recent assessment of their Glenfinnan Group metasedimentary precursors probably makes an isochemical process adequate explanation of the migmatisation process. The Caledonian metamorphism itself added to the complexity of the regional migmatisation pattern and there is a variety of late granitic and pegmatitic veins, lenses and sheets, the majority of which are clearly igneous, intrusive and appear to define a differentiation sequence. The relationship of these igneous bodies to larger masses of granite such as that at Loch Choire is not clear, though Barr (op. cit.) suggests a petrographic similarity.

The Strath Halladale section of the Sutherland migmatite complex is less well known but contains a greater abundance of granite as discrete sheets and also some larger intrusions, notably the Strath Halladale granite itself (Read 1931, McCourt 1980). This igneous body has been dated at 649±30 Ma (M. Brook quoted in Pankhurst 1982).

Older granites in the Moines

As described by McCourt (1980) the Strath Halladale granite postdates the peak of sillimanite grade metamorphism and migmatisation and is only partly affected by monoformal folds late in the local structural sequence. The age of the complex (649 ± 30 Ma; M. Brook op. cit.) confirms the pre-Caledonian timing of the main tectono–thermal events in this area of Sutherland.

The granite is locally foliated and is cut by a network of pegmatites and aplogranite veins as well as slightly younger, thick microgranite dykes which McCourt (op. cit.) suggests are the end products of the magmatic activity responsible for the granite itself. The granite has an igneous texture and is mineralogically simple with essential quartz–K-feldspar–oligoclase–biotite and rare hornblende.

According to Bailey (1960) it was Peach, in conversation with Crampton over hand specimens, who first suggested the pre-movement date of hornfelses around the Carn Chuinneag augen gneiss in Ross-shire. Subsequent to the initial mapping (Peach *et al.* 1912) and confirmation of Peach's suggestion, various subdivisions of the metamorphosed augen granite were established (Fig. 8.5) by Harker (1962), Shepherd (1973) and Wilson

and Shepherd (1979). The Inchbae rock is the commonest type of coarse-grained gneiss and consists of abundant phenocrysts of orthoclase in a matrix of quartz, K-feldspar, plagioclase and biotite. The Lochan a' Chairn rock has a finer grained groundmass and veins the Inchbae type. There is also a strongly foliated fine to medium-grained gneiss composed of microcline, plagioclase, quartz, riebeckite and some aegirine, the relationship of which to the other granite gneisses is not clear.

Detailed structural mapping by Shepherd and Wilson

Fig. 8.5. The Carn Chuinneag intrusion and its aureole (after Wilson and Shepherd 1979).

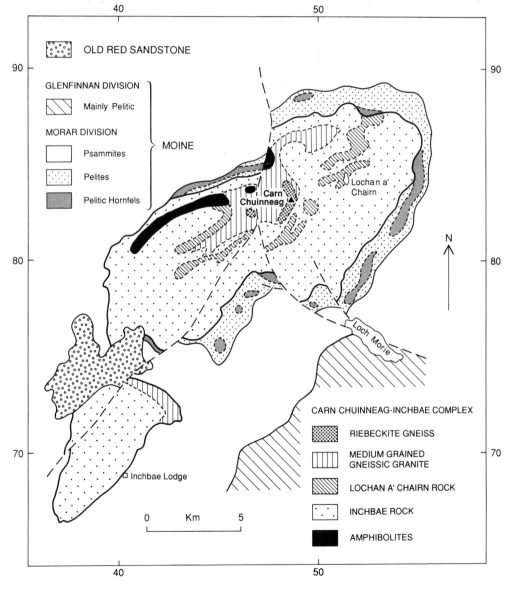

(*op. cit.*) has demonstrated a four-fold deformation sequence in which the emplacement of the granite took place after regional (D1) folding. Thus, although Peach believed the intrusion to be entirely pre-tectonic, it did in fact invade rocks which had attained garnet-grade metamorphism. The rocks of the intrusion became foliated in response to later (D2) regional deformation (Fig. 8.6) and in further (D3) deformation the intrusion came to occupy the core of a large synform plunging gently SW. Finally, D4 movement produced a major open SE-trending synform which affects the intrusion and its envelope and appears responsible for the acuate shape of the complex. Despite subsequent amphibolite facies metamorphism the 1 to 1·5 km wide contact aureole

contains pseudomorphs of cordierite and andalusite (Tilley 1935, Harker 1954).

A Rb–Sr whole-rock isochron age of 548 ± 10 Ma is interpreted as the age of intrusion of the Carn Chuinneag body (Long 1964) and is supported by a U-Pb zircon age of 560 ± 10 Ma (Pidgeon & Johnson 1974).

Early basic magmatism in the Dalradian

The earliest sign of basic igneous activity in the Dalradian is found as thin bands of pyroclastic or detrital volcanic material in the Ardrishaig Phyllite in the Argyll Group of the South-west Highlands (Borradaile 1973) and also as

Fig. 8.6. Generalised trend of D2 structures around the Carn Chuinneag intrusion (after Wilson and Shepherd 1979).

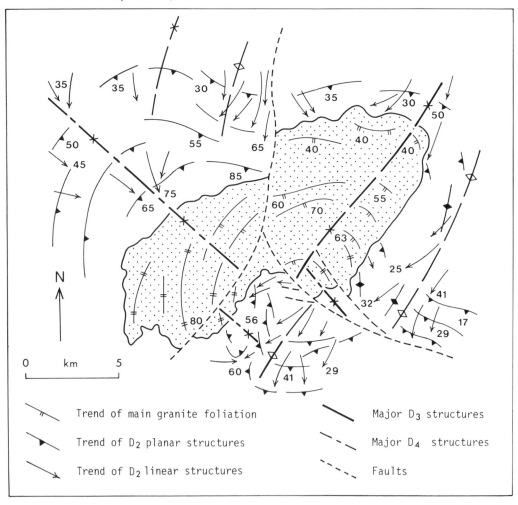

the approximately equivalent Farragon Green Beds which lie below the Ben Lui Schist in central Perthshire.

The main development of extrusive igneous rocks is at a higher stratigraphic level where in the South-west Highlands the Tayvallich and Loch Avich Volcanics occur as frequently pillowed lavas and intercalated tuffs in the Southern Highlands Group (Harris & Pitcher 1975). Other lesser occurrences of pillow lavas are found in the underlying Tayvallich Limestone at the top of the Argyll Group. Hornblende schists of probable volcanic origin in Perthshire and pillow lava at Ardwell Bridge in Banffshire (MacGregor & Roberts 1963) are at about the same stratigraphic level. In addition to these rocks volcanogenic sediments known as green beds occur, apparently at several stratigraphic levels, in the Southern Highlands Group. Harris and Pitcher (op. cit.) suggested that the formation specifically referred to as the Green beds should be restricted to rocks laterally equivalent to the Tayvallich Volcanics.

Table 8.1. Dalradian lithostratigraphic sequence and approximate thickness of igneous rocks in the South-west Highlands (Borradaile 1973).

Southern Highlands Group	⎧ Loch Avich Volcanics (0·3–0·5 km) Loch Avich Grits Tayvallich Volcanics ⎩ (2 km)	
Argyll Group	⎧ Tayvallich Slates/Limestones Crinan Grits – Erins Quartzite ⎨ Craignish – Ardrishaig ⎩ Phyllites Easdale Slates Scarba Conglomerate	Metadolerite Sill Complex (2 km)

In the South-west Highlands, where the pre-tectonic igneous activity was most voluminous, the extrusive rocks are accompanied by a broadly contemporaneous and comagmatic suite of dolerite and gabbro sills intruded into the underlying pile of wet unconsolidated sediments (Wilson & Leake 1972, Borradaile 1973, Graham 1976a). Massive metadolerite dykes in Jura and Islay (Graham & Borradaile 1984) have been shown to be likely feeders to the sills and volcanics.

In the South-west Highlands the low grade of metamorphism, up to epidote–amphibolite facies, has not affected igneous textures and structures in the centres of massive intrusions and flows and this has facilitated their interpretation (Graham 1976a, b). A portion of the suite, both sills and flows, underwent spilitic alteration prior to metamorphism, with notable adjustments in Ca and Na through the migration of heated pore fluids. The bulk of the metabasites comprise a suite of quartz and olivine – hypersthene normative tholeiites in which Graham and Bradbury (1981) identified as distinctive fractionation characteristics a strong iron enrichment trend, high concen-

trations of Zr, Y, P and Nb, plus extreme depletion in Ni and Cr.

Pre- or early tectonic basic intrusions known as the Older Basic Intrusions are also well developed elsewhere in the Dalradian. They occur mainly in the Argyll Group metasediments between the Central Highland Quartzite and the Loch Tay Limestone, or their lateral equivalents. This is at about the same general level as the Tayvallich sills.

In Perthshire the Ben Vrackie complex of metabasic sills (Pantin 1956) preserves evidence of pre-tectonic hornfels in the surrounding Ben Lawers and Ben Eagach Schists. Numerous basic horizons are also found in the overlying Ben Lui Schist and may possibly have been high-level feeders for extrusive activity in the Southern Highlands group.

In the north-east the Portsoy sill complex can be traced from the Banffshire coast southwards to Deeside as sheets and lenses of metagabbro, amphibolite and serpentine, again in the lateral equivalents of the Tayvallich Subgroup (Read 1961).

The tectonic implications of the early, widespread tholeiitic magmatism in the Dalradian have been reviewed by Graham and Bradbury (1981). On the Ti-Zr-Y discriminant plot (Pearce & Cann 1973) the SW Highland suite is transitional between 'within plate' and 'ocean floor' basalts, while the Central Highlands suite plots within the 'ocean-floor' field. The only comparable rock types to produce similar Fe and Ti enrichment trends are strongly differentiated tholeiitic ferrograbbros and ferrobasalts of ocean floor spreading centres and overall the Dalradian metabasaltic suite shows strong petrochemical affinity with tholeiites of accreting plate margins. The basic vulcanicity in the Dalradian occurred (Graham & Borradaile (op. cit.)) on the NW margin of an actively evolving Lower Cambrian ocean basin, or possibly a marginal basin to the Iapetus Ocean. More recently Graham (1986) has suggested an alternative model in which basaltic volcanism and sedimentation in the SW Highlands occurred in a quasi-oceanic trans-Caledonoid pull-apart basin bounded on its NE side by a major fault now identified as the Cruachan lineament (Fig. 8.7). In this model the Dalradian terrane may provide no evidence regarding the early history of the NW side of the Iapetus Ocean. The possible influence of trans-Caledonoid lineaments has been further extended by Fettes et al. (1986).

Dalradian 'Older' Granites and Migmatites

Granites which intrude previously deformed Dalradian metasediments, but which are themselves involved in subsequent regional deformation and metamorphism occur at Ben Vurich, Meall Gruaim, Glen Tilt and Dunfallandy

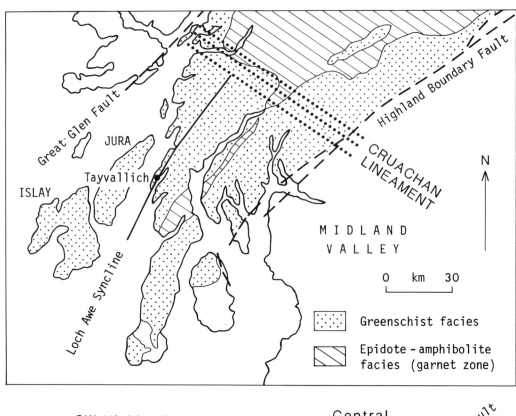

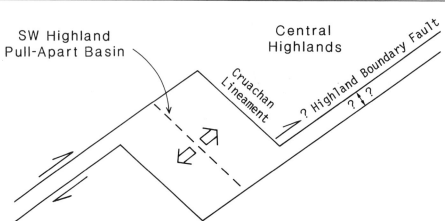

Fig. 8.7. The Cruachan lineament and the transcurrent pull-apart model for early Dalradian evolution (after Graham 1986).

Hill, all in the Pitlochry district of Perthshire (Fig. 8.8), where they form part of the 'older' Granite suite of Barrow (1893). The main intrusion is that at Ben Vurich which is composed of gneiss and foliated granite containing K-feldspar, oligoclase, biotite, and quartz, with abundant garnet, zircon and sphene. Similar granite occurs on Meall Gruaim. The Dunfallandy Hill granite occurs mainly as a set of sheet-like bodies consisting mainly of K-feldspar, quartz and muscovite with subsidiary plagioclase and garnet. The granite in Glen Tilt occurs as pink sheets and veins. All these occurrences post-date D2 and have tectonically imposed fabrics equivalent to S3 in the Dalradian country rocks (Bradbury *et al.* 1976). The granite magmatism is broadly coeval with regional migmatisation which occurred in the D2–D3 interval. A zircon U–Pb age for the Ben Vurich intrusion of $514 \pm 6/7$ Ma (lower intersection with concordia, Pankhurst & Pidgeon 1976; quoted as 519 ± 8 by Aftalion *et al.* 1984) has been regarded as the date of post-D2 but pre-D3 emplacement of the granite, while a Rb–Sr whole-rock isochron age of 481 ± 15 Ma for the Dunfallandy Hill gneiss (quoted as 506 ± 30 Ma by Aftalion *et al.* 1984) may date the D3 foliation. At the time of writing, however, a new high precision U–Pb age of 590 ± 2 Ma has been obtained for the Ben Vurich granite by Rogers *et al.* (*J. Geol. Soc.*, in the press). Rogers *et al.* point out that this new age has implications for the meaning of the term 'Grampian Orogeny'. As defined by Lambert & McKerrow (1976) the Grampian Orogeny is the set of events which caused the metamorphism and deformation of the Dalradian sediments in the Lower Ordovician. The new age for the Ben Vurich granite means, however, that the D1/D2 deformation of the Dalradian occurred before 590 Ma, and not in Lower Ordovician times. Rogers *et al.* suggest that the term Grampian Orogeny should be restricted to the D1/D2 (Precambrian) episode and that a new term 'Athollian' be used for events which occurred in the Lower Ordovician. This proposed usage would, however, conflict with the now established use of the term Grampian for Ordovician age events which included the intrusion of the Younger Basics, major granites and the climax of metamorphism in the Grampian area of NE Scotland. The best estimate of the age of these events is the whole-rock Rb–Sr age of *c.* 490 Ma obtained by Pankhurst (1970) for the Younger Basic intrusions (see below).

Traditionally included with the Older Granites are lensoid outcrops of augen gneiss at Portsoy in the north-east of the Dalradian outcrop. These granite gneisses occur in the Portsoy Group of metasediments and a whole-rock Rb–Sr isochron age of 655 ± 17 Ma was regarded by Pankhurst (1974) as the date of intrusion and hence the minimum age for sedimentation of the country rock. There is, however, no evidence preserved in the very poor outcrops of intrusive relationships and similar approximate 650 Ma ages on other migmatitic gneisses in the area have been interpreted by Sturt *et al.* (1977) as belonging to basement rocks tectonically emplaced in the Dalradian cover.

Migmatites feature strongly among the north-eastern

Fig. 8.8. Principal localities of 'Older' Granites in the Dalradian.

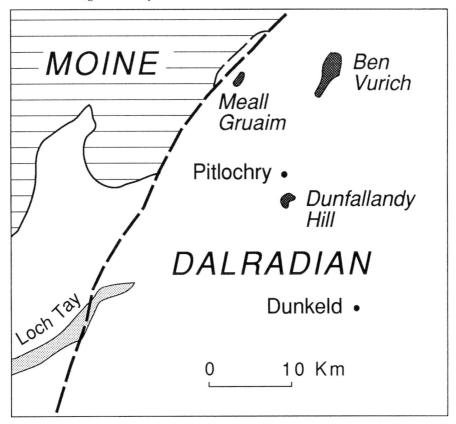

Dalradian rocks, occurring mainly around the horizon of the pelitic Ben Lui Schists in the Argyll Group. Atherton (1977) speculated that this localisation may be due to strong thermal anisotropy controlling preferential heat movement, and the possible focussing effect of large-scale structures in relation to the heat source responsible for the metamorphism (cf. Borradaile 1976). Not surprisingly the origin of the Dalradian migmatites is controversial and it seems highly likely that they are a polygenetic group of different ages ranging from association with the Older Granites to the Younger Gabbros. In the south-eastern Grampians, Barrow (1893, 1912) related the migmatites directly to intrusions of granite by a mechanism of filter-pressing under tectonic stress, the whole complex forming his suite of 'older granites', but in Barrow's type area of Glen Clova, Harry (1958b) found that the 'older granites' consist of two distinct phases with migmatitic quartz–oligoclase–mica gneiss separate from later discrete masses of microline granite.

Partial melting, or anatexis, is perhaps the most popular mechanism suggested for the origin of those migmatites, which are well developed in the sillimanite zone of metamorphism, an association regarded as an overprint, superimposed on the slightly earlier main metamorphic zonation due to an isobaric temperature increase which caused partial melting (Chinner 1966).

In the Glen Isla area (Fig. 8.9) McLennan (1983, 1984) has described two types of migmatite, the first with quartzose, stromatic leucosomes formed by metamorphic differentiation at staurolite–kyanite grade, and the second with trondhjemitic schlieren formed by anatexis at sillimanite grade.

In the Huntly–Portsoy area of the Buchan-type metamorphism the sillimanite overprint was attributed by Ashworth (1975) to thermal effects in the inner aureole of the Newer Basic intrusions. Ashworth (1976) also considered that partial melting was the main process in the formation of the migmatites which he subdivided into comparatively low temperature trondhjemitic and granitic types and higher temperature cordierite and ortho-pyroxene-bearing varieties. The last are widespread and clearly the result of partial melting of Dalradian metasediments by the basic magma of the Newer Gabbros (Gribble 1970) which were intruded into country rocks still at a high temperature following regional metamorphism. Ashworth's (1976) proposals of an anatectic origin for the leucosomes of the trondhjemitic and granitic migmatites were criticised by Yardley (1977) who, contrary to Ashworth, believed that comparison with experimental evidence on melts and plagioclase compositions in appropriate starting materials, made metamorphic segregation a more likely mechanism. These are just the sort of arguments to which this very difficult group of rocks are susceptible; in this case, however, both authors seem to agree that the process was not one involving overall changes in the bulk composition of the rock system.

Fig. 8.9. The Glen Clova and Glen Isla areas showing distribution of metamorphic differentiation migmatites and anatectic migmatites (after McLellan 1984).

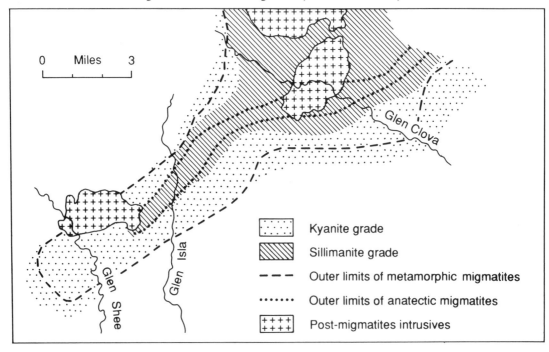

Highland Border Complex

The Highland Border Complex (HBC) is found as a number of fault bounded slivers along the Highland Border Fault (Fig. 8.10) and consists of the dismembered parts of an oceanic crustal sequence ranging in age from Early to pre-Arenig to Caradoc (Curry *et al.* 1984). Major components in the Complex are terrigenous-derived turbidites, a dismembered ophiolite, and ophiolite-derived sediments. Principal rock types are basaltic pillow lavas, serpentinites, black phyllites, cherts, limestones and greywackes.

On Arran and Bute the metamorphosed mafic rocks of the HBC are of two types (Ikin 1983): greenschist facies pillow lavas and epidote–amphibolite facies hornblende schists. Both types have been affected by post-magmatic alteration, including spilitisation. On the basis of 'immobile' element (Ti, Zr, Y, Nb) discriminant diagrams, Ikin (*op. cit.*) recognised two magmatic groups, one tholeiitic with similarities to ocean-floor or marginal-basin basalts, and the second within-plate alkali basaltic. One possible interpretation of this is that the HBC mafic

igneous rocks formed as tholeiitic oceanic-crust upon which alkaline ocean-island lavas erupted.

The same two magmatic groups were recognised by Robertson and Henderson (1984) who found mafic igneous rocks with tholeiitic MORB-like chemistry in outcrops the length of the Highland Boundary Fault and, more locally mafic rocks with an alkalic within-plate character.

Ultrabasic rocks in the HBC are generally pervasively altered to serpentinite, carbonate and quartz, probably with harzburgite as the dominant parental material (Henderson & Robertson 1982).

Suggestions that the HBC and Dalradian shared a common deformation history have been discounted by geochronological and palaeontological evidence that formation of the major nappe structures in the Dalradian preceded the deposition of the HBC. Harte *et al.* (1984) concluded that strike–slip motion probably along a boundary transform fault, in conjunction with major periodic uplift of the rocks on the northern side of the fault, offers the most acceptable explanation of the disposition of the HBC rocks along the Highland Border.

Fig. 8.10. The Highland Border Complex.

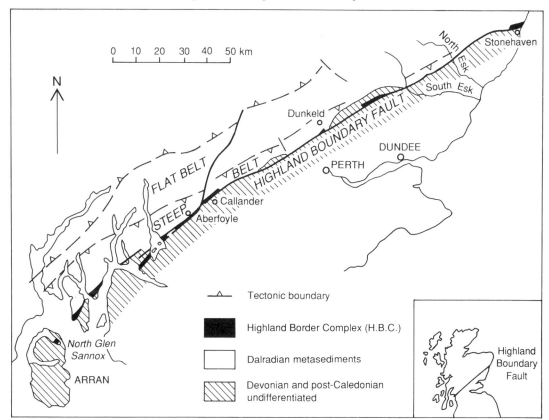

Hutton (1987) has summarised current thinking that the Highland Boundary Fault in Scotland was a major terrane boundary during the Caledonian and that movements along it have been large with final docking of the Highland Border Terrane not taking place until the Devonian.

Ordovician volcanics in the Southern Uplands

Volcanic rocks are found among the predominantly sedimentary Ordovician sequences which make up the Northern Belt of the Southern Uplands (Fig. 8.11). A few vestiges of basalts and oceanic chert of Arenig age found at the base of the greywacke sequence are considered to be slivers from the transform zone which partially floored an

extending Southern Uplands basin (Stone *et al.* 1987). Spilitic lavas also occur interbedded with greywackes of N. gracilis age at various localities (Williams 1962, Floyd 1982). These lavas occur as strips with a maximum thickness of *c.* 200 m, fault-bounded at the base and overlain by a generalised sequence of metalliferous sediment–chert–graptolitic mudstone–greywacke (Leggett 1980, Lambert *et al.* 1981). The outcrops are markedly lenticular with individual strips of volcanics only traceable for a maximum of 4 km along strike. The degree of secondary alteration of the lavas is extremely variable but some flows preserve evidence of original pyroxenes and feldspars.

Lambert *et al.* (*op. cit.*) considered the Northern Belt to be made up of several tectonic slices and mainly on the basis of 'immobile' element chemistry (Ti, Zr, Y, Nb) they

Fig. 8.11. The northern belt of the Southern Uplands showing the positions of the Bail Hill and Wrae Hill volcanics.

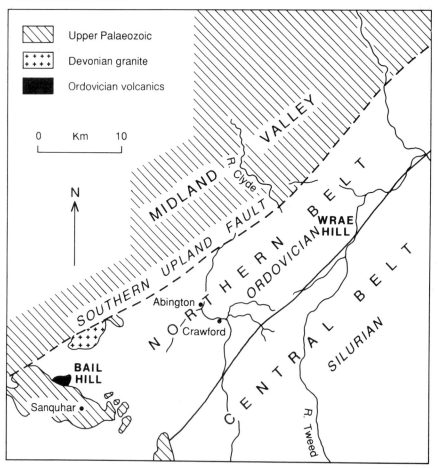

demonstrated a N–S range through successive slices from mildly alkaline to tholeiitic basalts. This could indicate early eruptions associated with rifting or fracture zone tectonics being replaced by later generation of oceanic crust.

The Bail Hill volcanics overlie black shales of N. gracilis age and constitute a range of mildly alkaline lavas and pyroclastics (Hepworth et al. 1982). Some basalts contain phenocrysts of clinopyroxene and bytownite and are locally vesicular and amygdaloidal, but more commonly have autobrecciated textures indicative of submarine eruption. Above the basalts are hawaiites, mugearites and pyroclastics containing phenocrysts of oligoclase-andesine, amphibole and biotite. The presence of gabbroic and dioritic xenoliths was suggested by Hepworth et al. (op. cit.) to indicate that a large magma body underlay the Bail Hill volcano in Ordovician times. Furthermore the presence of extrusive and shallow intrusive material now found only as clasts in a volcanic neck suggests that extrusive activity was lithologically more diverse than indicated by present outcrops. The Bail Hill volcano is believed to have originated as a mildly alkaline seamount which was accreted from the subducting Iapetus plate.

Alkaline lavas at Wrae Hill (Fig. 8.11) were interpreted by Leggett (1980) as allochthonous deposits, introduced along with limestone blocks to an active trench as debris flows produced by submarine sliding from a seamount. The lavas were described by Eckford and Ritchie (1931) and Lambert et al. (op. cit.) as soda–trachytes or quartz–keratophyres. They have been subjected to secondary alteration but some specimens contain groundmass riebeckite and on the basis of detailed chemistry Thirlwall (1981b) has demonstrated that they are in fact peralkaline rhyolites similar to pantellerites and comendites. Notably, the lavas have high concentrations of Zr, Y, Nb and rare-earth-elements, which along with a pronounced negative Eu anomaly are characteristic of peralkaline rhyolites. The interpretation of these rather extreme rock types is that they formed on an island or seamount volcano sited on the Iapetus Ocean plate.

Ballantrae Ophiolite

The Ballantrae Igneous Complex appears to have all the components of an ophiolite (Bluck 1978), with the possible exception of sheeted dykes, and has figured prominently in plate-tectonic reconstructions of the Iapetus Ocean. The complex (Fig. 8.12) consists of two major bodies of serpentinised ultramafic rock with minor gabbros and trondhjemites, pillow lavas and volcaniclastic sediments, radiolarian cherts, graptolite shales, and olistostromes containing blueschists. Amphibolites and schists beneath the serpentinites resemble aureoles found in ophiolites where obduction of hot oceanic lithosphere over pre-existing rocks has caused metamorphism (Church

& Gayer 1973, Treloar et al. 1980, Spray & Williams 1980).

Graptolite faunas from the Ballantrae Complex range from lowest to mid Arenig (Stone & Rushton 1983, Stone 1984). Radiometric ages (Bluck et al. 1980, Hamilton et al. 1984, Thirlwall & Bluck 1984) suggest a comparatively short time span of c. 30 m.y. for the generation of the igneous rocks and their obduction, with the age of the lavas lying between 505 ± 11 and 476 ± 14 Ma.

Major difficulties of interpretation are caused by the fact that although rocks which could be components of an ophiolite are present their original stratigraphic relationships are not preserved and internal contacts between rock units are tectonic.

A variety of environments have been postulated for the pre-obduction formation of the Ballantrae rocks, based mainly on a combination of geological observations and geochemistry.

From the 'immobile' element chemistry of spilitised pillow lavas (Fig. 8.13), Wilkinson and Cann (1974) deduced the presence of island arc, hotspot, and ocean floor basalts, and suggested that individual blocks of volcanic rocks represent juxtaposed material which may be unrelated in space and time. Jelínek et al. (1980) on the other hand concluded that the Ballantrae lavas have been affected by crustal contamination and furthermore that diagnostic elements (Zr, Y, Nb) used to identify tectonic environments had in fact been mobile during alteration, suggestions of which Thirlwall and Bluck (1984) were highly critical.

Bluck et al. (1980) pointed out that clastic rocks of volcanic provenance make up about 55 per cent of the volcanic pile and suggested that this is more indicative of an island arc environment than of ocean crust. It was suggested that the arc was south-west of a marginal basin from which the other elements of the complex were derived, notably the ultramafics which are stated by Jelínek et al. (1984) to have geochemical characteristics similar to those of harzburgites and lherzolites found in ophiolites elsewhere. The interpretation of Bluck et al. (op. cit.) was disputed by Barrett et al. (1982) who pointed out the absence of arc material along strike from Ballantrae and suggested instead an obducted seamount terrain on the northern margin of the Iapetus Ocean.

The most definitive statement to date on the isotopic and chemical characteristics of the Ballantrae lavas is by Thirlwall and Bluck (1984). They largely confirm the conclusion of Wilkinson and Cann (1974) that the basalts were generated in a wide variety of tectonic environments which have since become tectonically juxtaposed. A Ti-Y-Zr plot (Fig. 8.14) shows the affinities of the lavas and summarises the reasons why the Ballantrae Igneous Complex cannot be regarded as a single entity, produced in a single tectonic environment. These results favour the proposals of Bluck et al. (1980) for closure of a small marginal basin.

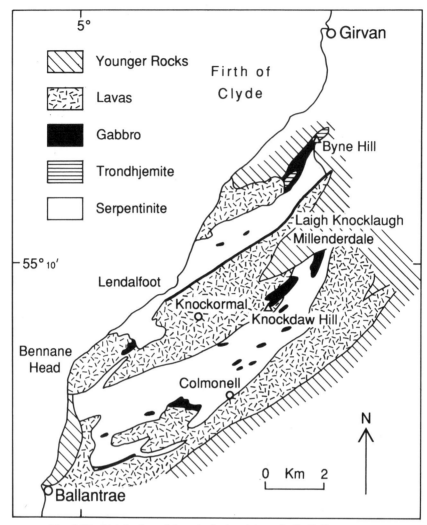

Fig. 8.12. Distribution of the principal rock types in the Ballantrae area.

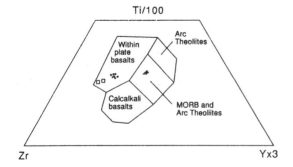

Fig. 8.14. Ti–Zr–Y plot showing affinities of Ballantrae lavas (after Thirwall and Bluck 1984).

Fig. 8.13. Pillow lavas, Downan Point. View onto bedding surface which dips to bottom right. Limestone, chert and tuff fill inter-pillow spaces. Photo B. Bluck.

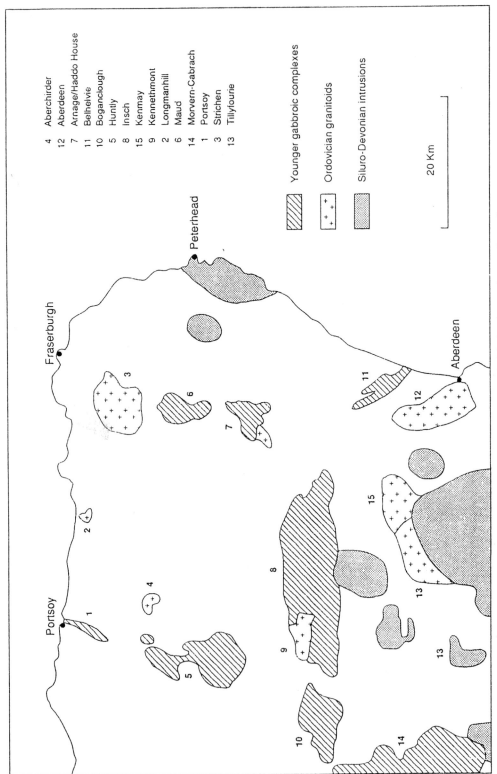

4   Aberchirder
12 Aberdeen
7   Arnage/Haddo House
11 Belhelvie
10 Boganclough
5   Huntly
8   Insch
15 Kenmay
9   Kennethmont
2   Longmanhill
6   Maud
14 Morvern-Cabrach
1   Portsoy
3   Strichen
13 Tillyfourie

Younger gabbroic complexes

Ordovician granitoids

Siluro-Devonian intrusions

20 Km

Fig. 8.15. Newer Gabbros and Ordovician age granitoids in the north-east Grampians. Siluro–Devonian (Newer Granites) are unnumbered.

Younger basic intrusions of the North-east Grampians

The term 'Younger Basic' was used by Read (1961) to distinguish a suite of substantial but very poorly exposed gabbroic intrusions in the NE of Scotland which, in contrast to earlier pre-metamorphic basic bodies, have not been pervasively deformed and metamorphosed. The suite consists of gabbro masses at Huntly, Boganclough, Insch, Haddo House, Arnage, Maud, Belhelvie and Morvern–Cabrach (Fig. 8.15) (*Scott. J. Geol.*, 1970, 6, special issue)

some of which display good igneous layering. The age of the Insch intrusion as deduced from a whole-rock Rb–Sr isochron is 489 ± 17 Ma (Pankhurst 1970) and the whole suite is believed to be of the same age. In fact Shackleton (in discussion of Read & Farquhar 1956) suggested that the various bodies represent a now disrupted single sheet folded after consolidation as a layered intrusion (see also Stewart & Johnson 1960, Fettes 1970), though arguments based on the attitude of supposed gravity-induced rhythmic layering (Stewart 1946, Shackleton 1948) are now less compelling in view of the known complexities of origin of igneous layering in general.

Fig. 8.16. The Belhevie intrusion (after Munro 1986).

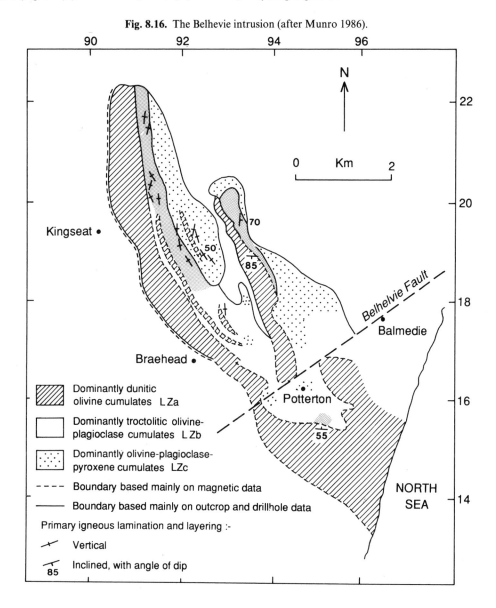

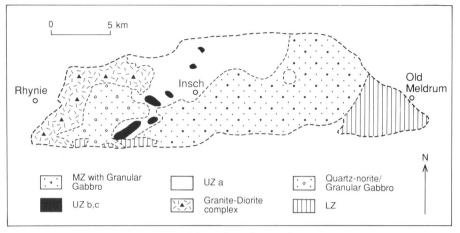

Fig. 8.17. The Insch intrusion (after Wadsworth 1982).

Table 8.2. Generalised differentiation sequence of Belhelvie, Insch and Huntly intrusive gabbros.

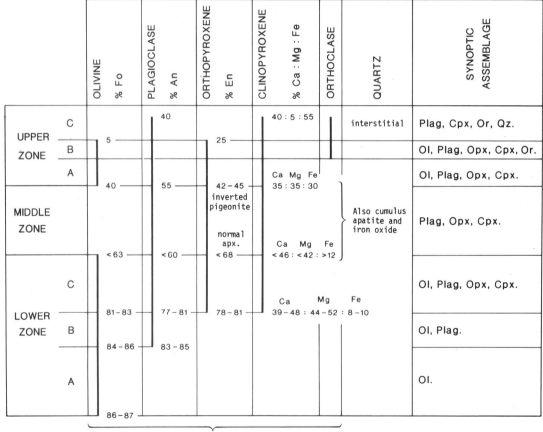

		OLIVINE % Fo	PLAGIOCLASE % An	ORTHOPYROXENE % En	CLINOPYROXENE % Ca : Mg : Fe	ORTHOCLASE	QUARTZ	SYNOPTIC ASSEMBLAGE
UPPER ZONE	C	5	40	25	40 : 5 : 55		interstitial	Plag, Cpx, Or, Qz.
	B							Ol, Plag, Opx, Cpx, Or.
	A	40	55	42–45 inverted pigeonite	Ca Mg Fe 35 : 35 : 30			Ol, Plag, Opx, Cpx.
MIDDLE ZONE		<63	<60	normal apx. <68	Ca Mg Fe <46 : <42 : >12		Also cumulus apatite and iron oxide	Plag, Opx, Cpx.
LOWER ZONE	C	81–83	77–81	78–81	Ca Mg Fe 39–48 : 44–52 : 8–10			Ol, Plag, Opx, Cpx.
	B	84–86	83–85					Ol, Plag.
	A	86–87						Ol.

CUMULUS PHASES

The gabbros were intruded into the Dalradian at about the peak of regional metamorphism, that is post D2 (Fettes 1970, Ashworth 1975, Kneller & Leslie 1984) and produced in places a sillimanite overprint of the regional metamorphic pattern, prior to themselves being involved in regional folding. During this F3 fold episode the gabbros were at temperatures above the Curie points of their ferromagnesian minerals and all retain approximately the same direction of remnant magnetism (Sallomy & Piper 1973). Droop and Charnley (1985) estimated that temperatures in the inner metamorphic aureoles of the gabbros were between 700°C to 850°C at the pressures between 4 kb and 5 kb.

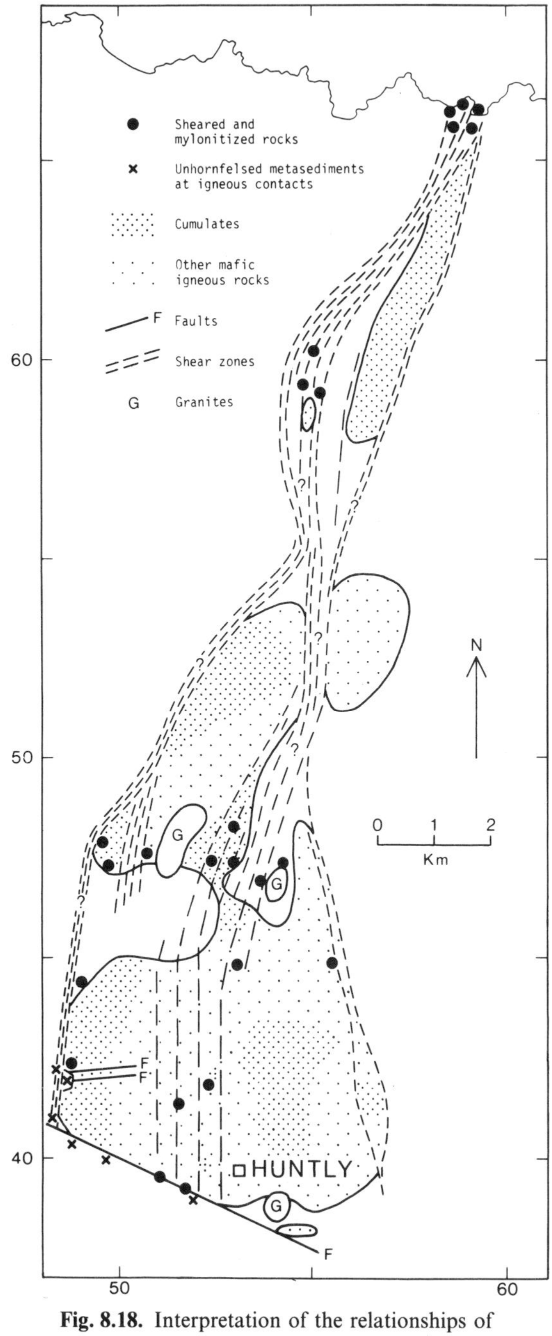

Cumulates are found in the Belhelvie, Insch and Boganclough, Huntly, and Morvern–Cabrach intrusions (Figs. 8.16, 8.17, 8.18, 8.19). Rhythmic, phase and cryptic layering are present (e.g. Wadsworth 1982, 1986, 1988; Munro 1984, 1986; Ashcroft & Boyd 1976), and Wadsworth (1982) has divided the cumulates into lower (LZ), middle (MZ) and upper (UZ) zones ranging in composition from peridotites (olivine cumulates) and troctolites (olivine–plagioclase cumulates), through gabbros (opx–cpx–plagioclase cumulates), to 'syenogabbros' (cpx–opx–plagioclase–alkali feldspar cumulates). This is a generalised fractionation

Fig. 8.18. Interpretation of the relationships of the gabbros in the Huntly–Portsoy area (after Munro and Gallagher 1984).

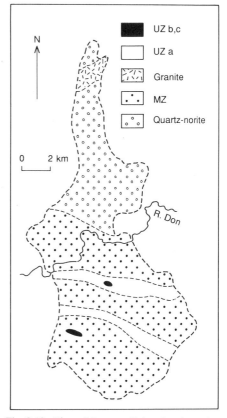

Fig. 8.19. The Morvern–Cabrach intrusion (after Wadsworth 1982).

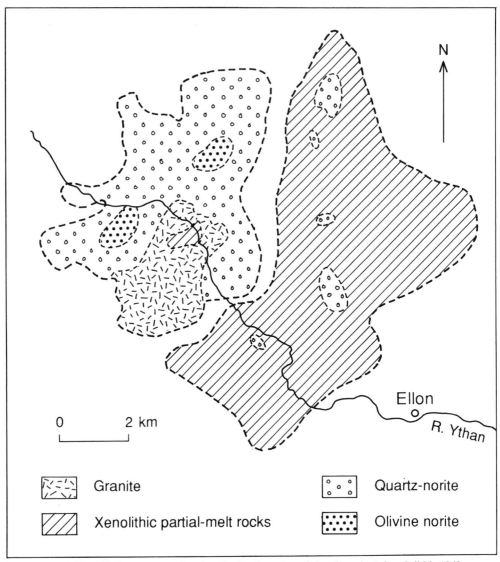

Fig. 8.20. The Arnage mass showing distribution of partial-melt rocks (after Gribble 1968 and Wadsworth 1982).

sequence, but the appearance and disappearance of cumulus phases differ in detail from one intrusion to another (Munro 1984, 1986; Ashcroft & Munro 1978) which suggests that separate batches of magma followed similar but differing fractionation paths (Wadsworth *op. cit.*). Based on compilation of evidence from Belhelvie, Insch and Huntly, this differentiation sequence is summarised in Table 8.2 (after Kneller 1987). The most complete sequence is found in the Insch intrusion and all three zones are present there.

The MZ of the Insch intrusion is further complicated by the intricate association of cumulates with texturally distinct fine-grained granular gabbros (FGG), a rock type also conspicuous in the Huntly mass. Wadsworth (1988) concluded that the MZ cumulates and FGG at Insch are essentially complementary, formed in different locations, and under slightly different conditions, but in the same magma chamber. The complex relationships between the two rock types and an apparently haphazard geographical variation in mineral compositions are enigmatic, but could be due to the foundering of large xenolithic rafts into the cumulus mush.

In addition to cumulates and granular gabbros Wadsworth (1982) recognised a norite group which differs texturally from the others and includes poikilitic plates of biotite. The majority of the norite group rocks are quartz-bearing. Quartz–biotite norites are found in the Insch and Morvern–Cabrach intrusions, and olivine–norites of the same textural group occur in the Haddo House–Arnage intrusion.

The Huntly and Arnage (Fig. 8.20) gabbros are associated with magmatic-looking xenolithic rocks referred to by Read (1935) as cordierite norites and which he suggested were the products of contamination of gabbro magma by pelitic metasediments. These rocks and their congeners throughout the Younger Basic Suite have been studied by Gribble (1966, 1967, 1968, 1970) and Gribble and O'Hara (1967) who showed that the cordierite-bearing xenolithic rocks and quartz–cordierite–norites were formed by partial melting of Dalradian country rocks in contact with the basic magma of the intrusions. Gribble pointed out that the distribution of these partial-melt cordierite-bearing rocks among the members of the basic suite is consistent with the view that they result from the intrusion of the same parent magma into country rocks which were still at relatively high temperature following regional metamorphism.

Post-consolidation deformation

Mention has been made of the effect of the regional F3 deformation on the gabbros, but shearing and mylonitisation are also known from all of the main basic intrusions. The incomplete or disrupted cumulate stratigraphy found in individual intrusions suggests that they are fragmentary and thermal aureoles are often attenuated or removed (Munro 1970, 1984, 1986; Ashcroft & Boyd 1976; Ashcroft & Munro 1978; Boyd & Munro 1978; Leslie 1984; Ashcroft et al. 1984). The shears responsible for this disruption are part of a major regional development of shear zones which commenced shortly after consolidation of the gabbro intrusions and during the peak and waning stages of regional metamorphism (Ashcroft et al., op. cit.).

Grampian (Ordovician) Granites in the North-east Grampians

Pankhurst (1974) used Rb–Sr whole-rock isochrons to define a group of granitic plutons in the north-east Highlands which are syn- to late-tectonic with respect to the Grampian orogenic episode. These Ordovician age granitic rocks (Fig. 8.15) were described by Pankhurst and Sutherland (1982) as Group One Newer Granites and by Brown (1983) as late Grampian Granites. As the Ordovician granites are compositionally quite distinct from the Siluro–Devonian Newer Granites, as well as being closely associated with the Younger Basic bodies, they are here

included in the Grampian plutonic suite. Thompson (1985) has suggested that this basic to acid suite may have volcanic arc affinities. An extensive subduction-related suite of Ordovician granites has also been postulated by van Breeman and Bluck (1981) on the basis of isotopic data from boulders of two-mica granite in the Lower Old Red Sandstone conglomerates along the Highland Boundary Fault.

The principal outcrops of Ordovician granites are at Aberchirder, Aberdeen, Auchedly, Kemnay, Kennethmont, Longmanhill, Strichen and Tilleyfourie (Fig. 8.15), but the true extent of the suite is ill-defined because of the poorly exposed terrain. Where emplaced into low-grade country rocks, as at Aberchirder, the granites are unfoliated and have contact aureoles in the surrounding schists. Where the grade of the country rocks is higher, the granites tend to be foliated and have migmatitic contacts, as at Aberdeen (Kneller 1987).

Kneller and Aftalion (1987) obtained an age from concordant U–Pb systems in monazite of 470 ± 1 Ma for the Aberdeen granite. This is in good agreement with a monazite age of 475 ± 5 Ma for the Strichen granite (Pidgeon & Aftalion 1978) and K–Ar muscovite age of 468 ± 10 Ma for the Auchedly granite (Pankhurst 1970). Kneller and Aftalion (op. cit.) suggested that ages younger than about 470 Ma on other members of the suite are largely reset (e.g. Pidgeon & Aftalion 1978, Pankhurst 1974, Bell 1968).

The Aberdeen granite passes outwards into a network of granitic sheets and veins (Fig. 8.21; Munro 1986) which are petrographically similar to the main body and chemically indistinguishable from it (Walsworth-Bell quoted by Kneller & Aftalion 1987). The veins are broadly syn-D3 in the regional deformation sequence and the 470 Ma age of the granite is regarded as a minimum age for this event, a maximum age being given by the pre-D3 Younger Basic intrusions dated at 489 ± 17 Ma. Foliation in the Aberdeen granite is defined by preferred orientation of micas (muscovite and biotite) and feldspar, and may be partly primary, though much of the granite shows evidence of recrystallisation.

Zones of ductile deformation are common among the granitic rocks of the Ordovician suite and T. N. Harrison (1987) has suggested that in the absence of clear field, petrographic or isotopic evidence, the presence of tectonic fabrics in granitic intrusives in the east Grampians can be taken as an indication of Ordovician age. Deformation occurred during the sequence of intrusion (Kneller & Leslie 1984, Boyd & Munro, 1978) which spanned the local D3 folding, whereas the Siluro–Devonian intrusions do not display any significant penetrative ductile strains.

The Ordovician suite comprises mainly muscovite–biotite granites and biotite granites, though there are associated dioritic rocks, notably in the Kennethmont granite–diorite complex (Busrewil et al. 1975). The granites have S-type features, including peraluminous chemistry, high initial $^{87}Sr/^{86}Sr$ (>0·71), high $^{18}O/^{16}O$ (>10 per ml and inherited zircons (Pankhurst 1974, Pidgeon & Aftalion 1978, Harmon 1983), all of which indicate a predominantly crustal origin.

For the Aberdeen granite isotopic analyses (Halliday 1984) give initial $^{87}Sr/^{86}Sr$ of 0·7119 and ϵNd of $-$ 12·45. The T_{chur} age of the Aberdeen granite is 1758 Ma and for the Strichen and Longmanhill granites 1406 Ma and 1608 Ma

Fig. 8.21. The Aberdeen granite and surrounding vein complex (after Kneller and Aftalion 1987).

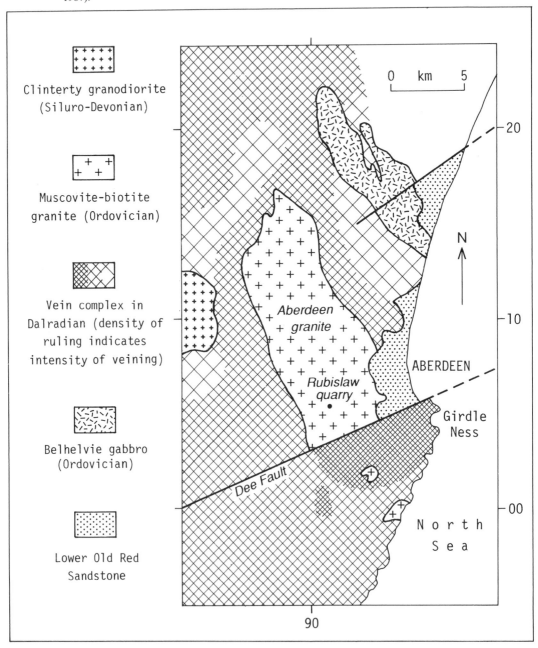

respectively (recalculated from Hamilton *et al.* 1980). Rocks with comparable isotopic characteristics form the Dalradian metasediments which are the country rocks to the granites (O'Nions *et al.* 1983) and for the Aberdeen granite in particular Kneller and Aftalion (*op. cit.*) suggest an origin by extensive melting at shallow depths below the level of emplacement (*c.* 17–20 km at the metamorphic climax).

Fig. 8.22. The Loch Borralan Intrusion (after Johnson and Parsons 1979).

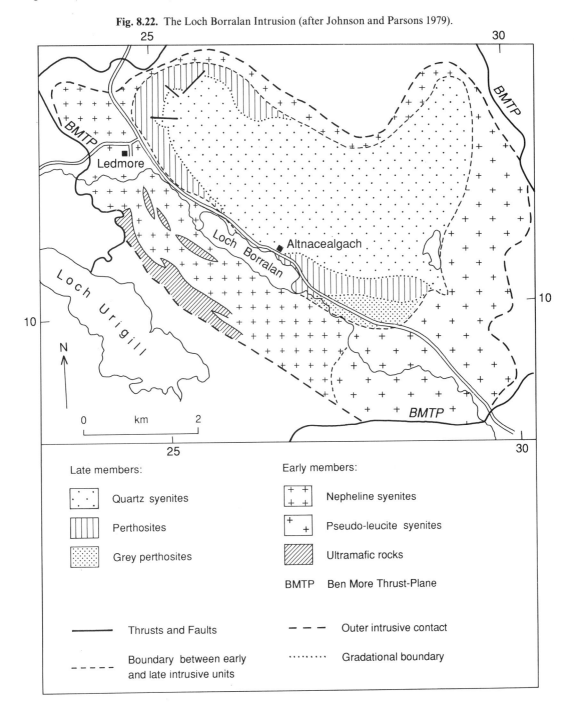

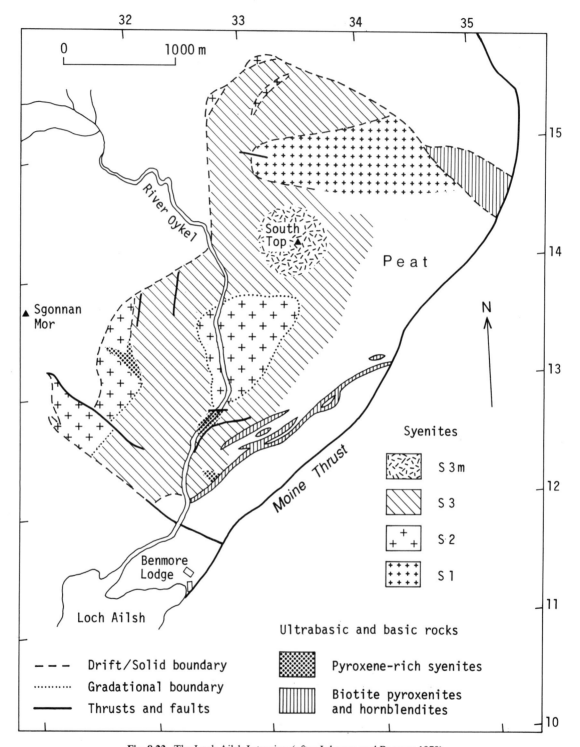

Fig. 8.23. The Loch Ailsh Intrusion (after Johnson and Parsons 1979).

Alkaline intrusions of the North West

In Assynt there are two large alkaline intrusions, at Loch Borralan and Loch Ailsh, and an associated suite of sills and dykes. The country rocks are Lewisian and Moine metamorphics and Cambro–Ordovician sediments, notably limestones and quartzites. The Ben Loyal syenite lies some 40 km north-west of Assynt in a region of Moine schists. Further to the south in western Inverness-shire is the Glen Dessary syenite which has undergone some of the deformation and metamorphism which affected the Moine country rocks.

The Loch Borralan intrusion

The Loch Borralan intrusion (Fig. 8.22) contains a great diversity of rock types including some which are highly potassic, with pseudoleucite and K_2O contents of up to 15 wt% (Wooley 1973). The intrusion was thought by Shand (1910, 1939) to be a stratified laccolith in which mafic nepheline syenites graded upwards into leucocratic syenite and quartz syenites. Explanation of the rock suite was sought in the Daly–Shand limestone assimilation hypothesis (Shand 1945), the suggestion being that after desilication by ingestion of Cambrian limestone the complex evolved by gravity differentiation *in situ*. In fact the leucosyenites are intrusive into the pyroxene–nepheline syenites and there are two separate suites of rocks involved (Woolley 1973). Geophysical surveys and drilling have also led to re-interpretation of the form of the intrusion (Parsons 1965a, Woolley 1970, Mathews & Woolley 1977). The early suite around the locality of Ledbeg appears to have a sheet-like form roofed and floored by limestone, while on the south-west flank of the intrusion there is a steeply inclined sheet of biotite–melanite–pyroxenite. Parsons (1979) suggested that these latter rocks are altered limestones, as are certain related diopside–forsterite–phlogopite rocks only found in boreholes. The later saturated and over-saturated syenites punch through these mafic rocks and probably have a plug-like form.

A limited outcrop of carbonatite has been found in association with the Borralan intrusion (B. Young. *pers. comm.* 1988). The carbonatite intrudes Durness Limestone on the shore of Loch Urigill and varies from early phlogopite–sovite containing xenoliths of nepheline syenite and pyroxenite, to late white sovite which forms the bulk of the outcrop and carries the earlier stages as xenoliths. The phlogopite–sovite contains diopside, iron sulphides and apatite and the white sovite carries in addition humite, K-richterite and Mg-rich ilmenite. Trace element, REE, and oxygen and carbon isotope analyses all confirm the identification of the carbonatite.

The Loch Ailsh intrusion

The Loch Ailsh intrusion (Fig. 8.23) also was originally described as a laccolith made up of successive intrusions of ultra-mafic, mesocratic and leucocratic syenites in upward sequence (Phemister 1926). Reinterpretation by Parsons (1965a, 1965b, 1968, 1972, 1979) has shown that the main rocks of the intrusion are rather sodic, saturated or over-saturated, leucosyenites in three units, S1 to S3, which form a regular chemical series. S1 (the pulaskite of Phemister) and S2 are not seen in contact, but predate S3 (perthosite). Pyroxenites and other ultramafic rocks do not form a lower zone but, on the eastern margin of the intrusion, occur as a near vertical sheet between leuco–syenite and limestone. Pyroxene syenites (shonkinites) occurring as xenoliths in leucosyenite are interpreted as remnants of a roof zone on S2 formed originally by contact alteration of limestones.

The Loch Loyal syenites

The Loch Loyal quartz syenites comprise the Ben Loyal intrusion and the two satellite bodies of Cnoc nan Cuillean and Ben Stumanadh (Fig. 8.24). The form of the Ben Loyal intrusion is that of a laccolith fed from the south-east flank and forcibly expanding into the Moinian country rock on the north-west (Robertson & Parsons 1974). The core and south-east part of the body are occupied by homogeneous and structureless syenite, but elsewhere the outer parts of the intrusion are formed by a somewhat variable foliated syenite in which the structures broadly conform to those in the forcibly deformed Moine envelope. The boundary between the core and marginal syenites is perfectly gradational.

The Ben Loyal and Cnoc nan Cuillean syenites were originally considered to be continuous (Read 1931) but later investigations have shown that the Cnoc nan Cuillean body is a separate stock-like intrusion separated from Ben Loyal by a strip of Moine schists. The Ben Stumanadh body is made up of irregular sheets, partly transgressing and partly concordant with the foliation of the Moine schists, the upper sheet being at least 400 m thick.

All three intrusions differ slightly in their chemistry and petrography, with the Ben Loyal leucosyenites the only members of the alkaline suite in the north-west Highlands to show consistent evidence of peralkalinity. The Cnoc nan Cuillean syenite, for which King (1942) postulated a metasomatic origin, is more basic and potassic than the patches of altered Moine schists which are concentrated in the diffuse margins of the body. The Ben Stumanadh syenite also has an extensive diffuse zone of hybridisation at the margins where feldspathised schists grade into normal syenite (Robertson & Parsons 1974).

The Glen Dessary syenite

The Glen Dessary intrusion consists of an inner felsic syenite rimmed by mafic syenite (Fig. 8.25), representing successive intrusions from a differentiating source (Lambert *et al.* 1964, Richardson 1968). The mineralogy is alkali–feldspar, aegirine–augite, biotite, magnetite, sphene, apatite, calcite, allanite and zircon. The distinction between the mafic and felsic syenites is made on the amount of feldspar relative to dark minerals. During subsequent metamorphism and deformation a fluid phase entered the igneous rocks giving rise to pyroxene–hornblende and pyroxene-free assemblages. At one time the complex, which is strongly deformed and contains inclusions of marble, was considered to be a Lewisian inlier (Harry 1951, 1952).

The Moine rocks at Glen Dessary had already been folded and metamorphosed before emplacement of the igneous complex. Of the three deformation episodes identified in the Moine metasediments only the latest (D3) is thought to have deformed the syenite (Fig. 8.26) which occupies the core of a large, intensely curvilinear, D3 synform (Roberts *et al.* 1984). The deformed igneous rocks have a strong linear fabric which in places develops into a steeply dipping L–S fabric. Prior to the work of Roberts *et*

al. (*op. cit.*) it was believed that the syenite had undergone both D2 and D3 deformation (van Breemen *et al.* 1979a).

Zircons from the Glen Dessary complex have yielded concordant U–Pb ages averaging 456 ± 5 Ma, which is believed to date the crystallisation of the syenites. Undeformed pegmatites which cut the syenites have been dated at *c.* 430 Ma (van Breemen *et al. op. cit.*) and this constrains the age of the D3 deformation and the formation of the Northern Highland steep belt (Roberts *et al. op. cit.*).

Hypabyssal intrusions

The Assynt hypabyssal intrusions are a petrographically diverse suite which Sabine (1953) grouped as aegerine felsites (grorudites), three varieties of quartz micro-syenite (Canisp porphyry, hornblende porphyries and nordmark-

Fig. 8.24. The Loch Loyal Syenites (after Robertson and Parsons 1974).

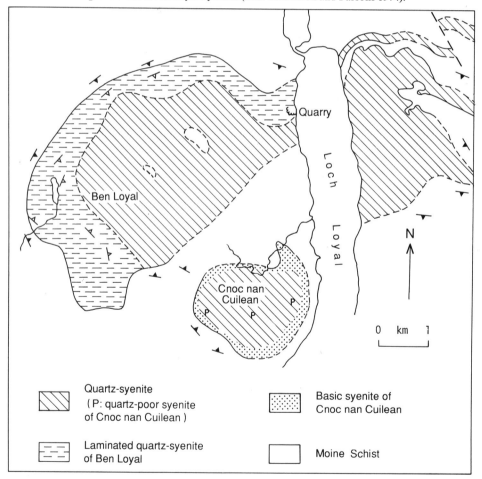

ites), hornblende lamprophyres (vogesites) and nepheline syenites (ledmorites and borolanites). Emplacement of the minor intrusions and the major syenites overlapped movements in the Moine thrust zone, and the spatial distribution of the various members of the suite enable the sequence of igneous and structural events to be determined (Sabine 1953; Woolley 1970; Parsons 1972, 1979; Halliday *et al.* 1987). In particular the most compelling evidence for post- Ben More Thrust displacements on the Moine Thrust is the complete absence, apart from a few deformed

Fig. 8.25. Geology and structure of the Glen Dessary Syenite (after Roberts *et al.* 1984).

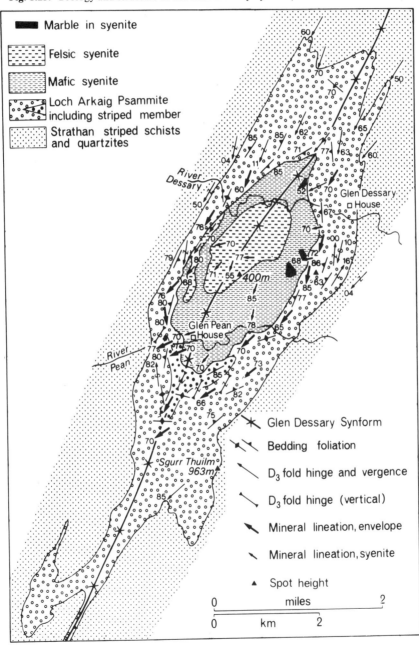

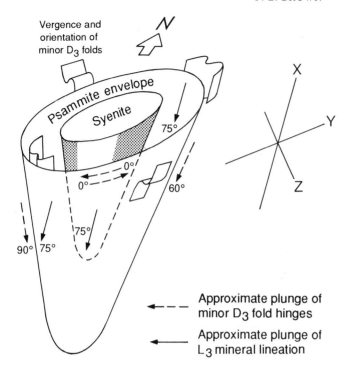

Fenites

Of uncertain lineage but fittingly described in this section dealing with alkaline rocks are certain sodic and ultrasodic rocks of metasomatic origin which crop cut at Glen Cannich and in a roughly north–south zone extending from Loch Hourn to the head of Loch Nevis (Tanner & Tobisch 1972, Peacock 1973). They occur in Moinian metasediments and post-date the regional metamorphism and folding. Feldspathised breccias and dykes occur in the Loch Hourn area but most of the sodic rocks appear to have resulted from the passage of albite-rich hydrous solutions along fractures in the country rocks. The

Fig. 8.26. The structural setting of the Glen Dessary Syenite (after Roberts *et al*. 1984).

nordmarkite sills near the thrust plane, of alkaline minor intrusions in the Moine nappe (Fig. 8.27).

The ages of the alkaline intrusions and events in the Moine Thrust zone

Zircon ages for the main intrusions are: Loch Ailsh 439 ± 4 Ma (Halliday *et al.* 1987), Loch Borralan 430 ± 4 Ma (van Breemen *et al.* 1979b), and for the Cnoc-nan-Cuilean syenite 426 ± 9 Ma (Halliday *et al. op. cit.*).

Table 8.3 (after Halliday *et al. op. cit.*) summarises the geological and radiometric information on relationships between the igneous rocks and movements in the thrust zone. The Borralan complex is thought to have been intruded entirely subsequent to the main movements on the Ben More thrust, but in contrast the Loch Ailsh complex is cut by grorudite dykes which occur only in the Ben More and Glencoul nappes (Sabine 1953), and was therefore emplaced prior to the Ben More thrust. The ages of the two intrusions thus bracket the main movements on the Ben More thrust.

Halliday *et al.* (*op. cit.*) suggested that the source of the alkaline magmas lay in the deep lithosphere rather than the asthenosphere. They linked the Assynt suite with the earlier Glen Dessary syenite (456 ± 5 Ma) and syenite components of the younger Ratagain complex as a linear belt of persistent alkaline magmatism formed during plate con-

vergence, subduction and crustal shortening. The alkaline rocks were seen as the product of a low degree of partial melting in a trace element and large-ionic–lithophile element enriched lithospheric mantle.

Thompson and Fowler (1986) likened the same suite of rocks to the fractional-crystallisation residua of potassic subduction-related magmas such as the shoshonitic series. A picritic shoshonitic parent was suggested for the Glen Dessary, Loch Borralan and Loch Ailsh syenites and a picritic ultrapotassic magma for the Loch Loyal and Ratagain syenites. These magmas were envisaged to have been generated in asthenospheric mantle, modified by infusion from melted subducted material and several hundred km behind a NW-directed subduction zone.

metasomatism resulted in the addition of Na and Al and loss of K, Ca, Mg and Fe, and in the formation of albitites with sodic pyroxenes and amphiboles. In Glen Cannich the metasomatism was controlled by lithological layering in the schists. Tanner and Tobisch found a strong similarity between these rocks and the outer zones of fenitisation around alkaline complexes and in particular with that around the Cnoc nan Cullean intrusion in Sutherland

Fig. 8.27. Geological map of the southern part of Assynt showing the relationship of the major plutons and important minor intrusives to the main structural features (after Halliday *et al.* 1987).

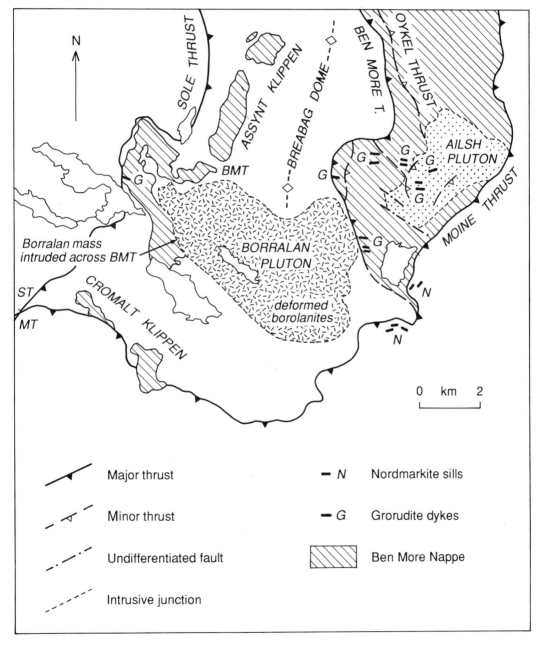

Table 8.3. Major tectonic and intrusive events in the Moine Thrust Zone.

FORELAND	SOLE THRUST SHEET	BEN MORE NAPPE	MOINE NAPPE	AGE
Peak of illite metamorphism in Foreland sediments				*c.* 408
		Ross of Mull Granite cuts Moine Thrust Plane		414 ± 4
	Late undeformed pegmatites in Loch Borralan		*Cnoc-nan-Cuilean complex*	426 ± 9
	Penetrative deformation of pseudoleucite rocks at Loch Borralan. Crush Zones in quartz syenites		*Final movements on the MTP*	
		Late crushing in Loch Ailsh		
		Nordmarkite sills near the MTP		
Nepheline syenite dykes	*Loch Borralan Complex*			430 ± 4
	Main movements on the STP, folding BMTP?			
		Main movements on the BMTP	Moine mylonites and 'D₁' Main movements on MTP	
		Grorudite dykes		
		Mylonites and greenschist-facies metamorphism in Loch Ailsh		
		Sgonnan Mor folds and fabric		
		Loch Ailsh pluton		439 ± 4
Canisp porphyry	Hornblende–porphyry and vogesite sills and dykes			
			'D₃' of Glen Dessary Moine Folding of syenite	
			Glen Dessary syenite	456 ± 5

Italic indicates events that are essentially synchronous.

(King 1942). This similarity leads to the suggestion of unexposed syenitic bodies at depth.

Fenite-type metasomatism has also been described by Deans *et al.* (1971) and Garson *et al.* (1984) from rocks cut by the Great Glen Fault in the Abriachan–Dochfur–Moniack, Rosemarkie and Foyers areas, where secondary crocidolite infills fractures along with aegerine and haematite. Areas of patchy replacement of country rock and cross-cutting breccia dykes also occur. The Newer Granites of Abriachan and Foyers are affected by the metasomatism. A tentative age of 394 ± 15 Ma is suggested from a K–Ar determination on crocidolite.

Newer Granites

The term Newer Granites will be used here for late Ordovician to Siluro–Devonian intrusions, excluding the early Ordovician suite intruded around *c.* 490 Ma to *c.* 470 Ma and referred to above as the Grampian granites. For a review of Caledonian granitoid plutonism across Europe the reader is referred to Stephens (1988). In Scotland Newer Granites occur widely in the metamorphic, or ortho-tectonic, Caledonides and also in the paratectonic Caledonides, notably in the slate belt of the Southern Uplands (Fig. 8.28). They are particularly abundant in a broad NE–SW belt through the Grampian Highlands. Though commonly referred to as granites a wide range of rock types is represented and granitic or granitoid is a more appropriate descriptive term for these intrusions. Diorite, tonalite, granodiorite and adamellite, as well as granite are all common.

Stratigraphic evidence for the age of the Newer Granites is sparse. In the Southern Uplands several important representatives of the Newer Granite suite cut Lower Palaeozoic strata, the youngest being of Wenlock age.

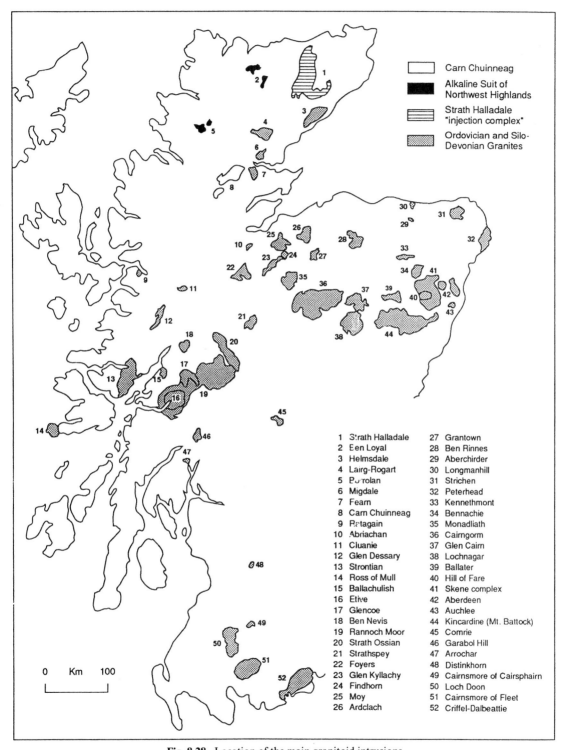

Carn Chuinneag

Alkaline Suit of Northwest Highlands

Strath Halladale "injection complex"

Ordovician and Silo-Devonian Granites

1 Strath Halladale
2 Een Loyal
3 Helmsdale
4 Lairg-Rogart
5 Purolan
6 Migdale
7 Fearn
8 Carn Chuinneag
9 Ratagain
10 Abriachan
11 Cluanie
12 Glen Dessary
13 Strontian
14 Ross of Mull
15 Ballachulish
16 Etive
17 Glencoe
18 Ben Nevis
19 Rannoch Moor
20 Strath Ossian
21 Strathspey
22 Foyers
23 Glen Kyllachy
24 Findhorn
25 Moy
26 Ardclach

27 Grantown
28 Ben Rinnes
29 Aberchirder
30 Longmanhill
31 Strichen
32 Peterhead
33 Kennethmont
34 Bennachie
35 Monadliath
36 Cairngorm
37 Glen Cairn
38 Lochnagar
39 Ballater
40 Hill of Fare
41 Skene complex
42 Aberdeen
43 Auchlee
44 Kincardine (Mt. Battock)
45 Comrie
46 Garabol Hill
47 Arrochar
48 Distinkhorn
49 Cairnsmore of Cairsphairn
50 Loch Doon
51 Cairnsmore of Fleet
52 Criffel-Dalbeattie

0   Km   100

Fig. 8.28. Location of the main granitoid intrusions.

However, the folding of these sediments preceded intrusion of the granites and is comparable with folding in the Lake District where uppermost Silurian strata are affected. The local upper age limit is indicated by detritus from the Criffel–Dalbeattie mass in nearby conglomerates and arkoses of Upper Old Red Sandstone age (Phillips 1956).

The Moor of Rannoch granite is believed to have contributed boulders to the basal Lower Old Red Sandstone conglomerates of Glencoe and is itself cut by the fault

Table 8.4. Localities numbered in Fig. 8.25 and radiometric ages where available.

1. Strath Halladale 649 ± 30 Rb–Sr, Brook quoted in Pankhurst 1982. **2.** Ben Loyal 426 ± 9 Zircon (Cnoc-nan-Cuilean), Halliday *et al.* 1987. **3.** Helmsdale *c.* 420 Zircon, Pidgeon & Aftalion 1978, 401 ± 14 k–Ar and 397 ± 14 K–Ar, Brown *et al.* 1968. **4.** Lairg-Rogart *c.* 420 K–Ar, Brown *et al.* 1968. **5.** Borralan 430 ± 4 Zircon, van Breemen *et al.* 1979b. **6.** Migdale *c.* 400 Zircon, Pidgeon & Aftalion 1978. **7.** Fearn *c.* 400 Zircon, Pidgeon & Aftalion 1978. **8.** Carn Chuinneag 555 ± 10 Zircon, Pidgeon & Johnson 1974. **9.** Ratagain 415 ± 5 Rb–Sr, Turnell 1982 (Ph.D. Thesis, Univ. Leeds, unpubl.). **10.** Abriachan –. **11.** Cluanie 425 ± 4 Rb–Sr, Brook (quoted in Brown *et al.* 1985) *c.* 417 Zircon, Pidgeon & Aftalion 1978. **12.** Glen Dessary 456 ± 5 Zircon, van Breemen *et al.* 1979. **13.** Strontian 435 ± 10 Zircon, Pidgeon & Aftalion 1978. **14.** Ross of Mull 414 ± 3 Rb–Sr, Halliday *et al.* 1979. **15.** Ballachulish *c.* 401 K–Ar, Brown *et al.* 1968. **16.** Etive Meall Odhar 401 ± 6 Rb–Sr, Starar 396 ± 12 Rb–Sr, Clayburn *et al.* 1983. **17.** Glen Coe –. **18.** Ben Nevis –. **19.** Rannoch Moor 402 ± 18 K–Ar, Brown *et al.* 1968. **20.** Strath Ossian 400 ± 10 Zircon, Pidgeon & Aftalion 1978. **21.** Strathspey *c.* 404 Rb–Sr, Brown, J. F. 1975, D.Phil. Thesis, Oxford Univ., unpubl. **22.** Foyers *c.* 385 Zircon, Pidgeon & Aftalion 1978. 409 ± 18 K–Ar, Brown *et al.* 1968. **23.** Glen Kyllachy 443 ± 5–15 Rb–Sr, van Breemen & Piasecki 1983. **24.** Findhorn 413 ± 5 Rb–Sr, van Breemen & Piasecki 1983. **25.** Moy 407 ± 5, Rb–Sr, Zaleski 1982. **26.** Ardclach –. **27.** Grantown –. **28.** Ben Rinnes 411 ± 3 Rb–Sr, Zaleski 1982. **29.** Aberchirder 449 ± 9 Rb–Sr, Pankhurst 1974. **30.** Longmanhill –. **31.** Strichen 475 ± 5 monazite, Pidgeon & Aftalion 1978. **32.** Peterhead –. **33.** Kennethmont 453 ± 4 Rb–Sr, Pankhurst 1974. **34.** Bennachie 404 ± 5 K–Ar, Brown *et al.* 1965. 401 ± 6, Darbyshire & Beer 1988. **35.** Monadliath 419 ± 5 Rb–Sr, Harrison 1987. **36.** Cairngorm 408 ± 3 Rb–Sr, Brook (quoted by Pankhurst 1982). **37.** Glen Gairn Phase II 404 ± 6 Rb–Sr, Harrison 1987. **38.** Lochnagar 415 ± 3 Rb–Sr, Halliday *et al.* 1979. **39.** Ballater –. **40.** Hill of Fare 413 ± 3 Rb–Sr, Halliday *et al.* 1979. **41.** Skene complex –. **42.** Aberdeen 470 ± 1 U–Pb monazite, Kneller & Aftalion 1987. **43.** Auchlee –. **44.** Kincardine 416 ± 4 Rb–Sr, Harrison 1987. **45.** Comrie 408 ± 5 Rb–Sr, Turnell 1982 (Ph.D. Thesis, Univ. Leeds, unpubl.). **46.** Garabal Hill 406 ± 4 Rb–Sr, Summerhayes 1966. **47.** Arrochar –. **48.** Distinkhorn 413 ± 6 Rb–Sr, Thirlwall 1988. **49.** Cairnsmore of Cairsphairn 410 ± 4 Rb–Sr, Thirlwall 1988. **50.** Loch Doon 408 ± 2 Rb–Sr, Stephens & Halliday 1979. **51.** Cairnsmore of Fleet 392 ± 2 Rb–Sr, Stephens & Halliday 1979. **52.** Criffel–Dalbeattie 397 ± 2 Rb–Sr, Stephens & Halliday 1979.

intrusion of the Glencoe Cauldron Subsidence which also cuts Lower Old Red Sandstone volcanics and sediments. It should be noted, however, that the age of these and other Old Red Sandstone volcanics has been shown to be late Silurian rather than Devonian (Thirlwall 1988).

A large body of isotopic age data produced by a variety of techniques is available for the Newer Granites and has been summarised by Brown *et al.* (1985). The ages span the interval 435 to 390 Ma and cluster strongly around 410 Ma (Table 8.4, Fig. 8.29). In referring isotopic ages to the stratigraphic time-scale use will be made of the summary of McKerrow *et al.* (1985) relating to the Ordovician, Silurian and Devonian periods, and in which the base of the Devonian is at 412 ± 3 Ma (an age identical with that proposed by Thirlwall (1988)) and the base of the Silurian at 435 ± 7 Ma.

A detailed isotopic chronological classification of the Newer Granites is, however, an as yet unattained ideal and for the purposes of this account precise stratigraphical subdivisions are not used (cf. the Groups 2 and 3 of Pankhurst and Sutherland (1982)). The number of reliable age determinations is however increasing rapidly and Thirlwall (1988) has claimed to recognise systematic differences in the age of the late Ordovician to Siluro–Devonian magmatic activity through the geological regions of Scotland from the Grampian Highlands to the Southern Uplands (see below).

Forceful and permitted intrusions and some problems of classification

For the Newer Granites Read (1961) suggested a subdivision on intrusive style into forceful and permitted granites and also broadly equated intrusive style with age. The older, forceful granites made their way in by shoving aside their country rocks, while the younger, permitted type arrived by the brittle method of cauldron subsidence. To these mechanisms should be added that of stoping.

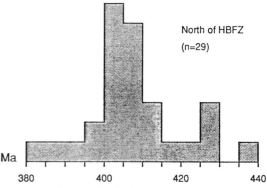

Fig. 8.29. Age distribution of Newer Granites north of the Highland Border fault (after Soper 1986 from data quoted in Brown *et al.* 1985).

It is now known that there is no simple correlation of intrusive style with age or depth. Structurally based studies of emplacement mechanisms of plutons within different tectonic settings (reviewed by Hutton 1988a) have shown that space for magma is usually created by the combination of tectonically created cavities and internal magma-related buoyancy. This occurs in transcurrent, extensional and compressive regimes and Hutton (op. cit.) concluded that transient and permanent space creation, such as may be exploited by magma, is a typical feature of tectonically stressed and deforming lithosphere. This, in combination with the buoyancy of ascending magmas, can generate the varied emplacement mechanisms observed in granitic masses. Thus the variety of emplacement mechanisms observed in the Donegal granites, from diapirism to cauldron subsidence can be related to crustal tectonics prevailing at the time of virtually synchronous intrusion of the different plutons at c. 400 Ma.

Detailed modern structural studies of the Scottish Newer Granites are few, though Hutton (1988b) has shown that the biotite granite of Strontian was emplaced in the termination of a major shear zone and also that the Ratagain complex was emplaced during sinistral shear (Hutton pers. comm.). Meanwhile the broad classification into forceful and permitted intrusions, if divorced from any time implications, remains valid, as does the observation by Read (1961) that the major classic examples of permitted intrusion by cauldron subsidence, referred to by him as the Last Granites, are in the main those which postdate the Lower Old Red Sandstone.

The intrusions – general comments

Space permits comment on only a limited number of individual intrusions illustrative of some of the salient features to be found in the Newer Granites.

Forceful intrusion is exemplified by the Rogart intrusion which consists of a zoned tonalite–granodiorite–adamellite body emplaced in Moine metasediments with a peripheral zone of migmatites (Read & Phemister 1925, Read et al. 1926). Although the migmatites are genetically related to the central igneous complex they are not a static contact effect. Soper (1963) suggested that the ascending body of magma was preceded by a front of migmatisation which it eventually overtook, then continued forceful intrusion of magma deformed the migmatites into a plastic envelope around the expanding magma body. Strong foliation in the outer zones of the intrusion is interpreted as a plane of flattening of semi-consolidated tonalite–granodiorite distended by continued ballooning due to central input of magma. Preferential horizontal distension produced a sub-horizontal lineation and as lateral distension exceeded upward movement a partly funnel-shaped body resulted. The outer tonalite grades rapidly into granodiorite but the adamellite crosscuts structures in the earlier members and

was emplaced under more brittle conditions. A K–Ar age of c. 420 Ma (Brown et al. 1968) is effectively the same as that of the regional country rocks showing that the pluton cooled through the argon blocking temperature along with the surrounding schists. An accurate emplacement age for the Rogart intrusion has yet to be determined.

The Strontian complex, like the Rogart intrusion, is a zoned body with outer tonalite grading into granodiorite and with a later central, cross-cutting biotite–granite (Sabine 1963; Munro 1965, 1973). The intrusion has been dated by U–Pb on zircons as intruded at 435 ± 10 Ma (Pidgeon & Aftalion 1978), which was at or close to the termination of regional deformation and metamorphism. According to Ashworth and Tyler (1983) the emplacement of the igneous rocks was at pressures of about 4 kbar into rocks which were probably still at amphibolite facies temperatures. In the northern part of intrusion (Fig. 8.30) the tonalite and granodiorite have a well-developed foliation broadly concordant with the Moine contact and defining a synformal pattern which extends southward of Loch Sunart. Hutton (1988b) has shown that the final, biotite–granite, member of the pluton was emplaced in the extensional termination of a dextral transcurrent shear zone which is itself a splay on the Great Glen Fault. The arcuate northern sheeted complex of biotite granite (Fig. 8.31) represents the extensional zone termination and the granites western contact is constrained to be long and straight by deformation in the shear zone. The main body of biotite–granite magma was emplaced sideways as a sheet into an extensional listric fault bounded cavity.

The Foyers intrusion also consists of tonalite, granodiorite and granite or adamellite (Mould 1946, Marston 1971), but with parts of the roof of country rock still in situ. The outer tonalitic mantle to the inclined funnel-shaped pluton forms a partial conformable envelope which was deformed by continued injection of granodiorite magma. Sediments in roof pendants at Foyers show marked thermal metamorphism with development of sillimanite, andalusite and cordierite and the country rocks are heavily injected by igneous material from the complex. Kennedy (1946) believed that similarities between the Foyers and Strontian intrusions were such that the two could be matched by allowing for a 105 km sinistral displacement on the Great Glen Fault. Marston (1971) suggested that such a comparison was made equivocal by the respectively high and low structural levels of the two intrusions. Detailed study of the thermal aureole of the Foyers intrusion (Tyler & Ashworth 1983) has, however, shown that the metamorphism took place at about 4 kbar pressure, similar to that at Strontian. The different metamorphic environment at Foyers can be explained by more prolonged cooling after regional metamorphism which, together with the fact that the metamorphic environment was drier at Foyers, may bear on the structural differences between the two areas. Rare-earth data and strontium isotope evidence strongly indicate lack of correlation between the two intrusions (Pankhurst

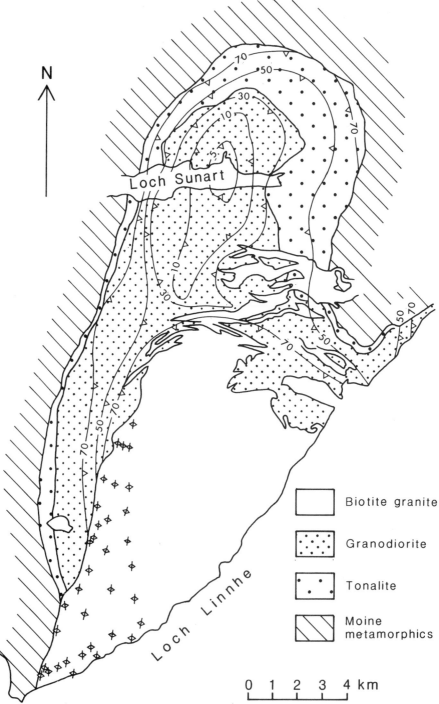

Fig. 8.30. Basic structure of the Strontian pluton. Generalised foliation trends are shown in the tonalite and granodiorite; single double barbed symbols are foliations in the biotite granite (after Hutton 1988a).

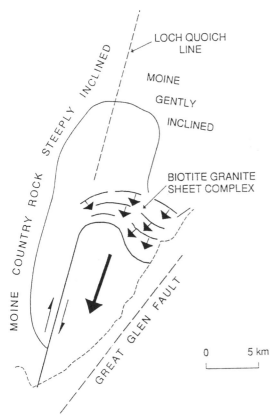

Fig. 8.31. Model for the emplacement of biotite granite in a dextral shear-zone termination related to the Great Glen fault (after Hutton 1988a).

1979), although isotopic evidence from zircons is not decisive (Pidgeon & Aftalion 1978, Halliday *et al.* 1979).

There are numerous other less well documented intrusions in the Highlands which appear to have been forcefully emplaced, such as the Cluanie pluton (Leedal 1952) which has a contact migmatitic aureole deflected into conformity with the margin of the intrusion.

The Lower Palaeozoic rocks of the Southern Uplands slate belt are cut by several concentrically zoned plutons which show forceful characteristics in deflection of regional cleavage trends and in some cases development of igneous foliations. The Loch Doon mass consists of a central granite surrounded by tonalite and some norite (Gardiner & Reynolds 1932, Brown 1979, Tindle & Pearce 1981), while the smaller Cairnsmore of Cairsphairn intrusion shows concentric zonation from a dioritic margin to a central granite core (Deer 1935, Tindle *et al.* 1988). Foliation in the Loch Doon mass was attributed by Oertel (1955) to forceful lateral expansion of the growing pluton.

The Criffel Pluton (Phillips 1956, Stephens & Halliday 1980, Phillips *et al.* 1981, Stephens *et al.* 1985) has an outer zone of clinopyroxene–hornblende–biotite granodiorite which surrounds a core of biotite granite, muscovite–biotite granite and biotite–muscovite granite. Foliation in the outer Criffel granodiorite was attributed by Phillips *et al.* (*op. cit.*) to both convection and uprise of the inner granite magma. This has been amplified by Courrioux (1987) who used analysis of strain trajectories, strain gradients and quartz C-axis fabrics to deduce a detailed emplacement history for the pluton. Most of the deformation in the granodiorite is attributed to diapiric emplacement of the inner granite with asymmetric ballooning resulting from an inclined intrusion direction (Figs. 8.33, 8.34).

The Cairnsmore of Fleet intrusion (Parslow 1968, 1971) is composed of an outer coarse granite with biotite and biotite–muscovite facies and an inner fine-grained biotite–muscovite granite. This is the most felsic pluton in the Southern Uplands and exhibits the smallest petrological range (Halliday *et al.* 1980). Like the other major intrusions it has deformed its envelope.

Plutonism in the Southern Uplands dates from about 410 ± 400 Ma except for the Cairnsmore of Fleet which was intruded at 391 ± 2 Ma (Halliday *et al.* 1980). Negative gravity anomalies running NE–SE parallel to the Caledonian trend along the Southern Uplands have been interpreted as an underlying granite batholith connecting the exposed plutons at depth (Lagios & Hipkin 1979, Parslow & Randle 1973).

Stoping although probably fairly common in the Newer Granites is not easily demonstrated. An example is seen in the Ross of Mull pluton (Riley 1966) which contains large free-swimming rafts of Moine metasediments not far removed from their place of origin and preserving their pre-intrusion stratigraphic relationships. This particular intrusion has one of the best developed contact aureoles among the Highland granites with cordierite, andalusite and sillimanite present (MacKenzie 1949). It also post-dates movements on the Moine Thrust and a Rb–Sr age of 414 ± 3 Ma by Halliday *et al.* (1979) gives a minimum age for the thrust movements.

Stoping has also been suggested as a major process in the emplacement of the large stock-like Cairngorm pluton (Harrison 1986) which was intruded at a high level (less than 6 km) into brittle crust with no evidence of diapirism. It is suggested that the partially crystalline magma behaved as a Bingham fluid and that the size of the majority of the stoped blocks was such that the critical yield strength of the magma was exceeded allowing the xenoliths to sink (cf. Marsh 1982). Such reasoning is no doubt applicable to numerous other high-level members of the Newer Granites (Harrison & Hutchinson 1987). Manganese-rich garnets in the margin of the Cairngorm granite are considered by Harrison (1988) to have formed due to ponding of Mn-rich fluids against wall of pluton and are not due to assimilation of metasedimentary material.

Fig. 8.32. View of the Rubislaw Quarry in the Aberdeen granite when working in the 1920s. B.G.S. photo.

Among the permitted intrusions those at Ben Nevis (Fig. 8.35), Glencoe and Glen Etive in the south-west Highlands are outstanding. The largest is the Etive complex (Clough *et al.* 1909, Bailey 1960, Anderson 1937) in which there were at least four successive pulses of magma related to repeated ring-fault subsidence of a central more-or-less cylindrical block. From study of the contact metamorphic aureole, Droop and Treloar (1981) estimated a depth of intrusion between 3 and 6 km. In the southern part of the complex diorite forms the outermost member and between it and the next innermost Cruachan granite is a screen of metamorphosed lavas. Each of the successive pulses of granite is identified by its own textural or mineralogical characteristics and usually by sharp junctions against the earlier arrivals. The Cruachan granite member is of particular interest in that the high-level Glencoe Cauldron Subsidence

foundered in to the Cruachan granite magma chamber, so that the Glencoe Cauldron is invaded by the granite which on the east side of the Cauldron merges with the ring-fault intrusion. On the west of the Cauldron, however, the ring-fault intrusion is apparently cut by the Cruachan granite.

The Etive complex has featured prominently in recent petrological studies. From Sr, O and Pb isotope data Clayburn *et al.* (1983) envisages a juvenile melt from a mantle source initiating melting within the lower crust. Mixing of these melts in a lower crustal reservoir resulted in a dry hybrid magma which was rapidly intruded by fracture-associated emplacement into the upper crust. A progressively increasing component of lower crust is found in the later intrusive units, implying that lower crustal melting continued throughout the igneous activity. Frost and O'Nions (1985) also found that the Etive complex had

Fig. 8.33. Foliation trajectories in the Criffel pluton (after Courrioux 1987).

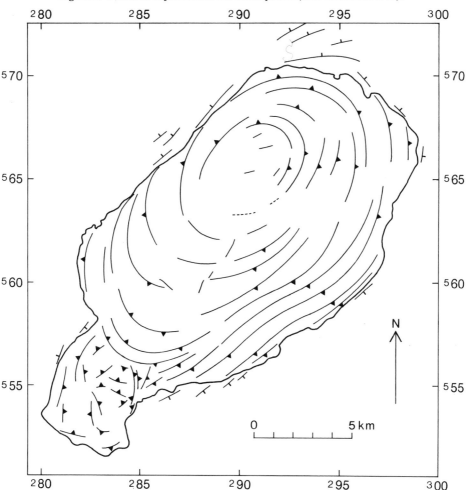

recycled material from the lower crust and lithospheric mantle to shallow levels, and that the last stages of evolution of the complex involved assimilation of up to 25 per cent of Dalradian metasediments.

According to Clough *et al.* (1909) and Bailey (1960) the Glencoe Cauldron Subsidence was formed by the two-stage subsidence on encircling ring-faults, of a cylindrical block of Lower Old Red Sandstone lavas and underlying schists. Each stage of cauldron subsidence led to the upwelling of

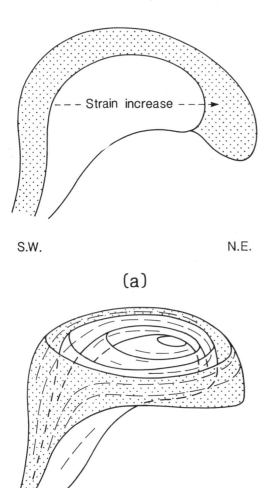

Fig. 8.34.
(a) Geometric relations between granite and granodiorite in the Criffel pluton and
(b) foliation and lineation trajectories (after Courrioux 1987).

granite magma along the ring fractures and this resulted in the early and late fault-intrusions. Seven groups of volcanics and minor sedimentary horizons, in all about 1 km thick, are present within the caldera.

A more sophisticated picture of the evolution of the Cauldron has been developed by Roberts (1963; 1966a, b; 1974); Taubeneck (1967) and Hardie (1968) who recognised that the two groups of flow-brecciated rhyolites mapped by the earlier investigators are, in fact, ignimbrites, often in the form of welded tuffs, and which can be correlated with two pyroclastic horizons in the Lorne Plateau lavas. The ring-faults are now seen as inward-dipping tensile fractures formed in response to excess pressure developed in a magma chamber at depth. Escape of magmatic material up the fractures allowed the central block to subside, thereby forming a surface caldera. The two episodes of caldera formation and fault intrusion can be correlated with the two surface eruptions of ignimbrites. Roberts (1974) has also shown that the first cycle of caldera formation was lop-sided with volcanic activity concentrated along the north-eastern margin and the caldera formed by flap-like sub-sidence of the floor.

The Ben Nevis complex (see Fig. 8.36) is another classic example of cauldron subsidence which consists of four concentrically arranged units. A narrow, discontinuous, outer rim of heterogeneous quartz–diorite was variously subdivided by Anderson (1935) and Haslam (1968), and grades rapidly inwards into hornblende–biotite adamellite which is known as the Porphyritic Outer Granite (Bailey 1960). Maufe (1910) visualised these rocks as filling a void created by sinking of a subterranean block into the underlying magma chamber. After dyke injection a further subsidence admitted the biotite granodiorite (the Inner Granite) into which there then subsided a block of schists overlain by rhyodacite lavas, against which the Inner Granite is chilled.

Further to the north-east in Aberdeenshire the Loch-nagar Granitic complex consists of nine concentric annular and crescentic masses of adamellite and granodiorite. The early members of the complex were forcibly intruded while the later central members were admitted by cauldron subsidence and the complex as a whole appears to combine aspects of both forceful and permitted mechanisms of emplacement (Oldershaw 1974).

Internal variation and magmatic evolution

Although zoned tonalite to granite plutons such as Rogart or Strontian are numerous among the Newer Granites, there are others (for example Helmsdale and Cairngorm) which consist almost entirely of granodiorite or adamellite. Yet others are more variable and range from ultramafic to granitic. Notable in the latter group are Comrie near the Highland Border and composed mostly of diorite; Glen Doll (south of Lochnagar), composed of olivine gabbro, pyroxene diorite and adamellite; Glen Tilt of diorite, tonalite and granite; and near the head of Loch Lomond where the intrusions of Arrochar and Garabal Hill display a remarkable array of ultramafic, gabbroic, dioritic and granitic rocks.

The Newer Granites as a whole display essentially calc–alkaline characteristics as shown by the plots of CaO and $Na_2O + K_2O$ Vs SiO_2 and the AFM diagram (Fig. 8.37). The order of emplacement is usually from mafic to felsic, and in individual complexes the major element variation usually displays characteristic smooth curves.

In his classic account of the Garabal Hill complex Nockolds (1941) concluded that the major process in the evolution of the complex was differentiation by fractional crystallisation. A parent pyoxene–mica diorite magma was postulated and more felsic rocks were considered from their smooth variation to represent the liquid line of descent. The more mafic members, such as pyroxenites and hornblendites were held to be cumulates. Differentiation took place beneath the present level of exposure and the various units of the complex arrived as separate pulses of magma.

Sharp intrusive contacts are not always displayed between members of this type of complex and some show progressive zonation inwards. The growth model for such plutons suggests that progressively less mafic magmas

Fig. 8.35. The cliffs of Ben Nevis formed predominantly of lavas within the Inner Granite. B.G.S. photo.

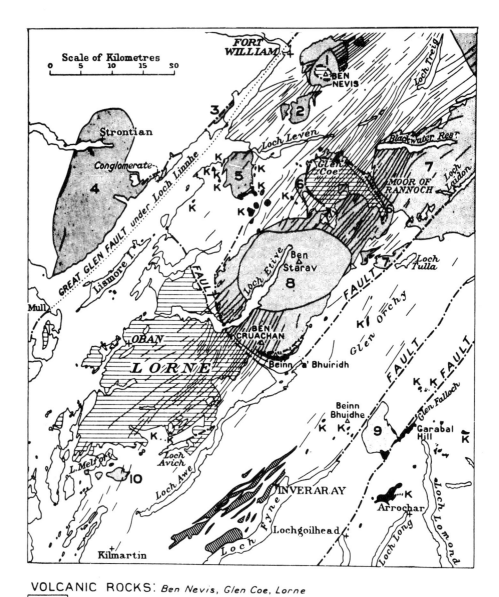

VOLCANIC ROCKS: *Ben Nevis, Glen Coe, Lorne*

BASALT, ANDESITE, RHYOLITE.

DYKES: *Mostly of Nevis and Etive Swarms.*

PORPHYRITE, MICRODIORITE, LAMPROPHYRE, PORPHYRY.

INCLINED SHEETS: *Loch Fyne.*

QUARTZ-PORPHYRY.

GRANITIC PLUTONS
GRANITE, QUARTZ-DIORITE: 1 *Ben Nevis*; 2 *Mullach nan Coirean*;
3 *Loch Linnhe*; 4 *Strontian*; 5 *Ballachulish*; 6 *Glen Coe*; 7 *Rannoch*; 8 *Etive*;
9 *Garabal*; 10 *Loch Melfort.*

BASIC and ULTRA-BASIC PLUTONS, *including a few giant xenoliths*

APPINITE SUITE including K = Kentallenite

Fig. 8.36. The igneous rocks of the South-west Highlands (after Bailey 1960).

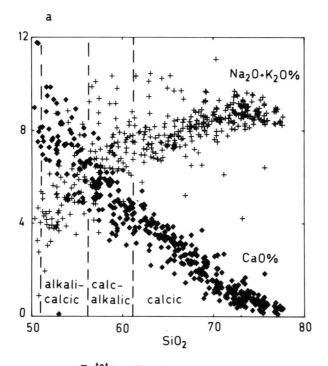

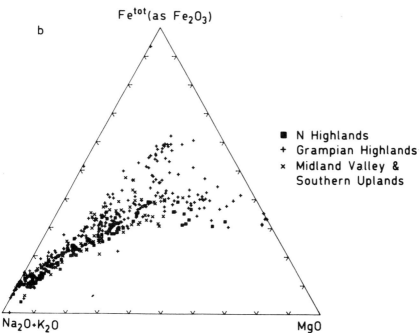

Fig. 8.37.
(a) Total alkalies and CaO versus SiO_2 in the Newer Granites.
(b) AFM plot showing the calc–alkaline nature of the Newer Granites (after Halliday *et al.*
1985).

arrived more or less continuously from below, though lack of internal contacts can also be seen as evidence of crystallisation differentiation *in situ* from the walls inwards. Even when merging rather than sharp internal contacts are present, however, the transition from one rock type to another is usually fairly rapid and reasonably explained by penecontemporaneous magma pulses rather than *in situ* differentiation.

Alternatively the concept of independent existence of felsic and mafic magmas and of hybridisation in the genesis of intermediate rock types has had many advocates. The Cairnsmore of Cairsphairn intrusion with its concentric zonation from dioritic margin to granite core was explained by Deer (1935) in this way. Hornblendic and pyroxene–biotite hybrids were produced at the present level of exposure by the interaction of a tonalite magma with semi-consolidated gabbro and the tonalite itself was closely followed by granite magma to produce a zone of acid hybrids. Deer also advocated a deeper-seated granite–gabbro hybrid origin for the main body of tonalite magma.

This model for the evolution of the Cairsphairn intrusion has been revised by Tindle *et al.* (1988) in the light of new comprehensive major and trace element data. This shows that hybridisation alone could not have produced the main volume of intermediate tonalite, although a form of hybridisation may account for transitional boundaries between the main rock types at the present level of exposure. The preferred model is one crystal fractionation of a dioritic magma, initially separating pyroxenes and plagioclase and subsequently biotite and plagioclase. A simple model of *in situ* crystallisation is, however, untenable. The locally sharp internal contacts and varying compositional gradients indicate that crystal fractionation occurred mainly below the level of emplacement and it is envisaged that the dioritic magma fractionated towards a granodiorite composition which broke through the partially crystallised roof to form the zone of granodiorite. The melt continued to fractionate and, after an interval, a granitic melt was intruded through the granodiorite to form the inner granite zone.

Crystal fractionation has also been proposed as the dominant mechanism of evolution of the Loch Doon intrusion (Brown *et al.* 1979, Tindle & Pearce 1981) with early crystallisation of the dioritic magma dominated by separation of pyroxene and plagioclase, which were later joined by biotite as granitic liquids developed.

In the Etive pluton, Batchelor (1987) suggested that two pulses of monzogranitic magma which formed the outer (Cruachan) component of the pluton resulted by separation of pyroxene, amphibole, plagioclase and accessory minerals from the dioritic magma. Four separate nested pulses which make up the central (Starav) component of the pluton successively tapped a magma reservoir which was fractionating clinopyroxene or amphibole and plagioclase. *In situ* crystallisation of plagioclase, K-feldspar, biotite and accessory minerals led to further variation in each pulse.

These few examples, which have the advantage of calling upon the armoury of modern analytical and computational techniques, are but some of the numerous studies illustrating the perceived importance of crystal fractionation in the evolution of the Newer Granites.

Fractional crystallisation is, however, but one of the many processes which operated during the evolution of the Newer Granites. In particular the concept of independent basic and felsic magmas and hybridisation has received support from isotope geochemistry. From isotope data on 25 plutons, ranging from Helmsdale in the north to Criffel in the south, Harmon and Halliday (1980) found a positive correlation between $\delta^{18}O$ and $^{87}Sr/^{86}Sr$ initial ratios, both of which increase sympathetically with progression from mafic to felsic rock compositions. This is interpreted as a result of the mixing of magmas derived by partial fusion of two isotopically distinct sources, one mantle-like (lower crust or continental lithospheric mantle) and the other crustal and relatively enriched in $\delta^{18}O$ and radiogenic Sr. No systematic distribution of Sr or O isotope variations was found with geographical position, age, or intrusive style.

This isotopic study was pursued in greater detail for the Doon (408 ± 2 Ma), Criffel (397 ± 2 Ma) and Fleet (393 ± 2 Ma) intrusions in the Southern Uplands (Halliday *et al.* 1980). This time-emplacement sequence was accompanied by a progression towards more siliceous compositions, higher $^{87}Sr/^{86}Sr$ initial ratios and higher $\delta^{18}O$ values. All three plutons are zoned and display an overall increase in initial $^{87}Sr/^{86}Sr$ and $\delta^{18}O$ inwards from the margins. These data, combined with isotopic studies of the surrounding Lower Palaeozoic sediments, common Pb and U–Pb (zircon) data and geophysical constraints, were interpreted by Halliday and his co-workers to indicate that the magmas were derived by melting of mantle and/or 'new' basic lower crust, plus metasediments. Incomplete hybridisation between magmas from these differing sources was responsible for the observed isotopic variations. Fractional crystallisation superimposed on the hybridisation process was also responsible for major variations in rock type in these intrusions.

The Criffel pluton has been the subject of further geochemical study by Stephens *et al.* (1985) in which it was shown that the rare earth elements (REE) are strongly correlated with Sr and O isotopes. There is progressive decrease in total REE with approach to the geochemical centre of the pluton along with change to more radiogenic Sr and more silicic and peraluminous compositions. From consideration of various petrogenetic models, including cumulate formation, restite separation, contamination by assimilation of wall rocks, thermogravitational diffusion and volatile mass transfer, it was concluded that a process of assimilation and fractional crystallisation (AFC) had been operative. The parental magma was an I-type crustally derived magma of granodiorite composition and contamination probably took place in the middle crust.

Although multiple magma sources are well accepted as having played a part in the genesis of the Newer Granites there is nevertheless much debate on the relative roles of continental crust and mantle, be it lithospheric mantle or asthenosphere. This is a facet of the major problem of generation and evolution of the continental crust to which granitoid intrusion is fundamental. It is not appropriate here to review the complex and at times convoluted arguments which centre largely upon the interpretation of isotopes of Sr, Nd and Pb (Halliday *et al.* 1979, O'Nions *et al.* 1983, Clayburn *et al.* 1983, Harmon *et al.* 1984, Halliday 1984, Halliday *et al.* 1985, Frost & O'Nions 1985, Clayburn 1988). The consensus opinion is that although the Lower Palaeozoic was a time during which a large amount of igneous material was introduced into the upper crust, it was not a major crust-forming period in that the Caledonian granites are dominated by recycled continental crust. It was a time when there was much melting near to the Moho. Thirlwall (1986) has, however, argued that an enriched mantle source can be identified as having contributed extensively to the incompatible element budget of the Caledonian granitoids, implying that there was in fact significant growth of the crust.

The degree of mantle involvement is debatable but the source of at least some of the chemical variation, the granitoid magmas, did lie in the lithospheric mantle. A significant study of Nd and Sr isotopes in microdiorite enclaves from the Criffel and Strontian intrusions enabled Holden *et al.* (1987) to show that the enclaves are representative of synplutonic mafic magma derived from mantle and injected into the host granite magma. This major input of melt and heat from the mantle substantially aided crustal melting and granitoid genesis.

Subduction, terranes and granite suites

Interest in the roles of mantle and deep crust in granitoid genesis is heightened because granitic intrusions are images of their source rocks and are to be seen as potential providers of information on subduction processes and terrane boundaries. Collision tectonics may bring together differing deep crustal and mantle lithologies with very varied magmatic products. With these factors in mind some of the salient compositional characteristics and regional variations of the Newer Granites will be described.

In terms of the major subdivision into I-type and S-type granites as recognised by Chappell and White (1974, 1984), the Newer Granites are predominantly I-type (however, see Plant *et al.* 1983, Atherton & Plant 1985). Very few are strongly peraluminous (A/CNK > 1·1), a notable exception being the Cairnsmore of Fleet intrusion which has both chemical and Sr and O isotope compositions tending towards S-type. Several authors (e.g. Stephens & Halliday 1984, Harrison & Hutchinson 1987) have also noted that some of the east Grampian granites have affinities with the A-type granites of SW Australia.

Like other calc–alkaline magmas intruded into continental crust the Newer Granites are a high-K suite. They are also relatively Na-rich and in some instances markedly so in Ba and Sr, which are considered to be provincial characteristics (Halliday & Stephens 1984). There are also consistent regional variations in chemistry across Scotland and in particular Na, Sr and Ba increase towards the north-west, whereas Th and Rb decrease (Halliday *et al.* 1985). In the extreme NW the Ratagain complex has up to 6,000 ppm Ba, 10,000 ppm Sr and > 7 per cent Na_2O.

Suites

Grouping of cognetic Newer Granite plutons into suites on the basis of petrographic, chemical and isotopic criteria will no doubt be developed as data accumulate. Chappell and Stephens (1988) recognise good trends on single plutons but not with groups of separate plutons. Suites are not clearly recognisable.

Nevertheless, within the regional variations observed, Stephens and Halliday (1984) provisionally recognised three suites of granitoids each with distinctive chemical characteristics. These are the Cairngorm, Argyll and South of Scotland suites, the distribution of which is shown in Fig. 8.38 and the essential features in Table 8.5.

The pyroxene–mica diorites which are associated with the plutons of the south of Scotland suite are comparatively rich in Ni and Cr and represent mantle-derived melts. The diorites associated with the Argyll suite, on the other hand, are thought to have a deep crustal source affected by mantle derived incompatible element-rich fluids. Some plutons in the Cairngorm suite (notably Cairngorm and Peterhead) are described as having characteristics transitional from I to A type granites.

The Newer Granites in the eastern Grampian region, in which most of the Cairngorm suite lie, have also been divided by Harrison and Hutchinson (1987) into an early group (Monadliath, Lochnagar, Hill of Fare, Kincardine), often associated with relatively abundant diorites, and a slightly later group not associated with diorites (Ben Rinnes, Moy, Cairngorm, Glen Gairn Phase II, Bennachie). Harrison and Hutchinson (*op. cit.*) suggested that the earlier group of intrusions resulted from subduction related melting of the lower crust (the diorites do not have an unequivocal mantle signature) whereas the later group, which have the incipient A-type chemistry, arose from melting induced by the introduction of mantle material into the lower crust along the root zones of major strike–slip faults.

Isotopic evidence

Relationships between systematic regional variations in the granitoids and spatial changes in the crust or mantle have been assessed using isotope data. Pidgeon and Aftalion (1978) demonstrated that granitoids from north of the Highland Boundary Fault contain zircons with a marked isotopic memory of old radiogenic Pb, whereas granitoids from south of the fault contain negligible memory of older zircon. This was interpreted as a fun-

damental change in the nature of the source regions for the granitoids, those from the north being largely derived from a continental basement of *c.* 1,600 Ma age, whereas those from the south were derived from Lower Palaeozoic

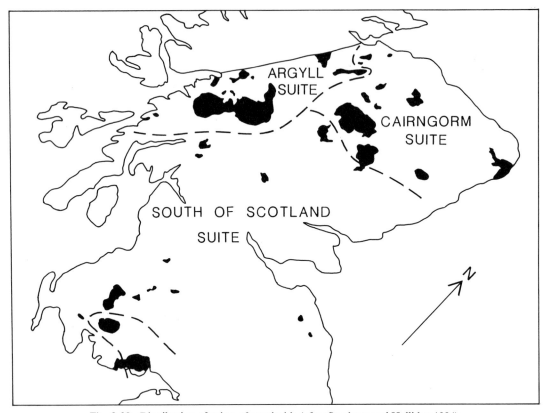

Fig. 8.38. Distribution of suites of granitoids (after Stephens and Halliday 1984).

Table 8.5. Characteristics of late granitoid suites.

	Cairngorm suite	*Argyll suite*	*S of Scotland suite*
Rock types	Mainly red biotite granites; intermediate types rare; few appinites	Common granodiorites and diorites; appinites abundant; hornblende characteristic of diorites	Commonly diorites and granodiorites; pyroxene typical of diorites; appinites in the N only
Major oxides	Highly silicic metaluminous compositions	Calc-alkalic high Na_2O compositions	Calc-alkaline
Trace elements	High Nb, Rb and Th; low Ba, Sr	Very high Sr and Ba; low Nb, Th, Rb	Low La, Ce, Ba, Sr
Age (Ma)	408–415	410–415	390–408
εSr	+24–+33	−7–+58	+1–+54
εNd	−8–−1	−10–+3	−4–+1
$\delta^{18}O$	8·2–11·1	7·2–10·7	7·9–10·4

sources. Subsequently Halliday *et al.* (1979) showed that the crustal metasediments to the north and south also have the same degree of U–Pb zircon memory and therefore the presence or absence of zircon memory can be explained in terms of melting and/or assimilation of the differing crustal rock types regardless of the ultimate source of the magma. There is, however, no doubt that the crust changes in terms of age from north to south and this is likely to be accompanied by changes in chemistry.

The initial ϵNd isotopic compositions of the Newer Granites also become more radiogenic southwards (Fig. 8.39) and very negative initial ϵNd values (less than 6) are restricted to the Highlands. A value of -6 ϵNd separates the granites with inherited zircons from those without (Halliday 1984). The ϵNd isotope systematics thus display parallels with common Pb and U–Pb zircon systematics in showing differences from north to south. As the inherited zircons are unequivocally derived from continental crust

Fig. 8.39. ϵNd$_t$ values and the position of the Mid-Grampian Line. Figures in brackets are numbers of samples (after Halliday 1984).

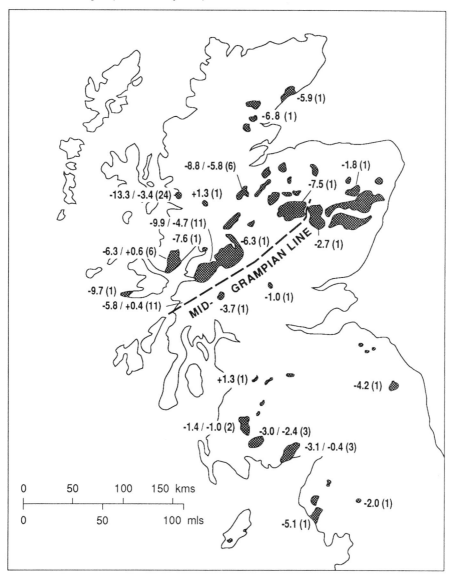

the lower ϵNd values are also believed to be a function of crustal contamination or melting (Halliday *op. cit.*). The geographical limit to the initial ϵNd values of less than 6 is shown on Fig. 8.39 as the Mid-Grampian Line. This is to the north of the Highland Boundary Fault but actually does not conflict with data on the distribution of inherited zircons. The eastward termination of the Mid-Grampian Line is not defined. Halliday (1984) concluded that there is a change in the deep basement about the middle of the Grampian Highlands. This is taken as the boundary between the Argyll and south of Scotland suites, as also positioned on trace element data.

The fact that the isotopic and chemical changes generally coincide geographically led Stephens and Halliday (1984) to conclude that as some of the changes originate in the deep continental crust and some in the mantle, both are probably zoned from N to S (but see Thirlwall 1982). The mantle in the north is probably old mantle attached to the base of old crust and enriched in elements such as K, Ba and Sr.

Terranes

There is increasing evidence that Scotland is made up of an uncertain number of diverse terranes brought together by strike–slip movements of great magnitude. Hutton (1987) regards the Caledonian rocks of Britain as a large number of disorganised terranes which originated to the southwest. These came together at the end of the Silurian period when the Caledonides became a major strike–slip zone which shredded the Ordovician palaeogeography of the Iapetus. Such a viewpoint has obvious implications for considerations of the source rocks of the Newer Granites and it is significant that many workers have advocated that granitoid emplacement was controlled by a major system of sinistral faults (e.g. Leake 1978, Hutton 1982, Pitcher 1982). Nevertheless the contribution of terrane geology to our understanding of the Newer Granites, or vice versa, is so far meagre. Stephens and Halliday (1984) concluded from their examination of the geochemistry and isotopic characteristics of the Newer Granites that there were fundamental differences in magma sources, including composition and age. The sources increase in age northwards towards the continental foreland and there are contemporaneous along-strike variations of considerable magnitude. There are, however, no obvious changes in granitoid chemistry across the major fault lines of the Great Glen, Highland Boundary and Southern Uplands, and the postulated deep crustal variations responsible for granitoid diversity do not correlate well with the fault traces.

Subduction

The Newer Granites are a calc–alkaline suite related to the Caledonian orogeny and comparisons have been drawn between their geochemical characteristics and those of modern magmatic arcs generated by subduction (e.g. Brown *et al.* 1984). The associated volcanic rocks (Thirlwall 1988) are also believed to be subduction-related. Nevertheless attempts to relate the granitoids to subduction and closure of the Iapetus Ocean about the Solway Line suture, have proved controversial. The plutons fail to show the progressive space–time changes in chemistry which are characteristic of magmatism at plate margins, where for example there is typically an increase in K with distance from the site of subduction. Also the Southern Uplands plutons occur too close to the Iapetus suture (Solway Line) and if the granites in England are included, actually straddle it. Most critically, the popular two-plate tectonic interpretations, in which closure of the Iapetus Ocean and cessation of subduction was complete by Ordovician or mid-Silurian times (e.g. Phillips *et al.* 1976, Watson 1984) can have no direct bearing on the subduction-related genesis of a Siluro–Devonian magmatic arc (Soper 1986, Thirlwall 1988). Rather, the granitoids have been regarded as being post-collision and a mechanism of magma genesis has been sought in melting induced by adiabatic decompression during rapid post-orogenic uplift (e.g. Simpson *et al.* 1979, Brown *et al.* 1981, Pitcher 1982, Zhou 1985). The problem with this, as pointed out by Soper (1986), is that the Highland metamorphic terrain was already deeply eroded before the emplacement of the Newer Granites and in the slate belts the crust was never sufficiently thickened to induce the rapid uplift and decompression required to produce widespread melting.

In fact the palaeotectonic situation is likely to have been much more complex than allowed for in a two-plate, mid-Silurian closure model. It has been demonstrated that final closure in the slate belts adjacent to the Solway Line did not take place until early Devonian times (Soper 1986, Soper *et al.* 1987, Thirlwall 1988) and in particular Soper and his co-authors have developed a three-plate collision model which in principle would allow the Newer Granites to be representative of a series of subduction-related magmatic arcs overlapping in space and time. The main features of this model, which is summarised in Fig. 8.40, are: convergence of Laurentia–Baltica with Ordovician to Early Silurian westward subduction beneath the Scottish sector of the Laurentian margin; northward Silurian to early Devonian subduction at the Solway Line (Iapetus suture) due to northward movement of the Cadomian terrane (Eastern Avalonia); northward accretion of Armorica leading to early Devonian collision with Avalonia at the Mid European Caledonide (Ardennes) suture.

On this palaeotectonic scenario the Newer Granites north of the Highland Border are related to northward Silurian to early Devonian subduction at the Solway Line, while the granites of the slate belts are associated with northward accretion of the Armorican terrane in the early Devonian. Soper *et al.* (1987) proposed adopting the term 'Acadian' for the late Caledonian deformation of the slate belts which took place synchronously with the Acadian orogeny of the Canadian Appalachians.

Thirlwall (1988) objected to some features of this model in that so far as the igneous rocks of the Southern Uplands were concerned, northward accretion of the Armorican terrane could neither explain a marked enrichment in K, Rb, Th and LREE from north to south, nor a shift in ages from 408 Ma in the north to 394 Ma in the south. He nevertheless found a clear genetic relationship to subduction for magmatism north of the Southern Uplands Fault where a limited number of radiometric ages considered by him to be 'precise and reliable' show that all magmatism

had terminated by 408 Ma, this being prior to closure of Iapetus and while subduction was still active.

In summary it must be said that despite rapid advances in both geochemistry and geotectonics there is no certainty about the origin of the Newer Granites. On the one hand, some 'mobilists' believe that there is unlikely to be a single Iapetus suture within the British Caledonides, there being many different terranes that represent broken and re-arranged fragments of Iapetus palaeogeography, so that the whole zone is the suture (Hutton 1987). On the other hand, geochemists are divided about the extent and meaning of regional variations in the granitoids. Nevertheless the

Fig. 8.40. Geotectonic model for early Devonian time, after collision between Baltica and Laurentia with Cadomian terranes in the process of accretion (after Soper *et al.* 1987).

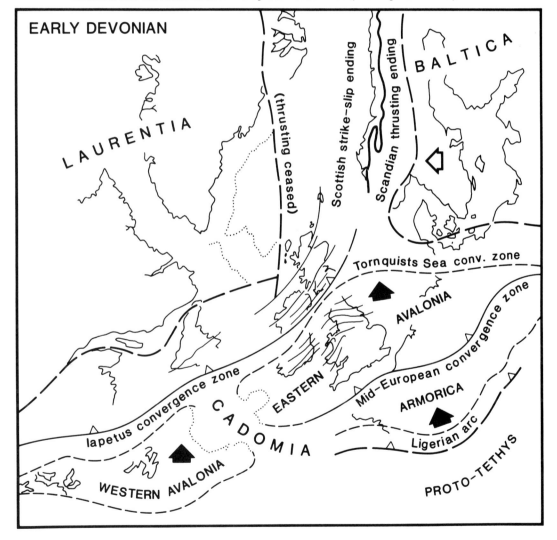

importance of subduction is strongly advocated and it is to be expected that as knowledge of the geochemistry and timing of magmatic activity is refined and the number and timing of terrane accretion events becomes better appreciated, so will the relationship between these processes and granitoid genesis be made clearer.

Old Red Sandstone volcanicity

Volcanic rocks of ORS age are found in the Shetland Isles, the cauldron subsidences of Glencoe and Ben Nevis, and also form the Lorne Plateau. Other extensive outcrops occur on both flanks of the Midland Valley: to the north in the Ochil and Sidlaw Hills and towards Stonehaven, and to the south in the Pentland Hills and Ayrshire (Fig. 8.41). Together, these outcrops constitute an extensive orogenic calc–alkaline suite of basalts, andesites, dacites, rhyolites and pyroclastic rocks, parts of which have a close relationship in space and time with the high-level permitted intrusions which make up Read's (1961) group of 'Last Granites'. It is unfortunate that secondary alteration renders the majority of these volcanic rocks difficult objects for petrographic study.

The lavas of the Lorne Plateau, which are predominantly basalts and andesites, cover about 300 km^2 and attain a thickness of about 800 m (Kynaston & Hill 1908, Lee & Bailey 1925, Groom & Hall 1974). Similar andesites to those of the Lorne Plateau are found in the Glencoe Cauldron Subsidence. Roberts (1974) concluded that these andesites were derived from outside the cauldron and Bailey (1960) thought they were probably originally contiguous with the Lorne Plateau. In contrast rhyolites and two thick ignimbrite horizons originated within the cauldron and can be related to two separate episodes of caldera formation. The Glencoe ignimbrites are correlated with two tuff horizons in the Lorne Plateau sequence which Kynaston (1900), writing before the recognition of nuées ardentes, described as the 'product of showers of minute glassy particles ejected from a volcanic orifice by discharge of a highly explosive character'.

The volcanic rocks which form the higher reaches of Ben Nevis are enclosed within the Inner Granite where cauldron subsidence preserved them as representatives of what, presumably, was a regional volcanic pile. The lavas rest on schist and are associated with conglomerates and sediments. Unlike the Glencoe caldera the range of composition is very limited and all the flows are hornblende or hornblende–biotite rhyodacites (Haslam 1968).

In the Sidlaw Hills a rather potassium-rich sequence of olivine basalts, basaltic andesites, andesites and dacites is interbedded with Old Red Sandstone sediments, the basic volcanics being the most voluminous (Harry 1956, 1958a; Gandy 1975). The lavas are generally porphyritic, the phenocryst phases being olivine, orthopyroxene, clinopyroxene, and plagioclase, but, apart from rare biotite,

primary hydrous minerals are absent. A similar sequence of porphyritic rocks, which the addition of felsic intrusions and pyroclastics, occurs in the nearby Ochil Hills (Francis et al. 1970). These rocks are somewhat more potassic than those of the Sidlaw Hills and in this respect similar to those of the Lorne Plateau (French et al. 1979).

In the Pentland Hills, intercalated with tuffs and occasional sediments, are olivine basalts, pyroxene and hornblende andesites and rhyolites, including some strikingly porphyritic rocks such as the Carnethy porphyry with fluxioned feldspar phenocrysts. There are also vents partly or wholly filled with agglomerate. In Ayrshire basic and intermediate lavas predominate and dacites and extrusive rhyolites are absent, although their equivalents appear as minor intrusions associated with the lavas (Richey et al. 1930, Eyles et al. 1949).

In Western Shetland basalts, pyroxene–andesite, ignimbrites and pyroclasts form a suite of extrusive rocks comparable to those of Argyll and the Midland Valley (Flinn et al. 1968, Mykura & Phemister 1976). The proportion of pyroclastic deposits is higher in Shetland than elsewhere and Mykura and Phemister speculated that the presence of tuff cones and extensive ignimbrites implies the one-time existence of calderas, though these can no longer be recognised.

Detailed petrochemical and isotopic studies have demonstrated that overall the Old Red Sandstone lavas of northern Britain have many similarities with calc–alkaline suites developed on continental margins (Thirlwall 1981a, 1982, 1983). Rocks rich in Ni and Cr are unusually abundant and have incompatible element concentrations little different from those of primary magmas. Such rocks show a marked increase in Sr, Ba, K, P and LREE in a north-west direction across Scotland, suggestive of subduction-related zonation. Volcanism in the Highlands occurred around 420 Ma and in the Midland Valley between 415 and 410 Ma, that is in Ludlow to earliest Gedinnian times (Thirlwall 1988), so that the traditional Lower Devonian age usually assigned to the whole suite is incorrect.

Dyke swarms and appinites

North of the Great Glen two major suites of dykes (disregarding early amphibolites) have been recognised (Smith 1979) – a microdiorite and a lamprophyre suite. In the former sheared microdiorites are abundant; these rocks do not have a cataclastic fabric and they show from NW to SE a facies variation from green-schist to amphibolite. This variation is attributed to the effects of a residual Caledonian regional temperature gradient at the time the microdiorite suite was intruded.

Representatives of the lamprophyre suite are particularly abundant in the vicinity of the Ratagain complex and around the Ross of Mull pluton where the suite ranges from

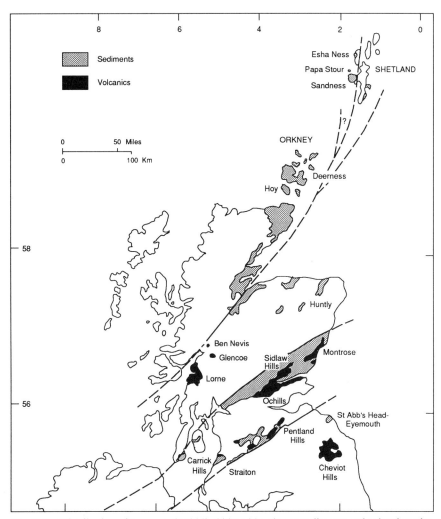

Fig. 8.41. Distribution of Lower and Middle Old Red Sandstone sediments and volcanic rocks (after Thirlwall 1981).

basic calc–alkaline lamprophyres, through microdioritic compositions (malchites) to microgranitic porphyries (Rock & Hunter 1987). The most primitive lamprophyres combine very high K, Rb, Th and Ba with contents of Cr, Ni, Ti, Y and Si which imply a substantial mantle-derived component, and in some cases near-primary magma status. The close association of lamprophyric and granitic magmatism is regarded by Rock and Hunter (*op. cit.*) as genetic and the lamprophyres are considered to be related to the genesis of the Ross of Mull pluton itself, representing mantle-derived magmas which had a dual role as parent magmas (undergoing crustal contamination) and as heat sources (facilitating crustal melting).

Extensive dyke swarms in the Grampians and South-west Highlands trend generally NNE (Richey 1938). In places the swarms are dense and of the 9 km long-axis of the Glencoe Cauldron Subsidence 4 km is elongation due to dyke intrusion (Bailey 1960). The dominant rock types are porphyrites and microdiorites and the main suite (Dearnley 1967) was emplaced after the intrusion of the Cruachan and Outer Nevis granites, but before the Starav and Inner Nevis granites.

In Southern Scotland the distributions of mafic (lamprophyre), intermediate (porphyrite) and felsic (porphyry) dykes are quite distinct (Barnes *et al.* 1986; Rock *et al.* 1986a, 1988). Biotite–lamprophyres are most abundant in a 10 km wide regional zone in the south, coincident with, but partly predating, granitic plutons, whereas hornblende–lamprophyres dominate two swarms around the Doon and Criffel plutons. Intermediate dykes are mostly centred on

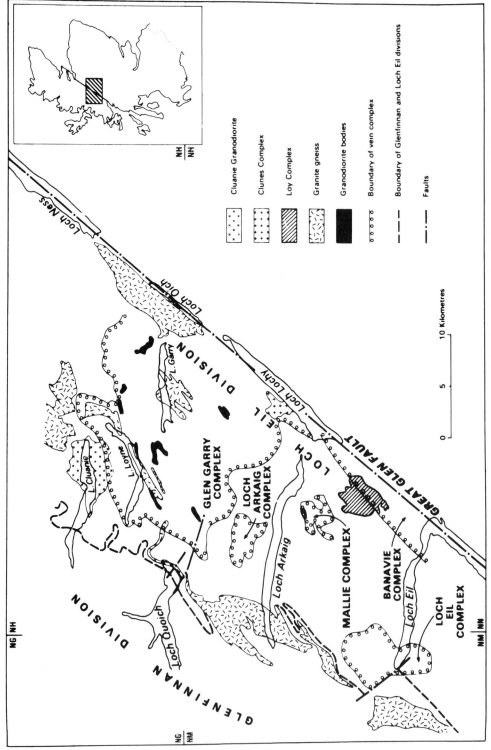

Fig. 8.42. The location of vein complexes north-west of the Great Glen (after Fettes and Macdonald 1978).

the three large granite plutons of Doon, Fleet and Criffel, and the felsic dykes are almost confined to a zone in the north-east which corresponds to the location of the concealed 'Tweeddale Batholith' (Lagios & Hipkin 1979). A petrological distinction between central (dominantly hornblendic) and regional (dominantly micaceous) lamprophyre swarms was made by Rock *et al.* (*op. cit.*), with the biotite lamprophyres having originated in a CO_2-dominated melting regime at greater depth than the hornblende lamprophyres, which are more closely associated with the 'wet' granites.

The Newmains dyke, near Dumfries, is the best exposed and largest of a local concentration of lamprophyres. It has received special attention because it exhibits a unique range of differentiation and contamination phenomena (Kennedy & Read 1936, Macdonald *et al.* 1986). Rock types found in the dyke range from primitive and at least partly mantle-derived lamprophyre with hornblende–pyroxene-rich cumulates, to quartz syenite and granite residua. The Newmains dyke appears, in miniature, to demonstrate petrogenetic processes widely applicable to the Newer Granites. These granites have high K, Sr and Ba and are considered to have developed from high Mg, high Ni, mantle-derived magmas which were affected to varying degrees by crustal contamination (see section on Newer Granites). According to Macdonald *et al.* (*op. cit.*) all these characteristics are shown by the Newmains lamprophyre.

Appinite was defined by Bailey (1916) as the plutonic equivalent of hornblende lamprophyres (vogesite and spessartite). It occurs normally in association with such other rocks as monzonite, augite–diorite, kentallenite and cortlandite (olivine–hornblende) and together they form the appinite suite (Bailey 1916). Their mode of occurrence is mainly as satellite pipes clustered around the Newer Granite plutons and on occasion included within them. They are widespread but are particularly common in the Appin district (Fig. 8.35) and in the vicinity of the Garabal

Hill and Arrochar complexes. Many of the appinite pipes are accompanied by intrusive breccias (Wright & Bowes 1968) produced by vertically streaming, high velocity gases which preceded the rise of magma. In the Kentallen area Bowes and Wright (1967) were able to show that the site of emplacement was closely controlled by steeply plunging, angular folds which provided steep shattered structures up which the gases and magma moved. Various phenomena attributable to fluidisation processes have been described by Platten (1982, 1984) and Platten and Money (1987) from pipes in the south-west Grampian Highlands. Other analogous subvolcanic vents are associated with lamprophyric activity in the Kirkcudbrightshire area of the Southern Uplands (Rock *et al.* 1986b).

Granitic vein complexes

Several late-Caledonian granitic vein complexes (Fig. 8.42) occur north-west of the Great Glen (Fettes & MacDonald 1978). Of these the Loch Eil, Loch Arkaig and Mallie complexes are cut by foliated members of the microdiorite suite and are apparently best regarded as major concentrations of the regionally developed belt of granitic pegmatites. The Banavie complex, adjacent to the Great Glen, appears to overlap the period of intrusion of the microdiorite suite. The youngest and largest vein complex at Glen Garry is only cut by some of the last felsic porphyrite representatives of the micro–diorite suite. The veins in the Glen Garry complex are predominantly granodiorite but range from quartz–diorite to leucogranite in a series which can adequately be related by fractional crystallisation. The regional significance of the Glen Garry vein complex in terms of the late-Caledonian magmatism is difficult to assess, but it lies in the area of possible high-heat flow associated with other late-stage intrusions (Fettes & MacDonald 1978).

REFERENCES

AFTALION, M. & VAN BREEMEN, O.	1980	U–Pb zircon, monazite and Rb–Sr whole rock systematics of granitic gneiss and psammitic to semi-pelitic host gneiss from Glenfinnan, Northwestern Scotland. *Contrib. Mineral. Petrol.,* **72**, 87–98.
AFTALION, M., VAN BREEMEN, O. & BOWES, D. R.	1984	Age constraints on basement of Midland Valley of Scotland. *Trans. R. Soc. Edinb., Ecrth Sci.,* **75**, 55–64.
ANDERSON, J. G. C.	1935	The marginal intrusions of Ben Nevis; the Coille Lianachain complex; and the Ben Nevis dyke swarm. *Trans. Geol. Soc. Glasg.,* **19**, 225–269.
	1937	The Etive complex. *Q. Jl. Geol. Soc. Lond.,* **93**, 487–533.
	1956	The Moinian and Dalradian rocks between Glen Roy and the Monadliath Mountains, Inverness-shire. *Trans. R. Soc. Edinb.,* **63**, 15–36.
ASHCROFT, W. A. & BOYD, R.	1976	The Belhelvie mafic igneous intrusion, Aberdeenshire – a re-investigation. *Scott. J. Geol.,* **2**, 1–14.

ASHCROFT, W. A. &   1978   The structure of the eastern part of the Insch mafic intrusion,
MUNRO, M.   Aberdeenshire. *Scott. J. Geol.*, **14**, 55–79.

ASHCROFT, W. A.,   1984   Major shear zones and autochthonous Dalradian in the north-
KNELLER, B. C.,   east Scottish Caledonides. *Nature*, **310**, 760–762.
LESLIE, A. G. &
MUNRO, M.

ASHWORTH, J. R.   1975   The sillimanite zones of the Huntly–Portsoy area in the north-
                             east Dalradian, Scotland. *Geol. Mag.*, **112**, 113–136.

                      1976   Petrogenesis of migmatites in the Huntly–Portsoy area, north-
                             east Scotland. *Mineral. Mag.*, **40**, 661–682.

ASHWORTH, J. R. &   1978   Co-existing garnet and cordierite in migmatites from the Scottish
CHINNER, G. A.   Caledonides. *Contrib. Mineral. Petrol.*, **65**, 379–394.

ASHWORTH, J. R. &   1983   The distribution of temperatures around the Strontian Grano-
TYLER, I. M.   diorite. *Geol. Mag.*, **120**, 281–290.

ATHERTON, M. P.   1977   The metamorphism of the Dalradian rocks of Scotland. *Scott. J.
                             Geol.*, **13**, 331–370.

ATHERTON, M. P. &   1985   High heat production granites and the evolution of the Andean
PLANT, J.   and Caledonian continental margins. *In* High heat production
                             granites, hydrothermal circulation and ore genesis. *Inst. Min.
                             Metall. London.*

BAILEY, E. B.   1960   Geology of Ben Nevis and Glencoe. *Mem. Geol. Surv. U.K.*
                             (2nd Ed.).

BAILEY, E. B. &   1916   The geology of Ben Nevis and Glen Coe. *Mem. Geol. Surv.
MAUFE, H. B.   Scotland.*

BARNES, R. P.,   1986   Late Caledonian dyke swarms in Southern Scotland: new field,
ROCK, N. M. S. &   petrological and geochemical data for the Wigtown Peninsula,
GASKARTH, J. W.   Galloway. *Geol. J.*, **21**, 101–125.

BARR, D.   1985   Migmatites in the Moines. *In* Ashworth, H. R. (Ed.)
                             Migmatites. Blackie & Son, Glasgow, 225–264.

BARR, D., ROBERTS, A. M., 1985   Structural setting and geochronological significance of the West
HIGHTON, A. J.,   Highland Granitic Gneiss, a deformed early granite within
PARSON, L. M. &   Proterozoic Moine rocks of NW Scotland. *Jl. Geol. Soc. Lond.*,
HARRIS, A. L.   **142**, 663–675.

BARRETT, T. J.,   1982   Comments and reply on 'Age and origin of the Ballantrae
JENKINS, H. C.,   ophiolite and its significance to the Caledonian orogeny and the
LEGGET, J. K. &   Ordovician time-scale'. *Geology*, **9**, 331–333.
ROBERTSON, A. H. E.

BARROW, G.   1893   On an intrusion of muscovite biotite gneiss in the south-east
                             Highlands of Scotland and its accompanying metamorphism. *Q.
                             Jl. Geol. Soc. Lond.*, **49**, 330–358.

                      1912   On the geology of lower Deeside and the southern Highland
                             Border. *Proc. Geol. Ass.*, **23**, 268–273.

BATCHELOR, R. A.   1987   Geochemical and petrological characteristics of the Etive
                             granitoid complex, Argyll. *Scott. J. Geol.*, **23**, 227–249.

BELL, K.   1968   Age relations and provenance of the Dalradian series in
                             Scotland. *Bull. Geol. Soc. Am.*, **79**, 1167–1194.

BLUCK, B. J.   1978   Geology of a continental margin. 1: the Ballantrae complex. *In*
                             'Crustal evolution in north-west Britain and adjacent regions'.
                             Geol. Journ. Special Issue No. 10, 151–162.

                      1982   Hyalotuff deltaic deposits in the Ballantrae ophiolite of SW
                             Scotland: evidence for crustal position of lava sequence. *Trans.
                             R. Soc. Edinb., Earth Sci.*, **72**, 217–228.

BLUCK, B. J.,   1980   Age and origin of the Ballantrae ophiolite and its significance to
HALLIDAY, A. N.,   the Caledonian orogeny and the Ordovician time scale. *Geology*,
AFTALION, M. &   **8**, 492–495.
MACINTYRE, R. M.

BORRADAILE, G. J.   1973   Dalradian structure and stratigraphy of the northern Loch Awe district, Argyllshire. *Trans. R. Soc. Edinb.*, **69**, 1–21.

       1976   Thermal anistropy – a factor contributing to the distribution of Caledonian metamorphic zones in the south-west Highlands. *Geologie Mijnb.*, **42**, 121–142.

BOWES, D. R. & WRIGHT, A. E.   1967   The explosion-breccia pipes near Kentallen, Scotland and their geological setting. *Trans. R. Soc. Edinb.*, **67**, 109–143.

BOYD, R. & MUNRO, M.   1978   Deformation of the Belhelvie mass, Aberdeenshire. *Scott. J. Geol.*, **14**, 29–44.

BRADBURY, H. J., SMITH, R. A. & HARRIS, A. L.   1976   'Older' granites as time-markers in Dalradian evolution. *Jl. Geol. Socl. Lond.*, **132**, 677–684.

BREWER, M. S., BROOK, M. & POWELL, D.   1979   Dating of the tectonometamorphic history of the SE Moine, Scotland. *In* Harris, A. L., Holland, C. H. and Leslie, B. E. (Eds.) The Caledonides of the British Isles Reviewed. *Spec. Publ. Geol. Soc., Lond.*, **8**, 129–137.

BROOK, M., POWELL, D. & BREWER, M. S.   1976   Grenville age for rocks in the Moine of north-western Scotland. *Nature*, **260**, 515–517.

BROWN, G. C.   1979   Geochemical and geophysical constraints on the origin and evolution of Caledonian granites. *In* Harris, A. L., Holland, C. H. and Leake, B. E. (Eds.) *Spec. Publ. Geol. Soc. Lond.*, **8**, 645–652.

BROWN, G. C., CASSIDAY, J., TINDLE, A. G. & HUGHES, D.   1979   The Loch Doon Granite: an example of granite petrogenesis in the British Caledonides. *J. Geol. Soc., Lond.*, **136**, 745–753.

BROWN, G. C., CASSIDAY, J., LOCKE, C. A., PLANT, J. A. & SIMPSON, P. S.   1981   Caledonian plutonism in Britain: a summary. *J. Geophys. Res.*, **86**, 10502–10514.

BROWN, G. C., THORPE, R. S. & WEBB, P. C.   1984   The geochemical evolution of constrasting arcs and comments on magma sources. *J. Geol. Soc. Lond.*, **141**, 413–426.

BROWN, G. C., FRANCIS, E. H., KENNAN, P. & STILLMAN, C. J.   .985   Caledonian igneous rocks of Britain and Ireland. *In* Harris, A. L. (Ed.) The nature and timing of orogenic activity in the Caledonian rocks of the British Isles. Mem. 9, *Geol. Soc. Lond.*

BROWN, P. E.   1967   Major element composition of the Loch Coire migmatite complex, Sutherland, Scotland. *Contrib. Mineral. Petrol.*, **14**, 1–26.

       1971   The origin of granitic sheets and veins in the Loch Coire migmatites, Scotland. *Mineral. Mag.*, **38**, 446–450.

       1983   Caledonian and earlier magmatism. *In* Craig, G. W. (Ed.) *Geology of Scotland.* Scottish Academic Press, 167–204.

BROWN, P. E., MILLER, J. E. & GRASTY, R. L.   1968   Isotopic ages of late Caledonian granitic intrusions in the British Isles, *Proc. Yorks, Geol. Soc.*, **36**, 251–276.

BUSREWIL, M. T., PANKHURST, R. J. & WADSWORTH, W. J.   1975   The origin of the Kennethmont granite–diorite series, Insch, Aberdeenshire. *Mineral. Mag.*, **40**, 363–376.

CHAPPEL, B. W. & STEPHENS, W. A.   1988   The origin of infracrustal (I-type) granite magmas. *Trans. Roy. Soc. Edinb. Earth Sci.*, **79**, 71–86.

CHAPPELL, B. W. & WHITE, A. J. W.   1974   Two contrasting granite types. *Pacific Geology*, **8**, 173–174.

       1984   I and S type granites in the Lochlan fold belt, southeastern Australia. *In* Xu Keqin & Tin Guangchi (Eds.) *Geology of granites and their metallogenic relations,* 87–101, Bejing: Science Press.

CHINNER, G. A.   1966   The distribution of temperature and pressure during Dalradian metamorphism. *Q. Jl. Geol. Soc. Lond.*, **122**, 159–186.

CHURCH, W. R. & GAYER, R. A.   1973   The Ballantrae ophiolite. *Geol. Mag.*, **110**, 497–510.

CLAYBURN, J. A. P.   1988   The crustal evolution of central Scotland and the nature of the lower crust: PB, Nd and Sr isotope evidence from Caledonian granites. *Earth Planet. Sci. Lett.*, **90**, 41–51.

CLAYBURN, J. A. P., HARMON, R. J., PANKHURST, R. J. & BROWN, J. F.   1983   Sr, O and Pb isotope evidence for origin and evolution of Etive Igneous complex, Scotland. *Nature*, **303**, 492–497.

CLOUGH, C. T., MAUFE, H. B. & BAILEY, E. B.   1909   The cauldron-subsidence of Glencoe and associated igneous phenomena. *Q. Jl. Geol. Soc. Lond.*, **65**, 611–676.

COURRIOUX, G.   1987   Oblique diapirism: the Criffel granodiorite/granite zoned pluton SW Scotland. *J. Struct. Geol.*, **9**, 313–330.

CURRY, G. B., BLUCK, B. J., BURTON, C. J., INGHAM, J. K., SIVETER, D. J. & WILLIAMS, A.   1984   Age, evolution and tectonic history of the Highland Border Complex, Scotland. *Trans. Roy. Soc. Edinb., Earth Sci.*, **75**, 113–134.

DALZIEL, I. W. D.   1963   Zircons from the granitic gneiss of western Ardgour, Argyll: their bearing on its origin. *Trans. Edinb. Geol. Soc.*, **19**, 349–362.

   1966   A structural study of the granitic gneiss of western Ardgour, Argyll and Inverness-shire. *Scott. J. Geol.*, **2**, 125–152.

DARBYSHIRE, D. P. F. & BEER, K. E.   1988   Rb–Sr age of the Bennachie and Middleton granites, Aberdeenshire. *Scott. J. Geol.*, **24**, 189–193.

DEANS, T., GARSON, M. S. & COATS, J. S.   1971   Fenite-type soda metasomatism in the Great Glen, Scotland. *Nature Phys. Sci.*, **234**, 145–147.

DEARNLEY, R.   1967   Metamorphism of minor intrusions associated with the Newer Granites of the western Highlands of Scotland. *Scott. J. Geol.*, **3**, 449–457.

DEER, W. A.   1935   The Cairnsmore of Cairsphairn igneous complex. *Q. Jl. Geol. Soc. Lond.*, **91**, 47–76.

DREVER, H.   1940   The geology of Ardgour, Argyllshire. *Trans. R. Soc. Edinb.*, **60**, 141–170.

DROOP, G. T. R. & TRELOAR, P. J.   1981   Pressure of metamorphism in the thermal aureole of the Etive Granite Complex. *Scott. J. Geol.*, **17**, 87–102.

DROOP, G. T. R. & CHARNLEY, N. R.   1985   Comparative geobarometry of pelitic hornfelses associated with the Newer Gabbros: a preliminary study. *J. Geol. Soc. Lond.*, **122**, 53–62.

ECKFORD, R. J. A. & RITCHIE, M.   1931   The lavas of Tweeddale and their position in the Caradocian sequence. Summary of progress. *Geol. Surv. Gt. Britain*, for 1930, 46–57.

EYLES, V. A., SIMPSON, J. B. & MACGREGOR, A. G.   1949   Geology of central Ayrshire. *Mem. Geol. Surv. U.K.*

FETTES, D. J.   1970   The structural and metamorphic state of the Dalradian rocks and their bearing on the age of emplacement of the basic sheet. *Scott. J. Geol.*, **6**, 108–118.

FETTES, D. J. & MACDONALD, R.   1978   The Glen Garry vein complex. *Scott J. Geol.*, **14**, 335–358.

FETTES, D. J., GRAHAM, C. M., HARTE, B. & PLANT, J. A.   1986   Lineaments and basement domains: an alternative view of Dalradian evolution. *J. Geol. Soc. Lond.*, **143**, 453–464.

FLINN, D., MILLER, J. A.,   1968   On the age of the sediments and contemporaneous volcanic
EVANS, A. L. &   rocks of western Shetland. *Scott. J. Geol.*, **4**, 10–19.
PRINGLE, I. R.

FLOYD, J. D.   1982   Stratigraphy of a flysch succession: the Ordovician of West
                     Nithdale, SW Scotland. *Trans. Roy. Soc. Edinb., Earth Sci.*, **73**,
                     1–9.

FLOYD, J. D. &   1983   Element mobility associated with meta-shear zones within the
WINCHESTER, J. A.   Ben Hope amphibolite suite, Scotland. *Chem. Geol.*, **39**, 1–15.

FRANCIS, E. H.,   1970   The geology of the Stirling district. *Mem. Geol. Surv. U.K.*
FORSYTH, I. H.,
READ, W. A. &
ARMSTRONG, M.

FRENCH, W. J.,   1979   The petrogenesis of Old Red Sandstone volcanic rocks of the
HASSAN, M. D. &   western Ochils, Stirlingshire. *In* Harris, A. L., Holland, C. H.
WESCOTT, J   and Leake, B. E. (Eds.) The Caledonides of the British Isles –
                     reviewed. *Spec. Publ. Geol. Soc. Lond. No. 8*, 635–642.

FROST, C. D. &   1985   Caledonian magma genesis and crustal recycling. *J. Petrol.*, **26**,
O'NIONS, R. K.   515–544.

GARDINER, C. I. &   1932   The Loch Doon granite area, Galloway. *Q. J. Geol. Soc. Lond.*,
REYNOLDS, S. H.   **83**, 1–34.

GANDY, M. K.   1975   The petrology of the Lower Old Red Sandstone lavas of the
                     eastern Sidlaw Hills, Perthshire, Scotland. *J. Petrology*, **16**, 189–
                     211.

GARSON, M. S., COATS, J. S.,   1984   Fenites, breccia dykes, albitites and carbonatitic veins near the
ROCK, N. M. S. &   Great Glen fault. *J. Geol. Soc. Lond.*, **141**, 711–732.
DEANS, T.

GRAHAM, C. M.   1976a   Petrochemistry and tectonic setting of Dalradian metabasaltic
                     rocks of the south-west Scottish Highlands. *J. Geol. Soc.*, **132**,
                     61–84.

                 1976b   Petrochemical affinities of Dalradian metabasaltic rocks: discus-
                     sion of paper by J. A. Winchester and P. A. Floyd. *Earth Plant.*
                     Sci. Lett., **32**, 210–212.

                 1986   The role of the Cruachan lineament during Dalradian evolution.
                     Scott. J. Geol., **22**, 257–270.

GRAHAM, C. M. &   1984   The petrology and structure of Dalradian metabasaltic dykes of
BORRADAILE, G. J.   Jura: implications for early Dalradian evolution. *Scott. J. Geol.*,
                     20, 257–270.

GRAHAM, C. M. &   1981   Cambrian and Late Cambrian basaltic activity in the Scottish
BRADBURY, H. J.   Dalradian: a review. *Geol. Mag.*, **118**, 27–39.

GRIBBLE, C. D.   1966   The thermal aureole of the Haddo House norite in Aberdeen-
                     shire. *Scott. J. Geol.*, **2**, 306–313.

                 1967   The basic intrusive rocks of Caledonian age of the Haddo and
                     Arnage districts, Aberdeenshire. *Scott. J. Geol.*, **3**, 125–136.

                 1968   The cordierite-bearing rocks of the Haddo House and Arnage
                     districts, Aberdeenshire. *Contr. Mineral. Petrol.*, **17**, 315–330.

                 1970   The role of partial fusion in the genesis of certain cordierite-
                     bearing rocks. *Scott. J. Geol.*, **6**, 75–82.

GRIBBLE, C. D. &   1967   Interaction of basic magma and pelitic materials. *Nature*, **214**,
O'HARA, M. J.   1198–1201.

GROOME, D. R. &   1974   The geochemistry of the Devonian lavas of the northern Lorne
HALL, A.   plateau, Scotland. *Mineral. Mag.*, **39**, 621–640.

HALLIDAY, A. N.   1984   Coupled Sm–Nd and U–Pb systematics in late Caledonian
                     granites and basement under northern Britain. *Nature*, **307**, 229–
                     233.

HALLIDAY, A. N.,   1979   Petrogenic significance of Rb–Sr and U–Pb isotopic systems in
  AFTALION, M.,   the c. 400 Ma old British Isles granitoids and their hosts. *In*
  VAN BREEMEN, O. &   Harris, A. L., Holland, C. H. and Leake, B. E. (Eds.) The
  JOCELYN, J.   Caledonides of the British Isles – reviewed. *Spec. Publ. Geol. Soc.
                                 Lond., No. 8*, 653–661.

HALLIDAY, A. N.,   1980   Rb–Sr and O isotopic relationships in three zoned Caledonian
  STEPHENS, W. E. &   plutons, Southern Uplands, Scotland: evidence for varied
  HARMON, R. S.   sources and hybridisation of magmas. *J. Geol. Soc. Lond.*, **137**,
                                 329–348.

HALLIDAY, A. N.,   1985   Isotopic and chemical constraints on the building of the deep
  STEPHEN, W. E.,   Scottish lithosphere. *Scott. J. Geol.*, **21**, 456–491.
  HUNTER, R. H.,
  MENZIES, M. A.,
  DICKIN, A. P. &
  HAMILTON, P. J.

HALLIDAY, A. N. &   1984   Crustal controls on the genesis of the 400 Ma Old Caledonian
  STEPHEN, W. E.   granites. *Phys. Earth Planet, Inter.*, **35**, 89–104.

HALLIDAY, A. N.,   1987   Syn-orogenic alkaline magmatism and its relationship to the
  AFTALION, M.,   Moine Thrust zone and the thermal state of the lithosphere in
  PARSONS, I.   NW Scotland. *J. Geol. Soc. Lond.*, **144**, 611–618.
  DICKIN, A. P. &
  JOHNSON, M. R. W.

HAMILTON, P. J.,   1980   Isotopic evidence for the provenance of some Caledonian
  O'NIONS, R. K. &   granites. *Nature*, **283**, 21–25.
  PANKHURST, R. J.

HAMILTON, P. K.,   1984   Sn–Nd ages from the Ballantrae complex, SW Scotland. *Trans.
  BLUCK, B. J. &   Roy. Soc. Edinb., Earth Sci.*, **75**, 183–187.
  HALLIDAY, A. N.

HARDIE, W. G.   1968   Volcanic breccia and the Lower Old Red Sandstone uncon-
                                 formity, Glencoe, Argyll. *Scott. J. Geol.*, **4**, 291–299.

HARKER, R. I.   1954   Further data on the petrology of the pelitic hornfelses of the
                                 Carn Chuinneag-Inchbae region, Ross-shire, with special
                                 reference to the status of almandine. *Geol. Mag.*, **91**, 445–462.

                          1962   The older ortho-gneisses of Carn Ghuinneag and Inchbae. *J.
                                 Petrology*, **3**, 215–237.

HARMON, R. S.   1983   Oxygen and strontium isotope evidence regarding the role of
                                 continental crust in the origin and evolution of the British
                                 Caledonian granites. *In* Atherton, M. P. & Gribble, C. D.
                                 (Eds.) Migmatites, melting and metamorphism. *Shiva Press*,
                                 62–79.

HARMON, R. S. &   1980   Oxygen and strontium isotope relationships in the British late
  HALLIDAY, A. N.   Caledonian granites. *Nature*, **283**, 21–25.

HARMON, R. S.,   1984   Chemical and isotope systematics of the Caledonian intrusions
  HALLIDAY, A. N.,   of Scotland and Northern England: a guide to magma source
  CLAYBURN, J. A. P. &   region and magma-crust interaction. *Philos. Trans. Roy. Soc.
  STEPHENS, W. E.   Lond. Ser. A.*, **310**, 709–742.

HARRIS, A. L. &   1975   The Dalradian Supergroup. *In* Harris, A. L. *et al.*, (Eds.) A
  PITCHER, W. S.   correlation of Precambrian rocks in the British Isles, 52–75,
                                 Spec. Rep. Geol. Soc. Lond. No. 6.

HARRIS, A. L.,   1978   Ensialic basic sedimentation: the Dalradian Supergroup. *In*
  BALDWIN, C. T.,   Bowes, D. R. and Leake, B. E. (Eds.). Crustal evolution in north-
  BRADBURY, H. J.,   west Britain and adjacent regions. *Geol. J. Spec. Issue*, **10**, 115–
  JOHNSON, H. D. &   138.
  SMITH, R. A.

HARRISON, T. N.   1986   The mode of employment of the Cairngorm granite. *Scott. J.
                                 Geol.*, **22**, 303–314.

HARRISON, T. N.   1987 The granitoids of eastern Aberdeenshire. *In* Trewin, N. H., Kneller, B. C. and Gillen C. G. (Eds.) *Excursion Guide to the geology of the Aberdeen area.* Scott. Academic Press, Edinb., 243–250.

1988 Magmatic garnets in the Cairngorm granite, Scotland. *Mineral. Mag.,* **52,** 659–667.

HARRISON, T. N. &   1987 The age and origin of the eastern Grampian Newer Granites.
  HUTCHINSON, J.   *Scott. J. Geol.,* **23,** 269–282.

HARRY, W. T.   1951 The Glen Dessary marble and its associated calc-silicate rocks. *Geol. Mag.,* **88,** 393–403.

1952 The migmatites and feldspar-porphyroblast rock of Glen Dessary, Inverness-shire. *Quart. J. Geol. Soc., Lond.,* **107,** 497–511.

1954 The composite granitic gneiss of Western Ardgour. *Quart. J. Geol. Soc., Lond.,* **109,** 285–309.

1956 The Old Red Sandstone lavas of the western Sidlaw Hills, Perthshire. *Geol. Mag.,* **93,** 43–56.

1958a The Old Red Sandstone lavas of the eastern Sidlaws. *Trans. Edinb. Geol. Soc.,* **17,** 105–112.

1958b A re-examination of Barrow's older granites in Glen Clova, Angus. *Trans. R. Soc. Edinb.,* **63,** 393–412.

HARTE, B., BOOTH, J. E.,   1984 Aspects of the post-depositional evolution of Dalradian and
  DEMPSTER, T. J.,   Highland Border Complex rocks in the Southern Highlands of
  FETTES, D. J.,   Scotland. *Trans. Roy. Soc. Edinb. Earth Sci.,* **75,** 151–163.
  MENDUM, J. R. & WATTS, D.

HASLAM, H. W.   1968 The crystallisation of intermediate and acid magmas at Ben Nevis, Scotland. *J. Petrology,* **9,** 84–104.

HENDERSON, W. G. &   1982 The Highland Border rocks and their relation to marginal basin
  ROBERTSON, A. H. G.   development in the Scottish Caledonides. *J. Geol. Soc. Lond.,* **139,** 433–450.

HEPWORTH, B. C.,   1982 Sedimentology, volcanism, structure and metamorphism of the
  OLIVER, G. J. H. &   Lower Palaeozoic accretionary complex; Bail Hill-Abington
  MCMURTY, N. J.   area of the Southern Uplands of Scotland. *In* Leggett, J. K. (Ed.) Trench–Forearc Geology. *Spec. Publ. Geol. Soc. Lond.,* **10,** 521–534.

HOLDEN, P.,   1987 Neodymium and strontium isotope content of microdiorite
  HALLIDAY, A. N. &   enclaves points to mantle input to granitoid production. *Nature,*
  STEPHEN, W. E.   **330,** 53–56.

HUTTON, D. H. W.   1982 A tectonic model for the emplacement of the Main Donegal Granite, NW Ireland. *J. Geol. Soc. Lond.,* **139,** 615–632.

1987 Strike–slip terranes and a model for the evolution of the British and Irish Caledonides. *Geol. Mag.,* **124,** 405–425.

1988a Igneous emplacement in a shear zone termination: the biotite granite at Strontian, Scotland. *Geol. Soc. Am. Bull.,* **100,** 1392–1399.

1988b Granite emplacement mechanisms and tectonic controls: inferences from deformation studies. *Trans. Roy. Soc. Edinb., Earth Sci.,* **79,** 245–255.

IKIN, N. P.   1983 Petrochemistry and tectonic significance of the Highland Border suite mafic rocks. *J. Geol. Soc., Lond.,* **140,** 267–268.

JELINEK, E., SOUCEK, J.,   1980 Nature and significance of beerbachites in the Ballantrae
  BLUCK, B. J.,   ophiolite, SW Scotland. *Trans. Roy. Soc. Edinb., Earth Sci.,* **71,**
  BOWES, D. R. &   159–179.
  TRELOAR, P. J.

JELINEK, E., SOUCEK, J.,   1984 Geochemistry of peridotites, gabbros and trondhjemites of the
  RANDA, Z., JAKES, P.,   Ballantrae complex, SW Scotland. *Trans. Roy. Soc. Edinb.,*
  BLUCK, B. J. & BOWES, D. R.   *Earth Sci.,* **75,** 193–210.

JOHNSON, M. R. W. & PARSONS, I. 1979 Geological Excursion Guide to the Assynt District of Sutherland. Edinb. Geol. Soc., 1–76.

JOHNSTONE, G. S., SMITH, D. I. & HARRIS, A. L. 1969 Moinian assemblage of Scotland. *Am. Ass. Petrol. Geol. Mem.*, **12**, 159–180.

KENNEDY, W. Q. 1946 The Great Glen fault. *Q. Jl. Geol. Soc. Lond.*, **102**, 41–76.

KENNEDY, W. Q. & READ, H. H. 1936 The differentiated dyke of Newmains, Dumfriesshire, and its contact and contamination phenomena. *Q. J. Geol. Soc. Lond.*, **92**, 116–145.

KING, B. C. 1942 The Choc nan Cuilean area of the Ben Loyal igneous complex. *Quart. J. Geol. Soc., Lond.*, **98**, 147–185.

KNELLER, B. C. 1987 A geological history of north-east Scotland. *In* Trewin, N. H., Kneller, B. C. and Gillen, C. (Eds.) *An excursion guide to the geology of the Aberdeen area*. Scott. Academic Press, Edinb., 1–50.

KNELLER, B. C. & AFTALION, M. 1987 The isotopic and structural age of the Aberdeen granite. *J. Geol. Soc. Lond.*, **144**, 717–722.

KNELLER, B. C. & LESLIE, A. G. 1984 Amphibolite facies metamorphism in shear zones in the Buchan area of NE Scotland. *J. Metam. Geol.*, **2**, 83–94.

KYNASTON, H. 1900 On some tuffs associated with the andesitic lavas of Lorne. *Trans. Edinb. Geol. Soc.*, **8**, 87.

KYNASTON, H. & HILL, J. B. 1908 The geology of Oban and Dalmally (Sheet 45). *Mem. Geol. Surv. U.K.*

LAGIOS, E. & HIPKIN, R. G. 1979 The Tweeddale granite a newly discovered batholith in the Southern Uplands. *Nature*, **280**, 672–675.

LAMBERT, R. St. J. 1969 Isotopic studies relating to the Precambrian history of the Moinian of Scotland. *Proc. Geol. Soc. Lond.*, **12**, 271–292.

LAMBERT, R. ST. J., HOLLAND, J. E. & LEGGETT, J. K. 1981 Petrology and tectonic setting of some Ordovician volcanic rocks from the Southern Uplands of Scotland. *J. Geol. Soc. Lond.*, **138**, 421–436.

LAMBERT, R. ST. J. & McKERROW, W. S. 1976 The Grampian orogeny. *Scott. J. Geol.*, **12**, 271–292.

LAMBERT, R. ST., J., POOLE, A. B., RICHARDSON, S. W., JOHNSTONE, G. S. & SMITH, D. I. 1964 The Glen Dessary syenite, Inverness-shire. *Nature*, **202**, 370–372.

LEAKE, B. E. 1978 Granite emplacement: the granites of Ireland and their origin. *In* Bowes, D. R. and Leake, B. E. (Eds.) Crustal evolution in north-western Britain and adjacent regions. *Geol. J. Spec. Issue*, **10**, 221–248.

LEE, G. W. & BAILEY, E. B 1925 Pre-Tertiary geology of Mull, Loch Aline and Oban. *Mem. Geol. Surv. U.K.*

LEEDAL, G. P. 1952 The Cluanie igneous intrusion, Inverness-shire and Ross-shire. *Quart. J. Geol. Soc. Lond.*, **108**, 35–63.

LEGGETT, J. K. 1980 Palaeogeographic setting of the Wrae Limestone: an Ordovician submarine-slide deposition Tweeddale. *Scott. J. Geol.*, **16**, 91–104.

LEGGETT, J. K., McKERROW, W. S. & EALES, M. H. 1979 The Southern Uplands of Scotland: a Lower Palaeozoic accretionary prism. *Jl. Geol. Soc. Lond.*, **136**, 755–770.

LESLIE, A. G. 1984 Field relations in the north-eastern part of the Insch mafic igneous mass, Aberdeenshire. *Scott. J. Geol.*, **20**, 215–235.

LONG, L. E. 1964 Rb–Sr chronology of the Carn Chuinneag intrusion, Ross-shire, Scotland. *J. Geophys. Res.*, **69**, 1589–1597.

LONG, L. E. & LAMBERT, R. ST. J. — 1963 — Rb–Sr isotope ages from the Moine Series. *In* Johnson, M. R. W. and Stewart, F. H. (Eds.) *The British Caledonides.* Oliver & Boyd, Edinburgh.

McCOURT, W. J. — 1980 — The geology of the Strath-Halladale–Altnabreac District. ENPU 80–81. *Rep. Inst. Geol. Sci.*, 1–18.

MACDONALD, R., ROCK, N. M. S., RUNDLE, C. C. & RUSSELL, O. J. — 1986 — Relationships between Caledonian lamprophyre and acidic magmas in a differentiated dyke. *Mineral. Mag.*, **50**, 547–557.

MACGREGOR, S. M. A. & ROBERTS, J. — 1963 — Dalradian Pillow Lavas, Ardwell Bridge, Banffshire. *Geol. Mag.*, **100**, 17–23.

McGREGOR, D. M. & WILSON, C. D. V. — 1967 — Gravity and magnetic surveys of the younger gabbros of Aberdeenshire. *Q. Jl. Geol. Soc. Lond.*, **123**, 99–123.

MACKENZIE, W. S. — 1949 — Kyanite-gneiss within a thermal aureole. *Geol. Mag.*, **86**, 251–254.

McKERROW, W. S., LAMBERT, R. ST. J. & COCKS, L. R. M. — 1985 — The Ordovician, Silurian and Devonian periods. *In* Snelling, N. J. (Ed.) The chronology of the geological record. *Geol. Soc. Lond., Mem.*, **10**, 73–80.

McLENNAN, E. L. — 1983 — Contrasting textures in metamorphic and anatectic migmatites: an example from the Scottish Caledonides. *J. Metam. Geol.*, **I**, 241–262.

1984 — Deformational behaviour of migmatites and problems of structural analysis in migmatite terrains. *Geol. Mag.*, **121**, 339–345.

MARSH, B. D. — 1982 — On the mechanics of the igneous diapirism, stoping and zone melting. *Am. J. Sci.*, **282**, 808–855.

MARSTON, R. J. — 1971 — The Foyers granite complex, Inverness-shire, Scotland. *Q. Jl. Geol. Soc. Lond.*, **126**, 331–368.

MATHEWS, D. W. & WOOLLEY, A. R. — 1977 — Layered ultramafic rocks with the Borralan complex, Scotland. *Scott. J. Geol.*, **13**, 223–236.

MAUFE, H. B. — 1910 — The geological structure of Ben Nevis. *Sum. Prog. Geol. Surv. Gt. Brit.* (for 1909), 80–89.

MERCY, E. L. P. — 1963 — The geochemistry of some Caledonian granitic and metasedimentary rocks. *In* Johnson, M. R. W. and Stewart, F. J. (Eds.) *The British Caledonides*, 229–267. Oliver & Boyd, Edinburgh.

MOORHOUSE, S. J. & MOOREHOUSE, V. E. — 1979 — The Moine amphibolite suites of central and northern Sutherland, Scotland. *Minerlog. Mag.*, **326**, 211–226.

MOULD, D. D. C. P. — 1946 — The geology of the Foyers granite and the surrounding country. *Geol. Mag.*, **83**, 249–265.

MUNRO, M. — 1965 — Some structural features of the Caledonian granitic complex at Strontian, Argyllshire. *Scott. J. Geol.*, **1**, 152–175.

1970 — A reassessment of the younger basic igneous rocks between Huntly and Portsoy based on new bore-hole evidence. *Scott. J. Geol.*, **6**, 41–52.

1973 — Structures in the south-eastern portion of the Strontian granitic complex, Argyllshire. *Scott. J. Geol.*, **9**, 99–108.

1984 — Cumulate relations in the 'Younger Basic' masses of the Huntly–Portsoy area, Grampian Region. *Scott. J. Geol.*, **20**, 343–359.

1986 — Geology of the country around Aberdeen. *Mem. Brit. Geol. Surv.*

MUNRO, M. & GALLAGHER, J. W. — 1984 — Disruption of the 'Younger Basic' masses in the Huntly–Portsoy area, Grampian Region. *Scott. J. Geol.*, **20**, 361–382.

MYKURA, W. & PHEMISTER, J. — 1976 — The geology of western Shetland. *Mem. Geol. Surv. U.K.*

NOCKOLDS, S. R. — 1941 — The Garabal Hill–Glen Fyne igneous complex. *Q. Jl. Geol. Soc. Lond.*, **96**, 451–511.

OERTEL, G.   1955 Der pluton von Loch Doon in, Südschottland. *Geotckt. Forsch.*, **11**, 1–83.

OLDERSHAW, W.   1974 The Lochnagar granitic ring complex, Aberdeenshire. *Scott. J. Geol.*, **10**, 297–310.

O'NIONS, R. K.   1983 A Nd isotope investigation of sediments related to crustal development in the British Isles. *Earth Planet, Sci. Lett.*, **63**, 229–240.

PANKHURST, R. J.   1970 The geochronology of the basic igneous complexes. *Scott. J. Geol.*, **6**, 83–107.

                                    1974 Rb–Sr whole-rock chronology of Caledonian events in northeast Scotland. *Bull. Geol. Soc. Am.*, **85**, 345–350.

                                    1979 Isotope and trace element evidence for the origin and evolution of Caledonian granites in the Scottish Highlands. *In* Atherton, M. P. and Tarney, J. (Eds.) *Origin of Granite Batholiths*, Shiva, 18–33.

PANKHURST, R. J. &   1976 Inherited isotope systems and the source region pre-history of
PIDGEON, R. T.   early Caledonian granites in the Dalradian series of Scotland. *Earth Planet. Sci. Lett.*, **31**, 55–68.

PANKHURST, R. J. &   1982 Caledonian granites and diorites of Scotland and Ireland. *In*
SUTHERLAND, D. S.   Sutherland, D. S. (Ed.) *Igneous Rocks of the British Isles*, Wiley, 141–190.

PANTIN, H. M.   1956 The petrology of the Ben Vrackie epidiorites and the contact rocks. *Trans. Geol. Soc. Glasg.*, **22**, 48–79.

PARSLOW, G. R.   1968 The physical and structural features of the Cairnsmore of Fleet granite and its aureole. *Scott. J. Geol.*, **4**, 91–108.

                                    1971 Variations in mineralogy and major elements in the Cairnsmore of Fleet granite, SW Scotland. *Lithos*, **4**, 43–55.

PARSLOW, G. R. &   1973 A gravity survey of the Cairnsmore of Fleet granite and its
RANDALL, B. A. O.   environs. *Scott. J. Geol.*, **9**, 219–232.

PARSONS, I.   1965a The feldspathic syenites of the Loch Ailsh intrusion, Assynt, Scotland. *J. Petrology*, **6**, 365–394.

                                    1965b The sub-surface shape of part of the Loch Ailsh intrusion, Assynt, as deduced from magnetic anomalies across the contact, with a note on transverses across the Loch Borralan complex. *Geol. Mag.*, **102**, 46–58.

                                    1968 The origin of the basic and ultrabasic rocks of the Loch Ailsh intrusion, Assynt. *Scott. J. Geol.*, **4**, 221–234.

                                    1972 Comparative petrology of the leucocratic syenites of the Northwest Highlands of Scotland. *Geol. Journ.*, **8**, 71–82.

                                    1979 The Assynt alkaline suite. *In* Harris, A. L., Holland, C. H. and Leake, B. E. (Eds.) The Caledonides of the British Isles – reviewed. *Spec. Publ. Geol. Soc. Lond., No. 8*, 677–681.

PEACH, B. N. & HORNE, J.   1899 The Silurian rocks of Great Britain 1, Scotland. *Mem. Geol. Surv. U.K.*

PEACH, B. N., GUNN, W.,   1912 The geology of Ben Wyvis, Carn Chuinneag, Inchbae and
CLOUGH, C. T.,   surrounding country. *Mem. Geol. Surv. U.K.*
HINXMAN, L. W.,
CRAMPTON, C. B.,
ANDERSON, E. M. &
FETTES, J. S.

PEACOCK, J. D.   1973 Sodic rocks of metasomatic origin in the Moine nappe. *Scott. J. Geol.*, **9**, 96–97.

PEARCE, J. A. &   1973 Tectonic setting of basic volcanic rocks determined using trace
CANN, J. R.   element analyses. *Earth Planet. Sci. Lett.*, **12**, 339–349.

PHEMISTER, J.   1926 The alkaline igneous rocks of the Loch Ailsh district. *In* Read, H. H. *et al.* The geology of Strath Oykell and Lower Loch Shin. *Mem.. Geol. Surv. U.K.*

PHILLIPS, W. J.   1956   The Criffel–Dalbeattie granodiorite complex. *Q. Jl. Geol. Soc. Lond.*, **112**, 221–239.

PHILLIPS, W. J., FUGE, R. &   1981   Convection and crystallisation in the Criffel Dalbeattie pluton. *J. Geol. Soc. Lond.*, **138**, 351–366.
   PHILLIPS, N.

PHILLIPS, W. E. S.,   1976   A Caledonian plate tectonic model. *J. Geol. Soc. Lond.*, **132**, 579–609.
   STILLMAN, C. J. &
   MURPHY, T.

PIDGEON, R. T. &   1978   Cognetic and inherited zircon U–Pb systems in granites: Palaeozoic granites of Scotland and England. *In* Crustal evolution in north-west Britain and adjacent regions. *Geol. J. Spec. Issue*, **10**, 183–220.
   AFTALION, M.

PIDGEON, R. T. &   1974   A comparison of zircon U–Pb and whole-rock Rb–Sr systems in three phases of the Carn Chuinneag granite, northern Scotland. *Earth Planet. Sci. Lett.*, **24**, 105–112.
   JOHNSON, M. R. W.

PITCHER, W. S.   1982   Granite type and tectonic environment. *In* Hsu, K. (Ed.) *Mountain building processes*. Academic Press, 19–40.

PLANT, J., SIMPSON, P. R.,   1983   Metalliferous and mineralised Caledonian granites in relation to regional metamorphism and fracture systems in northern Scotland. *Trans. Instn. Min. Metall.*, **B92**, 33–42.
   GREEN, P. M.,
   WATSON, J. V. &
   FOWLER, M. B.

PLATTEN, I. M.   1982   A late Caledonian breccia dyke swarm in Glen Creran, near Glencoe in the Grampian Highland. *Geol. Mag.*, **119**, 169–180.

                                 1984   Fluidised mixtures of magma and rock in a late Caledonian breccia dyke and associated breccia pipes in Appin, Scotland. *Geol. J.*, **19**, 209–226.

PLATTEN, I. M. &   1987   Formation of late Caledonian subvolcanic breccia pipes at Cruachan Cruinn, Grampian Highlands, Scotland. *Trans. Roy. Soc. Edinb., Earth Sci.*, **78**, 85–103.
   MONEY, M. S.

POWELL, D.   1964   The stratigraphical succession of the Moine schists around Loch Ailort (Inverness-shire) and its regional significance. *Proc. Geol. Ass.*, **75**, 223–250.

                                 1966   The structure of the south-eastern part of the Morar Antiform, Inverness-shire. *Proc. Geol. Ass.*, **77**, 79–100.

POWELL, D., BROOK, M. &   1983   Structural dating of a Precambrian pegmatite in Moine rocks of northern Scotland and its bearing on the status of the Morarian orogeny. *J. Geol. Soc. Lond.*, **140**, 813–824.
   BAIRD, A. W.

POWELL, D. &   1985   Time and deformation in the Caledonian orogen of Britain and Ireland. *In* Harris, A. L. (Ed.) The Nature and timing or orogenic activity in the Caledonian rocks of the British Isles. *Geol. Soc. Lond., Mem.*, **9**, 17–39.
   PHILLIPS, W. E. A.

READ, H. H.   1931   Geology of central Sutherland. *Mem. Geol. Surv. U.K.*

                                 1935   The gabbros and associated xenolithic complexes of the Haddo House district, Aberdeenshire. *Q. Jl. Geol. Soc. Lond.*, **91**, 591–638.

                                 1961   Aspects of the Caledonian magmatism in Britain. *Proc. Lpool. Manch. Geol. Soc.*, **2**, 653–683.

READ, H. H. &   1956   The Buchan anticline of the Banff nappe of Dalradian rocks in north-east Scotland. *Q. Jl. Geol. Soc. Lond.*, **112**, 131–154.
   FARQUHAR, O. C.

READ, H. H. &   1925   The geology of the country around Golspie, Sutherland. *Mem. Geol. Surv. U.K.*
   PHEMISTER, J.

READ, H. H.,   1926   The geology of Strath Oykell and Lower Loch Shin. *Mem. Geol. Surv. U.K.*
   PHEMISTER, J. & ROSS, G.

RICHARDSON, S. W.   1968   The petrology of the metamorphosed syenite in Glen Dessary, Inverness-shire. *Q. Jl. Geol. Soc. Lond.*, **124**, 9–51.

RICHEY, J. E.   1938   The dykes of Scotland. *Trans. Edinb. Geol. Soc. Surv. U.K.*

RICHEY, J. E.,   1930   The geology of north Ayrshire. *Mem. Geol. Surv. U.K.*
    ANDERSON, E. M. &
    MacGREGOR, A. G.

RILEY, P. J.   1966   *The geology of the Ross of Mull. Univ. Sheffield PhD thesis*
                                         (unpubl.).

ROBERTS, J. L.   1963   Source of the Glencoe ignimbrites. *Nature*, **199**, 901.

                                  1966a Ignimbrite eruptions in the volcanic history of the Glencoe
                                         cauldron subsidence. *Geol. J.*, **5**, 173–184.

                                  1966b The emplacement of the Main Glencoe Fault–Intrusion at Stob
                                         Mhic Mhartuin. *Geol. Mag.*, **103**, 299–316.

                                  1974   The evolution of the Glencoe cauldron. *Scott. J. Geol.*, **10**, 269–
                                         282.

ROBERTS, A. M.,   1984   The structural setting and tectonic significance of the Glen
    SMITH, D. I. &   Dessary syenite, Inverness-shire. *J. Geol. Soc. Lond.*, **141**, 1033–
    HARRIS, A. L.   1042.

ROBERTSON, A. H. &   1984   Geochemical evidence for the origins of igneous and sedimen-
    HENDERSON, W. G.   tary rocks of the Highland Border, Scotland. *Trans. Roy. Soc.
                                         Edinb. Earth Sci.*, **75**, 135–150.

ROBERTSON, R. C. R. &   1974   The Loch Loyal syenites. *Scott. J. Geol.*, **10**, 129–146.
    PARSONS, I.

ROCK, N. M. S.,   1985   Intrusive metabasite belts within the Moine assemblage, west of
    MacDONALD, R.,   Loch Ness, Scotland: evidence for metabasite modification by
    WALKER, B. H., MAY, F.,   country rock interactions. *J. Geol. Soc. Lond.*, **142**, 643–662.
    PEACOCK, J. D. &
    SCOTT, P.

ROCK, N. M. S.,   1986a Late Caledonian subvolcanic vents and associated dykes in the
    COOPER, C. &   Kirkcudbright area, Galloway, SE Scotland. *Proc. Yorks. Geol.
    GASKARTH, J. W.   Soc.*, **46**, 29–38.

ROCK, N. M. S.,   1986b Late Caledonian dyke swarms in southern Scotland: a regional
    GASKARTH, J. W. &   zone of primitive K-rich lamprophyres and associated vents. *J.
    RUNDLE, C. C.   Geol.*, **94**, 505–522.

ROCK, N. M. S. &   1987   Late Caledonian dyke swarms of northern Britain: spatial and
    HUNTER, R. H.   temporal intimacy between lamprophyric and granite magmat-
                                         ism around the Ross of Mull pluton, Inner Hebrides. *Geol.
                                         Rundsch.*, **76**, 805–826.

ROCK, N. M. S.,   1988   Late Caledonian dyke swarms of northern Britain: some
    GASKARTH, J. W.,   preliminary petrogenetic and tectonic implications of their
    HENNEY, P. J. &   province-wide distribution and chemical variation. *Canadian
    SHAND, P.   Mineral.*, 1988, 3–22.

SABINE, P. A.   1953   The petrography and geological significance of the post-Cam-
                                         brian minor intrusions of Assynt and the adjoining districts of
                                         north-west Scotland. *Q. Jl. Geol. Soc. Lond.*, **109**, 137–171.

                                  1963   The Strontian granite complex, Argyllshire. *Bull. Geol. Surv.
                                         G.B.*, **20**, 6–41.

SALLOMY, J. T. &   1973   Palaeomagnetic studies in the British Caledonide – II. The
    PIPER, J. D. A.   Younger Gabbros of Aberdeenshire, Scotland. *Geophys, J. R.
                                         Astr. Soc.*, **34**, 13–26.

SHACKLETON, R. M.   1948   Overturned rhythmic banding in the Huntly gabbro of Aber-
                                         deenshire. *Geol. Mag.*, **85**, 358–360.

SHAND, S. J.   1910   On Borolanite and its associates in Assynt. *Trans. Edinb. Geol.
                                         Soc.*, **9**, 202–215, 376–416.

                                  1939   Loch Borolan laccolith, north-west Scotland. *J. Geol.*, **47**, 408–
                                         420.

                                  1945   The present status of Daly's hypothesis of the alkaline rocks.
                                         Amer. J. Sci., **243A**, 495–507.

SHEPHERD, J. 1973 The structure and structural dating of the Carn Chinneag intrusion, Ross-shire. *Scott. J. Geol.*, **9**, 63–88.

SIMPSON, R. R., BROWN, G. C., PLANT, J. & OSTLE, D. 1979 Uranium mineralisation and granite magmatism in the British Isles. *Phil. Trans. R. Soc. Lond. A.*, **291**, 385–412.

SMITH, D. I. 1979 Caledonian minor intrusions of the Northern Highlands of Scotland. *In* Harris, A. L., Holland, C. H. and Leake, B. E. (Eds.) The Caledonides of the British Isles – reviewed. *Spec. Publ. Geol. Soc. Lond. No. 8*, 683–697.

SOPER, N. J. 1963 The Structure of the Rogart igneous complex, Sutherland, Scotland. *Q. Jl. Geol. Soc. Lond.*, **119**, 445–478.

1986 The Newer Granite problem: a geotectonic view. *Geol. Mag.*, **123**, 227–236.

SOPER, N. J. & ANDERTON, R. 1984 Did the Dalradian slides originate as extensional faults? *Nature*, **307**, 357–360.

SOPER, N. J. & BROWN, P. E. 1971 Relationship between metamorphism and migmatisation in the northern part of the Moine Nappe. *Scott. J. Geol.*, **7**, 305–326.

SOPER, N. J., WEBB, B. C. & WOODCOCK, N. H. 1987 Late Caledonian (Arcadian) transpression in north-west England: timing, geometry and geotectonic significance. *Proc. Yorks. Geol. Soc.*, **46**, 175–192.

SPRAY, J. G. & WILLIAMS, G. D. 1980 The subophiolite metamorphic rocks of the Ballantrae igneous complex, SW Scotland. *J. Geol. Soc. Lond.*, **137**, 359–368.

STEPHENS, W. E. 1988 Granitoid plutonism in the Caledonian orogen of Europe. *In* Harris, A. L. and Fettes, D. J. (Eds.) *Geol. Soc. Lond. Spec. Publ.*, **38**, 389–403.

STEPHENS, W. E. & HALLIDAY, A. N. 1980 Discontinuities in the composition of a zoned pluton, Criffel, Scotland. *Bull. Geol. Soc. Am.*, **91**, 165–170.

1984 Geochemical contrasts between late Caledonian granitoid plutons of northern, central and southern Scotland. *Trans. Roy. Soc. Edinb., Earth Sciences*, **75**, 259–273.

STEPHENS, W. E., WHITLEY, J. E. & THIRWALL, M. F. 1985 The Criffel zoned pluton: correlated behaviour of rare earth element abundancies with isotopic systems. *Contr. Miner. Petrol.*, **89**, 226–238.

STEVENSON, B. G. 1971 Chemical variability in some Moine rocks of Lochailort, Inverness-shire. *Scott. J. Geol.*, **7**, 51–60.

STEWART, F. H. 1946 The gabbroic complex of Belhelvie in Aberdeenshire. *Q. Jl. Geol. Soc. Lond.*, **102**, 465–498.

STEWART, F. H. & JOHNSON, M. R. W. 1960 The structural problem of the younger gabbros of north-east Scotland. *Trans. Edinb. Geol. Soc.*, **18**, 104–112.

STOKER, M. C. 1983 The stratigraphy and structure of the Moine rocks of eastern Ardgour. *Scott. J. Geol.*, **19**, 369–386.

STONE, P. 1984 Constraints on genetic models for the Ballantrae complex, SW Scotland. *Trans. Roy. Soc. Edinb., Earth Sci.*, **75**, 189–191.

STONE, P. & RUSTON, A. W. 1983 Graptolite faunas from the Ballantrae ophiolite complex and their structural implications. *Scott. J. Geol.*, **19**, 297–310.

STONE, P., FLOYD, J. D., BARNES, P. R. & LINTERN, B. C. 1987 A sequential back-arc and foreland basin thrust duplex model for the Southern Uplands of Scotland. *J. Geol. Soc. Lond.*, **144**, 753–764.

STUART, B. A., RAMSAY, D. M., PRINGLE, I. R. & TEGGIN, D. E. 1977 Precambrian gneisses in the Dalradian sequence of NE Scotland. *Jl. Geol. Soc. Lond.*, **134**, 41–44.

SUMMERHAYES, C. P. 1966 A geochronological and strontium isotope study of the Garabal Hill–Glen Fyne igneous complex, Scotland. *Geol. Mag.*, **103**, 155–165.

TANNER, P. W. G. & TOBISCH, O. T. 1972 Sodic and ultra-sodic rocks of metasomatic origin from part of the Moine Nappe. *Scott. J. Geol.*, **8**, 151–178.

TAUBENECK, W. H. 1967 Notes on the Glen Coe cauldron subsidence, Argyllshire, Scotland. *Geol. Soc. Am. Bull.*, **78**, 1295–1316.

THIRLWALL, M. F. 1981a Implications for Caledonian plate tectonic models of chemical data from volcanic rocks of the British Old Red Standstone. *J. Geol. Soc. Lond.*, **138**, 123–138.

1981b Peralkaline rhyolites from Ordovician Tweeddale lavas Peebles-shire, Scotland. *Geol. J.*, **16**, 41–44.

1982 Systematic variations in chemistry and Nd–Sr isotopes across a Caledonian calc–alkaline volcanic arc: implications for source materials. *Earth Planet. Sci. Lett.*, **58**, 25–50.

1983 Isotope geochemistry and origin of calc–alkaline lavas from a Caledonian continental margin volcanic arc. *J. Volc. Geotherm Res.*, **18**, 589–631.

1986 Lead isotope evidence for the nature of the mantle beneath Caledonian Scotland. *Earth Planet. Sci. Lett.*, **80**, 55–70.

1988 Geochronology of late Caledonian magmatism in northern Britain. *J. Geol. Soc., Lond.*, **145**, 951–968.

THIRLWALL, M. F. & BLUCK, B. J. 1984 Sr–Nd isotope and geochemical evidence that the Ballantrae "ophiolite", SW Scotland, is polygenetic. *In* Gass, I. G., Lipard, S. J. and Shelton, A. W. (Eds.) Ophiolites and oceanic litho-sphere. *Spec. Publ. Geol. Soc. Lond.*, **13**, 215–230.

THOMPSON, R. N. 1985 Model for Grampian tract evolution – comments on a paper by J. F. Dewey and R. M. Shackleton. *Nature*, **314**, 562.

THOMPSON, R. N. & FOWLER, M. B. 1986 Subduction-related shoshonitic and ultrapotassic magmatism: a study of Siluro–Ordovician syenites from the Scottish Caledon-ides. *Contr. Miner. Petrol.*, **94** 507–522.

TILLEY, C. E. 1935 The role of kyanite in the "hornfels" zone of the Carn Chuinneag granite (Ross-shire). *Mineral. Mag.*, **24**, 92–97.

TINDLE, A. G. & PEARCE, J. A. 1981 Petrogenetic modelling of *in situ* fractional crystallisation in the zoned Loch Doon pluton, Scotland. *Contr. Miner. Petrol.*, **78**, 196–207.

TINDLE, A. G., McGARVIE, D. W. & WEBB, P. C. 1988 The role of hybridisation and crystal fractionation in the evolution of the Cairnsmore of Cairsphairn intrusion, Southern Uplands of Scotland. *J. Geol. Soc. Lond.*, **145**, 11–22.

TRELOAR, P. J., BLUCK, B. J., BOWES, D. R. & DUDEK, A. 1980 Hornblende–garnet metapyroxenite beneath serpentinite in the Ballantrae complex of SW Scotland and its bearing on the depth provenance of obducted oceanic lithosphere. *Trans. Roy. Soc. Edinb., Earth Sci.*, **71**, 201–212.

TYLER, I. M. & ASHWORTH, J. R. 1983 The metamorphic environments of the Foyers granitic complex. *Scott. J. Geol.*, **19**, 271–285.

VAN BREEMEN, O., PIDGEON, R. T. & JOHNSON, M. R. W. 1974 Precambrian and Palaeozoic pegmatites in the Moines of northern Scotland. *Jl. Geol. Soc. Lond.*, **130**, 493–507.

VAN BREEMEN, O., HALLIDAY, A. N., JOHNSON, M. R. W. & BOWES, D. R. 1978 Crustal additions in late-Precambrian times. *In* Bowes, D. R. and Leake, B. E. (Eds.) Crustal evolution in northwestern Britain and adjacent regions. *Geol. J. Spec. Issue*, **10**, 81–106.

VAN BREEMEN, O., AFTALION, M., PANKHURST, R. J. & RICHARDSON, S. W. 1979a Age of the Glen Dessary syenite, Inverness-shire: diachronous Palaeozoic metamorphism across the Great Glen. *Scott. J. Geol.*, **15**, 49–62.

VAN BREEMEN, O., AFTALION, M. & JOHNSON, M. R. W. 1979b Age of the Loch Borrolan complex, Assynt, and late movements along the Moine thrust zone. *Jl. Geol. Soc. Lond.*, **136**, 489–496.

VAN BREEMEN, O. &   1981   Episodic granite plutonism in the Scottish Caledonides. *Nature*,
BLUCK, B. J.   **291**, 113–117.

VAN BREEMEN, O. &   1983   The Glen Kyllachy granite and its bearing on the nature of the
PIASECKI, M. A.   Caledonian orogeny in Scotland. *J. Geol. Soc. Lond.*, **140**, 47–62.

WADSWORTH, W. J.   1970   The Aberdeenshire layered intrusion of north-east Scotland.
                                   Spec. Publ. Geol. Soc. S.Afr., **1**, 565–575.

                            1982   The basic plutons. *In* Sutherland, D. (Ed.) *Igneous rocks of the
                                   British Isles.* Wiley, 135–148.

                            1986   Silicate mineralogy in the later fractionation stages of the Insch
                                   intrusion, NE Scotland. *Mineral. Mag.*, **50**, 583–595.

                            1988   Silicate mineralogy of the Middle Zone cumulates and
                                   associated gabbroic rocks from the Insch intrusion, NE
                                   Scotland. *Mineral. Mag.*, **52**, 309–322.

WATSON, J. V.   1984   The ending of the Caledonian orogeny in Scotland. *J. Geol. Soc.
                                   Lond.*, **141**, 193–214.

WILKINSON, M. J. &   1974   Trace elements and tectonic relationships of basaltic rocks in the
CANN, J. R.   Ballantrae igneous complex, Ayrshire. *Geol. Mag.*, **111**, 35–41.

WILLIAMS, A.   1962   The Barr and Lower Ardmillan Series (Caradoc) of the Girvan
                                   District, south Ayrshire. *Mem. Geol. Soc. Lond.*, **3**.

WILSON, J. R. & LEAKE, B. E. 1972 The petrochemistry of the epidiorites of the Tayvallich penin-
                                   sula, North Knapdale, Argyllshire. *Scott. J. Geol.*, **8**, 215–251.

WILSON, D. & SHEPHERD, J.   1979   The Carn Chuinneag granite and its aureole. *In* Harris, A. L.,
                                   Holland, C. H. and Leake, B. E. (Eds.) The Caledonides of the
                                   British Isles – reviewed. *Spec. Publ. Geol. Soc. Lond., No. 8*, 669–
                                   675.

WINCHESTER J. A.   1974   The zonal pattern of regional metamorphism in the Scottish
                                   Caledonides. *J. Geol. Soc. Lond.*, **130**, 509–524.

                            1976   Different Moinian amphibolite suites in northern Ross-shire.
                                   Scott. J. Geol., **12**, 187–204.

                            1984   The geochemistry of the Strathconon amphibolites, northern
                                   Scotland. *Scott. J. Geol.*, **20**, 37–52.

WINCHESTER, J. A. &   1984   The geochemistry of the Ben Hope sill, northern Scotland.
FLOYD, P. A.   *Chem. Geol.*, **43**, 49–75.

WOOLLEY, A. R.   1970   The structural relationships of the Loch Borrolan complex,
                                   Scotland. *Geol. J.*, **7**, 171–182.

                            1973   The pseudoleucite borolanites and associated rocks of the south-
                                   eastern tract of the Borralan complex, Scotland. *Bull. Br. Mus.
                                   Nat. Hist. (Miner.)*, **2**, 287–333.

WRIGHT, A. E. &   1968   Formation of explosion breccias. *Bull. Volc.*, **32**, 15–32.
BOWES, D. R.

                            1979   Geochemistry of the appinite suite. *In* Harris, A. L., Holland,
                                   C. H. and Leake, B. E. (Eds.) The Caledonides of the British Isles
                                   – reviewed. *Spec. Publ. Geol. Soc. Lond., No. 8*, 699–704.

YARDLEY, B. W. D.   1977   Petrogenesis of migmatites in the Huntly–Portsoy area, north-
                                   east Scotland – a discussion. *Mineralog. Mag.*, **41**, 292–294.

ZALESKI, E.   1982   *The geology of Speyside and lower Findhorn granitoids.* Univer-
                                   sity of St Andrews MSc thesis (unpubl.).

ZHOU, J.-X.   1985   The timing of calc–alkaline magmatism in parts of the Alpine–
                                   Himalayan collision zone and its relevance to the interpretation
                                   of Caledonian magmatism. *J. Geol. Soc. Lond.*, **142**, 309–317.

9

OLD RED SANDSTONE

W. Mykura

The term 'Old Red Sandstone' has been used in Britain since 1822 (Conybeare & Philips 1822) to denote the terrestrial (i.e. fluvial, aeolian and lacustrine) sediments which are roughly equivalent in age to the Devonian marine deposits in south-west England and continental Europe. Old Red Sandstone sediments intercalated with marine strata occur in south-west Ireland, South Wales, the Welsh Borderland, and in south-east and Midland England, but the Scottish Old Red Sandstone is entirely terrestrial (House *et al.* 1977). Boreholes in the northern North Sea have encountered mainly terrestrial sediments of 'Old Red Sandstone' type, but one bore in the Argyll Oil Field, 200 km south-east of Aberdeen has cut ostracod-bearing Middle-Devonian shales succeeded by dolomitic limestone with corals, overlain in turn, by terrestrial 'Upper Old Red Sandstone' (Pennington 1975). This record suggests that a bed of marine sediments must have extended southwards from the latitude of the Argyll Field to link with the marine Devonian in north-east France and North Germany.

The organic remains within the Old Red Sandstone consist of fish, rare arthropods, plants and mio- and micropores. Though great advances have been made in correlation by miospores, the correlation of the Old Red Sandstone subdivisions with stages of the Devonian is still very tentative and evidence from K–Ar ages has shown quite considerable disagreements. Many Old Red Sandstone lavas in the Midland Valley and the Highlands have yielded Silurian ages which are difficult to correlate with the fossil ages of the adjacent sediments (Thirlwall 1979).

The three-fold subdivision of the Scottish Red Sandstone into Lower, Middle and Upper Old Red was established by Murchison (1859) who believed that Lower Old Red Sandstone occurs only south of the Grampians, that the middle Old Red Sandstone is confined to outcrops around the Moray Firth, Caithness, Orkney and Shetland, and that the Upper Old Red Sandstone is present in all areas resting unconformably on older beds. Geikie (1878), however, suggested that Murchison's 'Middle' Old Red Sandstone,

though differing in facies, is of the same age as the Lower Old Red Sandstone in the Midland Valley. Detailed work on the fossil fish, mainly by Traquair over many years, has however verified the Middle Old Red Sandstone age of the northern sediments, though in recent years *some* strata of Lower Old Red Sandstone age have been recognised in the Moray Firth–Caithness area (Westoll 1951, 1977; Richardson 1967) and possibly also in Orkney and Shetland (Mykura & Phemister 1976). No fauna, flora, or micro-flora which could be ascribed to the Middle Old Sandstone has been found south of the Grampians, and it is almost certain that there are no sediments of this age in the southern outcrops.

The terrestrial sediments of the Old Red Sandstone have always been regarded as the consolidated molasse deposits of the emerging Caledonian mountains, which in the areas of the present Grampians may well have been of Alpine or even Himalayan proportions when the Lower and particularly the Middle Old Red Sandstone were being deposited. The gradual build-up of knowledge about the evolution of the Iapetus Ocean and the collision of the Canadian plate (Laurentia) with Fenno Scandia has greatly altered our ideas on the development of deposition of the Old Red, and at the moment ideas and theories are still somewhat in the melting pot.

The main contributor to this is Bluck (1983, 1984) who has studied the lithology of the boulders in the Old Red Sandstone conglomerates which are present in the Midland Valley, the Grampians, and the Southern Uplands, as well as the source directions of the Old Red Sandstone sediments. He has been able to demonstrate that whereas the Lower Old Red Sandstone conglomerates just north of the Highland Boundary Fault, which rest on eroded Dalradian, contain material derived directly from the Dalradian and have a dispersal to the SE, the conglomerates immediately south of the Highland Boundary Fault, have a dispersal to the NW, and contain no material which can be definitely shown to be derived from the

Dalradian. The first undoubted Dalradian clasts to be found south of the Highland Boundary fault are in the top of the Upper Old Red Sandstone which, in contrast to the older Old Red Sandstone also shows a dispersal to the south-east.

Bluck has concluded that throughout the Ordovician, Silurian and most of Old Red Sandstone times the Midland Valley was a major volcanic–plutonic arc massif trending ENE roughly parallel to the Ochil–Sidlaw anticline today. This massif shed sediment to a fore-arc basin to the south and a marginal basin to the north. Both of these basins extended well beyond the margins of the present Midland Valley, and the present boundaries of the Midland Valley are relatively late thrust features, the Highland Boundary Fault not being emplaced until just before the Carboniferous. Both thrusts resulted in the concealment of many important faults and fault blocks which have supplied sediment to the Midland Valley. Bluck also considers that the concealed faults may have had a considerable lateral movement. Thus the basin in the northern Midland Valley was being formed about 415 Ma to 380 Ma ago, a time when granite plutons were being emplaced in the Dalradian of the southern Highlands and the area was probably rising through buoyancy. If the basin had bordered that terrain, it would be expected to contain greenschist and higher grade metamorphic rocks, which are completely absent. As the material north of the Highland Boundary Fault is full of locally derived material and shows a south-eastward deposition there must have lain a considerable basin containing detritus full of Highland clasts to the south-east of the present HBF. It has also been shown by Bluck, that the Highland Border complex, which is Ordovician in age, contains no Highland detritus, even though the time of the deposition coincides with rapid uplift of the Dalradian, and that the tectonic history of the Highland Border rocks does not match the history of the adjacent Dalradian. This indicates that the Midland Valley was not adjacent to the Dalradian throughout the Lower Palaeozoic and did not come together till the late Upper Old Red Sandstone.

The juxtaposition of the Dalradian block with the Midland Valley thus occurred late in the Upper Old Red Sandstone times when boulders of Highland rock appear to the south of the fault for the first time. As only thin conglomerate was deposited then, followed by quartz-rich sandstone and caliche deposits, topographic change involved at the time was relatively minor. The Dalradian was therefore thrust to come in contact with Midland Valley terrane on a very low-angled thrust moving south-eastward. As its emplacement at the end of the Old Red Sandstone caused very little relief, it is suggested that the Dalradian was already in strike slip contact with the Midland Valley terrane, and considerably eroded when the final thrusting took place.

Another concept, which will alter our ideas of Old Red Sandstone palaeogeography, still not tested thoroughly on the ground, has been put forward by McClay et al. (1986)

and is based on ideas worked out in the Tertiary Basin and Range province of the western United States. There it has been shown that the mid-Tertiary sediments were laid down on the overthickened crustal belt of the North-American Orogen which then pulled apart to thin the sedimentary sequence by sinistral extension faults and low-angled extension faults which may have originated as thrusts. Norton (1986) has seen similar structures in the Old Red Sandstone in West Norway and Haller (1971) has described extension faults and Basin Range geometries in the Old Red Sandstone in East Greenland. Similar structures are now being investigated in the Old Red Sandstone in Scotland, but a lot of evidence has still to be found before a valid story emerges.

There is still much argument as to the exact age of the Scottish Old Red Sandstone in relation to the Devonian. The exact age of the Devonian itself is still not settled; the most recent figures (McKerrow et al. 1985) are: base of Lower Old Red Sandstone (base of Gedinnian) 400 Ma (395–410 Ma), base of Middle Old Red Sandstone 385 Ma (380–390 Ma), base of Upper Old Red Sandstone 375 Ma (370–380 Ma), top of Upper Old Red Sandstone 360 Ma (350–365 Ma). No exact agreement is, however, available. Thirlwall (1981) has carried out radio-metric measurements which suggest that the ages of the lavas of the Arbuthnott Group (p. 305) is probably close to 410 Ma, but this is not substantiated by fish (Westoll 1977) or spores (Richardson 1967).

Similarly, the palaeomagnetic evidence for the position of 'Scotland' during the Old Red Sandstone in relation to the equator is not clear-cut. According to Fuller & Briden (1978) 'Britain' lay between 20° and 30° south of the equator during the Lower Devonian moving slowly northward to between 10° and 20° S by the end of the Devonian. These figures are only approximate. Much work on palaeolatitudes in the North Scottish Old Red Sandstone has been carried out by Storetvedt and his students (Storetvedt 1974, Storetvedt & Carmichael 1979), and the results have been used by Van der Voo & Scotese (1981) to calculate the amount of horizontal movement along the Great Glen Fault and of associated faults (though their results are not generally accepted (p. 324)). Their main conclusions are that in Middle and Late Devonian time the Orcadian Basin lay, like the North American Craton, at 15°–20° North palaeolatitude whereas the readings from sediment the same age south of the Great Glen indicate a palaeolatitude of 15°–20° South. There is, however, general agreement that the climate throughout most of Old Red Sandstone time was warm to hot, generally semi-arid, at times arid, but with intermittent terrestrial downpours.

In *Lower Devonian* times the topography of the Scottish area was still controlled by the final closure of the Iapetus Ocean. Bluck has shown that between 460 and 415 Ma the Highlands were isostatically rising due to the emplacement of granites but from 410 Ma on the Southern Highlands had established the present level of erosion. The extrusion of lavas in the Highlands covered a larger area than the outcrops of the present Lorne Plateau lavas. The present Highlands were, at that time, well west of their present position in relation to the Midland Valley, not being finally emplaced in their present position till the end of the Devonian. The valleys in the Highlands were being filled with sediments of local derivation. As the sediments close to the Highland Boundary Fault, which rest directly on the Dalradian, show a south-easterly direction of deposition there must have been a trough filled with Highland clasts south-east of the present Highland Boundary Fault.

In the Strathmore Syncline, the lowest 1 km of sediments along the Highland Border show a south-eastwardly direction of source of clasts indicating that the main basin of deposition lay north-west of the present Highland Boundary Fault. The lava pebbles in the conglomerate may be derived from the Ochil lavas, and the quartzites, which are polyclic may come from the underlying basement. The higher part of the sequence in the Strathmore Syncline has sandstone in the centre which was laid down by a river flowing to the south-west. The higher conglomerates along the Highland Boundary Fault, however, have a palaeoflow to south-east. They contain pebbles of basic volcanic rock, together with greenschist, mica-schist and quartzite. None of the clasts, however, can be matched with the rocks now cropping out in the southern Highlands and they were probably derived from metamorphic rocks and lavas which are now overridden by the Grampian Dalradian.

On the south-east side of the Strathmore Syncline the sediments are replaced by the thick pile of lavas which now form the Ochil and Sidlaw Hills which probably extended south-westward to join with the lava outcrops south of Darvel and Ayr. South of these hills another valley, also flowing south-west, linked the present outcrops through Lanarkshire and Ayrshire. The age of these sediments in relation to the Strathmore Syncline is not known and may well be older. They contain volcanic clasts derived from the south-east, thus indicating that another volcanic field lay in the position of the present Southern Uplands (Bluck 1983, fig. 15). In the western and northern Highlands there were smaller fault-controlled land-locked basins.

The *post-Lower Old Red Sandstone earth movements* produced much faulting and relatively gentle folding in the Midland Valley where no Middle Old Red Sandstone is preserved and probably not much sediment was laid down at the time. In the Highlands the land was highest in the west and sloped generally to the north-east with large rivers flowing north-eastward to the alluvial flats around the present Moray Firth. North of these flats lay, in the area of Caithness and Orkney, the large shallow lake which accumulated a great thickness (4 km +) of rhythmically bedded, fine-grained, shallow-water lake sediments. In Middle Devonian times much of the drainage from the Scottish area may eventually have found its way into the arm of the sea which extended northwards along the middle of the present North Sea, at least to the latitude of

Edinburgh. The Upper Devonian earth movements were relatively minor, and Bluck (1984) has shown that the present Dalradian may have been thrust into its present position in relation to the Midland Valley toward the end of that period. The overall palaeoslope in the Southern Uplands and Midland Valley, as well as in the Northern Highlands in Upper Old Red Sandstone times was to the north-east. The widespread development of calcretes near the end of the period suggests relatively dry conditions with relatively long periods of non-deposition on the alluvial plains.

Lower Old Red Sandstone

Lower Old Red Sandstone sediments and lavas crop out along the northern and southern flanks of the Midland Valley graben. The extent of the northern outcrop is shown in Figure 9.1, which also shows that Lower Old Red Sandstone crosses the Highland Boundary Fault in the areas north of Crieff and north of Blairgowrie; though Bluck (1984) has now shown that the sediments north of the Fault were laid down on top of Dalradian and are of different composition from the Midland Valley rocks. The southern outcrop within the Midland Valley is smaller and discontinuous and is dissected by sub-parallel north-east-trending belts. It extends from the Pentland Hills via the Biggar and Lanark areas to central and west Ayrshire. South of the Southern Uplands Fault two north-north-west-trending belts of Lower Old Red Sandstone conglomerate extend south-south-eastward across the Lammermuir Hills and a small outcrop of volcanic rocks and sediments is developed around St Abb's Head and Eyemouth. Farther south in Roxburghshire Lower Old Red Sandstone lavas form part of the Cheviot Volcanic Series. The St Abb's Head and Cheviot Hills lavas differ from the Midland Valley lavas and were probably deposited slightly later (Thirlwall 1981).

North of the Highland Boundary Fault the largest outcrop of Lower Old Red Sandstone rocks forms the Lorne lava plateau between Loch Melfort and Loch Awe. Originally it may have extended north-eastward to Glen Coe and Ben Nevis, where remnants are preserved in areas of cauldron subsidence. The sediments underlying the lava pile are seen at Oban and on Kerrera. Further outcrops occur in Mull and on the Mull of Kintyre.

In the north-eastern province around the Moray Firth sediments ascribed to the Lower Old Red Sandstone occur in the outliers of Rhynie, Tomintoul and Cabrach within the Grampians, along the south shore of the Moray Firth at Pennan and possibly also at Buckie, and at intervals along the base of the Middle Old Red Sandstone outcrops extending from the Great Glen, northward via Easter Ross and East Sutherland to Caithness. There is one small outcrop of possible Lower Old Red Sandstone age on Orkney.

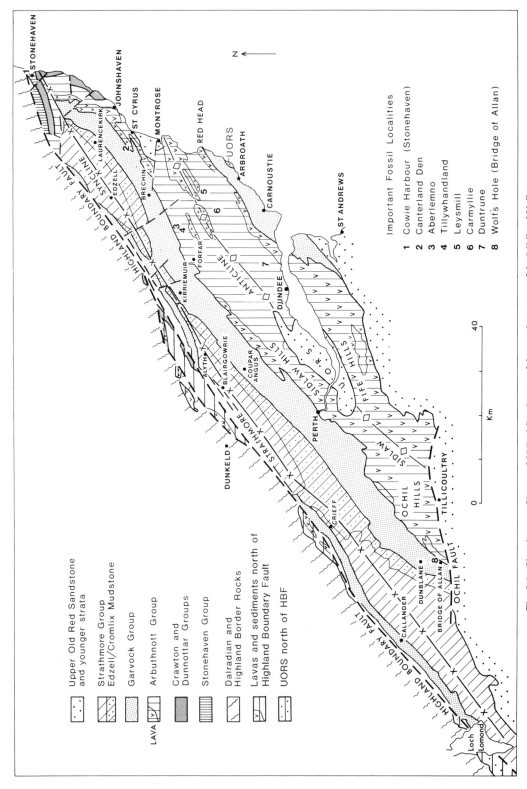

Fig. 9.1. Sketch map of Lower Old Red Sandstone of the northern part of the Midland Valley.

Southern outcrops

Northern Midland Valley

The Lower Old Red Sandstone rocks within the northern Midland Valley are folded into the asymmetrical Strathmore Syncline, and the parallel complementary Sidlaw Anticline (Fig. 9.1). The north-western limb of the former steepens towards the Highland Boundary Fault Zone and becomes vertical and locally overturned along the fault. Evidence for overthrusting towards the south-east along the fault zone is seen in the shattered conglomerates of the Loch Lomond area (Ramsay 1962). The south-eastern limb of the syncline is only gently inclined and the finest section of the Lower Old Red Sandstone within the Midland Valley is seen where this limb is exposed in the coastal cliffs between Stonehaven and St Cyrus. The Sidlaw Anticline is an open, symmetrical structure, the axial trace of which extends from Montrose south-westwards to the Ochil Hills. South-west of Dundee its north-west limb is formed by the lavas of the Sidlaw Hills, its south-east limb by the Fife Ochils, and its crest is downfaulted into a graben now filled by Upper Old Red Sandstone and Carboniferous

sediments. Farther south-west, in the Ochil Hills, the south-eastern limb of the anticline is, in part, cut out by the east–west-trending Ochil Fault.

The lavas and sediments within the northern belt have been extensively studied in the past and the most important earlier publications are by Jack & Etheridge (1877) [Callander area], du Toit (1905) [Balmaha], Hickling (1908, 1912) [mainly Angus], Campbell (1913) [Kincardineshire], and Allan (1928, 1940) [central Angus and Perthshire]. More recently the lavas of Eastern Forfarshire have been described by Robson (1948), those of the Sidlaw Hills by Harry (1956) and Gandy (1975), and the volcanic and sedimentary rocks of the Western Ochils (Fig. 9.2) and the Dunblane–Callander district by Francis et al. (1970). A recent synthesis and re-assessment of the stratigraphy of the area between the east coast and Callander by Armstrong & Paterson (1970) forms the stratigraphic basis of the present account and the palaeogeographic interpretation is largely based on papers by Bluck (1978, 1983, 1984), Morton (1979) and Wilson (1980).

The flagstones within the Strathmore succession have yielded a sparse but important assemblage of fossil fish and

Fig. 9.2. Escarpment of Ochil Hills showing near horizontal flows of basalt and andesite lava. The Ochil fault runs at the foot of the escarpment. (B.G.S. photo.)

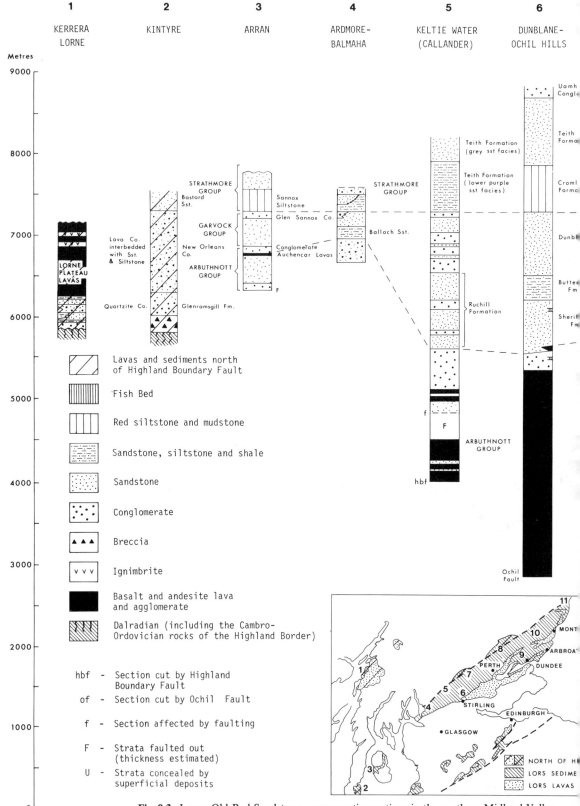

Fig. 9.3. Lower Old Red Sandstone – comparative sections in the northern Midland Valley and Argyll.

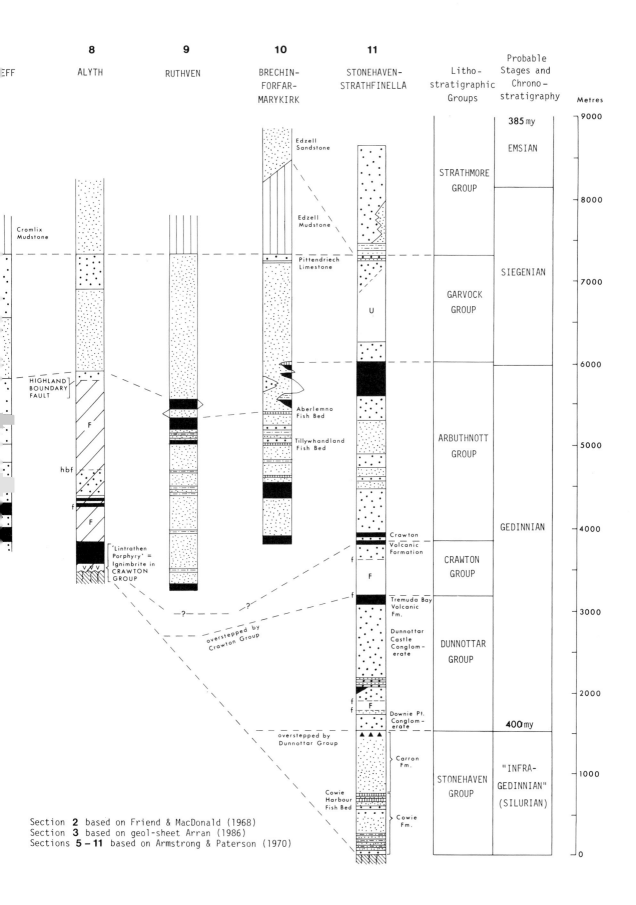

Section **2** based on Friend & MacDonald (1968)
Section **3** based on geol-sheet Arran (1986)
Sections **5 – 11** based on Armstrong & Paterson (1970)

plants. The fish remains have been described by Agassiz (1835), Hugh Miller (1841), Mitchell (1860, 1861), Powrie (1864) and Westoll (1945, 1951). Westoll has also been responsible for the stratigraphic interpretation of the faunal assemblages. The macroscopic plant fossils of the Strathmore Old Red Sandstone have been studied by Fleming (1831), Lyell (1865), Jack & Etheridge (1877), Kidston (1893), Reid & MacNair (1896), Reid *et al.* (1897), Kidston & Lang (1924) and Lang (1927, 1932). Richardson (1967) has described some of the spore assemblages and evaluated their stratigraphic significance. Rayner (1984) and Ford (1979) showed that macroflora and spore assemblages from the Strathmore Group suggest an Emsian Age. Thirlwall (1981) has disputed the ages based on macroscopic fauna and flora and microfauna, on the basis of radiometric age dating. He has shown that an intrusion which cuts the Arbuthnott Group is 407 + 5 Ma old which suggests that the Arbuthnott lavas are probably close to 410 Ma. If the base of the Lower Old Red Sandstone is taken at 400 Ma, that would put the rock of the Arbuthnott Groups and underlying Formations in the Silurian. As many of the best fossil beds (Fig. 9.3) are found in the Arbuthnott Group this problem is not resolved, and Thirlwall's contention (Thirlwall 1983), that early Devonian fossils evolved in the Scotland area 20 my before the European areas cannot be accepted.

The stratigraphic succession (Fig. 9.3) is based on the scheme set up by Campbell (1913) and modified and expanded by Armstrong & Paterson (1970). The subdivisions are of necessity lithostratigraphical and in some cases the boundaries between them are markedly diachronous. If Bluck's contention that the rocks north of the Highland Boundary Fault are quite unrelated to those in the Strathmore basin is accepted, sections 7 and 8 have to be re-interpreted and the lowest part with the Lintrathen Porphyry and the underlying Dalradian is not part of the section.

Stonehaven Group. The oldest rocks of the Old Red Sandstone within the Midland Valley form the Stonehaven Group which is probably Infra-Gedinnian (topmost Silurian) in age. They occupy a narrow belt of ground which extends for 13 km west of Stonehaven along the near-vertical north-western limb of the Strathmore (Mearns) Syncline (Fig. 9.1). They rest unconformably on the Cambro–Ordovician Margie Series and are succeeded, apparently without significant unconformity, by the conglomerates of the Dunnottar Group.

The Stonehaven Group is well exposed along the shore between the Highland Boundary Fault and Stonehaven Harbour. It is subdivided into the lower Cowie Formation (730 m) and the upper Carron Formation (820 m) (Fig. 9.3).

The *Cowie Formation* consists of a thin basal breccia with clasts derived mainly from the underlying Highland Border Series, overlain by dull red, grey and yellow, fine to medium grained sandstone with intercalations of red and grey mudstone. It contains a thin flow or intrusion of andesite

near the base and, higher up, a conglomerate with pebbles of acid volcanic rock, interbedded with tuffaceous sandstone. The highest beds are alternating grey sandstones and shales with *Dictyocaris*, which contain, near their base, grey shales with fish remains, known as the *Cowie Harbour Fish Bed*. These have yielded *Archidesmus*, *Kampecaris*, *Ceratiocaris*, *Dictyocaris slimoni*, *Hughmilleria norvegica*, *H. sp.*, *Pterygotus*, *Hemiteleaspis heintzi*, *Pterolepis?*, *Traquairaspis campbelli* and carbonised plant debris (Westoll 1945, 1951). This fauna suggests an age between that of the Downton Castle Formation and the base of the Ditton Group of the Welsh Borderland (Westoll 1977; Allen *in* House *et al.* 1977, fig. 7).

The *Carron Formation* comprises cross-bedded dull reddish-brown sandstones with a high proportion of volcanic debris. There are local conglomeratic lenses with pebbles of both igneous rocks and metasediment, and near the top there is a thick development of agglomerate with angular clasts of hypersthene- and biotite-andesite, set in a matrix of devitrified volcanic glass.

Armstrong *et al.* (*in* Friend & Williams 1978) suggest that the sediments of the Stonehaven Group were deposited largely by low-energy braided and meandering rivers, apparently flowing both north-eastward and south-eastward.

Dunnottar Group. The rocks of this group are exposed for 2 km on the coast south of Stonehaven and can be traced inland for 13 km west-south-westward along the northern limb of the Strathmore Syncline. The group was probably overlapped by the overlying Crawton Group not far west of its present outcrop. Along the coast it has the following sequence:

4. Tremuda Bay Volcanic Formation (olivine-basalts)	60 m +
5. Dunnottar Castle Conglomerate	1035 m
2. Strathlethan Formation (fine-grained sandstone and agglomerate)	360 m
1. Downie Point Conglomerate	560 m

The conglomerates contain rounded boulders, up to 1 m across, mainly of 'Highland' type (i.e. quartzite, schistose grit and subordinate Cambro–Ordovician lavas) and some of fine-grained igneous rock, including hypersthene- and biotite-andesite.

Crawton Group. The sediments and lavas of this group crop out almost continuously along the coast for 15 km south of Stonehaven and also appear on the north limb of the Strathmore Syncline for 13 km west of Stonehaven. The rocks taken to belong to the Crawton Group which crop out at intervals along the Highland Boundary Fault zone between Edzell and Dunkeld (Fig. 9.1) are north of the Highland Boundary Fault and are not connected. Along the coast the Crawton Group consists largely of conglomerate with 'Highland' type pebbles, including rounded boulders of quartzite and schist, interbedded with pebbly volcano–detrital sandstones. At the top of the group the

Crawton Volcanic Formation forms a series, up to 30 m thick, of distinctive 'Crawton type' macroporphyritic basic andesites and olivine basalts which are also present on the northern limb of the Strathmore Syncline.

Arbuthnott Group. The Arbuthnott Group was originally set up by Campbell to embrace the thick conglomeratic sequence in Kincardineshire above the Crawton Lavas. Armstrong & Paterson (1970) have suggested that this sequence can be correlated with the mixed volcanic sequence of the Sidlaw and Ochil Hills and with the lavas and conglomerates exposed along the Highland Boundary Fault zone between Kirriemuir and Dunkeld and between Crieff and Loch Lomond. The beds generally become coarser and more conglomeratic near the Highland Border. Along the Kincardineshire coast the group consists mainly of the Johnshaven Formation, which is composed largely of 'Highland' type conglomerates with some sandstones and a few thin lavas. Some of the conglomerates contain isolated boulders up to 3·7 m in diameter. South-west and west of Johnshaven a group of lavas termed the Montrose Volcanic Formation forms a thick series of flows in the upper part of the group. When traced south-westward into Angus the conglomerates die out and the sequence consists largely of cross-bedded drab-coloured sandstones (= Dundee Formation) with important intercalations of 'flagstone' up to 100 m thick, particularly near the middle and top. The flagstones are mainly thinly bedded (flaggy) sandstones and contain horizons of shale with some calcareous laminites of lacustrine origin. The flags and limy shales were much quarried in the past for paving stones and have yielded a sparse fauna, mainly of fish and arthopods, which is believed to be of basal Dittonian (= Gedinnian) age. The most important 'fish beds' are those of Tilly-whandland, Aberlemno, Leysmill, Carmyllie and Duntrune (Figs. 9.1, 9.3). The fauna comprises the myriapods *Archidesmus* and *Kampecaris*, a number of eurypterids, particularly *Pterygotus anglicus*, and the fish *Cephalaspis lyelli, C. pagei, Turinea sp. Climatius sp. Ischnacanthus sp.* and *Mesacanthus sp.* (Hickling 1912; Westoll 1951, 1977). The accompanying plant remains include *Parka decipiens, Pachytheca sp. Nematophyton sp.* and *Zosterophyllum sp.*, the first being particularly common (Hickling & Don 1915, Lang 1927). The spores so far examined suggest a somewhat younger (Siegenian) age for these strata than the fish and eurypterid remains (Westoll 1977).

West of Dundee the sediments pass by intercalation into volcanic rocks, and in the Sidlaw and Ochil Hills lavas, pyroclastic rocks and associated volcanic sediments form almost the entire thickness of the Arbuthnott Group. They reach 1500 m in the Sidlaws and at least 2400 m in the Ochils, being thickest in the vicinity of Tillicoultry. In the Ochils the lavas range in composition from rhyodacite through trachyandesite and hornblende-andesite to pyroxene-andesite and olivine-basalt, with the last three types predominant (Francis *et al.* 1970). In the Sidlaws they are more basic, being largely composed of basic andesites,

felspar–phyric olivine basalts, two-pyroxene basalts and andesite–basalts (Robson 1948; Harry 1956, 1958; Gandy 1975). Pyroclastic rocks are largely confined to the western Ochils and are thickest and most numerous in the south of these hills, alongside the Ochil Fault. The volcanic sediments are more widely distributed as intercalations in the lavas, and near Bridge of Allan thick lenses of conglomerate occupy deep valleys cut in the lavas (Armstrong *in* Francis *et al.* 1970).

Along the Highland Border the Arbuthnott Group consists largely of volcanic conglomerates with subordinate sandstones and lavas and a few thin shale beds. Just south of the Highland Boundary Fault zone the the beds are steeply dipping. Allan (1928, 1940) believed that the entire Kincardineshire sequence down to the Dunnottar Group is represented between Alyth and Crieff, but Armstrong & Paterson (1970) have suggested that west of Alyth the lowest beds seen are the topmost beds of the Crawton Group, and more generally the basal beds of the Arbuthnott Group. Farther west in the Keltie Water near Callander, basaltic and andesitic lavas with intercalated sandstones and conglomerates form most of the lower part of the Arbuthnott Group, overlain by massive 'volcanic' conglomerates with some basic lavas.

Garvock Group. The term Garvock Group was originally applied by Campbell to the mainly conglomeratic sequence of Kincardineshire, which lies between the lavas at the top of the Arbuthnott Group and the 'marls' of the Strathmore Group. Armstrong and Paterson have since extended the outcrop of the group to take in all the strata of roughly the same age throughout the northern Midland Valley. In Kincardineshire it consists mainly of 'Highland type' conglomerates with a few flows of olivine-basalt and andesite. Farther south-west cross-bedded sandstones with intraformational limestone detritus dominate. Between Auchterarder and Bridge of Allan there are thick intercalations of shale and fine-grained sandstone. Close to the top of the group there is a thin but persistent zone of lenticular bodies of concretionary limestone (cornstone). This can be traced at intervals from Brechin to Dunblane and is a useful marker horizon. It has local names, such as the '*Pittendriech Limestone*' at Brechin, and the '*Stanley Limestone*' near Perth. It is probably a fossil soil (caliche) which developed during a period of restricted sedimentation (see p. 299). The Garvock Group is distinguished from the underlying Arbuthnott Group by its lack of flagstones and shales and by the presence of intraformational conglomerates containing clasts and pebbles of limestone. The bright red-brown and grey colours of the Garvock Group also contrast with the drab colours of the underlying strata. These lithological changes are facies-tied and do not coincide with a stratigraphic horizon. There is a marked diachronism and intedigitation between the two groups, with strata of the same age ascribed to the Arbuthnott Group in Kincardineshire but to the Garvock Group in Angus (Fig. 9.3).

No fossil fish have been recorded within the group in the north-east of the outcrop and only *Pterapis mitchelli* in one locality east of Coupar Angus within the central area. Between Braco and Bridge of Allan in the south-west, however, there are a few fossiliferous localities at the base of the group. Of these the Wolf's Hole Quarry at Bridge of Allan has yielded *Pteraspis mitchelli, Cephalaspis scotica, C. lyelli, Securiapis caledonica* and *S. waterstoni*. These forms are similar to, but slightly younger than, the fish in the Arbuthnott Group (Westoll 1977).

Strata ascribed to the Garvock Group also crop out on the south-east limb of the Sidlaw Anticline where they comprise the Red Head Series, Auchmithie Conglomerate and Arbroath Sandstone of Hickling's (1908, 1912) and Robson's (1948) classification. They are intermittently exposed along the shore between Red Head, 10 km NE of Arbroath, and Tentsmuir in north-east Fife, and reach a thickness of over 1000 m. They consist mainly of cross-bedded sandstone with clasts of concretionary limestone and, near their base, of volcanic rock. The sandstone is bright red in the north-east around Arbroath and mainly grey in the south-west. It is interbedded with the thick Auchmithie conglomerate which contains pebbles of both Highland type and volcanic rocks.

On the north-west limb of the Strathmore Syncline the Garvock Group forms a narrow, almost continuous outcrop between Edzell and Loch Lomond. It is here composed mainly of conglomerate of mixed 'Highland' type and volcanic origin. There are also some thick beds of sandstone, particularly in the belt between Crieff and Loch Lomond. At Callander the lower part of the group (termed the *Ruchill Formation*) is composed mainly of sandstone, siltstone and shale, while the upper part is a coarse conglomerate. The only concretionary limestone horizon which could be correlated with the Pittendriech limestone of Strathmore is a 10 m thick sandstone full of calcareous concretions, exposed in the West Water near Edzell.

The *Strathmore Group* consists mainly of grey and purplish-grey sandstones with conglomerates, but also contains in its lower half much poorly bedded bright and dull red and green mottled calcareous mudstone. The mudstone is thickest in the south-east limb of the Strathmore Syncline and reaches 1200 m in the Brechin–Laurencekirk area (= Edzell Mudstone) and 760 m between Ruthven and Bridge of Allan (= *Cromlix Formation*). It is poorly sorted with some sand grains, poorly bedded and non-fissile and shows little evidence of cyclic or graded bedding. It is overlain by, and passes laterally into cross-bedded sandstone with pebbly lenses, and in places on the north-west limb of the Strathmore syncline almost the entire Strathmore Group consists of pebbly sandstones with thick conglomerates. Flaggy sandstones within the group have yielded the plant remains *Psilophyton princeps, P. goldschmidti* and *Arthrostigma gracile*. These suggest a late Siegenian date, but the spore content of the group favours a lower to middle Emsian age (Westoll 1977).

Intrusive Igneous Rocks. Minor intrusions are most common in the volcanic rocks of the Ochil and Sidlaw Hills and beneath the lavas in the Dundee area. They form dykes and a few sills, sheets and bosses and are mainly basalts, andesites, 'porphyrites', plagiophyres and quartz–albite porphyries. Along the Ochil Fault close to Tillicoultry the lowest Ochil lavas are intruded by a number of small stocks of diorite, which have produced a thermal aureole within the adjoining lavas. Dyke swarms (with preferred orientation along NNW and NNE lines) are centred on these stocks.

Outcrops north of Highland Boundary Fault. Outcrops north of this fault which at first sight are continuous with the outcrop in the Midland Valley occur north and west of Blairgowrie and north of Crieff. They rest directly on Dalradian rocks, and have a basal breccia full of local Dalradian clasts. Above this are sandstones and conglomerates with rounded quartzite and igneous clasts. Palaeoflow is towards the south-east. North and west of Blairgowrie the beds contain an ignimbrite which used to be known as the 'Lintrathen porphyry' (Paterson & Harris 1969) and has yielded a date of 411 ± 6 Ma (Thirlwall 1983). This ignimbrite originated north of the Highland Boundary Fault, but Paterson believes that it can also be found in the Crawton Group at Glenbervie west of Stonehaven well south of the Highland Boundary Fault. If this is the same flow as the ignimbrite north of Blairgowrie, Bluck's hypothesis that the sediments north and south of the Highland Boundary Fault are completely distinct cannot be accepted.

Loch Lomond area and Arran

The westward thinning of the Lower Old Red Sandstone in the northern Midland Valley belt continues rather more erratically between Loch Lomond and Antrim. Qureshi (1970) has estimated from gravity data that the maximum thickness of the Lower Old Red Sandstone around Loch Lomond is between 1500 m and 1800 m. In Arran it is possibly as much as 2500 m (Paterson 1987). Morton (1979) has shown that the sequence west of Loch Lomond can be correlated with the Garvock and Strathmore Groups. In the small outcrops of Lower Old Red Sandstone near Farland Head, North Ayrshire (Sandy's Creek Beds) Downie & Lister (1969) have recorded microspores of possible Late Downtonian or Dittonian type which would suggest correlation with the Stonehaven Group. In Arran the Arbuthnott, Garvock, and Strathmore Groups are represented. The Arbuthnott Group occurs only in the west with the basal breccias containing clasts of quartz and metasedimentary rocks and the thin Auchencar lava flow at the top. In Glen Rosa this group is completely overlapped by the Garvock Group, the basal division of which is breccia with clasts of 'Dalradian' type. The Strathmore Group crops out only around Glen Sannox, and is composed mainly of red-brown siltstones (the Sannox

Siltstones) with beds of fine-grained sandstone and quartz-conglomerate. At its base is a coarse conglomerate with large well-rounded quartzite boulders. The beds were laid down in braided rivers which flowed towards the SW. Alluvial fans entered this river from time to time along the flanks.

Kintyre, Argyll and Ben Nevis

North of the Highland Boundary Fault, sediments of Lower Old Red Sandstone age occur in Kintyre, on Kerrera and in a narrow coastal strip north and south of Oban. The sediments are overlain, interstratified with, and over-stepped both in the east and west, by lavas which extend from the Loch Don area in eastern Mull, eastward across the Lorne Plateau almost to the shores of Loch Awe and northwards to Loch Creran. Farther north lavas, which may originally have been continuous with the Lorne Plateau lavas, have been preserved by cauldron subsidence at Glen Coe (Fig. 9.4) and on Ben Nevis.

At the southern end of *Kintyre* there is a thick development of Lower Old Red Sandstone sediments. Friend &

MacDonald (1978, pp. 265–282) have established the following succession:

(3) *Bastard Sandstone*. Purple sandstone with red siltstones. Beds generally finely laminated, some cross-bedding, but more commonly flat-bedded. — 100 m

(2) *New Orleans Conglomerate*. Generally coarse with many large lava boulders, interbedded near top with reddish-purple sandstone and siltstone with some calcareous concretions. Local pumice lapilli. — 890 m

(1) *Glenramsgill Formation*
(B) Quartzite Conglomerate Member. 100 m of conglomerate with mainly quartzite clasts up to 1 m in size, overlain by 200 m of purple sandstone and siltstone. — 300 m
(A) Basal Breccia Member. Poorly bedded breccia full of Dalradian fragments, passing upwards into red sandstones and siltstones full of fine Dalradian detritus. Lava detritus comes in near top. — 150 m

Fig. 9.4. West face of Aonach Dubh and An't Sron (right), Glen Coe. Aonach Dubh is composed of gently inclined andesitic lavas (lower two-thirds), overlain by thick rhyolitic ignimbrites, which form the upper, slightly paler, cliffs. The gulley on the left edge of An't Sron marks the line of the Glen Coe Ring Fault, the rest of An't Sron is granite of the ring intrusion. (B.G.S. photo.)

At the south coast the top of the Glenramsgill and lower part of the New Orleans Formation is cut by three vents which were formed by gases pushing the underlying conglomerates and sandstones upwards. Lava was intruded in only one vent. Friend and MacDonald (1968) suggest that the sediments were derived from the north-west, with quartzite pebbles possibly coming from the Islay–Jura area and the lavas from a Devonian lava field similar to the Lorne Plateau that stretched across North Kintyre.

On *Kerrera* the sediments rest unconformably on an irregular topography of Dalradian slates. They consist of local basal breccia overlain by a sequence of conglomerates, sandstones, shales and siltstones up to 300 m thick, locally intercalated with lava. The lower 180 m of the succession is developed only in the south-eastern part of the island and contains a flaggy shale which has yielded *Cephalaspis lornensis, Kampecaris obanensis* and plant remains (Lee & Bailey 1925), and more recently a varied fauna of cephalaspids and an anaspid (Waterston 1965, p. 282). This fauna may be of Downtonian age, possibly equivalent to the Stonehaven Group of the Midland Valley. The suggestion by Tarlo & Gurr (1964) that the Lower Kerrera beds were laid down in a tidal cuvette has not been generally accepted. These basal beds are not present in the *Oban* area, where the sequence comprises a sandstone up to 30 m thick, overlain by conglomerate and then by grey shales interbedded with basalt lava. The conglomerate contains many boulders of andesite, indicating that volcanic activity had already begun in an adjoining area before the deposition of these sediments. The shales have yielded a 'Dittonian' fauna, including *Cephalaspis lornensis, Mesacanthus mitchelli, Thelodus sp. Pterygotus anglicus, Kampecaris forfarensis* and *K. obanesis* as well as ostracods and plants (Lee & Bailey 1925, Johnstone 1972). Morton (1979) has ascribed the Oban sediments and interbedded lavas to the Arbuthnott Group of the Midland Valley.

The succeeding volcanic succession covers about 300 km² and is up to 800 m thick. The lavas comprise basalts, hypersthene-, hornblende- and biotite-andesites and some small flows of dacite and rhyolite. The lower part of the sequence is composed mainly of basalts and basic andesites but higher in the succession there are two prominent tuff beds. The lower is an acid lithic tuff, and the upper a massive crystal tuff, which is in places 20 m thick (Kynaston & Hill 1908). Both tuffs have since been shown to be ignimbrites (Roberts 1966) with a chemical composition similar to that of rhyolite (Groome & Hall 1974). Thin intercalations of conglomerate, sandstone and shale, locally with thin concretionary limestones, are in places associated with the lower tuff, while the acid lavas are associated with the upper tuff. The last is overlain by thick flows of hypersthene-andesite which contain lenses of andesitic agglomerate. Brown (1975) has obtained an Rb–Sr isochron age of 415 ± 7 Ma and Clayburn *et al.* (1983) a date of *c.* 410 Ma. This would agree with a Downtonian age for the underlying sediments.

Lavas with ignimbrites and a few indurated sediments are preserved within the area of cauldron subsidence at *Glen Coe* and a small area of lavas (hornblende- and biotite-andesites) is similarly preserved on Ben Nevis (Bailey 1960). At Glen Coe the volcanic sequence is about 1200 m thick and the succession as established by Clough *et al.* (1909) and modified by Roberts (1966, 1974), is as follows:

Group		Approximate thickness (m)
7	Andesites and rhyolites interbedded with a thin ignimbrite horizon	100
6	Shales and grit (second hiatus in volcanic activity)	20
5	Ignimbrite (crystal tuff)	75
4	Hornblende-andesites, with some rhyolite lavas in the south	30
3	Breccias, grits and shales (first hiatus in volcanic activity)	80
2b	Ignimbrite (lithic tuff)	0–150 (in NE)
2a	Andesites and rhyolites with a thin ignimbrite near top	
1	Basalts and pyroxene-andesites	500

The source of the basalt and andesite lavas probably lay outside the Glencoe cauldron. Roberts (1966) has shown that the rhyolite lavas were derived from vents located along the western and north-western margins of the cauldron, and that the two major ignimbrite flows (groups 2b and 5) resulted from violent eruptions which were in each case followed by collapse and caldera formation. The ignimbrites were carried upwards along the Glen Coe ring intrusions emplaced respectively along the early and main boundary fault. Roberts suggests that the two Glen Coe ignimbrites can be equated with the two tuffs of the Lorne succession (p. 308). He also suggested (1974) that within the Glen Coe cauldron, subsidence and ignimbrite extrusion were concentrated along the north-east margin and that the caldera was asymmetric, with a high near-vertical north-eastern wall and a floor that rose continuously right up to its south-western rim. The sediments within the Glen Coe succession (groups 3 and 6) mark the periods when, after the violent eruption and caldera subsidence, volcanic activity had temporarily ceased. The shales have yielded some plant remains, including *Pachytheca fasciculata*.

Southern Midland Valley

The Lower Old Red Sandstone in the southern part of the Midland Valley consists predominantly of conglomerates and sandstones (lithic arenites) intercalated with a volcanic group. The conglomerates are thinner and less coarse than those in the northern Midland Valley. The clasts of the conglomerates below the volcanic group are mainly greywackes, shales, cherts and jasperised lavas from the Southern Uplands, but those above the lavas are, to a large extent, composed of volcanic rocks. The grains and matrix of the lithic arenites, both from above and below the volcanic group, are to a large extent of igneous origin. The present outcrops, which are much dissected by post-Lower Old Red Sandstone faulting, contain the remains of three major volcanic centres, which now form respectively the Pentland Hills, the hills of the Upper Clyde between West Linton and Douglas (the 'Biggar Centre' of Geikie 1897), and the Carrick and Dalmellington hills of Ayrshire.

In the *Pentland Hills* the Lower Old Red Sandstone rests on an eroded, gently undulating land surface of vertical Silurian rocks. In the south-western part of the range the lower 600 m of the succession comprises coarse pebbly grits and sandstones with lenses of greywacke–conglomerate which are coarsest at the base, and contain also pebbles of radiolarian chert, jasperised basic lava and rarely, fossiliferous Silurian limestone (Mitchell & Mykura 1962). The sediments interdigitate north-eastward with the Pentland Hills Volcanic Succession which has a maximum thickness of over 1800 m. It includes a basal cumulo-dome of felsite (the Black Hill Felsite), which rests directly on the floor of Silurian rocks, and contains a further ten groups of lava flows which include olivine-basalts, basic andesites, dacites and rhyolites, as well as acid tuffs and thin volcano–detrital sediments (Mykura 1960).

In *Lanarkshire* (Upper Clyde district) volcanic rocks occupy a large proportion of the south-west trending outcrop between the Southern Uplands Fault and the Carmichael Fault, which is the south-westward extension of the Pentland Fault (Read 1927). They form the rounded hills both north-east and south-west of Tinto Hill, which is a laccolithic intrusion of felsite, up to 1 km thick. The lavas are mainly basic pyroxene-andesites and olivine-basalts, though some trachytic lavas are present just north-east of Tinto and near Biggar. Intercalations of agglomerate and tuff are rare, but there are many thin bands of sediment. The sedimentary sequence below the lavas includes a thick basal greywacke-conglomerate, which also contains clasts of pebbly grit, chert, jasper and acid igneous rocks. The lavas are overlain by coarse green ashy sandstones and flaggy purple sandstones. The highest member of the succession is a coarse lava-conglomerate which forms the prominent Dungavel Hill, some 4 km SSW of Tinto.

North-west of the Carmichael Fault there are outcrops of Lower Old Red Sandstone sediments at Carmichael, Lanark, Lesmahagow and just west of Douglas. These contain no volcanic rocks and consist of a thin greywacke-conglomerate succeeded by mainly sandstone. Lavas were either never present in these areas or else all the exposed sediments are part of the sequence which underlies the volcanic series farther south-east. The whole succession is here relatively thin and in places Upper Old Red Sandstone rests directly on the Silurian. The sandstones of the lower sedimentary group, locally with some cornstone-like concretions, are seen again west of the Lesmahagow Inlier on the slopes of the Distinkhorn near Darvel. They are here

about 650 m thick and are partly within the aureole of the Distinkhorn plutonic complex. They have yielded cephalaspids including *C. traquairi* and ? *C. lyelli* and are overlain by volcanic rocks which consist of basalts and basic andesites with intercalated ashy sediments.

The Lower Old Red Sandstone of *West-central Ayrshire* has two outcrops, centred on Maybole and Dalmellington. In the Maybole area the lower sedimentary group is up to 1200 m thick and consists of a fairly uniform series of micaceous sandstone (mainly lithic arenite) with pebbly bands and marly partings. Conglomerates up to 20 m thick are developed near the top and there are also some thin tuffaceous beds in the upper part of the sequence. The latter may be derived from the large Lower Devonian volcanic vent at Mochrum Hill, 3 km west of Maybole. The overlying volcanic rocks form the Carrick Hills as well as two smaller coastal outcrops at Culzean Castle and Maidens (Geikie 1897, Smith 1909, Tyrrell 1914). In the Carrick Hills, the volcanic sequence is 300 m to 450 m thick and consists of olivine-basalts, augite- and hypersthene-andesites (Eyles *et al.* 1949). Vesicles filled with agate and calcite are common in both the Carrick Hill and coastal lavas (Smyth 1910). The tops of many lava flows are veined by sandstone, and sediments occur between the higher lava flows. They include lava-conglomerates, tuffs and fine volcano-detrital sediments, as well as thin bands of hard sandstone and sandy mudstone. In the last some well-preserved myriapod and other fossil trails have been found (Smyth 1909, Brade-Birks 1923).

The sequence in the Dalmellington area is similar to that of Maybole, but is much more disturbed by faulting. The sedimentary group may be up to 700 m thick and contains a thick basal conglomerate (up to 200 m), the clasts of which are composed chiefly of greywacke. A thinner conglomerate at the top of the sedimentary group contains, in addition to greywackes, clasts of acid intrusive rock and some chert and quartzite. The overlying volcanic group consists of nearly 600 m of olivine-basalt. It generally has a single olivine-rich flow at the base, overlain by a succession of feldspar–phyric flows.

A feature of the volcanic rocks within the southern Midland Valley is their association with an extensive suite of contemporaneous minor intrusions which include sheets and laccoliths of felsite and quartz porphyry, and sills and dykes of 'biotite–porphyrite', plagiophyre, kersantite, pyroxene–porphyrite, quartz dolerite and basalt. Garnets have been recorded in the Tinto felsite laccolith (Herriot 1956). The plutonic boss of granodiorite and diorite forming the Distinkhorn in east Ayrshire cuts the lower sediments and is probably of the same age as the volcanic group.

Southern Uplands

In the eastern part of the Southern Uplands two north to north-north-west-trending belts of poorly cemented and unsorted greywacke-conglomerate were formerly thought to be of Upper Old Red Sandstone age (Clough *et al.* 1910), but have recently been ascribed to the Lower Old Red Sandstone (Rock & Rundle 1986). One extends from the Southern Upland Fault at Dunbar southwards to the Dirrington Hills near Greenlaw, and the other occupies part of Lauderdale with a small outlier farther north-west on the plateau of the Lammermuir Hills. The conglomerates are evenly bedded and locally up to 600 m thick. They consist of mainly subangular clasts of locally derived greywacke, jasper and chert with substantial additions of granite clasts close to the outcrop of the Cockburnlaw granite. The matrix of the conglomerates is generally soft and ferruginous. In the Lauderdale outlier thin beds of sandstone are interdigitated with the conglomerate.

Lower Old Red Sandstone also crops out in the area extending inland for 10–15 km from the coast between Eyemouth and St Abbs. It consists of red feldspathic sandstones and conglomerates, with a few thin cornstones, some partings of red marl, and a volcanic succession of andesitic lavas and coarse tuff, which is at least 600 m thick (Greig 1971). Vents of andesitic agglomerate are exposed on the shore near Eyemouth and St Abbs. The Lower Devonian age of tne sediments and lavas of this area has been determined by the presence of *Pterygotus*.

Along the Northumberland border a considerable part of the *Cheviot Volcanic Series* lies within Scotland. The lavas are mainly augite-hypersthene-andesites but include some local acid flows of porphyritic rhyolite, known as 'mica-felsites'. The first stages of the volcanicity were explosive, producing coarse tuffs and agglomerates. There are also some intercalations of tuff between the lava flows, but there are few interbedded sediments. Some isolated patches of coarse breccia may mark the sites of volcanic vents. The lavas are intruded by the Cheviot Granite.

Conditions of deposition

Bluck (1978), Morton (1979) and Armstrong & Paterson (1970) have built up a picture (Fig. 9.5) of Lower Old Red

Fig. 9.5. Palaeogeography of the Midland Valley and south-west Highlands in Lower Old Red Sandstone times. Former interpretation after Bluck (1978) Morton (1979) and Armstrong & Patterson (1970 and 1978).

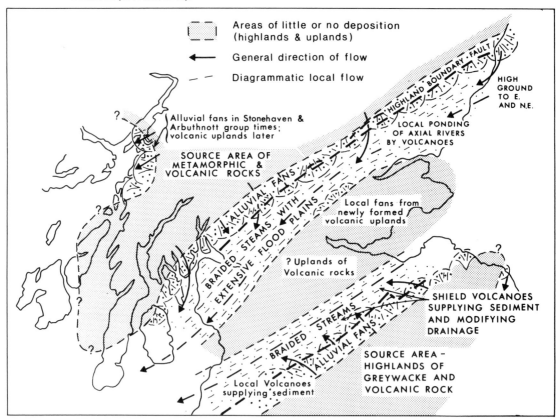

Sandstone palaeogeography of the Midland Valley which showed major flood plains of braided and meandering rivers flowing south-west on either side of volcanic uplands along the lines of the present Sidlaw and Ochil Hills extending south-westwards to Ayrshire. The northern alluvial plain was bounded by faults roughly parallel to the Highland Boundary Fault, along which alluvial fans produced most of the conglomerates found in the sediment close to the fault which form in places 56 per cent of the succession. The volcanic uplands to the south which in the Ochils are +3 km thick produced fans on the southern margin of the valley. In the southern plain fans of conglomerate formed along the southern margin, roughly along the Southern Upland Fault, and contain Lower Palaeozoic and volcanic rock. Another embayment occurred around Oban with material derived from the north-east forming the sediment in the Oban area and in Kerrera.

Since those ideas were set up, Bluck (1983, 1984) and his co-workers have studied the clasts in the Old Red Sandstone conglomerates and more closely considered the evidence from the Silurian in the Midland Valley and the Highland Border Rocks (p. 167). Bluck has shown that:

1. None of the pebbles in the Midland Valley conglomerates are derived from the Dalradian as at present seen north of the Highland Boundary Fault.
2. The lowest conglomerates south of the fault had a dispersal to the north-west, suggesting a basin to the north-west of the position of the Highland Boundary Fault.
3. The higher conglomerate seen south of the Fault is derived from the north-west but the boulders both of metamorphic rocks and volcanic rocks cannot be matched with rocks outcropping in Dalradian terrain as seen today.

The evolution of the northern Midland Valley as now envisaged by Bluck is illustrated by the two block diagrams (Fig. 9.6).

Northern area

Though it has long been suspected that deposits of Lower Old Red Sandstone age are present in the Highlands north of the Grampians (Westoll 1951) and in the Northern Isles, their existence has only recently been proved by spore determinations and by the discovery of diagnostic plants and rare fish remains. Though the outcrops with Lower Devonian fish are confined to Easter Ross and Caithness, evidence from spores and lithology suggests that several other deposits are likely to be of this age (Westoll 1977). On the mainland they include the outliers at Tomintoul, Cabrach and Rhynie in the northern Grampians, small outcrops at Aberdeen, the Crovie Group of the Turriff outlier, and a series of outcrops along the Great Glen, in the Black Isle and in East Sutherland.

The Tomintoul Outlier (Hinxman 1896) fills an irregular north-east-trending depression within the Dalradian basement. In the south-west it consists of a basal breccia overlain by a considerable, unknown, thickness of coarse breccio-conglomerates and conglomerates which are well seen in the Ailnack Gorge, and which contain mainly quartzite clasts, some up to 1m + in size, the conglomerates overlain in the north-east by red, medium-grained, immature sandstones. The Cabrach outlier, 12 km NE of Rhynie, consists of conglomerate overlain by soft grey micaceous sandstone intercalated with bands of coarse conglomerate. This outlier also contains andesite lava, similar to that at Rhynie.

The *Rhynie Outlier*, which forms a narrow, 21-km-long, NNE-trending outcrop, contains the following succession (Wilson & Hinxman 1890, Read 1923):

(5) Dryden Flags and Shales, containing the Rhynie Chert.
(4) Quarry Hill Sandstone: white, grey and pink sandstone with gritty, pebbly bands.
(3) Tillybrachty Sandstone: soft white to deep purple sandstone with conglomeratic lenses and a volcanic horizon containing at least one flow of slaggy andesite (the 'Cork Rock').
(2) Lower Shales: red and grey shales with beds of calcareous sandstone and flattened limestone concretions.
(1) Basal breccia and conglomerate.

Archer (1978) has shown that these deposits were laid down by northward-flowing palaeocurrents. The Lower Shales have yielded plant fragments, including cf. *Pachytheca*, and the beds of the Dryden group contain the famous *Rhynie Chert*, discovered by W. Mackie and believed to be a silicified peat deposit. This contains plants which grew close to a volcanic centre. The silica-rich waters from the volcano killed the plants quickly and prevented the microbial breakdown of the tissue so that the microscopic structure is perfectly preserved. They have been described by Kidston & Lang (1917–21). They include the psilophytes *Rhynia, Horneophyton* and *Asteroxylon* as well as fungi and blue-green algae. Other fossils from the chert include the crustacean *Lepidocaris rhyniensis*, the merostome *Palaeocharinus rhyniensis* and the insects *Rhyniella praecursor* and *Rhyniognatha hirsti*. Archer (1978) has recorded a collection of trace fossils including possible lungfish and annelid burrows. Westoll's (1951) suggestion that the Rhynie succession is of Lower Old Red Sandstone age has been confirmed by spores, which suggest a probable Siegenian age.

The lower of the two stratigraphic groups within the *Turriff basin*, termed the *Crovie Group* by Read (1932) is now considered, on the evidence from one spore assemblage, to be of high Lower Old Red Sandstone (possibly Siegenian–Emsian) age (Westoll 1977). The beds are well exposed at Gardenstown, Crovie, and New Aberdour, and

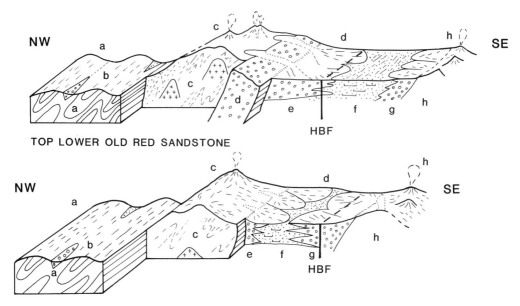

TOP LOWER OLD RED SANDSTONE

BASAL LOWER OLD RED SANDSTONE

Fig. 9.6. Evolution of NW Midland Valley in Lower Old Red Sandstone times according to Bluck (1985). Note that terranes (*a*) and (*b*) NW of the Highland Boundary Fault were covered by the Dalradian after the end of Upper Old Red Sandstone time. (*a*) Dalradian terrane eroded to present level at *c*. 410 Ma, with small internal basins (*b*) and volcanic–plutonic complexes. (*c*) Metamorphic terrane of uncertain affinites, probably part of the Midland Valley block, with plutonic (*c*. 460 Ma) and volcanic complexes. (*d*) Uplifted Old Red Sandstone feeding fans from N (*e*) and axial sediments (*f*), (*h*) Central Midland Valley Uplift (Sidlaw Hills–Heads of Ayr line) with adjacent fans (*g*) composed of volcanic rock and conglomerates with quartz clasts.

the unconformable junction between Lower and Middle Old Red Sandstone is seen near Pennan. The sequence is up to 500 m thick and Archer (1978) and Sweet (1985) have shown that it consists, above basal fan deposits, of a coarsening-upward sequence passing from low sinuosity, ephemeral stream deposits, through the flood plain deposits of meandering rivers into deposits of high velocity braided rivers, the last possibly laid down on an alluvial fan. The succession is as follows:

	metres
(6) Bright red sandstone, with trough cross-bedding and lag-conglomerates.	150–200
(5) Cyclic units of fine-grained cross-laminated sandstone units grading up into siltstone and mudstone.	32
(4) Thick siltstone units with interbedded thin sandstones and mudstones; many mudcracks.	95+
(3) Thick siltstone units, with rare thin sandstones; many mudcracks.	95+
(2) Interbedded thin sandstones and siltstones with many cornstone concretions (pedocals representing periods of reduced sedimentation).	45
(1) Conglomerate with thin sandstones (fanglomerate). The palaeocurrents suggest a north-westerly dispersion.	30+

At *Buckie* an andesite exposed in the Gollachy Burn has in the past been considered to be of Lower Old Red Sandstone age (Westoll 1951), but is now taken to be an intrusion (Peacock *et al.* 1968). Similarly, a possible Lower Old Red Sandstone age has been ascribed to the 'Buckie Formation' a 3 m thick sequence of limestones overlain by 2 m of flaggy sandstone. The age of this has also been questioned (Westoll 1977).

Great Glen. There are a number of narrow north-east-trending fault-bounded slices of probable Lower Old Red Sandstone along the Great Glen between Loch Lochy and Glen Urquhart.

In the stretch between *Loch Lochy* and Fort Augustus

there are three elongate outcrops of Old Red Sandstone sediment along the south-east side of the main Highland Boundary Fault (shown as a single outcrop of Middle Old Red Sandstone on the 10-mile Geological Map). They contain sediments composed of conglomerate, fine-grained flaggy sandstone and some shale (Eyles & MacGregor 1952), cut by a number of sub-parallel north-east-trending shear zones. Close to the Great Glen Fault the sediments have been affected by cataclasis and in some areas this has led to the development of secondary chlorite, calcite and sericite. Though no fossils have been found in these rocks, their similarity to the sediments of the Mealfuarvonie area suggests a Lower Old Red Sandstone age.

Mealfuarvonie. Along the north-west side of Loch Ness a fault-bounded outcrop of tightly folded and faulted sediments extends for 16 km south-west from Glen Urquhart (Mykura & Owens 1983). This contains lenticular masses of massive breccio-conglomerate and coarse arkosic grit, which give rise to prominent craggy hills and range in thickness from 50 m to 400 m. These are interbedded with a sequence of at least 1800 m of uniform, predominantly planar-bedded, reddish, fine- to medium-grained sandstone. Mudstones and siltstones generally form only thin films and intraformational clasts within the sandstone, but near the north-east end of the outcrop they become thicker and in the extreme north-east a number of pale green mudstones have yielded poorly preserved non-diagnostic plants as well as a rich microflora which suggests an Emsian or possibly Eifelian age. In the extreme south-west of the outcrop the conglomerates and arkosic grits contain many clasts of granite which must have been derived from a nearby outcrop on the south-east side of the fault. Their most likely source are the small granite outcrops now exposed on the east side of the Great Glen between Loch Lochy and Loch Oich, suggesting a post-Lower Devonian dextral displacement of the order of 25 km along the Great Glen fault (see p. 324).

Loch Linnhe. 17 km south-west of Fort William a small outcrop of Old Red Sandstone at Rubha na h-Earba comprises c. 12 m of coarse scree-breccia, overstepped by interdigitated conglomerate and sandstone (Stoker 1982). The beds have been folded, faulted and overthrust. Stoker believes them to be of Lower Old Red Sandstone age but the evidence for this is not conclusive.

Ross and Cromarty. The Old Red Sandstone forms an open south-west-trending syncline, the axial trace of which extends south-westward from Inver on the Dornoch Firth, via the centre of the Black Isle, to the vicinity of Beauly. Sediments of Lower Red Sandstone age form a narrow outcrop on the western margin of this syncline and may also be present on its eastern limb within the Black Isle.

The Lower Old Red Sandstone on the western limb of the syncline was first recognised by Armstrong (1964, 1977) who has termed it the *Struie Group.* Its contact with the underlying Moine is in part an unconformity and in part a fault, and on Struie Hill, at the extreme north of the

outcrop, strongly folded and faulted sediments of the Struie Group have been thrust over the Moine basement. The thrust plane contains crushed rock and some mylonite and forms a prominent feature on the western slope of Struie Hill. The basal part of the group contains a number of lenticular masses of breccio-conglomerate which form a series of small but prominent hills along the southern and south-western margin of the outcrop. They are well exposed in the Aigas Gorge, 6 km SW of Beauly, where the rugged pre-Old Red Sandstone topography of the Moinian basement can be seen (Smith 1963). They consist almost exclusively of locally derived clasts, in places up to 1 m in diameter, and appear to have formed in alluvial fans at the foot of actively moving fault scarps. Above them is a sequence of fine-grained sediments, mainly drab-coloured and dark grey siltstones, calcareous mudstones and rare flagstones. In the *Strathpeffer* area the following succession has been recorded (Horne & Hinxman 1914):

		metres
(5)	Red shales passing up into sandstones	200
(4)	Olive shales with shaly fetid limestones	85
(3)	Spa Beds – fetid bituminous and calcareous shales with thin limestones	30
(2)	Ord Beds – olive shales with fetid calcareous bands	250–300
(1)	Basal breccia-conglomerate	

The fetid beds, which give off a smell of hydrogen sulphide when split open, are the source of the 'sulphur waters' to which Strathpeffer owes its development as a Spa during the 19th century. Some of the fetid shales and limestones of the 'Ord beds' contain structures which resemble stromatolites and may be of algal origin and Parnell (1985) has recorded nests of calcite pseudomorphs after gypsum in several localities in this area. The olive shales (4) have yielded spores of Lower Devonian aspect, which include *Emphanisporites sp.* and pseudo-saccate spores suggesting a Siegenian or Emsian age (Richardson 1967). A find of Acanthodian spines south-west of Strathpeffer (Donovan *in* Friend & Williams 1978) confirms this age. The outcrop of the fetid calcareous shales can be traced southwards from Strathpeffer to Beauly and Kirkhill.

Along the south-east limb of the Black Isle Syncline the Old Red Sandstone rests on two Moinian/Lewisian inliers which crop out along the coast between hill of Nigg and Fortrose. It is probable that the northern inlier is directly overlain by Middle Old Red Sandstone, but in the area south-west of Fortrose Westoll (1977) has equated the Ord Hill Conglomerate, which crops out on the coast between Avoch and North Kessock, and the overlying sequence of sandstones, thin conglomerates and shales with limestone concretions, with the Strathpeffer succession. This correlation is, however, not supported by fossil evidence.

Western Outliers of Ross-shire. Small patches of breccia and conglomerate crop out on two hill tops north of Loch Luichart and on the hill slopes north and south of Loch

Garve. These were deposited on a mountainous landscape and in one section the junction with the underlying Moine is a near-vertical cliff. Farther north, the *Strath Rannoch Outlier*, situated 10 km NW of Ben Wyvis, fills a steep-sided pre-Old Red Sandstone depression, with the ORS–Moine junction rising in one place by over 400 m within a distance of less than 2 km. The sediments comprise coarse lenticular conglomerates which form prominent escarpments on the western margin of the outcrop, and fine-grained sandstones and mudstones with calcareous bands. No fossils have as yet been found and no order of succession has been established. As the interdigitation of coarse conglomerates with predominantly fine sediments is a characteristic feature of the Lower Old Red Sandstone of Easter Ross, the sediments of the western outliers have been provisionally ascribed to the Struie Group.

Sutherland. The Old Red Sandstone of Sutherland forms an irregular partly fault-bounded, outcrop extending north from the Dornoch Firth almost to Strath Helmsdale. The part west of the Helmsdale Fault is known as the *Brora Outlier* (Ross *in* Read *et al.* 1925) and its conglomerates and sandstones give rise to the prominent hills which bound the narrow coastal strip of Mesozoic strata between Loch Fleet and Helmsdale. The 'outlier' has the structure of a basin elongated in a north-easterly direction and contains two groups of strata (Table 9.1) which are in places separated by a marked angular unconformity. Though no fossils have yet been found the lower group is regarded as being of Lower Old Red Sandstone age, mainly on account of its lithological similarity to the proven Lower Old Red Sandstone, both south and north. Its basal conglomerate and breccia is thickest and coarsest in the south, where it contains boulders, up to 60 cm in diameter, set in a red gritty matrix and intercalated with impersistent sandstones. The boulders consist of schist and granite of the types which crop out immediately to the west and the breccia was probably deposited in fans and scree slopes at the foot of high hills which lay to the south-west. Above the fan-glomerates are pebbly sandstones which are, in turn, overlain by fine-grained red and purple ripple-marked sandstones and mudstones and, finally, by hard flaggy red sandstones. The lower sandstones and mudstones have sun-cracked surfaces and there are several bands of desiccation-breccia. The mudstones reach a thickness of 500 m at the north-east end of the outlier. They are the probable equivalents of the Ousdale/Braemore mudstones of southern Caithness and, like the latter, were probably laid down in a shallow lake.

Western outliers of Sutherland. In central Sutherland there are two isolated areas of Old Red Sandstone sediments, the outlier of Meall Odhar, 15 km north of Lairg and the three outliers of the Ben Griam hills, 20 km north-west of Helmsdale. The sediments of *Meall Odhar* (Read 1931, p. 207) comprise red and chocolate micaceous sandstones with ripple marks and worm casts, passing upwards by interdigitation into coarse conglomerates.

Their probable Lower Old Red Sandstone age is, again, based entirely on lithological comparison. The strata forming the three outliers of *Ben Griam More, Ben Griam Beg* and *Beinn a Mhadaidh* rest on a very uneven floor. The sequence, as seen in the largest outlier, is as follows (Read 1931, pp. 205–207):

(4) Arkose, red, cross-bedded with pebbly bands at base, but becoming yellow and massive at top.
(3) Conglomerate, with quartzite pebbles in arkosic matrix.
(2) Arkose, reddish, with conglomerate bands, and some thin red mudstones with sun-cracks near top.
(1) Basal conglomerate and breccio-conglomerate, locally with huge boulders of local origin.

No fossils have been obtained from these beds. However, formations 1 and 2 show lithological similarities to the Lower Old Red Sandstone of western Caithness and the Brora Outlier, and formations 3 and 4 may be a fluviatile facies of Middle Old Red Sandstone age, broadly similar to the Middle Old Red Sandstone of the Brora Outlier and Easter Ross (p. 321).

Caithness. The strata formerly ascribed to the 'Barren' or 'Basement' Group of the Middle Old Red Sandstone of Caithness (Crampton & Carruthers 1914) have now been shown, by both spore and fish evidence (Collins & Donovan 1977) to belong to the topmost Lower Old Red Sandstone. These beds form two main outcrops: a small coastal inlier which forms the Sarclet inticline 3 km to 6 km SSW of Wick, and the much more extensive western marginal belt which extends from south of Berriedale north-westward to near Reay and reappears in the northern coastal outliers as far west as Strathy.

The sequence at *Sarclet* is as follows (Donovan 1970):

		Thickness (m)
(4)	Ulbster/Ires Geo Sandstone Formation	107
(3)	Ulbster/Riera Geo Mudstone Formation	172
(2)	Sarclet Sandstone Formation	85
(1)	Sarclet Conglomerate Formation	70

The *Sarclet Conglomerate* consists of alternating lenses of conglomerate and pebbly sandstone and contains clasts of granite, schist, quartzite and abundant basalts, the latter implying the existence of a nearby area of possibly contemporaneous volcanism. The conglomerate and overlying *Sarclet Sandstone* were laid down by braided and, later meandering, rivers from the south-east, while the *Ulbster/Riera Geo Mudstone*, which is a rhythmic sequence dominated by green mudstone with some calcareous marls, appears to have been deposited in a shallow lake which periodically receded to give place to a delta plain environment. The *Ulbster/Ires Geo Sandstone* Formation has two thick fluvial sandstones interbedded with a grey mudstone of probable lacustrine origin. The latter has yielded mio-spores of probably Upper Emsian age. In this area the Lower Old Red Sandstone passes upwards without trace of non-sequence into the basal conglomerate (Ellen's Geo Conglomerate) of the Middle Old Red Sandstone (Table 9.1).

In *southern Caithness* the Lower Old Red Sandstone rests directly on the Helmsdale Granite (p. 261). Its basal member is the *Ousdale Arkose* which locally contains a basal conglomerate up to 50 m thick. The arkose is in places difficult to distinguish from the underlying granite and contains some local concentrations of uranium ore (Gallagher *et al.* 1971). The overlying *Ousdale Mudstone* comprises a sequence of purple, pale green and red siltstones and mudstones alternating with beds of locally pebbly sandstone. The sandy siltstones have yielded some large but non-diagnostic plant remains. Miospores, and a recent find of *porolepis* scales (Richardson 1967, Collins & Donovan 1977) suggest an Emsian or Lower Eifelian age for these beds. The Lower Old Red Sandstone was gently folded and eroded prior to the deposition of the basal Middle Old Red Sandstone *Badbea Breccia*, which in places oversteps both Lower Old Red Sandstone formations and rests directly on the basement.

Farther north Lower Old Red Sandstone forms an irregular outcrop along the western margin of the Caithness plain, with two westward projecting tongues in the Morven and Ben Alisky districts. In this belt there are a number of lenticular outcrops of basal arkosic conglomerate and breccia, the pebbles of which can be closely matched with the lithology of the local basement. Locally pebbles of probably penecontemporaneous andesite have been recorded in the conglomerate. Over a large part of the outcrop, however, fine-grained purplish flaggy sandstones, mudstones and subordinate siltstones (the *Braemore Mud-stones*) rest directly on the basement. The Lower Old Red Sandstone beds of this belt are generally concordant with the overlying Middle Old Red Sandstone. At its northern end around Reay, the belt becomes discontinuous, the isolated outcrops representing infilled north–south-trend-ing pre-Old Red Sandstone valleys (Peach *in* Crampton & Carruthers 1914, p. 69).

In the *Strathy outlier* on the north coast, conglomerates and arkoses of probable Lower Old Red Sandstone age occupy the southern part of the outcrop. Farther west small outliers of conglomerate, sandstone and subordinate silt-stone at *Kirtomy* and around *Tongue* have in the past been ascribed to the Old Red Sandstone, though their age was questioned by McIntyre *et al.* 1956). They have been described in greater detail by Blackbourn (1981) and O'Reilley (1983), both of whom regard them as Old Red Sandstone. There is, however, no reason why these rocks could not be New Red Sandstone.

Orkney. The only outcrop of Orkney sediments ascribed to the Lower Old Red Sandstone occurs at Yesnaby on the west shore of Orkney Mainland, 7 km NW of Stromness

(Fannin 1970, Mykura 1976). It contains two formations: the *Harra Ebb Formation* made up of planar-bedded sandstone and siltstone with breccia lenses, and the *Yesnaby Sandstone Formation* which contains a lower member of medium-grained yellowish sandstone with large-scale cross bedding, possibly of aeolian origin, and an upper fluvial sequence of pebbly sandstone, siltstone and shale. The Lower Old Red Sandstone age of those beds has not been confirmed by fossils. Farther south a borehole at Warebeth, 2 km west of Stromness, encountered 61 m of purple siltstone, sandstone and breccia below the basal conglomerate of the Middle Old Red Sandstone. These beds are similar to the Lower Old Red Sandstone red beds of north-west Caithness.

Conditions of deposition

The characteristic feature of nearly all the Lower Old Red Sandstone sediments north of the Grampians is the association of coarse, lenticular breccio-conglomerates with relatively fine-grained sediments comprising predominantly planar-bedded fine-grained sandstones, with in places a high proportion of siltstones and mud-stones. Fluvial sandstone with large-scale cross-bedding is virtually absent in all outcrops except Gardenstown (Turriff Basin). The areal distribution of the Lower Old Red Sandstone sediments is much more restricted than that of the succeeding Middle Old Red Sandstone which rests in places directly on the crystalline basement. This suggests that the former were laid down in isolated, mainly playa-filled, basins of limited extent, within the mountains of the Caledonian Chain. The local thick fanglomerates suggest that the margins of the basins were located along active fault scarps. Some of these basins contained lakes which received only fine sediment. The organic content of the fetid Strathpeffer lake must have been high, probably containing abundant blue-green algae and bacteria. In the outcrops south of the Great Glen the palaeoccurrent directions range from north-westward (Banff Coast) to north-eastward (Great Glen) suggesting that the highest ground lay close to the present Grampian watershed and that deeply incised ephemeral rivers flowed towards the present Moray Firth which probably contained a number of isolated basins. North of the Strathpeffer lake there is evidence for land-locked internal basins with relatively slow-flowing ephemeral streams and short-lived lakes as far north as southern Orkney. In eastern Caithness the palaeoccurrents are locally from the south-east, implying high ground in that direction while the presence of lava clasts suggests volcanic activity in the source area.

Another basin of Lower Old Red sediments has been described by Richards (1985b) in the Moray Firth. This was about 45 km NE of Lossiemouth and is possibly a continuation of the Turriff Basin outcrop.

Middle Old Red Sandstone

The most extensive development of the Middle Old Red Sandstone is in Orkney (Fig. 9.7) and Caithness, but its outcrop continues intermittently south-westwards through east Sutherland into Easter Ross, where it forms parts of the Tarbat Ness peninsula and most of the Black Isle. East of the Great Glen it forms a narrow outcrop along Loch Ness and the inner Moray Firth between Foyers and Nairn and reappears farther east in the Fochabers–Buckie area, south of Cullen, and in the Turriff Outlier. There is a marked change in facies from mainly lacustrine in the north to predominantly fluvial in the south. Extensive outcrops of Middle Old Red Sandstone are also found in Shetland, where three separate groups, which differ from each other in facies, and the extent of subsequent deformation have been brought into contact by later movement along two major dextral transcurrent faults.

Caithness and Orkney

Flagstone Succession. Large parts of Caithness and Orkney are underlain by flagstones which have been extensively quarried in the past, as paving slabs, building stones and roofing slates. The flagstone succession reaches a thickness of 4 km in Caithness and over 2 km in Orkney, and the major stratigraphic subdivisions (groups and sub-groups) in the two areas are shown in Table 9.1. The succession contains many fish-bearing beds, the most distinctive of which, both lithologically and faunally, is the Achanarras Limestone of Caithness which can be correlated with the Sandwick Fish Bed of Orkney and is also the probable equivalent of one of the Melby Fish Beds of west Shetland (p. 323), the Cromarty and Edderton fish beds of Easter Ross (p. 321) and probably also of the Nairnside, Lethan Bar, Tynet Burn and Gamrie fish beds of Moray and Banff (pp. 322–323).

Lithology and depositional environment. The base of the Middle Old Red Sandstone in the Orcadian basin is strongly diachronous, being oldest in Caithness (at Sarclet), where there is a conformable transition from the Lower Old Red Sandstone, and becoming progressively younger as the western margin of the outcrop is traced north-westwards from Berriedale towards Reay. Near Berriedale in the south there are nearly 2·5 km of sediment below the Achanarras Limestone, at Reay in the north-west the base is just below,

Fig. 9.7. Characteristic cliff scenery produced by the Middle Old Red Sandstone Flags of the Orcadian Basin, west coast of Orkney Mainland, just north of Yesnaby. (B.G.S. photo.)

and farther west at Strathy it is at or just above this horizon. In Orkney only about 215 m of strata are present below the Sandwick Fish Bed. The junction between the metamorphic basement and the basal beds of the Middle Old Red Sandstone is everywhere irregular, with small re-exhumed cliffs, knolls and gulleys often visible. The remains of small rocky islands within the Orcadian lake now form inliers at Dirlot and Red Point in Caithness and at Yesnaby, Stromness and Graemsay in Orkney. Basal breccias or breccio-conglomerates are usually thin and, in some areas, absent. In Orkney (Fannin 1969) and at Dirlot in Caithness (Donovan 1973) some of the pebbles of the basal con-

glomerate are coated with dolomitic stromatolites and along parts of the original basin margin marginal cherty limestones up to 5 km thick, pass laterally into finely laminated lake deposits. At Red Point, near Reay, the limestones alternate with thin fluvial breccio-conglomerates (Donovan 1975). In western Caithness the limits of the present outcrop correspond roughly with the western margin of the original lake basin.

The entire flagstone succession consists of well-defined rhythmic units or cycles generally 5 to 10 m thick but reaching 60 m in the Achanarras/Sandwick Fish Bed Cycle. Each cycle (Fig. 9.8) represents a sequence of events within

Fig. 9.8. Cyclic units within (*a*) Caithness Sandstone facies and (*b*) the Eday Flags.

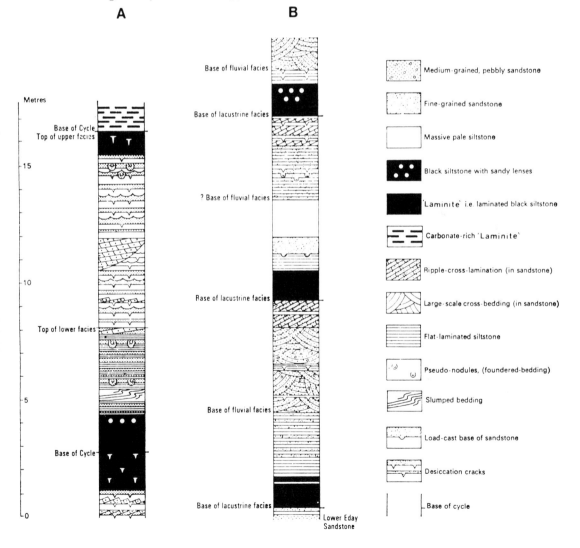

the shallow lake which was repeated with variations many times during the lake's existence (Fannin 1970, Donovan 1980, Trewin 1986). The most fully developed cycle commences with a bed of dark grey or black thinly laminated generally carbonate- and bitumen-rich siltstone ('laminate') which in many instances contains fish remains. This was laid down during the phase when the lake, though still shallow, was at its deepest, when its waters were not disturbed by wave action and when little or no coarse sediment entered the lake. It has been suggested that the individual laminae, which contain distinct carbonate- and bitumen-rich phases, may represent annual temperature-controlled 'non-glacial varves' (Rayner 1963). These result from an annual rhythm of fluctuating temperature gradients and biological activity within the lake, which may have been thermally stratified with a cool, dense, anaerobic hypolimnion and a warm algae-supporting epilimnion. The process of varve formation may have commenced with (i) a seasonal algal bloom, leading to increased photosynthesis, a rise in pH and the precipitation of carbonate (early summer), followed by (ii) algal death leading to

accumulation of organic matter on the lake floor (autumn) and (iii) increased accumulation of finely disseminated clastic material brought in from outside (winter). The 'laminate' passes upwards into thinly interbanded silty mudstones and siltstones with abundant subaqueous shrinkage cracks (syneresis cracks) (Donovan & Foster 1972) and with occasional beds of massive unlaminated siltstone. Slightly higher in the cycle sand- or silt-filled desiccation cracks (Fig. 9.9) appear and thin sandy beds become common. At this stage the lake had become so shallow that its floor was periodically exposed, giving rise to extensive mudflats. In most instances these beds are in turn overlain by thin beds of cross-laminated and frequently convoluted sandstone interbedded with sun-cracked siltstones. The sandstones are the deposits of river deltas spreading across the shallow lake floor. In the Clyth Subgroup of Caithness and in the basin-marginal deposits of the higher subgroups these delta-spread deposits are interbedded with, and channelled by, thicker lenticular, cross-bedded sandstones, which represent the filled-in channels of rivers crossing the mudflat, or of delta-distributaries in the shallow lake. The sandy phase is overlain by mudflat deposits composed of inter-banded siltstones and mudstones with sun-cracks and syneresis cracks, before the abrupt reversion to the dark, fish-bed facies ('laminite') of the next cycle.

The thicknesses of the lithological components within the cycle vary according to their stratigraphical position in the sequence and their geographical position within the original lake. Thus the sandstone phases are thick and laterally persistent in the Clyth and Mey subgroups of Caithness and in the lowest cycles of the Upper Stromness Flags in Orkney. The deep-water laminite facies is very carbonate-rich, approaching a dolomitic limestone in composition, in parts of the Clyth and Lybster subgroups of Caithness. Though it does not normally exceed 1 m in thickness, it is very thick in the Achanarras/Sandwick fish bed cycle in which two fish beds are developed within the laminite which reaches 6·6 m in Orkney (Trewin 1976). The colour of the weathered surface of the flags varies depending on the composition of the contained carbonate. Thus in Orkney the Lower and Upper Stromness Flags tend to be ochre or even orange weathering, due to the presence of dolomite, whereas the Rousay Flags weather grey (calcite). In Caithness, on the other hand, the flags below the Achanarras horizon weather to drab or buff (calcite) whereas those above are commonly ochre or bluish (dolomite and ferroan dolomite). A characteristic feature of the Stromness Flags of Orkney, which is less marked in other areas of the flagstone outcrop, is the presence of dolomitic (ochre weathering) algal stromatolites on the bedding planes of the shallow water and mud-flat facies of the cycle (Fannin 1969). They occur as sheets and as isolated mounds composed of convex-upward hemispheres, like the well-known 'Horse-tooth Stone of Yesnaby' (Mykura 1976, pl. XII).

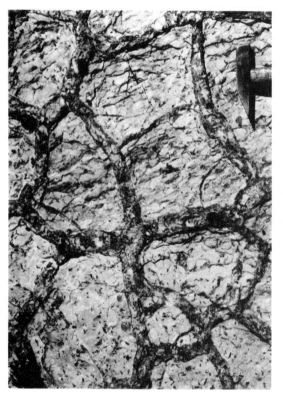

Fig. 9.9. Sand-filled polygonal sun cracks on bedding surface of Caithness Flags, Clairdon Head, north coast of Caithness. (B.G.S. photo.)

In the south of Caithness the lower part of the flagstone succession passes southwards into a sequence of red planar-bedded sandstones (the Berriedale Sandstone) overlain by red and grey rhythmically bedded flags with relatively thick fluvial sandstone phases (the Berriedale Flags). The southward transition from a lacustrine environment, via lake-marginal conditions, to a predominantly flood-plain environment is diachronous, with fluvial beds occupying only the basal part of the sequence in south Caithness, at least the lower 400 m in East Sutherland (see p. 321) and probably the whole succession in Easter Ross.

Fauna and Flora. The Caithness and Orkney flags are famous for their complete, well-preserved fossil fish (Fig. 9.10), which are normally found within the 'deep-water' laminate facies of the lacustrtine cycle. The distribution and range of diagnostic species in Caithness (Donovan *et al.* 1974), which also applies to the equivalent horizons in Orkney, is shown on Table 9.2. This table shows that *Coccosteus cuspidatus* is confined to the Lower Caithness Flagstone and Lower Stromness Flagstone Groups and that the small tadpole-like fish *Palaeospondylus gunni* has been recorded only at the Achanarras/Sandwick Fish Bed horizon. *Dickosteus threiplandi*, a species easily confused with *Coccosteus cuspidatus*, occurs only in the Latheron and Ham–Scarfskerry subgroups of Caithness and the

Upper Stromness Flags of Orkney and has not been found below the Achanarras horizon. The two species confined to the Mey Subgroup and Rousay Flags are *Millerosteus minor* and *Thursius pholidotus*. The only other fossil animal commonly recorded in the flags is the concostrachan branchiopod *Asmussia murchisoniana* (formerly *Estheria membranacea*), which occurs only in the Rousay Flags of Orkney and the equivalent subgroups in Caithness. Plant remains are scattered throughout the sequence, but these are not diagnostic for correlation. Similarly, miospores, which are present in many siltstones, have not yet proved useful as detailed stratigraphic indicators. The ecology and fauna of the Achanarras fish bed has been studied in detail by Trewin (1986).

John o'Groats and Eday Sandstone Groups. The lacustrine flagstones of Orkney and Caithness are succeeded, via a transitional 'Passage Beds' facies, by a predominantly fluvial sequence, which forms the John o'Groats Sandstone Group of Caithness and the Eday Beds of Orkney (Table 9.1). The *Lower Middle and Upper Eday Sandstones* and most of the *John o'Groats Sandstone* are composed of predominantly medium-grained, yellow and red sandstones with pebbly lenses, trough cross-bedding and abundant convolute bedding. Most sandstones appear to have been laid down by braided rivers, or as aeolian dunes and

Fig. 9.10. Middle Old Red Sandstone fish from the Orcadian Basin. *Thursius macrolepidotus* (Sedgwick and Murchison), Sandside Bay, Reay, Caithness. (B.G.S. photo.)

occasionally on lake beaches, while fining-upward cycles, with red overbank siltstones and mudstones, of the type formed on alluvial plains of meandering rivers, are only locally developed. Though palaeocurrent indicators suggest deposition by rivers flowing mainly from south-west to north-east there is a marked change of lithofacies from north to south, with the individual sandstone formations thickest, coarsest and most conglomeratic in the northern isles of Orkney. In the south-east of the archipelago they are finer grained, more clayey and locally interbedded with siltstones and marls. Astin (1985) has studied the Lower

Eday Sandstone, and concluded that it was laid down by a northern south-east-flowing alluvial fan, and a southern north-east-flowing braided and meandering river. The northern alluvial fan covered Eday, Sanday, Stronsay and Shapinsay. The deposits show a southward fining and interfere with the southern alluvial system at this margin. It is thought that the fan extended about 20 km into the basin. The southern river was much larger, and can be traced from Caithness, through South Ronaldsay and East Mainland to Shapinsay in a broad alluvial plain. Aeolian dunes with cross-bedding indicating winds from SSE are associated

Table 9.2 Faunal zones and ranges in Caithness Flags. *Dipterus* dominates the earliest and latest phases of the fish bed; *Coccosteus* is represented mainly by adult species (? bred outside the lake rivers); *Pterichthyodes* is present in all sizes and probably bred in the lake. Many of the lamination planes have concentrations of *Dipterus* and *Palaeospondylus–Mesacanthus* representing mass mortalities due to deoxygenation caused by algal blooms, water mixing in storms and lake overturn (Trewin 1986).

GROUP	SUBGROUP	ZONE	RANGE OF FOSSILS	FAUNA
UPPER CAITHNESS FLAGSTONE GROUP	MEY SUBGROUP	6	*Millerosteus minor*, *Thursius pholidotus*, *Asmussia murchisoniana*, *Dickosteus threiplandi* (ranges, ?)	*Millerosteus minor*, *Dipterus platycephalus*, *Thursius pholidotus*, *Glyptolepis paucidens*, *Osteolepis microlepidotus*, *Asmussia murchisonia[na]*, *Dipterus* sp., etc
	HAM-SCARFSKERRY SUBGROUP	5		*Asmussia murchisoniana*, *Dipterus valenciennesii*, *Homosteus milleri*, *Thursius* sp. (c.f. *pholidotus*), *Dickosteus threiplandi*, etc.
	LATHERON SUBGROUP	4		*Dickosteus threiplandi*, *Dipterus valenciennesii*, etc.
	ACHANARRAS LST	3		
LOWER CAITHNESS FLAGSTONE GROUP	ROBBERY HEAD SUBGROUP	2	*Palaeospondylus gunni*, *Coccosteus cuspidatus*	*Coccosteus cuspidatus*, *Thursius macrolepidotus*, *Dipterus valenciennes[ii]*
	LYBSTER SUBGROUP	1		*Thursius macrolepidotus*, *Dipterus valenciennesii*
	HILLHEAD RED BED SUBGROUP			
	CLYTH SUBGROUP	0	*Thursius macrolepidotus*	

From Donovan, Foster and Westoll 1974

The <u>Achanarras-Niandt</u> fauna: <u>Palaeospondylus gunni</u>, <u>Cheiracanthus murchisoni</u>, <u>Mesacanthus peachi</u>, <u>Coccosteus cuspidatus</u>, <u>Homosteus milleri</u>, <u>Pterichthys milleri</u>, <u>P. oblonqus</u>, <u>P. productus</u>, <u>Osteolepis macrolepidotus</u>, <u>Glyptolepis paucidens</u>, <u>Dipterus valenciennesii</u>, <u>Cheirolepis trailli</u> etc.

with both river systems, but they are thickest in western Eday. Lake depth at the time was nowhere more than 20 m and the beach deposits occur at the margins of alluvial deposits.

The Eday Flags are thickest in south-east Orkney (up to 150 m) and thin northwards by lateral interdigitation with the underlying sandstone to only 10 m in north Eday. They show a reversion to the cyclic sedimentation of the underlying Caithness Flags facies, but many of the cycles contain a relatively thick fluvial phase of red or yellow sandstone, which represents an infilled river channel or delta distributary on the lake floor (Fig. 9.8b). Dolosparite pseudomorphs after gypsum have been recorded at various localities in the Eday Marl by Parnell (1985). The deep-water laminite of the cycles contains fish remains which include *Microbrachius dicki, Pentlandia macroptera, Tristichopterus alatus* and *Watsonosteus fletti*. A similar assemblage is found in the John o'Groats fish beds of Caithness, which occur about 300 m above the base of the group and may be the most southerly representatives of the Eday Flags laminites. The Eday Flags contain some thin, laterally impersistent flows of olivine basalt associated with thin tuff beds and some minor intrusions. These volcanics contain analcime, natrolite and alkaline felspar, thus resembling the Carboniferous alkali basalts of the Midland Valley (Kellock 1969), though Thirlwall (1979) has shown that the analcime and zeolites are secondary and the basalts are basically calc–alkaline.

East Sutherland and Easter Ross

The Middle Old Red Sandstone of the *Brora Outlier* (Table 9.1) rests unconformably on beds ascribed to the Lower Old Red Sandstone, with a marked angular discordance seen along the margins of the basin. The sequence consists of locally thick basal conglomerates and pebbly sandstone overlain by 260 m + of mainly flaggy sandstone similar to the Berriedale Sandstone of Caithness (Read *et al.* 1925, 58–61). The presence of many blocks of Caithness flagstone in the Kimmeridge boulder beds (p. 137) exposed on the shore east of the Helmsdale Fault as far south as Lothmore, however, suggests that in the northern part of the outlier, the higher part of the Middle Old Red Sandstone sequence originally consisted of Caithness Flags.

The Middle Old Red Sandstone of Ross and Cromarty, termed the *Strath Rory Group* by Armstrong (1977), consists of fluvial sandstones with thick conglomerates, and subordinate calcareous fish-bearing shales. The lenticular basal conglomerates are up to 500 m thick and form conspicuous hills, and escarpments along the western limb of the Black Isle Syncline (p. 333). The pebbles of these conglomerates are mainly Moine gneiss and granulite but a number of clasts of Torridonian, and Cambrian rocks, which had their source to the west of the Moine Thrust have also been found (Peach *et al.* 1912). On the east limb of the syncline conglomerates reappear at the base of the

sequence. Westoll (1977) regards the lowest conglomerates of the southern Black Isle as Lower Old Red Sandstone and takes the Dumderfit Conglomerate as the basal unit of the Middle Old Red Sandstone. The Black Isle conglomerates contain, in addition to the Moine clasts, pebbles of biotite-granite and pink felsite. There are also some small quartzite pebbles, which may have been derived from the Dalradian quartzites of Banffshire.

The conglomerates of Easter Ross are interdigitated with and overlain by over 2·5 km of medium-grained yellow and red trough-cross-bedded fluvial sandstone, ascribed to the Strath Rory Group by Armstrong (1964, 1977) and termed the *Millbuie Sandstone* in the Black Isle (Horne & Hinxman 1914). In the Black Isle these sandstones contain pebbles of basic lava, suggesting contemporaneous volcanicity in the area. The sequence contains a number of thin grey, calcareous, locally slightly bituminous, shales and siltstones with concretionary nodules of limestone. The latter contain fish remains, and well-known fish beds are present at: (1) Black Park, 4 km WSW of Edderton, (2) the east coast of Tarbat Ness peninsula at and just north-east of Balintore, (3) Cromarty, and (4) the Killen Burn within the Black Isle. Species recorded in the Black Park (or Edderton) Fish Bed include *Diplacanthus striatus, Osteolepis macrolepidotus, Cheiracanthus murchisoni* and *Cheirolepis trailli*. A similar faunal assemblage has been recorded from the other fish beds. This assemblage has certain species in common with those of the Achanarras Limestone of Caithness (Westoll 1951, 1977), suggesting that the sandstones of Easter Ross can be roughly correlated with the middle part of the Caithness flagstone sequence. In the Tarbat Ness peninsula compound aeolian dunes and interdune sabka deposits (Hamilton & Trewin 1985) are found associated with pebbly sandstones of braided rivers.

Inverness, Moray and Banffshire

In the belt extending from Foyers north-eastwards to Nairn, the Middle Old Red Sandstone rests on an undulating basement of Moine gneiss and post-orogenic granites. South-west of Loch Duntelchaig (15 km south of Inverness) the sequence rests on a mountainous landscape (Fig. 9.11), with the pre-Old Red Sandstone palaeoslope locally as steep as 60°. In the extreme south-west, around Foyers and Inverfarigaig, where Old Red Sandstone in places overlies the Foyers Granite, the succession consists of lenticular granite-scree breccias, arkosic grits, Moine breccias and conglomerates, intercalated with and overlain by fine-grained sandstone, purple siltstone and shale (Stephenson 1972 *in* Gill 1977, Mykura 1982). These sediments were involved in contemporaneous landslips and affected by local thrusting. North-east of Inverfarigaig the locally derived intermontane fan and outwash deposits pass laterally into a sequence (up to 1·4 km thick south of Dores) of coarse breccio-conglomerates and conglomerates intercalated with arkosic sandstones (with NE directed palaeocurrents) and some purple shales. The conglomerates contain large (up to 1 m) boulders of biotite-granite and the basal breccio-conglomerates are rich in 'felsite' clasts.

North-east of Loch Duntelchaig the conglomerate sequence passes by intercalation, into a succession of sandstones, flagstones and shales with some nodular calcareous fish beds. The basal breccio-conglomerate is absent along some stretches of Strath Nairn but forms a number of lenticular fans which reach 75 m in thickness near Davoch and possibly 150 m at Cawdor. Though the area between Inverness and Nairn is poorly exposed, it has been possible to establish the following succession (Horne °: Hinxman 1914, pp. 66–68; Horne 1923, pp. 65–78):

Thickness (m)

(5) *Hillhead Group*: sandstone, flag-stones and shales.

(4) *Inshes and Holm Burn Flagstone Group*:   300
grey and purple flaggy micaceous sand-

stones locally pebbly and arkosic, with some dark calcareous flags and shales with limestone nodules.

(3) *Leanach and Dores Sandstone*: mainly   300
red sandstone, some flaggy sandstones and shales.

(2) *Nairnside Group*: mainly grey and   120
brown flagstones and shales with fish-bearing dark calcareous shales, containing some laminites and calcareous concretions. Includes Nairnside and Clava fish beds.

(1) Basal breccia and conglomerate with   0–150
some sandy beds.

The succession contains a number of important fish beds which consist of calcareous mudstone, occasionally laminated (e.g. Clava Fish Bed) but more commonly with calcareous concretions. The most important fish-bearing outcrops of Group 2 are those along the River Nairn at Easter Aultlugie and Clava (2 outcrops) along a southern

Fig. 9.11. Thick Middle Old Red Sandstone debris-flow breccia (forming the vertical cliffs) resting on a steeply undulating land surface of Moine basement. The vertical Moine Flags have been rotated to horizontal beneath the unconformity by Old Red Sandstone hill creep. Creag nan Clag, near Loch Duntelchaig, 15 km SSW of Inverness. (B.G.S. photo.)

tributary near Easter Town, at Knockloan, 5 km south of Nairn, and at several localities near Lethan Bar and Easter Clune, 8 km SE of Nairn. The fauna includes *Cheirolepis sp., Coccosteus cuspidatus, Dipterus valenciennesi, Mesacanthus sp.*, and *Osteolepis sp.* It has many elements of the Achanarras fauna, with which it has been equated (Westoll 1977). Only the Easter Town Fish bed appears to be below Achanarras. The presence of abundant *'Estheria'* just above the fish bed at Clava (Horne 1923, p. 67) casts doubt on the value of this crustacean as an index fossil elsewhere (see p. 319). Fish-bearing beds within Group 3 have been recorded at Leanach Quarry, just south of the Culloden Battle site (Horne 1923, p. 69), where the fauna includes *Pterichthyodes milleri, Homosteus sp., Glyptolepis sp.*, and *'Coccosteus decipiens'*. Group 4 contains some dark calcareous and slightly bituminous flags with calcareous concretions. Near Raigmore (Inverness) they have yielded *'Coccosteus decipiens' Osteolepis sp.*, as well as plant remains. The fossils from Group 5 were obtained from the Hillhead quarries, near Dalcross station, 12 km ENE of Inverness. They include *Millerosteus minor, Homosteus milleri* and *Coccosteus cf. decipiens*, an assemblage which suggests that the group may be of the same age as the Rousay Flags/Mey Subgroup beds of the Caithness–Orkney sequence.

There are a number of small outliers of Middle Old Red Sandstone between Forres and Elgin. At Altyre Burn, 5 km south of Forres the section was said to contain, above a basal breccia, a fish bed of shale with calcareous concretions which yielded Group 2 fauna similar to Lethen Bar. Andrews (1983) has however re-examined all the evidence about this area, and concludes that there are no Middle Old Red Sandstone outcrops in Altyre Burn, and that the fossils reputedly found here are from Lethen Bar. Farther east Middle Old Red Sandstone forms a large outcrop east of Elgin, between Rothes and Buckie. This contains a thin basal conglomerate overlain by a varied sequence of thin conglomerates, sandstones and red-green and purple shales. Calcareous concretions in the shales contain fish remains, and the largest fauna has been obtained from the Tynet Burn, 2 km SSW of Portgordon. It includes species of *Dipterus, Cheiracanthus, Osteolepis* and *Pterichthyodes* (Peacock *et al.* 1968, p. 37), which is a characteristic Achanarras type assemblage. Another fish bed crops out along the Spey at Dipple, near Fochabers Bridge. Its fauna includes *Dickosteus threiplandi* which indicates a higher horizon than the Tynet Burn Fish bed. The most easterly outcrop of Middle Old Red Sandstone occupies the greater part of the Turriff outlier, where it forms the *Findon Group* of Read (1923, pp. 168–173). A basal slate-breccia (up to 60 m thick) is here directly overlain by the Gamrie Fish Bed, which crops out in the Den of Findon and comprises 2 m of grey-red shales with limestone nodules full of fish remains, and lenticular shales with plants. This is, in turn, overlain by a sequence of breccias and conglomerates. The large fauna from the fish bed recorded by Read (1923,

p. 172) is, again, an Achanarras type assemblage (Westoll 1977).

Shetland

The two major north–south-trending transcurrent faults, which pass through Shetland Mainland separate three distinct sequences of Middle Old Red Sandstone (Mykura 1976, Mykura & Phemister 1976). Of these the *Walls Sandstone*, located between the Melby Fault and Walls Boundary Fault, was thought to be the oldest, as it is strongly deformed by the intrusion of granites and, near the base, contains poorly preserved plant remains which suggested a Lower Old Red Sandstone age. Astin (1982), however, has re-studied the whole Walls Formation and has shown that the Sulma Water Fault does not separate two distinct formations and that the whole sequence is much thinner than was formerly thought. He has shown that the formation in the north-east consists of an alluvial fan derived from the metamorphic rocks to the north, and this borders deposits which formed in a shallow lake and as beach ridges and now form most of the Walls Formation. Interbedded with the deposits of the alluvial fan are the Clousta Volcanic rocks which contain basaltic and andesitic lavas, rhyolitic ignimbrites, tuffs, and concordant intrusions of felsite. The tight folding in the Walls Formation is ascribed to the emplacement of the Sandsting Plutonic Complex (K–Ar dates 360±11 Ma) and the granites north of the basin which has produced a north-east-trending synclinorium, and a thermal aureole up to 2 km wide in the sediments.

There is, however, not universal agreement on the origin of the Walls Formation, and Melvin (1985) considers it to be made up entirely of river deposits, recognising the following environments: (1) high sinuosity (meandering) streams; (2) floodplain deposits; (3) braided streams; (4) sheet flood deposits.

The sequence west of the Melby Fault is termed the *Melby Formation*. Its lower portion consists of 600 m of mainly red and buff cross-bedded, locally pebbly, sandstone, laid down by rivers flowing to ESE. This is intercalated with two lacustrine fish-bearing beds, up to 17 m thick, of pale grey, in part, laminated siltstone and shale with bands and nodules rich in carbonate, and some sulphide nodules containing sulphides of Ca, Zn, Fe, and As (Hall & Donovan 1978). The fauna of the Melby Fish beds includes *Cheiracanthus sp., Coccosteus cuspidatus, Dipterus valenciennesi, Glyptolepis sp. Gyroptychius agassizi, Homostius milleri, Mesacanthus sp.* and *Pterichthyodes sp.* This assemblage is very similar to that of the Achannaras/Sandwick fish bed horizon and a direct correlation is probable. The sedimentary sequence is overlain by rhyolites which are probably contiguous with the volcanic succession of the neighbouring island of Papa Stour. The latter is up to 200 m thick and comprises a lower sequence of basalts overlain by two flows of spherulitic, flow-banded rhyolite, each directly overlain by rhyolitic (air fall) tuff. The probable northward continuation of this volcanic suite crops out at Esha Ness on north-west Mainland, where it contains, in addition to basalts and rhyolites, higher flows of mugearite and andesite, thick deposits of andesitic agglomerate, as well as a 'flow' of rhyolitic ignimbrite. The latter has yielded a Rb–Sr isochron age of 365 ± 2 Ma (recalculated from 373 ± 3 Ma) (Flinn *et al.* 1968). Thirlwall (1979) has shown that on geochemical grounds, the Shetland lavas have some characteristics transitional between calc–alkaline and tholeiitic types and that the lavas of Mainland, Papa Stour and Esha Ness are unconnected geochemically and of different ages.

Most of the island of *Foula*, 23 km west of Mainland, consists of 1800m of buff, locally pebbly, cross-bedded fluvial sandstones deposited by currents from the west-north-west. The sandstones are interbedded with two relatively thin sequences of siltstone and shale. Some of the latter contain plant remains which are not diagnostic, but their contained spore assemblage suggests a top Eifelian or low Givetian age (Donovan *et al.* 1978), which makes them roughly contemporaneous with the Melby Formation (Table 8.1) and the sediments of both areas roughly the same age as the Sandwick (Achanarras) fish bed.

The Middle Old Red Sandstone sediments of *Eastern Shetland* (Mykura 1976, Allen & Marshall 1981) crop out along the south-eastern coastal strip of Shetland Mainland and on the islands of Bressay Noss and Mousa. They rest on an undulating floor of metamorphic rock which has produced a diachronous base to the succession. The sediments were deposited in six major depositional environments: (1) the breccias of basement screes, which are well seen in the Quarff–Fladdabister region, (2) alluvial fan conglomerates seen on Rova head north of Lerwick, (3) the sandstones and conglomerates of braided streams seen at Lerwick, Northern Bressay as well as the Dunrossness and Scatsness peninsula in the south, (4) fine-grained alluvium of meandering streams well seen in the Cunningsburgh, Sandwick and Sumburgh peninsulas, (5) clastic and carbonate deposits of lakes, seen in two localities: (a) southern group which includes the Exnaboe Fish Bed (on the coast 2·3 km NE of Sumburgh Airport), the Sumburgh head 'limestones', the Hos Wick and Mousa fish beds: and (b) a northern group confined to the thinly bedded sandstones and siltstones at the top of the sequence, termed the *Bressay Flagstone facies*, which are best exposed on the Voe of Cullingsburgh in north-east Bressay. The southern group contains *Asterolepis thule, Coccosteus sp., Dipterus sp. Glyptolepis sp. Microbrachius dicki, Stegotrachelus finlayi, Tristichopteris cf. alatus* and at some horizons, numerous specimens of '*Estheria*'. This assemblage can be regarded as broadly contemporaneous with that of the John o'Groats/Eday Flags fauna of the Orcadian province, but the faunal differences are so great as to suggest development in a separate, unconnected basin. The Bressay fish beds have yielded *Asterolepis sp., Holenema ornatum* and *Glyotolepis*

paucidens. There is no comparable fish assemblage either in Orkney or on the Scottish Mainland and the age of the beds containing these fish may extend into the Upper Old Red Sandstone (p. 334).

(6) Sandstones of desert dunes are found in the Scatsness Peninsula and north of Exnaboe with large-scale cross-sets (up to 3 m).

The variations in facies of the East Shetland sediments from north to south is shown in Figure 9.12. Palaeocurrent data indicate a south-easterly dispersion except at Rova Head, where it is to the north-east. Wind directions are north to north-easterly.

The whole of the *Fair Isle* succession is of Middle Old Red Sandstone age (Marshall & Allen 1982). It is over 1·7 km thick and comprises conglomerate passing up into fluvial cross-bedded pebbly sandstone with a number of bands, up to 200 m thick, of ochre weathering dolomitic mudstone and siltstone (Mykura 1972). These are overlain by the deposits of meandering rivers flowing to the northeast and consisting of massive sandstones up to 9 m thick alternating with thinner beds of mudstone and siltstone. The middle of the succession has yielded the plants *Dawsonites roskiliensis, Hostimella*, and *Thursophyton milleri* (Challoner 1972) and at the top of the succession some fish remains have been found. Marshall has recorded a large microflora which proves conclusively the Middle Old Red Sandstone age of the whole succession and Astin (1982) has tentatively correlated it with the Walls succession.

Palaeogeography

Any attempt at reconstruction of the Middle Old Red Sandstone palaeogeography must take into account the subsequent movements along the major transcurrent faults which traverse the area, particularly the Great Glen Fault and Helmsdale Fault in northern Scotland and the Walls Boundary Fault and Melby Fault in Shetland. Most of these faults were already in existence and active before the deposition of the Old Red Sandstone and the Great Glen Fault appears to have formed a major valley at that time (Stephenson 1972, Smith 1977, Mykura 1982). The net post-Middle Devonian transcurrent movement along the faults appears to be as follows: Great Glen Fault: approximately 25–30 km dextral (Holgate 1969, Donovan *et al.* 1976, Speight & Mitchell 1979), though Armstrong (1977) favours a minimum dextral shift of 60 km; Helmsdale Fault: probably small dextral (Flinn (1969) suggests 50 km); Latheron Fault (Caithness): dextral (Donovan *et al.* 1974); Walls Boundary Fault: 60–80 km dextral (Flinn 1969; Mykura 1972, 1975, 1976; Astin 1982); Nesting Fault, Shetland: 16 km dextral (Flinn 1969); Melby Fault: 60–80 km dextral (Mykura 1976). These movements have been incorporated in the base map of Figure 9.13. This map suggests that the area occupied by the present Moray Firth as far north as Helmsdale was an alluvial plain, the western

N   S

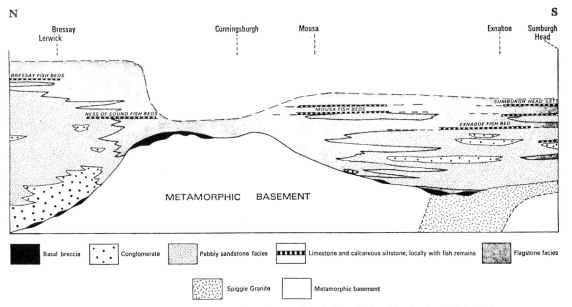

Fig. 9.12. Hypothetical section showing the possible facies relationships in the Middle Old Red Sandstone in east Shetland. After Mykura (1976, fig. 14).

and southern limits of which extended only slightly beyond the present Middle Old Red Sandstone outcrops. The plain was fed by major rivers derived from the west (palaeo-currents from the south-west to north-west) in the area north-west of the Great Glen fault, and from the south-west and south in the area south-east of the fault. Alluvial fans developed where the rivers debouched from the Highlands on to the plain, their position being probably controlled by contemporaneous faults sub-parallel to the Helmsdale–Great Glen Fault system. The Great Glen Fault formed a major valley which contained its own alluvial fans where tributaries descended into the valley.

It is likely that a ridge of high ground extended eastward from the present Scaraben ridge in southern Caithness via Berriedale into the present North Sea area. This ridge prevented the northward passage of coarse clastic sediments into the shallow Orcadian lake which received no coarse fluvial material throughout most of Eifelian and lower Givetian times. The lake extended from southern Caithness northwards beyond Orkney. It was bounded in the south and south-east by the Moray Firth alluvial plain. In the west its margin, though gradually transgressing north-westward, coincided roughly with the western limit of the present Old Red Sandstone outcrop. The fact that no major rivers entered the lake from the west throughout Caithness Flagstone times suggests that this margin was formed by a series of major north–south-trending ridges which formed an effective barrier to all rivers. The northern margin of the lake was probably bounded by the alluvial plain of a river flowing from the west-north-west, which

deposited the sediments of the Foula and Melby forma-tions. This plain was, in turn, bounded in the north-east by volcanic uplands. Nothing is known about the probable eastern margin of the Orcadian lake; a possible river connecting the lake with the tongue of open sea in the centre of the present North Sea (p. 297) may account for the lack of evaporites in the lake deposits. The maximum depth of the lake during the deep phase of each cycle (p. 317) was of the order 10 to 20 metres, but during the Achanarras/Sand-wick Fish Bed Cycle it was nearer 60 m (Fannin 1970). This was the period of maximum transgression when the lake extended northwards across the alluvial plain to Melby in Shetland, south-westwards into Easter Ross and south-wards across the alluvial plain into Moray and Banffshire to form the fish beds correlated with the Achanarras horizon. The early lake must have lain in south-eastern Caithness or farther east, transgressing from there gradually northward across Orkney and, later, westward towards Strathy and south-westwards across the Scaraben–Berriedale barrier to cover the fluvial sediments of, at least, the northern part of the Brora outlier.

Though there is evidence for an increased fluvial inflow from the west in Mey Subgroup times, the Orcadian lake was not finally filled till Eday/John o'Groats Group times, when major rivers from the south-west and north-west breached the western barrier ridge and their deposits filled the western part of the lake. A smaller lake must have remained for some time to the south-east of Orkney, as the cycles of the Eday Flags were formed by repeated transgres-sion of the lake north-westwards across the alluvial plain.

The volcanic uplands of Papa Stour and Esha Ness (Shetland) were probably not formed till upper Givetian times. Farther north, in the present Walls area, a subsiding basin with intense explosive volcanic activity may have developed at the same time, with a lake and lake margin deposits formed to the south and east of the alluvial fan. Towards the end of the Middle Devonian this basin was affected by a local but very intense orogenic episode, which produced a series of tight ENE folds and was probably

associated with the emplacement of the Sandsting and Northmaven Plutonic Complexes. Yet farther north the east Shetland sediments were laid down in late Givetian times on the eastern margin of a, probably north–south-trending, mountain range which exposed large areas of post-tectonic granite of the Northmaven type. The East Shetland sediments were laid down in fans and on alluvial plains bounded in the south-east by a shallow lake (p. 326).

The palaeogeography pictured above may have to be

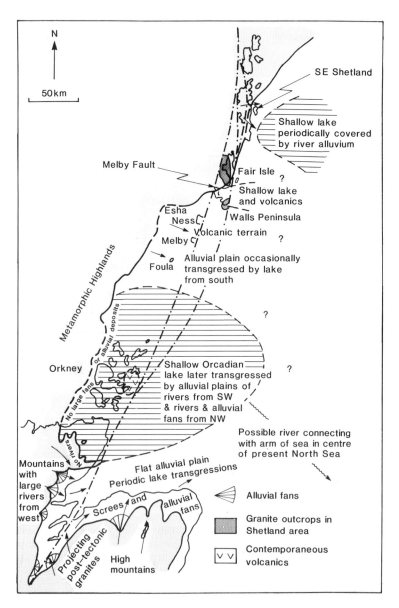

Fig. 9.13. Middle Old Red Sandstone palaeogeography in north-east Scotland and the Northern Isles.

modified if the concept of the collapse of the overthickened crustal welt formed during the development of the Caledonian orogen is proved. So far McClay *et al.* (1986) have propounded the theory and compared the Highlands with the Basin and Range area of the United States, but have not yet produced any concrete proof that their postulated structures occur in this area.

Pre-Upper Old Red Sandstone movements

The Upper Old Red Sandstone rests with marked unconformity on older rocks. South of the Grampians extensive earth movements took place between Lower and Upper Old Red Sandstone times. The final collision between the original Central Valley massif as visualised by Bluck (1983, 1984) and the Dalradian rocks of the Highlands took place, though it may not have been completed till nearly the end of Upper Old Red Sandstone times. The earth movements produced open folds and major faults. The ensuing uplift

and erosion not only removed all trace of Middle Old Red Sandstone sediments in the Midland Valley, if they were ever deposited, but eroded much of the Lower Old Red Sandstone, so that Upper Old Red Sandstone and locally basal Carboniferous now rests in places on Ordovician and Silurian rocks in the Southern Uplands and Midland Valley, and Dalradian rocks in the Southern Highlands. The resulting unconformity was first recognised by James Hutton at Lochranza (Arran), Jedburgh and Siccar Point (East Lothian).

Within the Midland Valley the Lower Old Red Sandstone and older formations were folded along the NE–SW axes into open anticlines and synclines. The effect of erosion of these open folds is well seen in Tayside, where the arch of the Sidlaw anticline was eroded and later covered by Upper Old Red Sandstone and Carboniferous sediments. A fine example of the intra-Old Red Sandstone unconformity is seen in the sea cliffs just north of Arbroath.

North of the Grampian Highlands the unconformity between the Middle and Upper Old Red Sandstone is not well seen. Horne (1923, p. 16) has suggested that in

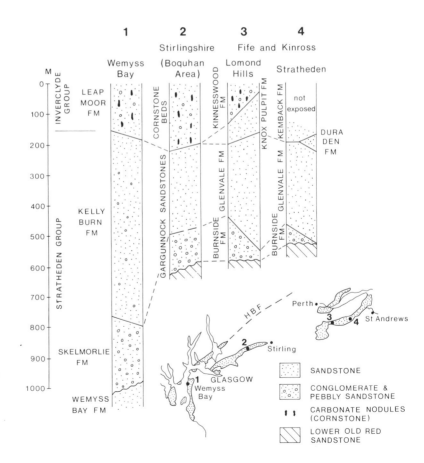

Fig. 9.14. Correlation of the Upper Old Red Sandstone in the Firth of Clyde, Stirlingshire, Kinross and Fife. After Chisholm & Dean (1974), Read & Johnson (1967), Bluck (1978) and Paterson & Hall (1986).

Morayshire the Upper Old Red Sandstone oversteps all the beds of the Middle Old Red Sandstone and rests directly on the Moine granulites south and east of Forres. This interpretation was, however, questioned by Black & Mackenzie (1957) who suggest that the junction between Moine and Upper Old Red Sandstone in this area is a fault. In Easter Ross and Caithness the pre-Upper Old Red Sandstone unconformity is nowhere exposed but in Hoy (Orkney) the basal Upper Old Red Sandstone volcanic rocks rest on an eroded surface of gently folded and faulted strata which range from Upper Stromness Flags to ?Lower Eday Sandstone.

Upper Old Red Sandstone

The Upper Old Red Sandstone consists mainly of red, yellow or buff fluvial sandstone, with local lenses of conglomerate which are thinner and less coarse than the conglomerates of the lower divisions. The sandstones contain fewer lithic clasts and a higher proportion of quartz grains than those in the Lower Old Red Sandstone, and in the upper part of the succession there are many horizons of cornstone, which indicate periods of reduced deposition during which caliche soils were able to develop. The upward passage into the Carboniferous is seen only in the Midland Valley and the Southern Uplands where it is everywhere transitional, with the topmost sandstones passing by upward alternation into the mudstones and shales of the Carboniferous. This upward passage is essentially a facies change which may be diachronous and usually takes place in the basal Carboniferous rather than in the Devonian. Volcanic rocks are absent in the Upper Old Red Sandstone of the Scottish mainland, the only sizeable volcanic horizon being at the base of the Hoy Sandstone in Orkney .

Midland Valley

Upper Old Red Sandstone forms two north-north-east-trending belts of outcrops within and just outside the Midland Valley. The northern outcrops (Fig. 9.14) include a cluster around the Firth of Clyde (Arran, Bute, Great Cumbrae and the west coast of North Ayrshire and Renfrewshire), a belt between Kilcreggan (west of Helensburgh) and Balmaha, a long narrow strip extending from Dumbarton on the Clyde to the vicinity of Stirling, two similar strips between Kinross and Tayport in Fife and between Bridge of Earn and Dundee in Tayside, and small coastal outcrops between Arbroath and Stonehaven in Angus. The outcrops in the southern belt are generally smaller and more broken up by faulting. They include several small outcrops in south and north-east Ayrshire and in Lanarkshire, and a larger outcrop which extends from Lanark north-eastwards to the southern Pentlands and reappears in Edinburgh. The drawing of a boundary between the Old Red Sandstone and Carboniferous has always been a problem and some of the strata taken in the past to be Upper Old Red Sandstone may well be Carboniferous. Paterson & Hall (1986) have recently set up a new subdivision of these beds as follows:

OLD NAMES	NEW NAMES		AGE
Calciferous Sandstone Measures	Clyde Plateau Volcanic Formation	} Strathclyde Group	Lower Carboniferous (Dinantian)
	Clyde Sandstone Formation	} Inverclyde Group	
	Ballagan Formation		
	Kinnesswood Formation		
Upper Old Red Sandstone	Knox Pulpit Formation	} Stratheden Group	Upper Devonian (Famennian)
	Dura Den Formation		
	Glenvale Formation		
	Burnside Formation		

LOWER DEVONIAN AND OLDER ROCKS

The Stratheden Group is regarded as entirely of Upper Devonian age, it comprises predominantly red-brown and yellow sandstones with subordinate amounts of red-brown siltstone and mudstone, and thick conglomerate formations in the west. Cornstone horizons are generally rare and poorly developed. The Inverclyde Group (Fig. 9.14) is believed to be mainly or completely of Carboniferous age. Its two arenaceous formations, the Kinnesswood and Clyde Sandstone divisions contain abundant beds and nodules of pedogenic limestone (cornstone) while the mudstone formation, the Ballagan Formation contains silty mudstone with thin ribs of dolomite or limestone and contains a marine fauna of Carboniferous.

In contrast to the Lower Old Red Sandstone, the succession is most complete and thickest in the west, where it may reach over 1000 m (Bluck 1978), thinning east-north-eastward to 600 m in Stirlingshire (Read & Johnstone 1967; Read *in* Francis *et al.* 1970, pp. 97–114), and 350 to 550 m in Fife and Kinross (Chisholm & Dean 1974). Within the southern outcrops the thickness ranges from 100 m to 425 m in Ayrshire (Eyles *et al.* 1949, pp. 46–50), 0 to 275 in Lanarkshire, and 0 to 300 m in the Pentland Hills, but reaches 640 m in Edinburgh (Mitchell & Mykura 1962, pp. 28–30). Parts of these successions, particularly in the southern belt, are now regarded by Paterson & Hall as Carboniferous.

Stratigraphy and palaeogeography

Northern Belt. The nomenclature and correlation of the Upper Old Red Sandstone formations within the northern belt are shown in Figure 9.14. The oldest sediments, which form the *Wemyss Bay Formation* (Bluck 1978) are seen only in the area around Wemyss Bay, where they were deposited in a restricted basin. They consist mainly of cross-bedded sandstones of fluvial origin, deposited by currents from the south-south-east, but may include some wind-deposited sediments. Their base is not seen. The overlying *Skelmorlie Formation* overlaps the Wemyss Bay Formation and consists of a number of lenticular conglomerates interbedded with conglomeratic sandstones (Bluck 1967). The con-

glomerates were laid down partly in alluvial fans and partly by braided rivers with a northerly to north-easterly direction of transport (Fig. 9.15) which swings to the south-east in the Loch Lomond area. The conglomerates die out southward, and in central and south Ayrshire almost the entire succession in sandstone with cornstones (p. 351). In North Ayrshire and south Bute there is evidence that the alluvial fans were formed at the foot of contemporaneous fault scarps oriented at right angles to the direction of sediment dispersal. The basin of sedimentation was probably enlarged by progressive marginal downfaulting, with the distal ends of the later alluvial fans burying the proximal portions of the preceding ones. The Skelmorlie Formation may be of the same general age as the basal pebbly and conglomeratic facies of the *Gargunnock Sandstone* of Stirlingshire (Read & Johnston 1967) and the *Burnside Formation* of Fife and Kinross. Both of these consist of brick red, purplish and white sandstones with rare thin siltstones and mudstone beds and with some pebbly sandstones and conglomerates. They are probably the deposits of braided rivers which in Fife and Kinross had an easterly flow.

The sandstones forming the *Kelly Burn Formation*, *Gargunnock Sandstone* (upper division), and *Glenvale Formation* are much finer than the underlying beds. They comprise brown, red and purple sandstones, which become

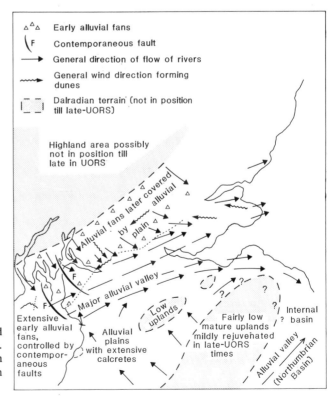

Fig. 9.15. Palaeogeography of the Midland Valley area in Upper Old Red Sandstone times. Based mainly on Bluck (1978), Chisholm & Dean (1974), Leeder (1973) and Hall & Chisholm (1984).

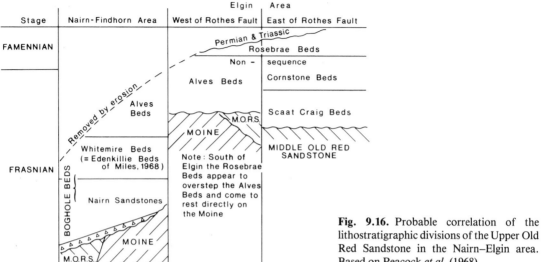

Fig. 9.16. Probable correlation of the lithostratigraphic divisions of the Upper Old Red Sandstone in the Nairn–Elgin area. Based on Peacock *et al.* (1968).

yellow and white in the Cupar area of Fife. They have rare extraformational pebbles, many intraformational clasts of red mudstone and siltstone, and occasional thin bands of siltstone. Cross-stratification is common, and the sequence is mainly braided stream in origin, with foresets indicating palaeocurrents (and palaeoslope) towards the east-north-east in the Firth of Clyde, to the south around Loch Lomond, to the south-east in Stirlingshire and to the east in Fife and Kinross (Fig. 9.15).

The *Knox Pulpit Formation* of the Lomond Hills area, which has been correlated with the *Kemback Formation* of Stratheden west of St Andrews, does not extend westwards beyond Kinross, though recently Monro (1982) has recorded coarse white sandstones (Fairlie Sandstone Formation) in the highest part of the Stratheden Group just east of the Largs Ruck. The rocks have many characteristics of the Knox Pulpit Formation and can probably be correlated with it. The Knox Pulpit Formation consists of a white or cream feldspathic sandstone which differs from the underlying fluvial sandstones in its lack of mudstone clasts, the marked variation of grain size in adjacent laminae, the presence of millet seed grains and well-rounded pebbles. It has a bipolar distribution of foreset azimuths with a marked preponderance of west-north-westward dips indicating deposition by currents or wind blowing mainly to the west-north-west. The sandstone also contains trace fossils consisting of vertical tubes ascribed to the ichnogenus *Skolithus*. Though Chisholm & Dean (1974) did not altogether rule out an aeolian origin for this formation, they believed that the evidence favours a shallow marine environment and ascribed its origin to a period of marine transgression and subsequent regression across an eastward sloping alluvial plain. Recently, however, Hall & Chisholm (1987) have re-examined the

Knox Pulpit Formation and the Kemback Formation and equivalent beds elsewhere in the Upper Old Red and concluded that most of it is of aeolian origin.

The *Dura Den Formation* which contains the famous fish assemblage (p. 331) is only 40 m thick and of limited extent. It lies between the Glenvale and Kemback Formations and contains a mixed lithology: in its lower part thin bands of fluvial sandstone of Glenvale facies are interbedded with fish-bearing layers of sun-cracked mudstone and siltstone and ripple-marked fine sandstone; in the top half the beds of fine sediment are interbedded with the sandstones of the Kemback Formation facies containing *Skolithus* tubes.

The highest sediments of the northern outcrops from west to east, the *Leap Moor Formation, Cornstone Beds, Powmill Formation* (in the small Powmill outcrop near Dollar, see Francis *et al.* 1970, p. 114) and *Kinnesswood Formation* (Fig. 9.14). They are all sandstones of fluvial origin with easterly palaeocurrents in all areas except Loch Lomond and Stirlingshire, where they indicate flow to the south-east. In the Cornstone Beds of Stirlingshire and to a less extent in the Kinnesswood Formation there are well-developed upward-fining cycles of the type formed by the channel and overbank deposits of meandering streams (Fig. 9.15). The sediments contain an appreciable proportion of pebbly sandstones and conglomerates, with clasts of quartzite, low-grade metamorphic rocks and lavas, suggesting a renewed uplift in the source area. Their most conspicuous feature, however, is the almost ubiquitous presence of bands of cornstone, which indicate long periods of reduced sedimentation (see below, p. 333).

Southern Outcrops. No consistent succession has as yet been recognised in the southern outcrops of the Midland Valley. The narrow, generally fault-bounded, outcrops in *South Ayrshire* (Eyles *et al.* 1949, Burgess 1961) fall into two groups: a western group extending from the coast at the Heads of Ayr south-eastwards to the Kerse Loch Fault and then south into the area around Straiton; and an eastern group around Muirkirk. The sediments thicken from 100 m in the north-west (near Ayr) to over 300 m near Straiton. They consist largely of pale pink, yellow or white calcareous sandstones, commonly arranged in upward-fining cycles, with sandstone units up to 5 m thick, grading up into red marly sandstone and mudstone. Pebbly grits and thin conglomerates are fairly common and the extraformational clasts are mainly greywacke, but include chert, jasper, quartzite and porphyrite. The lithology indicates a fluvial origin, with deposition in alluvial fans and on alluvial plains. Palaeocurrent indicators (Burgess 1961, fig. 1) suggest that the streams flowed mostly to the north-west from a source area beyond the Southern Upland Fault, which may at that time still have formed a degraded fault scarp (George 1960). There were also hills within the Midland Valley, such as the Carrick Hills (Lower Old Red Sandstone lavas) and the hills north of Muirkirk (lower Palaeozoic), which contributed both coarse and fine sediment. The most conspicuous feature of the Ayrshire Upper Old Red Sandstone is the abundance of cornstones (Burgess 1961) and they are regarded by Paterson & Hall (1986) as part of the Kinnesswood Formation. They contain bands of limestone, both massive and concretionary, ranging in thickness from 60 cm to 2 m which were quarried in the past as a source of lime. Several of the limestone horizons have well-marked profiles, consisting of a top layer of red-brown sandstone, up to 45 cm thick, which grades down into black marl of the same thickness with small nodules of limestone, underlain in turn by nodular or massive limestone. Burgess has shown that these sections are similar to profiles of the Pliocene caliche limestones of South–Central USA which are typical of soil profiles in semi-arid regions. In a few localities in Ayrshire (e.g. Girvan Water near Kirkmichael, and Blackhill just north of Sanquhar) a caliche soil profile is developed in the Lower Old Red Sandstone immediately beneath the sub-Upper Old Red Sandstone unconformity.

In *Lanarkshire* Upper Old Red Sandstone is absent in the area extending from Strathaven eastward beyond Lanark. Here strata of the basal Carboniferous facies rest directly on the Lower Old Red Sandstone. Farther south in the discontinuous much-faulted outcrops extending from Glenbuck to the southern Pentland Hills the succession thins both to the west and north-west. It consists largely of pale red, white and yellowish sandstone with gritty and pebbly lenses and some thin bands of mottled calcareous marls and shales. A coarse conglomerate, containing rounded clasts of greywacke and quartzite, forms the base of the sequence at the west end of the outcrop. Cornstone beds ranging from a few cm to 3 m are common throughout and the entire sequence of Upper Old Red Sandstone is now taken to be equivalent to the Kinnesswood Formation in the northern outcrops.

In the southern *Pentland Hills* the red sandstone facies is over 300 m thick and comprises pink cross-bedded quartz-rich sandstone with a calcareous cement, with thin conglomerates and bands of red marl, commonly with mudcracks. Cornstones are less abundant than in Ayrshire but the succession is regarded as equivalent to the Kinnesswood Formation of the Inverclyde Group. The sandstones are overlain by argillaceous beds of basal Carboniferous Ballagan Formation but there is a marked interdigitation of the two facies, with thick pink sandstones recurring throughout. The Old Red Sandstone facies thins and is locally absent on the north-west flanks of the Pentland Hills, but thickens to over 600 m in Edinburgh. In East Lothian both the Stratheden Group and a very thin development of the Inverclyde Group have been recorded.

The absence of Upper Old Red Sandstone over a large part of Lanarkshire and locally in the Pentland Hills suggests that some hills in the southern Midland Valley projected above the blanket of newly deposited sediment throughout Upper Old Red Sandstone times. The extreme alteration of the Lower Old Red Sandstone lavas of the Pentland Hills (Mykura 1960) is attributed to this protracted period of weathering. Jawad & Braithwaite (1977) have attributed the red coloration of the Old Red Sandstone sediments throughout the Midland Valley to two major weathering cycles, one within the Lower Old Red Sandstone and the other in the Upper.

Fossil fish are very rare in the western and southern outcrops, being confined to scales of *Holoptychius* and *Bothriolepis leptocheira*, the latter from Bracken Bay, Ayrshire. In Fife and Kinross fish fragments and complete *Holoptychius* have been recorded in the Glenvale Formation, but the most important assemblage is from the fish-bearing horizons in the Dura Den Formation (Waterston 1965), which includes: *Bothriolepis hydrophila, B. cristata, Phyllolepis woodwardi, Phaneropleuron andersoni, Glyptopomus minor, G. kinnairdi, Eusthenopteron dalgleissiensis, Holoptychius flemingi* and *H. nobilissimus. Phyllolepis*-bearing beds have also been found at Clashbennie, 10 km ESE of Perth. No diagnostic plant remains have been recorded.

Southern Uplands

Upper Old Red Sandstone crops out in the eastern Borders between Duns and Bonchester Bridge, and thence extends south-westward as a narrow, discontinuous belt via Langholm towards the Solway Firth. The beds rest on an irregular landscape of lower Palaeozoic and Lower Old Red Sandstone rocks. In the narrow outcrop of the southern Borders they range from 30 m to 200 m in thickness and consist of red pebbly sandstones, siltstones and minor conglomerates, which show an overall upward coarsening (Leeder 1973). They were laid down first by mainly meandering rivers and later by low-sinuosity braided rivers, all of which flowed to the north-east and had their provenance in the Galloway Uplands, where lower

Palaeozoic greywackes and late-Caledonian plutonic intrusions and dyke swarms provided the clasts. Cornstones, in places associated with chert (silcrete), are common in the upper part of the sequence. The succession is conformably overlain by the Carboniferous Birrenswark lavas.

In the large eastern outcrop the sediments are mainly soft, dark red-brown sandstones and mudstones (marls), with the sandstones becoming dominant towards the top, and along its northern and eastern borders conglomerates and pebbly sandstones appear. Palaeocurrent data by Smith (1967) have suggested that the streams which deposited the north Berwickshire sediments flowed towards the south-west. This has led Leeder (1973) to suggest that an interior drainage basin lay in the area west and northwest of the present Cheviot Hills. This concept has, however, been challenged by Paterson *et al.* (1976), who showed that cross-bedding measurements at Pease Bay, Kelso and Greenlaw indicate transport to the east and south-east, which is consistent with the overall easterly to south-easterly palaeoslope of a wider area as deduced from

the palaeocurrents in Upper Old Red sediments in central and southern Scotland (Fig. 9.15).

Fossil fish are usually scattered scales and plates of *Holoptychius* and *Bothriolepis*, and Waterston (1962) has recorded two very fossiliferous horizons near Duns. A block of sandstone from a volcanic vent in the Sundhope Burn, Roxburghshire, has yielded *Phyllolepis*, *Bothriolepis* and *Holoptychius*, which is an assemblage comparable to that from the Rosebrae Beds of Morayshire (p. 333) and a loose block from Pease Bay, Berwickshire, has yielded a plate of ?*Remigolepis sp.* which elsewhere occurs in strata of ?Famennian or Lower Carboniferous age (Andrews 1978). *Bothriolepis* has been found in Lauderdale (P. Aspen pers. comm.).

Morayshire and Nairnshire

Stratigraphy

Upper Old Red Sandstone occupies a belt of country extending westward from Spey Bay, east of Elgin to Fort George, west of Nairn. Though exposures are largely

Fig. 9.17. Old Man of Hoy, west coast of Hoy, Orkney, sea stack of red fluvial Upper Old Red Sandstone resting on sediment of basalt lava, which, in turn, rests unconformably on Upper Stromness Flags. The latter form part of the wave-swept foreshore. (B.G.S. photo.)

confined to two river gorges (River Findhorn and Muckle Burn) and a number of old quarries, the outcrop has the most important upper Old Red Sandstone succession in Scotland, as it has yielded fish faunas in which a well-defined stratigraphic zonation can be discerned. It is also important for the pioneering work on the heavy minerals of the sediments by Mackie (1923). Faunal and litho-stratigraphic zones were first established by Traquair (1896, 1897, 1905) and these were subsequently refined by Westoll (1951), Tarlo (1961) who undertook a systematic study of the psammosteids, and Miles (1968) who described the bothriolepids. The strata are mostly grey, buff and reddish brown cross-bedded sandstones, in places pebbly and conglomeratic, with subordinate marly beds and scattered thin partings of clay and shale and some outcrops of cornstone (Parnell 1983). Some beds contain mudstone chips with which are associated bony plates from the dermal armour of the fish. Some horizons, particularly in the Cornstone Beds (Fig. 9.16), are calcareous and contain numerous beds of cornstone with concretionary sandy limestone and some chert (Peacock *et al.* 1968, pp. 44–53). A thin basal breccia–conglomerate is seen in the River Findhorn and in the Muckle Burn (Horne 1923, pp. 78–82), and in the western part of the outcrop the lower junction of the Upper Old Red Sandstone may be a fault (Black & Mackenzie 1957).

Owing to the sparseness of exposure there has been some disagreement about the stratigraphic relationship of the various subdivisions. The recognition of facies changes across the Rothes Fault, which has a NNW trend and crosses the main outcrop 3 km west of Elgin, by Peacock *et al.* (1968) has, however, clarified some of the points at issue. The succession is diachronous from west to east with the lower formations only present in the west. The probable correlations within the outcrop are shown in Figure 9.16.

The lower subdivisions, up to the Alves Beds, are not separable on lithological criteria alone. The Cornstone Beds which are confined to the area east of the Rothes Fault are distinctive in that they are more calcareous and contain beds of sandy limestone and chert; the Rosebrae Beds, which range from 20 to 60 m in thickness, differ from the underlying sandstones in their yellow or yellow-brown colour and in the abundance of marly horizons, shale bands, and ships of red and green mudstone.

Palaeontology

The importance of the Moray–Nairn fish sequence lies in its comparability with the well-known sequence of the Baltic Province (Tarlo 1961) and also with the sequences of Greenland, Spitzbergen and Belgium (Miles 1968). The correlations suggested by Miles and the recorded species within the divisions are shown in Table 9.3. *Asterolepis maxima* is the diagnostic species of the *Nairn Sandstones* whick are also characterised by the absence of *Bothriolepis*. The *Boghole Beds* (Westoll 1951) are a transitional forma-tion containing fauna found in both the Nairn and Whitemire Beds. The *Whitemire Beds* see the incoming of the first Bothriolepis. The *Alves Beds* are characterised by the presence of both *Bothriolepis* and *Holoptychius*, and the *Scaat Craig Beds*, found only south-east of Elgin, are correlated with part of the Alves Beds and have yielded a similar fauna. The highest division, the *Rosebrae Beds*, is characterised by *Phyllolepis sp.* and *Glyptopomus sp.* There is probably a non-sequence beneath the latter division.

Conditions of deposition

The sandstones and conglomerates of the Nairn–Moray area were mostly laid down by braided and meandering rivers with the thin basal breccias and conglomerates perhaps representing small alluvial fans. Peacock *et al.* (1968) have suggested that the marly beds and pellet conglomerates may represent temporary lacustrine conditions, but these sediments could also have been laid down on alluvial plains. The presence of faceted pebbles suggests periods of re-working by wind, and the cornstones are caliche-type profiles formed during periods of reduced sedimentation. Palaeocurrent data indicate sediment transport to the north, ranging from north-west to north-east (Westoll 1977).

Easter Ross

Sediments of Upper Old Red Sandstone age, crop out in the axial region of the Inver–Black Isle Syncline within the Tarbat Ness Peninsula where they have been termed the *Balnagown Group* by Armstrong (1977). There is also a narrow outcrop along the coast north of Dornoch. The beds consist of yellow and red medium-grained cross-bedded sandstone with scattered pebbles. Isolated fish-scales have been found within the sandstone in the Balnagown River near Balnagown Castle, 6 km south of Tain, and on the shore at Embo, north of Dornoch. The fauna includes *Psammosteus taylori, P. megalopteryx, Asterolepis sp.,* and *Holoptychius sp.*, and the group may be equivalent in age to the section extending from the Boghole Beds to the Alves Beds in the Moray–Nairn area (Westoll 1977). The sandstones are of fluvial origin, laid down mainly by swift-flowing braided rivers flowing to the north-east.

Caithness and Orkney

The sandstones of Dunnet Head (Caithness) and Hoy (Fig. 9.17) were originally part of a single unit. They have now been dated as Upper Old Red Sandstone by the discovery of *Holoptychius sp.* at Dunnet Head. This age is supported by radiometric (stepwise degassing 40 Ar/39 Ar method) dating of the Hoy lavas as approximately 370 Ma (Halliday *et al.* 1977).

On Hoy the sandstones are underlain by volcanic rocks. The basal beds are partly air-deposited tuffs and tuffaceous sandstones which were laid down on an undulating land surface and vary in thickness from 0 to 15 m. The overlying olivine-basalts form five disconnected lenticular outcrops which may originally have been part of a single flow. Thirlwall (1979) has shown that the basalts contain primary analcime and are nepheline normative, and are the only alkaline basalts in the Scottish Old Red Sandstone. The Hoy and Dunnet Head sandstones comprise a succession of pink and red, medium-grained mainly trough-cross-bedded, locally pebbly sandstones with convolute bedding and some intraformational red mudstone clasts. Most of the sequence was laid down by probably braided rivers from the south-west, but some of the Hoy succession may be of aeolian origin (McAlpine 1979).

Shetland. The fish beds of Bressay, south-east Shetland (p. 324) contain two genera, *Asterolepis* and *Glyptolepis* which range from the Middle into the Upper Devonian. It is thus possible that the higher members of the Bressay Flagstone facies (p. 324), a sequence of fluvial sandstones and siltstones arranged in fining upward cycles, may be of Upper Devonian age.

Table 9.3. Probable correlation of the Upper Old Red Sandstone of the Moray–Nairn area with the succession of the Baltic Province, Greenland, Spitzbergen and Belgium. After Miles (1968, table 2).

GREENLAND	BELGIUM	BALTIC PROVINCE	MORAY-NAIRN AREA	FAUNA
Phyllolepis Series — Upper part	Condroz Sst.	"Post-Psammosteus-Stufe"	Rosebrae Beds	Phyllolepis sp., Bothriolepis alvesiensis, B. cristata B. laverocklochensis, Phaneropleuron sp., Rhynchodipterus elginensis, Conchodus sp., Eusthenopteron?, Glyptopomus elginensis, Holoptychius nobilissimus
Phyllolepis Series — Lower part		e Stage	Alves Beds	Psammosteus taylori, Bothriolepis alvesiensis, B. gigantea, Conchodus sp., Rhizodonts, Holoptychius nobilissimus, H. giganteus
			Scaat Craig Beds	Psammosteus cf. falcatus, Traquairosteus pustulatus, Bothriolepis paradoxa, Coccosteomorph arthrodire, Cosmacanthus malcolmsoni, Conchodus ostreiformis, Rhizodonts, Holoptychius nobilissimus, H. giganteus
		e-d Shelon-limen		
		b^1 Snetogor	Whitemire Beds (= Edenkillie Beds)	Bothriolepis taylori, Psammosteus taylori, "Coccosteus", Cosmacanthus sp., Conchodus ostreiformis, Holoptychius nobilissimus, H. giganteus
SPITSBERGEN		a^4 Amata	(Boghole Beds)	Psammosteids (indet), Asterolepis alta, Eusthenopteron traquairi, Plourdosteus magnus, Polyplocodus leptognathus, Holoptychius decoratus H. nobilissimus
? Fiskekløfta Formation ?		a^3 Gauja	Nairn Sandstones	Asterolepis maxima, Psammolepis tesselata, Plourdosteus magnus, Polyplocodus leptognathus, Holoptychius decoratus

REFERENCES

AGASSIZ, L.   1835 On the fossil fishes of Scotland. *Rep. Br. Ass. Advmt. Sci.*, **4**, 646–649.

ALLAN, D. A.   1928 The Geology of the Highland Border from Tayside to Noranside. *Trans. R. Soc. Edinb.*, **56**, 57–88.

                               1940 The Geology of the Highland Border from Glen Almond to Glen Artney. *Trans. R. Soc. Edinb.*, **60**, 171–193.

ALLEN, P. A.   1980 Sedimentological aspects of the Devonian strata of SE Shetland. *Ph.D. Cambridge* (Unpublished).

ALLEN & MARSHALL, J. E. A.   1981 Depositional environments and palynology of the Devonian East Shetland basin. *Scott. J. Geol.*, **17**, 257–273.

ANDREWS, S. M.   1978 A possible occurrence of *Remigolepis* in the topmost Old Red Sandstone of Berwickshire. *Scott. J. Geol.*, **14**, 311–315.

                               1983 Altyre and Lethen Burn, two Middle Old Red Sandstone fish localities? *Scott. J. Geol.*, **19**, 243–264.

ARCHER, R.   1978 The Old Red Sandstone outliers of Gamrie and Rhynie, Aberdeenshire. *Ph.D. Thesis, Newcastle upon Tyne* (Unpublished).

ARMSTRONG, M.   1964 The Geology of the Region between the Alness River and the Dornoch Firth. *Ph.D. Thesis, University of Newcastle upon Tyne* (Unpublished).

                               1977 The Old Red Sandstone of Easter Ross and the Black Isle. *In* Gill, G. (Ed.) *The Moray Firth Area – Geological Studies.* Inverness Field Club.

ARMSTRONG, M. &   1970 The Lower Old Red Sandstone of the Strathmore region. *Rep.*
  PATERSON, I. B.   *No. 70/12, Inst. Geol. Sci.*

ASTIN, T. R.   1982 The Devonian geology of the Walls peninsula, Shetland. *Ph.D. Cambridge* (Unpublished).

                               1985 The palaeogeography of the Middle Devonian Lower Eday Sandstone of Orkney. *Scott. J. Geol.*, **21**, 353–375.

BAILEY, E. B.   1960 The geology of Ben Nevis and Glen Coe 2nd Ed. *Mem. Geol. Surv. U.K.* (Sheet 53).

BLACK, G. P. &   1957 Supposed unconformities in the Old Red Sandstone of Western
  MACKENZIE, D. H.   Moray. *Geol. Mag.*, **94**, 170–171.

BLACKBOURN, G.   1981 Probable Old Red Sandstone conglomerates around Tongue and adjacent areas, north Scotland. *Scott. J. Geol.*, **17**, 103–118.

BLUCK, B. J.   1967 Deposition of some Upper Old Red Sandstone conglomerates in the Clyde area: a study in the significance of bedding. *Scott. J. Geol.*, **3**, 139–167.

                               1969 Old Red Sandstone and Palaeozoic Conglomerates of Scotland, pp. 711–723. *In* Kay, M. (Ed.) North Atlantic – geology and Continental Drift. *Amer. Assoc. Petroleum Geologists Memoir 12.*

                               1978 Sedimentation in a late orogenic basin: the Old Red Sandstone of the Midland Valley of Scotland. *In* Bowes, D. R. & Leake, B. E. (Eds.) *Crustal Evolution in North western Britain and adjacent Regions.* Geol. J. Special Issue 10.

                               1980a Structure, generation and preservation of upward fining, braided stream cycles in the Old Red Sandstone of Scotland. *Trans. R. Soc. Edinb.*, **17**, 29–46.

                               1980b Evolution of strike-slips fault-controlled basin, Upper Old Red Sandstone, Scotland. *In* Ballance, B. F. & Reading, H. G. (Ed.) Sedimentation in oblique-slip mobile zones. *Assoc. Sedimentology, Spec. Publ.*, **4**, 63–78.

BLUCK, B. J.   1983   Role of the Midland Valley of Scotland in the Caledonian Orogeny. *Trans. R. Soc. Edinb.*, **73**, 119–136.

                                1984   Pre-Carboniferous history of the Midland Valley of Scotland. *In* Bowes, D. R. (Ed.) The deep geology of the Midland Valley of Scotland. *Trans. R. Soc. Edinb.*, **75**, 275–295.

                                1986   Upward coarsening sedimentation units and facies lineages, Old Red Sandstone, Scotland. *Trans. R. Soc. Edinb.*, **77**, 251–264.

BRADE BIRKS, S. G.   1923   Notes on Myriapoda XXVIII, Kampecaris tuberculata n. sp. from the Old Red Sandstone of Ayrshire. *Proc. R. Phys. Soc. Edinb.*, **20**, 277–280.

BRAITHWAITE, C. T R. &   1978   Heavy mineral distribution in the Old Red Sandstone in the
    JAWADALI, A.   North-eastern Midland Valley of Scotland. *Scott. J. Geol.*, **14**, 273–288.

BROWN, J. F.   1975   Rb–Sr studies and related geochemistry on the Caledonian calc-alkaline igneous rocks of N.W. Argyllshire. *D.Phil. Thesis, univ. of Oxford* (Unpublished).

BROWNE, M. A. E.   1980   The upper Devonian and Lower Carboniferous (Dinantian) of the Firth of Tay, Scotland. *Rep. Inst. Geol. Sci.*, No 80/9.

BURGESS, I. C.   1961   Fossil soils of the Upper Old Red Sandstone of South Ayrshire. *Trans. geol. Soc. Glasg.*, **24**, 138–153.

CAMPBELL, R.   1913   The geology of south-eastern Kincardineshire. *Trans. R. Soc. Edinb.*, **48**, 923–960.

CHALLONER, W. G.   1972   Devonian plants from Fair Isle, Scotland. *Rev. Palaeobot. Palynol.*, **14**, 44–61.

CHISHOLM, J. I. &   1974   The upper Old Red Sandstone of Fife and Kinross: a fluviate
    DEAN, J. M.   sequence with evidence of marine incursion. *Scott. J. Geol.*, **10**, 1–30.

CLAYBURN, J. A. P. *et al.*   1983   Sr, O and Pb isotope evidence for origin and evolution of Etive Igneous Complex, Scotland. *Nature*, **303**, 492–497.

CLOUGH, C. T. *et al.*   1910   The Geology of East Lothian. *Mem. geol. Surv. U.K.* (Sheet 33).

CLOUGH, C. T.,   1909   The Cauldron-subsidence of Glen Coe and the associated
    MAUFE, H. B., &   igneous phenomena. *Q. J. geol. Soc. Lond.*, **65**, 611.
    BAILEY, E. B.

COLLINS, A. G. &   1977   The Age of two Old Red Sandstone sequences in southern
    DONOVAN, N. R.   Caithness. *Scott. J. Geol.*, **13**, 53–57.

CONYBEARE, W. D. &   1822   *Outlines of the Geology of England and Wales*. London.
    PHILLIPS, W.

CRAMPTON, C. B. &   1914   The Geology of Caithness. *Mem. Geol. Surv. U.K.* (Sheets 110
    CARRUTHERS, R. G.   and 116).

DAVIES, A. &   1986   Geology of the Dunbar district. *Mem. Br. geol. Surv.*
    McADAM, A. D.

DONOVAN, R. N.   1970   The Geology of the Coastal Tract near Wick, Caithness. *Ph.D. Thesis, Newcastle upon Tyne* (Unpublished).

                                1973   Basin margin deposits of the Middle Old Red Sandstone at Dirlot Caithness. *Scott. J. Geol.*, **9**, 203–211.

                                1975   Devonian lacustrine limestones at the margin of the Orcadian Basin, Scotland. *J. geol. Soc. Lond.*, **131**, 489–510.

                                1980   Lacustrine cycles, fish ecology and stratigraphic zonation in the Middle Devonian of Caithness. *Scott. J. Geol.*, **16**, 35–50.

                                1982   Devonian calcites (cornstones) near Tain. *Scott. J. Geol.*, **18**, 125–129.

DONOVAN, R. N.,   1976   Devonian Palaeogeography of the Orcadian Basin and the Great
    ARCHER, A.,   Glen Fault. *Nature*, **259**, 550–551.
    TURNER, P. &
    TARLING, D. H.

DONOVAN, R. N., &   1978   Mound structures from the Caithness Flagstones (Mid. Dev.),
  COLLINS, A.   northern Scotland. *J. Sedim. Petrol.*, **48**, 171–174.
DONOVAN, R. N.,   1978   The age of sediments on Foula, Shetland. *Scott. J. Geol.*, **14**,
  COLLINS, A.,   87–88.
  ROWLANDS, M. A., &
  ARCHER, R.
DONOVAN, R. N. &   1972   Subaqueous shrinkage cracks from the Caithness flagstone
  FOSTER, R. J.   Series (Middle Devonian) of Northeast Scotland. *J. sed. Pet.*, **42**,
                                    309–317.
DONOVAN, R. N.,   1974   A stratigraphical revision of the Old Red Sandstone of North-
  FOSTER, R. J., &   east Caithness. *Trans. R. Soc. Edinb.*, **69**, 167–201.
  WESTOLL, T. S.
DOWNIE, C. & LISTER, T. R. 1969   The Sandy's Creek beds (Devonian) of Farland Head, Ayrshire.
                                    Scott. J. Geol., **5**, 193–206.
DU TOIT, A. L.   1905   The Lower Old Red Sandstone Rock of the Balmaha Aberfoyle
                                    Region. *Trans. Edinb. geol. Soc.*, B 399–405.
EYLES, V. A. &   1952   The Great Glen Crush Belt. *Geol. Mag.*, **89**, 426–436.
  MACGREGOR, A. G.
EYLES, V. A.,   1949   The Geology of Central Ayrshire. *Mem. geol. Surv. U.K.* (Sheet
  SIMPSON, J. B., &   14).
  MACGREGOR, A. G.
FANNIN, N. G. T.   1969   Stromatolites from the Middle Old Red Sandstone of Western
                                    Orkney. *Geol. Mag.*, **106**, 77–88.
                           1970   The sedimentary environment of the Old Red Sandstone of
                                    Western Orkney. *Ph.D. Thesis, University of Reading* (Unpub-
                                    lished).
FLEMING, J.   1831   On the occurrence of the scales of vertebrated animals in the Old
                                    Red Sandstone of Fifeshire. *Edinb. J. nat. and geogr. Sci.*, **3**,
                                    81–86.
FLINN, D.   1969   A geological interpretation of the Aeromagnetic Maps of the
                                    Continental shelf around Orkney and Shetland. *Geol. J.*, **6**, 279–
                                    292.
FLINN, D., MILLER, J. A., 1968   On the age of the sediments and contemporaneous volcanic
  EVANS, A. L., &   rocks of western Shetland. *Scott. J. Geol.*, **4**, 10–19.
  PRINGLE, I. R.
FOSTER, R. J.   1972   The solid geology of North-east Caithness. *Ph.D. Thesis,
                                    Newcastle on Tyne* (Unpublished).
FRANCIS, E. H.,   1970   The Geology of the Stirling District. *Mem. Geol. Surv. U.K.*
  FORSYTH, I. H.,   (Sheet 39).
  READ, W. A., &
  ARMSTRONG, M.
                           1985   Molassic basins of Europe: a tectonic assessment. *Trans. R. Soc.
                                    Edinb.*, **76**, 451–462.
FRIEND, P. F.,   1963   The Old Red Sandstone and the Highland Boundary in Arran,
  HARLAND, W. D., &   Scotland. *Trans. Edinb. geol. Soc.*, **19**, 363–425.
  HUDSON, J. D.
FRIEND, P. F. &   1968   Volcanic sediments, stratigraphy and tectonic background of the
  MACDONALD, R.   Old Red Sandstone of Kintyre, W. Scotland. *Scott. J. Geol.*, **4**,
                                    265–282.
FRIEND, P. F. &   1978   A field guide to selected outcrop areas of the Devonian of
  WILLIAMS, B. P. J. (Eds.)   Scotland, the Welsh Borderland and South Wales. *Palaeon-
                                    tological Association.*
GALLAGHER, M. J.,   1971   New evidence of uranium mineralization in Scotland. *Trans.
  MICHIE, U. McL.,   Inst. Min. Met.*, **80**, B150–173.
  SMITH, R. T., &
  HAYNES, L.

GANDY, M. K.   1975   The Petrology of the Lower Old Red Sandstone lavas of the
                                       Eastern Sidlaw Hills, Perthshire, Scotland. *J. Petrol.*, **16**, 189–
                                       211.

GEIKIE, A.   1878   On the Old Red Sandstone of western Europe. *Trans. R. Soc.
                                       Edinb.*, **28**, 345–452.

                                1897   *The Ancient Volcanoes of Great Britain.* 2 vols. London.

GEORGE, T. N.   1960   The stratigraphic evolution of the Midland Valley. *Trans. geol.
                                       Soc. Glasgow*, **23**, 32–107.

GREIG, D. C.   1971   *British Regional Geology: The South of Scotland.*

GROOME, D. R. &   1974   The geochemistry of the Devonian lavas of the northern Lorne
   HALL, A.   plateau, Scotland. *Min. Mag.*, **39**, 621–640.

HALL, A. J. &   1978   Origin of complex sulphide nodules related to diagenesis of
   DONOVAN, R. N.   lacustrine sediments of Middle Devonian age from the Shetland
                                       Islands. *Scott. J. Geol.*, **14**, 289–299.

HALL, I. H. S. &   1987   Aeolian sediments in the late Devonian of the Scottish Midland
   CHISHOLM, J. I.   Valley. *Scott. J. Geol.*, **23**, 203–208.

HALLER, J.   1971   *Geology of the East Greenland Caledonides.* 413 pp. New York.

HALLIDAY, A. N.,   1977   The Age of the Hoy Lavas, Orkney. *Scott. J. Geol.*, **13**, 43–52.
   MCALPINE, A., &
   MITCHELL, J. G.

HAMILTON, R. F. M. &   1985   *Excursion Guide to the Devonian of Caithness.* P.E.S.B.
   TREWIN, N. F.   (Petroleum Exploration Society of G.B.), Aberdeen.

HARLAND, W. B.,   1987   A geologic time scale. *Cambridge University Press.*
   COX, A. V.,
   LLEWELLYN, P. G.,
   PICKTON, C. A. G.,
   SMITH, A. G., &
   WALTERS, R.

HARRY, W. T.   1956   The Old Red Sandstone Lavas of the Western Sidlaw Hills,
                                       Perthshire. *Geol. Mag.*, **93**, 43–56.

                                1958   The Old Red Sandstone Lavas of the Eastern Sidlaws. *Trans.
                                       Edinb. geol. Soc.*, **17**, 105–112.

HERRIOT, A.   1956   Notes on the occurrence of garnet in the felsite of Tinto,
                                       Lanarkshire. *Trans. geol. Soc. Glasgow*, **22**, 94–99.

HICKLING, G.   1908   **The Old Red Sandstone of Forfarshire. *Geol. Mag.*, 5, 396–
                                       408.**

                                1912   On the geology and palaeontology of Forfarshire. *Proc. Geol.
                                       Ass.*, **23**, 302–311.

HICKLING, G. &   1915   On *Parka decipiens. Q. J. geol. Soc. Lond.*, **72**, 648–663.
   DON, A. W. R.

HINXMAN, L. W.   1896   Explanation of Sheet 75 – West Aberdeenshire, Banffshire, parts
                                       of Elgin and Inverness. *Mem. Geol. Surv. Scotland.*

HINXMAN, L. W. &   1902   The Geology of Lower Strathspey. *Mem. Geol. Surv. U.K.* (Sheet
   WILSON, J. S. G.   85).

HIRST, S.   1923   On some Arachnid Remains from the Old Red Sandstone
                                       Rhynie Chert Bed, Aberdeenshire. *Ann. Mag. nat. Hist.*, **12**, 355–
                                       374.

HIRST, S. &   1926   On some Arthropod Remains from the Rhynie Chert (Old Red
   MAULIK, S.   Sandstone). *Geol. Mag.*, **63**, 69–71.

HOLGATE, N.   1969   Palaeozoic and Tertiary transcurrent movements on the Great
                                       Glen Fault. *Scott J. Geol.*, **5**, 97–139.

HORNE, J.   1923   The Geology of the Lower Findhorn and Lower Strath Nairn,
                                       including part of the Black Isle near Fortrose. *Mem. Geol. Surv.
                                       U.K.* (Sheet 84 and part of 94).

HORNE, J. &   1914   The Geology of the Country round Beauly and Inverness:
   HINXMAN, L. W.   including a part of the Black Isle. *Mem. Geol. Surv. U.K.* (Sheet
                                       83).

House, M. R.,
Richardson, J. B.,
Chaloner, W. G.,
Allen, J. R. L.,
Holland, C. H., &
Westoll, T. S.

1977 A correlation of Devonian rocks of the British Isles. *Geol. Soc. Lond. Special Report No. 7.*

Jack, R. L. &
Letheridge, R.

1877 On the discovery of plants in the Lower Old Red Sandstone of the neighbourhood of Callander. *Q. J. geol. Soc. Lond.*, **33**, 213–222.

Jarvik, E.

1948 On the morphology and taxonomy of Middle Devonian osteolepid fishes in Scotland. *K. svenska Vetenskaps Akad. Handl.*, (3) **25**, 1–301.

1950 On some osteolepiform crossypterygians from the Upper Old Red Sandstone of Scotland. *K. svenska Vetensk.-Akad. Handl.*, (4) **2**, 1–35.

Jawad Ali, A. &
Braithwaite, C. J. R.

1977 Penecontemporaneous weathering of the Old Red Sandstone of the Midland Valley of Scotland. *Scott. J. Geol.*, **13**, 305–312.

Johnstone, G. S.

1972 *British Regional Geology: The Grampian Highlands*, 3rd Edn.

Kellock, E.

1969 Alkaline basic igneous rocks in the Orkneys. *Scott. J. Geol.*, **5**, 140–153.

Kidston, R.

1893 On the occurrence of *Arthrostigma gracile* Dawson in the Lower Old Red Sandstone of Perthshire. *Proc. R. phys. Soc. Edinb.*, **12**, 102–111.

Kidston, R. &
Lang, W. H.

1917– On Old Red Sandstone plants showing Structure from the
1921 Rhynie Chert Bed, Aberdeenshire. Parts 1–5. *Trans. R. Soc. Edinb.*, **51** and **52**.

1924 Notes of fossil plants from the Old Red Sandstone of Scotland, III. *Trans. R. Soc. Edinb.*, **53**, 604.

Kynaston, H. &
Hill, J. B.

1908 The geology of Oban and Dalmally. *Mem. geol. Surv. U.K.* (Sheet 45).

Lang, W. H.

1925– Contributions to the study of the Old Red Sandstone Flora of
1932 Scotland. Parts I–VIII. *Trans. R. Soc. Edinb.*, **54–57**.

Lankester, E. R. &
Traquair, R. H.

1868– The fishes of the Old Red Sandstone of Britain. *Mon. Pal Soc.*
1914

Lee, G. W. &
Bailey, E. B.

1925 The pre-Tertiary Geology of Mull, Loch Aline and Oban. *Mem. Geol. Surv. U.K.*

Leeder, M. G.

1973 Sedimentology and Palaeogeography of the Upper Old Red Sandstone in the Scottish Border Basin. *Scott. J. Geol.*, **9**, 117–145.

1974 The origin of the Northumberland basin. *Scott. J. Geol.*, **10**, 283–296.

Leeder, M. G. &
Bridges, P. H.

1978 Upper Old Red Sandstone near Kirkbean, Dumfries and Galloway. *Scott. J. Geol.*, **14**, 267–272.

Lyell, C.

1865 *Elements of Geology*, 6th Edn., London.

McAlpine, A.

1979 The Upper Old Red Sandstone of Orkney, Caithness and neighbouring areas. *Ph.D. Newcastle* (Unpublished).

McClay, K. R.,
Norton, M. G.,
Cony, P., &
Davis, G. H.

1986 Collapse of the Caledonian Oregon and the Old Red Sandstone. *Nature*, **323**, 147–149.

McGiven, A.

1967 Sedimentation and provenance of some post-Valentian conglomerates, Midland Valley, Scotland. *Univ. Glasgow Ph.D. Thesis* (Unpublished).

McIntyre, D. B.,
Brown, W. L.,
Clarke, W. I., &
Mackenzie, D. H.

1956 On the conglomerates of supposed Old Red Sandstone age near Tongue, Sutherland. *Trans. geol. Soc. Glasg.*, **22**, 35–47.

McKERROW, W. S.,   1985   The Ordovician, Silurian and Devonian periods. *In* Snelling,
LAMBERT, R. ST. J., &   N. J. (Ed.) *The chronology of the Geological Record. Mem. geol.*
COCKS, L. R. M.   *Soc. Lond.*, **10**, 73–83.

MACKIE, W.   1923   The principles that regulate the distribution of heavy minerals in
                               sedimentary rocks, as illustrated by the sandstones of North-east
                               Scotland. *Trans. geol. Soc. Edinb.*, **11**, 138–164.

                       1923b The source of the purple zircons in the Sedimentary Rocks of
                               Scotland. *Trans. geol. Soc. Edinb.*, **11**, 200–213.

MARSHALL, J. E. A. &   1982   Devonian miospore assemblages from Fair Isle, Shetland.
ALLEN, K. C.   *Palaeontology*, **25**, 277–312.

MARSHALL, J. E. A.,   1985   Hydrocarbon source rock potential of the Devonian rocks of the
BROWN, J. F., &   Ordcadian Basin. *Scott. J. geol.*, **21**, 301–320.
HINDMARSH, J.

MELVIN, J.   1985   Walls Formation, Western Shetland, distal alluvial plain
                               deposits within a tectonically active Devonian Basin. *Geol.*
                               Journ. geol., **21**, 23–40.

MICHIE, U. McL. &   1979   Uranium in the Old Red Sandstone of Orkney. *Rep. No. 78/16,*
COOPER, U. C.   *Inst. geol. Sci.*

MILES, R. S.   1968   The Old Red Sandstone Antiarchs of Scotland: Family Both-
                               riolepidae. *Palaeontogr. Soc.* (Monogr), 552.

MILES, R. S. &   1963   Two new genera of Coccosteid Arthrodira from the Middle Old
WESTOLL, T. S.   Red Sandstone of Scotland and their stratigraphic distribution.
                               Trans. R. Soc. Edinb., **65**, 179–210.

MILES, R. S. &   1968   The placoderm fish *Coccosteus cuspidatus* Miller *ex* Agassiz
WESTOLL, T. S.   from the Middle Old Red Sandstone of Scotland. Part I. *Trans.*
                               R. Soc. Edinb., **67**, 373–476.

MILLER, HUGH   1841   *The Old Red Sandstone*. Edinburgh.

MITCHELL, G. H. &   1962   The Geology of the Neighbourhood of Edinburgh. *Mem. Geol.*
MYKURA, W.   *Surv. U.K.* (Sheet 32).

MITCHELL, H.   1860   Notice on new fossils from the Lower Old Red Sandstone. *The*
                               Geologist, **3**, 273–275.

                       1861   On the position of the beds of the Old Red Sandstone developed
                               in the counties of Forfar and Kincardine. *Q. J. geol. Soc. Lond.*,
                               17, 145–151.

MORTON, D. J.   1976   Lower Old Red Sandstone sedimentation in the north-west
                               Midland Valley and North Argyll areas of Scotland. *Univ.*
                               Glasgow Ph.D. Thesis (Unpublished).

                       1979   **Palaeogeographical evolution of the Lower Old Red Sand-**
                               stone basin in the Western Midland Valley. *Scott. J. Geol.*, **15**,
                               97–116.

MONRO, S. K.   1982   Sedimentation, stratigraphy and tectonics in the Dalry Basin,
                               Ayrshire. *Ph.D. Thesis, University of Edinburgh* (Unpublished).

MURCHISON, R. I.   1859   On the succession of the older rocks in the northernmost
                               counties of Scotland. *Q. J. Geol. Soc. Lond.*, **17**, 145–151.

MYKURA, W.   1960   The Lower Old Red Sandstone Igneous rocks of the Pentland
                               Hills. *Bull. geol. Surv. G.B.*, **16**, 131–155.

                       1972   The Old Red Sandstone sediments of Fair Isle, Shetland Islands.
                               Bull. geol. Surv. G.B., **41**, 1–31.

                       1975   Possible large-scale sinistral displacement along the Great Glen
                               Fault in Scotland. *Geol. Mag.*, **112**, 91–97.

                       1982   Old Red Sandstone east of Loch Ness, Inverness-shire. *Rep. Inst.*
                               geol. Sco., 82/13.

MYKURA, W. &   1976   The Geology of Western Shetland. *Mem. geol. Surv. U.K.*
PHEMISTER, J.

MYKURA, W. with   1976   *British Regional Geology: Orkney and Shetland.*
contributions by
FLINN, D. & MAY, F.

MYKURA, W. &   1983   The Old Red Sandstone of the Mealfuorvonie outlier, west of
  OWENS, B.   Loch Ness, Inverness-shire. *Report. Inst. Geol. Sci. 83/7.*

O'REILLY, K. J.   1983   Composition and age of the conglomerate outliers around the
                              Kyle of Tongue, north Sutherland, Scotland. *Proc. geol. Ass.*, **93**,
                              53–64.

PARNELL, J.   1981   Post-depositional processes in the Old Red Sandstone of the
                              Orcadian Basin, Scotland. *Ph.D. Thesis, University of London*
                              (Unpublished).

                     1983a   The Cothall Limestone. *Scott. J. Geol.*, **14**, 215–218.

                     1983b   The distribution of hydro-carbon minerals in the Orcadian
                              Basin. *Scott. J. Geol.*, **19**, 205–213.

                     1985a   Hydrocarbon source rocks, reservoir rocks and migration in the
                              Orcadian Basin. *Scott. J. Geol.*, **21**, 321–336.

                     1985b   Evidence for evaporites in the Old Red Sandstone of Northern
                              Scotland: replaced gypsum horizons in Easter Ross. *Scott. J.
                              Geol.*, **21**, 377–380.

PATERSON, I. B. &   1969   Lower Old Red Sandstone ignimbrites from Dunkeld, Perth-
  HARRIS, A. L.   shire. *Rep. 1969/7 Inst. Geol. Sci.*

PATERSON, I. B.,   1976   Upper Old Red Sandstone Palaeography. *Scott. J. Geol.*, **12**, 89–
  BROWNE, M. A. E., &   91.
  ARMSTRONG, M.

PATERSON, I. B. &   1986   Lithostratigraphy of the late Devonian and early Carboniferous
  HALL, I. S.   rocks of the Midland Valley of Scotland. *Rep. Br. Geol. Surv.*, **18**,
                              No. 3.

PATTERSON, E. M.   1949   The Lower Old Red Sandstone rocks of the West Kilbride–Largs
                              district, Ayrshire. *Trans. geol. Soc. Glasg.*, **21**, 207–236.

PEACH, B. N., GUNN, W., 1912 The Geology of Ben Wyvis, Carn Chuinneag, Inchbae and the
  CLOUGH, C. T.,   surrounding country. *Mem. Geol. Surv. U.K.* (Sheet 93).
  HINXMAN, L. W.,
  CRAMPTON, C. B., &
  ANDERSON, E. M.

PEACOCK, J. D.,   1968   The geology of the Elgin District. *Mem. Geol. Surv. U.K.* (Sheet
  BERRIDGE, N. G.,   95).
  HARRIS, A. L., & MAY, F.

PENNINGTON, J. J.   1975   The geology of the Argyll Field. *In* Woodland, A. W. (Ed.)
                              Petroleum and the Continental Shelf of Northwest Europe, 285–
                              291.

PLIMMER, R. S.   1974   The Sedimentology and Stratigraphy of the Middle Old Red
                              Sandstone Rousay Group of the Orkney Islands. *Univ. New-
                              castle Ph.D. Thesis* (Unpublished).

POWRIE, J.   1864   On the fossiliferous rocks of Forfarshire and their contents. *Q. J.
                              geol. Soc. Lond.*, **20**, 413–429.

QURESHI, I. R.   1970   A gravity survey in the region of the Highland Boundary Fault,
                              Scotland. *Q. J. geol. Soc. Lond.*, **125**, 481–502.

RAMOS, A. &   1982   Upper Old Red Sandstone sedimentation near the unconformity
  FRIEND, P. F.   at Arbroath. *Scott. J. Geol.*, **18**, 297–315.

RAMSAY, D. M.   1962   The Highland Boundary Fault: Reverse or Wrench Fault?
                              Nature, London, **195**, 1190.

RAYNER, D. H.   1963   The Achanarras Limestone of the Middle Old Red Sandstone,
                              Caithness, Scotland. *Proc. Yorks. Geol. Soc.*, **34**, 1–44.

                     1984   New finds of *Drepanophycus spineoformis* Goppert from the
                              Lower Devonian of Scotland. *Trans. R. Soc. Edinb.*, **75**, 353–363.

READ, H. H.   1923   The geology of the country around Banff, Huntly and Turriff.
                              Mem. Geol. Surv. U.K. (Sheets 86 and 96).

                     1927   The Tinto District. *Proc. geol. Ass. Lond.*, **38**, 499–504.

                     1931   The Geology of Central Sutherland. *Mem. Geol. Surv. U.K.*
                              (Sheets 108 and 109).

READ, H. H., ROSS, G., & PHEMISTER, J. — 1925 — The Geology of the Country around Golspie, Sutherlandshire. *Mem. Geol. Surv. U.K.* (Sheet 103).

READ, W. A., & JOHNSON, S. R. H. — 1967 — The sedimentology of sandstone formations within the Upper Old Red Sandstone and lowest Calciferous Sandstone measures west of Stirling. *Scott. J. Geol.*, **3**, 247–267.

REID, J. & McNAIR, P. — 1896 — On the Genera Lycopodites and Psilophyton of the Old Red Sandstone Formation of Scotland. *Trans. geol. Soc. Glasg.*, **10**, 323–330.

REID, J., GRAHAM, W., & McNAIR, P. — 1897 — Parka decipiens; its origin, affinities and distribution. *Trans. geol. Soc. Glasg.*, **10**, 323–330.

RICHARDS, P. C. — 1985a — Upper Old Red Sandstone sedimentation in the Buchan oilfield, North Sea. *Scott. J. Geol.*, **21**, 227–237.

1985b — A Lower Old Red Sandstone lake in the offshore Orcadian Basin. *Scott. J. Geol.*, **21**, 381–382.

RICHARDSON, J. B. — 1962 — Spores with bifurcate processes from the Middle Old Red Sandstone of Scotland. *Palaeontology*, **5**, 171–194.

1965 — Middle Old Red Sandstone spore assemblages from the Orcadian Basin, north-east Scotland. *Palaeontology*, **7**, 559–605.

1967 — Some British Lower Devonian spore assemblages and their stratigraphical significance. *Rev. Palaeobotan. Palynol.*, **1**, 111–129.

RIDGWAY, J. M. — 1974 — Sedimentology and Palaeogeography of the Eday Group, Middle Old Red Sandstone, Orkney. *Ph.D. Thesis, London University* (Unpublished).

ROBERTS, J. L. — 1966 — Ignimbrite eruptions in the volcanic history of the Glencoe cauldron subsidence. *Geol. J.*, **5**, 173–184.

1974 — The evolution of the Glencoe cauldron. *Scott. J. Geol.*, **10**, 269–282.

ROBINSON, M. A. — 1985 — Palaeomagnetism of volcanics and sediments of the Eday Group, Southern Orkney. *Scott. J. Geol.*, **21**, 285–300.

ROBSON, D. A. — 1948 — The Old Red Sandstone volcanic suite of Eastern Forfarshire. *Trans. Edinb. Geol. Soc.*, **14**, 128–140.

ROCK, N. M. S. & RUNDLE, C. C. — 1986 — Lower Devonian age for the 'Great (basal) Conglomerate', Scottish Borders. *Scott. J. Geol.*, **22**, 285–288.

SAXON, J. — 1975 — *The Fossil Fishes of the North of Scotland.* Thurso.

SCOURFIELD, D. V. — 1926 — On a new type of Crustacean from the Old Red Sandstone (Rhynie Chert Bed, Aberdeenshire). *Lepidocaris rhyniensis* gen. et. sp. nov. *Phil. Trans. R. Soc.*, B, **214**, 153–187.

SIMON, J. B. & BLUCK, B. J. — 1982 — Palaeodrainage of the Caledonian mountain chain in the northern British Isles. *Trans. R. Soc. Edinb.*, **73**, 11–15.

SMITH, D. I. — 1963 — Moine–Old Red Sandstone unconformity at Sugar Loaf Island, near Beauly, Inverness. *Bull. Geol. Surv. U.K.*, **20**, 1–5.

1977 — The Great Glen Fault. *In* Gill, G. (Ed.) *The Moray Firth area geological studies.* Inverness Field Club.

SMITH, J. — 1909 — *Upland Fauna of the Old Red Sandstone formation of Carrick, Ayrshire.* Kilwinning.

1910 — *Semi-precious Stones of Carrick.* Kilwinning.

SMITH, T. E. — 1967 — A preliminary study of sandstone sedimentation in the Carboniferous of Berwickshire. *Scott. J. Geol.*, **3**, 282–305.

SMITH-WOODWARD, A. & WHITE, E. I. — 1926 — The fossil fishes of the Old Red Sandstone of the Shetland Isles. *Trans. R. Soc. Edinb.*, **54**, 567–571.

SPEIGHT, J. M. & MITCHELL, J. G. — 1979 — The permo-Carboniferous dyke-swarm of northern Argyll and its bearing on dextral displacement on the Great Glen Fault. *J. geol. Soc. Lond.*, **136**, 1979.

STEPHENSON, D. — 1972 — Middle Old Red Sandstone alluvial fan and talus deposits at Foyers, Inverness-shire. *Scott. J. Geol.*, **8**, 121–127.

STEPHENSON, D.   1977   Intermontane Basin Deposits associated with an early Great
                                          Glen feature in the Old Red Sandstone of Inverness-shire. *In*
                                          Gill, G. (Ed.) *The Moray Firth area geological Studies.* Inverness
                                          Field Club.

STOKER, M. S.   1982   Old Red Sandstone sedimentation and deformation in the Great
                                          Glen Fault Zone, NW of Loch Linnhe. *Scot. J. Geol.*, **18**, 147–
                                          156.

STORHAUG, K. &   1985   Palaeomagnetism of the Sarclet Sandstone (Orcadian Basin), age
    STORETVEDT, K. M.   perspectives. *Scott. J. Geol.*, **21**, 275–284.

STORETVEDT, K. M. &   1985   Geological interpretation of palaeomagnetic results from
    MELAND, R. H.   Devonian rocks of Hoy, Orkney. *Scott. J. Geol.*, **21**, 337–352.

SWEET, I. P.   1985   Sedimentology of the Lower Old Red Sandstone near new
                                          Aberdour, Grampian Region. *Scott. J. Geol.*, **21**, 239–259.

TARLING, D. H.   1985   Palaeomagnetic studies of the Orcadian Basin. *Scott. J. Geol.*,
                                          21, 261–273.

TARLO, L. B. H.   1961   Psammosteids from the Middle and Upper Devonian of
                                          Scotland. *Q. J. geol. Soc. Lond.*, **117**, 193–1211.

TARLO, L. B. H. & GURR, P. R.   1964   The Lower Old Red Sandstone of the Lorne Couvette. *Advan.
                                          Sci.*, **20** (87), 446.

TAYLOR, D. M.   1972   The geochemistry and petrology of the andesites and associated
                                          igneous rocks of the area of the Ochil Hills, near Dunning,
                                          Perthshire. *Ph.D. Thesis, University of Nottingham* (Unpub-
                                          lished).

THIRLWALL, M. F.   1979   The Petrochemistry of the British Old Red Sandstone volcanic
                                          province. *Ph.D. Thesis, Edinburgh* (Unpublished).

                                   1981   Implications for Caledonian plate tectonics models of geo-
                                          chemical data from volcanic rocks of the British Old Red
                                          Sandstone. *J. geol. Soc. Lond.*, **138**, 123–138.

TILLYARD, R. J.   1928   Some remarks on the Devonian fossil insects from the Rhynie
                                          Chert beds, Old Red Sandstone. *Trans. ent. Soc. Lond.*, **76**, 65–
                                          71.

TRAQUAIR, R. H.   1890   On the fossil fishes found at Achanarras Quarry, Caithness. *Ann.
                                          Mag. Nat. Hist.*, **6**, 479–486.

                                   1894   On *Psammosteus taylori*, a new fossil fish from the Upper Old
                                          Red Sandstone of Morayshire. *Ann. Scot. nat. Hist.*, **3**, 225–226.

                                   1896   The extinct vertebrate fauna of the Moray Firth area. *In* Harvey-
                                          Brown, H. H. & Buckley, T. E. (Eds.) *Vertebrate Fauna of the
                                          Moray Firth.* Vol. 2, Edinburgh.

                                   1897   Additional notes on the fossil fishes of the Upper Old Red
                                          Sandstone of the Moray Firth area. *Proc. R. phys. Soc. Edinb.*,
                                          13, 376–385.

                                   1905   On the Fauna of the Upper Old Red Sandstone of the Moray
                                          Firth area. *Rep. Brit. Ass. Cambridge*, **1904**, 547.

                                   1908   On Fossil Fish Remains from the Old Red Sandstone of
                                          Shetland. *Trans. R. Soc. Edinb.*, **46**, 221–239.

TREWIN, N. H.   1976   Correlation of the Achanarras and Sandwick Fish Beds, Middle
                                          Old Red Sandstone, Scotland. *Scott. J. Geol.*, **12**, 205–208.

                                   1976   *Isopodichnus* in a trace fossil assemblage from the Old Red
                                          Sandstone. *Lethaia*, **9**, 29–37.

                                   1986   Palaeoecology and sedimentology of the Achanarras Fish Bed in
                                          the Middle Old Red Sandstone, Scotland. *Trans. R. Soc. Edinb.*,
                                          77, 21–46.

TYRRELL, G. W.   1914   A petrographic sketch of the Carrick Hills, Ayrshire. *Trans. geol.
                                          Soc. Glasg.*, **15**, 64–83.

VAN DER VOO, R. &   1981   Palaeomagnetic evidence for a large (c 2000 km) sinistral offset
    SCOTESE, C.   along the Great Glen Fault during Carboniferous times.
                                          Geology, **9**, 583–584.

WATERSTON, C. D.   1962   A new Upper Old Red Sandstone Fish locality in Scotland. *Nature, Lond.*, **196**, 263.

                           1965   The Old Red Sandstone. *In* Craig, G. Y. (Ed.) *The Geology of Scotland*. Edinburgh, pp. 270–310.

WATSON, D. M. S.   1908   Note on the occurrence of *Coccosteus minor*. *In* Miller, Hugh, The Old Red Sandstone of Dalcross, Inverness-shire. *Geol. Mag.*, **5**, 431.

                           1932   On three new species of fish from the Old Red Sandstone of Orkney and Shetland. *Mem. geol. Surv. Summ. Prog. for 1931*. Part 2, 157–166.

                           1934   Report on Fossil Fish from Sandness, Shetland. *Mem. geol. Surv. Summ. Prog. for 1933*. Part 1, 74–76.

                           1935   Fossil fishes of the Orcadian Old Red Sandstone. *In* Wilson, G. V. *et al*. The Geology of the Orkneys. *Mem. Geol. Surv. U.K.*

WATSON, D. M. S.,   1948   Guide to Excursions C16. *Internat. Geol. Congress, 18th Session, Great Britain*, London.
  WESTOLL, T. S.,
  WHITE, E. I., &
  TOOMBS, H.

WATT, J. F.   1963   Sedimentary studies of the Old Red Sandstone of Berwickshire and East Lothian. *Ph.D. Thesis, Edinburgh* (Unpublished).

WATTISON, A.   1958   Note on the occurrence of *Holoptychius nobilissimus* Ag. in Dura Den, Fife. *Trans. Edinb. geol. Soc.*, **17**, 179.

WESTOLL, T. S.   1937   Old Red Sandstone fishes of the North of Scotland, particularly of Orkney and Shetland. *Proc. Geol. Ass.*, **48**, 13–45.

                           1945   A new cephalaspid fish from the Downtonian of Scotland, with notes on the structure and classification of ostracoderms. *Trans. R. Soc. Edinb.*, **61**, 341–357.

                           1951   The vertebrate-bearing strata of Scotland. *Rep. XVIII Int. Geol. Congr.*, Part II, Great Britain, 1948, 5–21.

                           1977   Northern Britain. *In* House *et al*. A Correlation of Devonian rocks of the British Isles. *Geol. Soc. Lond. Special Report No. 7*.

WILSON, A. C.   1971   Lower Devonian Sedimentation in the north-west Midland Valley of Scotland. *Univ. Glasgow Ph.D. Thesis* (Unpublished).

                           1980   The Devonian sedimentation and tectonism of a rapidly subsiding semi-arid fluvial basin in the Midland Valley of Scotland. *Scott. J. Geol.*, **16**, 291–313.

WILSON, G. V.,   1935   The Geology of the Orkneys. *Mem. geol. Surv. U.K.*
  EDWARDS, W., KNOX, J.,
  JONES, R. C. B., &
  STEPHENS, J. V.

WILSON, I. S. G. &   1890   Explanation of Sheet 76 – Central Aberdeenshire. *Mem. geol. Surv. Scotland*.
  HINXMAN, L. W.

Addendum

Wally Mykura was tragically killed shortly after he had checked the first galley proofs of Chapter 9. Since then a number of new publications relevant to the Old Red Sandstone of Scotland have appeared. I am much indebted to *Matt Armstrong* for checking the page proofs of Chapter 9 and adding the following notes and references.—*Editor*

Notes on the Old Red Sandstone

The relations of the Lower Old Red Sandstone and the Dalradian terrane

At the time of the writing of Chapter 9 it was considered (p. 297) that the contrast between clasts in the Lower Devonian rocks south of the Highland Boundary Fault Zone and the lithologies of Dalradian rocks north of the fault zone was such as to indicate that these two areas must have been widely separated during the Lower Devonian period. Juxtaposition of the two areas by large-scale movements at the end of the Upper Devonian (Bluck 1983, 1984) was accordingly postulated, although the existence of apparently identical ignimbrites within the Lower Devonian sequences at Lintrathen and Glenbervie north and south respectively of the Highland Boundary Fault Zone (p. 306), was inconsistent with this hypothesis. The matter has excited considerable interest, and many data relating to clast petrography, geochemistry and radiometric dating are now available.

It has been shown (Haughton *et al.* 1990) that certain granitoid clasts of presumed northerly origin from the Crawton Group of Kincardineshire are chemically similar to, and share a common magmatic history with intrusions of Ordovician and Silurian age within the Dalradian rocks in part of north-east Scotland. This suggests that the Lower Devonian strata of Strathmore and the Dalradian rocks north of the Highland Border were in essentially their present relative positions before Lower Devonian times. In confirmation of this conclusion Haughton *et al.* (1990) have indicated that a proportion of the clasts in the Lower Devonian conglomerates resemble Dalradian rocks, and further has shown that the spatial variation in the composition of garnets within the Dalradian rocks is reflected by spatial variation in the composition of garnets contained in northerly derived Lower Devonian sediments south of the Highland Border.

Comparison of Apparent Polar Wander Paths, constructed on the basis of new and existing palaeomagnetic data relating to the Grampian Highlands, the Midland Valley and the Southern Uplands, indicates that these three terranes behaved as a single entity from Siluro–Devonian times (Trench *et al.* 1989). Finally the reported consistency of palaeolatitudes relating to the ignimbrites at Lintrathen and Glenbervie, together with similarities in the component structure of the rocks (Trench & Haughton 1988) support the correlation of these bodies. Relative translation or shortening across the Highland Border subsequent to the eruption of the ignimbrite cannot therefore exceed the former extent of this deposit.

Although these impressive new findings cannot be said to rule out some relative movement along the Highland Border after Lower Devonian times, the evidence does indicate that there has been no substantial post-Lower Devonian net displacement.

Age of the Lower Old Red Sandstone

Conflict between the apparent Silurian K–Ar ages of igneous rocks in the Lower Old Red Sandstone and the Lower Devonian ages indicated by associated fish-faunas and miospores (pp. 297–299, 304) hinges entirely on the assumed radiometric age for the base of the Devonian. Richardson *et al.* (1984), showing that miospore assemblages low in the Arbuthnott Group are of Devonian affinities, have resolved the problem with the suggestion that the base of the Devonian has a radiometric age of 410–415 Ma.

Matt Armstrong

REFERENCES TO NOTES TO CHAPTER 9

ARMSTRONG, M, PATERSON, I. B. & BROWNE, M. A. E.   1985   Geology of the Perth and Dundee district. *Mem. Br. Geol. Surv*. (Sheets 48E, 48W, 49).

GREIG, D. C. — 1988 — Geology of the Eyemouth district. *Mem. Br. Geol. Surv.* (Sheet 34).

HAUGHTON, P. D. W., ROGERS, G. & HALLIDAY, A. N. — 1990 — Provenance of Lower Old Red Sandstone conglomerates, S.E. Kincardineshire; evidence for the timing of Caledonian terrane accretion in central Scotland. *Journ. Geol. Soc. Lond.*, **147**, 104–120.

McADAM, A. D. & TULLOCH, W. — 1985 — Geology of the Haddington district. *Mem. Br. Geol. Surv.* (Sheets 33W, 41).

MUNRO, M. — 1986 — Geology of the country around Aberdeen. *Mem. Br. Geol. Surv.* (Sheet 77).

TRENCH, A., DENTITH, M. C., BLUCK, B. J., WATTS, D. R. & FLOYD, J. D. — 1989 — Short Paper. Palaeomagnetic constraints on the geological terrane models of the Scottish Caledonides. *Journ. Geol. Soc. Lond.*, **146**, 1–4.

TRENCH, A. & HAUGHTON, P. D. W. — 1988 — (Abs.). Palaeomagnetic study of the 'Lintrathen porphyry', Highland Border Region, Central Scotland: Terrane linkage of the Grampian and Midland Valley blocks in Late Silurian time? *Geophysical Journ.*, **93**, 554.

10

CARBONIFEROUS

E. H. Francis

The Carboniferous System in Scotland is formed mainly by sequences of sediments laterally varying in thickness and locally intercalated with penecontemporaneous volcanic rocks described in Chapter 11. The sediments are coals, limestones, mudstones, siltstones, sandstones, and fossil soils such as seatearths and seatclays. They are arranged in rhythmic sequences described as cyclothems up to 30 m or more thick but averaging about 10 m. The relative proportions of each rock type within a cyclothem vary from place to place and from one part of the succession to another so that, as demonstrated by Duff & Walton (1962), there is no such thing as an 'ideal' cyclothem. Nevertheless, as shown

graphically in Figure 10.1, the concept forms the basis of Scottish lithostratigraphical division and is thus of continuing practical value.

The cyclothems which are 'complete' point to widespread changes in sedimentary environment with time at any one locality and even incomplete sequences indicate alternation between two or three environments. Although some uncertainty remains as to the mechanics of this kind of sedimentation (Belt 1975), the current consensus view assumes an advance of deltas into lakes and lagoons, combining with fluctuating river and distributary channel positions. Added to these processes of sedimentation were local variations in

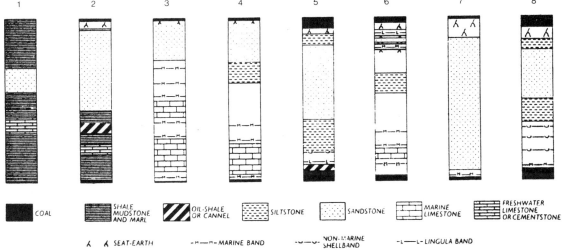

Fig. 10.1. Sections illustrating rhythmic sequences typical of some of the subdivisions of the Scottish Carboniferous (partly after Robertson 1948). 1 = Ballagan facies of Inverclyde Group; 2 = Oil-shale Sub-group; 3 = Lower and Upper Border groups; 4 = Lower Limestone Group and Liddesdale Group; 5 = Limestone Coal Group; 6 = Upper Limestone Group; 7 = Passage Group; 8 = Coal Measures.

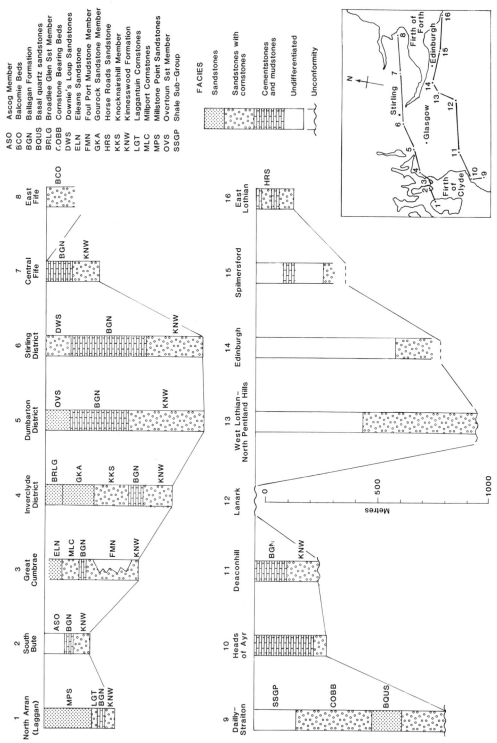

Fig. 10.2. Comparative vertical sections of the Inverclyde Group (partly after Paterson & Hall 1986).

FORMATIONAL NAMES

ASO Ascog Member
BCO Balcomie Beds
BGN Ballagan Formation
BQUS Basal quartz sandstones
BRLG Broadlee Glen Sst Member
COBB Cornstone Bearing Beds
DWS Downie's Loup Sandstones
ELN Eileans Sandstone
FMN Foul Port Mudstone Member
GKA Gourock Sandstone Member
HRS Horse Roads Sandstone
KKS Knocknairshill Member
KNW Kinnesswood Formation
LGT Laggantuin Cornstones
MLC Millport Cornstones
MPS Millstone Point Sandstones
OVS Overtoun Sst Member
SSGP Shale Sub-Group

FACIES

Sandstones

Sandstones with cornstones

Cementstones and mudstones

Undifferentiated

Unconformity

the amount of subsidence together with periodic rises in sea level giving rise to widespread marine incursions. Thus as quantified by Read & Dean (1967), though cycles tend to be thickest in areas of maximum subsidence, the number of cycles also tends to increase in direct proportion to the total amount of subsidence as measured by the thickness of strata between widespread marine bands.

The depth of water must have been relatively shallow even during marine transgressions, but by the end of the Carboniferous upwards of 3·5 km of sediments had accumulated in parts of Fife, the Lothians and Liddesdale. This implies an equivalent amount of subsidence broadly keeping pace with deposition throughout the period, with variations in that amount from place to place controlled by basement structures (pp. 376–381).

The two main areas of sedimentation were the Midland Valley and the Northumberland Basin, or Trough, wholly or partly separated by the Southern Uplands massif. Because of that separation the sequences in the two areas are different. They are therefore described here separately, the small Sanquhar and Thornhill basins being included with the Midland Valley. Described under a third regional heading are the Carboniferous rocks which extend from both areas out to the Scottish Continental Shelf, where they are mainly covered by younger sediments and are known in less detail than their counterparts on shore. The local classifications and correlations with successions elsewhere are discussed further on pp. 373–378.

Midland Valley

In the Midland Valley the subdivisions of the Carboniferous based on MacGregor (1960) for the Silesian and on Paterson & Hall (1986) for the Dinantian is shown relative to the main NW European classification as follows:

Table 10.1. Midland Valley lithostratigraphical divisions

	EUROPE	
	STAGE	SUB-SYSTEM
. Coal Measures	Westphalian	
Passage Group		
Upper Limestone Group	Namurian	Silesian
Limestone Coal Group		
Lower Limestone Group		
Strathclyde Group	Viséan	Dinantian
Inverclyde Group	Tournaisian	

Inverclyde Group

The definition of the base of the Carboniferous in the Midland Valley has always presented problems. In the absence of diagnostic fossils classification has depended on carbonate lithologies, with Upper Old Red Sandstone

facies assumed to have been characterised by nodular cornstones (pedogenic limestones) and the Calciferous Sandstone Measures of early Carboniferous age characterised by cementstones (argillaceous limestones or dolostones). On this basis, the boundary between the two lithostratigraphical groupings was variously drawn at the topmost cornstone or the lowest cementstone – a practice which has led to considerable discrepancy because in many places the two facies are interdigitated. The solution proposed by Paterson & Hall (1986) is to include both carbonate facies within a single Inverclyde Group (Fig. 10.2).

Taking the base of the Group at the lowest cornstone offers the additional advantage of seeming to offer a better equation with the Carboniferous–Devonian boundary than any hitherto used, especially as the early Tournaisian miospore zones are absent in Scotland (p. 374). The Group succeeds the Stratheden Group of the Upper Old Red Sandstone with conformity in most places, though in West Lothian and the Pentlands, where it reaches its maximum thickness of nearly 1 km, it rests unconformably on Lower Old Red Sandstone volcanics or even Silurian greywackes. The top of the Group is drawn at the base of the Clyde Plateau Volcanic Formation in the type area, at the base of approximately penecontemporaneous volcanics in Stirling and the Lothians, and at equivalent levels in the wholly sedimentary successions of East Fife and eastern East Lothian. In addition to the diachronism inherent in such a boundary, the onset of volcanism coincides with earth movements so that in places the overlying Strathclyde Group rests with unconformity either on some level within the Inverclyde Group or even, as in the Pentlands, directly on Upper Old Red Sandstone.

Although cornstones and cementstones are the dominant facies within the Inverclyde Group, subordinate facies are represented by non-calcareous channel sandstones and nondescript mudstone–siltstone–sandstone alternations. The cornstone facies comprises fining-upwards cycles (Fig. 10.3) with cross-bedded mainly white sandstones at the base passing upwards into reddish siltstones and mudstones (Read & Johnson 1967). The calcareous concretions are most common in the argillaceous beds though they also appear in the sandstones as well as in conglomeratic form; as noted by Burgess (1961) they are comparable with caliche formed in soil profiles during periods of low watertable in semi-arid climates. The sequences are generally described overall as fluviatile deposits of meandering streams having a current direction from the west or northwest (Read & Johnson 1967, Chisholm & Dean 1974).

The cementstone facies comprises sequences of grey, green, and red mudstones interbedded with argillaceous limestones or dolostones and ripple-marked sandstones (Fig. 10.4); they show evidence or aridity such as partings of gypsum, salt pseudomorphs, and desiccation breccias, some of them algal. With their limited fauna of *Spirorbis, Curvirimula, Naiadites*, ostracods and fish remains these

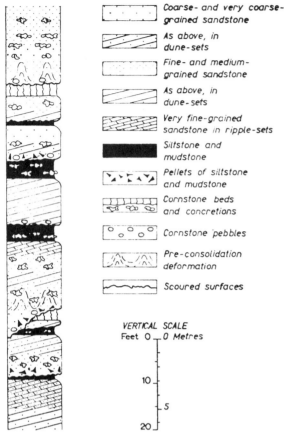

Coarse- and very coarse-
grained sandstone

As above, in
dune-sets

Fine- and medium-
grained sandstone

As above, in
dune-sets

Very fine-grained
sandstone in ripple-sets

Siltstone and
mudstone

Pellets of siltstone
and mudstone

Cornstone beds
and concretions

Cornstone pebbles

Pre-consolidation
deformation

Scoured surfaces

VERTICAL SCALE
Feet 0 __ 0 Metres

Fig. 10.3. Fluvial cyclothems in the Kinness-wood Formation, Gargunnock Burn, Stirling-shire (Read & Johnson 1968, fig. 3).

alternations are believed to be the deposits of isolated shallow-water basins undergoing changes of salinity (Belt *et al.* 1967). The isolation was effected in some places by barriers built by interdigitation with thick sandstones of fluviatile facies (Read & Johnson 1967).

In Fife, the two carbonate facies characteristic of the Inverclyde Group have been formalised in terms of a lower cornstone-bearing Kinneswood Formation (Chisholm & Dean 1974) and an overlying Ballagan Formation (Browne 1980a). These formational names have been extended by Paterson & Hall (1986) to most northern Midland Valley localities, including the Tayside region where their combined thickness is more than 270 m (Browne 1980b). Their absence in East Fife is explained by burial under younger strata. These include the Balcomie Beds correlated with the Downie's Loup Sandstone of Stirlingshire (Browne 1980a) as representing a reversion to a fluviatile pedogenic facies resembling that of the earlier Kinneswood Formation.

In addition to such interdigitation, there is clear indication that one carbonate facies may pass laterally into another (Read & Johnson 1967) or even into less distinctive clastic sequences (Fig. 10.5). Transitions of this kind are most pronounced in the southern part of the Midland Valley and there the Kinneswood–Ballagan nomenclature becomes less applicable with the whole of the group represented either by the pedogenic facies as in the southern Pentlands or, in contrast, by the equally thick but less distinctive sequences overlying the cornstones around and to the east of Edinburgh.

The Craiglockhart Lavas and Tuffs, 60 m thick, immediately above the cornstone facies in the Edinburgh district, are succeeded by some 700 m of siltstones and sandstones with occasional thin beds of carbonaceous shales containing plants, ostracods and fish scales as well as a few thin layers of cementstone and gypsum (Mitchell & Mykura 1962).

In East Lothian, too, the base of the Group is formed by cornstone facies of Kinneswood type, possibly as much as 100 m thick cropping out in several outliers. Volcanic rocks at the same horizon as those of Craiglockhart occur only at Oldhamstocks in the extreme east. North-west of the Dunbar–Gifford Fault the cornstones are overlain by alternations of sandstones, siltstones and shales, some of them carbonaceous in sequences of about 250 m and 350 m in the Spilmersford and East Linton boreholes (McAdam & Tulloch 1985, Davies *et al.* 1986). Both sequences are reddened in the upper part and both include a median band, about 50 m thick, of Ballagan-type cementstones and mudstones, with underlying tuffs. The latter thicken northwards to the coast and are also evident to the east of the Dunbar–Gifford fault where tuffaceous layers occur in Ballagan facies above and below the Horse Roads Sandstone.

Strathclyde Group

In the type area the base of Strathclyde Group is taken at the base of what earlier literature called the Clyde Plateau Lavas, but redesignated Clyde Plateau Volcanic Formation by Monro (1982). This can be identified eastwards as far as the Cleish Hills in north–central Fife to the north of the Forth; south of the Forth it approximates to the bases of other volcanic formations, namely the Torweaving Lavas of West Lothian, the Arthur's Seat Volcanic Formaton of the Edinburgh district and the Garleton Hills Volcanic Formation of East Lothian (Fig. 10.6). The top of the Group is defined with a great deal more precision at the base of the Hurlet Limestone and its equivalents representing the first marine transgression to have extended over the whole of the Midland Valley.

The contrasting volcanic and sedimentary patterns of deposition of the Group can be summarised as follows. The early and middle parts of the sequence are almost entirely volcanic both in the western Midland Valley and in East

Lothian whereas in the intervening area of West Lothian and Edinburgh, where only the basal beds are volcanic, great thicknesses of oil-shale-bearing cyclothems were deposited in a shallow, subsiding sedimentary basin surrounded everywhere except to north-east by the volcanic lands. The north-eastward outlet towards the sea is represented in East Fife by an equally thick group of cyclothems lacking oil-shales, but giving evidence instead of near-shore and deltaic accumulation (Greensmith 1965, Forsyth & Chisholm 1977). The uppermost part of the Group consists of Yoredale-type cyclothems with thin bioclastic limestone representing progressively more marine conditions preceding the more widespread marine transgression of Lower Limestone Group times beginning with deposition of the Hurlet Limestone. It followed the end of both Clyde and Garleton volcanic activity in the west and east respectively, but centrally, in West Lothian and south–central Fife new centres of activity became so established that virtually the whole of the upper part of the Group is there represented by basalt lavas.

Western and central areas

The Clyde Plateau Volcanic Formation forms the bulk of the succession in Machrihanish, Arran and most western and central parts of the Midland Valley.

The main mass of lavas (Chapter 11), consisting of olivine-basalts with subordinate trachytes, forms a horse-shoe-shaped outcrop extending from Strathaven in the south through the Ayrshire and Renfrewshire uplands before turning north-eastwards through the Kilpatrick and Campsie Hills. They are up to 1 km thick and have been proved by boring to continue eastward, at depth, beneath the Central Coalfield basin to join with the Bathgate Hills volcanic rocks which are mainly younger (Chapter 11). The lavas thus formed a barrier between the area in the west where the Lawmuir Formation (Monro 1982) was being laid down and the oil-shale basin of the Lothians farther east (Figs. 10.7, 11.16).

North and north-west of Ardrossan, there is no Lawmuir Formation above the lavas, but at Beith and Barrmill there

Fig. 10.4. Cementstones and mudstones ('marls') of Ballagan Formation, Ballagan Burn, Campsie Hills. Crown copyright photograph.

are about 45 m or these beds, including three marine limestones, resting on 15 m of volcanic detritus mantling the lavas. The limestones, named from the base upwards, Dykebar, Hollybush and Blackbyre (Wilson 1989) are a condensed equivalent of the limestone succession of the Glasgow district (Fig. 10.7). There the formation reaches its maximum thickness of 375 m near Paisley where the lower part includes some beds of volcanic detritus as well as the remarkable Quarrelton Thick Coal. This seam is locally as much as 15 m thick, and in one place even this thickness is doubled at the margin of a 'washout'. The succession does not include any oil-shales like those that were deposited farther east, but the Dykebar Marls, 45 m thick, represent a rock type common to both areas. They were probably derived from the finer elements of weathered lavas, such as boles, and may have been partly aeolian.

Along the northern outcrop, at Campsie, the lower part of the Lawmuir Formation is formed by the thick Craig-maddie Sandstone with a basal quartz-conglomerate. The upper part of the Group includes a more varied succession with thin coals, shales and limestones. Most of the shales and limestones are marine and are grouped as Craigenglen Beds and Balgrochan Beds. The highest limestone, however, the Baldernock White, is a freshwater bed rich in the ostracod *Carbonita fabulina*. North-eastwards towards Corrie and Bannockburn the formation becomes thinner

and all the sandstones are replaced largely by volcanic detritus of the Kirkwood Formation.

In the outlying Douglas Basin in the southern part of the Midland Valley the Clyde Plateau lavas are absent and the group is represented entirely by sediments which give evidence of approach to a Southern Uplands shoreline. Where the upper measures are correlatable with the Lawmuir Formation of the Central Basin, each limestone and marine band extends slightly farther than its predecessor. In addition penecontemporaneous movement along the Kennox Fault has caused such local attenuation that the combined Inverclyde and Strathclyde groups are represented by no more than 75 m of sandstones with conglomeratic beds, including one at the base which rests with angular unconformity on Lower Old Red Sandstone (Lumdsen 1967). Even greater attenuation is seen in the Sanquhar Basin, where some 10 m of marine sediments are the only representatives of the whole of the Lower Carboniferous.

West Lothian, Midlothian and South Fife

The Strathclyde Group attains maximum thickness in the formerly mined oil-shale fields of the Lothians and Fife, where up to 2·25 km of sediments overlie basal volcanics. These sediments have been traditionally subdivided at the

Fig. 10.5. Generalised sections along the northern limb of the Muirkirk syncline showing facies change in the Inverclyde Group (Davies 1972, fig. 1).

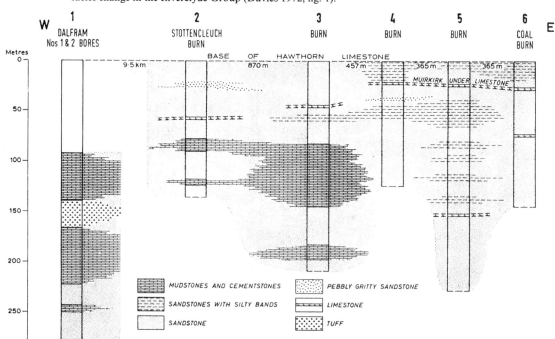

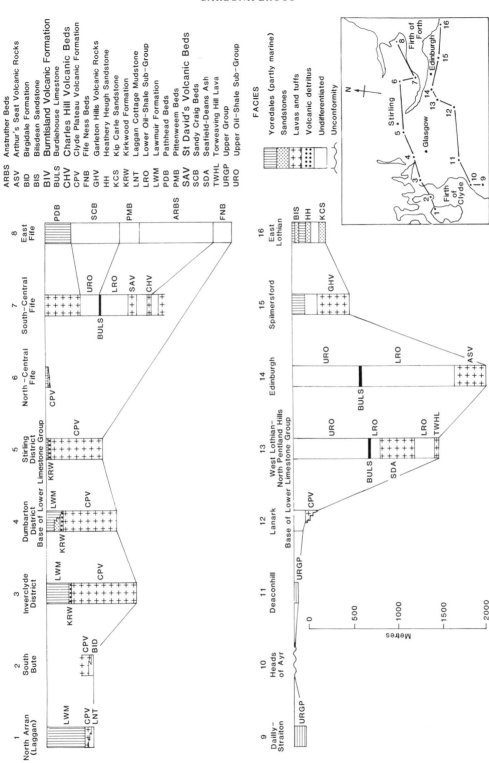

Fig. 10.6. Comparative vertical sections of the Strathclyde Group (partly after Paterson & Hall 1986).

FORMATIONAL NAMES

ARBS	Anstruther Beds
ASV	Arthur's Seat Volcanic Rocks
BID	Birgidale Formation
BIS	Bilsdean Sandstone
BIV	Burntisland Volcanic Formation
BULS	Burdiehouse Limestone
CHV	Charles Hill Volcanic Beds
CPV	Clyde Plateau Volcanic Formation
FNB	Fife Ness Beds
GHV	Garleton Hills Volcanic Rocks
HH	Heathery Heugh Sandstone
KCS	Kip Carle Sandstone
KRW	Kirkwood Formation
LNT	Laggan Cottage Mudstone
LRO	Lower Oil–Shale Sub–Group
LWM	Lawmuir Formation
PDB	Pathhead Beds
PMB	Pittenweem Beds
SAV	St David's Volcanic Beds
SCB	Sandy Craig Beds
SDA	Seafield–Deans Ash
TWHL	Torweaving Hill Lava
URGP	Upper Group
URO	Upper Oil–Shale Sub–Group

FACIES

Yoredales (partly marine)

Sandstones

Lavas and tuffs

Volcanic detritus

Undifferentiated

Unconformity

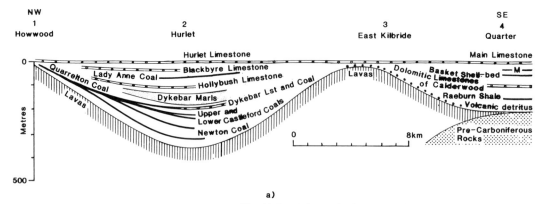

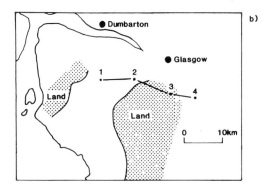

Fig. 10.7a. Relevant horizontal section 1–4 (see Fig. 10.7b).

Fig. 10.7b. Map showing land and sea areas near the close of Strathclyde Group times in North Ayrshire, Renfrewshire and Lanarkshire.

base of the marine Burdiehouse Limestone into Lower and Upper Oil-Shale 'Groups' – now redesignated Sub-groups.

In the Lower Oil-Shale Sub-group the type of rhythmic sedimentation shown in Figure 10.1 replaces that of the Ballagan Formation in such a gradual fashion that it is only towards the top of the Sub-group that workable seams such as the Pumpherston Shale are found. The succession is known in most detail in the Edinburgh district (Mitchell & Mykura 1962, p. 43; see also Fig. 10.8).

The principal faunas are ostracods and fish remains but marine beds are also recorded at six horizons. The lowest, in the shales between the Craigleith and Ravelston sand-stones, is the Granton 'Shrimp-Bed', the various layers of which contain, in addition to shrimp-like crustaceans, ostracods, bivalves, gastropods, nautiloid fragments and fish remains – all representing brief marine incursions into a stagnant lagoon (Briggs & Clarkson 1983, Cater 1987). Four other, less distinctive marine bands occur within the Wardie Shales and the sixth, the Pumpherston Shell-Bed, above the shales. There is a significant change in the non-marine bivalve faunas above the Pumpherston Shell-Bed where *Naiadites obesus* is replaced by *Curvirimula sp.* (Wilson 1974).

Most of this sequence is difficult to correlate with precision across the Forth into South Fife where, between

Rosyth and Burntisland the sediments are intercalated with three volcanic units, the base of the lowest not seen (Browne *et al.* 1987). The Burntisland Marine Band at the top of the sequence, however, can be equated with the Pumpherston Shell-Bed (Francis 1961). Moreover, between that marine horizon and the Burdiehouse Limestone a brecciated dolostone, representing a previously laminated stromatolitic mudflat deposit, provides a distinctive non-marine marker bed identified in three separate localities in Fife and West Lothian (Maddox & Andrews 1987).

The main outcrops of the Upper Oil-Shale Sub-group are in West Lothian. A maximum thickness of 850 m is recorded at West Calder, decreasing northwards to half that amount at Blackness on the Forth shore, and to some uncertain figure southwards. It is in this group that oil-shales of workable thickness and quality are most abundant (Cameron & McAdam 1978): their distribution within the succession is shown in Figure 10.9. Sandstones are generally thinner than in the Lower Oil-Shale Sub-group and unlike the latter this upper succession includes several beds of red 'marls' which are comparable with beds like the Dykebar Marl of the Glasgow area and are similarly derived, in all probability, from deeply laterised volcanic rocks. See p. 392 for reference to unique fossil fauna.

One source of these might be the Clyde Plateau Volcanic

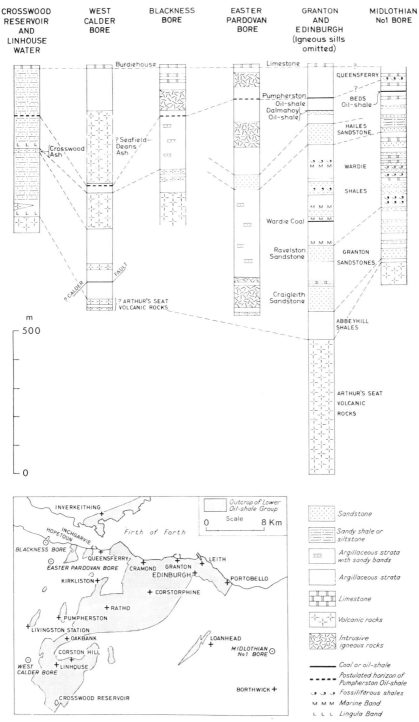

Fig. 10.8. Outcrops, successions and suggested correlations of the Lower Oil-Shale Sub-group (Mitchell & Mykura 1962, fig. 5).

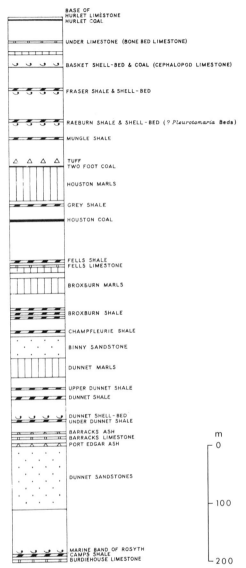

Fig. 10.9. Upper Oil-Shale Sub-group; generalised vertical section in West Lothian (Mitchell & Mykura 1962, fig. 16).

until at Seafield Colliery near Kirkcaldy more than 425 m of lavas, their base not proved, underlie the Lower Limestone Group (Francis 1961).

East Lothian

Where the Strathclyde Group reappears beyond the Midlothian Coalfield in East Lothian the lowest 600 m or so comprises the Garleton Hills Volcanic Formation which extends northwards from the Garleton Hills to the Forth and consists of trachytes, basalts and tuffs. Between the top of the volcanic rocks and the base of the Lower Limestone Group the Spilmersford Borehole proved a 265 m sequence including thick sandstones, thin coals and several marine bands. The lowest is the Dump Marine Band which is correlated (Cater 1987) with the 'Shrimp-beds' of Granton and of Cheese Bay, near Gullane, 15 m north of the borehole. The Cheese Bay faunas (Briggs & Clarkson 1985) occur within a 12 cm laminated bed interpreted by Hesselbro & Trewin (1984) as the deposits of a thermally stratified freshwater lake or brackish lagoon in which drainage was periodically interrupted by volcanism.

Higher marine horizons in the Spilmersford Borehole include the Macgregor Marine Bands equated by Wilson (1974) with the Pumpherston Shell-Bed to the west and the Cove Marine Beds to the east. They are overlain successively by the Bilsdean Sandstone and the Lower and Middle Longcraig limestones at the top of the Strathclyde Group.

South-east of the Lammermuir Fault the Garleton Hills volcanics are unrepresented except, perhaps, by the poorly exposed tuff and basalt at Fluke Dub, just east of Dunbar. Elsewhere the sedimentary sequence between the base of the Group (i.e. the top of the cementstone facies) and the Cove Marine Bands comprises two main divisions. The lowest, 90 m thick, is mainly carbonaceous, with thin impersistent coals; the upper is represented by the red 50 m Heathery Heugh Sandstone (Davies et al. 1986). The Kip Carle Sandstone of the coast section (Scott et al. 1984) can be equated with the base of the carbonaceous division.

North and East Fife

In northern Fife the Strathclyde Group is at its most attenuated. In the Cleish Hills a thin lava and tuff represent the 'feather edge' of the Clyde Plateau Volcanic Formation. Above this is a little-known succession of mainly arenaceous rocks, at least 226 m thick and much interrupted by intrusion; it is probably equivalent to the Lower and Upper Oil-Shale Sub-groups. There is further thinning eastward and north eastward until at Bishop and Lomond Hills a total thickness of no more than 20 to 30 m of partly marine strata (MacGregor 1968) probably represents only the topmost part of the group resting with non-sequence on the Kinnesswood Formation.

South-eastwards from there the succession thickens rapidly, measuring over 2 km in coastal exposures between St Monance and St Andrews where it consists dominantly

Formation, but more likely sources are to be found in penecontemporaneous volcanics. In West Lothian these include the Seafield–Deans Ash in the lower part of the Sub-group and the basalt lavas of the Bathgate Hills in the upper. In Fife too, above a once-mined extension of the Lothians oil-shale succession, the upper part of the Sub-group consists of olivine basalt lavas – the Burntisland Volcanic Formation. These take the place of successively lower sedimentary horizons towards the north and east, of sandstones up to 36 m thick with subordinate siltstones, thin dolomitic limestones, coals, seatearths and ironstones (Forsyth & Chisholm 1977). Environments of deposition probably ranged from fluviatile to deltaic (Greensmith 1961, 1965) with upwards-coarsening units representing the fillings of open stretches of water and the less-common upwards-fining units lying in either channel-shaped erosional hollows or planar erosional surfaces (Forsyth & Chisholm 1977).

This thick sequence is laterally equivalent to the Oil-Shale Sub-groups of the Lothians and many of the shales are carbonaceous, though few are true oil-shales in a commercial sense. As in the Lothians ostracods and fish faunas are commonly found in these shales and in associated non-marine limestones, but there are also several *Naiadites* beds (Bennison 1960, 1961) and a few marine bands, including the 'Encrinite Bed' equated with the Macgregor Marine bands and with the Cove Marine Band of East Lothian (Wilson 1974). Trace fossils occur throughout the sequence and give some indication as to the environment when the related body-fossils are missing (Chisholm 1968, 1970a, 1970b). As in other parts of the Midland Valley, marine sediments, such as the Upper and Lower Ardross limestones are more numerous in the upper part of the succession, where they are designated Pathhead Beds.

Lower Limestone Group

The Lower Limestone Group comprises, from base upwards, the strata from the bottom of the Hurlet to the top of the Top Hosie limestones of the Glasgow District and their equivalents elsewhere. The Hosie limestones contain prolific *Posidonia corrugata* and although the Top Hosie is, in places, impersistent, it is seldom difficult to define the top of the Group throughout the Midland Valley. The correlation of the Hurlet Limestone at the base is also now assured on palaeontological grounds after years of controversy caused, in part, by the number of very similar marine limestones of various thicknesses both below and above it. In particular, the distribution of two species of the algal genus *Calcifoleum* has enabled Burgess (1965) to distinguish between the Hurlet and overlying Blackhall Limestone, whilst the fauna of the Neilson Shell Bed, above the Blackhall and its equivalents include forms now recognised to be diagnostic and listed by Wilson (1966, 1974, 1979, 1989) as follows: *Microcyathus cyclostomus, Tornquistia youngi; Glabrocingulum atomarium, Straparollus (Euomaphalus) carbonarius, Tropidocyclus oldhami, Euchondria neilsoni, Pernopecten fragilis, Posidonia corrugata gigantea.*

The Group has a marine lithology which is distinguishable from the dominantly non-marine facies of the measures below and above, though marine bands (some of them with limestones) at the top of the Calciferous Sandstone Measures, suggest a transitional rather than abrupt change. The Group comprises several rhythmic sequences of Yoredale type each of which ideally includes a limestone near the base of shales containing marine shells and overlying a coal (Fig. 10.10). The coals are usually thin, though there are exceptions which have been exploited. They include the Wilsontown Main near the base of the succession at the southern end of the Central Basin, the Lillie's Shale Coal of Paisley, the various Radernie and Largoward coals of East Fife and the North Greens and Vexhim coals of Midlothian.

Isopachytes of the Group (Fig. 10.27) show the establishment of a pattern of sedimentation which was to continue throughout the deposition of the Limestone Coal Group and Upper Limestone Group. This pattern comprised a shelf area in Ayrshire marked by the least sedimentation, and several depositional basins including the large, complex synclinal area now forming the Central Coalfield (referred to throughout the rest of this chapter as the Central Basin), the large Fife–Midlothian Basin, and the small outlying basin of Douglas. In Ayrshire the Group is nowhere much more than 50 m thick and abrupt variations across lines of major north-easterly disturbance testify to contemporaneous movement (pp. 376–381). Local fault movement is also discernible in the sedimentary facies in East Fife (Fielding *et al.* 1988). In central and other eastern parts of the Midland Valley, however, thickness variations reflect contemporaneous downfolding which reached a maximum of 230 m in western Midlothian. Goodlet (1957) showed that these variations in thickness are matched by lateral variation in lithology, the ratio of sandstone to shale and limestone being greatest in the areas of greatest thickness.

The three main depositional regions were partially separated by areas of minimum subsidence which corresponded either to earlier or contemporaneous volcanism. Ayrshire and the Central Basin, for instance, were almost completely separated by an area of non-deposition (Figs. 10.7, 10.10) over a great thickness of earlier Clyde Plateau lavas. Similarly between the Central and Fife–Midlothian basins volcanism remained active into, and in places even throughout, the whole of Lower Limestone Group times. It greatly influenced the palaeoecology of local plant deposits (Rex & Scott 1987) and also provided a source of volcanic detritus which became inter-digitated with normal marine beds farther north in Fife.

Limestone Coal Group

The Limestone Coal Group extends from the top of the Top Hosie Limestone to the base of the Index Limestone and consists mainly of deltaic sandstones (fine-grained sheets and coarser transgressive channel-fills), siltstones and shales with many economically valuable coal seams. Limestones are unknown apart from rare thin freshwater varieties, but two important marine bands can be traced all over the Midland Valley. The lower is the Johnstone Shell-Bed about one-third the way up from the base of the Group, while the upper, known as the Black Metals, occurs midway

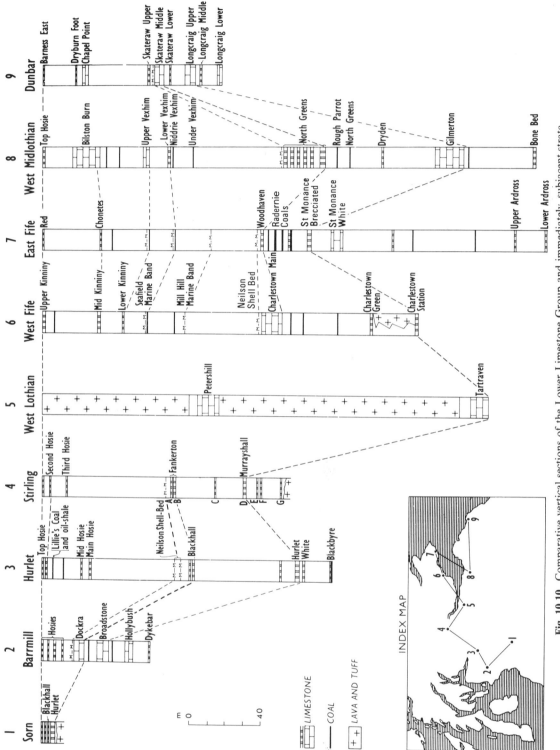

Fig. 10.10. Comparative vertical sections of the Lower Limestone Group and immediately subjacent strata.

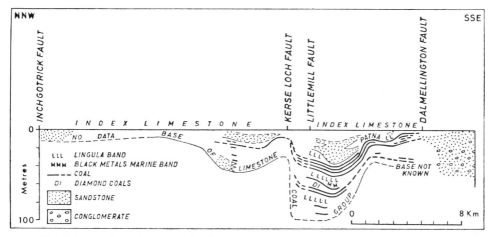

Fig. 10.11. Horizontal section showing variation in thickness and lithology within the Limestone Coal Group of SW Ayrshire (Mykura 1967, fig. 2b).

in the sequence. In attenuated successions both bands may be represented by *Lingula* alone, but in most areas they include more distinctive marine forms. The dominant forms of the Johnstone Shell-Bed are listed by Wilson (1967) as: *Campylites carbonarius*, '*Camarotoechia*' cf. *pleurodon*, *Palaeonilio luciniformis*, *P. mansoni* and *Streblopteria ornata*. He notes a less rich and less varied fauna in the Black Metals, where *Buxtonia* becomes more abundant than *Productus*.

In areas of maximum thickness the marine bands consist of up to 40 m of shales split by arenaceous beds into two or three leaves, each underlain by a thin coal or rooty under-bed and having a faunal sequence which passes up from marine through *Lingula* to non-marine and which may be repeated several times in one thick bed of shale. They represent eustatic marine transgressions modified by local subsidence and sedimentation, probably including the progressive building out and abandonment of local deltas (Read 1965, Read & Merriam 1966).

Apart from the two marine bands there are several

Lingula bands of proved persistence. None is distinctive on faunal grounds (Graham 1970) and so their value in correlation depends upon their frequency and spacing within local successions. The band beneath the Hartley Coal of Stirlingshire for instance, is useful since it is practically the only one in that part of the succession as it is traced eastwards to Fife and Midlothian. It is less useful farther west, for *Lingula* has now been recorded in the Glasgow district from no less than fifteen separate horizons between the Black Metals and the Index Lime-stone (Forsyth 1979). Non-marine bivalves ('mussels') such as *Curvirimula* and *Naiadites* occur in shales either alone, or overlying *Lingula*, or even in association with ostracods in cannels or cannelly shales. They do not, however, lend themselves to zonal work and are thus much less useful than their counterparts in the Coal Measures.

Isopachytes (Fig. 10.27) show that the pattern of thick-ness variation within the Group clearly follows that of the Lower Limestone Group, with marked variation over the contemporaneously active north-easterly faults (Fig. 10.11)

Fig. 10.12. Horizontal section showing variations in thicknesses of the Limestone Coal Group across Fife.

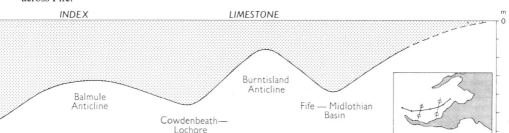

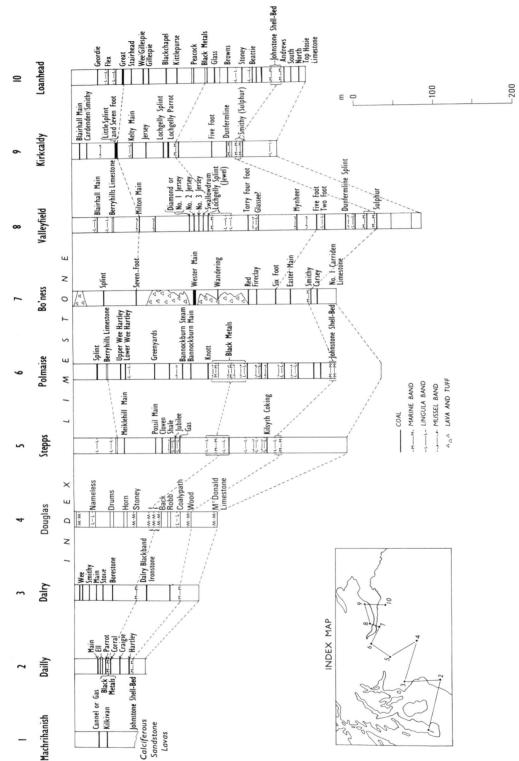

Fig. 10.13. Comparative vertical sections of the Limestone Coal Group.

and with maximum accumulation in the central parts of the main depositional troughs reaching as much as 550 m on the Fife–Clackmannan border in the northern part of the Central Basin. In the Limestone Coal Group, however, additional evidence shows that in the complex structural 'high' between the Central and Fife–Midlothian troughs, there are smaller north-easterly structures (the Lochore Syncline with 370 m of sediments may be mentioned especially) which were exerting similar effects on sedimentation (Fig. 10.12). The Group ranges in thickness from 60 to 170 m across a similar north-easterly axial 'high' in the Douglas basin and is virtually absent south of the penecontemporaneously active Kennox Faults (Lumsden 1964).

The main features of lateral variation within the Limestone Coal Group are shown in Figure 10.13. The lower part of the Group appears to remain constant over wide areas, reflecting stable conditions during deposition, but above the Black Metals these conditions continued only between Glasgow and Stirling (Read & Forsyth 1962).

Elsewhere abrupt lateral changes in lithology often make correlation difficult over quite small areas (Read 1961) – a problem that is the more acute because it is here that the richest seams are consistently found (Fig.10.14).

Towards the margins of the Midland Valley the Limestone Coal Group shows signs of attenuation. It is 77 m thick including two worked seams at Machrihanish (Manson 1953). At Sanquhar it has not been differentiated within the 30 m of sediments, presumed to incorporate at least two unconformities or non-sequences, and assigned to the Namurian (Davies 1970). In Arran the group is similarly undefined within a 'Carboniferous Limestone Series' 137–230 m thick (Macgregor 1965) lacking any economically workable coal.

Volcanism in the Limestone Coal Group is restricted to the Bo'ness district of West Lothian where earlier lava outpourings continued, and to localities in North Ayrshire and Fife where new, but relatively short-lived activity led to the formation of ash-rings (Chapter 11).

Fig. 10.14. Ribbon section showing seam-splitting in the Limestone Coal Group of the Muirkirk district (Davies 1972, fig. 3).

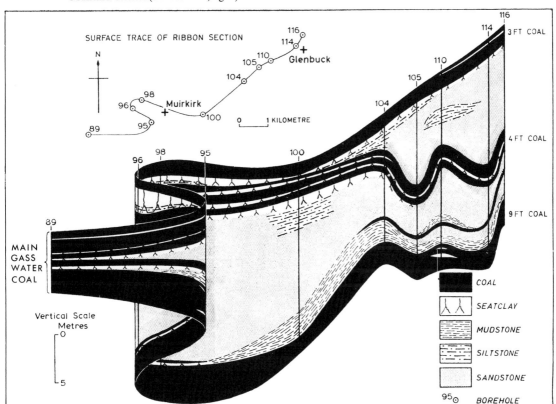

Upper Limestone Group

The Upper Limestone Group, which extends from the base of the Index Limestone to the top of the Castlecary Limestone, represents a reversion to the Yoredale type of rhythmic sequence, though the limestones are generally thinner and the marine faunas poorer than those of the Lower Limestone Group, particularly with regard to corals and bryozoans. Coals are usually thin, though some seams are thick enough to work, the most notable at present being the Upper Hirst, or Jenny Pate Coal of west Fife and Stirlingshire.

Correlation of the principal named limestones is precise in the Central, Fife and Midlothian basins (Fig. 10.15) though this is made possible more by lithological comparison of numerous boreholes than by recognisable faunas from specific horizons. These named limestones are, in ascending order, the Index, Lyoncross, Orchard, Calmy, Plean (1, 2 and 3) and Castlecary. In Ayrshire correlation with this standard sequence is made less certain by the greater proportion of limestone within the cyclothems – a factor indicative of the increasingly marine conditions as the group is traced from east to west. In east Fife and Midlothian by contrast, the Index and Orchard limestones are represented only by shales and as there are corresponding changes in the faunas (Wilson 1967) the influence of the Burntisland Anticline on sedimentation can be inferred.

In addition to the named horizons many other beds of shale, occasionally with thin limestone ribs and containing *Lingula* and marine bivalves, have been recorded. They are members of relatively thin rhythmic sequences interspersed within the framework of the thicker cyclothems. The number of these minor rhythms increases with the thickness of the group as a whole (Fig. 10.15). Many of the minor marine bands pass laterally first into *Lingula* bands, then into sandy beds apparently disturbed by the workings of marine organisms (cf. Goodlet 1959) before they wedge out altogether – in company with the other units of the same minor rhythm.

Gigantoproductus latissimus has long been regarded as characteristic of the Index Limestone, but as it is rare in Central Fife and Midlothian, Wilson (1967) suggests that *Meekospira sp.* is more diagnostic of that horizon. He notes the fauna of the Orchard Limestone to be the richest and most varied in the Scottish Namurian, naming *Antiquatonia costata* as diagnostic, with *Straparollus (Euomphalus) carbonarius* reaching its acme. The best-known form associated with the Calmy Limestone is *Edmondia punctatella* which is unusually well preserved in pyrite within a very thin layer of cannelly shale below the limestone over much of the Central Coalfield. As such it was once regarded as a precise marker band, but more diagnostic forms of this horizon (op. cit.) are *Pugnax* cf. *pugnus*, *Sinuatella* cf. *sinuata* and *Actinopteria regularis*.

The pattern of sedimentation within the Upper Limestone Group follows the same lines as those established for the Lower Limestone and Limestone Coal groups, with variation across the Ayrshire faults (Fig. 10.27) and thickening towards the centres of the sedimentary basins, a maximum of nearly 600 m being reached in the northern part of the Central Basin, in Clackmannan. In some of the areas of greatest thickness minor unconformities are recorded at several levels near the top of the Group: in places some of the Plean limestones as well as the Castlecary Limestone itself (Fig. 10.17) are absent owing to erosional episodes which are usually marked by pebbly or conglomeratic layers at the bases of overlying sandstones.

In Upper Limestone Group times volcanic activity further decreased and is represented only by small localised ash cones in Dalry, Ayrshire and in West Fife and by a few pillow lavas associated with hyaloclastites at the top of the Group at Westfield in Central Fife (Chapter 11).

Passage Group

The Passage Group comprises the strata from the top of the Castlecary Limestone to the base of the Coal Measures. The lower limit can be recognised readily in most places, but there is no single satisfactory criterion for identifying the upper. In Ayrshire and Douglas the top can be defined on faunal evidence to coincide reasonably closely to the Namurian–Westphalian boundary, even though the Subcrenatum Band itself has not been recognised. In most other regions the same equation can now be made using miospores (p. 376), but the group has not yet been formally redefined in those terms. The practice proposed by MacGregor (1960) is thus followed here, that is to use local lithologies to define the base of the Coal Measures. In those terms, the Passage Group is a lithological entity – dominantly arenaceous in contrast to the rocks below and above.

The sandstones are mainly massive and medium-grained orthoquartzites and protoquartzites with occasional coarse pebbly layers: colour is usually pale grey or white but there is some reddening. In places thick beds of clayrock occur, some of which appear to be structureless, while others containing roots are clearly seatclays. They are often variegated in red, lilac and yellow colours even where contiguous sandstones are white or grey, and Read (1959) has suggested that this may be due to partial oxidation under conditions of low water table produced by fluctuations of sea level relative to a delta surface. Valuable fireclays are included among the clayrocks (Read & Dean 1978) and are widely sought as refractories (Chapter 16). Marine bands occur within the succession, but are usually thin. Coal seams too are thin or even absent except in the small Westfield Coalfield of Central Fife, where sedimentation appears to have been aberrant (Fig. 10.18).

The Passage Group underlies and crops out as a fringe around the extensive areas of Coal Measures in the Ayrshire, central Fife and Midlothian coalfields and in the smaller outlying basins of Sanquhar, Douglas, and Westfield.

Ayrshire and the west

Discussion of the Passage Group in Ayrshire is complicated by volcanism. North-west of the Kerse Loch Fault the volcanic rocks consist mainly of lavas reaching a maximum thickness of over 150 m at Troon (Fig. 11.10). They are intercalated with two sedimentary sub-groups, the upper of which includes the valuable Ayrshire Bauxitic Clay (Monro *et al.* 1983) derived from the breakdown of the volcanic rocks. The lavas are overlapped by Lower Coal

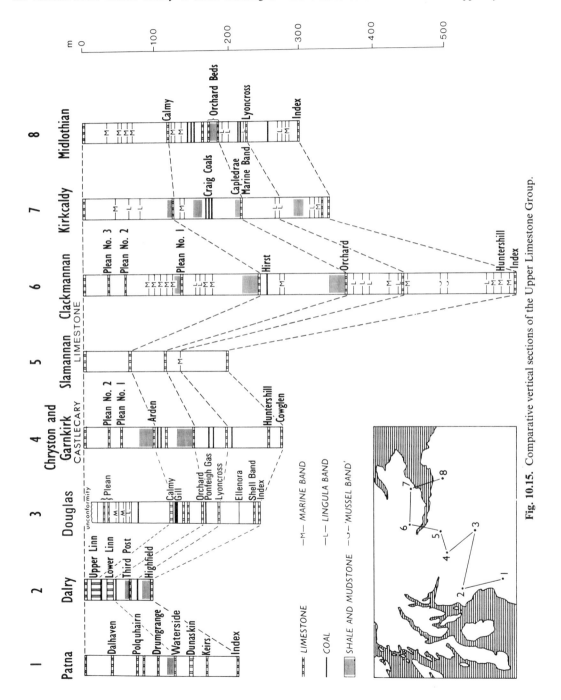

Fig. 10.15. Comparative vertical sections of the Upper Limestone Group.

Measures (Fig 10.21) and rest on Upper Limestone strata with similar angular disconformity (Richey *et al.* 1930). South-east of the Kerse Loch Fault the lavas give way to a sedimentary sequence 128 m thick originally termed 'Millstone Grit' by Eyles *et al.* (1949). However, most of the sequence belongs to the *A. lenisulcata* Zone of the Coal Measures and as the remainder is assigned to the Upper Limestone Group there appears to be no local representative of Passage Group sediments. As noted by Mykura (1967) this can be ascribed as much to correlation as to nonsequence.

The volcanic rocks extend westward into Arran and Kintyre. At Corrie Bay in Arran, Leitch (1941) records 'red ash' capped by bauxitic clay forming the lower part of a Passage Group sequence nearly 60 m thick and including sandstones and marls in the upper part. In the Machrihanish Coalfield there are 140 m of lavas with one marine band just below them and a second band contained in a sedimentary intercalation (Manson 1953).

Sanquhar and Thornhill

In Sanquhar the top of the Group is locally defined by the bottom of Tait's Marine Band, but the base, which rests in places directly on Ordovician, is difficult to date. With only 30 m of sediments to represent the Namurian (Davies 1970) it seems possible that the sandstones forming the upper part of the sequence may be the equivalent of the Passage Group. Similar lithologies extend to Thornhill where 20 m of sandstones overlie clays and shales with marine shells which rest unconformably on Ordovician greywackes in the right bank of the River Nith above Drumlanrig Bridge (Simpson & Richey 1936).

Central Coalfield

A type succession for the Central Coalfield Basin using a system of numbering the marine bands has been elaborated over the years to accommodate an increasing number of bands as they were discovered until now they total at least 17 (Fig. 10.16). In the lower bands fossils are abundant and well preserved and include the so-called 'Nebraskan' fauna which Wilson (1962) has shown to have affinities with Scottish Upper Limestone Group faunas only.

In the higher bands, particularly in Nos. 5 and 6 groups the faunas consist mainly of poorly preserved productoids and *Lingula*. Within the limits of the basin the individual bands or groups of bands can be recognised with remarkable accuracy even though there is no one locality at which all seventeen are present. Their impersistence is due to minor unconformities which affect different horizons at different places and are usually marked by coarse pebbly layers at the bases of overlying sandstones. The three main disconformities are above the Castlecary Limestone, above No. 2 Marine Band and above No. 3 Marine Band Group and the second of these is the probable equivalent of the

Mississippian–Pennsylvanian hiatus (Read 1981). The number of marine bands in the succession is further reduced towards the margins of the basin by attenuation like that affecting some of the bands in the Upper Limestone Group (p. 362).

The Passage Group reaches maximum thickness in the middle of the Central Basin between Airth and Clackmannan where it amounts to about 350 m. A continuous section through this thick succession is exposed along the north shore of the Forth between Kincardine and Culross. At the southern end of the basin there is marked unconformity (Fig. 10.17), the group being reduced to as little as 3 m and overstepping on to the Calmy Limestone.

Still farther south, in the outlying Douglas Basin, the group reappears, having a maximum thickness of *c.* 200 m (Lumsden 1967). The succession is very variable and locally includes two unconformities and several coals, one of which, the Manson Seam, is equated with the Netherwood coal of the Central Coalfield and is up to 2 m thick. At one locality Passage Group rocks overstep some of the older formations to rest on Lower Limestone Group, but they are themselves overstepped at another locality by Lower Coal Measures which rest directly on Plean limestones.

Westfield Basin

The Westfield Basin lies within the complex anticlinal area between the larger Central and Fife–Midlothian basins. Although the outcrop extends over only 3 km² the Passage Group here displays extraordinary variation in thickness and lithology as it is traced over a series of minor north-easterly fold axes (Figs. 10.18, 10.19). Where the Group is thickest it reaches 150 m of which 60 m (i.e. 40 per cent) is coal, concentrated mainly in the lower part of the sequence. This lithology would be unusual even for Coal Measures, but is so at variance with the Passage Group in other parts of Scotland that correlation was long debated. It was eventually only assured by the recognition of full Upper Limestone Group and Lower Coal Measures successions below and above, by the discovery of six marine bands intercalated among the coals (Francis 1961) and by miospores, which suggest that the Namurian–Westphalian boundary lies just above the Bogside Thick Group (Guienn *in* Brand *et al.* 1980). Virtually the whole of the basin has now been excavated by opencast mining.

Fife–Midlothian Basin

A maximum of 335 m of Passage Group strata is recorded off the Fife coast (Ewing & Francis 1960), but in Midlothian, where the Group is sometimes known as the Roslin Sandstone, it does not exceed 245 m in thickness (Tulloch & Walton 1958). On both sides of the Forth, sandstones with grit bands predominate and clayrocks tend to be concentrated in the middle of the succession rather than near top and base as they are in the Central Basin. In

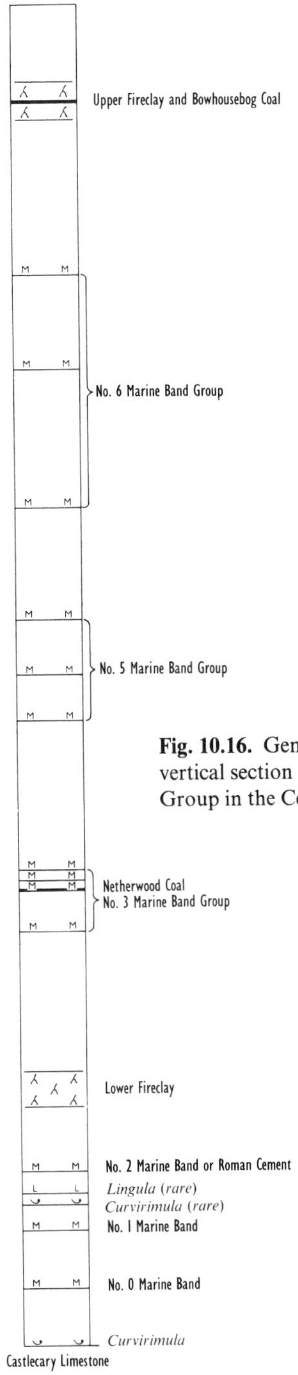

Upper Fireclay and Bowhousebog Coal

No. 6 Marine Band Group

Fig. 10.16. Generalised vertical section of the Passage Group in the Central Basin.

No. 5 Marine Band Group

Netherwood Coal
No. 3 Marine Band Group

Lower Fireclay

No. 2 Marine Band or Roman Cement
Lingula (rare)
Curvirimula (rare)
No. 1 Marine Band

No. 0 Marine Band

Curvirimula

Castlecary Limestone

further contrast there is widespread reddening which, in Fife at least, seems to have its upper limit at a constant horizon near the base of the Coal Measures, diminishing downwards over a depth of 300 m. A total of eight marine bands are recognised in Midlothian and four in Fife, but so far they have not proved to be of any correlative value either locally or with the Central Basin sequence of numbered bands. The best natural exposures of the group are along the Kirkcaldy shore in Fife and along the Joppa shore and in Bilston Burn in Midlothian. At the northern extremity of outcrop the Passage Group is represented by tuffs extending down to within 73 m of the Castlecary Limestone (Knox 1954, Forsyth & Chisholm 1977).

Coal Measures

The Scottish Coal Measures were traditionally subdivided into 'Productive Measures' below Skipsey's Marine Band and 'Barren Red Measures' above. By current practice, however (MacGregor 1960), the 'Barren Red' strata are redesignated Upper Coal Measures, while the 'Productive' beds are further subdivided at the Queenslie Marine Band (Manson 1957) into Lower and Middle Coal Measures.

So far, the Subcrenatum Marine Band which defines the base of the European Westphalian and of the Lower Coal Measures of England and Wales has not been recognised in Scotland. Hence the long-established practice of taking the base of the Coal Measures in each coalfield at an arbitrary horizon near the lowest worked seam. Such horizons include the Crofthead Slatyband Ironstone of the Central Coalfield, the Lower Dysart Coal of East Fife and the Seven Foot Coal of Midlothian. Since being so defined one or more marine or *Lingula* bands have been recognised just below most of those local bases. They can be collectively assigned, together with contiguous strata, to the *A. lenisulcata* Zone – the lowest of the Westphalian chronozones – so that, as implied by MacGregor's (1960) scheme, the Westphalian extends down into the Passage Group. Indeed, using miospores (p. 376) the base of the Westphalian has been located just above No. 6 Marine Band Group of the Central Coalfield sequence. The way is now open, therefore, for general adoption of the practice already established in Ayrshire (Mykura 1967) and in

Fig. 10.17. Horizontal section, north to south, across the Passage Group of the Central Basin.

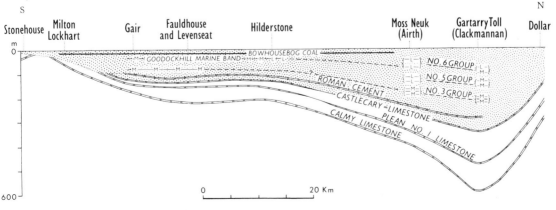

S
Stonehouse Milton Lockhart Gair Fauldhouse and Levenseat Hilderstone Moss Neuk (Airth) Gartarry Toll (Clackmannan) Dollar N

BOWHOUSEBOG COAL
GOODOCKHILL MARINE BAND
NO. 6 GROUP
NO. 5 GROUP
ROMAN CEMENT
NO. 3 GROUP
CASTLECARY LIMESTONE
PLEAN NO. 1 LIMESTONE
CALMY LIMESTONE

0 20 Km

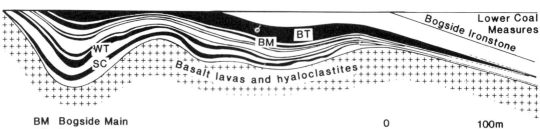

Fig. 10.18. Horizontal section showing lateral variation in the Passage Group at Westfield. Vertical scale *not* exaggerated.

BM Bogside Main
BT Bogside Thick
SC Shale coal
WT Westfield Thick

Douglas (Lumsden 1965) where bivalves diagnostic of the *A. lenisulcata* Zone have been used to redefine the base of the Coal Measures so as to correspond virtually to the base of the Westphalian.

At most localities the coal-bearing sequence follows conformably on Passage Group, but in Ayrshire it is unconformable on Passage Group lavas (Brand 1983) and at the southern margin of the Sanquhar Basin Coal Measures overlap onto Ordovician (Davies 1970). In Arran Lower *A. similis-pulchra* Zone beds overstep beds of *A. modiolaris* and possibly even *C. communis* Zone age to rest directly on the Passage Group (Leitch 1941).

Lower and Middle Coal Measures

The Lower and Middle Measures are mainly grey strata comprising rhythmic sequences of coal, shale, sandstone and seatclay. In central Ayrshire (Eyles *et al.* 1949) the

Fig. 10.19. North face in former Westfield Opencast Site, Fife (*c.* 1980), showing syncline of Coal Measures; bottom, Bogside Thick Coal. By courtesy of British Coal.

proportions of these rock types have been calculated as: coal 6 per cent, argillaceous beds 47 per cent, sandstone 47 per cent. Complete successions may contain as many as 20 coals of workable thickness as well as many others too thin

to have been named. Few of the individual seams can be correlated from one field to another and only some of the many names and local synonyms have been shown in Figure 10.20.

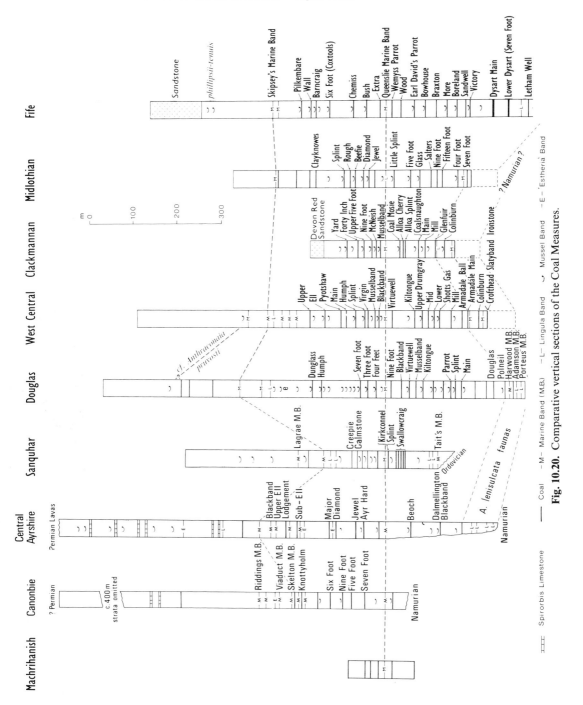

Fig. 10.20. Comparative vertical sections of the Coal Measures.

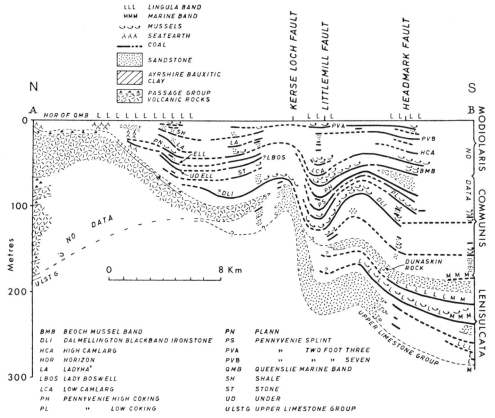

Fig. 10.21. Horizontal section showing variation in thickness and lithology within the Lower
Coal Measures and Passage Group in Ayrshire (Mykura 1967, fig. 7b).

Traced from west to east a gradual, but significant change in lithology takes place. Fossil beds become fewer and more impoverished so that there are, for instance, five marine bands, including Skipsey's, in the Lower *A. similis-pulchra* Zone in Ayrshire (Mykura 1960, 1967) as compared with two in Fife and Midlothian, where moreover, successive workers have commented on the relative impoverishment of the non-marine bivalve faunas (Knox 1954, Tulloch & Walton 1958). This change is not reflected in the distribution of the faunal facies of the only marine band (Queenslie) so far analysed in detail, for it shows a complex pattern apparently unrelated to the total thickness of the cycle and also to the marine portion of that cycle (Brand 1977).

The variation in thickness for the Lower and Middle Coal Measures is shown in Figure 10.20. Evidence of marked attenuation north-westwards is found in Machrihanish where a complete sequence, with faunas representing all the non-marine bivalve zones, is contained within a thickness of only 149 m (Manson 1953) as compared with over 500 m in parts of south-west Ayrshire (Mykura 1967).

The earlier variation across north-easterly faults in Ayrshire continued into the Coal Measures (Fig. 10.21), but in diminished degree. Similarly the thickness variations in the Clackmannan Coalfield (Francis 1956) show the Central Basin to have continued to function as a depositional trough throughout Lower and Middle Coal Measures times as did the northern part of the Fife–Midlothian Basin (Knox 1954), though only a short distance to the west there is an unusual superimposition of a Coal Measure depositional ridge on an older depositional trough.

In most coalfields reddening extends down irregularly into the Middle Coal Measures from some level of unconformity within Upper Coal Measures or beneath New Red Sandstone. The depth of reddening varies and with it the degree of oxidation of normally workable coals (Mykura 1960, Francis *et al.* 1970). An extreme is reached in Arran and in the small Thornhill Basin where attenuated successions of Lower and Middle Coal Measures, recognised by non-marine bivalve faunas, are reddened throughout and are completely devoid of coal.

Beneath the Firth of Forth the whole of the Lower Coal

Measures is represented by tuffs with a single flow of basalt lava (Fig. 11.11).

Upper Coal Measures

The main rock types of the Upper Coal Measures are red sandstones with subordinate structureless, variegated clayrocks which have sometimes been called 'fireclay' or 'marl'. Thin limestones are also recorded, some containing *Spirorbis* being primary, while others formed by oxidation of coal seams are secondary (Mykura 1960). As in the lower parts of the succession the facies vary markedly from west to east.

The thickest Midland Valley succession is found in Ayrshire where up to 550 m of Upper Coal Measures strata are capped by the Permian Mauchline Lavas. The sediments include some grey beds, with coals which have not been worked, and at least one marine band containing *Edmondia* and thus perhaps the equivalent of the *Edmondia* Band of the Upper *A. similis-pulchra* Zone of England. Higher beds contain faunas representative of the *A. phillipsii* and *A. tenuis* zones (Mykura 1967) and floras correspondingly assigned to the Westphalian C and D stages (Scott 1976).

There is no visible unconformity at the top of the sedimentary sequence but a Stephanian hiatus must be inferred because although plants in shales intercalated with the overlying lavas were originally assigned to the Westphalian D stage (Mykura 1965) both that flora (and hence the Mauchline Volcanic Formation itself) have since been reasssigned to the Permian (Wagner 1983).

As in Ayrshire the Upper Coal Measures in Douglas include some grey beds with thick, though unworked coals (Lumsden & Calver 1958). Little is known of the measures in the Central Basin, where they may be up to 300 m thick.

In Fife the Upper Coal Measures are over 300 m thick comprising 150 m of red sandstones resting on 45 m of grey beds – mostly shales with an *A. phillipsii-tenuis* Zone fauna – which lie, in turn, on 107 m of massive clayrocks with impersistent red sandstones (Francis & Ewing 1962). These lower 107 m, the approximate equivalent of the partly grey beds carrying Upper *A. similis-pulchra* Zone faunas farther west, are believed to be primary red deposits, and may include within them an unconformity equivalent to the 'Symon Fault' unconformity of parts of the English Midlands.

Southern Scotland

South of the Southern Uplands Massif, Carboniferous rocks reappear in a sedimentary basin – the Northumberland Trough. On the Scottish side of the Border the outcrops extend from the Kirkcudbrightshire shore of the Solway Firth north-eastwards through Annan and Langholm district to Liddesdale, thereafter being separated by

the Lower Old Red Sandstone volcanic pile of the Cheviots from a smaller area of Midland Valley type 'Cementstone Group' in the Tweed Basin. Only the lower part of the Carboniferous sequence has been preserved from erosion in the Solway and Tweed outcrops, but the Langholm–Liddesdale section (Fig. 10.22) is virtually as complete as that of the Midland Valley, approximating in total thickness to more than 3·5 km (Lumsden *et al.* 1967).

The lithostratigraphical divisions currently used for the Lower Carboniferous rocks of the region were defined by Day (1970) in the adjacent Bewcastle district of Northern England as follows:

Table 10.2. Northumberland Trough lithostratigraphical divisions

Coal Measures	Westphalian	
		Silesian
Millstone Grit Series	Namurian	
Liddesdale Group		
Upper Border Group	Viséan	Dinantian
Middle Border Group		
Lower Border Group	Tournaisian	

It can be seen from Figure 10.22, that these divisions are only recognisable with confidence as far west as the Kirkbean Outlier. Still farther west, around Rerrick, the Dinantian is represented by over 1000 m of sediments consisting mainly of alluvial fanglomerates and fluvial sandstones locally derived from the Southern Uplands (Deegan 1973).

Lower Border Group

As defined in the Bewcastle district, the Lower Border Group consists of four subdivisions. The Lynbank Beds, their base not seen, comprise 425 m of Yoredale type cyclic alternations of limestones, and shales containing abundant bivalves and less common brachiopods; a few thin sandstones occur in the lower part of the formation. The overlying Bewcastle Beds, 180 m thick, are also of Yoredale type, but with more sandstones and fewer fossils. They are succeeded by the Main Algal Beds, which include at least 14 distinctive algal bands in 85 m of alternating limestone and shales. The Cambeck Beds at the top of the group revert to cyclic sequences of limestone, shale and sandstone, altogether 200 m thick and including the distinctive Syringothyris Limestone in the upper part.

These subdivisions have not been recognised on the Scottish side of the Border, where an undifferentiated sequence of Yoredale-type cyclothems overlie the basal Carboniferous Birrenswark lavas (Chapter 11), which themselves rest upon, and are interbedded with, sediments of 'Old Red Sandstone' (in current terms Kinnesswood) facies (Lumsden *et al.* 1967). Equating the Harden Beds of the Newcastleton with the Main Algal Beds of Bewcastle, as

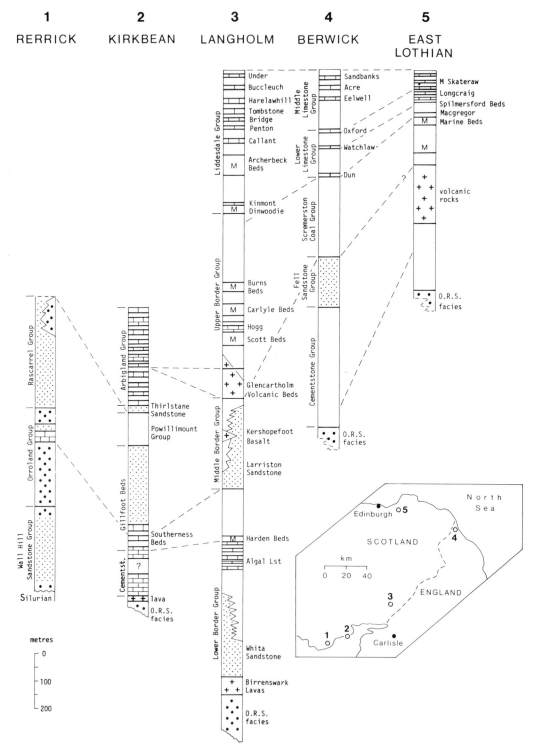

Fig. 10.22. Vertical sections to illustrate the Dinantian of the South of Scotland (mainly after George *et al.* 1976).

suggested by Day (op. cit.) and accepted by George *et al.* (1967) the estimates as to the thickness of the Group given by Lumsden *et al.* (1967) should be increased so as to include the presumed equivalents of the Cambeck Beds (i.e. the sediments above the Harden Beds); on that reading the group ranges in thickness from 400 m in the east of the Langholm district to 560 m in the west. The increase in thickness corresponds to an increase in sandstone, particularly the presence of the Whita Sandstone which reaches a maximum of 300 m before thinning out still farther to the west (Nairn 1956). It forms the Whita Formation of Leeder (1973), who resolves the problems of subdividing the non-arenaceous equivalents by identifying three other formations, one for each structural block.

The faunas of the group show progressively more marine influence with time, starting with only *Serpula, Modiolus latus* and ostracods in the Lynebank Beds, progressing to '*Camarotoechia*' *proava* and marine bivalves in the middle of the group and a richly varied fauna, including corals, polyzoa and brachiopods at the top. The Harden Beds are characterised by an abundance of *Syringothyris cuspidata* (Wilson *in* Lumsden *et al.* 1967).

At the north-eastern end of the Northumberland Trough, beyond the Cheviot Massif (which may have offered a barrier to sedimentation) the lowest part of the Dinantian shows more affinity to the Inverclyde Group of the Midland Valley than to the Lower Border Group of Liddesdale. In keeping with an earlier suggestion by Johnson (1967) some part of the Upper Old Red Sandstone might now be correlated with the cornstone-bearing Kinnesswood Formation. The overlying 'Cementstone Group', nearly 500 m thick at Burnmouth would then correspond to the Ballagan Formation. It consists of mudstones and siltstones with thin beds of argillaceous dolomitic micrite; in addition about 30 per cent of the sequence is formed by 13 beds of fluvial sandstone, individually ranging in thickness from 3 to 30 m (Smith 1967). Some of the nodular carbonates within this sequence are interpreted by Scott (1986) as diagenetic end-products of original nodular anhydrite developed in marginal sabkhas. Other nodules contain plant remains while faunas are of the same limited brackish-to-freshwater range as those of the Ballagan Formation. They include, most notably, the Foulden Fish Bed – the palaeontology and sedimentology of which has merited a series of papers collected in the covers of the *Transactions of the Royal Society of Edinburgh: Earth Sciences* (vol. 76, pt. 1, 1985).

Middle Border Group

The Middle Border Group, approximating to the Fell Sandstones and part of the Scremerston Coal Group of older classifications, is defined in the Bewcastle district as extending from the base of the Whitberry (*Chonetes cumbriense*) Band up to the base of the Clattering Band. Neither band is recognisable in Scotland. The use of the

Harden Beds as the lower limit now (p. 373) seems in need of revising upwards, approximately to the base of the Larriston Sandstone, diachronous though that must be. The Clattering Band horizon probably lies somewhere within the Glencartholm Volcanic Beds. On these assumptions, the group is about 350 m thick in the Langholm district reaching a maximum of 400 m + to the east, in the Larriston Fells, where the sandstone is thickest. The sequence includes intercalated thin cementstones and siltstones which are poorly fossiliferous like those at the base of the Lower Border Group. The Larriston Sandstone has not been traced west or south of Newcastleton, where mudstones and limestones occupy that part of the sequence. It is still uncertain whether the 30-m Kersehopefoot Basalt in the middle of the group is intrusive or extrusive, though Lumsden *et al.* (1967) favour the latter interpretation.

Upper Border Group

The Glencartholm Volcanic Beds at the base of the Upper Border Group are up to 150 m thick and consist of tuffs and tuffaceous sediments with basalt lavas included locally at the base. The beds have been noted for their fauna of arthropods and fish (Schramm 1983); they also contain foraminifera which facilitate correlation with the Clattering Band of Bewcastle defining the base of the Group. The Clattering Band macro-faunas, distinguished by *Lithostrotion martini, L. portlocki* and the gigantoproductid *Semiplanus* are also recognised, both in the Langholm district and in the Arbigland Group of Kirkbean (Craig 1956).

The Glencartholm Volcanic Beds are succeeded by 530 m of sediments extending to the top of the Group, taken in Scotland at the base of the Dinwoodie Beds. The lower part of this sequence is mainly marine, with limestones, up to 10 m thick, and contiguous mudstones and siltstones containing prolific faunas including the polyzoan *Dyscritella nana*, various species of *Lithostrotion*, common *Punctospirifer scabricosta* and bivalves diagnostic of the Upper Border Group such as *Prothyris breviformis, P. joblonga, Pteronites angustatus* and *Modiolus* cf. *oblongus* (Wilson *in* Lumsden *et al.* 1967). The upper part of the group, by contrast, is composed mainly of sandstones, with several thick coals (including the named Waverley Seam) and seatearths, though thin limestones and mudstones with rich faunas give evidence of continuing periodic marine incursions. These deltaic sediments become progressively more dominant as they are traced north-eastwards in the adjacent Bewcastle district and the group is reported to attain a thickness nearly 2000 m in the North Tyne catchment (Land & Mills *in* Day 1970), the equivalent Scremerston Coal group sequences being only slightly less thick in the Lewis Burn (Fowler 1966). Exposures are not good enough to establish whether or not there is similar north-easterly thickening in adjacent Scottish districts.

Liddesdale Group

This uppermost division of the Dinantian extends from the base of the Dinwoodie Beds to the base of the Catsbit (= Great) Limestone. It is most comprehensively known from the Archerbeck Borehole (Lumsden & Wilson 1961) where it is 532 m thick and can be subdivided into two almost equal parts in terms of distinctive lithologies. The Lower Liddesdale Group consists mainly of a heterogenous sequence of marine mudstones and limestones with a few thin sandstones and coals near the base. The Upper Liddesdale Group, by contrast is of typically Yoredale aspect, with each cycle represented by a limestone, up to 15 m thick, and shelly mudstone resting on thin coals and seatearths and overlain by siltstone and sandstone. The faunas, particularly those of the Lower Liddesdale Group, are the richest in the whole of the Carboniferous of the region. Species of the corals *Aulophyllum*, *Dibunophyllum* and *Lithostrotion* are prominent and productids are common, particularly *Eomarginifera* and Gigantoproductids.

Millstone Grit

The Millstone Grit sediments of Liddesdale are only poorly exposed and the 420 m succession has been established by drilling. The base, taken at the bottom of the Catsbit (= Great) Limestone, is unequivocal, but the top is even more uncertain than its equivalent in the Midland Valley, not only because the *Gastrioceras subcrenatum* Band is absent, but also because of the absence through uncon-formity, of most of the Lower Coal Measures (Lumsden *et al.* 1967).

The lowest 60 m of strata represent a continuation of the Yoredale-type sequences of the underlying Liddesdale Group. Together with the succeeding 210 m of sandstones, marine bands and the locally worked Penton Coals, they are equivalent to the Limestone Coal Group of the Midland Valley, though diagnostic faunas have not been recognised. The uppermost 150 m of the sequence are sandy with thin coals and several rootlet beds, but no marine bands comparable with those of the Midland Valley Upper Limestone Group and Passage Group are recorded. The uppermost of these beds are reddened for distances of as much as 120 m below the Coal Measures unconformity.

Coal Measures

Coal Measures crop out only over a small area of faulted ground around Canonbie and Rowanburn, but they have been established by drilling and geophysical surveys (Picken 1988) to extend southwards beneath New Red Sandstone to form a concealed coalfield over at least 70 km² (Fig. 10.23). The sequence is approximately 1 km thick with unconformities below, within and above, all of them related to westerly thinning and overstep on to a structural high. The Lower Coal Measures are most complete near the axis

of the syn-sedimentary basin, but its poor representation or absence elsewhere is exemplified by the angular uncon-formity exposed in the River Esk at Gilnockie Bridge (Lumsden *et al.* 1967). The main workable seams are in the Middle Coal Measures, which are about 200 m thick extending from the Queenslie Marine Band at the base to the Skelton (= Skipsey's) Marine Band at the top. The latter is cut out by unconformity near the base of the Upper Coal Measures in the north-west.

The Upper Coal Measures, nearly 800 m thick below the New Red Sandstone unconformity, are typically red like their Midland Valley equivalents. The lowest 100 m consists of grey siltstones and sandstone, with three main bands and several thin coals. All the succeeding beds are reddened by secondary alteration, so that original sedimentary struc-tures in the argillaceous rocks have become unrecognisable. In addition to freshwater limestones containing *Spirorbis* and ostracods, faunas representative of the *A. phillipsii* and *A. tenuis* zones have been found.

Continental Shelf

By means of geophysical surveys and drilling the Midland Valley Carboniferous rocks have been delineated offshore both westwards into the Firth of Clyde and eastwards into the Forth. They extend over a small area, partly covered by New Red Sandstone, east of Kintyre, forming a thin Namurian–Westphalian sequence comparable with that of Machrihanish (McLean & Deegan 1978). Sediments of various Carboniferous ages can also be traced outwards from the Ayrshire coast, their accumulation influenced, as on land, by the WSW Kersloch, Inchgotrick and Dusk Water faults. To the north of these faults, however, the synsedimentary controls, during late Carboniferous times at least, appear to have been NNW lineaments, particularly the Brodick Bay Fault (Fig. 10.28A). They contributed to the formation of the separate East Arran Basin in which 1·1 km of mainly Upper Carboniferous sediments rest on Clyde Plateau Lavas and are overlain by New Red Sandstone. Thicknesses in the basin are at variance with the attenuated sequences onshore in Arran to south-west and N Ayrshire to north-east.

In the Forth, Carboniferous sediments and volcanic rocks of various ages are recorded, though structures are too complex for divisions to be mapped. Those structures which have been delineated (Thomson 1977) include the presumed NE extension of the Pentland Fault, an unnamed E–W fracture following the buried channel of the Forth and separating the Fife and Midlothian coalfields, and several north-easterly fold axes (Fig. 10.28A). Farther offshore, seabed Lower Carboniferous outcrops link with those off the Tweed Basin, though showing marked attenuation over the buried continuation of the Southern Uplands Massif (Eden *et al.* 1979).

Lower Carboniferous sediments extend even farther

offshore into the Forth Approaches Basin and Moray Firth Basin (Fig. 10.24), where they are known from drilling through thick younger rocks to be at least 500 m thick and to consist mainly of interbedded sandstones, shales and coals with subordinate limestones; the lithology is reported to resemble that of the Limestone Coal Group of the Midland Valley and to contain microfloral assemblages diagnostic of Viséan to Namurian ages (Deegan & Scull 1977, Evans *et al*. 1982).

In the NE–SW Clair Basin, west of Shetland, more than 300 m of interbedded red sandstones and mudstones containing calcrete nodules are interpreted by Blackbourn (1987) as flood plain deposits of early Carboniferous age. His view is supported by the presence of an overlying Ballagan-type sequence up to 100 m thick, of fluvio-deltaic to lacustrine carbonaceous sands and dark grey organic-rich mudstones containing late Courceyan miospores.

Correlation

Dinantian

The Lower Carboniferous, or Dinantian sub-system as redefined by Conil *et al*. (1976) comprises a lower Tournaisian Series at the base and an upper Viséan Series. A further subdivision into stages is outlined in Table 10.3 though only the uppermost of these have been identified satisfactorily in the Midland Valley as yet. Modifications of the Vaughanian coral-brachiopod zonation have been particularly confusing and are omitted from Table 10.3, though some of the assemblages are of value in local correlation. Thus, the *Syringothyris* cf. *cuspidata* and associated faunas of the Harden Beds of Langholm and of the Syringothyris Limestone of Kirkbean have been used to delineate the base of the Chadian stage, while the base

Fig. 10.23. Canonbie Coalfield (after Picken 1988). A, Map showing area of outcrop and extent of Coal Measures beneath New Red Sandstone; B, Horizontal section along line shown in A; C, Generalised vertical section. Abbreviations: CB, Catsbit Limestone; LCM, Lower Coal Measures; MCM, Middle Coal Measures; MG, Millstone Grit; NRS, New Red Sandstone; OD, Ordnance Datum; Q, Queenslie Marine Band; SK, Skelton Marine Band; UCM, Upper Coal Measures; ULG, Upper Liddesdale Group.

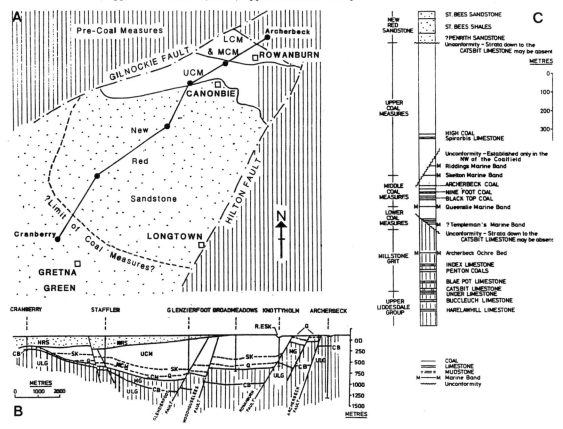

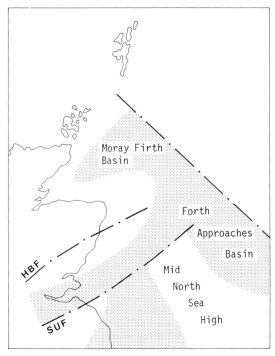

Fig. 10.24. Distribution of Carboniferous sediments offshore.

of the Arundian has been set 28 m higher in Kirkbean at a nodular algal band equated with the Glencartholm Volcanic Beds of Langholm and the Hillend Algal Band of Bewcastle (George *et al.* 1976). Similarly the base of the Asbian is located at the entry of a new and distinctive coral-brachiopod fauna at the base of the Upper Border Group.

Goniatites, which offer the potential of more precise zonation, are rare in Scotland and are, moreover, confined to higher levels of the Viséan. The lowest recorded goniatite is *Beyrichoceratoides redesdalensis* in the Cove Marine Band of West Lothian (Wilson 1952). Other specimens of *Beyrichoceratoides*, including *B.* cf. *truncatus* from the Pumpherston Shell-Bed of the Lothians, are suggestive of B_2 age and in terms of the Oil-shale Group succession the base of P_1 has been taken by Currie (1954) at the base of the Raeburn Shell-Bed. It is so indicated in Table 10.3, though it is based on a correlation of the Raeburn Shell-Bed with the Fordell Marine Band of Fife which is doubtful (Francis 1961). On grounds admitted to be arbitrary Currie (1954) takes the P_1–P_2 boundary in the Midland Valley at the Hurlet Limestone.

For the lower and middle parts of the Dinantian succession, therefore, the most useful means of correlation in the Midland Valley and Northumberland Trough is the zonal scheme based on miospores established by Neves *et al.* (1972, 1973) and modified by Clayton *et al.* (1978),

Clayton (1985), Scott *et al.* (1984) and Brindley & Spinner (1989). It seems that the lowest Tournaisian zones (VI to PC) of the type region are not represented in Scotland, where sediments of Upper Old Red Sandstone and Kinnesswood Formation facies are inimical to miospore preservation.

Namurian

The Viséan/Namurian junction is recognised with confidence both in the Midland Valley and Northumberland Trough. The base of the Pendleian (E_1) Stage is, by international agreement, marked by the entry of *Cravenoceras leion* and although the species has not yet been recorded in Scotland the presence of *Cravenoceras sp.* in the Top Hosie Limestone provides reasonable grounds for correlation with the Catsbit Limestone of Liddesdale and the Great Limestone of the base of the Northern England Namurian.

Goniatites remain scarce in younger strata, but the Limestone Coal Group must lie entirely within E_1, because the boundary with the overlying Arnsbergian (E_2) Stage is

Fig. 10.25. Characteristic Carboniferous miospores. All figured specimens are housed in the palynological collections of the British Geological Survey, Keyworth. Magnification × 500. By kind permission of Bernard Owens.

1. *Raistrickia nigra* Love; NC Zone (E_1), Namurian A, MPK 1360.
2. *Spelaeotriletes arenaceus* Neves & Owens KV Zone (R_1), Namurian B, MPK 1440.
3. *Mooreisporites trigallerus* Neves; KV Zone, (R_1) Namurian B, MPK 1428.
4. *Ibrahimispores magnificus* Neves; KV Zone (R_1), Namurian B, MPK 1473.
5. *Knoxisporites seniradiatus* Neves; KV Zone (R_1), Namurian B, MPK 1434.
6. *Cirratriradites rarus* (Ibrahim) Schopf, Wilson & Bentall; KV Zone (R_1), Namurian B, MPK 1436.
7. *Knoxisporites dissidius* Neves; TK Zone (E_2), Namurian A, MPK 1400.
8. *Cirratriradites saturni* (Ibrahim) Schopf, Wilson & Bentall; NJ Zone, Westphalian B, MPK 1107.
9. *Reinschospora triangularis* Kosanke; KV Zone (R_1), Namurian B, MPK 1426.
10. *Trinidulus diamphidos* Felix & Paden; FR Zone (G_1), Namurian C, MPK 1426.
11. *Densosporites pannosus* Knox; KV Zone (R_1), Namurian B, MPK 1438.
12. *Schulzospora campyloptera* (Waltz) Hoffmeister, Staplin & Malloy; NC Zone (E_1), Namurian A, MPK 1388.

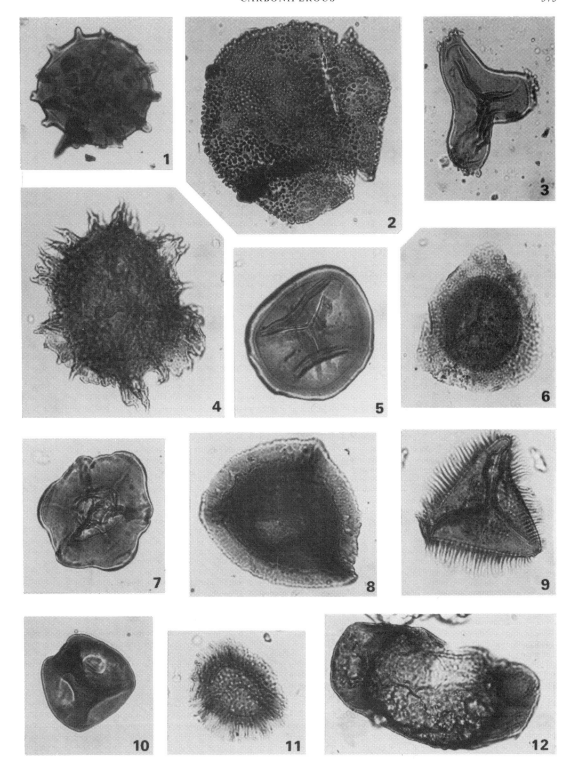

placed by Ramsbottom (1977) at the base of the Orchard Limestone following his recognition of *Eumorphoceras pseudobilingus* from the Index Limestone and his reassessment of Currie's (1954) determination of *E. bisulcatum grassingtonense* from the Orchard horizon. The Arnsbergian continues up into the lower part of the Passage Group for the E_2 goniatites *Anthracoceras glabrum* and *A. paucilobum*, plus nautiloid *Tylonautilus nodiferus*, are recorded from No. 1 Marine Band in association with E_2 miospores which also occur in No. 2 Marine Band (Neves *et al.* 1965).

There is no palaeontological evidence for the presence of the Chokerian (H_1) or Alportian (H_2) stages but they may be represented by the strata betrween No. 2 Marine Band and the base of No. 3 Marine Band Group for a specimen of *Homoceratoides* and associated spores (Fig. 10.25) show the latter to be late Alportian or early Kinderscoutian (R_1) in age (op. cit.). As younger Namurian goniatites have yet to be found, the strata from No. 3 Group up to the top of No. 5 Marine Band are assigned to the Kinderscoutian (R_1) and lower part of the Marsdenian (R_2) stages entirely on the basis of correlation with KV zone of miospores (Owens *et al.* 1977). The overlying FR miospore zone continues up to the top of the Namurian and includes the remainder of the Marsdenian as well as the Yeadonian (G_1) Stage. The assemblages suggest that the Namurian–Westphalian boundary might be drawn at the top of No. 6 Marine Band Group in terms of the Central Coalfield sequence. Hence, as implied by MacGregor's (1960) original classification the topmost beds of the Passage Group are really Westphalian in age. They correspond to the G_2 stage approximating to the *A. lenisulcata* Zone and including the marine bands, up to five in all, now widely recognised at the base of the Midland Valley Coal Measures. Where those marine bands are best seen in Ayrshire, the underlying Nos. 0–6 Marine Band sequence of the Central Coalfield is absent and unconformity, as established in Douglas, seems likely. However, the evidence outlined above suggests that there is unlikely to be any major break within the thicker Central Coalfield sequence.

Westphalian

Apart from the non-recognition of the Subcrenatum Band and hence the local inclusion of the lowest beds of the Series in the top of the Passage Group, the whole of the Westphalian can be equated with the Scottish Coal Measures. The highest, or Stephanian stage of the Carboniferous is not represented. Correlation with England and north-west Europe now presents relatively few problems (Calver 1969), the Vanderbeckei (Queenslie) Marine Band marking the boundary between the Westphalian A and B stages and the Aegiranum (Skipsey's) Marine Band separating Westphalian B and C being of particular international importance. The base of Westphalian D cannot be defined in terms of a widespread marine band; it

is taken in Britain to correspond with the base of the Zone of *Anthraconauta tenuis*. Further subdivision is facilitated by the non-marine bivalve zonal system first applied in Scotland by Weir & Leitch (1936) and subsequently refined by Calver (1969) and Brand (1983). Characteristics, range and relative abundance of some of the more important forms are shown relative to the Central Coalfield succession in Figure 10.26. Plant remains, well preserved in shales, have always been among the most spectacular Coal Measure fossils. But with the greater precision of faunal zoning, the use of macro-flora as tools of correlation has declined in all but the upper part of the Westphalian and interest in plants is currently focussed more on their value as indicators of Coal Measure palaeoecology (Scott 1977, 1979).

Miospores offer a more useful tool of zonation Clayton *et al.* (1977; Table 10.3) and have proved to be particularly useful for seam correlation (Smith & Butterworth 1967).

Structure

Penecontemporaneous structural controls

The Caledonian orogeny, commonly ascribed to the closure of a proto-Atlantic (Iapetus) Ocean had ended by Middle ORS times (Chapters 1 and 9) and by the beginning of the Carboniferous the region had become part of the southern margin of a new North American–North Europe continent. The cratonised sub-Carboniferous basement, however, remained unstable. Movement along fault systems extending upwards from basement into sedimentary cover exerted a profound influence on the pattern of deposition from early Dinantian to late Westphalian times. It is reflected in lateral variation in thickness of sediments corresponding to differential subsidence along syn-sedimentary hingelines.

Three components can be identified. The largest in scale is apparent in thickness variations shown by isopachs. In the Midland Valley (Fig. 10.27) these form a pattern which operated virtually throughout the Carboniferous with only minor changes in location of axes. In Strathclyde and Inverclyde Group times sedimentary basins aligned to NE and NNE are already defined in Fife and West Lothian though the pattern of sedimentation is distorted by volcanism in East Lothian and the western Midland Valley. Thereafter, the pattern consistently shows an Ayrshire shelf cut off by the Clyde Plateau from two major NNE basins, one now comprising the Central–Stirling–Clackmannan Coalfields (bifurcated by a ridge formed by thick Bathgate lavas), the other comprising the East Fife–Midlothian Coalfields. Separate from these are smaller basins which also have NNE alignment as well as further basins such as Douglas in the south-west which are aligned NE parallel to the Southern Uplands Fault.

The second component of synsedimentary control consists of NE fractures with strike–slip elements (Figs. 10.28A, 10.30). Ayrshire members such as the Kerse Loch,

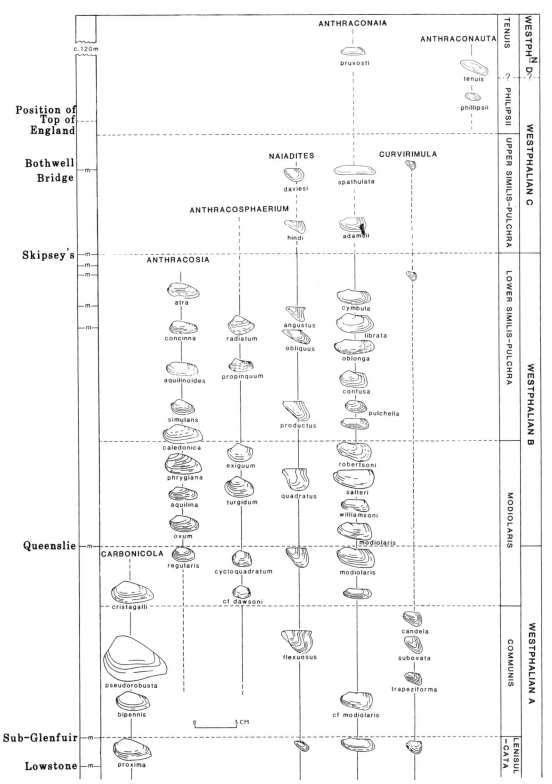

Fig. 10.26. Characteristic non-marine lamellibranchs in the Scottish Coal Measures and their approximate zonal position (after Calver 1956, figs. 3, 4 and 5).

Table 10.3. Carboniferous correlation chart.

MIDLAND VALLEY — AYRSHIRE & CENTRAL	EAST FIFE	LOTHIANS	WESTERN NORTHUMBERLAND TROUGH	Miospore Zone	Goniatite Zone	STAGE	SERIES	SUB-SYSTEM
Upper Coal Measures				OT		D	WESTPHALIAN	SILESIAN
Skipsey's MB			Skelton MB	SL		C		
Middle Coal Measures / Queenslie MB				NJ		B		
Lower Coal Measures			SS →	RA	G_2	A		
Passage Group			Millstone Grit Series	FR	G_1	Yeadonian	NAMURIAN	
				KV	R_2	Marsdenian		
					R_1	Kinderscoutian		
				SO	H_2	Alportian		
					H_1	Chokerian		
Castlecary Lst				TK	E_2	Arnsbergian		
Upper Limestone Group		Orchard						
Index								
Limestone Coal Group			Catsbit	NC	E_1	Pendleian		
Lower Limestone Group / Hurlet		Top Hosie			P_2	Brigantian	VISÉAN	DINANTIAN
Lawmuir Formation	Pathhead Beds	Upper Oil-Shale	Liddesdale Group	VF	P_1			
	Sandy Craig Beds	Two Ft / Sub-Group / Burdiehouse		NM		Asbian		
Clyde Plateau Volcanic Formation	Pittenweem Beds	Pumpherston Shell Bed	Dinwoodie Beds					
			Upper Border Group					
Strathclyde Group	Anstruther Beds	Lower Oil-Shale Sub-Group		TC	B	Holkerian Arundian Chadian		
			Glencartholm Volc					
	Fife Ness Beds		Middle Border Group	TS				
				Pu				
Clyde Sst Fm	Balcomie Beds		Lower Border Group ?	CM		Courceyan	TOURNAISIAN	
Ballagan Formation		Inverclyde Group		PC ?				
Inverclyde Group	Kinnesswood Formation			BP ?				
				HD ?				
				VI ?				

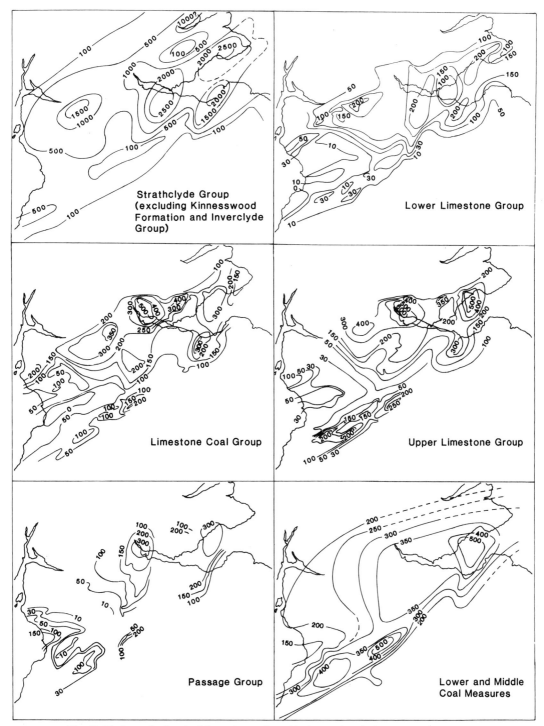

Fig. 10.27. Isopachs (in m) of the main stratigraphical groupings in the Midland Valley (from Browne *et al.* 1985).

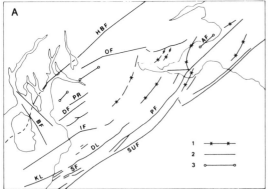

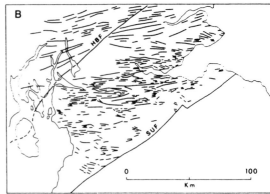

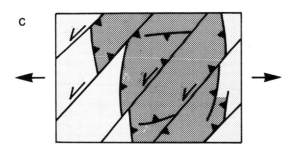

**Composite basin of strike–slip and
domino faults
NB Fault throws shown for carapace**

Fig. 10.28. Synsedimentary structural components and their resolution.

A. Axes of syndepositional basins (first com-
ponent; lines with crosses) and main strike
slip faults (second component: lines only).
AF, Ardross Fault; BP, Brodick Bay Fault;
DF, Dusk Water Fault; DL, Dron Line;
HBF, Highland Boundary Fault; IF, Inch-
gotrick Fault; KL, Kerseloch Fault; OF,
Ochil Fault; PF, Pentland Fault; PR, Paisley
Ruck; SF, Straiton Fault; SUF, Southern
Uplands Fault.

B. East–west fractures (third component).

C. Diagram illustrating interrelationship of the
three components (Gibbs 1987).

Inchgotrick and Duskwater faults have long been recognised as responsible for spectacular thickness variations
indicative of progressive activity (Figs. 10.11, 10.21, 10.30).
The Highland Boundary and Southern Upland faults,
already defining the Midland Valley rift (George 1960) as
well as the Ardross and Pentland faults are probably in the
same category.

The third component is represented by the system of E–
W fractures (Fig. 10.28B). With the exception of the largest
and northernmost – the Ochil Fault – these have been
traditionally interpreted as late- or post-Carboniferous in
age because they displace the youngest of Coal Measures.

However, it is now clear from the behaviour of dolerite
intrusions (Chapter 11) that these too were synsedimentary
and active throughout the Carboniferous.

Superimposed on these variations there is a general
increase in the thickness towards the north-east from which
Kennedy (1958) inferred a general downwards movement
of the Midland Valley floor which was greater in the north-
east than in the south-west (see also Fig. 10.31). Additional
to this is a northward thinning which is apparent particularly in Fife and Clackmannan, throughout the period.

Thickness data of Midland Valley quality are lacking for
the Northumberland Trough, though it is clear that its NE-

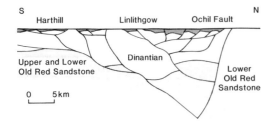

Fig. 10.29. Horizontal N–S section across the
centre of the Midland Valley to show intermediate and upper fault leaves developed as a
'carapace' to the simpler basement strike–slip
shears (modified from Gibbs 1987).

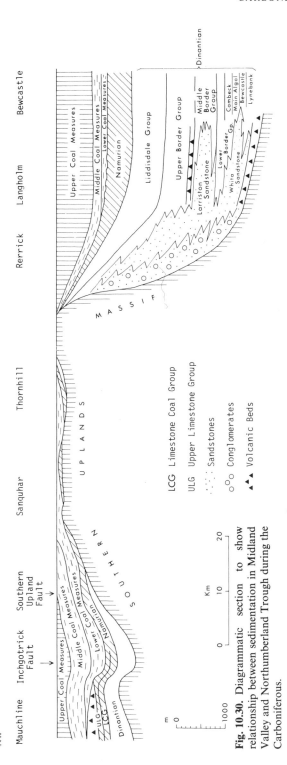

Fig. 10.30. Diagrammatic section to show relationship between sedimentation in Midland Valley and Northumberland Trough during the Carboniferous.

aligned margins are comparable hinge–line controls of sedimentation (Leeder 1974b). Local variations in the early Westphalian times are indicated by unconformity in the Canonbie coalfield.

The mechanics of development of the Scottish Carboniferous basins is a subject of intense research at the time of writing. Earlier models (Bott 1976, Leeder 1976) postulated extra buoyancy from Caledonian granites emplaced in massifs such as the Southern Uplands, coupled with mantle flow, lower crustal creep and partial crustal melting.

These models have evolved to encompass basin formation by extensional tectonics and lithospheric stretching (Dewey 1982, Leeder 1982). More specific to the Midland Valley, Gibbs (1987) and Read (1988, 1989) have developed the concept in a way which encompasses all three of the components described above, describing them as a combination of strike–slip and dip–slip arrays formed by mixed-mode transpression and transtension (Figs. 10.28C, 10.29).

Late Carboniferous and early Permian structures

A change in the synsedimentary pattern appears to have been signalled in late Carboniferous times by an extensional episode which accentuated movement along the existing E–W faults. The episode is dated (at *c.* 295 Ma) by the suite of tholeiitic dykes emplaced along many such faults not only within the Carboniferous basins but also in the Highlands (Fig. 10.28B).

Subsequent rifting and related sedimentation formed a new pattern dominated by NW–SE lineaments (Fig. 10.32). Mykura (1967) believes that this pattern exerted control over Permian volcanism and sedimentation in the Mauchline Basin. Hall (1974) has traced the system south-eastwards through Sanquhar and Thornhill to the Vale of Eden, and McLean (1978) includes it, together with the offshore East Arran Basin, and a further Stranraer–Liverpool Bay–Cheshire Basin linear system in a 'Clyde Belt'. Magmatic associations (Chapter 11) support Mykura's view that the system is mainly Permian in age, but Hall (op. cit.) suggests that it began to operate as a synsedimentary control during the late Carboniferous. Moreover, he follows Russell (1972) in believing that the fractures are related to the initiation of Atlantic rifting. Although this view has not met with universal acceptance, further arguments related to the tholeiitic intrusions have been made in its support and are discussed in the next chapter.

Palaeogeography and conditions of deposition

By the start of the Carboniferous the region had become part of the southern marginal shelf of the North America–North Europe craton flanked to the north-west by the partly eroded remains of the Caledonian mountains and

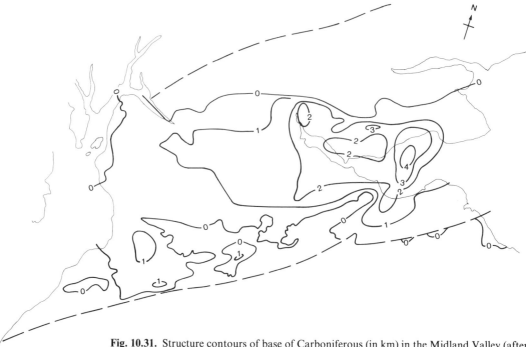

Fig. 10.31. Structure contours of base of Carboniferous (in km) in the Midland Valley (after Browne *et al.* 1985, fig. 18).

with a Hercynian ocean some uncertain distance to the south (Fig. 10.33).

The mountains continued to supply the bulk of the sediments to the variable subsiding environments of deposition which ranged in time and space from alluvial plains through deltas to shallow seas encroaching from the south. A subsidiary source of sediments was the Southern Uplands which then, as now, formed a barrier between the Midland Valley and the Northumberland Trough (Fig. 10.35). Although it was occasionally breached, as at Loch Ryan. Sanquhar–Thornhill and Cockburnspath, the

barrier was frequently renewed by uplift and denudation and it is questionable whether it ever became completely submerged even in Westphalian times. Similar periodic and localised rift-controlled breaching of the Highlands might offer a more satisfactory explanation of the isolated, imperfectly dated Carboniferous outliers at Inninmore and Bridge of Awe (Johnstone 1966) than the Highland Basin proposed by George (1960).

In the Midland Valley additional barriers to sedimentation were erected by the great outpourings of lava in the Clyde and Forth areas. Their effects were the greatest in the

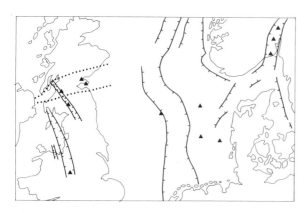

Fig. 10.32. Map showing Permian rifting. Generalised location of volcanic rocks shown by black triangles (Francis 1978, fig. 9).

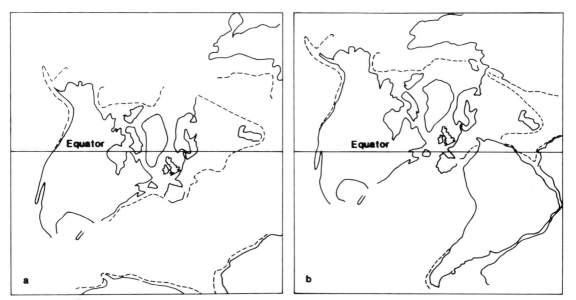

Fig. 10.33. Palaeogeographic plate reconstructions, mainly after Smith *et al.* (1981). a, Early Carboniferous; b, early Permian.

early part of the period when they virtually broke down the area of deposition into isolated basins, the most important being that of the Lothians oil-shales (Fig. 11.16). Even after they were buried by younger beds these lava piles continued to influence sedimentation by virtue of differential compaction; this and some of the other localised reactions between volcanism and sedimentation are discussed further in the next chapter.

To the various possible mechanisms governing differential hinge–line subsidence, mentioned above, must be added the continuing debate as to the cause of cyclicity in the Carboniferous. Eustatic marine transgression has been invoked by Ramsbottom (1973, 1977) to explain the rhythmic pattern of sedimentation in the Dinantian and Namurian. The Mississipian–Pennsylvanian hiatus is similarly singled out by Read (1988) as relating to a great eustatic fall in sea level in Passage Group times. George (1978), though primarily concerned to dissociate this concept from the definition and recognition of named stages, offers regional pulsed diastrophism as better fitting the evidence. In detail, the pattern of cyclicity supports this view. On the other hand, the European-wide distribution of marine bands (Calver 1971) in an otherwise prolonged period of on-delta paralic sedimentation suggests that in Westphalian times, at least, eustatic movement did give rise to some marine incursions.

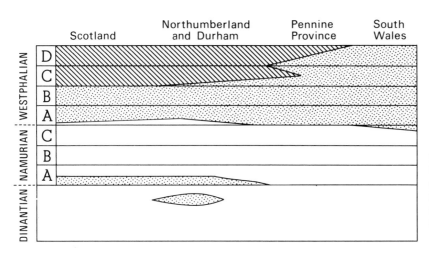

Fig. 10.34. Relationship of Coal Measures facies (stippled) and red bed facies (oblique lines) to stratomeric divisions of the Carboniferous (Francis 1979, fig. 2).

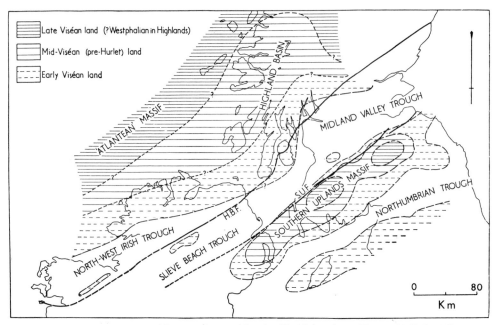

Fig. 10.35. Palaeogeographic map of central Scotland in Viséan times (George 1960, fig. 15).

A progressive change in climate is indicated by other aspects of sedimentation. The red facies and cornstones of the Upper Old Red Sandstone and earliest Dinantian indicate semi-arid conditions suitable for the formation of calcrete. Later shelf limestones and coals point to tropical environments having a high humidity before a return was made to the semi-arid conditions of the late Westphalian. The progression fits well with the palaeo-magnetic evidence (Smith *et al.* 1973, Turner & Tarling 1975) (Fig. 10.33) and is further supported by the earlier onset of Upper Carboniferous red-bed facies in Scotland than in England (Fig. 10.34).

The red facies of the Upper Old Red Sandstone and its interdigitation with earliest grey Dinantian facies indicates a semi-arid fluviatile (alluvial fan and flood-plain) environment which was only intermittently inundated. The cementstone facies of the Ballagan Formation in the central and south-eastern part of the Midland Valley and as far around the eastern end of the Southern Uplands massif as the Tweed Basin show these conditions to have given way to lacustrine or estuarine environments. Salt pseudomorphs and desiccation breccias, layers of gypsum and stunted faunas all testify to high ratio of evaporation and salinity; some of the cementstone themselves were probably formed as chemical precipitates.

The Cheviots separated this environment from the shallow gulf sea which occupied the western part of the Northumberland Trough and in which the Dinantian carbonate lithofacies were deposited (Leeder 1974a). It also received fluviatile and deltaic input from at least three systems draining the broad area to the north (Fig. 10.36) including the unroofing of the Criffel pluton to form the alluvial-fan sandstones and conglomerates deposited on a constantly renewed steep slope (Deegan 1973). Some of these river systems declined by the end of Lower Border Group times, but sheet-sands deposited from braided river systems continued around the Cheviots to form the Fell Sandstones (Hodgson 1978).

In the Midland Valley, Viséan sedimentation continued in two different environments. The Oil Shale sub-groups, deposited in an enclosed basin, surrounded by Clyde and Forth lavas to north, west and east (providing detritus in the form of red 'marls') and by Southern Uplands to the south, include desiccation breccias and freshwater limestones; the oil-shales themselves provide evidence of stagnant anoxic waters in which abundant algae accumulated and putrified. The basin was periodically invaded by deltas advancing from the other major Viséan environment to the north-east, in Fife, where thick sequences incorporated

Fig. 10.36. Tournaisian palaeogeography and summary graphical logs (fining-upward cycles) of some penecontemporaneous fluvial and deltaic sediments in the Northumberland Trough (Leeder 1974a, figs. 9 and 17).

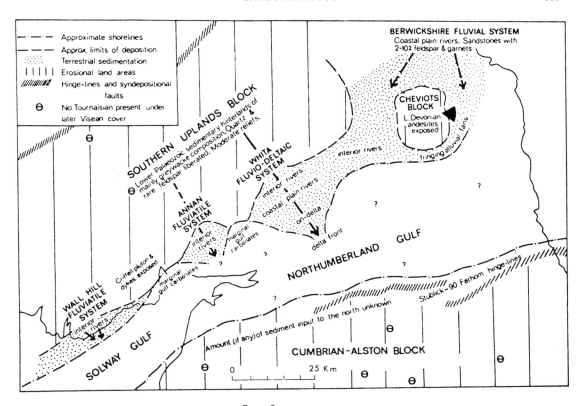

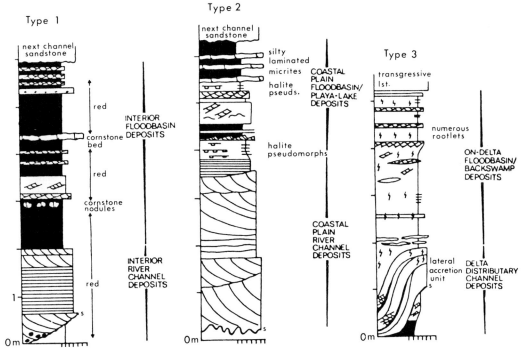

debris derived from the erosion of Lower Old Red Sand-stone still farther to the north-east (Greensmith 1961, 1965).

Sediments of Yoredale facies accumulated in late Viséan times indicate a replacement of local environments by more widespread shallow seas in which shelf limestones were deposited in alternation with deltaic sandstones and on-delta coals so as to form the Lower Limestone Group. Like the earlier Scremerston Coal Group of Northumberland the Limestone Coal Group of the Midland Valley is essentially a fluvio-deltaic clastic facies representing a temporary early Namurian shallowing before the extensive

re-establishment of Upper Limestone Group Yoredale conditions. The latest Namurian Passage Group was a time of further shallowing, giving rise to the deposition of coarse fluvial sediments derived from slightly different sources than those of the Upper Limestone Group and Coal Measures (Muir 1963); there were also periodic marine incursions which continued into the Westphalian when, in addition, peat accumulated to form coals in on-delta surfaces, their deposition being interrupted by fluvial and distributary-channel sands and muds. Conditions finally returned to fluvial late in the Westphalian so as to give rise to the 'Barren Red' measures.

REFERENCES

BELT, E. S.   1975 Scottish Carboniferous cyclothem patterns and their paleo-environmental significance. *In* Broussard, M. L. (Ed.) *Deltas: models for exploration*. Houston Geological Society, 427–449.

BELT, E. S. *et al*.   1967 Sedimentology of Carboniferous cementstone facies, British Isles and eastern Canada. *J. Geol.*, **75**, 711–721.

BENNISON, S. M.   1960 Lower Carboniferous non-marine lamellibranchs from east Fife, Scotland. *Palaeontol.*, **3**, 137–152.

                                   1961 Small *Naiadites obesus* from the Calciferous Sandstone Series (Lower Carboniferous) of Fife. *Palaeontol.*, **4**, 300–311.

BLACKBOURN, G. A.   1987 Sedimentary environments and stratigraphy of the Late Devon-ian–Early Carboniferous Clair Basin, west of Shetland. *In* Miller, J., Adams, A. E., & Wright, V. P. (Eds.) *European Dinantian Environments. Geol. J. Spec. Issue*, No. 12, 75–91. John Wiley & Sons, 402 pp.

BOTT, M. H. P.   1976 Formation of sedimentary basins of graben type by extension of continental crust. *Tectonophysics*, **36**, 77–86.

BRAND, P. J.   1977 The fauna and distribution of the Queenslie Marine Band (Westphalian) in Scotland. *Rep. Inst. geol. Sci.*, 77/18, 9 pp.

                                   1983 Stratigraphical palaeontology of the Ayrshire Coalfield, Scotland. *Trans. roy. Soc. Edinb.: Earth Sci.*, **73**, 173–190.

BRAND, P. J. *et al*.   1980 The Carboniferous strata at the Westfield Opencast Site, Fife, Scotland. *Rep. Inst. geol. Sci.*, 79/11, 26 pp.

BRIGGS, D. A. G. &   1983 The Lower Carboniferous Granton 'Shrimp-Bed', Edinburgh.
  CLARKSON, E. N. K.   *Spec. Pap. Palaeontol.*, **30**, 161–177.

                                   1985 The Lower Carboniferous Shrimp *Tealliocaris* from Gullane, East Lothian, Scotland. *Trans. roy. Soc. Edinb.: Earth Sci.*, **76**, 173–201.

BRINDLEY, S. &   1989 Palynological assemblages from Lower Carboniferous
  SPINNER, E.   deposits, Burntisland district, Fife, Scotland. *Proc. Yorks. geol. Soc.*, **47**, 215–231.

BROWNE, M. A. E.   1980a Stratigraphy of lower Calciferous Sandstone Measures in Fife. *Scott. J. Geol.*, **16**, 321–328.

                                   1980b The Upper Devonian and Lower Carboniferous (Dinantian) of the Firth of Tay, Scotland. *Rep. Inst. geol. Sci.* 80/9, 13 pp.

BROWNE, M. A. E. *et al*.   1985 The Upper Palaeozoic basins of the Midland Valley of Scotland. *Invest. Geotherm. Potent. U.K. Br. Geol. Surv.*, 48 pp.

                                   1987 The Upper Devonian and Carboniferous sandstones of the Midland Valley of Scotland. *Invest. Geotherm. Potent. U.K. Br. Geol. Surv.*, 55 pp.

BURGESS, I. C.   1961   Fossil soils of the Upper Old Red Sandstone of Ayrshire. *Trans. geol. Soc. Glasg.*, **24**, 138–153.

                                1965   *Calcifoleum* (Codiaceae) from the upper Viséan of Scotland. *Palaeontol.*, **8**, 192–198.

CALVER, M. A.   1956   Die stratigraphische Verbreitung der nichtmarinen Muscheln in den penninschen Kolenfeldern Englands. *Z. dtsch geol. Ges.*, **107**, 26–39.

                                1969   Westphalian of Britain. *C. R. 6me Congr. Int. Strat. Geol. Carb.* (Sheffield 1967), **1**, 233–254.

                                1971   *Westphalian marine faunas in northern England and adjoining areas. PhD Thesis, Univ. Reading (Unpubl.).*

CAMERON, I. B. &   1978   The oil-shales of the Lothians, Scotland: present resources and
   MCADAM, A. D.   former workings. *Rep. Inst. Geol Surv.*, 78/28, 21 pp.

CATER, J. M. L.   1987   Sedimentology of the Lower Oil-Shale Group (Dinantian) sequence at Granton, Edinburgh. *Trans. Roy. Soc. Edinb.: Earth Sci.*, **78**, 29–40.

CHISHOLM, J.   1968   Trace-fossils from the Geological Survey boreholes in east Fife 1963–4). *Bull. geol. Surv. G.B.*, **28**, 103–119.

                                1970a Lower Carboniferous trace-fossils from the Geological Survey boreholes in West Fife (1965–6). *Bull. geol. Surv. G.B.*, **31**, 19–35.

                                1970b *Teichichnus* and related trace fossils in the Lower Carboniferous at St Monance, Scotland. *Bull. geol. Surv. G.B.*, **32**, 21–51.

CHISHOLM, J. I. &   1974   The Upper Old Red Sandstone of Fife and Kinross: a fluviatile
   DEAN, J. M.   sequence with evidence of marine incursion. *Scott. J. Geol.*, **10**, 1–30.

CLAYTON, G.   1985   Dinantian miospores and intercontinental correlation. *C. R. 10me Congr. Int. Strat. geol. Carbonif.* Madrid 1983, **4**, 9–23.

CLAYTON, G. *et al.*   1977   Carboniferous miospores of western Europe: illustration and zonation. *Mededelingen. Rijks. geol. Dicnst.*, **29**, 1–71.

                                1978   Correlation of palynological zonation of the Dinantian of the British Isles. *Coloq. Int. Palinol. Léon*, **1**, 137–147.

CONIL, R. *et al.*   1977   Nouvelle charte stratigraphique du Dinantian type de la Belgique. *Ann. Soc. géol. Nord*, **96**, 363–371.

CRAIG, G. Y.   1956   The Lower Carboniferous outlier of Kirkbean, Kirkcudbrightshire. *Trans. geol. Soc. Glasg.*, **22**, 113–132.

CURRIE, E. D.   1954   Scottish Carboniferous goniatites. *Trans. roy. Soc. Edinb.*, **62**, 527–602.

DAVIES, A.   1970   Carboniferous rocks of the Sanquhar Outlier. *Bull. geol. Surv. G.B.*, **31**, 37–57.

                                1972   Carboniferous rocks of the Muirkirk, Gass Water and Glenmuir areas of Ayrshire. *Bull. geol. Surv. G.B.*, **40**, 1–49.

DAVIES, A. *et al.*   1986   Geology of the Dunbar district. *Mem. Br. geol. Surv.*, 69 pp.

DAY, J. B. W.   1970   Geology of the country around Bewcastle. *Mem. geol. Surv. U.K.*, 357 pp.

DEEGAN, C. E.   1973   Tectonic control of sedimentation at the margin of a Carboniferous depositional basin, Kirkcudbrightshire. *Scott. J. Geol.*, **9**, 1–28.

DEEGAN, C. E. &   1977   A standard lithostratigraphic nomenclature for the central and
   SCULL, B. J.   northern North Sea. *Rep. Inst. geol. Sci.*, No. 77/25, 36 pp.

DEWEY, J. F.   1982   Plate tectonics and the evolution of the British Isles. *J. geol. Soc. Lond.*, **139**, 371–412.

DUFF, P. McL. D. &   1962   Statistical basis for cyclothems; a quantitative study of the
   WALTON, E. K.   sedimentary succession in the East Pennine Coalfield. *Sedimentology*, **1**, 235–255.

EDEN, R. A. *et al.*   1969   Submarine examination of Lower Carboniferous strata on inshore regions of the continental shelf of south-east Scotland. *Marine Geol.*, **7**, 235–251.

EVANS, D. *et al.*   1982 The offshore geology of Scotland in relation to the IGS shallow
                                        drilling programme, 1970–1978. *Rep. Inst. geol. Sci.*, 81/12,
                                        36 pp.

EWING, C. J. &   1960 Nos. 1 and 2 off-shore borings in the Firth of Forth (1955–1956).
  FRANCIS, E. H.   *Bull. geol. Surv. G.B.*, **16**, 1–47.

EYLES, V. A. *et al.*   1949 Geology of Central Ayrshire. *Mem. geol. Surv. U.K.*, 160 pp.

FIELDING, C. R. *et al.*   1988 Deltaic sedimentation in an unstable tectonic environment – the
                                        Lower Limestone Group (Lower Carboniferous) of East Fife,
                                        Scotland. *Geol. Mag.*, **125**, 241–255.

FORSYTH, I. H.   1979 The *Lingula* bands in the upper part of the Limestone Coal group
                                        (E_1 Stage of the Namurian) in the Glasgow district. *Rep. Inst.
                                        geol. Sci.*, 79/16, 15 pp.

FORSYTH, I. H. &   1977 The geology of east Fife. *Mem. geol. Surv. U.K.*, 284 pp.
  CHISHOLM, J. I.

FOWLER, A.   1966 The stratigraphy of the North Tyne Basin around Kielder and
                                        Falstone. *Bull. geol. Surv. G.B.*, **24**, 57–104.

FRANCIS, E. H.   1956 The economic geology of the Stirling and Clackmannan Coal-
                                        field, Scotland: Area north of the River Forth. *Coalfield Papers
                                        geol. Surv. G.B.*, No. 1, 53 pp.

                                 1961 Economic geology of the Fife Coalfields, Area II (2nd Ed.).
                                        Mem. geol. Surv. U.K., 152 pp.

                                 1978 The Midland Valley as a rift, seen in connection with the late
                                        Palaeozoic European rift system. *In* Ramberg, I. B. & Newman,
                                        E.-R. (Eds.) *Tectonics and geophysics of continental rifts*. Reidel,
                                        Dordrecht, Holland, 133–147.

                                 1979 British Coalfields. *Sci. Prog. Oxf.*, **66**, 1–23.

FRANCIS, E. H. &   1962 Skipsey's Marine Band and red coal measures in Fife. *Geol.
  EWING, C. J. C.   Mag.*, **99**, 145–152.

FRANCIS, E. H. *et al.*   1970 Geology of the Stirling district. *Mem. geol. Surv. U.K.*, 357 pp.

GEORGE, T. N.   1960 The stratigraphical evolution of the Midland Valley. *Trans. geol.
                                        Soc. Glasg.*, **24**, 32–107.

                                 1978 Eustasy and tectonics: sedimentary rhythms and stratigraphical
                                        units in British Dinantian correlation. *Proc. Yorks. geol. Soc.*,
                                        42, 229–254.

GEORGE, T. N. *et al.*   1976 A correlation of Dinantian rocks in the British Isles. *Geol. Soc.
                                        Lond. Spec. Rep.*, No. 7, 87 pp.

GIBBS, A.   1987 Development of extension and mixed-mode sedimentary basins.
                                        In Coward, M. P. *et al.* (Eds.) Continental Extensional Tecton-
                                        ics, *Geol. Soc. Spec. Publ.*, No. 28, 19–33.

GOODLET, G. A.   1957 Lithological variations in the Lower Limestone Group in the
                                        Midland Valley of Scotland. *Bull. geol. Surv. G.B.*, **12**, 52–65.

                                 1959 Mid-Carboniferous sedimentation in the Midland Valley of
                                        Scotland. *Trans. Edinb. geol. Soc.*, **17**, 217–240.

GRAHAM, D. K.   1970 Scottish Carboniferous Lingulacea. *Bull. geol. Surv. G.B.*, **31**,
                                        139–184.

GREENSMITH, J. T.   1961 The petrology of Oil-Shale group sandstones of West Lothian
                                        and southern Fifeshire. *Proc. geol. Assoc., Lond.*, **72**, 49–71.

                                 1965 Calciferous Sandstone Series sedimentation in the eastern end of
                                        the Midland Valley of Scotland. *Jl. Sed. Petrol.*, **35**, 223–242.

HALL, J.   1974 A seismic reflection survey of the Clyde Plateau Lavas in North
                                        Ayrshire and Renfrewshire. *Scott. J. Geol.*, **9**, 253–279.

HESSELBRO, S. P. &   1984 Deposition, diagenesis and structures of the Cheese Bay Shrimp
  TREWIN, N. H.   Bed, Lower Carboniferous, East Lothian. *Scott. J. Geol.*, **20**,
                                        281–296.

HODGSON, A. V.   1978 Braided river bedforms and related sedimentary structures in the
                                        Fell Sandstone Group (Lower Carboniferous) of North North-
                                        umberland. *Proc. Yorks. geol. Soc.*, **41**, 509–532.

| JOHNSON, G. A. L. | 1967 | Basement control of Carboniferous sedimentation in northern England. *Proc. Yorks. geol. Soc.*, **36**, 175–194. |

JOHNSTONE, G. S. — 1966 — The Grampian Highlands. *Br. Reg. Geol.* (3rd Edit.) HMSO, 107 pp.

KENNEDY, W. Q. — 1958 — The tectonic evolution of the Midland Valley of Scotland. *Trans. geol. Soc. Glasg.*, **23**, 106–133.

KNOX, J. — 1954 — The economic geology of the Fife coalfields, Area III. *Mem. geol. Surv. U.K.*, 134 pp.

LEEDER, M. R. — 1973 — Sedimentology and palaeogeography of the Upper Old Red Sandstone in the Scottish Border Basin. *Scott. J. Geol.*, **9**, 117–144.

1974a — Lower Border Group (Tournaisian) fluviodeltaic sedimentation and the palaeogeography of the Northumberland Basin. *Proc. Yorks. geol. Soc.*, **40**, 129–180.

1974b — Origin of the Northumberland Basin. *Scott. J. Geol.*, **10**, 283–296.

1976 — Sedimentary facies and the origin of basin subsidence along the northern margin of the supposed Hercynian ocean. *Tectonophysics*, 167–179.

1982 — Upper Palaeozoic basins of the British Isles – Caledonide inheritance versus Hercynian plate margin processes. *J. geol.Soc. Lond.*, **139**, 479–491.

LEITCH, D. — 1941 — The Upper Carboniferous rocks of Arran. *Trans. geol. Soc. Glasg.*, **20**, 141–154.

LUMSDEN, G. I. — 1964 — The Limestone Coal Group of the Douglas Coalfield, Lanarkshire. *Bull. geol. Surv. G.B.*, **21**, 37–71.

1965 — The base of the Coal Measures in the Douglas Coalfield, Lanarkshire. *Bull. geol. Surv. G.B.*, **22**, 80–91.

1967 — The Upper Limestone Group and Passage Group of Douglas, Lanarkshire. *Bull. geol. Surv. G.B.*, **27**, 17–48.

LUMSDEN, G. I. & CALVER, M. A. — 1958 — The stratigraphy and palaeontology of the Douglas Coalfield. *Bull. geol. Surv. G.B.*, **15**, 32–70.

LUMSDEN, G. I. *et al.* — 1967 — The geology of the neighbourhood of Langholm. *Mem. geol. Surv. U.K.*, 255 pp.

McADAM, A. D. & TULLOCH, W. — 1985 — Geology of the Haddington district. *Mem. Br. geol. Surv.*, 99 pp.

MACGREGOR, A. G. — 1960 — Division of the Carboniferous on Geological Survey Scottish maps. *Bull. geol. Surv. G.B.*, **16**, 127–130.

MACGREGOR, A. R. — 1968 — *Fife and Angus Geology.* Blackwood, Edinburgh & London, 266 pp.

MACGREGOR, M. — 1965 — Excursion guide to the geology of Arran. *Geol. Soc. Glasg.*, 192 pp.

McLEAN, A. C. — 1978 — Evolution of fault-controlled ensialic basins in north-western Britain. *In* Bowes, D. R. & Leake, B. E. (Eds.) *Crustal evolution in northwestern Britain and adjacent regions, Geol. J. Spec. Issue,* No. 10, 325–346.

McLEAN, A. C. & DEEGAN, C. E. — 1978 — The solid geology of the Clyde sheet. *Rep. Inst. geol. Sci.*, No. 78/9, 114 pp.

MADDOX, S. J. & ANDREWS, J. E. — 1987 — Lithofacies and stratigraphy of a Dinantian non-marine dolostone from the Lower Oil-Shale Group of Fife and West Lothian. *Scott. J. Geol.*, **23**, 129–147.

MANSON, W. — 1953 — *In* Sum. Prog. geol. Surv. UK, 1951.

1957 — On the occurrence of a marine band in the *Anthraconaia modiolaris* Zone of the Scottish Coal Measures. *Bull geol. Surv. G.B.*, **12**, 66–86.

MITCHELL, G. H. & MYKURA, W. — 1962 — The geology of the neighbourhood of Edinburgh. *Mem. geol. Surv. U.K.*, 159 pp.

MONRO, S. K.   1982   Sedimentation, stratigraphy and tectonics in the Dalry Basin, Ayrshire. *Univ. Edinburgh Ph.D. Thesis* (unpublished).

MONRO, S. K. *et al.*   1983   The Ayrshire Bauxitic Clay: an allochthonous deposit? *In* Wilson, R. C. L. (Ed.) *Residual Deposits, Spec. Publ. Geol. Soc. Lond.*, No. 11, 47–58.

MUIR, R. O.   1963   Petrography and provenance of the Millstone Grit of central Scotland. *Trans. Edinb. geol. Soc.*, **19**, 439–485.

MYKURA, W.   1960   The replacement of coal by limestone and the reddening of coal measures in the Ayrshire Coalfield. *Bull. geol. Surv. G.B.*, **16**, 69–109.

                                1965   The age of the lower part of the New Red Sandstone of south-west Scotland. *Scott. J. Geol.*, **1**, 8–18.

                                1967   The Upper Carboniferous rocks of south-west Ayrshire. *Bull. geol. Surv. G.B.*, **26**, 23–98.

NAIRN, A. E. M.   1956   The Lower Carboniferous rocks between the River Esk and Annan, Dumfriesshire. *Trans. geol. Soc. Glasg.*, **22**, 80–93.

NEVES, R. *et al.*   1965   Note on the recent spore and goniatite evidence from the Passage Group of the Scottish Upper Carboniferous Succession. *Scott. J. Geol.*, **1**, 185–188.

NEVES, R. *et al.*   1972   A scheme of miospore zones for the British Dinantian. *C. R. 7me. Congr. Int. Strat. geol. Carb.* Krefeld 1971, **1**, 347–353.

NEVES, R. *et al.*   1973   Palynological correlations within the Lower Carboniferous of Scotland and northern England. *Trans. roy. Soc. Edinb.*, **69**, 23–70.

OWENS, B. *et al.*   1977   Palynological division of the Namurian of northern England and Scotland. *Proc. Yorks geol. Soc.*, **41**, 381–398.

PATERSON, I. B. &   1986   Lithostratigraphy of the late Devonian and early Carboniferous
    HALL, I. H. S.   rocks in the Midland Valley of Scotland. *Rep. Br. Geol. Surv.*, v, 18 [3], 14 pp.

PICKEN, G. S.   1988   The concealed coalfield at Canonbie: an interpretation based on boreholes and seismic surveys. *Scott. J. Geol.*, **24**, 61–71.

RAMSBOTTOM, W. H. C.   1973   Transgressions and regressions in the Dinantian: a new synthesis of British Dinantian stratigraphy. *Proc. Yorks. geol. Soc.*, **39**, 567–607.

                                1977   Major cycles of transgression and regression (Mesothems) in the Namurian. *Proc. Yorks geol. Soc.*, **41**, 261–291.

READ, W. A.   1959   The economic geology of the Stirling and Clackmannan Coalfield; Area south of the River Forth. *Coalfield Papers geol. Surv. G.B.*, No. 2, 73 pp.

                                1961   Aberrant cyclic sedimentation in the Limestone Coal Group of the Stirling Coalfield. *Trans. Edinb. geol. Soc.*, **18**, 271–292.

                                1965   Shoreward facies changes and their relation to cyclical sedimentation in part of the Namurian east of Stirling, Scotland. *Scott. J. Geol.*, **1**, 69–92.

                                1981   Facies breaks in the Scottish Passage Group and their possible correlation with the Mississippean–Pennsylvanian hiatus. *Scott. J. Geol.*, **17**, 295–300.

                                1988   Controls on Silesian sedimentation in the Midland Valley of Scotland. *In* Besly, B. M. & Kelling, G. (Eds.) *Sedimentation in a synorogenic basin complex: the Upper Carboniferous of Northwest Europe.* Blackie, London, 222–241.

                                1989   Sedimentological evidence for a major subsurface fracture system linking the eastern Campsie and the eastern Ochil faults. *Scott. J. Geol.*, **25**, 187–200.

READ, W. A. &   1967   A quantitative study of a sequence of coal-bearing cycles in the
    DEAN, J. M.   Namurian of Central Scotland. *Sedimentology*, **9**, 137–156.

READ, W. A. &
DEAN, J. M. — 1978 — High-alumina fireclays in the Passage Group of the Clackmannan Syncline, Scotland. *Rep. Inst., Geol. Soc.*, 78/12, 21 pp.

READ, W. A. &
FORSYTH, I. H. — 1962 — The correlation of the Limestone Coal Group above the Kilsyth Coking Coal in the Glasgow–Stirling region. *Bull. geol. Surv. G.B.*, **19**, 29–52.

READ, W. A. &
JOHNSON, S. R. H. — 1967 — The sedimentology of sandstone formations within the Upper Old Red Sandstone and lowest Calciferous Sandstone Measures west of Stirling, Scotland. *Scott. J. Geol.*, **3**, 242–267.

READ, W. A. &
MERRIAM, D. F. — 1966 — Trend-surface analysis of stratigraphic thickness data from some Namurian rocks east of Stirling, Scotland. *Scott. J. Geol.*, **2**, 96–100.

REX, G. M. & SCOTT, A. C. — 1987 — The sedimentology, palaeoecology and preservation of the Lower Carboniferous plant deposits at Pettycur, Fife, Scotland. *Geol. Mag.*, **124**, 43–66.

RICHEY, J. E. *et al.* — 1930 — The geology of North Ayrshire. *Mem. geol. Surv. U.K.*, 417 pp.

ROBERTSON, T. — 1948 — Rhythm in sedimentation and its interpretation: with particular reference to the Carboniferous sequence. *Trans. Edinb. geol. Soc.*, **14**, 141–175.

RUSSELL, M. J. — 1972 — North–south geofractures in Scotland and Ireland. *Scott. J. Geol.*, **8**, 75–84.

SCHRAMM, F. R. — 1983 — Lower Carboniferous biota of Glencartholm, Eskdale, Dumfriesshire. *Scott. J. Geol.*, **19**, 1–15.

SCOTT, A. C. — 1976 — Fossil plants from the Barren Red Measures, near Ochiltree, Ayrshire. *Proc. geol. Soc. Glasg.*, **117**, 9–12.

— 1977 — A review of the ecology of Upper Carboniferous plant assemblages, with new data from Strathclyde. *Palaeontol.*, **20**, 447–473.

— 1979 — The ecology of Coal Measure floras from northern Britain. *Proc. geol. Assoc.*, **90**, 97–116.

SCOTT, A. C. *et al.* — 1984 — Distribution of anatomically preserved floras in the Lower Carboniferous in Western Europe. *Trans. roy. Soc. Edinb.: Earth Sci.*, **75**, 311–340.

SCOTT, W. B. — 1986 — Nodular carbonates in the Lower Carboniferous Cementstone Group of the Tweed Embayment, Berwickshire; evidence for a former sulphate evaporite facies. *Scott. J. Geol.*, **22**, 325–345.

SIMPSON, J. B. &
RICHEY, J. E. — 1936 — The geology of the Sanquhar Coalfield and the adjacent basin of Thornhill. *Mem. geol. Surv. U.K.*, 97 pp.

SMITH, A. G. *et al.* — 1981 — *Phanerozoic palaeocontinental world maps.* Cambridge University Press, 102 pp.

SMITH, A. H. V. &
BUTTERWORTH, M. — 1967 — Miospores in the coal seams of the Carboniferous of Great Britain. *Spec. Pap. Palaeont.*, **1**, 324 pp.

SMITH, T. E. — 1967 — A preliminary study of sandstone sedimentation in the Lower Carboniferous of the Tweed Basin. *Scott. J. Geol.*, **3**, 282–305.

THOMSON, M. E. — 1977 — The geology of the Firth of Forth and its approaches. *Rep. Inst. geol. Sci.*, No. 77/17, 56 pp.

TULLOCH, W. &
WALTON, H. S. — 1958 — The geology of the Midlothian Coalfield. *Mem. geol. Surv. U.K.*, 157 pp.

TURNER, P. &
TARLING, D. H. — 1975 — Implications of new palaeomagnetic results from the Carboniferous of Britain. *Jl. geol. Soc. Lond.*, **131**, 469–488.

WAGNER, R. H. — 1983 — A lower Rotliegend Flora from Ayrshire. *Scott. J. Geol.*, **19**, 135–155.

WEIR, J. & LEITCH, D. — 1936 — The zonal distributions of non-marine lamellibranchs in the Coal Measures of Scotland. *Trans. roy. Soc. Edinb.*, **57**, 697–751.

WILSON, H. H. — 1952 — The Cove Marine Bands in East Lothian and their relation to the ironstone shale and limestone of Redesdale. *Geol. Mag.*, **89**, 305–319.

WILSON, R. B.   1962   A review of the evidence for a 'Nebraskan' fauna in the Scottish
                                         Carboniferous. *Palaeontol., Lond.*, **4**, 507–519.

                                 1966   A study of the Neilson Shell Bed, a Scottish Lower Carbon-
                                         iferous marine shale. *Bull. geol. Surv. G.B.*, **24**, 105–130.

                                 1967   A study of some Namurian faunas in central Scotland. *Trans.
                                         roy Soc. Edinb.*, **66**, 445–490.

                                 1974   A study of the Dinantian marine faunas of south-east Scotland.
                                         Bull. geol. Surv. G.B., **46**, 35–65.

                                 1979   The base of the Lower Limestone Group (Viséan) in North
                                         Ayrshire. *Scott. J. Geol.*, **15**, 313–319.

                                 1989   A study of the Dinantian marine macrofossils of central
                                         Scotland. *Trans. roy. Soc. Edinb.: Earth Sci.*, **80**, 91–126.

Note: (p. 354) The East Kirkton Limestone underlying the Hurlet Limestone (p. 357 and Fig. 10.9) has recently yielded a unique terrestrial fauna that includes the earliest known reptile.

SMITHSON, T. R.   1989   The earliest known reptile. *Nature*, **342**, 676–678.

11

CARBONIFEROUS–PERMIAN IGNEOUS ROCKS
E. H. Francis

It will have been apparent from Chapter 10 that a considerable proportion of the Scottish Carboniferous sequence is formed of lavas and tuffs. Most of these are alkaline in composition and affinity, as are the minor intrusions of basic dykes and sills forming three distinct groupings. One comprises swarms of olivine–dolerite, camptonite and monchiquite dykes emplaced during mid-Carboniferous to late Permian pulses mainly into the crystalline basement of the Highlands where extrusive equivalents, if any, are not preserved. A second group forms alkali–dolerite sills emplaced, unaccompanied by dykes, in the sedimentary basins of the Midland Valley where they are almost certainly comagmatic with the extrusive rocks, particularly the pyroclastics, and thus span a wide age-range. The third consists of tholeiitic sills (the Midland Valley Sill complex) and dykes, apparently emplaced during a single late Carboniferous magmatic pulse in a separate stress regime; it contrasts with the alkali–dolerite sills in lacking any known extrusive equivalents.

The distribution of erupted rocks shows a broad change with time. The greatest volume – c. 6,000 km³ according to Tomkeieff (1937) – was produced during the Dinantian in the form of lavas, of which c. 85 per cent were mildly undersaturated basalts, the remainder consisting of under-saturated and saturated trachytic and rhyolitic differentiates (Fig. 11.1). Although further lavas were emitted during Namurian and early Permian times, the bulk of the post-Dinantian activity was phreatomagmatic and gave rise to pyroclastic rocks smaller in total volume than the lavas, but from a large number of centres, the magmas consisting entirely of basalts and basanites with no felsic differentiates.

Interpretations of the morphology of the Dinantian volcanoes differ. A 27-km line of vents trending NE–SW between Dumbarton and Fintry (White & MacDonald 1974) and one of similar trend along the line of the Campsie Fault (Craig & Hall 1975) are assumed to have fed large central volcanoes. By contrast, the lateral continuity (up to 6 km) of most of the Campsie flows is taken by Read (*in* Francis *et al.* 1970) to suggest fissure eruption. A shield volcano of Hawaiian type might resolve the differences indicated by the Campsie evidence, but at least one strato-volcano is likely to have existed there on the evidence of the caldera identified by Craig (1980; Fig. 11.4). Other strato-volcanoes may have formed elsewhere, judging from the relative abundance of acid rocks farther south and in the Renfrewshire Hills (Johnstone 1965). Indeed, the welded trachytic ashflow tuffs found in East Lothian (Upton 1982) suggest both eruption from such a central structure and also the likelihood of caldera collapse, though the focus is not yet located.

The classification of the Carboniferous–Permian basalts, traditionally based on phenocryst content, was standardised by MacGregor (1928) so as to provide a sixfold grouping which continues to form an invaluable basis for field description. Its relationship to current international usage has been outlined by Upton (1969) and Macdonald (1975) and is reproduced in Table 11.1. Flows are commonly composite, showing lateral variation (e.g. from mugearite to Markle type or Markle type to Dunsapie type) indicative of changes in composition during eruption (Kennedy 1931; MacDonald 1967) which may reflect, in turn, differentiation either in a magma chamber or during ascent.

Volcanic stratigraphy

The time–space distribution of volcanism originally shown graphically as Fig. 10.1 of the first edition of this book has progressed through several modifications (e.g. Upton 1969, De Souza 1979, Cameron & Stephenson 1985) and the version here forming Fig. 11.2 is no more than the best reconciliation possible at the time of writing between the sometimes conflicting evidence provided by stratigraphical relationships and radiometric age-dating.

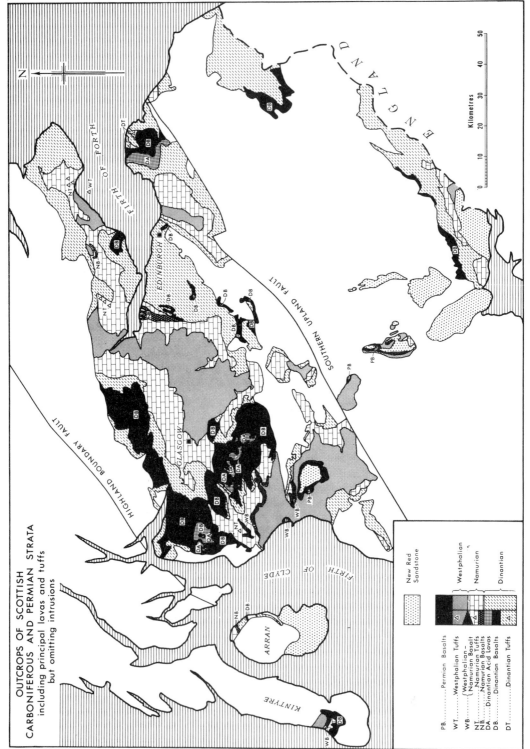

Fig. 11.1. Outcrops of Carboniferous and Permian volcanic rocks in the Midland Valley and Borders.

Table 11.1

	Basalt type of MacGregor, 1928	Phenocrysts		
		Essential	Sometimes present	
Microphyric (< 2 mm)	Jedburgh	Plag, Ol		Microphyric Ol-plag ± Fe-oxides-phyric (basaltic) hawaiites (occasionally basalt)
	Dalmeny	Ol	Cpx, Plag,	Microphyric Ol– or Ol + cpx -phyric basalt
	Hillhouse	Ol	Cpx	Microphyric Ol ± cpx-phyric basalt
Macroporphyritic (> 2 mm)	Markle	Plag, Ol		Macroporphyritic Ol-plag ± Fe-oxides-phyric basalts, basaltic hawaiites or hawaiites
	Dunsapie	Plag, Ol Cpx		Macroporphyritic Ol-cpx-pl -Fe-oxides-phyric basaltic hawaiites, or Ol-cpx-pl- phyric basalts
	Craiglockhart	Ol + Cpx		Ankaramite

Ol = olivine, cpx = clinopyroxene, plag. = plagioclase

Evidence of interbedding is uncontroversial only in the late Viséan, Namurian and early Westphalian where the stratigraphy of the sediments is closely controlled. The problem of dating volcanics within the earlier and later, mainly continental sedimentary sequences has been considerably resolved by miospore dating and whilst some anomalies remain (Paterson & Hall 1986) they are ignored in Fig. 11.2. Radiometric ages obtained by whole-rock K–Ar analyses (De Souza 1979, 1982) are mostly internally consistent, but the time-scale based on them is challenged (Lippolt *et al.* 1984) by $^{40}Ar/^{39}Ar$ determinations on single minerals – a technique which also brings into question the whole-rock dates themselves as well as long-held views on petrographic affinites of some of the Scottish suites (Henderson *et al.* 1987). The unsuitability for whole-rock K–Ar dating of many rock types must also be stressed. For instance, pyroclastic rocks which were the main mid-Carboniferous extrusions, are not suitable for analysis or age dating and consequently tend to be under-represented in graphs and histograms designed to depict petrochemical trends or peaks of activity. In east Fife, for example, bedded tuffs indicate almost continuous phreatomagmatic activity from one locality or another almost through the Namurian, but only the sills give corresponding dates (Macintyre *et al.* 1981); all the necks which contain intrusions fresh enough for isotopic dating appear to be Stephanian or early Permian (Forsyth & Rundle 1978).

Dinantian extrusive activity

Midland Valley

Activity in the Midland Valley during Tournaisian times is represented by the Craiglockhart Volcanic Rocks, comprising 60 m of basic tuffs overlain by 30 m of basaltic lavas. They lie at or near the base of the former 'Cementstone Group' (Mitchell & Mykura 1962), that is in the middle of the Inverclyde Group.

It is now accepted, however, that the bulk of the outpourings in the Midland Valley were Viséan. Although details of timing remain unresolved, a case can be made for assuming the onset of activity to have been approximately simultaneous, early in the Viséan, across the region (Fig. 11.3). In the west the case depends on the new classification (Chapter 10) which takes the base of the Clyde Plateau Volcanic Formation as approximating to the Tournaisian–Viséan boundary; De Souza's 1982 misgivings over his own 1979 dating of the earliest of these lavas can be offset by evidence of early Viséan miospores *above* the base of the lavas in some places (Paterson & Hall 1986). In the south–central and eastern parts of the Midland Valley the presence of PU zone miospores respectively in or just below the lowest of the Arthur's Seat and Garleton Hills volcanic formation similarly points to early Viséan start of activity.

Some degree of diachronism at the base of the lavas is nevertheless to be expected as well as unconformities, though it must be recalled that unconformities in volcanic terrains tend to be more angular and represent relatively shorter time spans than they do in other depositional environments.

Diachronism is most apparent in the west, for basalt flows and pyroclasts of the Clyde Plateau rest on the Tournaisian Ballagan Formation in Stirlingshire and on Old Red Sandstone sediments in Ayrshire. Moreover, it seems possible (De Souza 1979) that the earliest Viséan eruptions equivalent to the Arthur's Seat rocks are preserved only locally in the west, where they include the fault-bounded macrophyric hawaiites of the Beith Hills in Ayrshire. De Souza (op. cit.) suggests that these early eruptions were followed by a long period of uplift and extensive erosion preceding the main outpourings of lava forming the Clyde Plateau in late Viséan times. The lava formation once covered an area of at least 3,000 km² to a thickness of up to 1 km. It now forms a 'C'-shaped outcrop in the western part of the Midland Valley (Fig. 11.1),

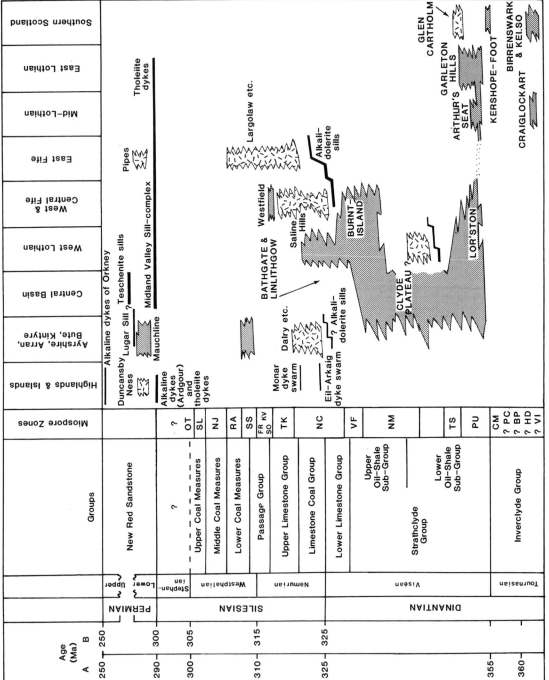

Fig. 11.2. Range and distribution of Carboniferous and Permian extrusive rocks in Scotland. Intrusions (dykes and sills) black; mainly lavas stippled; mainly volcaniclastics black triangles. Timescale Column A after Harland *et al.* (1982) and Forster & Warrington (1985); Column B after Lippolt *et al.* (1984).

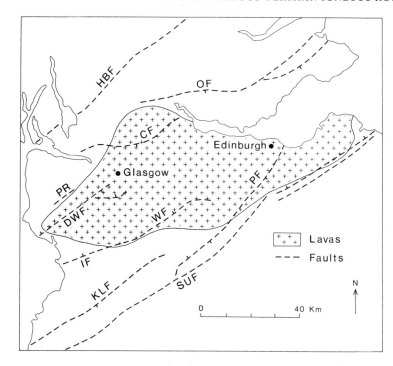

Fig. 11.3. Possible extent of early Viséan lavas in the Midland Valley, Major faults active at that time are also shown. CF, Campsie Fault; DWF, Dusk Water Fault; HBF, Highland Boundary Fault; IF, Inchgotrick Fault; KLF, Kerse Loch Fault (Dron line); PF, Pentland Fault; PR, Paisley Ruck; SUF, Southern Uplands Fault; WF, Wilsontown Fault (De Souza 1979).

extending eastward beneath younger sediments. It also extends eastward and northward in the form of sporadic outliers. To the west the formation is reduced to 135 m of basalts overlying 15 m of agglomerates in Arran, while still farther west, in Kintyre, an uncertain thickness of basalts, hawaiites, mugearites, benmoreites, trachytes and rhyolites (Macdonald 1975) suggests a source or sources quite separate from those of the mainland.

The northern arc of the 'C'-shaped outcrop is formed by the Fintry, Gargunnock and Touch Hills, where an early group of macroporphyritic hawaiites are overlain discontinuously by nearly 50 m of volcaniclastic beds. These are the Slackgun Interbasaltic Beds of Read (*in* Francis *et al.* 1970), reinterpreted by Craig (1980) as the degraded remains of a line of ash-cones. They are overlain by nearly 400 m of lavas in a sequence comprising a lower group of microphyric hawaiites and a thicker upper group of macrophyric hawaiites with subordinate microphyric basalts, mugearites and trachybasalts. This broad twofold subdivision continues westward into the Campsie Fells and Kilsyth Hills (Whyte & MacDonald 1974, Craig & Hall 1975), though the lower is locally faulted out or subordinated by locally derived macrophyric hawaiites. The sequence continues above a thick basal accumulation of tuffs even farther west in the Kilpatrick Hills, where early and late basalts of more basic composition (including ankaramites) are additionally recorded (Hamilton 1956). This northern part of the Clyde Plateau includes two

prominent NE–SW lines of necks, one extending 27 km from Fintry to Dumbarton (Whyte & MacDonald 1974), the other approximating to the line of the Campsie Fault (Craig & Hall 1975). One of the former, the Meikle Binn, contains trachytic agglomerate, and though no lavas of such composition have been preserved, field evidence supported by geophysical survey (Cotton 1968) indicates that later in the volcanic history a major central volcano was focused here or in the Waterhead area nearby, where Craig (1980) has identified the remains of a caldera (Fig. 11.4).

Beyond Stirling this northern edge of the Clyde Plateau Volcanic Formation is obscured beneath younger sediments of the Kincardine Basin, but it reappears first at Monks Grave, near Dollar, in the form of basalts and a felsite formerly assigned to the Lower Old Red Sandstone (Browne & Thirlwall 1981), and finally in the Cleish Hills of Fife, where *c.* 30 m from the base of an attenuated Dinantian succession, 30 m of tuffs are intercalated with a lens of felsite below and impersistent basalts above.

Outcrops south of the Clyde form a suite of rocks which Macdonald (1975) recognises as an entirely different volcanic association from those to the north and which, moreover, appears to have had different volcano–tectonic controls. Although the lavas show rapid variations of thickness along NE–SW fractures, such as the Duskwater and Inchgotrick faults, as well as minor NE folds (Richey *et al.* 1930; McLean 1966), there is seismic evidence of original

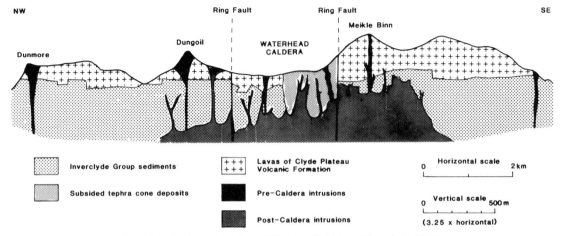

Fig. 11.4. Horizontal section of Waterhead Caldera (after Craig 1980).

thicknesses having reached their maxima (up to nearly 1 km) along NW–SE lines accentuated by later folding (Hall 1974). The lavas comprise a wider range of alkaline types than those to the north and include macrophyric and microphyric basalts, hawaiites, mugearites, benmoreites, trachytes and rhyolites. An original twofold basic-acid sequence advanced by Richey *et al.* (1930) for the Renfrewshire, Kilbirnie and Eaglesham Hills has been substantially modified by Johnstone (1965) who recognises two trachytic phases within a dominantly basaltic succession. In general, however, the more evolved lavas appear to have been erupted about midway in the pile. The ESE line of trachytic necks extending from Irish Law through Misty Law to Loudoun Hill appears to have been the main source of the rocks. The even more extensive acid lavas cropping out farther east in the Lanarkshire Block, between Lugton and Strathaven must also have been derived from local vents, but these are now concealed by younger flows.

In the middle and to the east of the 'C'-shaped outcrop, the volcanic formation continues at least as far as the axis of the Central Coalfield basin, as established by the microphyric olivine-basalts and hawaiites cored in the Rashiehill Borehole at Slammanan (Anderson 1963), but still farther east all except perhaps the lowest lavas give way abruptly to the thick sedimentary sequence of the oil-shale field of West Lothian. The transition is marked by tuffs, particularly in the Lower Oil-Shale Sub-group near West Calder where the Crosswood Ash and overlying Seafield-Deans Ash are *c.* 100 and 250 m thick respectively (Mitchell & Mykura 1962). Several thinner, but widespread layers, including the Port Edgar, Barracks, Two Foot and Hopefield 'ashes' also appear in the Upper Oil-Shale Sub-group farther north. Other records in this area (op. cit.) show that the source of the pyroclastic rocks must be in the transition zone and it is tempting to suggest that that source may be preserved, at depth, in the form of a thick volcaniclastic pile analogous to

those of the Saline Hills and Largo Law in Fife (p. 408 and Figs. 11.11, 11.15). This might account for the Bathgate gravity and magnetic anomaly (Davidson *et al.* 1984) as well as reconciling that anomaly with the absence of any crystalline intrusion structure discernible by seismic traverse (Conway *et al.* 1987).

Lithological and mapping evidence suggest that the lowest basaltic lavas of the Clyde Plateau Volcanic Formation continue impersistently around the southern rim of the oil-shale field to link with the Arthur's Seat Volcanic Rocks of Edinburgh. They can also be identified at depth beneath the oil-shales in boreholes at West Calder and Blackness.

The main outcrop of the Arthur's Seat Volcanic Rocks, in Edinburgh, marks the site of a separate centre, with agglomerate necks and basaltic plugs flanked by between 400 and 500 m of lavas, including 13 flows (Fig 11.5) ranging in composition from ankaramite to mugearite (Upton 1969). Some indication that these rocks continue eastwards to link with the Garleton Hills Volcanic Formation of East Lothian is provided by two boreholes sunk through the sedimentary cover in the intervening coalfields. Midlothian No. 1 Borehole, nearest to Edinburgh (12 km) encountered a similar though thinner sequence to that of Queen's Park, comprising 75 m of basalts and mugearites at the same stratigraphic position (Mitchell & Mykura 1962). Basic rocks also occur at the same horizon farther east in the Spilmersford Borehole (McAdam 1974) where 68 m of basaltic tuffs and agglomerates are overlain by 153 m of hawaiites (15 flows ranging from 3 to 16 m in thickness) intercalated with basic tuffs; here, however, more acid rocks appear in the form of 36 m of trachytic tuffs overlying the hawaiites. These must be the distal equivalent of the upper part of the Garleton Hills Volcanic Formation of East Lothian, which at its maximum 32 km E of Edinburgh, comprises 200 m of basaltic tuffs overlain by 160 m of hawaiite lavas, with subordinate mugearites, ankaramites and kulaites, capped in turn by 160 m of trachytic lavas and tuffs (Upton 1969). The ages of the acid rocks (De Souza 1979) relate them not only to the Eildon Hills episode farther south, but also to the well-known phonolitic intrusions of East Lothian, such as Traprain Law, North Berwick Law and the Bass Rock (Fig. 11.6) – all representing further and approximately penecontemporaneous centres of activity.

Towards the end of the Viséan times activity shifted north-eastwards to become focused around the Bathgate Hills of Linlithgow and the Burntisland Anticline of Fife (Allan 1924). The lavas of both areas are similar and contrast with the earlier Clyde Plateau flows in being more restricted in composition – mainly microphyric olivine- and pyroxene-basalts with subordinate ankaramites. There appears to have been physical overlap between the Clyde and Bathgate formations at Slamannan (Fig. 11.7). The Bathgate lavas, with subordinate tuffs, locally represent the highest part of the Upper Oil-Shale Sub-group and most of the Lower Limestone Group. There and at Burntisland, boles and intercalations of marine limestones give evidence of alternating subaerial weathering and inundation by the sea. The Burntisland pile appears to thicken eastwards, reaching a known maximum of 430 m without reaching base, in Seafield Colliery at Kirkcaldy (Francis 1961a) and it crops out farther east on the island of Inchkeith where it is at least 150 m thick (Davies 1936).

Fig. 11.5. Queen's Park, Edinburgh, Lion's Head and Haunch necks in background; Salisbury Crags teschenite sill to right; basalt lavas in left middle distance; Holyrood Palace in foreground (Planair, Edinburgh copyright).

Borders

The Birrenswark Lavas, cropping out discontinuously north of the Solway (Fig. 11.8) form a pile up to 90 m thick of short interdigitated flows of transitional (hawaiites) to alkaline basalts with thin intercalated red sediments (Pallister 1952, Elliott 1960, Lumsden *et al.* 1967). The presence of thick overlying Tournaisian sediments establishes this as the earliest phase of Carboniferous volcanism in Scotland – penecontemporaneous with the petrologically similar Kelso Traps, some 120 m thick, forming a 'C'-shaped outcrop of *c.* 160 km² in the Tweed Basin (Eckford & Ritchie 1939; Tomkeieff 1945, 1953). The two lava fields were not continuous, though both emanated from necks aligned north-eastwards along the northern margin of the Northumberland Trough (Leeder 1974).

Some of these necks are the sites of intermittent later Viséan activity on a smaller scale. It was represented first by the 60 m Kershopefoot Basalt of Liddesdale, then by the Glencartholm Volcanic Beds, consisting of basaltic and trachytic tuffs and tuffaceous sediments with a few thin basalt lavas (Lumsden *et al.* 1967). Isolated lavas of uncertain stratigraphical position near Newcastleton, Carter Fell and Dunbar are also probably Viséan (Greig 1971). The trachytic fragments in the Glencartholm Volcanic Beds can be matched by the composition of some of the necks as well as by the larger intrusions of trachytic riebeckite microgranite and phonolite in the Eildon Hills (McRobert 1914, Irving 1930).

Namurian extrusive activity

In the Bathgate–Hilderston area, microphyric basalts continued to be erupted to a thickness of over 200 m,

Fig. 11.6. The Bass Rock (phonolitic trachyte plug). The background is the East Lothian shore of the Firth of Forth and is formed by Dinantian volcanic rocks including the phonolitic trachyte plug of North Berwick Law to the left (Planair, Edinburgh copyright).

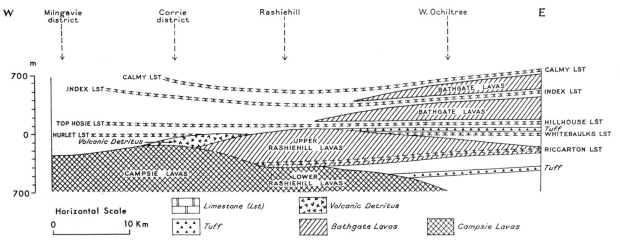

Fig. 11.7. Horizontal section to show relationship between lavas of Clyde Plateau and of Bathgate Hills (Anderson 1963, fig. 3).

representing much of the early Namurian Limestone Coal Group and interfingering with sediments northwards and westwards. Elsewhere, however, the beginning of the Namurian coincided with a distinct change in the pattern of activity. It continued intermittently both in Ayrshire and in Fife, but on a much smaller scale, and it was represented by phreatomagmatic eruptions from multiple, generally short-lived, centres. In Ayrshire the tuffs appear at several levels in the Limestone Coal Group and extend up into the Upper Limestone Group as high as the Calmy Limestone (Richey *et al.* 1930). The thicker beds occupy the positions of coal seams and the Dalry Blackband Ironstone. The source of

Fig. 11.8. Map to show generalised distribution of Lower Carboniferous volcanics in the Northumberland Trough and isopachs of the Birrenswark Lavas (after Leeder 1974).

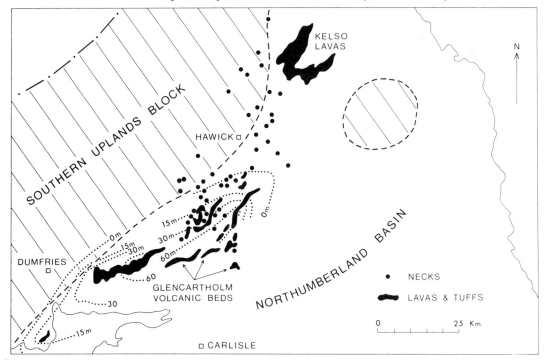

the tuffs is believed to lie farther west, where several necks are seen to cut older strata. The tuffs consist of decomposed green basaltic material with much sedimentary debris.

In Fife, tuffs of Namurian age are scattered over a much wider area. The main centre during Limestone Coal Group times was in the west, around Saline, where almost throughout the period thick tuffs with only a few patches of lava were accumulated from several necks (Francis 1961b). The tuffs, like those of Ayrshire, are admixtures, in various proportions, of altered basalt and comminuted sedimentary rock. Farther east, between Kirkcaldy and Elie, several separate, though minor accumulations of similar pyroclastic rock are recognised and activity of this kind continued both in west and east Fife throughout Upper Limestone Group times, culminating at Westfield with an episode which produced five flows of pillow lava associated with tuffs and hyaloclastites.

Fig. 11.9a. Graded kaolinised tuff in mudstone, near base of Namurian, Kinneil Colliery, Bo'ness.

Fig. 11.9b. Tonstein-like kaolinised tuff in the "Jersey Yellowstone" of the mid-Fife Limestone Coal Group.

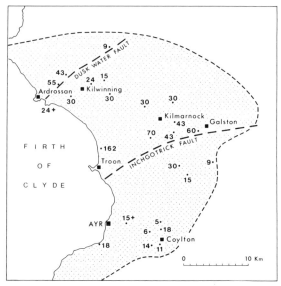

Fig. 11.10. Map showing original extent of Passage Group lavas in Ayrshire. Figures give thickness in metres (after Richey *et al.* 1930, fig. 26).

Although many of the centres of explosive activity have not been located at outcrop or by mining activity, their presence can be inferred from the occurrence of thin layers of graded tuffs (Fig. 11.9) or tuffaceous siltstones representing ash and dust fall-out from eruption clouds extending from distances of up to 32 km from the originating vents (Francis 1961c, 1968a). Many layers contain a high proportion of sand grains derived from the coring out of the vents through sediments (p. 412), but some consist mainly of basaltic debris and this is invariably altered by diagenesis to kaolinite, especially in or near the coal seams (Fig. 11.9). Such kaolinised tuffs are akin to the *tonsteins* of NW European Coalfields (Price & Duff 1969; Francis 1969; Huff & Spears 1989) and are of correlation value. Moreover they can be regarded as time-planes and as such they throw light on the nature of cyclic sedimentation in coal-bearing sequences. For instance, at the one time represented by one of the late Namurian ash-falls of W Fife (Francis 1961c, figs. 5, 6) different elements of the sedimentary cycle, such as coal and the sediments either below or above it, were being formed at different places.

Phreatic activity continued on a small scale in Fife until late Namurian (Passage Group) times, though the principal centre of activity was then in Ayrshire, where a series of microphyric olivine–clinopyroxene basalts with intercalated sediments lie between Namurian and Westphalian sediments; they are thus perhaps partly Westphalian in age (Mykura 1967). They reach a maximum thickness of 162 m near Troon (Fig. 11.10) and are represented in Arran by 'red ash' (Leitch 1941). The lavas were subjected to subaerial weathering before burial by younger sediments and this weathering gave rise to the Ayrshire Bauxitic Clays (Monro *et al*. 1983).

Westphalian extrusive activity

The main area of Westphalian volcanism extends southwards from Largo Law in East Fife beneath the Firth of Forth. Largo Law, which resembles the older Saline Hills in being formed mainly of pyroclastic rocks with only a few localised patches of lava, may be partly late Namurian in age. Westphalian activity is, however, clearly evidenced by the Westphalian A spores in the tuffaceous sediments which

have subsided into the neighbouring Viewforth Neck (Forsyth & Chisholm 1977) and by drilling offshore, where most of Westphalian A is represented by a single basalt lava capping 162 m of tuffs and agglomerates, their base not proved (Fig. 11.11).

Westphalian activity ended with the intrusion of a tholeiitic suite of dykes and a sill-complex (p. 407)–a single magmatic pulse initiated during an entirely new, short-lived stress regime and lacking any known extrusive expression.

Permian extrusive activity

The NW–SE to N–S rifting which appears to have influenced Permian sedimentation (McLean 1978) is also

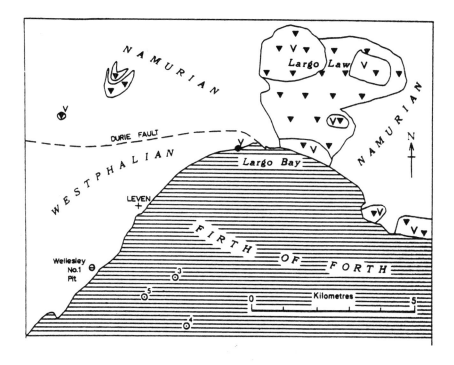

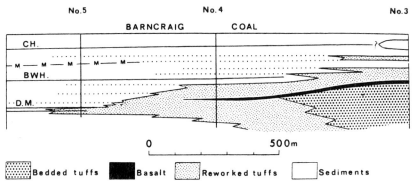

Fig. 11.11. Map and section to show distribution of Westphalian volcaniclastic and volcanic rocks around the Fife coast of the Forth. M, *A. vanderbeckei* Marine Band; CH, Chemiss Coal; BWH, Bowhouse Coal; D.M., Dysart Main Coal.

reflected in the distribution of volcanic rocks – all now reassigned from Stephanian to Permian by revisions of timescale as well as review of the related flora (Chapter 10). The westernmost rift, tangential to the Ulster coast at Larne, contains more than 616 m of basalts and tuffs (Penn *et al.* 1983). The next rift to the east is the Islay–Machrihanish–Stranraer line, where, on the island of Glas Eilean in the sound of Jura, some 120 m of alkali–olivine-basalts formerly assumed to be Carboniferous have now been dated and reinterpreted as the products of an isolated Permian volcano (Upton *et al.* 1987).

Perhaps the most extreme outpourings of this time, however, are contained within the next rift of the east, which contains the main Mauchline Basin of Ayrshire and the outlying basins of Sanquhar and Thornhill. In Ayrshire, the Mauchline Volcanic Formation comprises olivine-clinopyroxene–phyric basalts with a few basanites and with intercalated beds of tuff, agglomerate, desert sandstone and mudstone collectively reaching a maximum thickness of 238 m (Mykura 1967). More than 60 necks which probably acted as feeders to the volcanics are located within a radius of 30 km around the outcrop of the Formation. Further necks are recorded at Sanquhar, but here and farther south-east in Thornhill, the extrusive basalt thickness is no more than 50 m (Simpson & Richey 1936).

Fracture control over volcanism is further evidenced by the Permian subaerial basaltic flows of the Central North Sea Graben (Dixon *et al.* 1981) and the edge of the Magnus Trough, north of Shetland (Hitchen & Ritchie 1987). It is, however, less apparent in the intervening ground, particularly in Fife, Duncansby Head (Caithness) and the Orkneys where pipes now radiometrically dated as Permian (Forsyth & Rundle 1978; Macintyre *et al.* 1981; Mykura 1976) are randomly distributed; if they had any extrusive products (presumably tuffs), moreover, they have since been removed by erosion.

Intrusions

Minor intrusions commonly form integral parts of the extrusive activity described above. They include plugs and apophyses of olivine-rich basalts, basanites and moni-chiquites, with subordinate phonolites, occurring in or near necks of various ages as well as small-scale dyke swarms related to eruptive centres such as Misty Law, in Renfrew-shire, and Meikle Binn in Stirlingshire (Richey 1939). More widespread and voluminous than these, however, are dyke swarms and sill complexes of various Carboniferous and Permian ages some of which are apparently related, while others are not, to the processes of extrusion.

Highland dyke swarms

Distinguishing Carboniferous–Permian dykes from those of Caledonian and Tertiary ages has been a problem of

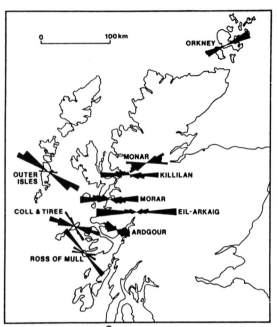

Fig. 11.12. Location and azimuth distribution of the main alkali lamprophyre dyke swarms of the Highlands and Islands (after Rock 1983 and Baxter 1987).

long-standing and is still not readily resolved in the field. However, on petrographic characteristics, Rock (1983) recognises over 3,000 dykes grouped into 9 swarms (Fig. 11.12) consisting of alkaline lamprophyres (camptonites and monchiquites) with subordinate varieties such as basanite and basalt. K–Ar dating by Speight & Mitchell (1979) and Baxter & Mitchell (1984) indicates three main intrusive pulses, during late Viséan, early Permian and late Permian. Their proposal that these times can be correlated with dyke trends (E–W Viséan, NW–SE early Permian, NE–SW late Permian) appears to conform with general Scottish volcano–tectonic associations described elsewhere in this chapter. However, trend alone is evidently an unsuitable criterion for dating emplacement judging from the discovery by Morrison *et al.* (1987) that many of the Carboniferous–Permian E–W dykes of the Ardgour Swarm were emplaced along Caledonian fractures. This leaves open to question the age of the 20 or so agglomeratic vents and bosses of Lochaber grouped with the Ardgour Swarm (Rock 1983).

Carboniferous alkali–dolerite sills

In contrast to the crystalline basement terrain of the Highlands, where dyke-swarms are typical minor intrusions, the syn-sedimentary basins of the Midland

Valley are characterised by dolerite sills or sill-complexes; dykes which might have acted as feeders to those sills are unknown. The sills are not found in the deeper parts of the syn-sedimentary basins, being restricted to the same hinge-line localities as the necks, with which they are closely related petrochemically as well as geographically.

In Ayrshire most of the sills are assumed to be post-Carboniferous, though by analogy with other areas some of the southern sills might well be contemporaneous with the Namurian phreatomagmatic activity around Dalry. It is in the central part of the Midland Valley where the sills can be placed most readily in context; West Lothian examples intruded into Dinantian oil-shales give ages comparable with the Namurian Bathgate volcanics (De Souza 1982) while those farther north, in west and central Fife give evidence of having been emplaced into Namurian sediments barely older than the sills themselves (Francis & Walker 1987).

These intrusions are up to 120 m thick, ranging in composition from olivine–dolerite to teschenite and basanite with indications of gravitative sinking in west Fife (Flett 1931) and also near Burntisland, where the Braefoot Outer Sill comprises, from the base upwards, basalt 6 m, picroteschenite 15 m, teschenite becoming coarser and olivine-poorer upwards 30–45 m dolerite–pegmatite 21 m, and dolerite and basalt 15 m (Campbell *et al.* 1932, 1934). Between these geographical extremes the sill complex of west and central Fife comprises 9 components; some of these now separate may once have been physically joined, though on geochemical grounds at least three pulses of injection can be recognised – the earliest consanguineous with the immediately underlying late Viséan–early Namurian Burntisland lavas (Walker 1987). The shapes of the components indicate control by syn-sedimentary structures, with restored sill geometries showing thickening towards the bottoms of saucer-shapes mimicking sediment disposition at time of emplacement and thus implying flow downdip (pp. 410–412). Changes of horizon along fracture lines are taken to indicate that these are growth faults (Francis & Walker 1987).

The olivine–dolerite and teschenite sill-complex of east Fife is similarly assumed to be Namurian or early Westphalian in age (Forsyth & Rundle 1978) though in view of the lack of stratigraphical control it remains possible that some of the younger radiometric ages determined from them are genuinely Permian and contemporaneous with the known pipes.

Permian alkali–dolerite sills

East Fife is not alone in lacking adequate three-dimensional control of sill geometry and the Permian age of sills elsewhere is thus essentially dependent on radiometric age-dating. The group includes the teschenites of the Glasgow–Paisley area, dated at 276–279 Ma by De Souza (1979) as

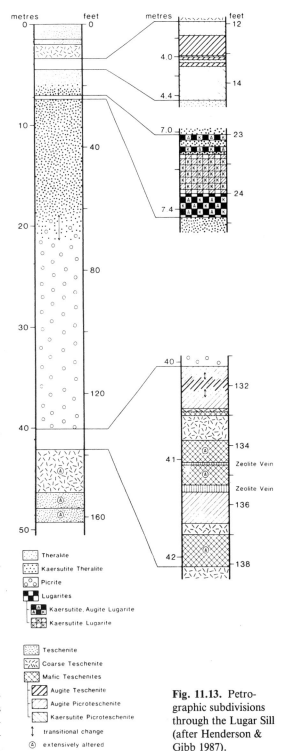

Theralite
Kaersutite Theralite
Picrite
Lugarites
Kaersutite, Augite Lugarite
Kaersutite Lugarite

Teschenite
Coarse Teschenite
Mafic Teschenites
Augite Teschenite
Augite Picroteschenite
Kaersutite Picroteschenite
transitional change
extensively altered

Fig. 11.13. Petrographic subdivisions through the Lugar Sill (after Henderson & Gibb 1987).

Fig. 11.14a

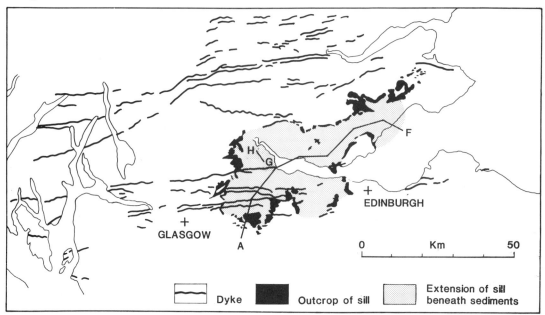

Fig. 11.14b

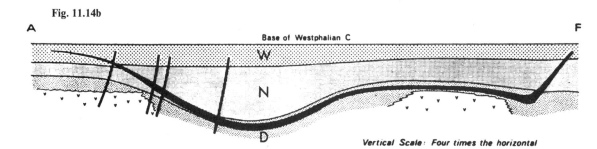

Fig. 11.14c

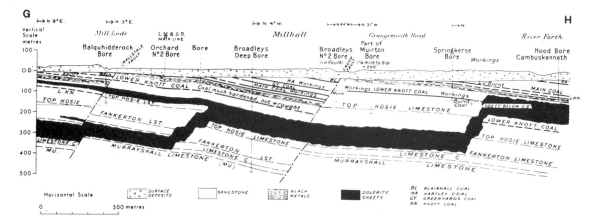

well as the various kylite and teschenite sills of Ayrshire. Using the incremental heating $^{40}Ar/^{39}Ar$ method, Henderson *et al.* (1987) have dated one of the latter – the Lugar Sill – to be 268±6 Ma; by expressing doubt over the reliability of previous K–Ar whole-rock ages they now argue that there is no reason why many of the various petrological sill types in W Scotland could not have been emplaced over a much shorter time-span than previously assumed, thereby forming a single petrological suite comagmatic with the early Permian Mauchline Volcanic Formation.

The Lugar Sill, a classic text-book example of differentiated basic sill since its investigation by Tyrell (1917, 1948, 1952), has been re-examined by Henderson & Gibb (1987). They conclude that it was formed by multiple injection from the outside inwards, of continuously less evolved teschenitic magma followed by a large pulse of olivine-rich theralitic magma which, by differentiation *in situ* and upward enrichment in residual liquid and volatiles, gave rise to lugarites (Fig. 11.13).

Carboniferous–Permian tholeiites and quartz–dolerites

The dominantly alkali–basaltic magmatism which continued throughout most of the Carboniferous and the early Permian appears to have been interrupted by a phase of

Fig. 11.14a. Distribution of quartz–dolerite and tholeiite dykes and outcrops and underground continuation of the Midland Valley Quartz–dolerite Sill.

Fig. 11.14b. Horizontal section along line A–F of Fig. 11.14a drawn palinspastically relative to the base of Westphalian C (Skipsey's Marine Band) as datum. Quartz–dolerite in black ('step-and-stair' smoothed off); sediments stippled Dinantian (D), Namurian (N), Westphalian (W) (after Francis 1982).

Fig. 11.14c. Horizontal section along G–H of Fig. 11.14a showing detail of 'step-and-stair' and relationship of horizon-change to both penecontemporaneous and growth faulting (Dinham & Haldane 1932, fig. 10).

tholeiitic intrusion. Depending on the timescale used (Fig. 11.2) and bearing in mind the likely validity of its whole-rock K–Ar age of *c.* 296 Ma its time of emplacement would appear to be latest Carboniferous or earliest Permian; its cross-cutting relationships certainly shows it to be post-Westphalian B.

The suite consists of interrelated dykes and sills. The dykes were emplaced along E–W fractures and are up to 300 km long, emplaced *en echelon* and are usually 3 to 75 m wide; they form a belt some 200 km wide which extends across Scotland and Northern England and continues eastward to Sweden. Like the alkali–dolerites, therefore, they are restricted to dyke-like form at levels traversing the crystalline basement.

A further resemblance with the alkaline suite is that in the sedimentary basins of the Midland Valley, sills are volumetrically by far the dominant form of intrusion. However, in any discussion of their mode of emplacement, there are also important differences to note. One is that the alkali–dolerite sills are associated with extrusion conduits represented by necks and lack any connection with dykes in the basins, whereas the tholeiites lack any known extrusive equivalents and are directly fed by dykes. The other is that the tholeiites appear to have been intruded in a single pulse rather than several (Macdonald *et al.* 1981).

The main manifestations of that pulse is the Midland Valley Sill-complex of quartz–dolerite which extends over an area of 1,600 km² and is up to 150 m thick (Fig. 11.14a). It consists essentially of basic plagioclase with clino- and ortho-pyroxenes, and interstitial micropegmatite or quartz; tholeiitic varieties have a glassy mesostasis (Walker 1935, 1965). Where the sill-complex is more than 50 m or so thick it displays a consistent variation in grain size from top to bottom as first described by Robertson (*in* Robertson & Haldane 1937) wherein a coarse pegmatitic zone occurs about one-third the distance below the top of the sill and grades upwards and downwards through thick medium-grained zones towards chilled margins.

The structure of the sill has been uncovered in the course of mining which shows it to follow some stratigraphical horizons for long distances, but also to change horizon along high-angled 'risers', some of which are faults (Fig. 11.14c). However, structure contours and isopachs show that its shape, which includes a general thickening towards basin bottoms (Fig. 11.14b) is clearly related to, and possibly dictated by, the shape of the syn-sedimentary basins at the time of intrusion and a possible mechanism to explain this is outlined on pp. 410–412.

Preservation within the stratigraphical column

The Carboniferous–Permian expanses of shallow water, where accumulation of sediments broadly kept pace with subsidence, provided an environment in which ash-rings or outpourings of lavas rapidly become partly subaerial. Thus the occasional records of pillow lavas or chloritised palagonitic tuffs which suggest an eruption of magmatic material directly into the sea are balanced by the occurrence of red boles testifying to conditions of subaerial lateritisation between flows. Once raised above sea level the accumulations of lava and tuff were subjected to erosion and they remained subaerial only if eruption kept pace with erosion and regional subsidence. Thus the final relationship between sedimentary and volcanic rocks which is preserved in the stratigraphic record is governed by the dominance of one or other of those three factors.

Three cases of entombed ash-rings may be quoted in illustration. One at Dalry in Ayrshire exemplifies the case where volcanism was brief as compared with the erosion and sedimentation; in consequence, the younger sediments are banked with overlap, against the upper surface of a cone which has been partly remoulded by erosion. The position is reversed in the Saline district of Fife (Fig. 11.15) where the growth of a cone by eruption kept ahead of the inner-ring slumping and subsidence described on p. 413, as well as of external erosion and sedimentation, until volcanism ceased. Rapid planation and burial by sediments then gave rise to a body of volcaniclastics, that is ash partly *in situ*, and partly redeposited, which is concordant with the sediments at its top and transgresses them below. The third case concerns volcanic rocks of Westphalian age in the former Wellesley Colliery, off the Fife Coast (Fig. 11.11) where, in contrast to Dalry and Saline, records are modern enough for distinction to be made between pyroclastic

rocks *in situ* and similar rocks redistributed by sedimentary processes. The distribution of these rocks within the sedimentary sequence is partly reminiscent of Dalry and partly of Saline, from which it may be deduced that the balance between eruption, erosion and deposition changed from time to time. In an earlier section (p. 398) it is suggested that the Bathgate geophysical anomaly may reflect the presence of a comparable volcaniclastic body beneath younger sediments in West Lothian.

It is of interest to note that the volcanic detritus of Wellesley forms rhythmic sequences analogous to those of normal Coal Measures sediments, including thin, impure coals. Some of these beds are reddened in a fashion which indicated that the eroded surface of the ash was lateritised: other reddening suggests subaerial oxidation of beds after they had been deposited within the rhythmic sequences (Francis & Ewing 1961). It is reasonable to assume that similar factors apply to the burial of lavas. The thick red 'marls' interbedded with the Upper Oil-Shale Sub-group of the Central Basin and Lothians, for instance, can be taken as evidence of penecontemporaneous erosion of the surrounding Dinantian lavas. The inter-digitation of sediments, some of them marine, among lavas of all ages testifies to the periodic failure of eruption to keep pace with this erosion and with regional subsidence. One of the best examples of this is at Rashiehill (Fig. 11.7), where marine beds occur more than 300 m above the base of a lava-bole sequence and are themselves overlain by further mainly subaerial lavas.

Episodes of lava formation seem to have been followed

Fig. 11.15. Horizontal section showing lateral variation in lithology associated with volcanism in the Limestone Coal Group of the Saline district, Fife (Francis 1961a, fig. 2).

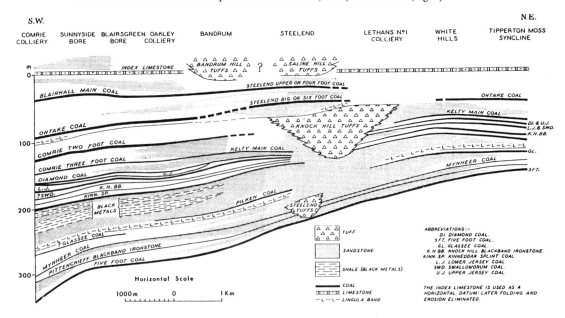

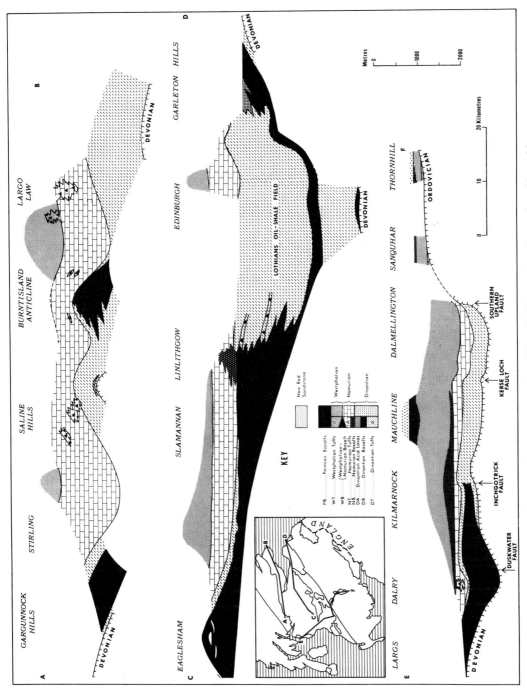

Fig. 11.16. Horizontal sections across the Midland Valley to show the stratigraphic position of the principal lavas and tuffs.

by relatively rapid planation in the course of which volcanic detritus was washed into hollows at the tops of the lava-remnants and spread around their flanks. Even so, there appear to have been instances when this process was overtaken by the deposition of post-volcanic sediments which overlapped the lavas. There is also a question of differential compaction. For instance, coals and shales are well known to thin as they are traced above beds of sandstone and similar effects have been demonstrated to occur above bodies of comparably incompactable volcanic rocks. The attenuation is sometimes increased by the presence of thick sandstones around the peripheries of volcanic rocks where they originally accumulated as shoals of sand. An extreme example from Saline (Fig. 11.15) shows the replacement of virtually the whole of the Limestone Coal Group by such sandstones and by the associated tuffs.

Variations in thickness of sediments above some of the lava sequences have received less detailed study, but differential compaction above some of the Clyde Plateau Lavas is probable (Fig. 11.16) and there may be a similar correlation between the attenuated sequences of Namurian rocks above the thick lavas in the area around the Rashiehill Bore and along the axis of the Burntisland Anticline.

Volcanic mechanisms

Crustal structural controls

Remarking on the universal association between alkaline volcanism and rift valleys, Holmes (1965) noted that activity is generally expressed along fractures associated with folds inside the grabens than along the marginal fractures. The Scottish Carboniferous–Permian volcanic rocks in the Midland Valley graben and Northumberland Trough half-graben fit this pattern. The north-easterly fractures and basin-swell axes which were inherited from the Caledonian orogeny as well as related E–W growth faults which influenced sedimentation throughout the period (Chapter 10), seem equally to have controlled the ascent of alkaline basalt magmas. Hence the absence of volcanism from the centres of major sedimentary troughs, its location at crests or hinges of syn-sedimentary upfolds and the prominent north-easterly neck alignments such as the Fintry–Dumbarton line, the Campsie Faults and Ardross Fault (Fig. 11.17).

Ascent of magma

MacGregor's (1948) suggestion that magma rising along fissures became checked 2 to 3 km from the surface, there to form several co-existing magma chambers, is supported by the pattern of extrusion of differentiated lavas during the Dinantian. For instance Upton (1982), noting that the Garleton Hills lie over the subsurface extension of the

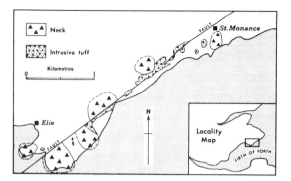

Fig. 11.17. Sketch-map showing line of necks and related intrusive tuffs associated with the Ardross Fault – a north-easterly strike–slip fault active intermittently during the Carboniferous (Francis & Hopgood 1970).

Southern Uplands Fault, suggests the formation, along the zone of weakness, of a magma chamber in which basalts fractionated to trachytes and phonolites.

Ascent by upwards wedging along fissures so as to form dykes, some of which gave way at the surface to pipe-breccias, conforms with the pattern of alkaline activity in the Highlands. It also conforms to MacGregor's (1948) vision of how the Carboniferous and Permian necks were generated in the sedimentary basins of the Midland Valley and Borders and whilst it is clearly supported by some it may not be true for all. Those for which the process seems undisputed are the structures containing xenoliths and xenocrysts from mantle peridotite (Hunter *et al.* 1984, Upton *et al.* 1984, Alexander *et al.* 1986). Such direct mantle source implies magma columns in the order of 70 km high with initial temperatures assumed by Chapman (1976) to be in the range of 1,300–1,450°C, though from a recalculation of the kinetics of pyrope megacrysts Donaldson (1984) concludes that either the initial temperatures were much less or that the magma cooled very rapidly during ascent.

Sill emplacement

Most, if not all, of the alkaline dolerites as well as the tholeiite complex are post-Viséan in age; the mode of intrusion did not begin, therefore, until after the broad change from voluminous lava effusion to more sporadic phreatomagmatic activity. One of the controlling factors was almost certainly the progressively increasing thickness of water-bearing sediments in the basins; these would have been too low in density always to support the weight of magma columns which would thus spread laterally in the form of sills.

Analyses by means of palinspastic structure contours and isopachs reveal a close correspondence between sill structure and the shape of the host sediments at the time of

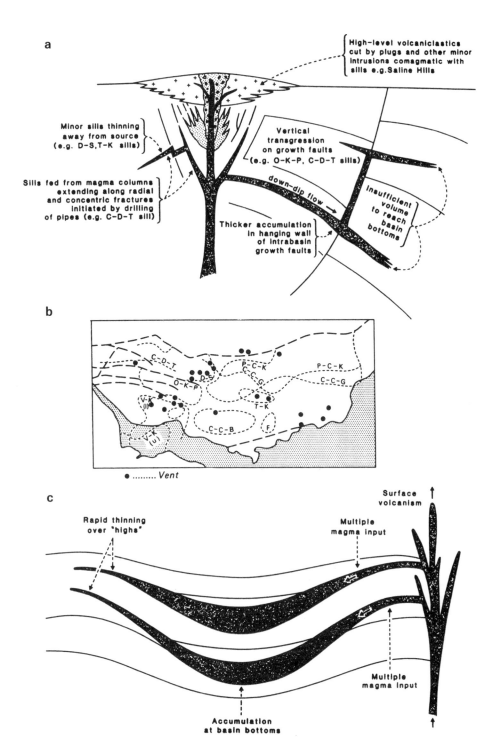

Fig. 11.18. Diagrammatic representation of relationships between alkali–dolerite sills and necks in west-central Fife: **(a)** magma pathways to minor sill formation along small radial and concentric fractures formed during neck emplacement and directly from neck walls; **(b)** sketch map showing location of principal necks relative to sill components (C–C–B, Crombie–Cairneyhill–Bellknowes; C–C–G, Craigluscar–Cluny–Glenrothes; C–D–T, Cairnfolds–Dollar–Tillycoultry; D–S, Dunnygask–Steelend; F, Fordell; O–K–P, Oakley–Kinneddar–Parklands; P–C–K, Parkhill–Cowdenbeath–Kinglassie; T–K, Townhill–Kingseat; V–K, Valleyfield–Kinneil); **(c)** large-volume multi-leaf sills fed from basin hinge-line pipes, infrabasinal fracturing omitted (Francis & Walker 1987, fig. 11).

intrusion, whether that was Namurian as in the case of the west–central Fife alkali–dolerites (Francis & Walker 1987) or Carboniferous–Permian for the tholeiites (Francis 1982). As, at the time of intrusion, the syn-sedimentary bedding was already dipping at angles of up to 5° and as, moreover, the sills are thickest at the bottoms of basins (Fig. 11.14b, c) the mechanics of emplacement are inferred to include some element of gravitational flow under the pressure of head of magma. In the tholeiite complex the dykes are presumed to have provided the necessary head, for they extend above sill levels without reaching the contemporaneous surface (judging from the absence of extrusives).

Relationship between alkaline sills and volcanic necks

Apart from their small extent, the alkali–dolerites differ from the tholeiites in being associated with volcanic necks rather than with dykes. Dyke-like elements such as minor apophyses and horizon changes along E–W faults are important only in showing the faults to be penecontemporaneous growth structures rather than the entirely post-Carboniferous structures as once believed; such elements are not large enough to have acted as heads from which magma could have flowed down-dip in the pattern indicated by the contours. The associated volcanic pipes offer alternative forms of heads (Francis & Walker 1987). It is inferred that magma columns ascending through the basement interacted phreatomagmatically with wet sediments in the basins so as to drill pipes which served as conduits for surface eruption, and that when, periodically, these became choked or plugged, degassed magma broke through their walls along radial and concentric minor fractures to begin flow down-dip (Fig. 11.18).

This assumed mechanism differs from earlier postulates (e.g. Francis 1968b) that the sills may have acted as high-level magma chambers generating pipes from their roofs. Interaction between such magma and wet sediments at the advancing propagating edges of the sills certainly seems to have given rise to bodies of fluidised tuffisite identical with

that of some pipe fillings, but they are lateral rather than vertical (Walker & Francis 1987) directed in part by the particularly intense reaction between the magma and seams of coal or (as it then might have been) peat (Fig. 11.19).

Volcanic pipe (neck) formation

Most of the several hundred known pipes are rounded or oval in plan and funnel-shaped, with inwardly inclined margins in section. Although some contain central plugs or minor impersistent dykes the majority consist of tuff or agglomerate alone. These rocks are made up of varying proportions of parent basaltic magma and of the country rock traversed by the pipes. In most cases the parent magma corresponded to the predominating Carboniferous lava type: thus in Ayrshire and east Fife, where many of the basalts are monochiquitic, the tuffs commonly contain xenocrysts of alkali–feldspar, augite, hornblende and biotite and xenoliths of peridotite. The composition of the country-rock debris in the tuffs is more varied. Xenoliths of granulite are significant indicators of the underlying crust (Upton et al. 1976, 1984; Hunter et al. 1984), but they are rare; most debris is of recognisably high-level derivation. Among the plateau lavas, fragments from older flows are abundant, but elsewhere the wall rock is formed by sediments which have been broken down to large fragments, to grains of sand, or to shale and coal dust (Fig. 11.20). Some of the short-lived volcanoes are now represented by vents filled with sedimentary debris alone, but it is more usual to find such debris mixed with pale green, ragged-edged lapilli or decomposed parent basalt.

The initial pipe-drilling process is inferred from exposures and drill cores at various levels below the penecontemporaneous surface. It assumes up-doming associated with minor radial and concentric fracturing (Barnett 1985) followed by gas-fluxion and wall-rock stoping to explain the concentration of the larger blocks of country rock at (and aligned parallel to) the neck margins as compared with the finer-grained, better-mixed and more homogeneous tuffs at the centre. It is consistent too with the frequent lack of any evidence of significant thermal

Fig. 11.19. Diagrammatic representation of coal-magma distribution and the formation of tuffisite (Walker & Francis 1987, fig. 9).

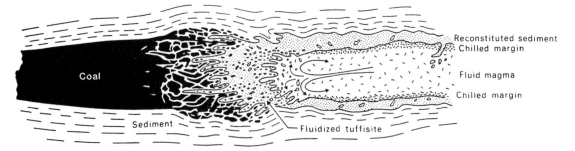

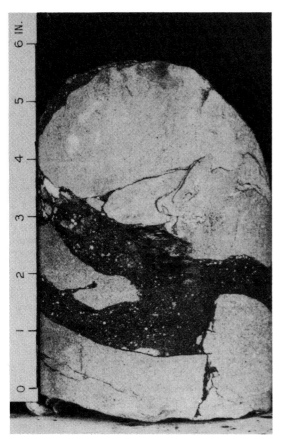

Fig. 11.20. Intrusion into neck-marginal sandstone of tuffisite which is composed mainly of sedimentary debris. Boring near Culross, Fife.

alteration to coaly and other debris both in the tuffs and in the walls of the vents. Crypto-volcanic structures and dykes and sills of intrusive tuffs (tuffisites) emplaced along the radial and concentric fractures (Fig. 11.20) are associated effects.

Some further explanation is needed for pipes filled with bedded tuffs and agglomerates; these are most common among the larger pipes and dips are usually inclined inwards. Included lenses of sediment and fragments of fossil wood show these rocks to have been subaerially accumulated and to have reached their present position in the necks as a result of subsidence. The amount of subsidence is calculated to be at least 500 m in some of the Fife structures; the neck-margins are thus effectively ring-faults against which the wall-rock sediments have been dragged down (Francis 1970). The lithologies and textures, together with base-surge cross-bedding as well as slumped sheets, conform to tuffs found in modern Surtseyan ash-rings (Leys 1982) formed by basaltic eruptions in shallow

water. The modern deposits are usually palagonitised and it is reasonable to suppose the same of many of the Carboniferous tuffs, though the unstable palagonite has long since altered to chlorite.

Recognition of the Surtseyan nature of the bedded material in the necks has cast new light on models reconstructing activity and subsidence. Water from the sediments or in the shallow seas and lagoons would have given rise to phreatomagmatic eruption, readily comminuting the sediments into the sand, mud and coal dust, which, with basaltic lapilli, are major matrix constituents of the tuffs. By modern analogy, wide-diameter, low-altitude tuff-rings would have been built at the surface, and as long as water was able to gain access, by breaching or other means, Surtseyan activity would have continued intermittently. The ring fracturing and subsidence are most satisfactorily explained by reference to a model proposed by Lorenz (1973). This suggests initiation of ring-fracturing during the first formation of the pipes above the magma sources with each eruptive pulse thereafter followed by some measure of subsidence. Thus, as the earliest bedded tuffs descended progressively further, they were pierced by later conduits filled with basaltic debris (Fig. 11.21).

Petrogenesis

In a review of the petrochemistry of the Dinantian lavas, Macdonald (1975) has shown that the sequence in each geographical area has a distinctive magmatic lineage recognisable in terms of silica undersaturation, Fe/Mg ratio, Na_2O/K_2O ratios and TiO_2 and P_2O_3 content (Fig. 11.22). In addition he recognises three main associations; one ranging from ankaramitic basalt to trachyte and rhyolite: a second

Fig. 11.21. Schematic diagram illustrating some of the processes involved in the emplacement of volcanic rocks in a sedimentary pile and, particularly, the intermittent subsidence of bedded proclastic rocks from the inner flanks of superficial ash-rings (Forsyth & Chisholm 1977, fig. 17, after Lorenz 1973 and Francis 1970).

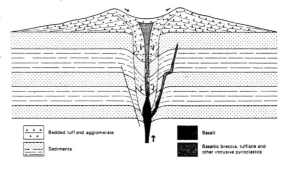

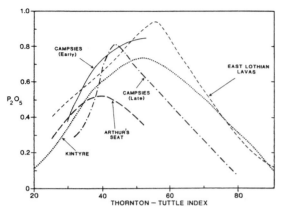

Fig. 11.22. Differentiation index against P_2O_5 for various Dinantian lava suites (Macdonald 1975, fig. 15).

dominated by plagioclase–phyric hawaiites and basalts with subordinate mugearites and local trachytes or rhyolites: and a third restricted to olivine–clinopyroxene

microphyric basalts with rare ankaramites. The more acid variants in these associations are assumed to be differentiates formed in high level magma chambers.

Subsequent work (Macdonald 1980) on the Carboniferous–Permian suite as a whole has established that each area retained its own distinctive incompatible trace element chemistry over periods of up to 50 Ma. This evidence discounts the possibility of crustal contamination of the ascending magmas and implies, instead, a long-lasting heterogeneity of the underlying mantle from which the basaltic magmas were derived by variable degrees of closed-system partial melting over a limited range of pressures; the melting may have ranged from 20 per cent for the hypersthene-normative magmas to 7 per cent for the basanites – at depths ranging from 100 to 50 km respectively.

Belief in mantle heterogeneity has since been stressed by petrochemical work on dykes in the Highlands (Baxter 1987) and xenocrysts in the Lowlands (Hunter *et al.* 1984), while Smedley (1988) cites similarities with modern ocean island basalts as evidence of a sub-lithospheric mantle source for the Scottish Carboniferous lavas. Nevertheless, an important consideration in such modelling is whether there was a broad change with time in the silica-saturation

Fig. 11.23. Degree of silica-saturation of Carboniferous–Permian alkaline basic rocks (100 An/An + Ab 45) plotted as a function of stratigraphical age. Circles are lavas; triangles are radiometrically dated intrusions, 'Q' = normative quartz + quartz in normative hypersthene. The trend lines indicate the inferred changes of basic magma compositions during each thermal cycle (Macdonald *et al.* 1977, fig. 1).

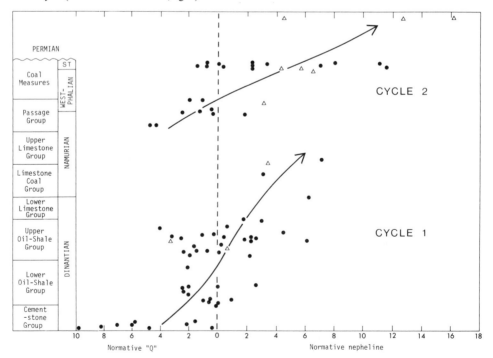

of the suite as a whole. MacGregor (1948) asserted that the later Dinantian basalts were progressively more undersaturated than the earlier and Macdonald *et al.* (1977) extend this by proposing that there were two thermal cycles, each represented by an overall reduction in saturation in respect to silica, one Dinantian to early Namurian, the other mid-Namurian to Permian in age. It can be seen from Fig. 11.23 that this view is coloured by the paucity of Namurian rocks suitable for analysis (as distinct from tuffs established by stratigraphy to be of that age). The second 'cycle', moreover, would have been interrupted by the end-Carboniferous intrusive episode, when large volumes of tholeiitic magma were presumed to have been derived from partial melting at possibly higher levels in the mantle and under a different stress regime.

Balancing structural control with the chemical evidence for the period as a whole, the petrogenetic history might be synthesised as follows. The crustal fractures which acted as hinge-lines determining the location of syn-sedimentary basins and swells (Chapter 10) are presumed to have extended downward to provide access for the rising alkali-basaltic magma from early Dinantian to mid-Westphalian times. Hence the NE alignment of many of the eruptive centres, the location of activity at swells and corresponding absence of volcanic rocks both in the centres of the major depositional basins and in the surrounding Highlands and Southern Uplands areas of dyke injection.

Seismic traverse (Bamford *et al.* 1976) and the presence of granulitic xenoliths in pipes in the Midland Valley and Southern Uplands (Upton *et al.* 1984) establish a continuous underlying sialic crust. The volcanism of Carboniferous times is thus within-plate and rift-related continental in character owing little or nothing to subduction of the oceanic crust from the marginal ocean basin of the Rheno–Hercynian Zone (Francis 1978, 1988), or a Tethys still farther south (Anderton *et al.* 1979). Generation by mantle plume of 'hot spot' is no less difficult to reconcile with northern drift of a continental plate through 15° of latitude (Fig. 10.35, Irvine 1977) or, for that matter, with the geochemistry (Smedley 1986, 1988).

The end-Carboniferous, and apparently short-lived episode of voluminous tholeiitic intrusion appears to have been an interruption to this pattern. It is explained by Russell & Smythe (1983) and Haszeldine (1984) as the first manifestation of Atlantic rifting – a view which has not found universal favour because the timing seems too early. In favour of the concept, however, is that all the great tholeiite sill-complexes of the world appear to be related to continental rifting, the emplacement of the Karoo–Tasmania–Antarctica sill-complex during the break-up of Gondwanaland being the outstanding example.

Resumption of continental rift-related alkali volcanism during the Permian – albeit along new NW–SE to NE–SW lines – has also been accommodated within the framework of Atlantic opening (op. cit.) though the reduction in scale and succeeding (early Mesozoic) quiescence suggest that if so it must have been an incipient opening which went temporarily into abeyance.

REFERENCES

ALEXANDER, R. W. S. *et al.*   1986   The megacryst and inclusion assemblage from the Black Rock Vent, Ayrshire. *Scott. J. Geol.*, **72**, 203–212.

ALLAN, D. A.   1924   The igneous geology of the Burntisland district. *Trans. roy. Soc. Edinb.*, **53**, 479–501.

ANDERSON, F. W.   1963   The Geological Survey bore at Rashiehill, Stirlingshire (1951). *Bull. geol. Surv. G.B.*, **20**, 43–106.

ANDERTON, R. *et al.*   1979   *A dynamic stratigraphy of the British Isles.* Allen & Unwin, London, 301 pp.

BAMFORD, D. *et al.*   1976   A lithospheric seismic profile in Britain – I. Preliminary results. *Geophys. J. R. astron. Soc.*, **44**, 145–160.

BARNETT, J. A. M.   1985   Fracture patterns related to volcanic necks and pipes in an Upper Limestone Group (Namurian) coal seam in the Kincardine Basin, West Fife, Scotland. *Proc. Yorks. geol. Soc.*, **45**, 249–259.

BAXTER, A. N.   1987   Petrochemistry of late Palaeozoic alkali lamphrophyre dykes from N Scotland. *Trans. roy. Soc. Edinb.: Earth Sci.*, **77** (for 1986), 267–277.

BAXTER, A. N. &   1984   Camptonite–monchiquite dyke swarms of Northern Scotland; MITCHELL, J. G.   age relationships and their implications. *Scott. J. Geol.*, **20**, 297–308.

BROWNE, M. A. E. &   1981   An occurrence of Lower Carboniferous lavas at Monksgrave
  THIRLWALL, M. F.   (Powmill), near Dollar. *Scott. J. Geol.*, **17**, 275–279.
CAMERON, I. B. &   1985   The Midland Valley of Scotland. *Brit. reg. Geol. Br. Geol. Surv.*,
  STEPHENSON, D.   172 pp.
CAMPBELL, R. *et al.*   1932   The Braefoot Outer Sill, Fife: Part I. *Trans. Edinb. geol. Soc.*, **12**,
                                   342–375.
CAMPBELL, R. *et al.*   1934   The Braefoot Outer Sill, Fife: Part II. *Trans. Edinb. geol. Soc.*,
                                   13, 148–173.
CHAPMAN, N. A.   1976   Inclusions and megacrysts from undersaturated tuffs and basan-
                                   ites, East Fife, Scotland. *J. Petrol.*, **17**, 472–498.
CONWAY, A. *et al.*   1987   Preliminary interpretation of upper crustal structure across the
                                   Midland Valley of Scotland from two east–west seismic refrac-
                                   tion profiles. *J. geol. Soc. Lond.*, **144**, 865–870.
COTTON, W. R.   1968   A geophysical survey of the Clyde Plateau Lavas. *PhD Thesis,
                                   Univ. Glasgow* (Unpublished).
CRAIG, P. M.   1980   Volcanic geology of the Campsie Fells area, Stirlingshire. *PhD
                                   Thesis, Univ. Lancaster* (Unpublished).
CRAIG, P. M. &   1975   The Lower Carboniferous rocks of the Campsie–Kilpatrick
  HALL, I. H. S.   area. *Scott. J. Geol.*, **11**, 171–174.
DAVIDSON, K. A. S. *et al.* 1984 Geophysical model for the Midland Valley of Scotland. *Trans.
                                   roy. Soc. Edinb.: Earth Sci.*, **75**, 175–181.
DAVIES, L. H.   1936   The Geology of Inchkeith. *Trans. roy. Soc. Edinb.*, **58**, 753–786.
DE SOUZA, H. A. F.   1979   The geochronology of Scottish Carboniferous volcanism. *PhD
                                   Thesis, Univ. Edinburgh* (Unpublished).
                         1982   Age data from Scotland and the Carboniferous time scale. *In*
                                   Odin, G. S. (Ed.) *Numerical dating in stratigraphy.* John Wiley
                                   and Sons Ltd., 456–465.
DINHAM, C. H. &   1932   The economic geology of the Stirling and Clackmannan Coal-
  HALDANE, D.   field. *Mem. geol. Surv. U.K.*, 242 pp.
DIXON, J. E. *et al.*   1981   The tectonic significance of post-Carboniferous igneous activity
                                   in the North Sea Basin. *In* Illing, L. V. & Hobson, G. D. (Eds.)
                                   *Petroleum Geology of the Continental Shelf of North-West
                                   Europe.* Institute of Petroleum, London, 121–137.
DONALDSON, C. H.   1984   Kinetics of pyrope megacryst reactions in ascending basaltic
                                   magma – relevance to high-pressure magmatic crystallization at
                                   Elie Ness, Fife. *Geol. Mag.*, **121**, 615–620.
ECKFORD, R. J. A. &   1939   The igneous rocks of the Kelso district. *Trans. Edinb. geol. Soc.*,
  RITCHIE, M.   **13**, 464–472.
ELLIOTT, R. B.   1960   The Carboniferous volcanic rocks of the Langholm district.
                                   Proc. geol. Assoc., Lond., **71**, 1–24.
FLETT, J. S.   1931   The Saline No. 1 Teschenite. *Summ. Prog. Geol. Surv.* for 1930,
                                   pt. 2, 44–51.
FORSTER, S. C. &   1985   Geochronology of the Carboniferous, Permian and Triassic. *In*
  WARRINGTON, G.   Snelling, N. J. (Ed.) *The Chronology of the Geological Record.*
                                   Geol. Soc. Lond. Mem., No. 10, 99–113.
FORSYTH, I. H. &   1977   The geology of East Fife. *Mem. geol. Surv. U.K.*, 284 pp.
  CHISHOLM, J. I.
FORSYTH, I. H. &   1978   The age of the volcanic and hypabyssal rocks of east Fife. *Bull.
  RUNDLE, C. C.   geol. Surv. G.B.*, **60**, 23–29.
FRANCIS, E. H.   1961a The economic geology of the Fife coalfields, Area II (2nd Edit.).
                                   Mem. geol. Surv. U.K., 152 pp.
                         1961b Volcanism in relation to sedimentation in the Carboniferous
                                   rocks of the Saline district, Fife. *Bull. geol. Surv. G.B.*, **17**, 116–
                                   144.
                         1961c Thin beds of graded tuffs and tuffaceous siltstone in the
                                   Carboniferous of Fife. *Bull. geol. Surv. G.B.*, **17**, 191–215.

FRANCIS, E. H.   1968a Pyroclastic and related rocks of the Geological Survey boreholes in East Fife. *Bull. geol. Surv. G.B.*, **28**, 121–135.

1968b Effect of sedimentation on volcanic processes, including neck-sill relationships in the British Carboniferous. *Rep. 23rd Int. geol. Congr.*, Prague, **2**, 163–174.

1969 Les tonstein du Royaume-Uni. *Ann. Soc. geol. Nord.* Lille, **89**, 209–214.

1970 Bedding in Scottish (Fifeshire) tuff-pipes and its relevance to maars and calderas. *Bull. volcanol.*, **34**, 697–712.

1978 Igneous activity in a fractured craton: Carboniferous volcanism in northern Britain. *In* Bowes, D. R. & Leake, B. E. (Eds.) *Crustal evolution in northwestern Britain and adjacent regions. Geol. J. Spec. Issue*, No. 10, 279–296.

1982 Emplacement mechanism of late Carboniferous tholeiite sills in north Britain. *Jl. geol. Soc. Lond.*, **139**, 1–20.

1988 Mid-Devonian to early Permian volcanism: Old World. *In* Harris, A. L. & Fettes, D. J. (Eds) *The Caledonian–Appalachian Orogen. Spec. Publ. Geol. Soc. Lond.,* **38**, 573–584.

FRANCIS, E. H. &   1961 Coal measures and volcanism off the Fife coast. *Geol. Mag.*, **98**,
EWING, C. J. C.   501–510

FRANCIS, E. H. &   1970 Volcanism and the Ardross Fault, Fife, Scotland. *Scott. J. Geol.*,
HOPGOOD, A. M.   **6**, 162–185.

FRANCIS, E. H. &   1987 Emplacement of alkali-dolerite sills relative to extrusive volcan-
WALKER, B. H.   ism and sedimentary basins in the Carboniferous of Fife, Scotland. *Trans. roy. Soc. Edinb.: Earth Sci.*, **77** (for 1986), 309–323.

FRANCIS, E. H. *et al.*   1970 The geology of the Stirling district. *Mem. geol. Surv. U.K.*, 357 pp.

GREIG, D. C.   1971 The south of Scotland. *Br. Reg. Geol.* (3rd Edit.). HMSO, Edinburgh, 125 pp.

HALL, J.   1974 A seismic reflection survey of the Clyde Plateau lavas in north Ayrshire and Renfrewshire. *Scott. J. Geol.*, **9**, 253–279.

HAMILTON, J.   1956 Mineralogy of basalts from the western Kilpatrick Hills and its bearing on the petrogenesis of Scottish Carboniferous olivine-basalts. *Trans. Edinb. geol. Soc.*, **16**, 280–298.

HARLAND, W. B. *et al.*   1982 *A geologic timescale.* Cambridge Univ. Press, 131 pp.

HASZELDINE, R. S.   1984 Carboniferous North Atlantic palaeogeography: stratigraphic evidence for rifting not megashear or subduction. *Geol. Mag.*, **121**, 443–463.

HENDERSON, C. B. M. &   1987 The petrology of the Lugar Sill, S.W. Scotland. *Trans. roy. Soc.*
GIBB, F. G. F.   *Edinb.: Earth Sci.*, **77** (for 1986), 325–347.

HENDERSON, C. B. M. *et al.*   1987 The age of the Lugar Sill and a discussion of the Late Carboniferous–Early Permian sill complex of S.W. Scotland. *Geol. J.*, **22**, 43–52.

HITCHEN, K. &   1987 Geological review of the West Shetland area. *In* Brooks, J. &
RITCHIE, J. D.   Glennie, K. (Eds.) *Petroleum Geology of Northwest Europe.* Graham & Trotman, 737–749.

HOLMES, A.   1965 *Principles of physical geology* (2nd Edit.). Nelson, Edinburgh, 1288 pp.

HUFF, W. D. &   1989 A tonstein from the Lower Limestone Group of Arran. *Scot.*
SPEARS, D. A.   *J. Geol.*, **25**, 161–172.

HUNTER, R. H. *et al.*   1984 Meta-igneous granulite and ultramafic xenoliths from basalts of the Midland Valley of Scotland: petrology and mineralogy of the lower crust and upper mantle. *Trans. roy. Soc. Edinb.: Earth Sci.*, **75**, 75–84.

IRVING, E.   1977 Drift of the major continental blocks since the Devonian. *Nature Lond.*, **270**, 304–309.

IRVING, J.   1930   Four 'felstone' intrusions in central Berwickshire. *Geol. Mag.*, **67**, 529–541.

JOHNSTONE, G. S.   1965   The volcanic rocks of the Misty Law–Knockside Hills district, Renfrewshire. *Bull. geol. Surv. G.B.*, **22**, 53–64.

KENNEDY, W. Q.   1931   On composite lava flows. *Geol. Mag.*, **68**, 166–181.

LEEDER, M. R.   1974   The origin of the Northumberland basin. *Scott. J. Geol.*, **10**, 283–296.

LEITCH, D.   1941   The Upper Carboniferous rocks of Arran. *Trans. geol. Soc. Glasg.*, **20**, 141–154.

LEYS, C. A.   1982   Volcanic and sedimentary process in phreatomagmatic volcanoes. *PhD Thesis, Univ. Leeds* (Unpublished).

LIPPOLT, VON H. J. *et al.*   1984   Isotopische Alter von pyroklastischen Korrelationsmarken für das mitteleuropäische Oberkarbon. *Fortschr. Geol. Rheinld. u. Westf.*, **32**, 119–150.

LORENZ, V.   1973   On the formation of maars. *Bull. volcanol.*, **37**, 183–204.

LUMSDEN, G. I. *et al.*   1967   The geology of the neighbourhood of Langholm. *Mem. geol. Surv. U.K.*, 255 pp.

McADAM, A. D.   1974   The petrography of the igneous rocks in the Lower Carboniferous (Dinantian) at Spilmersford, East Lothian, Scotland. *Bull. geol. Surv. G.B.*, **45**, 39–46.

MacDONALD, J. G.   1967   Variations within a Scottish Lower Carboniferous lava flow. *Scott. J. Geol.*, **3**, 34–45.

MACDONALD, R.   1975   Petrochemistry of the early Carboniferous (Dinantian) lavas of Scotland. *Scott. J. Geol.*, **11**, 269–314.

                        1980   Trace element evidence for mantle heterogeneity beneath the Scottish Midland Valley in the Carboniferous and Permian. *Phil. Trans. roy. Soc. Lond.*, **280**, 111–123.

MACDONALD, R. *et al.*   1977   Variations in basalt chemistry with time in the Midland Valley province during the Carboniferous and Permian. *Scott. J. Geol.*, **13**, 11–22.

                        1981   Geochemistry of a continental tholeiite suite: late Palaeozoic quartz dolerite dykes of Scotland. *Trans. roy. Soc. Edinb.*, **72**, 57–74.

MacGREGOR, A. G.   1928   The classification of the Scottish Carboniferous olivine-basalts and mugearites. *Trans. geol. Soc. Glasg.*, **18**, 324–360.

                        1948   Problems of Carboniferous–Permian volcanicity in Scotland. *Q. J. geol. Soc. Lond.*, **104**, 133–153.

MACINTYRE, R. M. *et al.*   1981   Geochronological evidence for phased volcanic activity in Fife and Caithness necks, Scotland. *Trans. roy. Soc. Edinb.*, **72**, 1–8.

McLEAN, A. C.   1966   A gravity survey in Ayrshire and its geological interpretation. *Trans. roy. Soc. Edinb.*, **66**, 239–265.

                        1978   Evolution of fault-controlled ensialic basins in northwestern Britain. *In* Bowes, D. R. & Leake, B. E. (Eds.) *Crustal evolution in northwestern Britain and adjacent regions. Geol. J. Spec. Issue*, No. 10, 325–346.

McROBERT, R. W.   1914   Acid and intermediate intrusions and associated ash necks in the neighbourhood of Melrose (Roxburghshire). *Q. J. geol. Soc. Lond.*, **70**, 303–315.

MITCHELL, G. H. &   1962   The geology of the neighbourhood of Edinburgh. *Mem. geol. MYKURA, W.   Surv. U.K.*, 159 pp.

MONRO, S. K. *et al.*   1983   The Ayrshire Bauxitic Clay: an allochthonous deposit? *In* Wilson, R. C. L. (Ed.) *Residual Deposits. Spec. Publ. Geol. Soc. Lond.*, No. 11, 47–58.

MORRISON, M. A. *et al.*   1987   Regional and tectonic implications of parallel Caledonian and Permo–Carboniferous lamprophyre dyke swarms from Lismore, Ardgour. *Trans. roy. Soc. Edinb.: Earth Sci.*, **77** (for 1986), 279–288.

MYKURA, W.   1967   The Upper Carboniferous rocks of south-west Ayrshire. *Bull geol. Surv. G.B.*, **26**, 23–98.

MYKURA, W.   1976   Orkney and Scotland. *Br. Reg. Geol.* HMSO, Edinburgh, 149 pp.

PALLISTER, J. W.   1952   The Birrenswark Lavas, Dumfries-shire. *Trans. Edinb. geol. Soc.*, **14**, 336–348.

PATERSON, I. B. &   1986   Lithostratigraphy of the late Devonian and early Carboniferous
   HALL, I. H. S.   rocks in the Midland Valley of Scotland. *Rep. Br. Geol. Surv.*, v. 18 [3], 14 pp.

PENN, I. E. *et al.*   1983   The Larne No. 2 Borehole: discovery of a new Permian volcanic centre. *Scott. J. Geol.*, **19**, 333–346.

PRICE, N. R. &   1969   Mineralogy and chemistry of tonsteins from Carboniferous
   DUFF, P. McL. D.   sequences in Great Britain. *Sedimentology*, **13**, 45–69.

RICHEY, J. E.   1939   The dykes of Scotland. *Trans. Edinb. geol. Soc.*, **13**, 393–435.

RICHEY, J. E. *et al.*   1930   The geology of North Ayrshire. *Mem. geol. Surv. U.K.*, 417 pp.

ROBERTSON, T. &   1937   The economic geology of the Central coalfields, Area I. *Mem.*
   HALDANE, D.   *geol. Surv. U.K.*, 169 pp.

ROCK, N. M. S.   1983   The Permo–Carboniferous camptonite–monchiquite dyke-suite of the Scottish Highlands and Islands: distribution, field and petrological aspects. *Rep. Inst. geol. Sci.*, **82/14**, 31 pp.

RUSSELL, M. J. &   1983   Origin of the Oslo Graben in relation to the Hercynian–Allegh-
   SMYTHE, D. K.   enian orogeny and rifting of the North Atlantic. *Tectonophysics*, **94**, 457–472.

SIMPSON, J. B. &   1936   The geology of the Sanquhar Coalfield and the adjacent basin of
   RICHEY, J. E.   Thornhill. *Mem. geol. Surv. U.K.*, 97 pp.

SMEDLEY, P. L.   1986   The relationship between calc-alkaline volcanism and within-plate continental rift volcanism; evidence from Scottish Palaeozoic lavas. *Earth Planet. Sci. Lett.*, **77**, 113–128.

                            1988   Trace element and isotopic variations in Scottish and Irish Dinantian volcanism; evidence for an OIB-like mantle source. *J. Petrol.*, **29**, 413–443.

SPEIGHT, I. M. &   1979   The Permo–Carboniferous dyke-swarm of northern Argyll and
   MITCHELL, J. G.   its bearing on the dextral displacement of the Great Glen Fault. *Jl. geol. Soc. Lond.*, **136**, 3–11.

TOMKEIEFF, S. I.   1937   Petrochemistry of the Scottish Carboniferous–Permian igneous rocks. *Bull. Volcanol.*, **1**, 59–87.

                            1945   Petrology of the Carboniferous igneous rocks of the Tweed basin. *Trans. Edinb. geol. Soc.*, **14**, 53–75.

                            1953   The Carboniferous igneous rocks of the Kelso district. *Proc. Univ. Durham Phil. Soc.*, **11**, 95–101.

TYRRELL, G. W.   1917   The picrite–teschenite sill of Lugar (Ayrshire). *Q. J. geol. Soc. Lond.*, **72**, 84–131.

                            1948   A boring through the Lugar Sill. *Trans. geol. Soc. Glasg.*, **21**, 157–202.

                            1952   A second boring through the Lugar Sill. *Trans. Edinb. geol. Soc.*, **15**, 374–392.

UPTON, B. G. J.   1969   *Field excursion guide to the Carboniferous rocks of the Midland Valley of Scotland.* Edinburgh.

                            1982   Carboniferous volcanism. *In* Sutherland, D. (Ed.) *Igneous rocks of the British Isles.* Wiley, London, 255–275.

UPTON, B. G. J. *et al.*   1976   Pre-Palaeozoic basement of the Midland Valley. *Nature Lond.*, **260**, 517–518.

                            1984   Xenoliths and their implications for the deep geology of the Midland Valley of Scotland and adjacent regions. *Trans. roy. Soc. Edinb.: Earth Sci.*, **75**, 65–70.

                            1987   The Glas Eilean lavas: evidence of a Lower Permian volcano–tectonic basin between Islay and Jura, Inner Hebrides. *Trans. roy. Soc. Edinb.: Earth Sci.*, **77** (for 1986), 289–293.

WALKER, B. H.   1987   Emplacement mechanism of high-level dolerite sills and related
                                         eruptions in sedimentary basins, Fife, Scotland. *PhD Thesis,
                                         Univ. Leeds* (Unpublished).

WALKER, B. H. &   1987   High-level emplacement of an olivine–dolerite sill into
    FRANCIS, E. H.   Namurian sediments near Cardenden, Fife. *Trans. roy. Soc.
                                         Edinb.: Earth Sci.*, **77** (for 1986), 295–307.

WALKER, F.   1935   The late Palaeozoic quartz–dolerites and tholeiites of Scotland.
                                         Mineral. Mag., **24**, 131–159.

                                 1965   The part played by tholeiitic magma in the Carbo–Permian
                                         vulcanicity of Central Scotland. *Mineral. Mag.*, **34**, 498–516.

WHYTE, F. &   1974   Lower Carboniferous vulcanicity in the northern part of the
    MACDONALD, J. G.   Clyde plateau. *Scott. J. Geol.*, **10**, 187–198.

12

PERMIAN AND TRIASSIC
J. P. B. Lovell

The Permo-Triassic rocks of Scotland were formed at a time of major regression. The withdrawal to the south of the late Palaeozoic sea begun in the Carboniferous was completed by the beginning of the Permian; only at the end of the Triassic were there indications of the first of the major transgressions of the Mesozoic.

Offshore work, much of it in search of oil and gas, has led to many discoveries of Permian and Triassic rocks below the continental shelf around Scotland. These studies have had great value in setting the isolated and largely unfossiliferous outcrops on land (Fig. 12.1) into a more comprehensive regional framework (Fig. 12.2) than has ever been possible before. This has in turn led to more confident reconstruction of palaeogeography and structural setting.

There has also been much work on land since the first edition of this book was published in 1965. Although this has been limited as before by the lack of good stratigraphical control, considerable progress has been made in facies analysis in many areas. Figure 12.3 reflects the difficulties in zoning and correlation between basins.

This account emphasises work carried out since 1965. An outline of earlier work is recorded in Craig (1965). The Permian and Triassic rocks are here considered together, because of the uncertainty about their age that arises from the lack of fossils. In the sections on stratigraphy that follow, offshore and onshore localities are discussed together, in the three main areas of Hebrides, SW Scotland, and the Moray Firth Basin and northern North Sea.

Hebrides

Seven main Hebridean basins of Permo-Triassic age have been recognised (Fig. 12.2). The existence of the 'Outer Hebrides Basin' remains uncertain; proponents of early continental separation to the north-west of the British Isles would not be surprised at the discovery of a thick Permo-Triassic sequence on the floor of the Rockall Trough (Jones 1978, Russell 1976, Smythe, Kenolty, and Russell 1978). Along strike to the north-east, it is believed that over 1,000 m of Permo-Triassic sediments were formed in the fault-bounded West Shetland Basin (Ridd 1981).

North Minch Basin

The Stornoway Formation of north-east Lewis is c. 4,000 m thick, unconformable on, or faulted against, Lewisian, and is mainly composed of conglomerates. It has been variously assigned Torridonian, Devonian or Permo-Triassic ages; though there is no palaeontological evidence, the Stornoway Formation is most probably late Permo-Triassic (Steel 1971, Steel and Wilson 1975, Storetvedt and Steel 1977, compare with Smith 1976). It apparently lies on the western margin of a North Minch Basin; small outliers of unconformable Permo-Triassic in Wester Ross, such as those at Udrigle, Isle of Ewe and Camas Mor (Fig. 12.4) lie on the south-eastern margin of this basin, in the centre of which over 4 km of sediment may have accumulated in Permo-Triassic times. The great contrast in thickness between the Stornoway Formation in the west and the thin successions in the east indicates a marked asymmetry in the development of the basin.

Sea of the Hebrides Basin

Samples of late Permian and probable Rhaetian age have been reported from just east of Benbecula, on the western margin of a basin that may contain 2 km of Permo-Triassic sediment (Binns, McQuillin, and Kenolty 1974, Smythe et al. 1972, Steel 1977). On land there is only the evidence of small patches of mainly thin Permo-Triassic sediments forming unconformable outliers along the eastern margin of the basin, from Gairloch through Applecross and Skye to Rhum. There are 75–90 m of probable Triassic conglomerates, sandstones and carbonates on Raasay and Scalpay, that pass up into marine Lower Jurassic (Hettangian) sediments on Raasay (Bruck,

Dedman, and Wilson 1967). Probable Late Triassic ostracod and fish and plant remains occur in the highest part of the Rhum succession (Bailey 1944). The age of the outcrops on land of Permo-Triassic rocks in this basin has for some time been taken as Triassic (Richey 1961).

Inner Hebrides Basin and Great Glen Basin

The main evidence for the existence of these basins is reported by Binns, McQuillin, and Kenolty (1974) and

McQuillin and Binns (1973). The Inner Hebrides Basin may contain over 2 km of sediments (including Permo-Triassic) banked up against the Camasunary Fault scarp east of Coll and Tiree. To the east, in the west of Mull, a thin (c. 12 m) sequence of marine Late Triassic carbonates lies conformably on non-marine probable earlier Triassic sediments (Richey 1961) of probable Rhaetian age (Warrington and Pollard 1985). The Permo-Triassic of south-east Skye is predominantly coarse-grained (Nicholson 1978, Steel 1971, 1974a, b, and Steel, Nichol-

Fig. 12.1. Permo-Triassic outcrop in Scotland. After Craig (1965), Smith *et al.* (1974) and Steel (1974a).

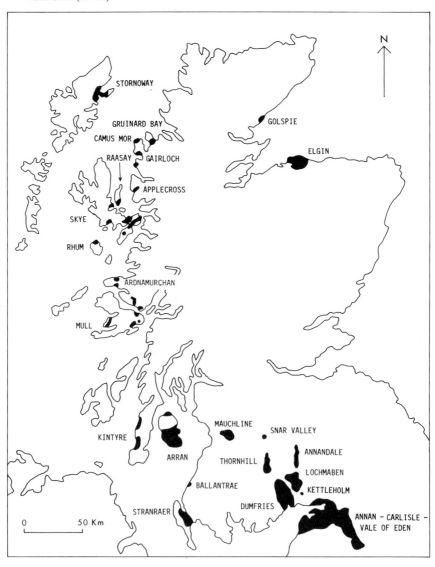

son, and Kalander 1975). Otherwise land outcrops in this area are restricted to unconformable, unfossiliferous outliers of non-marine sediments in north Mull, Morven and Ardnamurchan.

The Great Glen Basin was a deep (more than 1 km) but very narrow trough along the south-eastern margin of an active Great Glen Fault. Direct evidence for this is a reputedly thick (up to 2,000 m?) sequence of probable Triassic conglomerates and sandstones reported from south-eastern Mull by Rast, Diggens and Rast (1968). 'New Red Sandstone' rocks are reported from 11 km north-east of Colonsay in what may be a south-western extension of the Great Glen Basin (Institute of Geological Sciences 1974).

Arran Basin and Rathlin Basin

Thin, unconformable outliers of non-marine Permo-Triassic on the southern tip of Islay and along part of the west coast of Kintyre provide the only land-based evidence concerning the Rathlin Basin, which offshore may contain a relatively thick Permo-Triassic sequence in part dated as Triassic (Institute of Geological Sciences 1976, McLean and Deegan 1978, Steel 1977, Warrington *in* Owens and Marshall 1978).

The Arran Basin was probably closely connected with the Rathlin Basin, at least in the later stages of their development (Steel 1977). The extent of probable Permo-Triassic rocks offshore in the Arran Basin, and the poss-

Fig. 12.2. Permo-Triassic basins of Scotland. After Kent (1975), Steel (1977) and Fisher (1984).

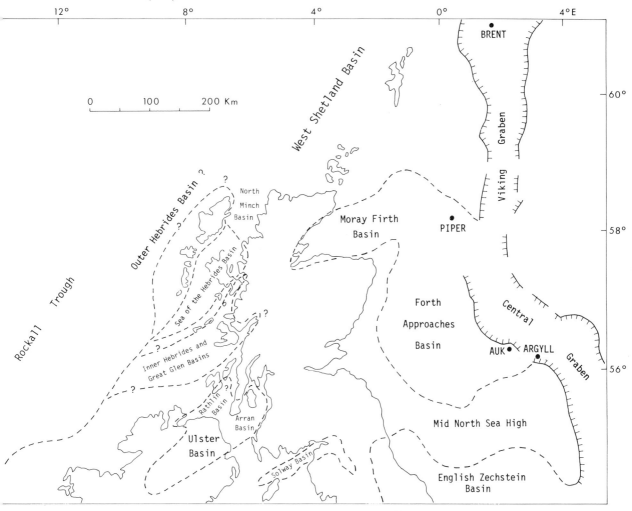

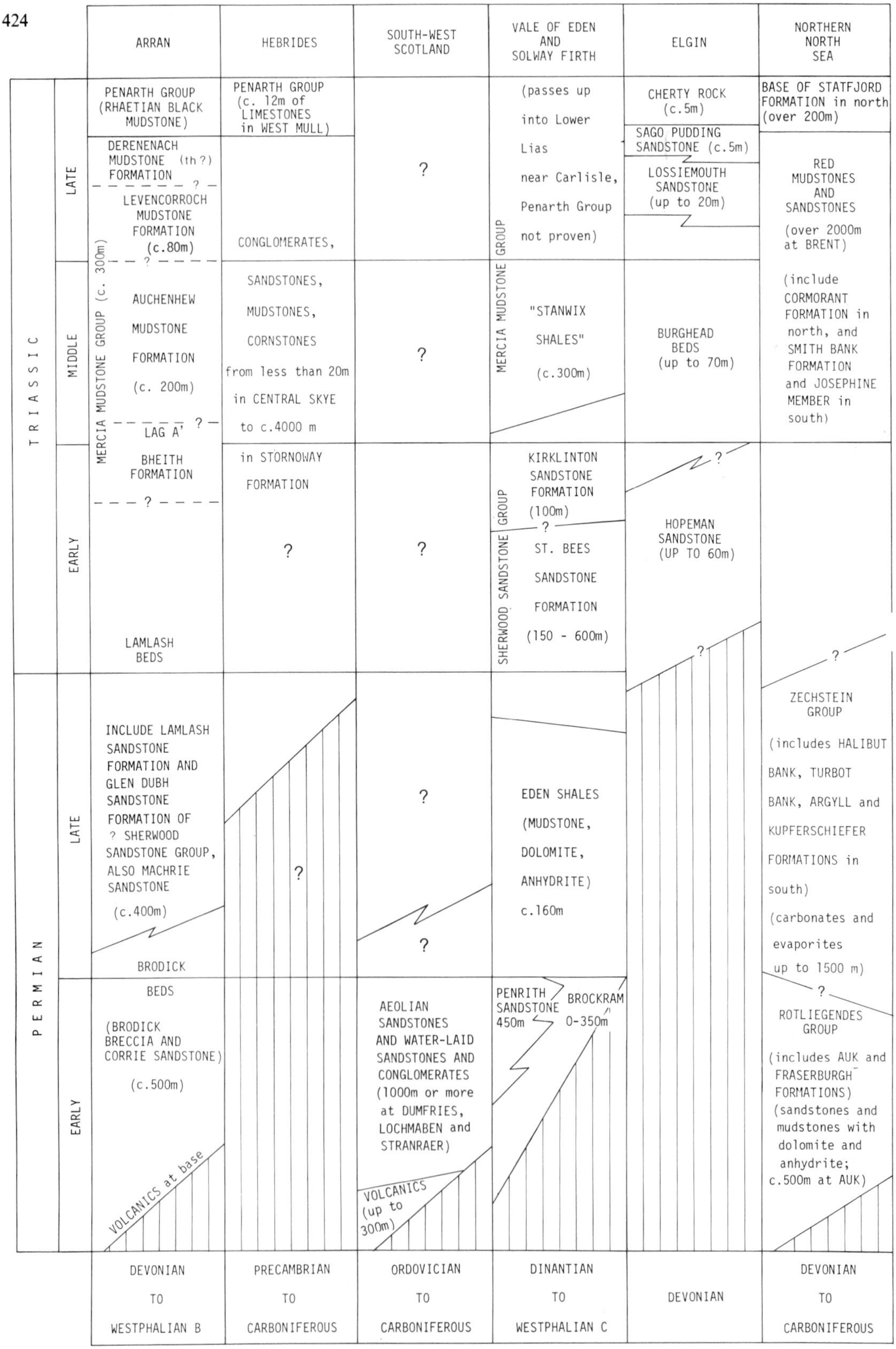

		ARRAN	HEBRIDES	SOUTH-WEST SCOTLAND	VALE OF EDEN AND SOLWAY FIRTH	ELGIN	NORTHERN NORTH SEA
T R I A S S I C	LATE	PENARTH GROUP (RHAETIAN BLACK MUDSTONE) / DERENENACH MUDSTONE (th?) FORMATION — ? — / LEVENCORROCH MUDSTONE FORMATION (c.80m) — ? —	PENARTH GROUP (c. 12m of LIMESTONES in WEST MULL) / CONGLOMERATES,	?	(passes up into Lower Lias near Carlisle, Penarth Group not proven)	CHERTY ROCK (c.5m) / SAGO PUDDING SANDSTONE (c.5m) / LOSSIEMOUTH SANDSTONE (up to 20m)	BASE OF STATFJORD FORMATION in north (over 200m) / RED MUDSTONES AND SANDSTONES (over 2000m at BRENT)
	MIDDLE	AUCHENHEW MUDSTONE FORMATION (c. 200m) — LAG A' — ? —	SANDSTONES, MUDSTONES, CORNSTONES from less than 20m in CENTRAL SKYE to c.4000 m	?	"STANWIX SHALES" (c.300m)	BURGHEAD BEDS (up to 70m)	(include CORMORANT FORMATION in north, and SMITH BANK FORMATION and JOSEPHINE MEMBER in south)
	EARLY	BHEITH FORMATION — ? — / LAMLASH BEDS	in STORNOWAY FORMATION / ?	?	KIRKLINTON SANDSTONE FORMATION (100m) — ? — ST. BEES SANDSTONE FORMATION (150 - 600m)	HOPEMAN SANDSTONE (UP TO 60m) ?	
P E R M I A N	LATE	INCLUDE LAMLASH SANDSTONE FORMATION AND GLEN DUBH SANDSTONE FORMATION OF ? SHERWOOD SANDSTONE GROUP, ALSO MACHRIE SANDSTONE (c.400m) / BRODICK	?	? / ?	EDEN SHALES (MUDSTONE, DOLOMITE, ANHYDRITE) c.160m		ZECHSTEIN GROUP (includes HALIBUT BANK, TURBOT BANK, ARGYLL and KUPFERSCHIEFER FORMATIONS in south) (carbonates and evaporites up to 1500 m) ?
	EARLY	BEDS (BRODICK BRECCIA AND CORRIE SANDSTONE) (c.500m) VOLCANICS at base	?	AEOLIAN SANDSTONES AND WATER-LAID SANDSTONES AND CONGLOMERATES (1000m or more at DUMFRIES, LOCHMABEN and STRANRAER) VOLCANICS (up to 300m)	PENRITH SANDSTONE 450m / BROCKRAM 0-350m		ROTLIEGENDES GROUP (includes AUK and FRASERBURGH FORMATIONS) (sandstones and mudstones with dolomite and anhydrite; c.500m at AUK)
		DEVONIAN TO WESTPHALIAN B	PRECAMBRIAN TO CARBONIFEROUS	ORDOVICIAN TO CARBONIFEROUS	DINANTIAN TO WESTPHALIAN C	DEVONIAN	DEVONIAN TO CARBONIFEROUS

Notes appearing within table spines: MERCIA MUDSTONE GROUP (c. 300m); MERCIA MUDSTONE GROUP; SHERWOOD SANDSTONE GROUP.

Fig. 12.3

Fig. 12.4. Coarse-grained Triassic (?) braided-stream deposits at Camas Mór, Wester Ross. Exposure *c.* 70 m across. Photograph by R. J. Steel.

ible links with the sequences in Northern Ireland and south-west Scotland, are discussed by McLean and Deegan (1978). The sedimentary sequence of Permo-Triassic on Arran itself is over 1,000 m thick. It passes into marine Rhaetian (on the evidence of fragments of country rock preserved in the Tertiary igneous Central Ring Complex), and includes Triassic fossils and a great variety of facies.

A simple classification of the Arran rocks (Rhaetic, Auchenhew Beds, Lamlash Beds, Brodick Beds) is adopted by Craig (1965). This is designed to express the many variations in facies in the succession; Craig (pp. 387–388) recognises that this 'must be a temporary

Fig. 12.3 (facing page)
Correlation of Permo-Triassic rocks of Scotland. After Abrahamsen (1979), Brookfield (1978), Craig (1965), Deegan & Scull (1977), Peacock *et al.* (1968), Richey (1961), Smith *et al.* (1974), Steel (1977), Warrington (1973), Warrington *et al.* (1980), Williams (1973) and Woodland (1975). See text for discussion of uncertainties involved.

measure until this important New Red Sandstone succession in Arran is revised'. Formal units for the upper part of the sequence are proposed by Warrington *et al.* (1980) (Fig. 12.3) on the basis of Tyrrell's (1928) work. The main limits of the age of the Permo-Triassic rocks of Arran remain the Rhaetian fossils (including *Rhaetavicula contorta, Schizodus* and *Protocardia*) in the block in the Central Ring Complex, and the *similis-pulchra* zone mussels, recorded from small, faulted and intruded exposures about 7 km south-west of Lamlash, that may indicate a Late Carboniferous age for the very base of the succession (Craig 1965, Leitch 1941).

Abrahamsen (1979) seeks to integrate various studies, and to correlate the sequences at Brodick, Cock of Arran and Machrie (Fig. 12.5). Interbedded aeolian and water-deposited sediments form the basal Brodick Beds, which contain fulgurites ('fossil' lightning strikes) (Harland and Hacker 1966, Piper 1970). Astin and Macdonald (1983) have described in detail local tectonic control of sedimentation in the Brodick Beds. It may be possible to correlate the overlying Lamlash Beds across Arran, using a possible time plane defined by an influx of detrital agates in water-laid sandstones and conglomerates that now form a belt running from Corrygills and Lamlash in the east to Machrie in the west. The Lag a' Bheith Formation (Mercia

Mudstone Group) in south-east Arran is dated as Triassic (late Scythian or Anisian). Stratigraphically non-diagnostic trace fossils, sedimentary features and detailed rock sequences have been described from the Mercia Mudstone Group and Lamlash Beds in west Arran.

South-west Scotland

The only good direct fossil evidence of the age of the Permo-Triassic rocks in this area comes from the base of the sequence in the Mauchline Basin, where a series of

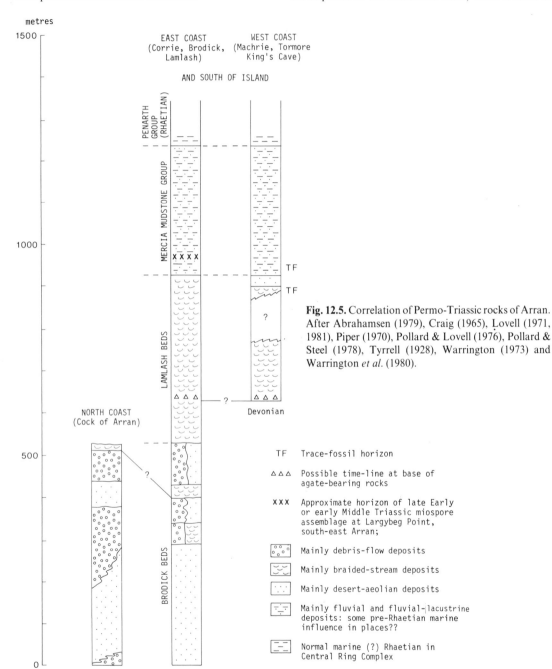

Fig. 12.5. Correlation of Permo-Triassic rocks of Arran. After Abrahamsen (1979), Craig (1965), Lovell (1971, 1981), Piper (1970), Pollard & Lovell (1976), Pollard & Steel (1978), Tyrrell (1928), Warrington (1973) and Warrington *et al.* (1980).

TF   Trace-fossil horizon

△ △ △   Possible time-line at base of agate-bearing rocks

x x x   Approximate horizon of late Early or early Middle Triassic miospore assemblage at Largybeg Point, south-east Arran;

          Mainly debris-flow deposits

          Mainly braided-stream deposits

          Mainly desert-aeolian deposits

          Mainly fluvial and fluvial-lacustrine deposits: some pre-Rhaetian marine influence in places??

          Normal marine (?) Rhaetian in Central Ring Complex

interbedded sediments and volcanics has yielded a sparse flora, tentatively interpreted as Early Permian (Mykura 1965, Wagner 1966, Smith *et al.* 1974). This suggests an Early Permian age for the basal Permo-Triassic rocks of south-west Scotland. It is tempting to seek links between the scattered outcrops of south-west Scotland, and the Permo-Triassic of northern England, and offshore areas to the south (Colter and Barr 1974), but lithological correlation between the scattered outcrops themselves is difficult enough. Local correlation has been attempted in some detail by Smith *et al.* (1974), and in considerable detail by Brookfield (1978); Brookfield's stratigraphy is followed in most of this account.

Mauchline Basin

This sequence provides the fossil flora referred to above. The large-scale cross-bedded well-sorted aeolian Mauchline Sandstone (over 450 m thick) overlying the basal volcanics (up to 300 m thick) is barren, but may be assigned to the Early Permian on the strength of the (?)

Early Permian age of the Mauchline Volcanic Group (Figs. 12.6 and 12.7). It is suggested that the volcanic group rests on Westphalian rocks and that 'the field appearance of conformity is probably misleading, and . . . a hiatus is present which may represent much of Stephanian time' (Smith *et al.* 1974, p. 23).

Thornhill Basin

A thin (*c.* 20 m) Carron Basalt Formation, unconformable on Ordovician to Westphalian rocks, passes up through a thickness of up to 50 m of the Durisdeer and Locherben Breccias to the aeolian sandstones of the Thornhill Formation (over 70 m thick).

Lochmaben Basin

The maximum thickness of the sequence is about 1,000 m (Bott and Masson-Smith 1960). The basal volcanics of Mauchline and Thornhill are absent; the Hartfield and Lockerbie Breccias rest directly on Silurian rocks and pass

Fig. 12.6. Large-scale aeolian cross-bedding in Mauchline Sandstone. IGS photo.

up into several hundred metres of aeolian sandstone (Cornockle Sandstone).

The sandstone quarries at Cornockle have provided many fossil reptilian footprints; footprints have also been noted in the Dumfries/Locharbriggs Sandstone. These prints were much studied in the last century (see Sarjeant 1974); they have been discussed by Delair (1966, 1967 and 1969), and by Haubold (1971), who suggests that some of the footprints were made by caseasaurs (edaphosaurs) which reached their peak during the Late Carboniferous and Early Permian.

A 50 m thickness of Kettleholm Breccias, in a small outlier to the south of the basin may be rather younger than the main Lochmaben sequence; like the isolated Snar and Ballantrae Breccias to the north and west, these rocks may be of Late Permian or even Early Triassic age.

Dumfries Basin

The statigraphy has been considerably simplified by Brookfield (1978), who recognises just two formations, with a total thickness of 1,000 m or more, unconformable on rocks of Silurian to Dinantian age (Bott and Masson-Smith 1960). The Dumfries Sandstone is renamed the Locharbriggs Sandstone, after the town north-east of Dumfries, where up to 25 m thickness of aeolian sand-

stone is exposed in a series of large quarries (Brookfield 1977, 1979). The Doweel Breccia, best seen to the west of Dumfries, is partly contemporaneous with and partly younger than the Locharbriggs Sandstone.

Stranraer Basin

Geophysical work indicates that the basin closes northwards, and is asymmetrical, with a thickness of up to 1,200 m of sediment at its margin in the south-east (Mansfield and Kennet 1963). The Loch Ryan Breccias in the west are unconformable on Ordovician to Carboniferous rocks; the eastern margin of the basin is faulted. Offsets of structural belts in lower Palaeozoic rocks on either side of Loch Ryan may be used to calculate a downthrow of about 1,500 m to the west on this eastern boundary fault (Kelling and Welsh 1970). Sandstone has been found beneath the thick sequence of breccias at Stranraer; its extent is uncertain.

Other areas

The possibility that the isolated Snar and (locally derived) Ballantrae Breccias may be younger than the main body of (?) Early Permian rocks of south-west Scotland has been discussed above in connection with the age of the outlier

Fig. 12.7. Tuff and overlying red sandstone, River Ayr near Howford Bridge, Mauchline. Contact is about half-way up the cliff face.

of Kettleholm Breccias south of the Lochmaben Basin. Arguments of age based on roundness of clasts and inferred climatic change are necessarily highly tentative (Brookfield 1978, Smith *et al.* 1974).

Exposure in the narrow Moffat Basin is poor (Fig. 12.8). The sequence is partly unconformable on lower Palaeozoic rocks, and partly faulted against them. It includes breccias of the Auchencat and Bellcraig Formations up to

150 m thick, and about 40 m of aeolian sandstones called the Corehead Formation. The sequence in the Annan district is a northern extension into Scotland of the important sequence of the Carlisle Basin and Vale of Eden (Fig. 12.3). The poorly exposed unfossiliferous sequence of conglomerates and sandstones in the Annan area is unconformable on Carboniferous rocks. It is assigned to the Permian by Smith *et al.* (1974), who suggest that some of

Fig. 12.8. The River Annan Valley, north of Moffat. View towards the north-west, into Devil's Beef Tub. Poorly exposed Permian (?) rocks on the floor of the valley lie unconformably on lower Palaeozoic rocks that form the surrounding hills. Photograph by Aerofilms Limited.

the water-laid footprint-bearing sandstones and mudstones found higher in the sequence may be of Late
Permian age.

Moray Firth Basin and Northern North Sea

These are the areas in which there has been the most
spectacular growth of geological knowledge since 1965.
The well-known outliers of Permo-Triassic rocks on the
shores of the Moray Firth will be discussed here with
evidence from offshore. Information from elsewhere in
the North Sea will be considered in relation to work in the
region of the Brent, Piper, Argyll and Auk oil fields.
Relevant background information on the Permo-Triassic
rocks offshore is to be found in Brooks and Glennie
(1987), Glennie (1984a), Illing and Hobson (1981) and
Ziegler (1982). The work of Taylor (1981 and 1984),
Glennie (1984b) and Fisher (1984) gives particularly
detailed coverage.

Moray Firth Basin

The Permo-Triassic rocks on the shores of the Moray
Firth are either unconformable on, or faulted against,
Devonian rocks. The reptilian fauna from the Elgin area,
first made famous by Huxley (1877), has been restudied by
Walker (1961, 1964, 1973). The older fossils come from the
aeolian Cuttie's Hillock (now Hopeman) Sandstone west
of Elgin. The remains are nearly all referable to two
groups, the dicynodonts or anomodonts and the
pareiasaurs. They are of Late Permian to Early Triassic
age, more probably the latter. With the aeolian Hopeman
Sandstone to the north they are taken to be the lowest
beds in the succession. The unfossiliferous Burghead
Beds are faulted against these aeolian sandstones, and
are thought to be younger (Peacock et al. 1968). The
aeolian Lossiemouth Sandstone contains a reptilian fauna
of probable Late Triassic age. Overlying the Lossiemouth
Sandstone are the thin Sago Pudding Sandstone and
Cherty Rock (Williams 1973). At Golspie on the north-
western shore of the Moray Firth, Triassic calcretes and
silcretes pass upwards unconformably into (?) latest
Rhaetian to Hettangian freshwater conglomerates (Batten,
Trewin and Tudhope 1986).

A borehole at Lossiemouth in 1964 encountered early
Lias rocks (Peacock et al. 1968), providing a hint of the
sequence offshore (Institute of Geological Sciences 1977).
The Permo-Triassic sandstones offshore are 300 to 500 m
thick south-east of the Great Glen Fault; they are unconformable on Devonian rocks, and are themselves unconformably overlain by Jurassic rocks. Marine Zechstein is
present in the Moray Firth Basin, at the base of a sequence
or pre-Upper Cretaceous sedimentary rocks that has a
total thickness of about 2,000 m (Kent 1975). Up to 450 m
of Triassic has been drilled, with greater thicknesses indicated by seismic evidence. Even thicker (over 1,000 m)

Triassic sequences occur farther offshore, in the northern
Viking Graben and the Central North Sea (Brennand
1975).

A particular feature of the Hopeman Sandstone that may
be of regional significance is the presence of structureless
and distorted sequences up to 200 m or more in thickness.
Such rocks are also found in the offshore Permian;
according to Glennie and Buller (1983) they result from a
very rapid rise in water level associated with the Zechstein
transgression.

Triassic sedimentation in the Inner Moray Firth has
been compared to that in the Tertiary to Recent rifts of
East Africa. Frostick et al. (1988) suggest that in the
Triassic the area of the Firth was a simple half-graben,
with major tectonic control exercised by dip-slip on the
Great Glen Fault to the north-west. The coarsest
sediments are found on the unfaulted margin to the south,
around Burghead. As in the East African rifts, deposits
both thicken and get finer towards the main fault, in this
case towards the north-west.

Brent Oil Field

This lies in the northern part of the U.K. sector of the North
Sea. The lower of the two main reservoirs is of Early
Jurassic to Rhaetian age (Bowen 1975), this paralic Statfjord Formation (over 200 m thick) overlies a thickness of
over 2,000 m of Triassic sandstones and red mudstones, the
upper part of which is Rhaetian to Norian in age. The latter
sequence may be correlated with Triassic rocks found
elsewhere in the northern Viking Graben, at the Cormorant
and Dunlin oil fields (Brennand 1975).

Piper Oil Field

This lies at the eastern end of the Moray Firth Basin.
About 120 m of anhydrite with thin-bedded dolomite and
mudstone, containing Late Permian palynomorphs, pass
up into a thickness of over 50 m of red probable Triassic
mudstones that are overlain unconformably by non-
marine Middle Jurassic rocks. The Permian evaporites are
a Zechstein facies; they unconformably overlie Carboniferous carbonaceous sandstones (Williams, Conner, and
Peterson 1975).

Argyll Oil Field

This field is 320 km east-south-east of Aberdeen. Oil has
been found in Late Permian Zechstein dolomite and (?)
Early Permian Rotliegend-facies sandstones at depths
between 2,500 and 3,000 m (Pennington 1975). The
Rotliegend–facies consists of a thickness of several
hundred metres of supposed fluvial and aeolian sandstones, with some highly altered basalt, lying unconformably on a Devonian sequence that includes marine Middle
Devonian. The Zechstein is mainly dolomite that varies
from 10 to 30 m in thickness, and has been dated as Late

Permian on floral evidence. The Rotliegend is largely unfossiliferous and its age is uncertain. The conventional view would place it in the Early Permian (Deegan and Scull 1977), but it may be Late Permian (see discussion in Pennington 1975). The Zechstein passes up through a thickness of about 40 m of red probable Triassic mudstones into probable Jurassic sandstones. Cretaceous rocks overlie this sequence unconformably.

Permian anhydrite is found in several places well to the west of the Argyll Field; for example, 11 km south-south-east of Aberdeen and 56 km east-north-east of Dunbar (Institute of Geological Sciences 1974).

Auk Oil Field

This lies in the south-eastern part of the Forth Approaches Basin, about 50 km north-west of the Argyll Field. A thickness of several hundred metres of Rotliegend-facies sandstones unconformably overlies (?) Devonian sandstones, and passes up into oil-bearing Zechstein carbonates up to 40 m thick in places (Brennand and Van Veen 1975). These in turn pass up unconformably into mudstones, dated palynologically as Early Triassic (Scythian). These mudstones are over 30 m thick in places, but are cut out over part of the area by the sub-Cretaceous unconformity.

There are great variations in the thickness of the Triassic from near zero to over 1,000 m in the Central

Graben and Forth Approaches area (Brennand 1975). The upper boundary is everywhere erosional or non-sequential; the red Triassic mudstones lie directly on Permian Zechstein of both basin and shelf facies without themselves showing any variation in facies. A thickness of 900 m of Zechstein basin facies is found north-west of Auk; the widespread Zechstein in the northern North Sea contrasts with the variable Rotliegend sequence, which includes some igneous extrusives (Ziegler, W. H. 1975).

Palaeogeography

Permian

Evidence of (?) Early Permian volcanic activity is seen both at the base of the sequence in south-west Scotland and also in the Rotliegend (Early Permian) of the northern North Sea and the area of the Mid North Sea High. The Rotliegend aeolian sandstones of the North Sea are well represented in the supposed Early Permian rocks of south-west Scotland and Arran that overlie the volcanics. Wind directions (Fig. 12.9) were from east and north-east, in sympathy with the trend for the Permian Rotliegend of north-west Europe. Contrary directions in the northern North Sea Auk Formation are reported by Glennie (1984b); he suggests that a barometric high may have existed over the Mid North Sea structural high.

Fig. 12.9. Early Permian palaeogeography of Scotland. After Abrahamsen (1979), Brookfield (1977, 1978, 1979, 1980), Craig (1965), Glennie (1972, 1984b), Lovell (1977), Piper (1970), Smith (1976), Waugh (1970), and W. H. Ziegler (1975).

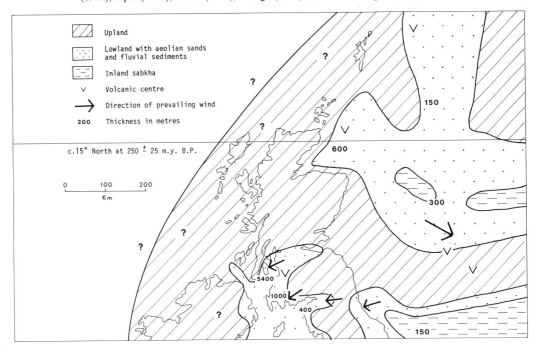

These Early Permian sand-dune fields were flanked by wetter, upland areas intermittently supplying coarse-grained water-laid sediment. Glennie (1984b) suggests a measure of fault-control of these deposits in the northern North Sea. The aeolian sandstones at Locharbriggs near Dumfries may have formed in a sand sea in which no bedrock was exposed, and into which no wadis led. Elsewhere, as in Arran, there is clear evidence of deposition at the margin of a desert basin, where water-deposited breccias and sands were in places reworked by wind. In the area of the North Sea, for example at the Argyll oil field, there is evidence of alternating fluviatile and aeolian deposition in Rotliegend facies.

There the resemblance ceases between conditions in the North Sea area and conditions over the area of Scotland to the west (Fig. 12.10). In the North Sea area marine influence becomes obvious for a while during the formation of the thick sequence of Late Permian Zechstein facies of carbonates and evaporites. Farther west, over Scotland, Permo-Triassic stratigraphy is at its haziest. Some of the breccias and aeolian sandstones of south-west Scotland, the Hebrides and the Moray Firth may be Late Permian; water-laid deposits may be linked with the Late Permian climatic amelioration inferred from elsewhere. What is known is that the Late Permian marine transgression brought epicontinental seas close to the present-day coast-line of south-west and east Scotland; Glennie (1984b) gives a graphic and quantitative account of a postulated rapid flooding by the Zechstein Sea of a Rotliegend desert lying below the level of the open ocean.

Triassic

In the west, part of the Arran sequence is dated as late Early or early Middle Triassic, and in the east the lower part of the Moray Firth sequence is probably Early Triassic. Figure 12.11 therefore includes the suggestion that the Arran rocks are fluvial-lacustrine, or possibly even near-marine, shows the persistence into the Early Triassic of Permian trends with north-east winds in the Moray Firth area, and includes the suggestion that transport of water-laid sediments in that area may have been partly from the west.

In both the Hebridean and Moray Firth areas the Late Triassic to Early Jurassic transgression is recorded in places (Fig. 12.12). Evidence of this transgression may also be seen in the Viking Graben to the north but not at Piper and farther south.

The Hebridean Permo-Triassic has traditionally been assigned to the Triassic (Fig. 12.11). The Stornoway Formation consists of a conglomeratic, alluvial-fan facies and an interfingering floodplain facies. Similar facies may be

Fig. 12.10. Late Permian palaeogeography of Scotland. After Colter & Barr (1975), Pattison, Smith & Warrington (1973), Wilson (1972), and W. H. Ziegler (1975).

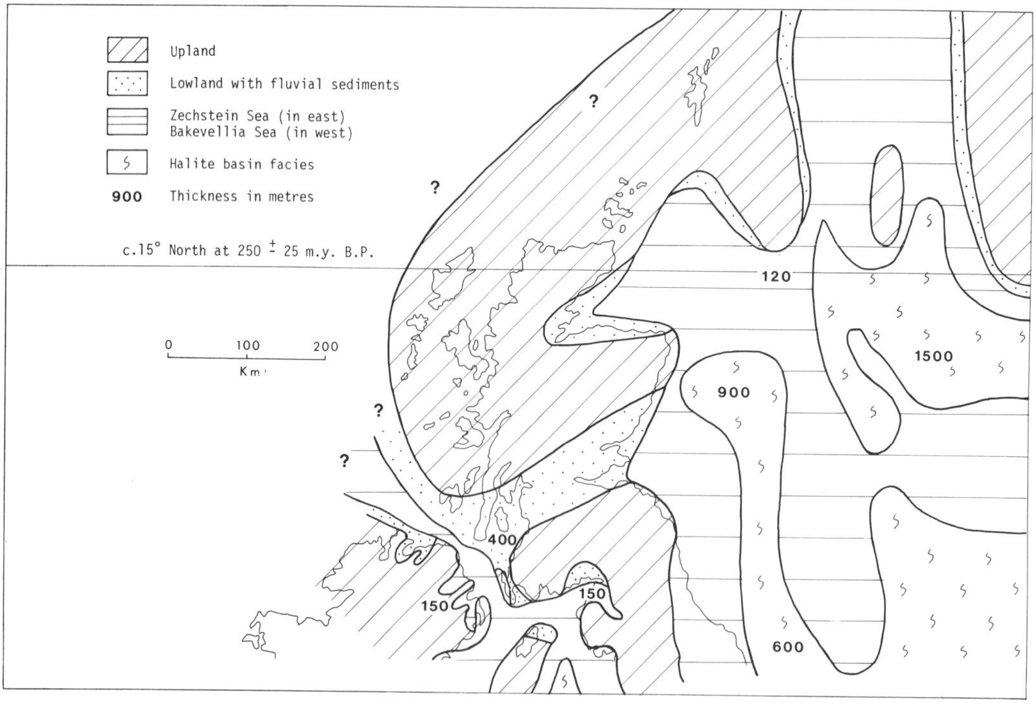

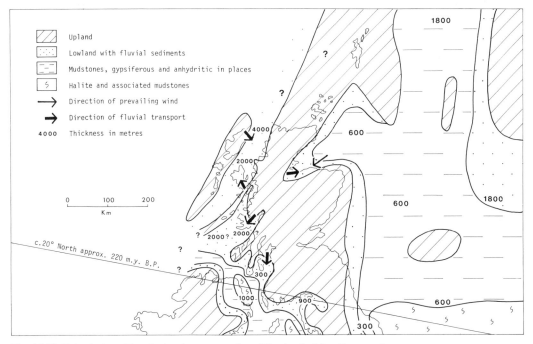

Fig. 12.11. Triassic (pre-Rhaetian) palaeogeography of Scotland. After Brennand (1975), Pattison, Smith & Warrington (1973), Peacock *et al.* (1968), Steel (1977), Warrington (1974), Williams (1973), Wilson (1972) and W. H. Ziegler (1975).

Fig. 12.12. Rhaetian palaeogeography of Scotland. After Audley-Charles (1970), Brennand (1975), Hallam & Sellwood (1976), Steel (1977), Williams (1973) and Wilson (1972).

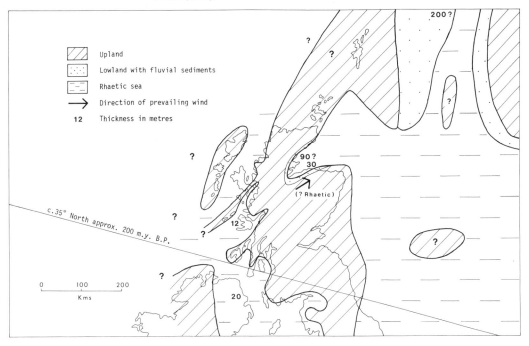

seen in Central Skye and elsewhere in the Hebridean Province; the floodplain deposits contain well developed cornstone (fossil caliche), which is normally formed in semi-arid regions.

Trends in sedimentation through time may be recognised in the characteristic Hebridean shallow, eastern basin-margin sequences. The alluvial fan sequences show a change from mudflow through stream-flood to braided-stream deposits. In laterally equivalent floodplain sequences the comparable trend is from low to high sinuosity and less ephemeral stream systems. These two trends probably indicate a combination of general lowering of relief and a long-term climatic change to less arid conditions. Comparable trends are also apparent in the Arran sequence. Such a pattern fits well with the known Late Triassic marine transgression in the area. The deeper basin-margin sequences found along the great faults to the north and west show an opposite trend, a record of vigorous and repeated progradation (Steel 1977). The facies contrasts across the Hebridean basins are typical of half-grabens. Thickness variations in the Triassic sediments in the Viking Graben in the northern North Sea are also compatible with fault-control of deposition (Fisher 1984), as are those in the Triassic of the inner Moray Firth Basin (Frostick et al. 1988).

The palaeoclimatological and palaeontological evidence of both the Permian and the Triassic rocks of Scotland is consistent with movement from a position just north of the Equator in the Permian to c. 30° N by the Late Triassic (Smith and Briden 1977, Smith, Briden, and

Drewry 1973). So is the colour that first led Sedgwick and Murchison (1828) to give them the name New Red Sandstone.

Structural evolution

A Permo-Triassic 'intracratonic' stage may be recognised in the tectonic evolution of north-western Europe, followed by a Late Triassic to early Tertiary rifting or 'taphrogenic' stage (Ziegler, P. A. 1975). The Scottish Permo-Triassic provides some of the evidence supporting these ideas. In the broadest terms, the Rotliegend and Zechstein facies of the northern North Sea indicate a subsiding, intracratonic basin during the Permian; the later faulting can be shown to have controlled much of the supposed Triassic sedimentation in the Hebridean province to the west. Just how this Triassic faulting is related to the opening of the North Atlantic remains an open question, the resolution of which must await drilling of the largely unexplored offshore area north-west of the British Isles.

In face of the welter of information that is coming to hand concerning the geological evolution of offshore areas during the Permo-Triassic, it must be remembered that the long-established ideas of the structure of the mainland are largely unchanged. Throughout the Permo-Triassic Scotland remained a mostly stable, upland area, a structural high never to be fully submerged by the sea from then to the present day.

REFERENCES

ABRAHAMSEN, K. A.   1979   *Aspects of sedimentology and palaeogeography of the lower New Red Sandstone (? Permian) of Arran, western Scotland.* Thesis, Cand. Real., Univ. Bergen. (Unpublished).

ASTIN, T. R. &   1983   Syn-depositional faulting and valley-fill breccias in the Permo-
  MACDONALD, D. I. M.   Triassic of Arran. *Scott. J. Geol.*, **19**, 47–58.

AUDLEY-CHARLES, M. G.   1970   Triassic palaeogeography of the British Isles. *Q. J. geol. Soc. London*, **126**, 49–90.

BAILEY, E. B.   1944   Tertiary igneous tectonics of Rhum (Inner Hebrides). *Q. J. geol. Soc. London*, **100**, 165–191.

BATTEN, D. J.,   1986   The Triassic–Jurassic junction at Golspie, inner Moray Firth
  TREWIN, N. H. &   Basin. *Scott. J. Geol.*, **22**, 85–98.
  TUDHOPE, A. W.

BINNS, P. E.,   1974   The geology of the Sea of the Hebrides. *Rep. Inst. geol. Sci.*
  McQUILLIN, R. &   *London*, **73/14**, 44 pp.
  KENOLTY, N.

BOTT, M. H. P. &   1960   A gravity survey of the Criffel Granodiorite and the New Red
  MASSON-SMITH, D.   Sandstone deposits near Dumfries. *Proc. Yorkshire geol. Soc.*, **32**, 317–332.

BOWEN, J. M.   1975   The Brent Oil-Field. *In* Woodland, A. W. (Ed.), q.v. 353–361.

BRENNAND, T. P.   1975   The Triassic of the North Sea. *In* Woodland, A. W. (Ed.), q.v., 295–310.

BRENNAND, T. P. & VAN VEEN, F. R.
1975 The Auk Oil-Field. *In* Woodland, A. W. (Ed.), q.v., 275–283.

BROOKFIELD, M. E.
1977 The origin of bounding surfaces in ancient aeolian sandstones. *Sedimentology,* **24**, 303–332.

1978 Revision of the stratigraphy of Permian and supposed Permian rocks of southern Scotland. *Geol. Rdsch.,* **67**, 110–149.

1979 Anatomy of a Lower Permian aeolian sandstone complex, Southern Scotland. *Scott. J. Geol.,* **15**, 81–96.

1980 Permian intermontane basin sedimentation in southern Scotland. *Sediment. Geol.,* **27**, 167–194.

BROOKS, J. & GLENNIE, K. W.
1987 ***Petroleum Geology of North West Europe.*** **Graham & Trotman, London, 1219 pp.**

BRUCK, P. M., DEDMAN, R. E. & WILSON, R. C. L.
1967 The New Red Sandstone of Raasay and Scalpay, Inner Hebrides. *Scott. J. Geol.,* **3**, 168–180.

COLTER, V. S. & BARR, K. W.
1975 Recent developments in the geology of the Irish Sea. *In* Woodland, A. W. (Ed.), q.v., 61–73.

CRAIG, G. Y.
1965 Permian and Triassic. *In* Craig, G. Y. (Ed.). *The Geology of Scotland.* Oliver & Boyd, Edinburgh, 383–400.

DEEGAN, C. E. & SCULL, B. J.
1977 A standard lithostratigraphic nomenclature for the Central and Northern North Sea. *Rep. Inst. geol. Sci. London,* **77/25**; *Bull. Norw. Petrol. Direct* No. 1; 36 pp.

DELAIR, J. B.
1966 Fossil footprints from Dumfriesshire with descriptions of new forms. *Trans. J. Dumfries Galloway nat. Hist. Antiq. Soc.,* **43**, 14–30.

1967 Additional records of British Permian footprints. *Trans. J. Dumfries Galloway nat. Hist. Antiq. Soc.,* **44**, 1–5.

1969 Preliminary notice of vertebrate footprints from the Triassic of Dumfriesshire. *Trans. J. Dumfries Galloway nat. Hist. Antiq. Soc.,* **46**, 178–179.

EYLES, V. A., SIMPSON, J. B. & MACGREGOR, A. G.
1949 Geology of Central Ayrshire. *Mem. geol. Surv. U.K.,* 160 pp.

FISHER, M. J.
1984 Triassic. *In* Glennie, K. W. (Ed.), q.v., 85–101.

FROSTICK, L., REID, I., JARVIS, J. & EARDLEY, H.
1988 Triassic sediments of the Inner Moray Firth, Scotland: early rift deposits. *J. geol. Soc. London,* **145**, 235–248.

GLENNIE, K. W.
1972 Permian Rotliegendes of northwest Europe interpreted in light of modern desert sedimentation studies. *Bull. Am. Assoc. Petrol. Geol.,* **56**, 1048–1071.

1984a *Introduction to the Petroleum Geology of the North Sea.* Blackwell Scientific Publications, Oxford, 236 pp.

1984b Early Permian-Rotliegend. *In* Glennie, K. W. (Ed.), q.v., 41–60.

GLENNIE, K. W. & BULLER, A. T.
1983 The Permian Weissliegend of NW Europe: the partial deformation of aeolian dune sands caused by the Zechstein transgression. *Sedimentary Geology,* **35**, 43–81.

HALLAM, A. & SELLWOOD, B. W.
1976 Middle Mesozoic sedimentation in relation to tectonics in the British area. *J. Geol. Chicago,* **84**, 301–321.

HARLAND, W. B. & HACKER, J. F.
1966 Fossil lightning strikes 250 million years ago. *Rep. Br. Assoc. Advmt. Sci.,* **22**, 663–671.

HAUBOLD, H.
1971 Ichnia Amphibiorum et Reptiliorum fossilium. *In* Kuhn, O. (Ed.). *Handbuch der Palaeoherpetologie 18.* Fischer, Stuttgart, 124 pp.

HUXLEY, T. H.
1877 The crocodilian remains found in the Elgin sandstones, with remarks on the ichnites of Cummingstone. *Mem. geol. Surv. U.K.,* Monograph 3.

| ILLING, L. V. & HOBSON, G. D. | 1981 | *Petroleum geology of the continental shelf of North-West Europe.* Heyden and Son, London, 521 pp. |

ILLING, L. V. & HOBSON, G. D. — 1981 — *Petroleum geology of the continental shelf of North-West Europe.* Heyden and Son, London, 521 pp.

INSTITUTE OF GEOLOGICAL SCIENCES — 1974 — IGS boreholes 1973. *Rep. Inst. geol. Sci. London,* **74/7**, 23 pp.

— 1976 — IGS boreholes 1975. *Rep. Inst. geol. Sci. London,* **76/10**, 47 pp.

— 1977 — Moray-Buchan: Sheet 57N 04W (1:250 000). Ordnance Survey, Southampton.

JONES, E. J. W. — 1978 — Seismic evidence for sedimentary troughs of Mesozoic age on the Hebridean continental margin. *Nature. London,* **272**, 789–792.

KELLING, G. & WELSH, W. — 1970 — The Loch Ryan Fault. *Scott. J. Geol.,* **6**, 266–271.

KENT, P. E. — 1975 — The tectonic development of Great Britain and the surrounding seas. *In* Woodland, A. W. (Ed.), q.v., 3–28.

LEITCH, D. — 1941 — The Upper Carboniferous rocks of Arran. *Trans. geol. Soc. Glasgow,* **20**, 141–154.

LOVELL, J. P. B. — 1971 — Petrography and correlation of sandstones in the New Red Sandstone (Permo-Triassic) of Arran. *Scott. J. Geol.,* **7**, 162–169.

— 1977 — *The British Isles through geological time: a northward drift.* George Allen & Unwin, London, 40 pp.

— 1981 — Intertidal sediments in the Auchenhew (Triassic) of Arran: discussion of paper by J. E. Pollard and R. J. Steel. *Scott. J. Geol.,* **17**, 223—224.

McLEAN, A. C. & DEEGAN, C. E. (Eds.) — 1978 — The solid geology of the Clyde Sheet (55° N/6W). *Rep. Inst. geol. Sci. London,* **78/9**, 114 pp.

McQUILLIN, R. & BINNS, P. E. — 1973 — Geological structure in the Sea of the Hebrides. *Nature (phys. Sci.),* **241**, 2–4.

MANSFIELD, J. & KENNET, P. — 1963 — A gravity survey of the Stranraer sedimentary basin. *Proc. Yorkshire geol. Soc.,* **34**, 139–151.

MYKURA, W. — 1965 — The age of the lower part of the New Red Sandstone in southwest Scotland. *Scott. J. Geol.,* **1**, 9–18.

NICHOLSON, R. — 1978 — The Camas Malag Formation: an interbedded rhythmite/conglomerate sequence of probable Triassic age, Loch Slapin, Isle of Skye. *Scott. J. Geol.,* **14**, 301–309.

OWENS, B. & MARSHALL, J. (Compilers) — 1978 — Micropalaeontological biostratigraphy of samples from around the coasts of Scotland. *Rep. Inst. geol. Sci. London,* **78/20**, 35 pp.

PATTISON, J., SMITH, D. B. & WARRINGTON, G. — 1973 — A review of late Permian and early Triassic biostratigraphy in the British Isles. *In* Logan, A. V. and Mills, L. V. (Eds.). *The Permian and Triassic Systems and their mutual boundary. Mem. Can. Soc. Petrol. Geol.,* **2**, 220–260.

PEACOCK, J. D., BERRIDGE, N. G., HARRIS, A. L. & MAY, F. — 1968 — The geology of the Elgin District. *Mem. geol. Surv. Scotland,* 165 pp.

PENNINGTON, J. J. — 1975 — The geology of the Argyll Field. *In* Woodland A. W. (Ed.), q.v., 285–291.

PIPER, D. J. W. — 1970 — Eolian sediments in basal New Red Sandstone, Arran, Scotland. *Scott. J. Geol.,* **6**, 295–308.

POLLARD, J. E. & LOVELL, J. P. B. — 1976 — Trace fossils from the Permo–Triassic of Arran. *Scott. J. Geol.,* **12**, 209–225.

POLLARD, J. E. & STEEL, R. J. — 1978 — Intertidal sediments in the Auchenhew Beds (Triassic) of Arran. *Scott. J. Geol.,* **14**, 317–328.

RAST, N., DIGGENS, J. N. & RAST, D. E. — 1968 — Triassic rocks of the Isle of Mull, their sedimentation, facies, structure and relationship to the Great Glen Fault and the Mull caldera. *Proc. geol. Soc. London,* **1645**, 299–305.

RICHEY, J. E.   1961 *Scotland: The Tertiary Volcanic Districts.* British Regional
                       Geology (3rd Edition). H.M.S.O., Edinburgh, 120 pp.

RIDD, M. F.   1981 Petroleum geology west of the Shetlands. *In* Illing, L. V. and
                       Hobson, G. D. (Eds.), q.v., 414–425.

RUSSELL, M. J.   1976 A possible Lower Permian age for the onset of ocean floor
                       spreading in the northern North Atlantic. *Scott. J. Geol.,* **12,**
                       315–323.

SARJEANT, W. A. S. 1974 A history and bibliography of the study of fossil vertebrate
                       footprints in the British Isles. *Palaeogeogr. Palaeoclimatol.*
                       Palaeoecol., **16,** 265–378.

SEDGWICK, A. &   1828 On the geological relations of the secondary strata of the Isle of
   MURCHISON, R. I.   Arran. *Proc. geol. Soc. London,* **1,** 41–42.

SMITH, A. G. &   1977 *Mesozoic and Cenozoic Paleo-continental Maps.* Cambridge
   BRIDEN, J. C.   University Press, Cambridge, 64 pp.

SMITH, A. G.,   1973 Phanerozoic world maps. *In* Hughes N. F. (Ed.). *Organisms*
   BRIDEN, J. C. &   *and continents through time. Spec. Pap. Palaeontol. London,*
   DREWRY, G. E.   **12,** 1–42.

SMITH, D. B.   1976 A review of the Lower Permian in and around the British
                       Isles. *In* Falke, H. (Ed.). *The continental Permian in Central,*
                       West and South Europe. D. Reidel, Dordrecht, 14–22.

SMITH, D. B.,   1974 A correlation of the Permian rocks in the British Isles. *Spec.*
   BRUNSTROM, R. G. W.,   *Rep. geol. Soc. London,* **5,** 45 pp.
   MANNING, P. I.,
   SIMPSON, S. &
   SHOTTON, F. W.

SMYTHE, D. K.,   1972 Deep sedimentary basin below Northern Skye and the Little
   SOWERBUTTS, W. T. C.,   Minch. *Nature (phys. Sci.),* **236,** 87–89.
   BACON, M. &
   MCQUILLIN, R.

SMYTHE, D. K.,   1978 Seismic evidence for Mesozoic sedimentary troughs on the
   KENOLTY, N. &   Hebridean continental margin. *Nature. London,* **276,** 420.
   RUSSELL, M. J.

STEEL, R. J.   1971 New Red Sandstone movement on the Minch Fault. *Nature*
                       (phys. Sci.), **234,** 158–159.

                1974a New Red Sandstone floodplain and piedmont sedimentation
                       in the Hebridean province, Scotland. *J. sediment. Petrol.,* **44,**
                       336–357.

                1974b Cornstone (fossil caliche) – its origin, stratigraphic, and sedi-
                       mentological importance in the New Red Sandstone, Western
                       Scotland. *J. Geol. Chicago,* **82,** 351–369.

                 1977 Triassic rift basins of northwest Scotland – their configuration,
                       infilling and development. *In* Finstad, K. G. and Selley, R. C.
                       (Eds.). *Proceedings: Mesozoic Northern North Sea Symposium*
                       1977. Norwegian Petroleum Society, Stavanger, Paper 7, 18 pp.

STEEL, R. J.,   1975 Triassic sedimentation and palaeogeography in Central Skye.
   NICHOLSON, R. &   *Scott. J. Geol.,* **11,** 1–13.
   KALANDER, L.

STEEL, R. J. &   1975 Sedimentation and tectonism (?Permo-Triassic) on the margin
   WILSON, A. C.   of the North Minch Basin, Lewis. *J. geol. Soc. London,* **131,**
                       183–202.

STORETVEDT, K. M. & 1977 Palaeomagnetic evidence for the age of the Stornoway Forma-
   STEEL, R. J.   tion. *Scott. J. Geol.,* **13,** 263–269.

TAYLOR, J. C. M. 1981 Zechstein facies and petroleum prospects in the central and
                       northern North Sea. *In* Illing, L. V. and Hobson, G. D. (Eds.),
                       q.v., 176–185.

                 1984 Late Permian-Zechstein. *In* Glennie, K. W. (Ed.), q.v., 61–83.

TYRRELL, G. W.   1928   The geology of Arran. *Mem. geol. Surv. U.K.,* 292 pp.

WAGNER, R. H.   1966   On the presence of probable Upper Stephanian Beds in Ayr-
                                                  shire, Scotland. *Scott. J. Geol.,* **2**, 122–123.

WALKER, A. D.   1961   Triassic reptiles from the Elgin area: Stagonolepis, Dasyg-
                                                  nathus and their allies. *Philos. Trans. R. Soc. London (B),*
                                                  244, 103–204.

                                          1964   Triassic reptiles from the Elgin area: Ornithosuchus and the
                                                  origin of Carnosaurs. *Philos. Trans. R. Soc. London (B),* **248**,
                                                  53–134.

                                          1973   The age of the Cuttie's Hillock Sandstone (Permo-Triassic) of
                                                  of the Elgin area. *Scott. J. Geol.,* **9**, 177–183.

WAUGH, B.   1970   Petrology, provenance and silica diagenesis of the Penrith
                                                  Sandstone (Lower Permian) of northwest England. *J. sediment.
                                                  Petrol.,* **40**, 1226–1240.

WARRINGTON, G.   1973   Miospores of Triassic age and organic-walled microplankton
                                                  from the Auchenhew Beds, southeast Arran. *Scott. J. Geol.,*
                                                  9, 109–116.

                                          1974   Les évaporites du Trias britannique. *Bull. Soc. géol. Fr.,* **16**,
                                                  708–723.

WARRINGTON, G.,   1980   A correlation of Triassic rocks in the British Isles. *Spec. Rep.
  AUDLEY-CHARLES, M. G.,   geol. Soc. London,* **13**, 78 pp.
  ELLIOTT, R. E.,
  EVANS, W. B.,
  IVIMEY-COOK, H. C.,
  KENT, PETER, ROBINSON,
  PAMELA, L.,
  SHOTTON, F. W. &
  TAYLOR, F. M.
WARRINGTON, G. &   1985   Late Triassic miospores from Gribun, Western Mull. *Scott. J.
  POLLARD, J. E.   Geol.,* **21**, 218–221.

WILLIAMS, D.   1973   *The sedimentology and petrology of the New Red Sandstone of
                                                  the Elgin Basin, northeast Scotland.* Thesis, Ph.D., Univ. Hull.
                                                  (Unpublished).

WILLIAMS, J. J.,   1975   The Piper Oil-Field, U.K. North Sea: a fault-block structure
  CONNER, D. C. &   with Upper Jurassic beach-bar reservoir sands. *In* Woodland,
  PETERSON, K. E.   A. W. (Ed.), q.v., 363–377.
WILSON, H. E.   1972   *Regional geology of Northern Ireland.* H.M.S.O., Belfast,
                                                  115 pp.

WOODLAND, A. W. (Ed.)   1975   *Petroleum and the continental shelf of northwest Europe,
                                                  Volume 1, Geology.* Applied Sci. Publ., Barking, 501 pp.

ZIEGLER, P. A.   1975   North Sea basin history in the tectonic framework of north-
                                                  western Europe. *In* Woodland, A. W. (Ed.), q.v., 131–148.

                                          1982   *Geological atlas of western and central Europe.* Shell Inter-
                                                  nationale Petroleum Maatschappi B.V., distributed by Elsevier,
                                                  Amsterdam, 130 pp.

ZIEGLER, W. H.   1975   Outline of the geological history of the North Sea. *In* Wood-
                                                  land, A. W. (Ed.), q.v., 165–190.

13

JURASSIC, CRETACEOUS AND TERTIARY SEDIMENTS

A. Hallam

Until recently, the only known Scottish Jurassic, Cretaceous and Tertiary sediments were small outcrops mainly confined to the north-west and north-east of the country. The most important zone of outcrops is in the Inner Hebrides from Skye to Mull and the second most important occurs on the east coast of Sutherland between Golspie and Helmsdale. Minute exposures also occur in the Shiant Isles and the Isle of Arran (where they are preserved as fragments in a Tertiary volcanic vent); also on

the mainland bordering the Hebrides of Applecross and the east coast of Ross-shire, at Balintore, Port an Righ and Eathie (Fig. 13.1).

Since the late 1960's, however, drilling for oil in the central and northern sectors of the North Sea has revealed a thick series of sediments, with some intercalated volcanics, ranging back from the Quaternary to the pre-Jurassic. Analysis of North Sea geology has greatly amplified our knowledge and transformed ideas about

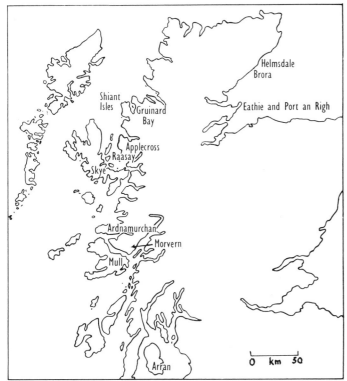

Fig. 13.1. Locality map of Jurassic and Cretaceous outcrops.

contemporary environments and tectonics (Woodland 1975, Illing & Hobson 1981). Offshore geophysical and geological investigations have also been carried out in the Hebridean seas and west of the Shetlands. While it has become evident that substantial Mesozoic and Tertiary deposits occur in these regions little of the relevant work has been published.

Among the earlier geologists who have helped to unravel the stratigraphy of the deposits exposed on land three should be singled out for special mention: MacCulloch the pioneer (1819), Judd, whose great comprehensive work (1873–78) forms the basis of our present detailed knowledge, and Lee, the most notable of several Geological Survey workers whose careful fossil collecting enabled the Scottish Jurassic ammonite succession to be tied in accurately with those of the type European areas. Good reviews by Lee and Pringle (1932) and Arkell (1933) obviate the need to give a comprehensive list of references to older work at the end of this chapter. Besides the references cited in the text below, valuable sources of information on the regional stratigraphy are provided by a number of Geological Survey memoirs: Anderson and Dunham (1966), Bailey *et al.* (1924), Harker (1908), Lee (1920), Lee and Bailey (1925), Peach *et al.* (1910), Read *et al.* (1925), Richey *et al.* (1930) and Tyrell (1928).

The stratigraphy will be dealt with in terms of the two main regions of the Hebrides and north-east Scotland extending into the North Sea. With regard to the Jurassic, there has been dispute about whether to place the Callovian stage in the Middle or Upper Jurassic. While, following Arkell and others, a consensus favours the former assignation, it will be convenient here because of different facies developments to treat the Callovian as the basal stage of the Upper Jurassic

Hebridean Region

Lower Jurassic

The thickest and best exposed Lias occurs in Skye and Raasay, where an almost complete succession of ammonite zones has been recognised. It is divided into a number of marine formations which will be dealt with briefly in turn (Fig. 13.2). The *Broadford Beds* range from the Angulata to the Turneri Zone (Hettangian–Lower Sinemurian) reaching a maximum thickness of about 140 m, and rest on an irregular substratum of Triassic and older rocks. Hallam (1959) has divided the formation into two units, the Lower Broadford Beds, consisting of sandstones, mudstones and micritic, oolitic and coral-bearing

Fig. 13.2. Correlation chart for the Jurassic.

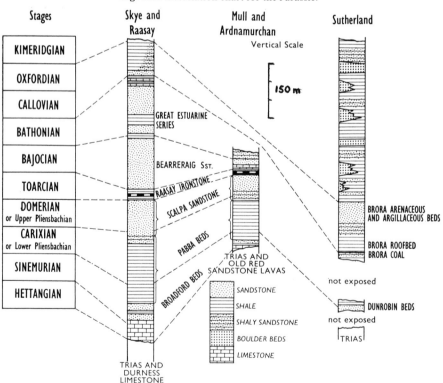

limestones, which reach their maximum development at Applecross and thin south-westwards to zero on the shores of Loch Slapin in Skye, and the more argillaceous and ammonitiferous Upper Broadford Beds, which thicken towards the south. The overlying *Pabba Beds* (Obtusum to Davoei Zones; Sinemurian and Carixian (Lower Pliensbachian)) are a 200 m thick homogenous group of somewhat silty and sandy micaceous mudstones. Both formations are substantially thinner in the Mull–Morvern region further south, and the Broadford Beds equivalent lacks sand, and has a facies of alternating limestones and mudstones comparable to the contemporary Blue Lias of southern England.

Oates (1978) has recently made a number of important discoveries of ammonites not previously found in Scotland and has thereby been able to demonstrate that the Lower Lias has an even more complete zonal sequence than recognised hitherto. He has also been able to add further precision to the stratigraphy and has established the existence in Mull of a notable erosional break within the Upper Broadford Beds equivalent (Fig. 13.3). His use of the term 'Blue Lias' for the Broadford Beds equivalent in Mull is questionable, however. In accordance with currently accepted procedure in stratigraphic nomenclature, a regional name ought to be used instead.

The Pabba (or Pabay) Beds pass up gradually into the Scalpa Sandstone of the Middle Lias, best developed in Raasay, where it is 90 m thick. Howarth (1956) gives full details of the ammonite succession. The remaining three formations belong to the Upper Lias (Toarcian). The *Portree Shales*, varying from 1 to 25 m in thickness, are succeeded by the *Raasay Ironstone*, a thin chamosite oolite formerly worked for ore in that island. The iron-stone is overlain non-sequentially by the *Dun Caan Shales* (8–21 m) of the highest Toarcian. All these formations are represented further south in Mull, Morvern and Ardnamurchan, where they are attenuated compared with Skye and Raasay (Fig. 13.2).

Middle Jurassic

Rocks of the Aalenian and Bajocian stages in Scotland consist predominantly of coarse marine clastics and have been termed the Bearreraig Sandstone Series (Morton 1965, 1976). Morton recognised two main basins in the northern area, in northern Skye (Trotternish) and in Raasay (Fig. 13.4) and Strathaird. The Strathaird coastal exposures exhibit spectacular large scale cross bedding. The maximum thickness of 530 m in Strathaird is the thickest Aalenian–Bajocian sequence known in north-west Europe. Three major sedimentary cycles of shale coarsening up into sandstone are recognisable, whose boundaries correspond with ammonite zones. The most striking of the sudden reversions up the sequence into shale occur at the Lower–Upper Bajocian boundary.

A comparable but much thinner sequence occurs in Mull and Ardnamurchan, but the Aalenian is represented by a rather condensed limestone group and the Upper Bajocian consists of sandstones indistinguishable from those below.

Resting conformably upon the Bajocian is one of the most interesting rock groups in the Scottish Jurassic, the Great Estuarine Group (Harris & Hudson 1980). This lacks normal marine fossils but may be dated by the marine strata above and below as being essentially Bathonian in age. The lithology is highly variable, with

Fig. 13.3. Facies variations in the Lower Lias formations of the Inner Hebrides. After Oates 1978, Fig. 2.

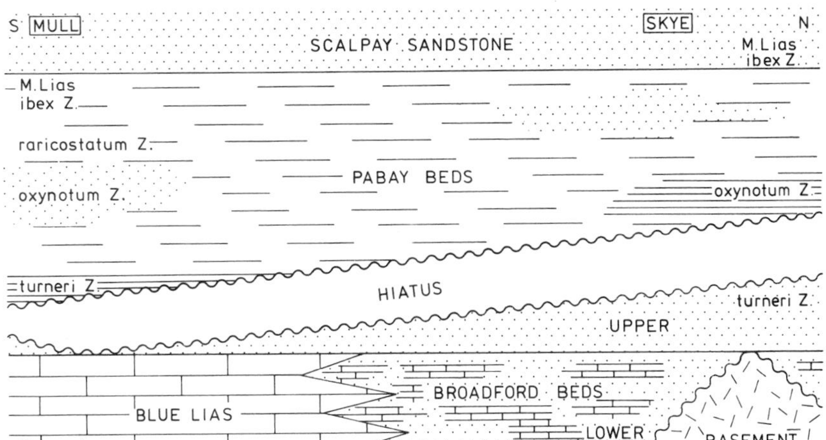

sandstones, shales containing ostracods and estheriids, a variety of limestones including beds crowded with brackish and fresh-water bivalves, oolites, dolomites and algal beds (Anderson 1948, Hudson 1962, 1970). Though lateral variation is considerable, a few thin horizons including the basal oil shale and certain algal limestones have a remarkably wide lateral extent and prove very useful in correlation. The type succession, about 250 m thick, is taken in Trotternish, northern Skye, where the following formations are recognised (Harris and Hudson 1980).

> Skudiburgh
> Kilmaluag
> Duntulm
> Valtos Sandstone
> Lealt Shales
> Elgol Sandstone
> Cullaidh Shale

Some of the formations may be traced far afield, to Raasay, Strathaird (Skye), Eigg, Muck, Ardnamurchan and Mull, but the more southerly successions tend to be much reduced because of later erosion. The supposed Upper Bathonian 'Cornbrash' of Raasay has been shown to be of late Bajocian age (Bradshaw and Fenton 1982).

	Thickness in metres
(7) Mottled Clays	15
(6) Ostracod Limestones	30
(5) Lower Ostrea Beds	21
(4) Concretionary Sandstone Beds	80
(3) Estheria Shales	39
(2) White Sandstone	10 to 32
(1) Basal Oil Shale	3
	Total 198 to 220

The supposed Upper Bathonian 'Cornbrash' of Raasay has recently been shown to be of late Bajocian age (Bradshaw & Fenton 1982).

Fig. 13.4. View on east side of Raasay looking northwards from Rudha' na Leac. Sandstones of the Broadford Beds form the small promontory in the foreground and Bajocian sandstones the main coastal cliff in the background. Dun Caan, on the left, consists of Tertiary dolerite overlying shales of the Great Estuarine Series. BGS photo. Crown Copyright.

Upper Jurassic (including Callovian)

From the point of view of the ammonite stratigrapher, by far the best succession occurs in northern Skye, where nearly all the rocks belong to argillaceous marine series, termed the Staffin Shales by Turner (1966), which ranges in age up to the Mutabilis Zone of the Lower Kimmeridgian (Wright 1973).

Sykes (1975) gives a comprehensive account of the Callovian and Oxfordian stratigraphy (Fig. 13.5). At the base is the Staffin Bay Formation (18 m) of Lower Callovian (Macrocephalus Zone) age, consisting of the so-called Upper Ostrea Shales overlain by the Belemnite Sands. The overlying Staffin Shale Formation (over 117 m thick) ranges up from the Middle Callovian into the Kimmeridgian without a significant break. Bituminous shales of the Jason Zone, containing thin glauconitic silt horizons, pass up into light-coloured clays and subordinate siltstone. In south eastern Skye (Strathaird) the Lower Callovian Carn Mor Sandstone (9 m) is overlain disconformably by 150 m of siltstones and sandstones ranging in age from late Middle Callovian (Coronatum Zone) to Upper Oxfordian. Other Callovian and Oxfordian rocks occur in Scalpay, Strollamus and Eigg. The only Upper Jurassic known in the Mull area is a minute down-faulted patch of Lower Kimmeridgian sandstone near Craignure (Arkell 1933, p. 371).

Cretaceous and Tertiary

In comparison with the Jurassic, younger sediments occur onshore in only negligible quantities. Thin representatives of marine Upper Cretaceous are widespread in the west of Scotland and are without doubt merely the isolated remnants of a much more extensive series that has been largely removed by Tertiary and Quaternary erosion. They must also have been preceded by a prolonged period of erosion because they rest non-sequentially on Triassic and Jurassic beds of a variety of ages. They are best developed in Mull and Morvern, where generally some 13 m of glauconitic sandstone dated as Cenomanian are overlain by the so-called White Sandstone, about 8 m thick and of extraordinary purity at Loch Aline, where it has been worked for glass sand. Above this comes a few cm of clay that has yielded Cenomanian fossils, followed by a thin bed of silicified white chalk, of Senonian age. (The Turonian seems to be missing everywhere). The Cenomanian overlies Middle Lias at Carsaig and Bathonian near Loch Don, in Mull. Small patches of Upper Cretaceous sandstone and sandy limestone, prob-

Fig. 13.5. Correlation of the major Callovian–Oxfordian sections in Scotland. Simplified from Sykes 1975, Fig. 7.

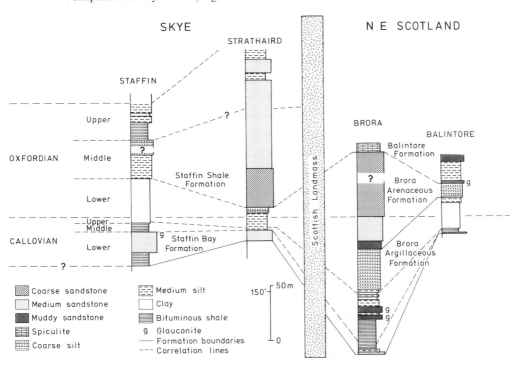

ably Cenomanian in age, are also found in Skye, Raasay and Eigg, where the so-called Laig Gorge Beds have been shown to be Cretaceous not Jurassic as previously assumed (Hudson 1960). Fragments of hardened chalk occur in a volcanic vent in central Arran.

Subsequent to the deposition of the Upper Cretaceous in the Inner Hebrides land was uplifted and eroded so that the Lower Tertiary volcanic pile rests upon an irregular surface of rocks ranging in age back to the Precambrian. Sediments are confined to thin and local terrestrial accumulations below and within the lava sequence. They are best seen in Mull, where up to 7 m of conglomerates, sandstones, lignite and a very constant mudstone underlie the lavas. Locally lignite and sandstone are found interbedded within the volcanics. Various estimates have been made of their age on the basis of included plant remains. Older work on the macroflora had suggested an Eocene age to most workers (Seward and Holttum 1924) but a palynological analysis led Simpson (1961) to propose an appreciably younger age. He argued that the pre-basaltic horizon could not be older than Oligocene and might even be Miocene, and the interbasaltic horizon could be either Miocene or Lower Pliocene. However, potassium-argon dating carried out at Cambridge indicates that the plateau basalts of Mull are Palaeocene in age (Miller and Harland 1963, Evans et al. 1973).

North-east Scotland and the North Sea

Lower Jurassic

A small thickness of Lower Lias is exposed on the shore near Golspie, below Dunrobin Castle on the east coast of Sutherland (Read et al. 1925, Neves and Selley 1975). Some 20 m of non marine carbonaceous siltstone and mudstone passing up into white sandstone containing drifted plant remains are overlain by about 16 m of marine micaceous shales and shaly sandstones containing a marine fauna of molluscs and brachiopods. Ammonites indicate the presence of the upper Raricostatum and lower Jamesoni zones (Berridge and Ivimey-Cook 1967). The rock sequence has been assigned by Neves and Selley (1975) to the Dunrobin Bay Formation. On the opposite side of the Moray Firth, rocks of similar age and facies have been penetrated in a borehole at Lossiemouth (Berridge and Ivimey-Cook 1967). 75 m of sandstones and mudstones, some of them richly kaolinitic, are predominantly non-marine, containing only estheriids and plant fragments, but a shelly horizon with ammonites and a diverse bivalve fauna also occurs; its age is the same as in the Dunrobin Bay Formation.

Lias is missing over the Mid North Sea High, along with the rest of the Jurassic, and in at least part of the eastern Moray Firth Basin, but occurs in quantity in the Viking Graben east of the Shetlands. Thus in the Brent Oilfield (Bowen 1975) the non-marine Statfjord Sand Formation

is about 200 m thick and is assigned tentatively in the absence of precise palaeontological dating to the Rhaetian–Sinemurian interval. It is overlain by the Dunlin Formation, about 180 m thick, consisting of shale and siltstone with subordinate sand. Much of this is apparently marine Toarcian, a stage not represented on land.

Middle Jurassic

Rocks of this age are almost completely absent in the section on the east Sutherland coast, the only deposit of note being the thin Brora Coal (Hurst 1981), which has been worked in a small mine at Fascally, mainly as fuel for the neighbouring brickworks. Both it and the underlying non-marine sandstones and mudstones have been long assigned to the Bathonian because of the immediately overlying Lower Callovian marine deposits. This age has now been confirmed by palynological evidence (Lam and Porter 1977).

The situation is very different offshore, where a thick Middle Jurassic sequence contains some of the principal oil reservoirs discovered in the North Sea. In the Brent field, for instance, some 150 m of strata have been dated as Bajocian–Bathonian and termed the Brent Sand Formation (Bowen 1975). This consists of massive coarse- to fine-grained non marine sandstone overlain by mixed shales, sandstones and thin coals, with an upper unit of bioturbated marine sands. Boreal Bathonian ammonites of the genera *Arcticoceras*, *Crenocephalites* and *Kepplerites* have been encountered in borehole cores, and exhibit a close resemblance to the East Greenland faunas (Callomon 1975).

In the Piper oilfield of the east Moray Firth Basin a lithologically comparable non marine Middle Jurassic sequence directly overlies red Triassic shales. Of especial interest is the occurrence of interbedded basaltic lava, agglomerate and tuff, which compares with the much more substantial Middle Jurassic volcanic deposits discovered in the Forties field of the Central Graben, which will be discussed later (Williams et al. 1975).

Upper Jurassic (including Callovian)

An excellent sequence of Upper Jurassic rocks can be observed on and near the coast of east Sutherland between Brora and Helmsdale. Sykes (1975) gives a thorough description of the Callovian and Oxfordian, and creates three new formation names. The oldest, Brora Argillaceous, formation (30 m) starts with a shelly sandstone, the Brora Roofbed, of Lower Callovian (Calloviense Zone) age. It is overlain by Middle Callovian bituminous shales with thin horizons of glauconitic sand, which pass up into Upper Callovian glauconitic sandstone, brick clay and siltstone. The overlying Brora Arenaceous Formation (more than 56 m) contains both bioturbated and cross bedded sandstones. It lacks ammonites but is tentatively dated as Lower Oxfordian. The Balintore Formation above (more than 12 m) corresponds with the earlier-named Ardassie Limestone and is a spiculite similar to the contemporary Middle Oxfordian Arngrove Stone of Oxfordshire. It consists of calcitised spicules of the siliceous sponge *Rhaxella*. Further south, in the Balintore region of eastern Rosshire, the Callovian and Oxfordian deposits are more condensed and argillaceous (Fig. 13.5).

Above these deposits comes a celebrated sequence of Kimmeridgian bituminous marine shales with inter-calated boulder beds, the boulders consisting of Middle Old Red Sandstone. Derived shallow water fossils including corals occur in these boulder beds and contrast markedly with the ammonites and small bivalves characteristic of the shales.

Lam and Porter (1977) have studied the palynology of the shales and, in conjunction with the earlier work on ammonites, propose the following litho- and biostrati-graphic sequence:

	Ammonite Zone	Thickness in m
Helmsdale Boulder Beds	Eudoxus–?	
	Pallasioides	527
Loth River Shales	Mutabilis	30
Allt na Cuile Sandstones	Cymodoce	122
Kintradwell Boulder Beds	Cymodoce	58

Contrary to what had been claimed earlier, an essentially complete Kimmeridgian sequence appears to be present. It is of great tectonic interest, as will be discussed later.

A small outcrop of Kimmeridgian shales, lacking boulder beds, also occurs at Eathie in Rosshire (Waterston 1951).

Kimmeridgian bituminous shales are now known to be very widely distributed in the North Sea and are generally thought to be the petroleum source rock. In the eastern Moray Firth Basin they are underlain some tens of metres of Upper Oxfordian by marine sandstones with *Amoebo-ceras* resting discordantly on Middle Jurassic, which are the reservoir rock of the Piper field (Williams *et al.* 1975). A sub-Oxfordian unconformity is also recognised in the Brent field (Bowen 1975). The youngest, Volgian, stage is generally absent in the North Sea, so that Lower Cretaceous rests non-sequentially on Kimmeridgian shales (Johnson 1975).

Cretaceous

Glacial erratics of Lower Cretaceous age have long been known from the eastern mainland. Thus a block of sandstone so huge that it has been quarried occurs near Wick, Caithness and contains ammonites that date it as Valanginian or Hauterivian (Lee and Pringle 1932). Richly fossiliferous cherty sandstone boulders from Moreseat and

Fraserburgh, Aberdeenshire, range in age from Ryazanian to perhaps basal Barremian (Cumming and Bate 1933).

These erratics must have come from the North Sea region and hence it is somewhat surprising that pre-Barremian sediments are generally missing from the oil company boreholes (Johnson 1975). The facies of the off-shore Lower Cretaceous is usually non radioactive (and hence non bituminous) shales and marls.

Upper Cretaceous (Albian–Maastrichtian) occurs everywhere in the North Sea and progressively oversteps older Mesozoic deposits and the ancient massif of the Mid North Sea High. Above the Albian the deposits east of Scotland consist predominantly of chalk of normal English facies, reaching a maximum thickness of over 1,400 m in the Central Graben. Northwards, however, the deposits become progressively more argillaceous until in the northern part of the Viking Graben chalk is completely absent (Hancock and Scholle 1975).

Tertiary

It is now recognised that the North Sea was a major Tertiary basin, with over 3,000 m of deposits occurring in the most rapidly subsiding part east of Scotland (Parker 1975). Two features are of especial interest. Several hundred metres of Palaeocene sands provide the major oil reservoir of the Forties and Montrose fields (Fowler 1975, Walmsley 1975). These are capped by a widespread volcanic ash horizon. Palaeocene and early Eocene tuffs having been recognised over an area of more than 400,000 sq km (Jacqué and Thouvenin 1975). South of the Viking Graben the underlying Danian is represented by argillaceous and sandy chalk, while the Eocene and younger deposits consist of a thick, monotonous series of shales and mudstones.

Tectonics and Volcanism

Though it was not to be appreciated until very recently, the famous Kimmeridgian boulder beds of the Helmsdale region provide significant clues to the tectonic history of a much larger region both east and west of the Scottish mainland. In their classic study many years ago, Bailey and Weir (1932) worked out a complex tectonic history involving repeated movements along a large submarine fault scarp (Figs 13.6 and 13.7). These movements were associated with the detachment and transport of blocks of Middle Old Red Sandstone, ranging in size from a few cm across to the so-called 'Fallen Stack' of Portgower (32 × 29 × 6 m), which were periodically carried into deep water by the action of submarine currents or tidal waves and embedded in the midst of black ammonite- and plant-bearing shales (Fig. 13.8). The sandstone matrix of these boulder beds contains abundant shallow-water fossils including thick-shelled bivalves, echinoids and reef corals and was derived from a shelf zone to the north-west, on the upthrown side of the fault. Post-Jurassic movement (in the same sense) along this, the Brora-Helmsdale fault, has brought the Kimmeridgian beds into contact with the Helmsdale granite, which stratigraphically underlies the Middle Old Red Sandstone and cannot have been exposed at the time of the Kimmeridgian movements.

A more recent study of directional structures in the boulder beds and associated sandstones has enabled Crowell (1961) to confirm an essential point in Bailey and Weir's interpretation, that the sediments were derived mainly from the north-west. Crowell argued that the main breccia beds formed when scree at the foot of the fault scarp became unstable as a result of tsunamis triggered off by faulting. The finer sediments and shallow-water fossils were probably carried seaward through steep canyons cut

in the fault scarp (Fig. 13.7). A recent sedimentological analysis of the Helmsdale deposits (Pickering 1984) suggests that they may be part of an ancient fan delta, with various mass flow sediments including debris flow, rockfall, turbidity current and storm surge washover deposits. Redeposition by soft-sediment sliding has also taken place.

Unpublished oil company data indicate that similar late Jurassic breccia beds occur offshore to the east, and extensive seismic and drilling work demonstrate very clearly that the Viking Graben (Fig. 9A) is underlain by a series of (mainly) westerly tilted fault blocks (Bowen 1975). These structures and the associated facies bear an extraordinarily close resemblance to approximately end-Jurassic features in East Greenland (Neves and Selley 1975, Surlyk 1978).

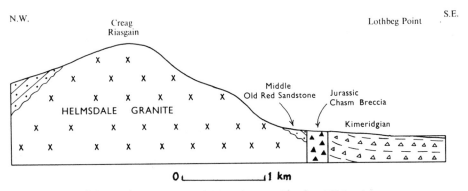

Fig. 13.6. Section across the coastal region midway between Brora and Helmsdale.

Fig. 13.7. Diagrammatic reconstruction of the Kimeridgian environment in the Helmsdale region.

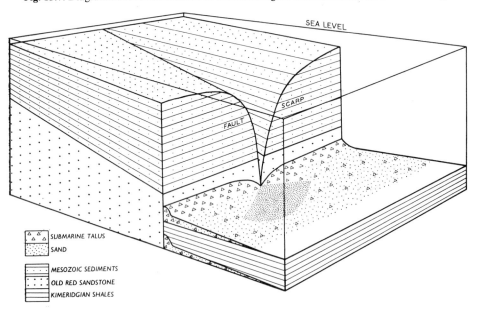

Fig. 13.8. Kimeridgian boulder-bed, Sutherland. BGS photo. Crown Copyright.

The North Sea has evidently had a long and complex taphrogenic history dating back probably into pre-Mesozoic time (Kent 1975) but major tectonic activity involving the creation of tilted fault blocks must have taken place in the axial graben zone in late Middle Jurassic time, as evidenced by the widespread overstep of Upper Jurassic sediments. Another important tectonic phase has been dated by Johnson (1975) as Middle Volgian by comparison with neighbouring regions and because of the widespread absence of strata between the Kimmeridgian and Barremian. After a further Aptian-Albian faulting episode the tectonic style changed to one of axial down-sagging, but a final phase of minor rifting took place in the early Palaeocene, with the reactivation of pre-existing fault systems (Parker 1975).

With regard to the Hebridean region, it has long been recognised that the Camasunary Fault of Skye (Fig. 13.10) is an intra-Mesozoic feature, with an easterly downthrow of more than 650 m dateable as post late Oxfordian-pre Cenomanian (Peach *et al.* 1910, Binns *et al.* 1975). Off-shore work by the I.G.S. has now demonstrated that the Camasunary Fault can be traced south-westwards beyond Coll and Tiree and that Mesozoic sediments thin eastwards from the fault. The Inner Hebrides, Sea of the Hebrides and North Minch Basins can all be interpreted as features which became established in late Jurassic or early Cretaceous times as a result of the creation and westerly tilting of fault blocks (Binns *et al.* 1975) (see Fig. 9B). The close resemblance to both the North Sea and East Greenland in both character and age of these structures and the sedimentary sequence can hardly be coincidental. Taking into account the conventional pre-drift North Atlantic reconstruction, it suggests that a rift zone extended southwards from East Greenland to be split into eastern and western branches by the ancient massif of the Shetlands and Scottish mainland.

One of the most spectacular discoveries of North Sea oil exploration is that part of the Central Graben and eastern Moray Firth Basin was a major volcanic province in the Middle Jurassic, reinforcing the structural/stratigraphic evidence of a significant rift zone (Howitt *et al.* 1975, Gibb and Kanaris-Sotiriou 1976). The maximum development is in the Forties field, where several hundred metres of porphyritic olivine basalts and tuffs have been encoun-

tered in a number of boreholes. Radiometric dating and
spores from the intercalated sediments indicate a Bathonian age. Persistence of volcanic activity into the late
Jurassic is indicated by thin tuff bands in the Brent field
(probably Kimmeridgian). Knox (1977) has recognised
thin bands of altered tuff in the Staffin Shales of Skye,
ranging in age from the Middle Callovian to the Lower
Kimmeridgian. He infers a peak of activity in the late
Callovian and Oxfordian and speculates that eruptions in
the newly opening Rockall Trough might have been
responsible.

A significantly younger episode of volcanism is indicated by the widespread Palaeocene-early Eocene basaltic
tuff horizons in the North Sea. Since the age range of these
tuffs corresponds closely with that of the Hebridean
volcanic province a westerly source is considered the most
probable by Jacqué and Thouvenin (1975).

Since montmorillonite is normally indicative of breakdown of volcanic ash it is worth recording that both
Middle Jurassic and basal Liassic shales in Skye are rich in
this mineral (Knox 1977, Amiri-Garroussi 1977).

Palaeogeography and depositional environments

The abundance of coarse clastic sediments in the Jurassic
and Palaeocene indicate the proximity of one or more
major landmasses, of which the most important was
probably the Scottish mainland. Such a source was inferred by Hudson (1964) in a heavy mineral study of sandstones in the Great Estuarine Group. It is confirmed by
the westerly diminution of sand in the Broadford Beds of
the Skye area (Amiri-Garroussi 1978) and the inferred
westerly provenance of Palaeocene sands in the North Sea
(Parker 1975). More generally, the abundant mica in
Jurassic argillaceous sediments in Scotland points to a
source in eroded Moine Schists, while the occurrence of
abundant fresh feldspar and metamorphic rock fragments
in the Statfjord Sandstone of the Brent Province indicates
rapid erosion and deposition from a nearby uplifted orogenic belt.

Some lateral facies changes suggest that the predominant direction of sediment transport was not simply east–

Fig. 13.9. A. Schematic cross section across the Brent Oilfield. Simplified from Bowen 1975, Fig. 7.

B. Schematic cross section across the North Minch Basin, based on seismic profiles. Simplified from Binns *et al.* 1975, Fig. 7.

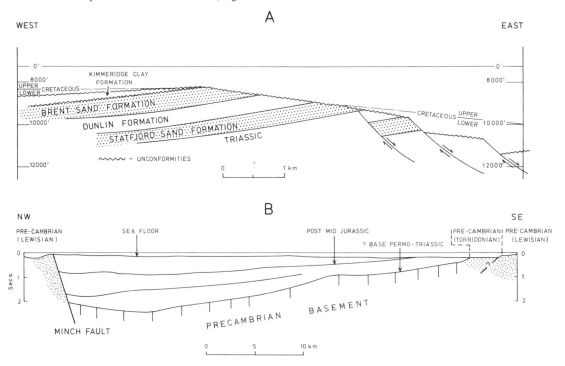

west. Thus the sand content of the Broadford Beds diminishes sharply southwards towards Mull from the Skye area, and that of the Oxfordian of Skye diminishes northwards. In north-east Scotland the lateral passage of late Jurassic sandy into shaly rocks south-eastwards confirms the Bailey and Weir interpretation of a land source to the north-west.

The extent to which the Scottish and neighbouring landmasses were inundated by the sea fluctuated considerably through time (Hudson 1983). The earliest transgression was early in the Lias. The sea reached Mull as early as Planorbis Zone times but there is no evidence in the Skye area of marine deposits older than Late Hettangian (Angulate Zone), while locally in Skye the transgression can be dated precisely as Semicostatum Zone.

After an extensive Bathonian regression (Bajocian in the North Sea) the sea returned almost everywhere in the early Callovian. Another major transgressive episode in the North Sea region took place in the Upper Oxfordian-Kimmeridgian. By far the biggest transgression, however, as elsewhere in the world, occurred in the Upper Cretaceous, when the sea returned to the Hebridean region after a long interval and inundated the Mid North Sea High and other ancient massifs. By Campanian–early Maastrichtian times land areas must have been very limited in extent and subdued in relief, because hardly any clastic sediment reached the sea except north-east of the Shetlands. Greenland is considered the likely source for these terrigenous deposits (Hancock and Scholle 1975).

With regard to climate, Upper Jurassic reef corals signify warm waters, as does the mineralogy of a Middle Jurassic mytilid bivalve from Eigg (Hudson 1968). A warm and fairly humid climate is indicated by plants from the Palaeocene Ardtun leaf bed of Mull (Bailey et al. 1924). Clay mineral analysis of Jurassic rocks in north east Scotland (Hurst 1985) suggests a subtropical-type climate, with abundant kaolinite associated with mixed layer illite/smectite indicating varying, possibly seasonal humidity.

Many years ago Bailey (1924) proposed the existence of a desert shoreline in late Cretaceous times. This was based primarily on the presence of exceptionally well-rounded grains of quartz, especially in the White Sandstone of Loch Aline. This interesting proposal receives, however,

no support from Humphries (1961) detailed study of the Loch Aline Sandstone (see also Hancock 1975).

Apart from the Chalk, whose depositional environments in Great Britain are thoroughly discussed by Hancock (1976), and the rudaceous deposits of Helmsdale, the vast majority of the Scottish Jurassic and younger deposits are arenaceous or argillaceous.

The sandstones range from fine to coarse-grained and fall into two distinct categories, non-marine and marine. Non-marine sandstones occur in the Hettangian and Sinemurian of north-east Scotland, the Bajocian-Bathonian of the North Sea, the Bathonian of the Inner Hebrides and the Palaeocene of the North Sea. Characteristic features include trough cross bedding, channelling and intercalated carbonaceous shales and lignitic or coal beds. They are most reasonably interpreted as the deposits of coastal plains, migrating river channels and deltas (e.g. Bowen 1975, Neves and Selley 1975, Parker 1975).

The more abundant marine sandstones are characteristically bioturbated and hence may contain an admixture of clay. Also, for this reason, inorganic sedimentary structures are not often preserved. Some deposits, however, exhibit cross bedding, most notably the Bearreraig Sandstone of Strathaird, Skye (Morton 1965). The fauna is dominated by bivalves and such trace fossils as *Rhizocorallium* and *Thalassinoides*. There is little doubt that the great majority of these sandstones were laid down in a wave – rather than tide – dominated shoreface or shallow offshore environment (e.g. Williams et al. 1975, Amiri-Garroussi 1978). Parker (1975), has described deep water graded turbidite sands with characteristic sole marks from the Palaeocene of the North Sea, and proposed a progradational model whereby basinal turbiditic sandstones and shales were successively replaced by argillaceous slope and sandy delta plain deposits.

The argillaceous deposits, which are usually somewhat silty, may variously be termed shales, mudstones or clays depending on the degree of fissility or plasticity. Most such deposits are marine, as shown by abundant ammonites and a wide diversity of bivalves together with subordinate brachiopods, echinoderms etc. The most detailed palaeoecological research has nevertheless been

Fig. 13.10. Section across Strathaird, Skye (redrawn from figure 3, Peach *et al.* 1910).

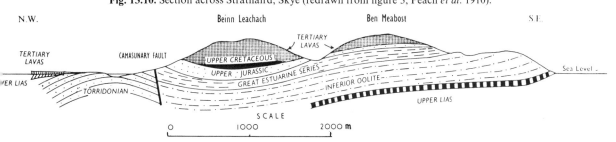

conducted on non marine shales and associated beds in the Great Estuarine Group by Hudson (1963, 1966). He has recognised a series of faunas characterised by low diversity and high density, including estheriids, ostracods and the molluscs *Neomiodon, Liostrea* and *Viviparus*, and inferred considerable fluctuations of salinity, subsequently confirmed by carbon and oxygen isotope analysis (Tan and Hudson 1974). The closest modern analogues to the deposits of the Great Estuarine Series appear to be the coastal lagoons of Texas and the bays of the Florida Everglades. A recent facies analysis of the top two formations of the Great Estuarine Group (Andrews 1985) identifies a range of environments reflecting a late Bathonian regression.

Particular interest attaches to the laminated bituminous shales strongly developed in the Kimmeridgian and Middle Callovian (as also in southern England). In contrast to the normal non-bituminous shales these lack bioturbation and trace fossils, and benthonic fauna is either rare or of small individual size and low diversity. Variable degrees of bottom stagnation can be inferred from this distinctive facies. While such deposits are characteristically basinal and were probably laid down offshore of normal shales and mudstones, no great depth of water need be envisaged. While the bituminous development of the Kimmeridgian is almost certainly bound up to some extent with transgression because of rising sea level, restriction of free water circulation as a result of fault-block collapse in the late Jurassic probably also played a part (Hallam and Sellwood 1976).

REFERENCES

AMIRI-GARROUSSI, K.	1977	Origin of montmorillonite in the early Jurassic shales of NW Scotland. *Geol. Mag.*, **114**, 281–290.
	1978	Sedimentological and palaeoenvironmental studies in the Broadford Beds (Hettangian and Sinemurian) of north west Scotland. *Unpubl. D.Phil. thesis, Univ. Oxford.*
ANDERSON, F. W.	1948	Algal beds in the Great Estuarine Series of Skye. *Proc. roy. phys. Soc. Edinb.*, **23**, 123–142.
ANDERSON, F. W. & DUNHAM, K. C.	1966	The geology of North Skye. *Mem. geol. Surv.*
ANDREWS, J. E.	1985	The sedimentary facies of a late Bathonian regressive episode: the Kilmaluag and Skudiburgh Formations of the Great Estuarine Group, Inner Hebrides, Scotland. *J. geol. Sci.*, **142**, 1119–1138.
ARKELL, W. J.	1933	*The Jurassic System in Great Britain.* Oxford Univ. Press.
BAILEY, E. B.	1924	The desert shores of the Chalk seas. *Geol. Mag.*, **61**, 102–116.
BAILEY, E. B. *et al.*	1924	Tertiary and post-Tertiary geology of Mull, Loch Aline and Oban. *Mem. geol. Surv.*
BAILEY, E. G. & WEIR, J.	1932	Submarine faulting in Kimmeridgian times: East Sutherland. *Trans. roy. Soc. Edinb.*, **47**, 431–467.
BERRIDGE, N. G. & IVIMEY-COOK, H. C.	1967	The geology of a Geological Survey borehole at Lossiemouth, Morayshire. *Bull. geol. Surv. G.B.*, No. 27, 155–169.
BINNS, P. E. *et al.*	1975	Structure and stratigraphy of sedimentary basins in the Sea of the Hebrides and the Minches. *In* Woodland, A. W. (Ed.). *Petroleum and the continental Shelf of north-west Europe*, vol. 1, 93–102. Applied Sci. Publishers.
BOWEN, J. M.	1975	The Brent Oil-Field. *In* Woodland, A. W. (Ed.). *Petroleum and the continental shelf of north-west europe*, vol. 1, 353–360. Applied Sci. Publishers.
BRADSHAW, M. J. & FENTON, J. P. G.	1982	The Bajocian 'Cornbrash' of Raasay, Inner Hebrides: palynology, facies analysis and a revised geological map. *Scott. J. Geol.*, **18**, 131–145.
CALLOMON, J. H.	1975	Jurassic ammonites from the northern North Sea. *Norsk Geol. Tidsskr.*, **55**, 373–386.
CROWELL, J. C.	1961	Depositional structures from Jurassic boulder beds, east Sutherland. *Trans. Edinb. geol. Soc.*, **18**, 202–220.

CUMMING, G. A. & BATE, P. A.	1933	The Lower Cretaceous erratics of the Fraserburgh district, Aberdeenshire. *Geol. Mag.,* **70**, 397–413.
EVANS, A. L. *et al.*	1973	Potassium-argon age determination on some British Tertiary igneous rocks. *Jl. geol. Soc. Lond.,* **129**, 419–443.
FOWLER, C.	1975	The geology of the Montrose Field. *In* Woodland, A. W. (Ed.). *Petroleum and the continental shelf of north-west Europe,* vol. 1, 467–476. Applied Sci. Publishers.
GIBB, F. G. F. & KANARIS-SOTIRIOU, R.	1976	Jurassic igneous rocks of the Forties Field. *Nature,* **260**, 23–25.
HALLAM, A.	1959	Stratigraphy of the Broadford Beds of Skye, Raasay and Applecross. *Proc. Yorks. geol. Soc.,* **32**, 165–184.
HALLAM, A. & SELLWOOD, B. W.	1976	Middle Mesozoic sedimentation in relation to tectonics in the British area. *J. Geol.,* **84**, 302–321.
HANCOCK, J. M.	1975	The sequence of facies in the Upper Cretaceous of northern Europe compared with that in the Western Interior. *Geol. Ass. Canada Spec. Paper* no. 13, 83–118.
	1976	The petrology of the Chalk. *Proc. Geol. Ass. Lond.,* **86**, 499–435.
HANCOCK, J. M. & SCHOLLE, P. A.	1975	Chalk of the North Sea. *In* Woodland, A. W. (Ed.). *Petroleum and the continental shelf of north-west Europe,* Vol. 1, pp. 413–425. Applied Sci. Publishers.
HARKER, A.	1908	Geology of the small isles of Invernesshire. *Mem. geol. Surv.*
HARRIS, J. P. & HUDSON, J. D.	1980	Lithostratigraphy of the Great Estuarine Group (Middle Jurassic), Inner Hebrides. *Scott. J. Geol.,* **16**, 231–250.
HOWARTH, M. K.	1956	The Scalpa Sandstone of the Isle of Raasay, Inner Hebrides. *Proc. Yorks. geol. Soc.,* **30**, 353–370.
HOWITT, F. *et al.*	1975	The occurrence of Jurassic volcanics in the North Sea. *In* Woodland, A. W. (Ed.). *Petroleum and the continental shelf of northwest Europe,* Vol. 1, pp. 379–386. Applied Sci. Publishers.
HUDSON, J. D.	1960	The Laig Gorge Beds, Isle of Eigg. *Geol. Mag.,* **97**, 313–325.
	1962	The stratigraphy of the Great Estuarine Series (Middle Jurassic) of the Inner Hebrides. *Trans. Edinb. geol. Soc.,* **19**, 139–165.
	1963	The recognition of salinity – controlled mollusc assemblages in the Great Estuarine Series of the Inner Hebrides. *Palaeontology,* **6**, 318–326.
	1964	The petrology of the sandstones of the Great Estuarine Series, and the Jurassic palaeogeography of Scotland. *Proc. Geol. Ass. Lond.,* **75**, 499–528.
	1966	Hugh Miller's reptile bed and the Mytilus Shales, Middle Jurassic, Isle of Eigg, Scotland. *Scott. J. Geol.,* **2**, 265–281.
	1968	The microstructure and mineralogy of the shell of a Jurassic mytilid (Bivalvia), *Palaeontology,* **11**, 163–182.
	1970	Algal limestones with pseudomorphs after gypsum from the Middle Jurassic of Scotland. *Lethaia,* **3**, 11–60.
	1983	Mesozoic sedimentation and sedimentary rocks in the Inner Hebrides. *Proc. roy. Soc. Edinb.,* **83B**, 47–63.
HUMPHRIES, D. W.	1961	The Upper Cretaceous White Sandstone of Loch Aline, Argyll, Scotland. *Proc. Yorks. geol. Soc.,* **33**, 47–76.
HURST, A.	1981	Mid Jurassic stratigraphy and facies at Brora, Sutherland. *Scott. J. Geol.,* **17**, 169–177.
	1985	The implications of clay mineralogy to palaeoclimate and provenance during the Jurassic in NE Scotland. *Scott. J. Geol.,* **21**, 143–160.
ILLING, L. V. & HOBSON, G. D. (Eds)	1981	*Petroleum geology of the continental shelf of north-west Europe.* Heyden.

JACQUÉ, M. &   1975   Lower Tertiary tuffs and volcanic activity in the North Sea.
  THOUVENIN, J.   *In* Woodland, A. W. (Ed.). *Petroleum and the continental
                              shelf of north-west Europe*, Vol. 1, pp. 455–465. Applied Sci.
                              Publishers.

JOHNSON, R. J.   1975   The base of the Cretaceous: a discussion. *In* Woodland, A. W.
                              (Ed.). *Petroleum and the continental shelf of north-west Europe*,
                              Vol. 1, pp. 389–399. Applied Sci. Publishers.

JUDD, J. W.   1873– The Secondary rocks of Scotland. *Quart. Jl. geol. Soc. Lond.*,
                       8   **29**, 97–197; **30**, 220–301; **34**, 660–743.

KNOX, R. W. O'B.   1977   Upper Jurassic pyroclastic rocks in Skye, West Scotland.
                              Nature, **265**, 323–324.

LAM, K. A. & PORTER, R. 1977 The distribution of palynomorphs in the Jurassic rocks of the
                              Brora outlier, N.E. Scotland. *Jl. geol. Soc. Lond.*, **134**, 44–55.

LEE, G. W.   1920   The Mesozoic rocks of Applecross, Raasay and north-east
                              Skye. *Mem. geol. Surv.*

LEE, G. W. &   1925   The pre-Tertiary geology of Mull, Loch Aline and Oban.
  BAILEY, E. B.   *Mem. geol. Surv.*

LEE, G. W. & PRINGLE, J. 1932 A synopsis of the Mesozoic rocks of Scotland. *Trans. geol.
                              Soc. Glasg.*, **19**, 158–224.

MACCULLOCH, J.   1819   *A description of the Western Isles of Scotland, including the
                              Isle of Man.* London.

MILLER, J. A. &   1963   Ages of some Tertiary intensive rocks in Arran. *Miner. Mag.*,
  HARLAND, W. B.   **33**, 521–523.

MORTON, N.   1965   The stratigraphy of the Bearreraig Sandstone Series of Skye
                              and Raasay. *Scott. J. Geol.*, **1**, 189–216.

                       1976   Bajocian (Jurassic) stratigraphy in Skye, western Scotland.
                              Scott. J. Geol., **12**, 23–33.

NEVES, R. & SELLEY, R. C. 1975 A review of the Jurassic rocks of north-east Scotland. *In*
                              Finstad, K. G. and Selley, R. C. (Eds.). Proceedings, Jurassic
                              North Sea Symposium. Stavanger, Sept. 1975. JNNSS/5, 1–29.
                              Norsk Petroleumforening.

OATES, M. J.   1978   A revised statigraphy for the western Scottish Lower Lias.
                              Proc. Yorks. geol. Soc., **42**, 143–156.

PARKER, J. R.   1975   Lower Tertiary sand development in the Central North Sea.
                              In Woodland, A. W. (Ed.). *Petroleum and the continental shelf
                              of north-west Europe*, Vol. 1, 447–452. Applied Sci. Publishers.

PEACH, B. N. *et al.* 1910   Geology of Glenelg, Lochalsh and south-east Skye. *Mem. geol.
                              Surv.*

PICKERING, K. T.   1984   The Upper Jurassic 'Boulder Beds' and related deposits: a fault-
                              controlled submarine slope, NE Scotland. *J. geol. Soc.*, **141**, 357–
                              374.

READ, H. H. *et al.*   1925   Geology of the country around Golspie, Sutherlandshire. *Mem.
                              geol. Surv.*

RICHEY, J. E. *et al.* 1930   The geology of Ardnamurchan, north-west Mull and Coll.
                              Mem. geol. Surv.

SEWARD, A. C. &   1924   See *Mull Tertiary Memoir.*
  HOLTTUM, R. E.

SIMPSON, J. B.   1961   The Tertiary pollen-flora of Mull and Ardnamurchan. *Trans.
                              roy. Soc. Edinb.*, **64**, 421–468.

SURLYK, F.   1978   Submarine fan sedimentation along fault scarps on tilted fault
                              blocks (Jurassic-Cretaceous boundary, East Greenland). *Grøn.
                              geol. Undersøg.* Bull. no. 128.

SYKES, R. M.   1975   The stratigraphy of the Callovian and Oxfordian stages
                              (Middle–Upper Jurassic) in northern Scotland. *Scott. J. Geol.*,
                              11, 51–78.

TAN, F. C. & HUDSON, J. D. 1974 Isotopic studies of the palaeoecology and diagenesis of the Great Estuarine Series (Jurassic) of Scotland. *Scott. J. Geol.*, **10**, 91–128.

TURNER, J. A. 1966 The Oxford Clay of Skye, Scalpay and Eigg. *Scott. J. Geol.*, **6**, 371–378.

TYRRELL, G. W. 1928 The geology of Arran. *Mem. geol. Surv.*

WALMSLEY, P. J. 1975 The Forties Field. *In* Woodland, A. W. (Ed.). *Petroleum and the continental shelf of north-west Europe*, Vol. 1, pp. 477–484. Applied Sci. Publishers.

WATERSTON, C. D. 1951 The stratigraphy and palaeontology of the Jurassic rocks of Eathie (Cromarty). *Trans. roy. Soc. Edinb.*, **62**, 33–51.

WILLIAMS, J. J. *et al.* 1975 The Piper Oil-Field, UK North Sea: a fault-block structure with Upper Jurassic beach-bar reservoir sands. *In* Woodland, A. W. (Ed.). *Petroleum and the continental shelf of north-west Europe*, Vol. 1, pp. 363–377. Applied Sci. Publishers.

WOODLAND, A. W. (Ed.) 1975 *Petroleum and the continental shelf of north-west Europe*, vol. I. Applied Sci. Publishers.

WRIGHT, J. K. 1973 The Middle and Upper Oxfordian and Kimmeridgian Staffin Shales at Staffin, Isle of Skye. *Proc. Geol. Ass. Lond.*, **84**, 447–457.

14

TERTIARY IGNEOUS ACTIVITY
C. H. Emeleus

Extensive, largely subaerial volcanism marked the start of the Tertiary era in the NW of the British Isles in contrast to the marine and deltaic sedimentation which was taking place in the North Sea and other adjoining basins. In Scotland, where the igneous activity was spread over about eleven million years (from c. 63 Ma to c. 52 Ma, Table 14.3), the accumulation of thick sequences of basaltic lavas to some extent overlapped with, but was largely followed by, the establishment of major central volcanoes, the remains of which are now found on Skye, St. Kilda, Rhum, Ardnamurchan, Mull, Arran and several submarine sites. The intrusion of extensive dyke swarms occurred throughout much of the activity (Fig. 14.1).

The Tertiary igneous rocks, which form some of the most spectacular scenery in NW Scotland (Fig. 14.12), have attracted geologists and other visitors for the past two centuries. While the igneous origins of many of the rocks were recognised in the 18th century their Tertiary age was not established until the systematic investigations of Geikie (1871), Judd (1874) and others, summarised in Geikie's (1897) comprehensive review. These early studies were followed by a succession of detailed maps and memoirs of the Geological Survey of Great Britain, which, together with the Survey's Summaries of Progress, contain a wealth of factual information of enduring value. The Memoirs cover Skye (Harker 1904, Anderson and Dunham 1966),

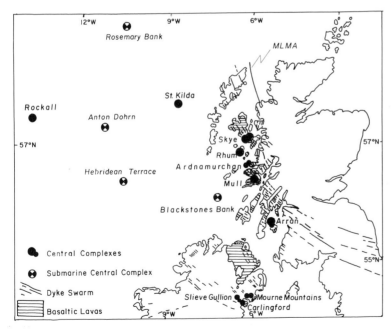

Fig. 14.1. Sketch map showing the positions of the Tertiary central complexes, lava fields and dyke swarms of Scotland and north Ireland. The positions of submarine central complexes and the Minches Linear Magnetic Anomaly (MLMA) are also indicated.

455

Fig. 14.2. Sketch map of the Sea of the Hebrides and adjoining land areas. The Central Complexes are numbered: 1, Skye; 2, Rhum; 3, Ardnamurchan; 4, Mull; 5, Blackstones Bank (after Binns *et al.* 1974, fig. 2, reproduced by permission of H.M.S.O.)

Rhum (Harker 1908), Mull (Bailey *et al.* 1924), Arran (Tyrrell 1928a) and Ardnamurchan (Richey and Thomas 1930) and, more recently (in the British Geological Survey's Report Series): Rockall (Sabine 1960), and St. Kilda (Harding *et al.* 1984). These investigations laid the foundations for numerous theoretical studies on the emplacement of igneous rocks (Richey 1932; Anderson 1936, 1951) and on magma associations and genesis (Bowen 1928, Kennedy and Anderson 1938).

The intensive research interest in these rocks since 1945 has continued with the emphasis on laboratory investigations. Increasingly sophisticated chemical and physical analytical techniques have led to a much fuller understanding of the probable origins of the magmas and the extent to which they may be of crustal or mantle derivation, or mixtures from both sources (e.g. Carter *et al.* 1978, Thompson 1982, Thompson and Morrison 1988). Significant new data are now available from radiometric age determinations (e.g. Dickin 1981) and combined radiometric age determinations and palaeomagnetic studies are providing promising results about the duration of igneous activity and the relative ages of central complexes, lava fields and dyke swarms (e.g. Mussett 1984, Dagley and Mussett 1986, Mussett *et al.* 1988). Geophysical investigations have largely concentrated on offshore investigations (e.g. Binns *et al.* 1974), but new approaches have been tried on some of the problems of the central complexes (e.g. Bott and Tantrigoda 1987). Reassessment of the field evidence from several of the central complexes indicates that pyroclastic acid magmatism was an important feature and was probably intimately connected with caldera formation (Williams 1985, Bell and Emeleus 1988), while combined field, petrographic, geochemical and fluid dynamic investigations have resulted in new theories on the origins of layered mafic rocks present in several of the complexes (e.g. Huppert and Sparks 1980, Bedard *et al.* 1988, Young *et al.* 1988).

The Plateau Lavas

Thick successions of predominantly basaltic lavas cover northern Skye, western Mull and western Morven (Fig. 14.4, Table 14.1). The lavas have submarine extensions: marine geophysical surveys (Binns *et al.* 1974) show that the lavas of Sanday and Canna lie on a basalt ridge extending about 40 km SSW of Skye, while to the east of the Camasunary–Skerryvore Fault the Mull–Morvern lava field may be joined to the lavas of Ardnamurchan, Muck and Eigg (Fig. 14.2). The lavas on Raasay are outliers of the Skye field while those of NW Rhum were once joined to Sanday and Canna. To the south, nearly equidistant from the lavas of Mull and Antrim, remnants of basaltic lava flows of presumed Tertiary age occur along with blocks of Mesozoic sediments within the Arran Central Complex (Fig. 14.25).

Thick successions of basaltic lavas and tuffaceous rocks have been identified in the NE Atlantic off Scotland, using geophysical techniques and by drilling. The seismic reflection characteristics of Tertiary lavas in the Rockall Trough have been interpreted to indicate the presence of subaerial flows, tuffs and hyaloclastite deposits (Wood *et al.* 1988); sediments beneath the lavas appear to have been intruded by sills. To the NE of Rosemary Bank (Fig. 14.1), drilling penetrated over 500 m of basalt lavas overlying cordierite–phyric dacites (Morton *et al.* 1988a). Wells drilled in sedimentary successions around the Shetland Isles encountered volcanic debris and ashes and some of the sequences have been intruded by dolerite sills (Morton *et al.* 1988b, Fitch *et al.* 1988). The amount of volcanism in these areas has been considerable, probably much in excess of that preserved onshore in the British Tertiary Volcanic Province (BTVP).

The majority of the lavas were erupted subaerially. Red, oxidised and lateritised flow tops are common outside the central complex aureoles and thin sedimentary horizons which underlie or are interbedded with the lavas contain plant remains, some in original growth positions (Bailey *et al.* 1924, frontispiece). Lava effusion appears to have been quiet; flows followed each other after short periods of largely *in situ* weathering, to build up essentially flat-lying successions (Fig. 14.4). However, a most convincing example of violent explosive basaltic volcanism remote from a central complex occurs towards the base of the lava succession on Canna where thick agglomerates contain fragments of basalt, dolerite and (Torridonian) sandstone (Geikie 1897, Harker 1908). Elsewhere, ash bands testify to occasional explosive eruptions while the well-known tuff horizons in the Tertiary sediments of the North Sea probably originated in part from the Hebridean volcanoes

Table 14.1. Distribution of basaltic lavas

	Approximate area of outcrop, in square kilometres	Approximate thicknesses, in metres
Skye and Raasay	1090	600
Small Isles	46	300–600
Ardnamurchan	15	90
Mull	720	1800
Morven	116	460
Arran	Blocks and thin flows in caldera	
Submarine:*		
Canna ridge, SW Skye W & NW of Mull. to	700	1000
Eigg	630	?
N.E. Ireland	3900	800

(Based on Richey (1961) with additional information from Binns *et al.* (1974), Smythe & Kenolty (1975) and Wilson (1972)).

* There are also extensive lava fields in the Rockall Trough and other parts of the NE Atlantic off Scotland (see text).

although these were predominantly of Faeroe–Greenland provenance (Knox and Morton 1988).

The pre-lava landscapes were hilly and dissected by river valleys. On Eigg, the thick pitchstone forming the Sgurr (Fig. 14.3) fills part of a system of steep-sided valleys (Bailey 1914, Allwright 1980) and in SW Skye the distinctive tholeiitic basalts of Preshal Mhor and Preshal Beg occupy a valley floored with clastic and pyroclastic debris (Williamson 1979). On Rhum, the lavas ponded in valleys carved from high ground formed by the central complex.

Fig. 14.3. The Sgurr of Eigg: a massive pitchstone flow filling a valley eroded in basalt lavas seen in the foreground and on the lower slopes of the Sgurr. The base of the flow, which is marked by a zone of inclined columns, climbs steeply from left to right against the side of the Tertiary river valley (C. H. Emeleus).

The conglomerates underlying and interbedded with the lavas of Rhum, Canna, Sanday, Eigg and Skye are important since their clasts aid palaeogeographic interpretations and help establish an inter-island stratigraphic sequence. Thick, valley-fill conglomerates underlie three of the Rhum lava groups. They contain coarse clasts of the local Torridonian arkosic sediments, Lewisian-type gneisses, representatives of almost all the igneous members of the Rhum central complex, and vesicular and nonvesicular basalts some of which were derived from lava flows not now found on or near Rhum. Clearly, the Rhum central complex was unroofed and undergoing active erosion during the accumulation of the lavas, and from the interrelationships of the distinctive lava groups it is apparent that new river valleys were formed and old ones re-excavated between successive major volcanic episodes (Emeleus 1985). Pebbles within the Sanday and Canna inter-lava conglomerates include granophyre and felsite almost certainly derived from Rhum (Emeleus 1973) and on Skye conglomerates interbedded with lavas SW of Glenbrittle contain, in addition to Torridonian fragments, pieces of granophyre also closely similar to the distinctive Rhum Western Granite (Williamson 1979, Meighan et al. 1982). Since the Skye lava succession is intruded by the Skye central complex, the clasts from these conglomerates provide crucial evidence suggesting that the Rhum centre pre-dates central intrusive activity on Skye. On Eigg, conglomerates underlying the Sgurr pitchstone contain Torridonian-type sandstone clasts which must have been derived from a source to the west of the Camasunary Fault. The common occurrence of arkosic sandstones and grits in these Tertiary sediments clearly indicates that the Torridonian rocks formed high ground during the Palaeocene, as they do now.

Pillow lavas have long been known from the early caldera on Mull (Bailey et al. 1924), and pillow lavas and hyaloclastite deposits crop out at the base of the Skye lava succession (Anderson and Dunham 1966), on Canna (Allwright 1980) and in NW Rhum (Emeleus 1985). The presence of these deposits demonstrates that the lavas occasionally erupted into standing water, probably in shallow lakes.

The lava successions include a wide range of compositions. Two of the major divisions originally distinguished on Mull, the Plateau Basalts and the Non-Porphyritic Central type, are now recognised as representatives of the world-wide alkali olivine basalt and tholeiitic basalt types; furthermore the commonly associated, more-fractionated lava suites (hawaiite–mugearite–benmoreite–trachytes and basaltic andesite–icelandite–rhyolite, respectively) also occur in the Hebrides. The two suites are not mutually exclusive and lavas of transitional type between alkali olivine basalt and tholeiite are common. In Skye, the earlier flows are predominantly alkali and transitional basalts but more fractionated varieties (hawaiite, benmoreite, mugearite, trachyte) become common in the higher parts of the succession. The youngest flows are formed by distinctive high-lime, low-alkali tholeiites termed the Preshal Mhor Series (Mattey *et al.* 1977). On Mull the early eruptions were of alkali and transitional basalts, including a number of feldsparphyric basaltic flows. These were followed by tholeiites, now largely confined to the central complex or its surroundings. A considerable variety of compositions is found in the relatively restricted lava outcrops of NW Rhum. Early tholeiitic flows, now removed by erosion, contributed to the clasts in the oldest sub-lava conglomerate. These were followed by alkali basalts, basaltic andesites, icelandites, and basaltic hawaiites and hawaiites in a succession of eruptive episodes separated by erosion and conglomerate accumulation. Rhyolitic lavas, ignimbrites and other acid pyroclastic deposits occur within several of the central complexes, but salic products are uncommon away from the igneous centres. Trachytes are present in the Skye and Mull plateau lava successions but the only acidic flow is the Sgurr of Eigg pitchstone. This flow, which is petrographically identical with the Hyskeir pitchstone 20 km to the west, has textural features, such as fiamme, at its base which indicate that it is probably an ash flow rather than a lava.

The originally vesicular, now amygdaloidal lavas contain a wide variety of zeolites along with calcite and silica minerals. Walker (1971) showed that the zeolites of Mull and Morvern occurred in flat-lying zones (Fig. 14.5) which corresponded to the deeper levels of the more complete succession in the younger lavas of eastern Iceland. By comparison with Iceland, it was suggested that the Mull flows represent the lower part of a lava succession originally up to 2,200 m in thickness. On Mull, the zeolite zones are overprinted by thermal effects associated with the central complex (Fig. 14.5). In the inner, high-temperature contact aureoles of gabbroic intrusions on Mull and Skye, where basalts are recrystallised to fine-grained two-pyroxene granulites, the zeolites are changed to aggregates of calcic plagioclase (Harker 1904, Bailey *et al.* 1924, LeBas 1955, Almond 1964).

Individual lava flows range in thickness from less than a metre to over fifty metres and, under favourable conditions such as the cliffs of Mull, northern Skye or Canna, they may sometimes be traced for several kilometres. Actual connections between flows and feeders are rarely preserved and the modes of eruption have caused controversy. Judd (1874) favoured central volcanic sources as witnessed by the close spatial connection between intrusions, caldera formation and lava effusion in the early stages of the Mull central complex (Bailey *et al.* 1924), whereas Geikie (1888, 1897)

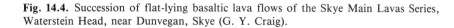

Fig. 14.4. Succession of flat-lying basaltic lava flows of the Skye Main Lavas Series, Waterstein Head, near Dunvegan, Skye (G. Y. Craig).

was of the opinion that the flows were predominantly fissure fed: in support of this view he cited the evidence from recent Icelandic fissure eruptions and the numerous dykes associated with the lava fields of the British Tertiary Volcanic Province (BTVP). Olivine dolerite plugs with high-temperature aureoles occur on Mull, Morvern and Ardnamurchan (Cann 1965, Butler 1961) and, farther afield at Sithean Slugaich on the E of Loch Fyne (Smith 1965, 1969). These intrusions may have been feeders for surface flows (cf. Walker 1959; Preston 1963, 1982), although it is obviously difficult to match plugs with particular flows since plugs which cut lavas at the present day are likely to have fed flows long since eroded. However, Mattey *et al.* (1977) found a compositional correspondence between dykes and flows on Skye although the proportions of the distinctive basalt types differed in each. Anderson and Dunham's (1966) studies on the northern Skye lavas lend weight to the view that these flows were fed from several sources, since they identified a number of geographically restricted groups, separated by erosion intervals. Furthermore, individual flows in the BTVP are laterally impersistent which is also a pointer towards the existence of numerous, short-lived feeders: today, Geikie's

view that the plateau lavas were predominantly fissure-fed is generally accepted.

The central complexes

The Scottish central complexes generally occur on ridges of pre-Mesozoic and pre-Cambrian rocks whereas the thicker lava successions are associated with Mesozoic basins (Fig. 14.2; Walker 1979). The centres often show a close spatial association with major faults (Fig. 14.2; Skye, Camasunary–Skerryvore; Rhum, Camasunary–Skerryvore; Ardnamurchan, Strathconon(?); Mull, Great Glen; Arran, Highland Boundary), furthermore the Tertiary volcanism in general appears to have been located in areas of somewhat thinned crust (cf. Meissner *et al.* 1986, fig. 1b).

The igneous complexes are made up of suites of arcuate intrusions that define one or more centres of igneous activity each up to 15 km in diameter. The classical concept that a centre consists of groups of steeply outward dipping ring-dykes and centrally dipping cone-sheets arose from the investigations of E. B. Bailey, J. E. Richey and others, and was based largely on mapping carried out in Scotland and

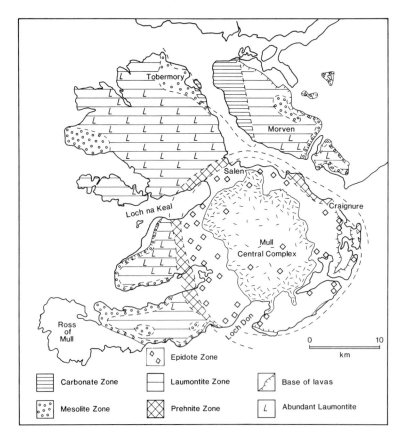

Fig. 14.5. Zeolite zones in the basalt lavas of Mull and Morven. Epidote-dominated alteration characterises much of the area of the Central Complex, a steep-sided zone with prehnite marks and outer limits of the area affected by the Central Complex. Beyond this, the flat-lying, pre-Central Complex zeolite zones show a laumontite zone sandwiched between zones characterised by mesolite. The deepest zone exposed is carbonate-rich (based on Walker 1971, fig. 2).

NE Ireland. Over the years the importance of ring-dykes in these centres has been modified: several of the granitic centres (e.g. Skye Western and Eastern Red Hills) are recognised as nested plutons comprising successions of small, stock-like intrusions which young progressively towards the core of a centre (Fig. 14.14) although true ring-dykes or partial ring-dykes do occur (e.g. Loch Ba and Knock on Mull, Fig. 14.15). Similarly, several of the arcuate gabbroic intrusions in Ardnamurchan which were mapped as ring-dykes are now regarded as parts of layered intrusions cut by a few ring-dykes (Richey and Thomas 1930, cf. Gribble 1976). Large numbers of basaltic cone sheets form important constituents in several of the centres in Skye, Ardnamurchan and Mull, and occur elsewhere. Although individual sheets are rarely more than a few metres thick, their large numbers may provide a significant volume of intrusive material (Figs. 14.15, 14.16, 14.19).

The central complexes are dominated by intrusions, of which many are large and coarse-grained, and resulted from slow cooling and crystallisation of magmas in chambers at depths of from 1 to 4 km. There is also much evidence for near-surface and subaerial activity: on Mull, a caldera structure is filled by basaltic lavas while on Rhum and in the Eastern Red Hills of Skye volcanic breccias are accompanied by air fall and ash flow tuffs both of which may have accumulated subaerially. The surface volcanicity usually appears to have occurred early in the development of a centre, and tends to be fragmentary and often overprinted or largely obliterated by later intrusions.

Emplacement of the central complexes almost invariably caused structural changes in the surrounding country rock. The effects are most obvious in the Mull complex, where a

series of annular folds occurs (Fig. 14.15), and in the distortions of the country rocks around the North Arran Granites (Fig. 14.24a, b), the Rhum centre (Fig. 14.9) and the Skye Red Hills (Fig. 14.14). These disturbances can frequently be related to emplacement of acid magmas (Walker 1975, England 1988). Pronounced thermal metamorphism is present around many of the central complexes, especially where the intrusions are gabbroic although the effects are limited to the immediate contact zones. Much more widespread is the hydrothermal altera- tion both around and within the centres, for example, the well-defined zeolite zones of the lava piles are overprinted for several kilometres distance from central complex con- tacts on Mull (Fig. 14.5) and Skye. Oxygen isotope studies by Taylor and Forester (1971) show that there must have been substantial convective circulation of meteoric waters through both the complexes and their surrounding country rocks (Fig. 14.6). The movement of these heated waters, derived largely from groundwater within the lava piles caused considerable hydrothermal metamorphism of the lavas and the intrusive rocks of the central complexes. Initially, it was thought that the alteration had been so pervasive that it must have significantly modified the trace element and isotopic chemistry of the intrusive rocks, however, examination of granitic rocks in Mull showed that this need not be so since the alteration was essentially channelled along joints and fissures, leaving the more massive parts of intrusions unaffected.

The central complexes are all characterised by positive Bouguer gravity anomalies which may be large (Table 14.2 and Fig. 14.7a, b). The anomalies, which are very localised and have steep gradients, are attributed to the presence of steep-sided bodies of dense rock (olivine-rich gabbro or

Fig. 14.6. Diagram showing circulation of heated groundwaters in and adjacent to a Tertiary gabbro–granite central complex (after Taylor and Forester 1971, fig. 5, reproduced by permission of the Oxford University Press).

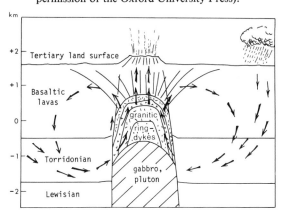

Table 14.2. Bouguer anomalies of the Tertiary central complexes

	Maximum anomalies mgal	Regional background mgal	Approximate amplitude of anomalies mgal
Skye	73	20	53
Rhum	76	10	66
Ardnamurchan	45	20	25
Mull	72	20	52
St. Kilda (Free air anomaly)	125	50	75
Blackstone Bank	135	50	85
Arran	40	25	15

Data are from the following sources:
Skye, Ardnamurchan and Mull from Bott and Tuson (1973); Rhum and Arran from McQuillin and Tuson (1963; see also McLean and Deegan (1978) for further data on seas around Arran); the Blackstones Bank from McQuillin et al. (1975); St. Kilda from Himsworth (1973, the maximum Bouguer anomaly is slightly more than the free-air anomaly, the amplitudes are similar).

feldspathic peridotite) extending to depths of 15 km or
more beneath the central complexes (McQuillin and Tuson
1963, Bott and Tuson 1973). From the gravity data it is thus
evident that mafic and ultramafic rocks dominate these
central complexes and that the granitic rocks which crop
out so extensively are probably very thin; thus on Mull,
Bott and Tantrigoda (1987) estimate that the granitic rocks
form less than 10 per cent by volume of the complex and
that granites in the Loch Ba area (Fig. 14.7b) are probably
only 2 km thick, overlying mafic rocks. The central
complexes are also marked by strong magnetic anomalies;
however, these lack the simplicity of the gravity anomalies
(Bott and Tantrigoda 1987, figs. 1 and 5). Clearly the dense
mafic rocks which dominate the centres may or may not
contain significant modal amounts of magnetic minerals, as
would be expected from surface outcrops where gabbroic
rocks with high modal amounts of Fe–Ti oxides occur
together with dense feldspathic peridotites generally
deficient in strongly magnetic minerals.

St. Kilda

The precipitous islands of the St. Kilda group are situated
on the continental margin about 80 km west of the Outer
Hebrides (Fig. 14.1). The islands, which consist entirely of
intrusive rocks, were first mapped in detail by Cockburn
(1935) and subsequently investigated by Harding (1966,
1967). More recently, the British Geological Survey has
published a well-illustrated account of the geology and a
new map which includes notes on the submarine geology
(Harding *et al.* 1984).

The earliest member of the complex is the layered
Western Gabbro on Hirta and Dunn (Fig. 14.8) focussed
on a centre east of Hirta. The Western Gabbro contributes
blocks to a remarkable breccia of gabbro and dolerite
which forms the islands of Boreray, Soay, and Levenish and
the north of Hirta (Fig. 14.8). The matrices of the breccias
are basaltic: sometimes they show chilled, glassy selvedges
against blocks, in other places they are more crystalline

Fig. 14.7a. Simplified Bouguer gravity anomaly map of Mull and adjacent areas.
Contours are at 10 mGal intervals. Asterisks denote the intrusive centres: 1, Glen More;
2, Beinn Chaisgidle; 3, Loch Ba. The A–A' marks the section shown in Fig. 14.7b.
The line with dots marks the margin of the main subsurface body, granites in the Central
Complex are shown with random dash ornament. The Caledonian Ross of Mull granite (R)
is marked by crosses (based on Bott and Tantrigoda 1987, fig. 1, gravity data based on
measurements by the British Geological Survey, reproduced by permission of
H.M.S.O.).

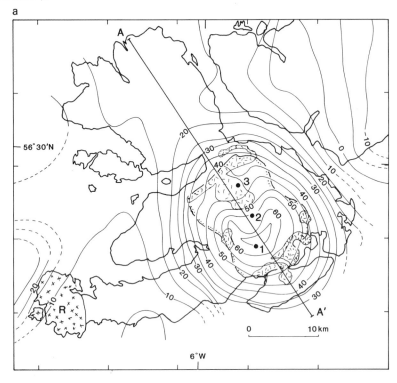

where they came into contact with hot gabbro, and they may also form micro-breccias. Harding *et al.* suggested that formation of the breccias was relatively near-surface (< 5 km of cover) and connected with surface volcanism. The gabbroic and granitic rocks exposed in Glen Bay followed formation of the breccias and are themselves intruded by the extensive Mullach Sgar Complex. This complex consists of an intimate mixture of acid and basic rocks, together with some of intermediate compositions and rocks interpreted as hybrids. The field relations of the acid and basic members led to the suggestion that basic magma had chilled against acid magma (Wager and Bailey 1953), and the complex, lobate contacts between the contrasted types provide excellent examples of 'liquid–liquid' contacts (Blake *et al.* 1965). The youngest major intrusion is the Conachair Granite of Hirta. The granite is compositionally

comparable to the other siliceous and alkali-rich granites of the Province. Harding *et al.* (1984) suggest on geochemical grounds that it originated from fractionation of a mafic magma. The granite has been dated at about 55 Ma (Table 14.3).

All the major intrusions of St. Kilda are cut by several generations of doleritic, acidic and composite dykes and sheets. These are magnificently exposed in the numerous cliff sections and are especially conspicuous on the coast below Conachair and Oiseval (Harding *et al.* 1984, figs. 24C and 25C).

Rhum

The Rhum central complex was emplaced in a thick succession of Torridonian sandstones which were originally

Fig. 14.7b. Interpretation of the gravity profile A–A' on Fig. 14.7a. The densities are shown in kg m^{-3}. Sediments are stippled, mafic igneous rocks have diagonal shading, granite has random dashes. If a denser root (say 3,100 kg m^{-3}) were assumed, the mafic cylinder underlying Mull would extend to only 7 km depth but the thickness of the granites would remain at about 2 km (based on Bott and Tantrigoda 1987, fig. 3a).

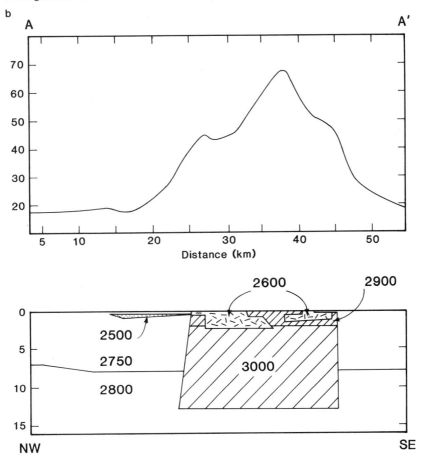

covered by Mesozoic sediments and early Tertiary lavas. The complex developed in two stages: the first dominated by acid magmatism involved both intrusive and extrusive activity accompanied by ring-faulting, the second saw intrusion of mafic magmas which gave rise to the layered ultrabasic rocks for which Rhum is especially noted. Subsequently, but still in the early Tertiary, the central complex was deeply eroded and lavas were ponded against hilly ground formed by the Western Granite.

Harker (1908) provided the first comprehensive description and map of Rhum. Bailey (1945, 1956) reinterpreted much of Harker's data, demonstrating the presence of the Main Ring Fault which showed early central Tertiary uplift of as much as 2 km. The crucial evidence for uplift comes from the restriction of Lewisian gneisses (cf. Tilley 1944) and basal members of the Torridonian succession to within the ring fault. (See for reviews Dunham and Emeleus (1967), Emeleus and Forster (1979), *Geological Magazine*, vol. 122, No. 5, 1985.)

Stage 1

This gave rise to the granophyre and microgranite of western Rhum (Fig. 14.9) (Black 1954), together with the porphyritic felsite, volcanic breccias, bedded tuffs and associated Jurassic sediments, the lowermost members of the Torridonian succession and Lewisian gneisses of SE Rhum (Hughes 1960, Smith 1985) and the northern margins of the central complex (Dunham 1968, Williams 1985). Outside the Main Ring Fault (Fig. 14.9), the regional strike of the Torridonian sediments is deflected suggesting that development of the Stage 1 rocks was accompanied by doming and possibly circumferential folding of the country rocks during central uplift. Smith (1985) confirmed the presence of Jurassic sediments within the Main Ring Fault near Allt nam Ba (Fig. 14.9) and showed that the limestones, shales and sandstones are of Lower Lias age overlain by amygdaloidal basaltic lavas similar to the Tertiary lavas on Eigg. Thus, it is clear that the core of the Main Ring Fault also subsided and it seems likely that a complex history of repeated subsidence and uplift along the line of this fault has been preserved in south-east Rhum (Emeleus et al. 1985).

Rhum contains excellent examples of the association of porphyritic felsites and volcanic breccias often found in the central complexes of the BTVP (cf. Richey 1940). On Rhum, and elsewhere, the felsites have been considered

Fig. 14.8. Geological sketch map of the St Kilda group (based on the British Geological Survey special sheet of St Kilda in Harding *et al.* 1984, by permission of H.M.S.O.).

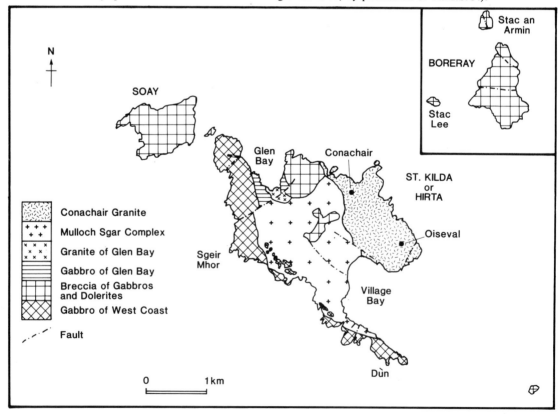

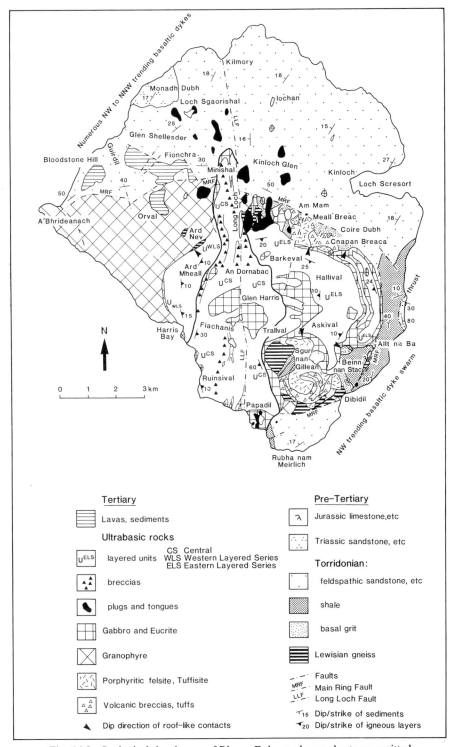

Fig. 14.9. Geological sketch map of Rhum. Dykes and cone-sheets are omitted.

intrusive whereas breccias resulted from shattering of country rocks during the explosive release of gases from crystallizing acid magmas (Hughes 1960, Dunham 1968). Recent re-examination of these rocks has shown that the breccias are associated with well-bedded, tuffaceous horizons and the overlying felsite contains relics of shards and fiamme, exhibiting the eutaxitic structures of welded tuffs (Williams 1985). Some, at least, of these fragmental rocks formed subaerially, possibly within a caldera associated with the Main Ring Fault (Emeleus et al. 1985, Bell and Emeleus 1988).

The granitic rocks of western Rhum are bounded either by the Main Ring Fault or later intrusions (Fig. 14.9), except on Ard Nev where areas of Lewisian gneiss form relics of the original roof. The contact between granophyre and Torridonian sediments at Minishal has been shown to be thermally metamorphosed fault breccia on the Main Ring Fault (Hughes et al. 1957) and not a metasomatic transition zone (Black 1954).

Stage 2

The Rhum ultrabasic rocks are the most extensive and best exposed in the BTVP and the layered series of eastern Rhum is one of the most notable geological features of the island (Figs. 14.10, 14.11). Sixteen alternating layers, each tens of metres in thickness, of easily weathered feldspathic peridotite and more resistant troctolite, troctolitic gabbro, feldspathic gabbro and anorthosite (= Harker's (1908) 'allivalite') form Hallival, Askival and other peaks of the Rhum Cuillin. The peridotites consist of olivine and chromite crystals poikilitically enclosed by calcic plagioclase and diopsidic augite, the more feldspathic rocks have a similar mineralogy but feldspar frequently shows good igneous lamination and the textures may not immediately suggest a crystallisation sequence. Thin (1–2 mm) chromite-rich layers often separate major thicknesses of allivalitic rocks from overlying peridotite. These layered rocks are regarded as classic examples of igneous rocks formed by crystal settling under gravity (Brown 1956, Wadsworth 1961, Wager and Brown 1968) and their textures interpreted accordingly by Wager et al. (1960) who used them as type examples of igneous cumulates. However, the present textures are probably also the result of considerable post-consolidation recrystallisation, analogous to diagenetic processes in sediments, since they frequently appear to have attained high degrees of textural equilibration (Hunter 1987). Layering on the scale of

Fig. 14.10. Layering in the ultrabasic rocks of Hallival, Rhum. The resistant layers are feldspar-rich allivalite, etc., feldspathic peridotites occupy the intervening slack ground (C. H. Emeleus).

centimetres to metres is also found in the feldspathic peridotites and gabbroic rocks of SW Rhum (Wadsworth 1961) where the layered structures are often defined by textural features rather than modal variation in mineral proportions. Elongate, tabular crystals of olivine, frequently 0·5 m in length, appear to have grown upwards as coral-like structures which nucleated on planar surfaces of normal granular-textured peridotite. These distinctive rocks are the harrisites first described by Harker (1908): they formed from olivine-saturated magmas by upwards-growth of olivine from the contemporaneous floor of the intrusion, under tranquil conditions (Wadsworth 1961), or from melt ponded beneath impermeable parts of the layered rocks (Donaldson 1982).

The external margins of the ultrabasic complex are well defined. The mafic rocks were emplaced into a variety of relatively silicic country rocks (Fig. 14.9) all of which have been thermally metamorphosed and frequently melted, giving rise to small amounts of rheomorphic acid magmas. The results of partial melting and the generation of rheomorphic magmas are best seen in the marginal zones of intrusion breccia where blocks of dolerite, gabbro and sometimes peridotite are embedded in light-coloured acid

and hybrid matrices. Hybridisation in and near the Rhum intrusion breccia zones was initially recognised by Dunham (1964) on Cnapan Breaca and Meall Breac. More recently Greenwood (1988) has found that marginal modifications to the Rhum layered complex, mainly in the area of Brown's (1956) fine-grained marginal gabbro, resulted from interaction of the rheomorphic acid magmas with the ultrabasic rocks.

The Rhum ultrabasic complex is shaped like a steep-sided cylinder with a domed roof. The steep contacts were at least in part determined by structural weakness along the Main Ring Fault. The outward-dipping, roof-like contacts occur at several places, most obviously on Beinn nan Stac (Fig. 14.9; Young et al. 1988, fig. 1; Emeleus 1987, figs. 1 and 2). The overall geometry of the complex and in particular the relationship of the layering to the contacts, argues against the suggestion that the layered rocks were emplaced upwards as an essentially solid block (Brown 1956); the contacts are now considered to be intrusive rather than tectonic and thus the complex is preserved *in situ* (e.g. Young et al. 1988). Opinion has also moved away from a tholeiitic basaltic parent magma for the ultrabasic rocks (e.g. Wager and Brown 1968) in favour of magmas of

Fig. 14.11. Gravity-stratified layering in the ultrabasic rocks of the Central Series, Long Loch, Rhum. A late basaltic dyke cuts the exposure (C. H. Emeleus).

mildly alkaline picritic or ultrabasic composition. Direct support for this view comes from a number of features including the occurrence of highly magnesian aphyric minor intrusions in the eastern layered rocks (Forster 1980, McClurg 1982; see also Gibb 1976 and Hutchison and Bevan 1977) and the occurrence of quenched ultrabasic rocks at certain contacts. The layered rocks of eastern and south-western Rhum are cut by a later, strongly transgressive Central Series of ultrabasic rocks (Fig. 14.9; Volker 1983) which are occasionally layered but are characterised by extensive areas of feldspathic peridotite breccias (Wadsworth 1961, Donaldson 1975).

In attempting to explain the origin of the layering in eastern Rhum, much has been made of the relationships between peridotite layers and rocks of the allivalite suite. Boundary structures suggest replacement of allivalite by peridotite (see Butcher *et al.* 1985, figs. 5, 8 and 11). Some of the peridotites appear to be intrusive sheets into allivalites (Renner and Palacz 1987, Bedard *et al.* 1988), representing a return to a model similar to that originally propounded by Harker (1908); other models involve, for example, formation of the layering by contemporaneous crystallisation

Fig. 14.12. The rugged main Cuillin Ridge and Loch Coruisk. Exposures of gabbro and ultrabasic rocks are cut by numerous dykes and cone-sheets which have weathered to form gullies. Cuillin centre, Skye (C. H. Emeleus).

Fig. 14.13. Beinn na Caillich, formed by smooth-weathering Inner Granite of the Eastern Redhills centre, Skye (C. H. Emeleus).

from stratified picrite and basalt magmas (Young *et al.* 1988).

Skye

The rough-weathering gabbros and peridotites of the Cuillin ridge form some of the finest mountain scenery in the British Isles (Fig. 14.12), contrasting strongly with the smooth-weathering granites of the adjoining Red Hills (Fig. 14.13). The complex contains several centres, the earliest is formed by the Cuillin mafic rocks, which are cut by the granites of the small Srath na Creitheach Centre and by the later, Western Red Hills Centre. The Eastern Red Hills Centre is the youngest, flanked by gabbros of Creag Strollamus and hybrids, breccias and tuffs at Kilchrist. The Tertiary igneous geology of Skye is summarised and discussed by J. D. Bell (1976) and Gass and Thorpe (1976), and an up-to-date review is given by B. R. Bell and Harris (1986). Fig. 14.14 summarises the geology of the complex.

Cuillin Centre

The centre consists of a succession of gabbroic and ultrabasic intrusions which young progressively in an ENE direction towards a focus beneath Meall Dearg (Fig. 14.14). Many of the intrusions show well-defined mineral layering and lamination (e.g. Geikie and Teall 1894, plates 24 and 25) and are cut by numerous cone-sheets which also focus beneath Meall Dearg.

The contacts between the Cuillin centre and the country rocks dip steeply inwards, as is seen either side of Loch Scavaig on the shoulders of Gars Bheinn and Sgurr na Stri. On the north-west side of Sgurr nan Gillean and in Upper Glen Brittle, interleaved sheets of gabbro and metamorphosed basalt lavas dip to the south-east, lending support to Walker's suggestion (1975) that the gabbros were emplaced as a confluent sheet system. Intense thermal metamorphism occurs around the centre. Where lavas adjoin gabbro, high-temperature pyroxene hornfels assemblages formed (Almond 1964) and amygdales were recrystallised to plagioclase aggregates (Harker 1904) but where the gabbros cut less refractory rocks, such as the Torridonian sandstones at Camasunary, partial melting of the acidic rocks occurred and spectacular intrusion breccias were formed.

The Cuillin mafic rocks may be divided into three major groups (cf. Bell and Harris 1986, pp. 45–46), (1) The earliest intrusions form a group of generally structureless gabbros extending in an arc from Loch Scavaig to near Sgurr nan Gillean (Fig. 14.12; Fig. 14.14, OG). Thermal metamorphism by later mafic intrusions has changed the Outer Gabbros to dark, matt-grey rocks on the inner side of the group. (2) The next group of intrusions comprises a Border Group, in which a feldspathic gabbroic rock termed the 'White Allivalite' (Hutchison 1968), is a major component accompanied by layered peridotites, allivalites and eucrites, termed the Outer Layered Series (OLS, Fig. 14.14; Wads-

worth 1982). Weedon (1961, 1965) and Hutchison (1968) have demonstrated that members of the group have the structures, textural features and cryptic variation characteristic of layered intrusions. (3) The OLS and the Outer Gabbros are cut by the strongly transgressive Druim nan Ramh Ring Eucrite (DR, Fig. 14.14) within which lies a further Inner Layered Series of allivalites, eucrites and gabbros (ILS, Fig. 14.14; Wadsworth 1982).

Irregular areas of basalt and breccias with basaltic matrices occur within the Outer Gabbros and the later Border Group and OLS rocks. Harker (1904) regarded these as inclusions of basalt lavas and volcanic agglomerate but Hutchison (1966) has reinterpreted them as tholeiitic intrusions which possibly formed late in the Cuillin centre (cf. B. R. Bell and Harris 1986).

The structures, textures and cryptic variation of the ultrabasic and basic rocks of the Cuillin Centre strongly suggest that we are dealing with the remains of a mafic, layered intrusion. Zinovieff (1958) considered that the OLS and ILS were originally part of the same body, separated by movement associated with emplacement of the Druim nan Ramh Ring Eucrite whereas Wager and Brown (1968) suggested that the OLS and ILS were parts of separate layered intrusions. The exact relationships remain to be resolved.

The thin, impersistent Coire Uaigneich Granophyre on the east and south-east side of the Cuillin Centre relates to this centre rather than the later granitic centres, although it is younger than cone-sheets cutting the gabbros. The granophyre is well exposed at the west end of Camasunary (Fig. 14.14) where relic xenoliths of partly digested Torridonian Sandstone are visible. From considerations of the mineralogy and bulk composition (Wager *et al.* 1953, Brown 1963) it was suggested that the granophyre resulted from partial melting of Torridonian sandstone under low pressures. Subsequently, however Dickin and Exley (1981) concluded that contributions came both from this source and from fractionation of mafic magmas.

The granite centres

The smallest and earliest of the granite centres is Srath na Creitheach (Fig. 14.14). It consists of three intrusions which extend under the Cuillin gabbros on the north-west side of Blaven. On Meall Dearg a chilled contact of spherulitic granophyre truncates structures in the ILS gabbros and fragmental rocks of the Srath na Creitheach vent. To the north of Srath na Creitheach the Western Red Hills Centre cuts the Srath na Creitheach granites although exposure is poor and in part the two centres may be separated by screens of gabbro.

The Western Red Hills Centre (WRHC) consists of several arcuate granite intrusions (Fig. 14.14), an early gabbroic member on the summit of Marsco and a late, thin arcuate intrusion of acid and mafic rocks (the Marscoite Suite, Fig. 14.14). Steep dips in the Torridonian, Triassic and Jurassic sediments and to a lesser extent Tertiary

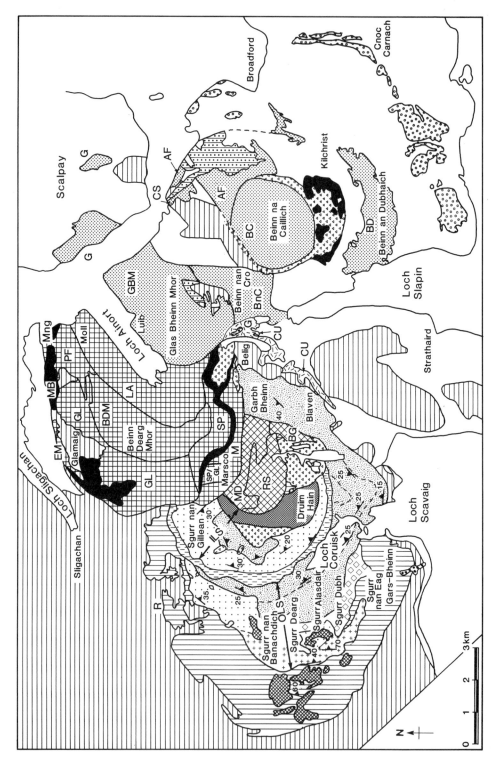

Granite intrusions referred to by initials

Eastern Redhills	Western Redhills	Strath na Creitheach
CS Creag Strollamus	M Marsco	MD Meall Dearg
BC Beinn na Caillich	MB Meall Buidhe (in M)	RS Ruadh Stac
BD Beinn an Dubhaich	SP Southern porphyritic	BG Blaven
AF Allt Fearna	PF Porphyritic felsite	
BnC Beinn na Cro	LA Loch Ainort	
GBM Glas Bheinn Mhor	BDM Beinn Dearg Mhor	
	EM Eas Mhor	
	MnG Maol na Gainmich	
	GL Glamaig	

CU Coire Uaigneich
R Silicified trachyte flows, etc. of Fionn Choire
G Other granite intrusions
ILS, OLS Inner and Outer Layered Series (see text)
⌐ Dip of layering

Composite and acid sills
Hybrids (Marscoite Suite and Kilchrist Vent)
Granites of the Eastern Redhills Centre
Granites of the Western Redhills Centre
Granites of the Strath na Creitheach Centre
Vent agglomerates, explosion breccias (various ages)
Coire Uaigneich and related granites
Intrusive tholeiites
Drum nan Ramh Eucrite
Gabbro Series
Eucrite Series (dotted lines indicate zone boundaries)
Sgurr Dubh peridotite
Allivalite Series
Border Group, including White Allivalite
Gars-bheinn sill
Ring Eucrite and Outer Gabbros, dots indicate other gabbros
Tertiary lavas
Pre-Tertiary rocks

Cuillin Centre

Fig. 14.14. Geological sketch map of the Cuillin and Redhills centres, Skye (after Bell 1976, figs. 2 and 3).

basalts on the north side of the centre suggest the presence of annular folds similar to those of Mull. On the summit of Glamaig and to the east, cappings of Tertiary basalt lavas are relics of country-rock roof. The earliest major intrusion is the Glamaig granite (GL, Fig. 14.14), characterised by very numerous small clots of mafic minerals. It becomes progressively more mafic towards the Marsco summit gabbro with which it may have been thoroughly mixed (Thompson 1969). Much larger (up to metre-sized) rounded, lobate fine-grained acid inclusions also occur. These are structurally and texturally similar to inclusions in the Beinn an Dubhaich granite and may be cognate, derived from early chilled margins of the granite, or possibly from partially melted and mobilised pre-existing acid rocks. The Marscoite Suite ring intrusions formed late in the centre. It is intruded by the Marsco Granite (M, Fig. 14.14) but the presence of 'liquid–liquid' type contacts against the Southern Porphyritic Felsite on Marsco indicates that it probably only just post-dated that intrusion. The Marscoite Suite includes a range of compositions from ferrodiorite to porphyritic felsite with intermediate types (glamaigite and marscoite) representing various degrees of mixing and homogenisation of the two extremes. Taken overall, this centre presents unequivocal evidence for the co-existence, and mixing, of acid and basic magmas (Wager et al. 1965, Thompson 1980, Vogel et al. 1984).

The Glas Bheinn Mhor granite of the Eastern Red Hills Centre (ERHC) truncates WRHC granites and the Marscoite Suite members of the WRHC south of Loch Ainort (Fig. 14.14). The granite is roofed by basalt cut by gabbroic sheets on Beinn na Cro and it is cut by the later Beinn na Cro Granite to the east. B. R. Bell and Harris (1986) group the granites of Beinn na Cro, Allt Fearna and Beinn an Dubhaich into an Outer Granite, peripheral to the Beinn na Caillich Granite (BC, Fig. 14.14; Fig. 14.13) which they term the Inner Granite. The Beinn an Dubhaich Granite is the most extensively studied member of the Outer Granite. It intrudes and metamorphoses dolomitic siliceous Cambro–Ordovician limestones forming a wide range of calc–silicate minerals and small skarn deposits (Tilley 1951). The intrusion figured prominently in controversies surrounding the origin of granite: by analysing the quartz and feldspar, Tuttle and Keith (1954) demonstrated that it had characteristics which linked rhyolites of undoubted magmatic origins, with the then more problematical, deep-seated granites. Opinions have differed over its form. King (1960) and Whitten (1961) considered that the complicated relationships between the granite and limestone at its east end resulted from erosion through to the floor of a granite sheet whereas Harker (1904), Stewart (1965) and Raybould (1973) interpreted the limestone areas as roof pendants and Raybould demonstrated that the granite was a multiple intrusion. Investigations of the Inner Granite by B. R. Bell (1985) show that this has a thin, well-chilled marginal facies carrying fayalite and hedenbergite and a coarser amphibole and biotite bearing central facies (B. R. Bell and Harris

1986). It is a steep-sided stock which intrudes Mesozoic sediments, earlier granites and subaerial acid and basic lava flows, and dominantly silicic pyroclastic rocks (B. R. Bell 1985). The majority of pre- and early Tertiary rocks surrounding the Inner and Outer granites of the ERHC are fairly flat lying but considerable tectonic upheaval is suggested by, for example, the juxtaposition of Lewisian gneisses, Mesozoic sediments and Tertiary lavas to the NW of Beinn na Caillich (Fig. 14.14).

Good evidence for magma mixing in the ERHC comes from the hybrid rocks which intrude and envelop the pyroclastic rocks of the Kilchrist vent (see below). Further east, the close association of acid and basic magmas is also found in the suite of composite sills and sheets which extends in an arc around the east of the granites from Scalpay and Broadford to Loch Slapin (Fig. 14.14). This evidence, together with that provided by the Glamaig granite and the Marscoite suite of the WRHC strongly suggests that basic magmas had an important role even in these granite-dominated areas.

Volcanic vents

Several distinct phases of volcanic vent formation have been recognised in the Skye complex (J. D. Bell 1976). The earliest vents cut the gabbros between Sgurr nan Gillean and Druim Hain. The later, large Srath na Creitheach vent was formed in the interval between emplacement of the ILS gabbros of the Cuillin Centre and the Srath na Creitheach granites. The bedded tuffs of subaerial origin contained within this vent (Jassim and Gass 1970) demonstrate that high-level, surface activity occurred between two plutonic events. Gabbro cut by cone-sheets forms large inclusions within these deposits, which possibly represent megablocks derived by collapse of a caldera wall. Vents north of Belig contain a variety of rocks including clasts of early granites and of gneisses, the latter presumably derived from Lewisian gneisses beneath the complex (J. D. Bell 1966). The Inner Granite of the ERHC is surrounded for much of its margin by pyroclastic rocks and lava flows. Ray (1960) demonstrated the presence of ignimbrites and these have been confirmed by B. R. Bell (1985) who has shown that the pyroclastic rocks around Beinn na Caillich and at Kilchrist were in large part accumulated subaerially and sub-aqueously. The country rocks around the Inner Granite include bedded tuffs, rhyolite lava flows, ignimbrites and a considerable thickness of coarse agglomerates with basalt and granite, Cambrian limestone, quartzite, and Jurassic sedimentary fragments. These surface deposits owe their preservation to downfaulting within a ring-fault which was subsequently intruded by a hybrid magma. It is of note that, despite the presence of a large area of Lewisian gneiss north-west of Beinn na Caillich, gneiss clasts are not found in these pyroclastic deposits indicating that explosive activity occurred above the level of the gneiss–sediment or lava unconformity.

Mull

The Mull central complex consists of three major foci of activity, The Glen More Centre, the Ben Chaisgidle Centre and the Loch Ba Centre or Late Caldera (centres 1, 2 and 3 respectively, Fig. 14.15). Later centres intrude and over-print earlier centres and this, together with the exceptionally large number of major and minor intrusions makes Mull the most complicated area of igneous rocks in the BTVP, if not the British Isles. The complex was comprehensively described in the Mull Memoir (Bailey et al. 1924), in which the authors' formulated concepts that were to strongly influence igneous petrogenetic theory for several decades. More recently, many studies have drawn on the Mull Tertiary rocks. These have led to revision of hypotheses regarding the origins of basaltic and granitic magmas and the extent to which they have been modified through reaction with crustal sources and with each other. By contrast, comparatively few new data have appeared on the field relations and structure of the Mull intrusions.

The Glen More Centre (Centre 1)

The Mull complex is later than most of the thick succession of Tertiary basalt lavas which covers much of the island (Figs. 14.5, 14.15) but there is evidence that the latest tholeiitic lavas were ponded within a caldera which marked the start of Centre 1. Two early intrusions co-focal with the caldera, the Glass Bheinn and Derrynaculean granophyres, form partial ring dykes and it is thought that the marked circumferential folding around Centre 1 was contemporaneous with, or slightly earlier than the acid intrusions (Fig. 14.15; Rast et al. 1968). A series of vents margining the Early Caldera shattered the granophyres and earlier rocks. In these, basalt clasts predominate within the caldera but outside it fragments of Moinian metasediments are common, suggesting that the explosive activity occurred at a high structural level, well above the basalt–Moine unconformity within the caldera limits. The major mafic intrusions of Ben Buie and Ben Bheag were emplaced towards the end of Centre 1 activity, possibly as parts of a much larger layered intrusion. They consist of gabbro, eucrite, allivalite and feldspathic peridotite which often develop mineral layering and have igneous cumulate textures (Lobjoit 1959, Wager and Brown 1968, Skelhorn 1969).

Numerous cone-sheets injected immediately before and after the plutonic mafic rocks are mainly basaltic in composition although intermediate and acid types occur. They are termed the Early Basic Cone Sheets and the Early Acid and Intermediate Cone-Sheets (Bailey et al. 1924). Those cutting the mafic plutonic rocks are considered to mark the start of Centre 2 activity (Skelhorn 1969).

The Ben Chaisgidle Centre (Centre 2)

After the emplacement of the early cone sheets, the Corra Bheinn Gabbro and the Gaodhial and An Cruachan augite diorites, activity in the centre was dominated by the intrusion of acid and basic ring-dykes, as exposed in stream sections in Glen More, most notably the Allt Molach (Bailey *et al*, 1924, fig. 52; Skelhorn 1969, fig. 3). Re-investigation of the Allt Molach section by Skelhorn has resulted in a revision of the relationships established by Bailey. The key to the intrusive sequence lies in the recognition that many of the thin basaltic bodies previously interpreted as screens between acid ring dykes are in fact later basic intrusions which have intruded the acid rocks and caused rheomorphic melting along the contacts. Back-veining of the basic rocks by rheomorphic acid magma gave rise to confusing field relationships which were initially misinterpreted here, as in many other localities in the BTVP.

Intrusion of a dense suite of basalt cone-sheets (the Later Basic Cone Sheets) overlapped intrusion of the acid and basic ring-dykes and continued intermittently into Centre 3 (Skelhorn 1969). However, a large number of these sheets pre-date intrusion of the next major intrusion, the Glen

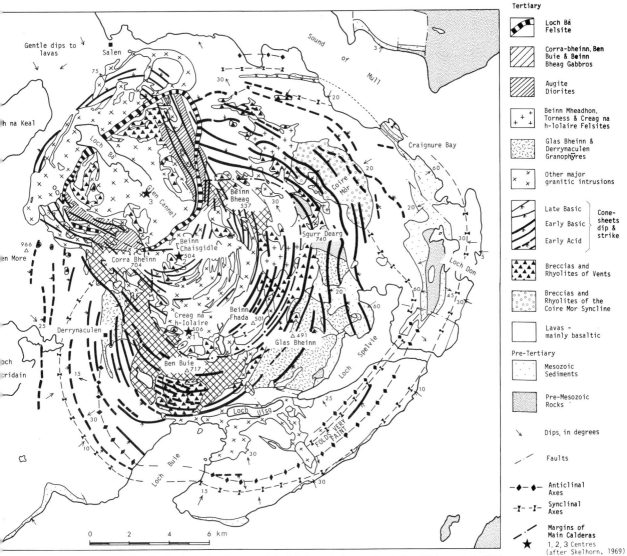

Fig. 14.15. Geological sketch map of the Mull central complex (after Richey *et al*. 1961, Plate IV, by permission of H.M.S.O.). Asterisks denote the igneous centres; 1, Glenmore; 2, Beinn Chaisgidle; 3, Loch Ba (after Skelhorn 1969).

More Ring Dyke, which cuts across cone-sheets on Maol nam Fiadh. This ring-dyke is well exposed on Cruach Choiredail (Skelhorn 1969, fig. 8) where it appears to grade from olivine and quartz gabbros in the lower parts through intermediate rocks to granophyre in the highest exposures. The variation has been cited as an example of either the generation of granitic magma through crystal fractionation and *in situ* differentiation (Bailey *et al.* 1924, Koomans and Kuenen 1938), or the mixing of contrasted basic and acid magmas to form intermediate rocks by hybridisation (Holmes 1936, Fenner 1937). The latter view is now generally favoured.

The Loch Ba Centre (Centre 3)

The Late Basic Cone Sheets continued into Centre 3 where they cut both the early granophyre on Beinn a' Chraig and the Glen Cannel Granophyre. The Glen Cannel

Granophyre has an upper domed surface and covers a considerable area inside and a small area outside the Loch Ba ring dyke. It overlies the centre of the Mull positive Bouguer gravity anomaly (Fig. 14.7, Table 14.2) and despite its considerable outcrop, it is probably thin (Fig. 14.7b).

The last major igneous event in Centre 3, and in the Mull Complex, was the formation of the Loch Ba Felsite (Fig. 14.15). This intrusion occupies a cylindrical fracture which dips steeply south-east and it has been suggested that space for the intrusion was made by enlargement of the fracture by explosively escaping gases eroding its sides: explosion breccias are preserved at several points along the ring fracture (Bailey *et al.* 1924; Lewis 1968; cf. Richey 1928, fig. 7). The Loch Ba Felsite is a fine-grained, in places glassy, acid intrusion charged with mafic inclusions of basaltic andesite and andesite compositions. Their shapes and fine-

Fig. 14.16. Geological sketch map of the Ardnamurchan central complex (based on Richey and Thomas 1930, Plate II, with minor additions and alterations, by permission of H.M.S.O.).

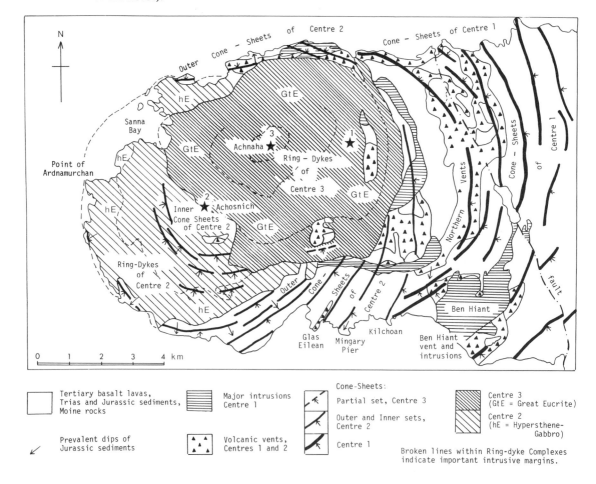

grained to semi-vitreous character indicate that they represent quenched liquids; the ring-dyke is thus a further example of a mixed-magma intrusion (Blake *et al.* 1965, Thompson 1980, Marshall and Sparks 1984), in this instance quenching on emplacement preserved the identity of the components to a much greater extent than in more slowly cooled bodies where rocks of a hybrid character formed. From a study of the inclusions, Sparks (1988) deduced that they had been derived from a zoned magma chamber overlain by rhyolitic magma.

Ardnamurchan

The Ardnamurchan central complex is noted for its numerous arcuate gabbroic intrusions and basic cone sheets (Richey and Thomas 1930). Despite the dominance of mafic rocks in the three centres composing the complex (Fig. 14.16), it has a much smaller positive Bouguer gravity anomaly than Skye, Rhum or Mull (Table 14.2). The geology of Ardnamurchan has been summarised by Gribble (1976) and specific topics were reviewed by Wadsworth and the writer (*in* Sutherland 1982).

Fig. 14.17. Coarse, unbedded agglomerate in the Ben Hiant volcanic, near Maclean's Nose, Ardnamurchan (C. H. Emeleus).

The Ben Hiant volcano and Centre 1

The Ben Hiant (Fig. 14.18) volcano was the earliest major igneous event on Ardnamurchan. Much of the activity was highly explosive, involving the formation of coarse agglomerates (Fig. 14.17) together with bedded pyroclastic rock. The character of these deposits indicates subaerial accumulation, as do volcanic bombs and the occurrence of pitchstone lavas interbedded with tuffs. Later massive intrusions of dolerite form the prominent ridges on the NW of Ben Hiant (Fig. 14.18).

To the north, an ill-exposed belt of agglomerates, termed the Northern Vents, extends across the eastern end of the complex (Fig. 14.16). The agglomerates are closely associated with numerous large areas of basalt lavas and Mesozoic sediments which may be subsided masses similar to those found within the Central Complex of Arran. The basalts are relics derived from the plateau lavas now found in and near the Ardnamurchan centre and which formed part of the lava field of Mull and Morvern. In both the Northern Vents and the Ben Hiant pyroclastic deposits, the clasts include rhyolite and dacite ignimbrite fragments suggesting, as elsewhere in the BTVP, an association of acid magmatism with explosive volcanism. Both the Northern Vents and Ben Hiant are cut by numerous dolerite cone-sheets which form the principal intrusions of Centre 1.

Centre 2

The cone-sheets exposed on the southern edge of the Centre provide some of the most accessible, and arguably the best exposures of these intrusions in the BTVP (Fig. 14.19). Termed the Outer Cone-Sheets since they largely pre-date the major Hypersthene Gabbro and other gabbros of Centre 2, which are cut by an Inner Cone-Sheet system, these intrusions are compositionally indistinguishable from those of Centre 1 (Holland and Brown 1972) but converge on a different focus (Fig. 14.16). A thin, elongate volcanic vent cuts Outer Cone-Sheets at Glas Eilean and is the only such occurrence in the centre, indicating a marked change from Centre 1 conditions.

The Hypersthene Gabbro is the earliest and most extensive of the dozen or so major intrusions assigned to Centre 2 (Richey and Thomas 1930). In common with most of major intrusions in Ardnamurchan, this was regarded as a ring dyke but it is more probably the remnant of a large layered intrusion cored out by later gabbros (Wells 1954a, Skelhorn and Elwell 1971). High-grade thermal metamorphism, including the formation of sapphires in baked bole horizons in adjoining lava successions, occurs on its margins and there is good, centrally inclined igneous layering at several localities (Wells 1954a). Extensive partial melting and mobilisation of Mesozoic sediments and Moine rocks has occurred along the margins of the Hypersthene Gabbro, where distinctive rheomorphic magmas were developed from the different source rocks (Day 1989).

Centre 2 contains extremely clear examples of the complex relationships frequently developed between acid and basic rocks in the BTVP. Coast exposures near the Point of Ardnamurchan show a typical range: angular and sub-angular blocks of dolerite and basalt are embedded in a fine to medium grained acid matrix (Fig. 14.20) providing classic examples of net veining and intrusion breccias; basic inclusions with fine-grained chilled, lobate and crenulated

Fig. 14.18. Ben Hiant and Mingary Castle, Ardnamurchan. The rough ground of Ben Hiant is formed by quartz dolerite intrusions, Mingary Castle rests on a composite sill but the reefs here and towards the shore beneath Ben Hiant are formed by resistant dolerite cone sheets cutting Mesozoic and Moinian rocks. Photograph by © Aerofilms Ltd.

Fig. 14.19. Cone-sheets west of Mingary Pier, Ardnamurchan (C. H. Emeleus).

Fig. 14.20. Net-veining and intrusion breccia near Eilean Carrach, north of the Point of Ardnamurchan. Net-veining is a marked feature of the Granophyric Quartz Dolerite of Centre 2, Ardnamurchan (C. H. Emeleus).

margins are enclosed in unchilled acid material; and blocks and vague areas of dark rock are set in acidic matrices which themselves appear contaminated by mafic material. The exposures provide field, petrographic, mineralogical and chemical evidence for conditions ranging from the invasion of cold brittle basic rocks by acid magma to the mixing and hybridisation of these contrasting magmas and also for the co-existence of acidic and basic liquids (Richey and Thomas 1930, Wells 1954b, Skelhorn and Elwell 1966, Blake *et al.* 1965, Vogel 1982).

Centre 3

This centre contains some of the most striking and complete annular intrusions in the BTVP. The Great Eucrite, the most notable, forms a nearly complete ring of hills rising to *c.* 250 m altitude (GtE, Fig. 14.16; Stewart 1965, Fig. 13.15; and Gribble 1976, cover illustration).

Richey and Thomas (1930) interpreted the Centre 3 intrusions as ring dykes but subsequent unpublished work (quoted in Gribble 1976) suggests that the Great Eucrite and other gabbroic intrusions, including fluxion gabbros with pronounced alignment of plagioclase feldspars, form part of a layered mafic pluton cut by quartz dolerite ring dykes. The youngest intermediate and acid intrusions of the centre are probably of hybrid origin.

Blackstones Bank

This submarine complex is of late Cretaceous or early Tertiary age. The centre is the site of a strong magnetic anomaly and was delimited during geophysical surveys of the Sea of the Hebrides (Binns *et al.* 1974). It has the largest positive Bouguer gravity anomaly in the British Isles (Table 14.2), probably caused by a large cylindrical mass of ultrabasic rock extending to the base of the crust. Unlike

Fig. 14.21. Geological sketch map of the Blackstones Bank submarine central complex (based on Durant *et al.* 1982, fig. 4, by permission of H.M.S.O.).

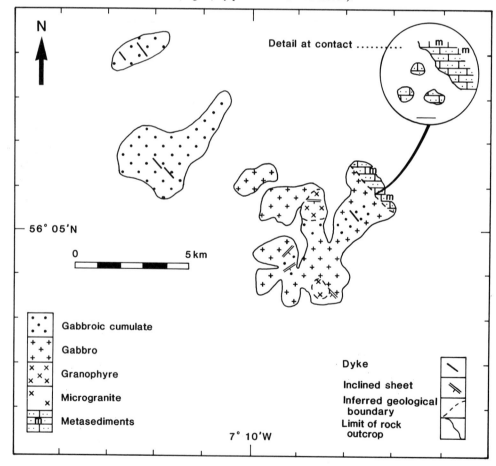

the subaerial centres, the Blackstones Bank Centre lies within a Mesozoic basin (Fig. 14.2).

A large number of *in situ* samples have been recovered by scuba diving, enabling the construction of a geological sketch map (Fig. 14.21; Durant *et al*. 1982). Gabbros, often with cumulate textures, are the most abundant rocks sampled, although granophyre and microgranite also occur. Original calcareous sedimentary country rocks are probably represented by high-temperature calc–silicate hornfelses, which crop out on the east of the site. Basaltic dykes and inclined sheets were found on most sites, their differential erosion contributes to the rugged (submarine) topography.

Rockall

The peralkaline granite forming this islet is probably part of a larger igneous complex emplaced in sedimentary and metamorphic rocks (Roberts 1975). Sheets of rockallite, a rock rich in alkali pyroxene and amphibole cut the granite which has an unusually varied mineralogy (Sabine 1960, Harrrison 1975).

Igneous rocks have been dredged from features identified as late Cretaceous–early Tertiary centres in the Atlantic west of Scotland (Fig. 14.1; Scrutton 1971; Jones *et al*. 1974; Roberts 1975) and a further centre, the Erlend Complex (Fig. 14.28) has been identified near Shetland (Chalmers and Western 1979, Gatliff *et al*. 1984).

Arran

The Northern Granite and the Central Igneous Complex (Fig. 14.22) form high ground in the north and centre of Arran respectively (Fig. 14.23). The near-circular Northern Granite is the older of the two complexes, consisting of two major granitic intrusions, the earlier of which was forcibly emplaced and caused considerable distortion to the country rocks (Fig. 14.24). The later fine-grained inner granite, by contrast, was passively emplaced. Both granites are biotite-bearing and the outer coarse granite has drusy facies with topaz and beryl sometimes present in the cavities.

The Central Igneous Complex (Fig. 14.25) represents a shallower level of igneous activity than the Northern Granite which it post dates. It contains granites, gabbros and hybrid rocks together with acid and intermediate lavas, tuffs and pyroclastic breccias and an important suite of large remaniée blocks of Mesozoic sediments and Tertiary olivine basalt lavas. The sediments include fossiliferous Rhaetic and Lower Lias rocks together with sandstone and chalk of probable Cretaceous age. Skarns have been identified at the edge of some of the chalk inclusions (Cressey 1987). The remaniée masses provide convincing evidence that Arran was once covered by Jurassic and Cretaceous sediments and Tertiary plateau basalts although the nearest representatives of these formations on land are now found in Antrim and Mull. The western half of the Complex was shown by King (1955) to have been a caldera in which he distinguished three volcanic cones in the

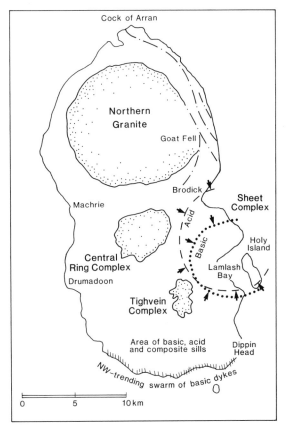

Fig. 14.22. Sketch map of Arran showing the principal elements of the Tertiary igneous geology (based on Tomkeieff 1961, fig. 2, with some alterations).

vicinity of Ard Bheinn (Fig. 14.25b). Granites and gabbros were intruded more or less in conformity with the caldera margins; in places the contrasted magmas must have existed more or less contemporaneously to judge from the numerous occurrences of hybrid rocks and the mixed-magma relationships found in the east of the complex.

A further small centre may exist at Tighvein (Fig. 14.22) where felsite, microgranite, hybrid rocks and dolerite crop out in ill-exposed ground (Herriot 1975).

Ailsa Craig

Ailsa Craig is a boss of Tertiary granite rising out of the Clyde estuary 20 km south of Arran (Fig. 14.1). The rock is a distinctive blue-grey microgranite containing alkali amphibole, zirconium-rich alkali pyroxene, aenigmatite and glauconite–celadonite (Howie and Walsh 1981, Harding 1983, Harrison *et al*. 1987). Its thoroughly peralkaline character sets it apart from the majority of BTVP granitic rocks. It is a very distinctive rock which has proved a valuable indicator in tracing the flow direction of Quaternary ice sheets (Charlesworth 1957). It has also been quarried extensively as a source of the stones used in the game of curling (termed 'Ailsas').

Sill complexes

Alkaline dolerite sills and sheets occur widely in the Hebrides and Arran. They are particularly abundant in the Mesozoic sediments beneath the plateau lavas and in the basin structures. Individual sills may be over 100 m in thickness and their resistance to weathering makes them stand out as prominent headlands and cliffs. Particularly good examples crop out around and south of Lamlash Bay on Arran where several sheets are gently transgressive towards Permian sediments and appear to converge on a focus beneath Holy Island (Fig. 14.22). Possibly the best development of dolerite sills is found on the Trotternish peninsula of northern Skye where a sill complex is responsible for the spectacular cliff scenery from the vicinity of Portree Harbour north to Rubha Hunish and Duntulm (Fig. 14.26). This system of sills extends well beyond the coastline where some of the seismic 'dipping reflectors' are probably sills in Mesozoic sediments, occasionally the

intrusions break surface to form islets as in the Shiant Isles between Skye and Lewis. The thicker sills show marked textural and modal mineral variations. At Rubha Huinish in N. Skye the basal part of the sill is strongly olivine enriched (up to 40 wt. per cent) whereas the mineral is virtually absent in the top of the sill (Anderson and Dunham 1966, fig. 15) and similar relationships are found in the Oskaig sill on Raasay. Olivine concentration occurred by settling under gravity or through flow differentiation (Simkin 1967). Segregation of alkali-rich residual liquid in these sills may, in extreme instances, give rise to pegmatitic rocks with syenite or nepheline syenite compositions, one of the best examples of these unusual rocks occurs in the Shiant Isles sill (Walker 1932, Gibb and Henderson 1984). The dolerites are usually analcite-bearing dolerites (crinanites) but occasionally fresh nepheline is present or relics of nepheline are found in analcite, as in the Dippin Sill of Arran (Gibb and Henderson 1978), attesting to the initially alkaline, undersaturated character of the magmas involved. On the mainland, the Howford Bridge sill in Ayrshire is also Tertiary (de Souza 1979) and contains beautiful analcime syenite segregations (Tyrrell 1928b, Cameron and Stephenson 1985). On Skye the sills on Trotternish and near Portree intrude the basal Tertiary sediments and are therefore younger than at least the lower plateau lavas. A large sill complex has been identified in a

Fig. 14.23. Coarse granite forming Cir Mhor and other peaks in the Northern Granite, Arran. Strong jointing of the granite and a number of easily weathered dykes give rise to the slabs and gullies (C. H. Emeleus).

thick succession of Mesozoic rocks to the north of the Shetland Isles, on the edge of the Faeroe–Shetland trough (Gibb *et al.* 1986, Gibb and Kanaris-Sotiriou 1988). Encountered in two wells, this complex may extend for over 130 km; unlike the mildly alkaline basaltic rocks of the Hebridean sill complexes, these intrusions are of olivine tholeiite (MORB) basalt composition.

The sills and sheets near the central complexes show a much greater compositional variety, ranging from felsites and granophyres to intermediate and quartz dolerite varieties and, quite frequently, including composite examples. Quartz dolerites are well represented in southern Arran along with felsite and medium-grained quartz porphyry sheets, the latter providing extensive outcrops at Brown Head. SE of Brodick, the Corrygills Pitchstone sheet is a fine example of a natural acid glass. Composite sills are particularly well represented on Arran. In the Whiting Bay area the Glenashdale sill is an example of one group where the margins are of felsite or craignurite which grade without obvious break into a central quartz dolerite. Another group has basaltic or quartz doleritic margins which enclose a relatively thick core of quartz porphyry as at Drumadoon and Bennan Head. These latter sills are similar to the sills south-east of Broadford on Skye. In the two Arran examples, lobate, rounded and sometimes marginally chilled basalt xenoliths occur in the base of the quartz porphyry member which itself is in sharp, unchilled contact with the dolerite margins and sometimes shows slight contamination by the mafic material. Some of the chilled, lobate margins to basic rock in contact with unchilled acid material suggest magma mixing and 'liquid–liquid' contacts. A further ubiquitous feature is the occurrence of quartz and alkali–feldspar xenocrysts in the quartz dolerites, indicating that there has been mixing of basaltic and porphyritic acid magmas prior to intrusion. These complex relationships are particularly well displayed in the Drumadoon Sill and in associated composite intrusions on the west coast of Arran (Kanaris-Sotiriou and Gibb 1985).

A swarm of basalt, basaltic andesite and more acidic sheets cuts the lavas around Loch Scridain in SW Mull and extends into the Moine metasediments on the Ross of Mull. These intrusions are remarkable for their cognate, coarse-grained gabbroic inclusions and xenoliths of partially fused aluminous rocks which contain, *inter alia*, sapphires (Thomas 1922). These sheets characteristically have chilled, semi-vitreous margins but one on the south coast of the Ross of Mull lacks chilled edges. This sheet, originally described in detail by Cunningham Craig (1911) appears to have differentially thermally eroded the pelitic schists in a quartzite–pelite succession (Kille *et al.* 1986, figs. 2 and 3), suggesting that the magma underwent local turbulence maintaining high marginal temperatures which permitted melting of the relatively fusible pelitic schist, but not the quartzites. These sheets may have acted as feeders for surface lava flows.

Minor ultrabasic intrusions occur around some of the central complexes, notably Rhum where there are numerous peridotite plugs and Skye where peridotite (and allivalitic) dykes penetrate far into the country rocks (Bowen 1928, Gibb 1968, Donaldson 1977). Peridotite plugs and sheets are present on the southern side of the Cuillin centre where they cut the Tertiary lavas; picrite sills which fuse the Torridonian of Soay may be related to the Cuillin feldspathic peridotites and provide evidence of picritic liquids (Wyllie and Drever 1963). An ultrabasic sheet intruding basalts on the south side of Gars-bheinn, Skye shows a striking development of alternating layers of feldspathic peridotite and coarse, pegmatitic feldspar-rich bands which increase in number towards the top of the intrusion (Weedon 1960). The structures were considered by Weedon to have developed in place through a complex differentiation process. Although generally flat-lying and parallel, the pegmatitic feldspathic layers show strongly transgressive features at several places which led Bevan and Hutchinson (1984) to conclude that the pegmatitic rocks represented intrusion of a liquid unrelated to the peridotite; in their view, one consequence of this would be that the feldspathic peridotite crystallised from a thoroughly ultrabasic magma. The Gars-bheinn sill has a remarkable feeder dyke which shows a strongly chilled marginal facies in which forsteritic olivine phenocrysts (up to 30 modal per cent) lie in a variolitic matrix.

The dyke swarms

Regional dykes swarms of lower Tertiary age extend from NW Scotland and the north of Ireland into southern Scotland, northern England, Wales and the Midlands (Fig. 14.1). Geophysical investigations of the surrounding seas show that swarms are also present offshore (Kirton and Donato 1985), two pronounced linear magnetic anomalies in the North Minch and off the Butt of Lewis have been interpreted as thick (1·1 km and 0·8 km respectively) dykes probably of Tertiary age (Ofoegbu and Bott 1985). As with the sill complexes, the dykes are mainly of basaltic compositions except in the vicinity of central complexes where intermediate, acid and composite examples occur. The dykes generally trend in a NW–SE direction but there are marked regional and local perturbations and these are indicated on Fig. 14.27 where the individual dyke swarms are shown.

The southern shore of the Isle of Arran provides a classic section through a dyke swarm. Over 300 dykes are exposed in a 3 km section east of Bennan Head with an aggregate crustal extension of about 10 per cent although locally much greater; Richey (1961) has estimated an overall 7 per cent extension in a 20 km section across this swarm. The dykes cut red, brown and pink Triassic sandstones, usually with very limited thermal effects. However, a few dykes are associated with strong thermal metamorphism, including the formation of quartz paramorphs after tridymite and it is

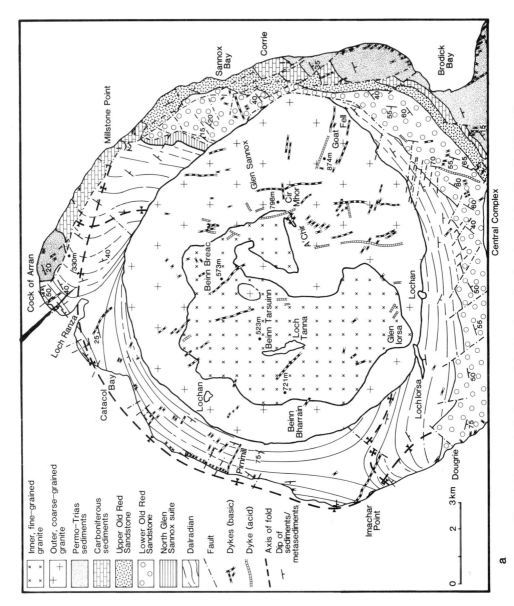

Fig. 14.24a. Geological map of the Northern Granite, Arran (after Tyrrell, 1928a, Plate III, by permission of H.M.S.O.; Friend et al. 1964 and England, personal communication).

a

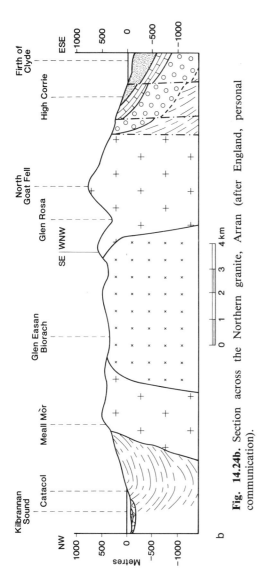

Fig. 14.24b. Section across the Northern granite, Arran (after England, personal communication).

likely that they functioned as conduits for magma which fed surface flows. Halsall (1978) has shown that the southern part of the Arran swarm contains a wide variety of compositions including alkaline olivine dolerite, mildly alkaline and transitional olivine dolerite, tholeiitic olivine dolerite and quartz dolerite. In addition the island has classic exposures of composite pitchstone–dolerite dykes, some of which intrude the Northern Granite (Judd 1893).

Mattey *et al.* (1977) studied the geochemistry of the dykes of the Skye swarm in four transects between Harris in the Outer Hebrides and Arisaig on the mainland. Dykes of a high calcium, low alkali magma type similar to the very late lavas of Preshal Mhor formed 70 per cent of the samples and are restricted to a narrow central zone along the axis of maximum dilation of the swarm. Another group closely resembles the basalt lavas of the Skye Main Lava series, while a third group, the Fairy Bridge magma type, which is typified by flat chondrite-normalised rare earth element patterns, is not represented amongst the lavas. Production of the Preshal Mhor type magmas, which involved high degrees of partial melting of their source area, was considered to be associated with the Skye central complex and this activity certainly continued to a late stage since some of these dykes cut the Skye granites. The Mull centre has long been considered the focus of the dykes of NE England and Southern Scotland. However, it has been uncertain whether the magma was derived from Mull or from a source or sources beneath the course of the dyke. Macdonald *et al.* (1988) showed that the Cleveland dyke has close chemical similarities with tholeiitic rocks in Mull but that it is not comagmatic with dykes in the western part of the Scottish Midland Valley with which it had been linked. The magma for the Cleveland dyke is now considered to have come from the major magma chamber beneath Mull and to have been emplaced laterally as a single pulse in a matter of days.

In a comprehensive review of the Scottish Tertiary dyke swarms, Speight *et al.* (1982) subdivided the regional linear swarms and distinguished major swarms and secondary swarms; groups of dykes which departed from the normal directions of the regional linear swarms are termed subswarms. They recognised the close correspondence between the swarms, their maximum dilation and the central complexes and pointed out that the manner in which different swarms link indicates a broad contemporaneity. These authors argued strongly against extensive lateral flow during dyke emplacement and considered that the swarms were fed from more or less vertical ridge-like magma chambers whose crests are indicated by the positions of the dilation axes (Fig. 14.27). The pattern of the swarms and their dilation axes is very similar to the well-known tension-gash patterns and the authors suggested that shearing stresses associated with the early opening of the North Atlantic were responsible for the initiation and development of the magma ridges and the dyke swarms. From an examination of the early Tertiary stress field in the BTVP, England (1988) concluded that a stress field with the

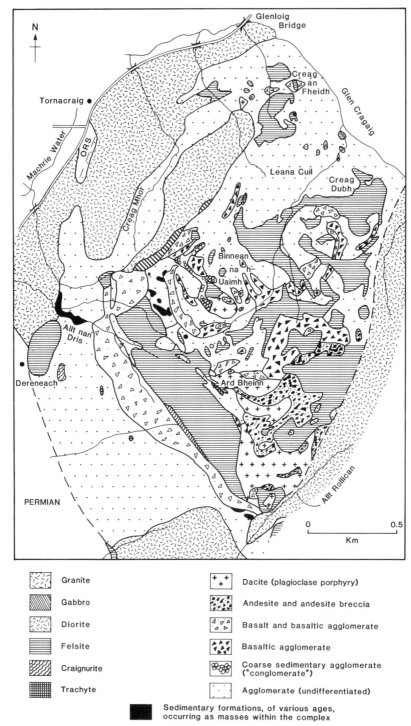

Fig. 14.25. Geological sketch map of the Central Igneous Complex, Arran (after Macgregor 1983, fig. 18).

Fig. 14.26. The Kilt Rock, Trotternish, north Skye. A thick, columnar jointed olivine dolerite sill overlies Jurassic strata while a second, smaller sill crops out near sea level. The cliffs are *c.* 60 m in height (C. H. Emeleus).

compositions. Basaltic rocks occur throughout the province as dykes, sills, plugs and lava flows as well forming gabbroic and other intrusions within the central complexes, strongly contrasting with the distribution of granitic rocks which are almost wholly restricted to the central complexes or their immediate surroundings. Silicic extrusives rarely occur except close to central complexes. Within the central complexes, granitic rocks are often the dominant type in terms of surface outcrop but they frequently exhibit close spatial relationship to the basaltic rocks and are generally bracketed in time by basaltic intrusions. These diverse features must be considered in any scheme of magma genesis and are of particular importance when considering the origins of the granitic rocks.

The granitic rocks

Granite and granophyre stocks and ring dykes coalesce to give large outcrops in all of the Scottish central complexes with the exception of Ardnamurchan. In addition effusive silicic rocks, in the form of lava flows, domes, pyroclastic deposits are found in the Central Igneous Complex of Arran, in Mull, at an early stage on Rhum, and around the Eastern Red Hills of Skye. The isolated alkali granite boss of Ailsa Craig and very young pitchstone flow of the Sgurr of Eigg, with its possible continuation on Hyskeir (Fig. 14.2), are the only notable acidic bodies in Scotland outside the immediate environs of the central complexes.

Despite their considerable outcrop area, modelling of the pronounced positive gravity anomalies over the centres indicates that the granites are quite thin, probably not often more than 2 km in thickness (e.g. Bott and Tantrigoda 1987). The underlying dense rocks, which are responsible for the gravity highs, are considered to be either olivine-rich gabbros or feldspathic peridotites and it is likely that Rhum provides an example where erosion has cut down to the dense, mafic root of a central complex.

There are often clear indications of the former co-existence of granitic and basaltic magmas. This is dramatically demonstrated by the intricate liquid–liquid contacts found between the contrasted types in net-veined and pillowed outcrops of Ardnamurchan, St. Kilda and other centres, and in the intimate association of basic and acid rocks in composite sills and in intrusions such as the Marscoite Suite and Skye and Loch Ba ring-dyke of Mull. Mafic magmas have dominated the central complexes, whatever the apparent evidence of surface outcrop, and were closely associated with silicic magmas in both time and space. In the BTVP the field evidence shows unequivocally that mafic intrusions can melt and mobilise a variety of less refractory country rocks. However, we also know from Iceland and other areas where continental crust is lacking, that mafic magmas are also fully capable of generating magmas of granitic composition through fractionation processes. The dilemma is that in the BTVP the granitic rocks could easily have been formed by either fractionation of basaltic magmas or by melting or partial melting of

dominant NE–SW extension acted through the crust and controlled the siting of the intrusions which fed the NW–SE trending dyke swarms. A slight but persistent clockwise offset of the dilation axes of the swarms relative to the swarm axes was attributed to minor dextral shear within the regional stress field. Except possibly in the Arran area and farther south, upper crustal structures were not considered to have been of major importance in determining the orientation of the swarms. England (1988) also pointed out that although NW–SE directed extension in association with continental splitting became important during and after the late Eocene, the dominant pattern over all the BTVP and the North Sea was very different in the early Eocene which argues for a stress field influenced by tectonic events in NW Europe as well as in the NE Atlantic area.

The Tertiary magmas

Basaltic and granitic rocks form the majority of the rocks of the BTVP with remarkably few examples of intermediate

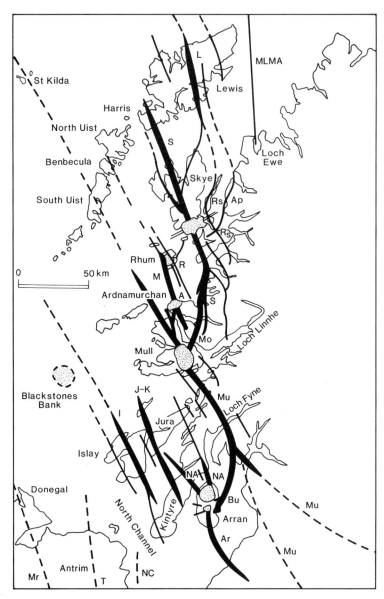

Fig. 14.27. Sketch map showing the dilation axes of the Tertiary dyke swarms in western Scotland and the north of Ireland. Major swarms are shown by thick lines and by broken lines where the axes are less certain. Stippled areas mark the central complexes. Thicknesses are approximately proportional to the concentrations of dykes in the swarms. Regional linear swarms are: L, Lewis; S, Skye; M, Muck; Mu, Mull; A, Ardnamurchan; Mo, Morven; J–K, Jura–Kintyre; I, Islay; NA, North Arran; Ar, South Arran; NC, North Channel; T, Tardee; Mr, Mourne. Subswarms are: Rs, Raasay; Ap, Applecross; R, Rhum; Bu, Bute (after Speight *et al.* 1982, fig. 33.5, by permission of John Wiley & Sons). MLMA = Minches Linear Magnetic Anomaly = dyke, or dyke swarm (after Ofoegbu and Bott 1985).

crustal rocks, or by a combination of these processes. Much of the research in the Province over the last four decades has been directed towards a solution of these problems (see Emeleus 1983, and the thorough reviews of Gass and Thorpe (1976), J. D. Bell (1982) and Thompson (1982)).

To attempt a solution to the problem of the origins of the granites it has been necessary to draw upon very diverse evidence provided by field observations, geophysical measurements and elemental and isotopic analyses of the igneous rocks and the associated crustal rocks. It has become apparent from studies on Skye and Mull that the granites always contain a major contribution from the fractionation of mantle-derived basalts and a significant, but lesser component from the partial melting of crustal rocks. For example, the investigations of the granites from the three Mull centres by Walsh et al. (1979) show that the early, Centre 1 granites have a considerable crustal component whereas the granites of Centre 2 have significantly less and the granites of Centre 3 have slightly more crustal material than those of Centre 2 but still less than Centre 1. Viewed broadly, this may be interpreted to show that when the crustal rocks beneath Mull were initially invaded by rising basaltic magma they underwent significantly more partial melting than when subsequent batches of mafic magma were emplaced; clearly, if the same crustal sources are involved as time went on they would become progressively depleted in low-melting point constituents, and become less capable of augmenting the silicic residua from basalt fractionation. Examination of the Skye Red Hills granites by Dickin et al. (1984) showed that these too have received varying contributions from mantle, lower crustal and upper crustal sources. In this instance, successive intrusions show progressively more contamination by upper crustal material, suggesting that the magma chambers beneath the Western and Eastern Red Hills granites were long-lived allowing a build-up of crust contaminants, with time. Taking a broad view of the Skye Red Hills granites, Dickin et al. (1984) suggested that they are 'little more than late-stage high-level differentiation products of crustally contaminated Preshal Mhor basic magmas'.

The basaltic rocks

The Tertiary lavas of NW Scotland comprise two major groups. The most abundant form the Skye Main Lava Series (SMLS; Thompson et al. 1972) and the Mull Plateau Group (MPG; of Beckinsale et al. 1978), these include mildly alkaline to transitional basalts and more fractionated hawaiites, mugearites and benmoreites. The magmas of this group equilibrated at the base of the crust (Fig. 14.29; Thompson 1982). The other, less abundant group includes the classic flow at Fingal's Cave, Staffa and is also represented by other flows at the base of the Mull and Skye succession. Further flows may also be present in NW Rhum and Canna. This group, the Staffa Magma Type (SMT) of Bailey et al. (1924), equilibrated within the upper third

of the crust (Morrison et al. 1985, fig. 1; see also Fig. 14.28).

In their consideration of the genesis of the early Tertiary lava pile of Mull, Beckinsale et al. (1978) suggested that the compositional variability between lavas was principally the result of differential partial melting of a vertically inhomogeneous mantle. Crystal fractionation in high-level magma chambers was thought to have been of minor importance. A radically different interpretation has been variously proposed by Thompson, Dickin and Morrison (Thompson 1982, Morrison et al. 1985, Thompson et al. 1986, Dickin et al. 1987). They start with the assumption that the Palaeocene mantle beneath NW Scotland was depleted in incompatible elements and volatile constituents during a small-scale partial melting event in the Permian, as evidenced by widespread Permian minor intrusions of nephelinitic compositions, with their suites of mantle-derived xenoliths (see Chapter 11). Partial melting of this same mantle source during the early Tertiary gave rise to mafic magmas which were both incompatible-element depleted and dry. The dense, dry picritic magmas experienced difficulty in rising through the density trap at the Moho where they ponded, fractionating to form basalts and more evolved compositions. Further density contrast barriers and possibly variable rheological properties in the overlying crust caused the ascending magmas to pond for varying lengths of time in granulite-facies Lewisian gneisses close to the base of the less dense amphibolite-facies gneisses (Fig. 14.29). Depending on how long they were held at this level, they underwent varying amounts of contamination through the assimilation of liquids derived by the partial melting of acid sheets in the lower crustal gneisses while at the same time probably experiencing further crystal fractionation. The melts from this stage then rose into the upper crust where there were renewed possibilities of contamination by other partial melts with their distinctive signatures, the degree of contamination once more depending very much on the residence time in the amphibolite-facies upper crustal gneisses. Specific suites from the BTVP were closely examined (Morrison et al. 1985, Thompson et al. 1986) and it was concluded that the south-west Mull SMT lavas must have undergone a complex series of events between the base of the crust and final extrusion; a distinct granulite facies gneiss signature is present in most flows indicating a period of residence in the lower crust and in addition the flows show variable degrees of contamination by Moine metasediments implying that the magmas for some of the flows resided in quite near-surface reservoirs: their varying progresses are charted in Fig. 14.30. By contrast, the SMLS lavas and the MPG flows do not show appreciable evidence of contamination by acidic melts derived from upper crustal sources; however, they usually do exhibit elemental and isotopic features which indicate up to 10 per cent contamination by acid melts derived by large-scale partial melting of granitic sheets in the granulite-facies Lewisian gneisses.

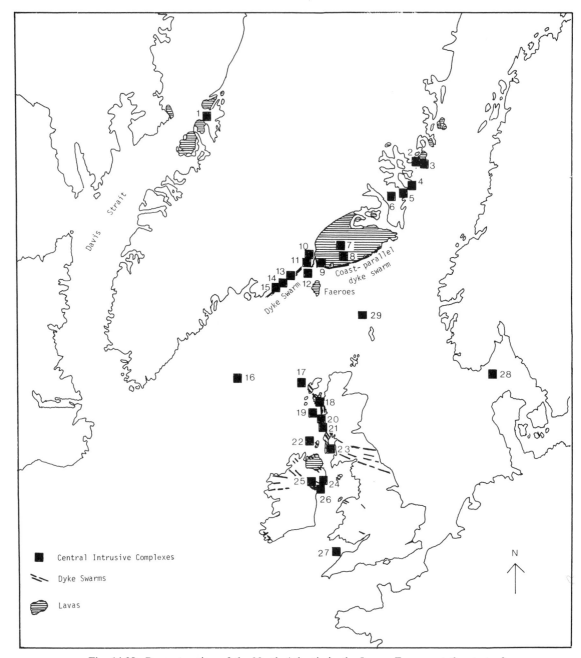

Fig. 14.28. Reconstruction of the North Atlantic in the Lower Eocene, at the start of Tertiary volcanism in Britain and Ireland (after Bell 1976, fig. 6). The central intrusive complexes numbered 1–29 are: 1, Ubekendt Island; 2, Myggebukta; 3, Kap Broer Ruys; 4, Kap Parry; 5, Kap Simpson; 6, Werner Bjerge; 7, Borgtinderne; 8, Lilleoise; 9, Skaergaard; 10, Gardiner's Plateau, Kaerven and Kangerdlugssuaq; 11, Kap Edvard Holm; 12, Nordre Apuiteq; 13, Nugaliq; 14, Kialineq; 15, Kap Gustav Holm; 16, Rockall and Helen's Reef; 17, St Kilda; 18, Skye; 19, Rhum; 20, Ardnamurchan; 21, Mull; 22, Blackstones; 23, Arran; 24, Mourne Mountains; 25, Slieve Gullion; 26, Carlingford; 27, Lundy; 28, Skagerrak; 29, Erland complex.

Crustal contamination of the mantle-derived magmas can overprint and effectively obscure characteristics of the basalts which may relate directly to their mantle sources. However, amongst the lower Tertiary lavas and minor intrusions of Skye, Thompson and Morrison (1988) have recognised rocks in which crustal contamination is not significant. Using these uncontaminated samples, they have suggested that the BTVP mafic magmas are likely ultimately to have come from asthenospheric sources but that the larger magma batches became ponded in the lithosphere where melting and assimilation occurred. Some smaller batches, however, traversed the lithosphere without significant contamination.

From earlier discussions of the gravity anomalies associated with the BTVP central complexes it is established that mafic rocks must be dominant. Examination of the central complexes shows that in addition to gabbro, dolerite and basalt, the mafic rocks include a variety of thoroughly ultrabasic rocks of which feldspathic peridotites are the most abundant; on Rhum they are the dominant mafic rock. Modelling of the gravity anomalies strongly suggests that ultrabasic rocks must be important constituents of the dense, mafic cylindrical masses beneath the complexes (e.g. Fig. 14.7). Field and laboratory evidence, principally from Rhum and Skye, points strongly to the participation of highly magnesian picritic or ultrabasic magmas in the growth of the rocks of the complex. Perhaps the intense igneous activity in these central complexes creates thermal conditions which allow

high temperature magnesian magmas, which would normally pond at the base of the crust and/or undergo significant crystallisation while travelling through it, to rise to near surface levels, although not to the point where they form lava flows.

The almost complete absence of peridotite and other xenoliths of mantle or deep crustal origin from the BTVP lavas and intrusions contrasts strongly with their common occurrence in Permian and Carboniferous alkaline basaltic lavas and intrusions (cf. Chapter 11), and is further evidence that the Tertiary magmas were ponded at various stages on their way to the surface. One exception is known at Loch Roag in the Outer Hebrides where a monchiquite dyke of Tertiary age contains xenoliths of ultramafic xenoliths, fragments of mafic granulites and a very varied megacryst suite (Hunter and Upton 1987, Menzies et al. 1987, Menzies and Halliday 1988). The compositions of the xenoliths indicate that under Lewis the lithosphere was chemically enriched, in marked contrast to that postulated for much of the rest of the subjacent BTVP mantle.

The timing, duration and location of Tertiary igneous activity

The range of radiometric age determinations on rocks from the BTVP (Table 14.3) indicates that the igneous activity spanned about 11 million years (Mussett et al. 1988). However, most took place between about 63 Ma and 57 Ma with the majority of determinations on the plateau lavas between 63 and 59 Ma and those on the central complexes somewhat younger, between about 60 and 57 Ma. Several acid bodies have yielded significantly younger ages, notably the Beinn an Dubhaich granite and the Sgurr of Eigg pitchstone flow. Activity within the BTVP was thus rather earlier than that in East Greenland (Soper et al. 1976).

The accuracy of the age determinations is rarely sufficient to discriminate between intrusive events within a centre, the determinations generally fail to provide satisfactory answers to questions such as what are the relative ages of the different centres or of groups of flows in different parts of the BTVP. They do, however, clearly demonstrate that individual centres must have had extremely limited life spans, of the order of a few million years at the most, and that the lava fields were built up over equally short intervals; thus, the bulk of the very widespread igneous activity in the BTVP occurred within a fairly restricted time span.

A promising approach to the problem of fine-tuning the dating of events in the BTVP combines radiometric age determinations, measurements of the rocks' remnant magnetisations and the data available from field observations (Dagley and Mussett 1981, 1986; Dagley et al. 1984, 1987; Mussett 1984, 1986; Mussett et al. 1987, 1988). Systematic measurements of the magnetisation of the plateau lavas show that virtually all have reversed polarities ('R') imply-

Fig. 14.29. The relationship of Skye and Mull lavas to the 1 atmosphere (1 atm) and 9 kilobar (9 + 1·5 kb = base of continental crust) cotectics plotted in terms of normative olivine (Ol), hypersthene (Hy), diopside (Di), nepheline (Ne) and quartz (Qz). Circles: basal basaltic and tholeiitic andesite lavas from SW Mull; squares: inclined dolerite sheets cutting lavas, SW Mull; triangles: basal pillow basalts and hyalocastite tuff, Skye (after Morrison et al. 1985, fig. 1).

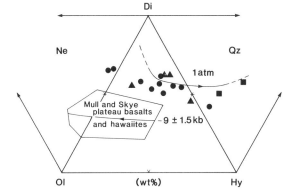

ing that in any given sequence eruption almost certainly occurred within one of the periods of reversed polarity of the Earth's magnetic field during Lower Tertiary times. Many of the central complex intrusions and members of the regional dyke swarms also show reversed polarities but some groups of major intrusions and dykes have normal polarities ('N'); thus, R–N–R and other sequences may be recognised in the central complexes which, when combined with specific age determinations and an agreed polarity time scale (Harland et al. 1982), set maximum time limits to the duration of igneous activity. Mussett (1986) applied these procedures to the Mull centre where he argued for an R–N–R sequence, starting with the plateau lavas (60 + 0·5

Ma; R), including the Loch Uisg Granophyre (58·1 + 1·6 Ma; R) acid intrusions of Centre 1 (c. 58–57 Ma; N) and ending with late basic dykes cutting the Centre 3 intrusions (dykes = R). The R–N–R events are correlated with chrons 26r, 26 and 25r, respectively, on the polarity time scale (Harland et al. 1982, but see Mussett et al. 1988, p. 339) from which Mussett estimates that the *maximum* duration for the Mull igneous activity was 4·7 million years and the *minimum* 0·7 million years. Elsewhere, field-based evidence demonstrates the very early plateau lavas of Eigg and Muck (R) to be earlier than the Rhum central complex (R) which was unroofed and eroded during eruption of the lavas of NW Rhum, Canna and south-west Skye (R). These lavas

Table 14.3. Tertiary igneous activity: selected age determinations

	Age (Ma)	Method	Source
SKYE:			
lavas	59	K/Ar	Brown and Mussett (1976)
Coire Uaigneich granite	59·3 ± 0·7	Rb/Sr isochron	Dickin (1981)
Beinn an Dubhaich granite	53·5 ± 0·4	Rb/Sr isochron	Dickin (1981)
Composite sill (late)	54·9 ± 0·6	$^{40}Ar/^{39}Ar$	Mussett et al. (1988)
RHUM:			
lavas	60·1 ± 0·5	$^{40}Ar/^{39}Ar$	Mussett (1984)
Western Granite	58·4 ± 0·4	$^{40}Ar/^{39}Ar$	Mussett (1984)
EIGG and MUCK:			
lava (N. Eigg)	63·2 ± 2·0	$^{40}Ar/^{39}Ar$	Dagley & Mussett (1986)
lava (Muck)	63·0 ± 3·4	$^{40}Ar/^{39}Ar$	Dagley & Mussett (1986)
Camas Mhor dyke	c.63	$^{40}Ar/^{39}Ar$	Dagley & Mussett (1986)
Sgurr pitchstone	52·1 ± 0·5	Rb/Sr isochron	Dickin & Jones (1983)
MULL:			
lavas	59·4 to 61·1	$^{40}Ar/^{39}Ar$	Mussett (1986)
acid rocks:			
Loch Uisg	58·1 ± 1·6	$^{40}Ar/^{39}Ar$	Mussett (1986)
Beinn a'Ghraig	57·0 ± 0·8	$^{40}Ar/^{39}Ar$	Mussett (1986)
Loch Ba felsite	56·5 ± 1·0	$^{40}Ar/^{39}Ar$	Mussett (1986)
dyke (Fishnish)	57·5 ± 1·1	$^{40}Ar/^{39}Ar$	Mussett (1986)
Centre 3 granites	58·2 ± 1·3	Rb/Sr isochron	Walsh et al. (1979)
ARDNAMURCHAN:			
Various plutonic intrusions	59 – 61	K/Ar	Mitchell & Reen (1973)
Centre 3 intrusions	60·0 ± 1·7	Rb/Sr isochron	Beckinsale & Walsh in Dagley et al. (1984)
ARRAN:			
Northern Granite	58·8 ± 0·6	$^{40}Ar/^{39}Ar$	Evans et al. (1973)
Northern Granite	60·3 ± 0·8	Rb/Sr isochron	Dickin et al. (1981)
Central Complex	58·3 ± 2·2	$^{40}Ar/^{39}Ar$	Evans et al. (1973)
Quartz Porphyry (Drumadoon)	58·5 ± 0·8	$^{40}Ar/^{39}Ar$	Mussett et al. (1987)
AILSA CRAIG:			
microgranite	61·5 ± 0·5	Rb/Sr isochron	Harrison et al. (1987)
ST. KILDA:			
gabbros etc.	60	K/Ar	Miller and Mhor (1965)
Conachair granite	55·0 ± 1·0	Rb/Sr isochron	Brook, in Harding et al. (1984)
ROCKALL:			
alkali granite	52·0 ± 0·8	Rb/Sr	Harrison (1975)

were intruded by the Cuillin gabbros (R), followed by the Western Red Hills granites (N) and the Eastern Redhills granites (R) (Meighan *et al.* 1982; Dagley and Mussett 1986, especially table 1). Despite the complexity of these events, this history of activity in the Small Isles and Skye seems also to have been accomplished within much the same time span as that for Mull, except that the youngest Eastern Red Hills granite was intruded appreciably after the last event on Mull for which age data are available (Table 14.3).

The problem of the siting of the central complexes is far from resolved and the writer would reiterate the view expressed in the second edition of this book that studies on central complexes in other areas, as well as in the Hebrides, point inescapably to pre-existing, major structures having played important roles in siting the BTVP central complexes (cf. Pitcher and Bussell 1977). In considering possible factors influencing the siting of the centres, significance must attach to their location on ridges of Precambrian rocks (with the exception of the submarine Blackstones centre which lies within a flanking Mesozoic basin), their fairly regular spacing in a NNW–SSE trending zone extending from Skye to Arran, NE Ireland and Lundy, and their coincidence with major dyke swarms (Fig. 14.2). The centres also occur within somewhat thinned crust (< 30 km thick) which extends along the west of Scotland into the Irish Sea area (Meissner *et al.* 1986). There is often a close spatial association with major pre-Tertiary faults which intersect the ridges at or near the central complexes (Fig. 14.2) which was considered by Richey (1937) to be a possible control on siting; the role of these faults was also emphasised by Vann (1978) who tentatively suggested that there may have been high heat flow where the dykes intersected the ridges, leading to the formation of granites. The diapiric uprise of the granites (Walker 1975) initiated the central complexes and provided foci at which subsequent basaltic magmatism was concentrated, building up the massive, cylindrical bodies of high density, mafic rocks which underlie each centre. Some authors (e.g. Harker 1904) considered that the dyke swarms originated from the central complexes and this is certainly true on a small scale (e.g. Vann 1978). However it is generally accepted that many of the dykes of the regional swarms were earlier than the central complexes and fed the plateau lavas which are frequently intruded by the centres. Nevertheless, the investigation of the Cleveland dyke of north-east England by Macdonald *et al.* (1988) demonstrates that far-travelled, regional dykes can originate in centres.

The BTVP is one of several regions around the North Atlantic where extensive volcanism occurred during the early Tertiary (Fig. 14.28, Upton 1988) caused possibly by the uprise of a narrow, hot, mantle plume, centred near East Greenland, which spread laterally beneath the overlying plate to form a mushroom-shaped head about 2,000 km in diameter (White 1988). The two other principal areas are the East Coast of Greenland (Deer 1976, Noe-Nygaard

1976) and the Faeroes (Noe-Nygaard and Rasmussen 1968) with substantial submarine occurrences of (apparently sub-aerial) basalt lavas in, for example, the Rockall Trough (Roberts 1975, Morton *et al.* 1988b, Wood *et al.* 1988). Although all three are underlain by continental crust (cf. Bott *et al.* 1974), the magmatism differs between them, as does the relationship of the areas to opening of the North Atlantic: East Greenland and the Faeroes are both characterised by massive successions of tholeiitic basaltic lavas and adjoin zones of successful sea-floor spreading whereas the BTVP remains well within an area of continental crust, albeit somewhat thinned. In the BTVP there is strong evidence of contributions by crust to the acid rocks of the

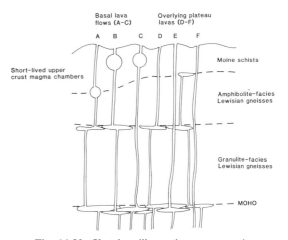

Fig. 14.30. Sketch illustrating magmatic plumbing within the continental crust beneath SW Mull during the extrusion of the Palaeocene flood basalts. The magmas resided for varying lengths of time in reservoirs below and within the continental crust; during residence they became contaminated by wall-rock material, thus acquiring distinctive geochemical signatures. In this diagram the magma reservoirs (staging posts!) are shown as flat lenses or circles. All lavas except F have been contaminated between initial generation and surface effusion and all others except E and C show distinctive granulite-facies Lewisian signatures. The others show additional contributions as follows: A, diffuse amphibolite/Moine schists signature; B, signature of derived from Moine schists; C is as B except it lacks input from granulite facies Lewisian; D came through the upper part of the continental crust without contamination (based on Morrison *et al.* 1985, fig. 4).

Province. Comparable rocks are lacking in the Faeroes and although there are strong similarities between the central complexes of the BTVP and those post-dating the East Greenland lavas, there is a marked tendency towards alkaline, syenitic and nepheline syenitic intrusions in East Greenland, suggesting a different late magmatic history which possibly involved less interaction with the crust than in the BTVP.

REFERENCES

ALMOND, D. C. — 1964 — Metamorphism of Tertiary lavas in Strathaird, Skye. *Trans. roy. Soc. Edinb.*, **65**, 413–434.

ALLWRIGHT, A. E. — 1980 — The structure and petrology of the Tertiary volcanic rocks of Eigg, Muck and Canna, N.W. Scotland. *M.Sc. Thesis, University of Durham* (unpublished).

ANDERSON, E. M. — 1936 — Dynamics of cone-sheets, ring-dykes and cauldron subsidences. *Proc. roy. Soc. Edinb.*, **56**, 128–157.

1951 — *The Dynamics of Faulting and Dyke Formation with Applications to Britain* (2nd Edition). Oliver and Boyd, Edinburgh.

ANDERSON, F. W. & DUNHAM, K. C. — 1966 — The geology of northern Skye. *Mem. geol. Surv. G.B.*, Scotland.

BAILEY, E. B. — 1914 — The Sgurr of Eigg. *Geol. Mag.*, **51**, 296–305.

1945 — Tertiary igneous tectonics of Rhum (Inner Hebrides). *Quart. J. geol. Soc. Lond.*, **100**, 165–188.

1956 — Hebridean notes: Rhum and Skye. *Lpool. Manch. geol. Jl.*, **1**, 420–426.

BAILEY E. B., CLOUGH, C. T., WRIGHT, W. B., RICHEY, J. E. & WILSON, G. V. — 1924 — Tertiary and post-Tertiary geology of Mull, Loch Aline and Oban. *Mem. geol. Surv. G.B.*, Scotland.

BECKINSALE, R. D., PANKHURST, R. J., SKELHORN, R. R. & WALSH, J. N. — 1978 — Geochemistry and petrogenesis of the Early Tertiary lava pile of the Isle of Mull, Scotland. *Contr. Mineral. and Petrol.*, **66**, 415–427.

BEDARD, J. H., SPARKS, R. S. J., RENNER, R., CHEADLE, M. J. & HALLWORTH, M. A. — 1988 — Peridotite sills and metasomatic gabbros in the Eastern Layered Series of the Rhum complex. *J. geol. Soc.*, **145**, 207–224.

BELL, B. R. — 1985 — The pyroclastic rocks and rhyolitic lavas of the Eastern Red Hills District, Isle of Skye. *Scott. J. Geol.*, **21**, 57–70.

BELL, B. R. & HARRIS, J. W. — 1986 — An excursion guide to the Geology of the Isle of Skye. *Geol. Soc. Glasgow*.

BELL, B. R. & EMELEUS, C. H. — 1988 — A review of silicic pyroclastic rocks of the British Tertiary Volcanic Province. *In* Morton, A. C. and Parsons, L. M. (Eds.) (1988), *Early Tertiary Volcanism*. Geol. Soc. Spec. Publ. **39**, 365–379.

BELL, J. D. — 1966 — Granites and associated rocks of the eastern part of the Western Redhills Complex, Isle of Skye. *Trans. roy. Soc. Edin.*, **66**, 307–343.

1976 — The Tertiary intrusive complex on the Isle of Skye. *Proc. Geol. Ass.*, **87**, 247–271.

1982 — Acid intrusions. *In* Sutherland, D. S. (Ed.) *Igneous Rocks of the British Isles*. Wiley, Chichester.

BEVAN, J. C. & HUTCHINSON, R. — 1984 — Layering in the Gars-bheinn ultrabasic sill, Isle of Skye: a new interpretation, and its implications. *Scott. J. Geol.*, **20**, 329–342.

BINNS, P. E., McQUILLIN, R. & KENOLTY, N. 1974 The geology of the Sea of the Hebrides. *Rep. Inst. geol. Sci.*, 73/14.

BLACK, G. P. 1954 The Acid Rocks of Western Rhum. *Geol. Mag.*, **91**, 257–272.

BLAKE, D. H., ELWELL, R. D. W., GIBSON, I. L. SKELHORN, R. R. & WALKER, G. P. L. 1965 Some relationships resulting from the intimate association of acid and basic magmas. *Quart. J. Geol. Soc. Lond.*, **121**, 31–49.

BOTT, M. H. P. & TUSON, J. 1973 Deep structure beneath the Tertiary volcanic regions of Skye, Mull and Ardnamurchan, north-west Scotland. *Nature (Physical Sciences)*, **242**, 114–116.

BOTT, M. H. P., SUTHERLAND, J. & SMITH, P. J. 1974 Evidence for continental crust beneath the Faeroe Islands. *Nature*, **248**, 202–204.

BOTT, M. H. P. & TANTRIGODA, D. A. 1987 Interpretation of the gravity and magnetic anomalies over the Mull Tertiary intrusive complex, NW Scotland. *J. geol. Soc.*, **144**, 17–28.

BOWEN, N. L. 1928 *The evolution of the igneous rocks.* Dover Press, New York (1956 reprint).

BROWN, G. M. 1956 The layered ultrabasic rocks of Rhum, Inner Hebrides. *Phil. Trans. roy. Soc.*, **A240**, 1–53.

BROWN, G. C. & MUSSETT, A. E. 1976 Evidence for two discrete centres in Skye. *Nature*, **261**, 218–220.

BUTCHER, A. R., YOUNG, I. M. & FAITHFULL, J. W. 1985 Finger structures in the Rhum Complex. *Geol. Mag.*, **122**, 491–502.

BUTLER, B. C. M. 1961 Metamorphism and metasomatism of rocks of the Moine Series by a dolerite plug in Glenmore, Ardnamurchan. *Mineral. Mag.*, **32**, 866–897.

CAMERON, I. B. & STEPHENSON, D. 1985 British Regional Geology: the Midland Valley of Scotland (3rd Ed.). HMSO, London.

CANN, J. R. 1965 The metamorphism of amygdales at 'S Airde Beinn, northern Mull. *Mineral, Mag.*, **33**, 533–562.

CARTER, S. R., EVENSEN, N. M., HAMILTON, P. J. & O'NIONS, R. K. 1978 Nd- and Sr- isotopic evidence for crustal contamination of continental volcanics. *Science N.Y.*, **202**, 743–747.

CHALMERS, J. A. & WESTERN, P. G. 1979 A Tertiary igneous centre north of the Shetland Isles. *Scott. J. Geol.*, **15**, 333–342.

CHARLESWORTH, J. K. 1957 *The Quaternary Era*, vol. I, 368–369. Edward Arnold, London.

COCKBURN, A. M. 1935 The geology of St. Kilda. *Trans. roy. Soc. Edin.*, **63**, 511–547.

CRAIG, G. Y. 1983 *Geology of Scotland* (2nd Edition). Oliver & Boyd, Edinburgh.

CRESSEY, G. 1987 Skarn formation between metachalk and agglomerate in the Central Ring Complex, Isle of Arran, Scotland. *Mineral., Mag.*, **51**, 231–246.

CUNNINGHAM CRAIG, E. H., WRIGHT, W. B. & BAILEY, E. B. 1911 The geology of Colonsay and Oronsay, with part of the Ross of Mull (Explanation of Sheet 35, with part of 27). *Mem. geol. Surv. G.B., Scotland.*

DAGLEY, P. & MUSSETT, A. E. 1981 Palaeomagnetism of the British Tertiary igneous province: Rhum and Canna. *Geophys. J.R. Astr. Soc.*, **65**, 475–491.

1986 Palaeomagnetism and radiometric dating of the British Tertiary Volcanic Province: Muck and Eigg. *Geophys. J.R. Astr. Soc.*, **85**, 221–242.

DAGLEY, P., MUSSETT, A. E. & SKELHORN, R. R. 1984 The palaeomagnetism of the Tertiary igneous complex of Ardnamurchan. *Geophys. J.R. astr. Soc.*, **79**, 911–922.

DAGLEY, P.,
MUSSETT, A. E. &
SKELHORN, R. R.
1987   Polarity, stratigraphy and duration of the Tertiary igneous activity of Mull, Scotland. *J. geol. Soc.*, **144**, 985–996.

DAY, S. J.
1989   Structural, thermal and geochemical evolution of the hypersthene-gabbro of Ardnamurchan, Argyll. *Ph.D. Thesis, University of Durham* (unpublished).

DEER, W. A.
1976   Tertiary igneous rocks between Scoresby Sund and Kap Gustav Holm, East Greenland. *In* Escher, A. and Watt, W. S. (Eds.) *Geology of Greenland.* Geol. Surv. Greenland, Copenhagen.

DE SOUZA, H. A. F.
1979   The Geochronology of Scottish Carboniferous volcanism. *PhD. Thesis, University of Edinburgh* (unpublished).

DICKIN, A. P.
1981   Isotope geochemistry of Tertiary igneous rocks from the Isle of Skye. *J. Petrol.*, **22**, 155–190.

DICKIN, A. P. &
EXLEY, R. A.
1981   Isotopic and geochemical evidence for magma mixing in the petrogenesis of the Coire Uaigneich granophyre, Isle of Skye, N.W. Scotland. *Contr. Mineral. and Petrol.*, **76**, 98–108.

DICKIN, A. P.,
MOORBATH, S. &
WELKE, H. J.
1981   Isotope, trace element and major geochemistry of Tertiary igneous rocks, Isle of Arran, Scotland. *Trans. roy Soc. Edin.*: Earth Sciences, **72**, 159–170.

DICKIN, A. P. &
JONES, N. W.
1983   Isotopic evidence for the age and origin of pitchstones and felsites, Isle of Eigg, NW Scotland. *J. geol. Soc.*, **140**, 691–700.

DICKIN, A. P., BROWN, J. L.,
THOMPSON, R. N.,
HALLIDAY, A. N. &
MORRISON, M. A.
1984   Crustal contamination and the granite problem in the British Tertiary Volcanic Province. *Phil. Trans. roy. Soc. Lond.*, **A310**, 755–780.

DICKIN, A. P., JONES, N. W.,
THIRLWALL, M. F. &
THOMPSON, R. N.
1987   A Ce/Nd isotope study of crustal contamination processes affecting Palaeocene magmas in Skye, Northwest Scotland. *Contr. Mineral. and Petrol.* **96**, 455–464.

DONALDSON, C. H.
1975   Ultrabasic breccias in layered intrusions – the Rhum complex. *J. Geol.*, **83**, 33–45.
1977   Petrology of anorthite-bearing gabbroic anorthosite dykes in north-west Skye. *J. Petrol.*, **18**, 595–620
1982   Origin of some of the Rhum harrisite by segregation of intercumulus liquid. *Miner. Mag.*, **45**, 201–209.

DUNHAM, A. C.
1964   A petrographic and geochemical study of back-veining and hybridization at a gabbro-felsite contact in Coire Dubh, Rhum, Inverness-shire. *Miner. Mag.*, **33**, 887–902.
1968   The felsites, granophyre, explosion breccias and tuffisites of the north-eastern margin of the Tertiary igneous complex of Rhum, Inverness-shire. *Quart. J. geol. Soc. Lond.*, **123**, 327–350.

DUNHAM, A. C. &
EMELEUS, C. H.
1967   The Tertiary geology of Rhum, Inner Hebrides. *Proc. Geol. Ass.*, **78**, 391–418.

DURANT, G. P.,
KOKELAAR, B. P. &
WHITTINGTON, R. J.
1982   The Blackstones Bank Igneous Centre, Western Scotland. *In* Blanchard, J., Mair, J. and Morrison, I. (Eds.), Proc. 6th Symn. Coned. Mondiale des Activities subaquatique, Sept. 1980. Publ. by Nat. Env. Res. Council, Swindon, pp. 297–308.

EMELEUS, C. H.
1973   Granophyre pebbles in Tertiary conglomerates on the Isle of Canna, Inverness-shire. *Scott. J. Geol.*, **9**, 157–159.
1983   Tertiary igneous activity. *In* Craig, G. Y. (Ed.) *Geology of Scotland* (Second Edition), 357–398. Scottish Academic Press, Edinburgh, Scotland.
1985   The Tertiary lavas and sediments of northwest Rhum, Inner Hebrides. *Geol. Mag.*, **122**, 419–437.
1987   The Rhum layered complex, Inner Hebrides, Scotland. *In* Parsons, I. (Ed.) Origins of Igneous Layering. *NATO ASI Series, Series C: Mathematical and Physical Sciences*, **196**, 263–286. Reidel, Dordrecht.

EMELEUS, C. H. & FORSTER, R. M.	1979	Field Guide to the Tertiary Igneous Rocks of Rhum. *Nature Conservancy Council, London.*
EMELEUS, C. H., WADSWORTH, W. J. & SMITH, N. J.	1985	The early igneous and tectonic history of the Rhum Tertiary Volcanic Centre. *Geol. Mag.*, **122**, 451–457.
ENGLAND, R. W.	1988	The early Tertiary stress regime in NW Britain: evidence from the patterns of volcanic activity. *In* Morton, A. C. and Parsons, L. M. (Eds.) *Early Tertiary volcanism.* Geol. Soc. Spec. Publ., **39**, 381–390.
EVANS, A. L., FITCH, G. J. & MILLER, J. A.	1973	Potassium-argon age determinations and some British Tertiary Igneous rocks. *J. geol. Soc.*, **129**, 419–443.
FENNER, C. N.	1937	A view of Magmatic differentiation. *J. Geol.*, **45**, 158–168.
FITCH, F. J., HEARD, G. & MILLER., J. A.	1988	Basaltic magmatism of late Cretaceous and Palaeogene age recorded in wells NNE of the Shetlands. *In* Morton, A. C. and Parson, L. M. (Eds.) *Early Tertiary Volcanism and the Opening of the NE Atlantic.* Geol. Soc. Spec. Publ., **39**, 253–262.
FORSTER, R. M.	1980	A geochemical and petrological study of the Tertiary minor intrusions of Rhum, Northwest Scotland. *Ph.D. Thesis, University of Durham* (unpublished).
FRIEND, P. F., HARLAND, W. B. & HUDSON, J. D.	1964	The Old Red Sandstone and the Highland Boundary in Arran. *Trans. Edinbr. geol. Soc.*, **19**, 363–425.
GASS, I. G. & THORPE, R. S.	1976	Igneous case study: The Tertiary igneous rocks of Skye, N.W. Scotland. Open University Press, Milton Keynes.
GATLIFF, R. W., HITCHEN, K., RITCHIE, J. D. & SMYTHE, D. K.	1984	Internal structure of the Erlend Tertiary volcanic centre, north of Shetland, revealed by seismic reflection. *J. geol. Soc.*, **141**, 555–562.
GEIKIE, A.	1871	On the Tertiary Volcanic Rocks of the British Islands. First Paper. *Quart. J. geol. Soc. Lond.*, **27**, 279–311.
	1888	The history of volcanic action during the Tertiary period in the British Isles. *Trans. roy. Soc. Edinb.*, **35**, 21–184.
	1897	*The Ancient Volcanoes of Great Britain.* Vol. 2. Macmillan, London.
GEIKIE, A. & TEALL, J. J. H.	1894	On the banded structure of some Tertiary gabbros in the Isle of Skye. *Quart. J. Geol. Soc. Lond.*, **50**, 645–659.
GIBB, F. G. F.	1968	Flow differentiation in the xenolithic ultrabasic dykes of the Cuillins and the Strathaird Peninsula, Isle of Skye. *J. Petrol.*, **9**, 411–433.
	1976	Ultrabasic rocks of Rhum and Skye: the nature of the parent magma. *J. geol. Soc.*, **132**, 209–222.
GIBB, F. G. F. & HENDERSON, C. M. B.	1978	The petrology of the Dippin sill, Isle of Arran. *Scott. J. Geol.*, **14**, 1–27.
	1984	The structure of the Shiant Isles sill complex, Outer Hebrides. *Scott. J. Geol.*, **20**, 21–29.
GIBB, F. G. F., KANARIS-SOTIRIOU, R. & NEVES, R	1986	A new Tertiary sill complex of mid-ocean ridge basalt type NNE of the Shetland Isles: a preliminary report. *Trans. roy. Soc. Edinb.: Earth Sci.*, **77**, 223–230.
GIBB, F. G. F. & KANARIS-SOTIRIOU, R.	1988	The geochemistry and origin of the Faeroe–Shetland sill complex. *In* Morton, A. V. and Parson, L. M. (Eds.) *Early Tertiary Volcanism and the Opening of the NE Atlantic.* Geol. Soc. Spec. Publ., **39**, 241–252.
GREENWOOD, R. C.	1988	Geology and petrology of the margin of the Rhum ultrabasic intrusion, Inner Hebrides. *Ph.D. Thesis, University of St. Andrews* (unpublished).
GRIBBLE, C. D.	1976	Ardnamurchan: a guide to geological excursions. *Geol. Soc. Glasgow.*

HALSALL, T. J. — 1978 The emplacement of the Tertiary dykes of the Kildonan shore, south Arran (abstract). *J. geol. Soc.*, **135**, 142.

HARDING, R. R. — 1966 The Mullach Sgar complex, St. Kilda, Outer Hebrides. *Scott. J. Geol.*, **2**, 165–178.

1967 The major ultrabasic and basic intrusions of St. Kilda, Outer Hebrides. *Trans. roy. Soc. Edin.*, **66**, 419–444.

1983 Zr-rich pyroxenes and glauconitic minerals in the Tertiary alkali granite of Ailsa Craig. *Scott. J. Geol.*, **19**, 219–227.

HARDING, R. R., MERRIMAN, R. J. & NANCARROW, P. H. A. — 1984 St. Kilda: an illustrated account of the geology. Report of the British Geological Survey, **16**, 46 pp + 1:25,000 map. H.M.S.O. London.

HARKER, A. — 1904 The Tertiary igneous rocks of Skye. *Mem. geol. Surv. G.B.*, Scotland, H.M.S.O., Edinburgh.

1908 The Geology of the Small Isles of Inverness-shire. *Mem. geol. Surv. G.B.*, Scotland, H.M.S.O., Edinburgh.

HARLAND, W. B., COX, A. V, LLEWELLYN, P. G., PICKTON, C. A. G., SMITH, A. G. & WALTERS, R. — 1982 *Geological Time Scale*. Cambridge University Press.

HARRISON, R. K. — 1975 Expeditions to Rockall 1971–1972. *Rep. Inst. Geol. Sci.* 75/1.

HARRISON, R. K., STONE, P., CAMERON, I. B., ELLIOT, R. W. & HARDING, R. R. — 1987 Geology, petrology and geochemistry of Ailsa Craig, Ayrshire. *Brit. geol. Surv. Rep.* **16**, No. 9, 29 pp. H.M.S.O. London.

HERRIOT, A. — 1975 Observation on the Tighvein 'complex' Arran. *Proc. geol. Soc. Glasg.*, **115 & 116**, 7–11.

HIMSWORTH, E. M. — 1973 Marine geophysical studies between northwest Scotland and the Faeroe plateau. *Ph.D. Thesis, University of Durham* (unpublished).

HOLLAND, J. G. & BROWN, G. M. — 1972 Hebridean tholeiitic magmas: a geochemical study of the Ardnamurchan cone sheets. *Contr. Mineral. and Petrol.*, **37**, 139–160.

HOLMES, A. — 1936 The idea of contrasted differentiation. *Geol. Mag.*, **73**, 228–238.

HOWIE, R. A. & WALSH, J. N. — 1981 Riebeckite, arfvedsonite and aenigmatite from the Ailsa Craig microgranite. *Scott. J. Geol.*, **17**, 123–128.

HUGHES, C. J. — 1960 The Southern Mountains Igneous Complex, Isle of Rhum. *Quart. J. geol. Soc. Lond.*, **114**, 111–138.

HUGHES, C. J., WADSWORTH, W. J. & EMELEUS, C. H. — 1957 The contact between Tertiary Granophyre and Torridonian Arkose on Minishal, Isle of Rhum. *Geol. Mag.*, **94**, 37–339.

HUNTER, R. H. — 1987 Textural equilibrium in layered igneous rocks. *Origins of igneous layering*, Parsons, I. (Ed.) NATO ASI Series, *Series C: Mathematical and Physical Sciences*, **196**, 473–504.

HUNTER, R. H. & UPTON, B. G. J. — 1987 The British Isles – a Palaeozoic mantle sample. *In* Nixon, P. H. (Ed.) *Mantle Xenoliths*. Wiley, New York, 107–118.

HUPPERT, H. E. & SPARKS, R. S. J. — 1980 The fluid dynamics of a basaltic magma chamber replenished by influx of hot, dense, ultrabasic magma. *Contr. Mineral. and Petrol.*, **75**, 279–289.

HUTCHISON, R. — 1966 Intrusive tholeiites of western Cuillin, Isle of Skye. *Geol. Mag.*, **103**, 352–363.

1968 Origin of White Allivalite, Western Cuillin, Isle of Skye. *Geol. Mag.*, **105**, 338–347.

HUTCHISON, R. & BEVAN, J. C.   1977   The Cuillin layered igneous complex – evidence for multiple intrusion and former presence of a picritic liquid. *Scott. J. Geol.*, **13**, 197–210.

JASSIM, S. Z. & GASS, I. G.   1970   The Loch na Creitheach volcanic vent, Isle of Skye. *Scott. J. Geol.*, **6**, 285–294.

JONES, E. J. W., RAMSAY, A. T. S., PRESTON, N. J. & SMITH, A. C. S.   1974   A Cretaceous guyot in the Rockall Trough. *Nature*, **251**, 129.

JUDD, J. W.   1874   The Secondary Rocks of Scotland. Second Paper. On the ancient volcanoes of the Highlands and relations of their products to the Mesozoic strata. *Quart. J. geol. Soc. Lond.*, **30**, 220–301.

        1893   On composite dykes in Arran. *Quart. J. geol. Soc. Lond.*, **49**, 536–565.

KANARIS-SOTIRIOU, R. & GIBB, F. G. F.   1985   Hybridization and the petrogenesis of composite intrusions: the dyke at An Cumhann, Isle of Arran, Scotland. *Geol. Mag.*, **122**, 361–372.

KENNEDY, W. Q. & ANDERSON, E. M.   1938   Crustal layers and the origins of magmas. *Bull. Volcanol.*, Ser. ii, vol. iii, 23–41.

KILLE, J. C., THOMPSON, R. N. MORRISON, M. A. & THOMPSON, R. F.   1986   Field evidence for turbulence during flow of basalt magma through conduits from southwest Mull. *Geol. Mag.*, **123**, 693–697.

KING, B. C.   1955   The Ard Bheinn area of the central igneous complex of Arran. *Quart. J. geol. Soc. Lond.*, **110**, 323–355.

        1960   The form of the Beinn an Dubhaich granite, Skye. *Geol. Mag.*, **97**, 326–333.

KOOMANS, C. & KUENEN, P. H.   1938   On the differentiation of the Glen More ring-dyke, Mull. *Geol. Mag.*, **75**, 145–160.

KIRTON, S. R. & DONATO, J. A.   1985   Some buried Tertiary dykes of Britain and surrounding waters deduced by magnetic modelling and seismic reflection methods. *J. geol. Soc.*, **142**, 1047–1058.

KNOX, R. W. O'B. & MORTON, A. C.   1988   The record of early Tertiary N. Atlantic volcanism in sediments of the North Sea Basin. *In* Morton, A. C. and Parson, L. M. (Eds.) *Early Tertiary Volcanism and the Opening of the NE Atlantic*. Geol. Soc. Spec. Publ., **39**, 407–420.

LE BAS, M. J.   1955   Magmatic and amygdaloidal plagioclases. *Geol. Mag.*, **92**, 291–296.

LEWIS, J. D.   1968   Form and structure of the Loch Ba ring-dyke, Isle of Mull. *Proc. geol. Soc. Lond.*, **1649**, 110–111.

LOBJOIT, W. M.   1959   On the form and mode of emplacement of the Ben Buie Intrusion, Isle of Mull, Argyllshire. *Geol. Mag.*, **96**, 393–402.

McCLURG, J.   1982   Geology and structure of the northern part of the Rhum ultrabasic complex. *Ph.D. Thesis, University of Edinburgh* (unpublished).

MACDONALD, R., WILSON, L., THORPE, R. S. & MARTIN, A.   1988   Emplacement of the Cleveland Dyke: evidence from geochemistry, mineralogy, and physical modelling. *J. Petrol.*, **29**.

MACGREGOR, A. G.   1983   Excursion Guide to the Geology of Arran. *Geological Society of Glasgow* (3rd Edition).

McLEAN, A. C. & DEEGAN, C. E. (Eds.)   1978   The solid geology of the Clyde sheet (55 N/6 W) Rep. Inst. Geol. Sci., **78/9**.

McQUILLIN, J. & TUSON, J.   1963   Gravity measurements over the Rhum Tertiary plutonic complex. *Nature*, **199**, 1276–1277.

McQuillin, R.,   1975   The Blackstones Tertiary igneous complex. *Scott. J. Geol.*, **11**,
  Bacon, M. &   179–192.
  Binns, P. E.

Marshall, L. A. &   1984   Origins of some mixed-magma and net-veined ring intrusions. *J.
  Sparks, R. S. J.   Geol. Soc.*, **141**, 171–182.

Mattey, D. P.,   1977   The diagnostic geochemistry, relative abundance and spatial
  Gibson, I. S.,   distribution of high-calcium, low-alkali olivine tholeiite dykes in
  Marriner, G. F. &   the Lower Tertiary regional swarm of the Isle of Skye, N.W.
  Thompson, R. N.   Scotland. *Min. Mag.*, **41**, 273–285.

Meighan, I. G.,   1982   Geological evidence for the different relative ages of the Rhum
  Hutchinson, R.,   and Skye Tertiary central complexes. *J. geol. Soc. Lond.*, **139**,
  Williamson, I. T. &   659.
  Macintyre, R. M.

Meissner, R.,   1986   The 'Moho' in and around Great Britain. *Annales Geophysicae*,
  Matthews, D. &   **4, B.6**, 659–664.
  Wever, Th.

Menzies, A. M.,   1987   Evidence from mantle xenoliths for an enriched lithospheric keel
  Halliday, A. N.,   under the Outer Hebrides. *Nature*, **235**, 44–47.
  Palacz, Z.,
  Hunter, R. H.,
  Upton, B. G. J.,
  Aspen, P. &
  Hawksworth, C. J.

Menzies, M. A. &   1988   Lithospheric Mantle Domains beneath the Archaean and
  Halliday, A.   Proterozoic crust of Scotland. *In* Menzies, M. A. and Cox, K. G.
                                (Eds.) *Oceanic and Continental Lithosphere: Similarities and
                                Differences*. J. Petrol. Spec. Vol., 275–302.

Miller, J. A. &   1965   Potassium–argon age determination on rocks from St. Kilda and
  Mohr, P. A.   Rockall. *Scott. J. Geol.*, **1**, 93–99.

Mitchell, J. G. &   1973   Potassium–argon ages from the Tertiary ring complexes of the
  Reen, K. P   Ardnamurchan peninsula, western Scotland. *Geol. Mag.*, **110**,
                                331–340.

Morrison, M. A.,   1985   Geochemical evidence for complex magmatic plumbing during
  Thompson, R. N. &   development of a continental volcanic center. *Geol.*, **13**, 581–584.
  Dickin, A. P.

Morton, A. C. &   1988   Early Tertiary Volcanism and the opening of the NE Atlantic.
  Parson, L. M   *Geol. Soc. Lond. Spec. Publ.*, **39**.

Morton, A. C.,   1988a Volcanic ash in a cored borehole W of the Shetland Islands:
  Evans, D.,   evidence for Selandian (late Palaeocene) volcanism in the
  Harnald, R.,   Faeroes region. *In* Morton, A. C. and Parson, L. M. (Eds.) *Early
  King, C. &   Tertiary Volcanism and the Opening of NE Atlantic*. Geol. Soc.
  Ritchie, D. K.   Spec. Publ., **39**, 263–270.

Morton, A. C.,   1988b Early Tertiary rocks in Well 163/6-1A, Rockall Trough. *In*
  Dixon, J. E.,   Morton, A. C. and Parson, I. M. (Eds.) *Early Tertiary Volcanism
  Fitton, J. G.,   and the Opening of the NE Atlantic*. Geol. Soc. Spec. Publ., **39**,
  Macintyre, R. M.,   293–308.
  Smythe, D. K. &
  Taylor, P. N.

Mussett, A. E.   1984   Time and duration of Tertiary igneous activity of Rhum and
                                adjacent areas. *Scott. J. Geol.*, **20**, 273–279.

                         1986 ^{49}Ar–^{39}Ar step-heating ages of the Tertiary igneous rocks of
                                Mull, Scotland. *J. geol. Soc.*, **143**, 887–896.

Mussett, A. E., Dagley, P., 1987 Palaeomagnetism and age of quartz porphyry intrusions, Isle of
  Hodgson, B. &   Arran. *Scott. J. Geol.*, **23**, 9–22.
  Skelhorn, R. R.

MUSSETT, A. E., DAGLEY, P. & SKELHORN, R. R.	1988	Time and duration of activity in the British Tertiary Igneous Province. *In* Morton, A. C. and Parson, L. M. (Eds.) *Early Tertiary Volcanism and the opening of the NE Atlantic.* Geol. Soc. Spec. Publ., **39**, 337–348.
NOE-NYGAARD, A.	1976	Tertiary igneous rocks between Shannon and Scoresby Sund, East Greenland. *In* Escher, A. and Watt, W. S. (Eds.) *Geology of Greenland*, Geol. Surv. Greenland, Copenhagen.
NOE-NYGAARD, A. & RASMUSSEN, J.	1968	Petrology of a 3,000 metre sequence of basaltic lavas in the Faeroe Islands. *Lithos*, **1**, 286–304.
OFOEGBU, C. O. & BOTT, M. H. P.	1985	Interpretation of the Minch linear magnetic anomaly and a similar feature on the shelf north of Lewis by non-linear optimization. *J. geol. Soc.*, **142**, 1077–1088.
PARSONS, I.	1987	*Origins of Igneous Layering.* NATO ASI Series, Series C: Mathematical and Physical Sciences, Vol. **196**, Reidel, Dordrecht.
PITCHER, W. S. & BUSSELL, M. A.	1977	Structural control of batholith emplacement in Peru: a review. *J. geol. Soc.*, **133**, 249–255.
PRESTON, J.	1963	The dolerite plug at Slemish, County Antrim, Ireland. *Lpool., Manchr. geol. J.*, **3**, 301–314.
	1982	Eruptive volcanism. *In* Sutherland, D. (Ed.) *Igneous rocks of the British Isles.* Wylie, Chichester.
RAST, D. E.	1968	Age relationships and geometry of the Knock and Beinn a Ghraig Granophyres, Isle of Mull. *Proc. geol. Soc. Lond.*, **1649**, 114–115.
RAST, N., DIGGENS, J. N. & RAST, D. E.	1968	Triassic rocks of the Isle of Mull; their sedimentation, facies, structure and relationship to the Great Glen Fault and the Mull caldera. *Proc. geol. Soc. Lond.*, **1645**, 299–304.
RAY, P. S.	1960	Ignimbrite in the Kilchrist vent, Skye. *Geol. Mag.*, **97**, 229–238.
RAYBOULD, J. G.	1973	The form of the Beinn an Dubhaich granite, Skye, Scotland. *Geol. Mag.*, **110**, 314–350.
RENNER, R. & PALACZ, Z.	1987	Basaltic replenishment of the Rhum magma chamber. *J. geol. Soc.*, **144**, 961–970.
RICHEY, J. E.	1928	The structural relations of the Mourne granites (Northern Ireland). *Quart. J. geol. Soc. Lond.*, **83** (for 1927), 653–688.
	1932	Tertiary ring structures in Britain. *Trans. geol. Soc. Glasg.*, **19**, 42–140.
	1937	Some features of Tertiary volcanicity in Scotland and Ireland. *Bull. Volc. Ser. 2*, **1**, 13–34.
	1940	Association of explosive brecciation and plutonic intrusion in the British Tertiary igneous province. *Bull. Volc., Ser. 2*, **6**, 157–175.
	1961	*Scotland: The Tertiary Volcanic Districts.* British Regional Geology, H.M.S.O., London.
RICHEY, J. E. & THOMAS, H. H.	1930	The geology of Ardnamurchan, Northwest Mull and Coll. *Mem. geol. Surv. G.B.*, Scotland, H.M.S.O., Edinburgh.
ROBERTS, D. G.	1975	Marine Geology of the Rockall Plateau and Trough. *Phil. Trans. roy. Soc.*, **278A**, 447–509.
SABINE, P. A.	1960	The geology of Rockall, North Atlantic. *Bull. geol. Surv. G.B.*, **16**, 156–178.
SCRUTTON, R. A.	1971	Gravity and magnetic interpretation of Rosemary Bank, North-East Atlantic. *Geophys. J.R. astr. Soc.*, **24**, 51–58.
SIMKIN, T.	1967	Flow differentiation in the picrite sills of North Skye. *In* Wyllie, P. J. (Ed.) *Ultramafic Rocks*, pp. 64–69. John Wiley, New York.
SKELHORN, R. R.	1969	The Tertiary igneous geology of the Isle of Mull. *Geol. Ass. Guides*, **20**.

SKELHORN, R. R. & ELWELL, R. W. D. 1966 The structure and form of the granophyric quartz-dolerite intrusion, Centre II, Ardnamurchan, Argyllshire. *Trans. roy. Soc. Edin.*, **66**, 285–306.

1971 Central subsidence in the layered hypersthene-gabbro of Centre II, Ardnamurchan, Argyllshire. *J. geol. Soc.*, **127**, 535–551.

SMITH, D. G. W. 1965 The chemistry and mineralogy of some emery-like rocks from Sithean Sluaigh, Strachur, Argyllshire. *Amer. Mineral.*, **50**, 1982–2022.

1969 Pyrometamorphism of phyllites by a dolerite plug. *J. Petrol.*, **10**, 20–55.

SMITH, N. J. 1985 The age and structural setting of limestones and basalts on the Main Ring Fault in southeast Rhum. *Geol. Mag.*, **122**, 439–445.

SMYTHE, D. K. & KENOLTY, N. 1975 Tertiary sediments in the Sea of the Hebrides. *J. geol. Soc.*, **131**, 227–233.

SOPER, N. J., HIGGINS, A. C., DOWNIE, C., MATTHEWS, D. W. & BROWN, P. E. 1976 Late Cretaceous–Early Tertiary stratigraphy of the Kangerdlugssuaq area, East Greenland, and the age of opening of the north-east Atlantic. *J. geol. Soc.*, **132**, 85–104.

SPARKS, R. S. J. 1988 Petrology and geochemistry of the Loch Ba ring-dyke, Mull (N.W. Scotland): an example of the extreme differentiation of tholeiitic magmas. *Contrib. Mineral. and Petrol.*, **100**, 446–461.

SPEIGHT, J. M., SKELHORN, R. R., SLOAN, T. & KNAAP, R. J. 1982 The dyke swarms of Scotland. *In* Sutherland, D. (Ed.) *Igneous Rocks of the British Isles*, pp. 449–459. Wiley, Chichester.

SUTHERLAND, D. 1982 *Igneous Rocks of the British Isles*. Wiley, Chichester.

STEWART, F. H. 1965 Tertiary Igneous Activity. *In* Craig, G. Y. (Ed.) *The Geology of Scotland*. Oliver and Boyd, Edinburgh.

TAYLOR, H. P. & FORESTER, R. W. 1971 Low-^{18}O igneous rocks from the intrusive complexes of Skye, Mull and Ardnamurchan, western Scotland. *J. Petrol.*, **12**, 465–497.

THOMAS, H. H. 1922 On certain xenclithic minor intrusions in the Island of Mull (Argyllshire); with chemical analyses by E. G. Radley. *Quart. J. geol. Soc. Lond.*, **78**, 219–260.

THOMPSON, R. N. 1969 Tertiary granites and associated rocks of the Marsco area, Isle of Skye. *Quart. J. geol. soc. Lond.*, **124**, 349–385.

1980 Askja 1875, Skye 56 Ma: basalt triggered, Plinian, mixed-magma eruptions during the emplacement of the Western Redhills granites, Isle of Skye, Scotland. *Geol. Rundsc.*, **69**, 245–262.

1982 Magmatism of the British Tertiary Volcanic Province. *Scott. J. Geol.*, **18**, 49–107.

THOMPSON, R. N., ESSON, J. & DUNHAM, A. C. 1972 Major element chemical variation in the Eocene lavas of the Isle of Skye, Scotland. *J. Petrol.*, **13**, 219–253.

THOMPSON, R. N., MORRISON, M. A., DICKIN, A. P., GIBSON, I. L. & HARMON, R. S. 1986 Two contrasted styles of interaction between basic magmas and continental crust in the British Tertiary Volcanic Province. *J. geophys. Res.*, **91**, 5985–5997.

THOMPSON, R. N. & MORRISON, M. A. 1988 Asthenospheric and lower lithospheric mantle contributions to continental extensional magmatism: an example from the British Tertiary Province. *Chem. Geol.*, **68**, 1–15.

TILLEY, C. E. 1944 A note on the gneisses of Rhum. *Geol. Mag.*, **81**, 129–131.

1951 The zoned contact skarns of the Broadford area of Skye; a study of boron-fluorine metasomatism in dolomites. *Mineral. Mag.*, **29**, 621–666.

TOMKEIEFF, S. I.   1961   Isle of Arran. No. 32 in Excursion Guide Series. *Geol. Ass. Guide*
                               32.
TUTTLE, O. F. &   1954   The granite problem: evidence from the quartz and feldspar of
    KEITH, M. L.   Tertiary granite. *Geol. Mag.*, **91**, 61–72.
TYRRELL, G. W.   1928a The Geology of Arran. *Mem. geol. Surv. G.B.*, Scotland,
                               H.M.S.O., Edinburgh.
                        1928b Dolerite sills containing analcite syenite in central Ayrshire.
                               Quart. J. geol. Soc. Lond, **84**, 540–569.
UPTON, B. G. J.   1988   History of Tertiary igneous activity in the N. Atlantic border-
                               lands. *In* Morton, A. C. and Parson, L. M. (Eds.) *Early Tertiary*
                               Volcanism and the Opening of the NE Atlantic. Geol. Soc. Spec.
                               Publ., **39**, 429–454.
VANN, I. R.   1978   The siting of Tertiary vulcanicity. *In* Bowes, D. R. and Leake,
                               B. E. (Eds.) *Crustal evolution in northwestern Britain and adjacent*
                               regions, 393–414. Seel House Press, Liverpool.
VOGEL, T. A.   1982   Magma mixing in the acidic-basic complex of Ardnamurchan:
                               implications on the evolution of shallow magma chambers.
                               Contrib. Mineral. and Petrol., **79**, 411–423.
VOGEL, T. A.,   1984   Magma mixing: the Marsco suite, Isle of Skye, Scotland.
    YOUNKER, L. W.,   *Contrib. Mineral. and Petrol.*, **87**, 231–241.
    WILBAND, J. T. &
    KAMPMUELLER, E.
VOLKER, J. A.   1983   The geology of the Trallval area, Rhum, Inner Hebrides. *Ph.D.*
                               Thesis, University of Edinburgh (unpublished).
WADSWORTH, W. J.   1961   The layered ultrabasic rocks of South-west Rhum, Inner
                               Hebrides. *Phil. Trans. roy. Soc. Lond.*, **B.244**, 21–64.
                        1982   The major basic intrusions. *In* Sutherland, D. S. (Ed.) *Igneous*
                               Rocks of the British Isles, pp. 416–426. Wiley, Chichester.
WAGER, L. R. &   1953   Basic magma chilled against acid magma. *Nature*, **172**, 68.
    BAILEY, E. B.
WAGER, L. R.,   1953   A granophyre from Coire Uaigneich, Isle of Skye, containing
    WEEDON, D. S. &   quartz paramorphs after tridymite. *Min. Mag.*, **30**, 263–276.
    VINCENT, E. A.
WAGER, L. R.,   1960   Types of Igneous Cumulates. *J. Petrol.*, **1**, 73–85.
    BROWN, G. M. &
    WADSWORTH, W. J.
WAGER, L. R.,   1965   Marscoite and related rocks of the Western Red Hills complex,
    VINCENT, E. A.,   Isle of Skye. *Phil. Trans. R. Soc. Lond.*, **A 257**, 273–307.
    BROWN, G. M. &
    BELL, J. D.
WAGER, L. R. &   1968   *Layered Igneous rocks*. Oliver and Boyd, Edinburgh.
    BROWN, G. M.
WALKER, F.   1932   Differentiation of the sills of northern Trotternish (Skye). *Trans.*
                               roy. Soc. Edin., **57**, 241–257.
WALKER, G. P. L.   1959   Some observations on the Antrim basalts and associated dolerite
                               intrusions. *Proc. Geol. Ass.*, **70**, 179–205.
                        1971   The distribution of amygdale minerals in Mull and Morvern
                               (Western Scotland). *In* Murty, T. V. V. G. R. K. and Rao, S. S.
                               (Eds.) *Studies in Earth Sciences, West Commemoration Volume.*
                               pp. 181–194.
                        1975   A new concept of the evolution of the British Tertiary intrusive
                               centres. *J. geol. Soc.*, **131**, 121–141.
                        1979   The environment of Tertiary igneous activity in the British Isles.
                               Bull. Geol. Surv. G.B., **70**, 5–6 (abstract).
WALSH, J. N.,   1979   Geochemistry and petrogenesis of Tertiary granitic rocks from
    BECKINSALE, R. D.,   the Island of Mull, Northwest Scotland. *Contrib. Mineral. and*
    SKELHORN, R. R. &   *Petrol.*, **71**, 99–116.
    THORPE, R. S

WEEDON, D. S.   1960   The Gars-Bheinn ultrabasic sill, Isle of Skye. *Quart. J. geol. Soc. Lond.*, **116**, 37–54.

                             1961   Basic igneous rocks of the Southern Cuillin, Isle of Skye. *Trans. geol. Soc. Glasg.*, **24**, 190–212.

                             1965   The layered ultrabasic rocks of Sgurr Dubh, Isle of Skye. *Scott. J. Geol.*, **1**, 41–68.

WELLS, M. K.   1954a The structure and petrology of the hypersthene-gabbro intrusion, Ardnamurchan, Argyllshire. *Quart. J. geol. Soc. Lond.*, **109**, 367–397.

                             1954b The structure of the granophyric quartz-dolerite intrusion of Centre 2, Ardnamurchan, and the problem of net-veining. *Geol. Mag.*, **91**, 293–307.

WHITE, R. S.   1988   A hot-spot model for early Tertiary volcanism in the N. Atlantic. *In* Morton, A. C. and Parson, L. M. (Eds.) *Early Tertiary Volcanism and the Opening of the NE Atlantic*. Geol. Soc. Spec. Publ., **39**, 393–414.

WHITTEN, E. T. H.   1961   Modal variation and the form of the Beinn an Dubhaich granite, Skye. *Geol. Mag.*, **98**, 467–472.

WILLIAMS, P. J.   1985   Pyroclastic rocks in the Cnapan Breaca felsite, Rhum. *Geol. Mag.*, **122**, 447–450.

WILLIAMSON, I. T.   1979   The petrology and structure of the Tertiary volcanic rocks of west-central Skye, N.W. Scotland. *Ph.D. Thesis, University of Durham* (unpublished).

WILSON, H. E.   1972   Regional Geology of Northern Ireland. *Geol. Surv. N. Ireland*, H.M.S.O., Belfast, 115 pp.

WOOD, M. V., HALL, J. &   1988   Distribution of Early Tertiary lavas in the NE Rockall Trough. *In* Morton, A. C. and Parson, L. M. (Eds.) *Early Tertiary Volcanism and the Opening of the NE Atlantic*. Geol. Soc. Spec. Publ., **39**, 283–292.
  DOODY, J. J.

WYLLIE, P. J. &   1963   The petrology of picritic rocks in minor intrusion – a picritic sill on the Island of Soay (Hebrides). *Trans. R. Soc. Edin.*, **65**, 155–177.
  DREVER, H. I.

YOUNG, I. M.,   1988   Formation of the Eastern Layered Series of the Rhum complex, northwest Scotland. *Canad. Mineral.*, **26**, 225–233.
  GREENWOOD, R. C. &
  DONALDSON, C. H.

ZINOVIEFF, P.   1958   The basic layered intrusions and the associated igneous rocks of the central and eastern Cuillin Hills, Isle of Skye. *D.Phil. Thesis, University of Oxford* (unpublished).

15

QUATERNARY

G. S. Boulton, J. D. Peacock and D. G. Sutherland

From time to time in its history, large parts of the Earth's surface have been covered by great ice domes analogous to the modern ice sheets of Antarctica and Greenland. The Ice Age which culminated during the Quaternary, the last 1·8 million years of geological time, began to develop about 35–40 million years ago at the Eocene–Oligocene boundary. The Ice Age intensified during the Middle Miocene by the growth of the Antarctic ice sheet (about 14 Ma) and still further by the growth of the Greenland ice sheet by 3 million years ago. During the last 0·75 million years it has been characterised by the rhythmic growth and decay of large ice sheets in the middle latitudes of North America and Europe. Periods of mid-latitude ice sheet growth in *glacial periods*, have been separated by relatively short *interglacials*, at intervals of about 100,000 years, during which climatic conditions have been very similar to those of the present day. These glacial/interglacial oscillations have involved exchanges of up to 50,000,000 km³ of water mass between the oceans and centres of ice sheet growth in Europe and North America, where ice sheets up to 3–4 km thick have formed, and which have cyclically loaded the lithosphere with excess pressures of up to 0·3–0·4 kilobars. Recent glacial periods have been accompanied by lowering of global sea levels by up to 150 m, major shifts in the global distributions of animals and plants and fundamental changes in atmospheric and oceanic circulation patterns.

The changes appear to be driven by predictable changes in the intensity of solar radiation reaching the Earth as a consequence of geometrical variations in the Earth's orbit around the sun. The modern ice sheets of Antarctica and Greenland represent successive layers of frozen, fossil atmosphere, the lowest layers of which are as much as 0·25 million years old. Trapped air bubbles sampled in cores from boreholes through these ice sheets reveal that 'greenhouse gas' (e.g. carbon dioxide, methane) concen-trations were low during glacial periods and high during interglacials, and thus that a natural 'greenhouse effect' may play a role in governing the magnitude of change between glacials and interglacials. Extrapolation of the climatic cycles of the recent past into the future suggests that the present interglacial would naturally come to an end within the next 5,000 years, unless a man-made greenhouse effect prevents or delays it.

The base of the Quaternary is now taken at 1·8 Ma, at the boundary between the Olduvai and Gilsa geomagnetic events (Fig. 15.1). As little species-evolution has occurred during this period, climatic change, which has had a dominant influence on sedimentation, is used to define a series of climato-stratigraphic stages, in contrast to the bio-stratigraphically defined stages of earlier geological periods, which are based on evolutionary changes. An alternating sequence of glacial and interglacial stages has now been defined which, in lowland Scotland, include both glacial and tundra environments during the glacial stages, and boreal (equivalent to the modern pine/spruce forest areas of central Scandinavia) and temperate environments during the interglacial stages.

The long-term tempo of these alternating glacial and interglacial stages is best given by cores from the deep ocean which preserve a record of changing oceanic water temperature in changing microfaunal assemblages (Fig. 15.1), and a record of global ice sheet volume in the changing oxygen isotopic composition preserved in the skeletons of these microfaunas. During glacial periods water which is lost from the oceans to form growing ice sheets tends to be relatively enriched in the light isotope of oxygen, ^{16}O, causing the oceans to become enriched in the heavy isotope, ^{18}O. The oscillating ^{18}O content of ocean water through time can thus be used as an index of ice sheet growth and decay (Fig. 15.1).

The Scottish Quaternary event which has left the

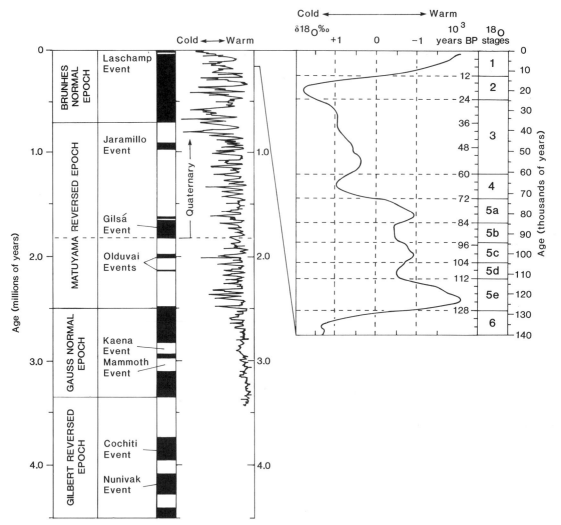

Fig. 15.1. Late Tertiary and Quaternary timescale and indices of global environmental change. On the left are shown the principal geomagnetic epochs. Periods of normal magnetisation are shown black and reversed magnetisation white. The curve of global climate change is based on variations in carbonate content of deep-ocean sediments (Ruddiman & Raymo 1988). A major step in the amplitude of climatic variation, reflecting more intense cold phases, occurs at 2·4 Ma before present. At 0·75 Ma before present both the intensity and length of cold periods increases. The climate of the last glacial cycle is reflected on the right in O-isotope variations in oceanic faunas (Imbrie *et al.* 1984).

strongest mark on the landscape and in the terrestrial sedimentary record was the last major ice sheet expansion in mid-latitudes, when glaciers formed in and expanded from the Highlands to form an ice sheet which by 18,000 years ago covered most of Britain (Fig. 15.17) and was over a kilometre thick in the Midland Valley of Scotland.

It flowed radially outwards from centres in the Highlands and Southern Uplands and was a powerful agent of erosion, moulding the uplands, removing earlier sediments from the lowlands, locally depositing great thicknesses of till directly from the ice and depositing sand and gravel from meltwater rivers. Consequently very little sedimentary evidence of events prior to this last major ice sheet expansion has survived. Those fragments of the earlier Quaternary stratigraphic record which have survived are isolated in time and space and difficult to re-integrate to give a coherent picture of the varied sequence of environments which must have characterised Quaternary Scotland.

Pre-Late Devensian history

The last glacial stage, the Devensian, began about 120,000 years ago (Fig. 15.2). It culminated 18,000 years ago in development of an ice sheet which covered all of Scotland and most of Britain. Evidence of pre-Late Devensian Quaternary events comes from three sources:

— large-scale erosional features on land;
— sporadic sedimentary units on land;
— sedimentary sequences in the offshore zone.

Large-scale erosional features on land

Major structural blocks in western Britain underwent several phases of strong uplift through the Cenozoic, in contrast to parts of the North Sea where subsiding basins have permitted almost continuous Cenozoic sedimentation. In Scotland, there has been general relative uplift in the west and relative down-warping in the east. Successive phases of erosional planation of the landscape have been identified (Geikie 1901, Peach & Horne 1930, Hollingworth 1938, George 1955, Goddard 1965, Sissons 1967) which some have regarded as a product of marine erosion (George 1955) others as subaerial (Fig. 15.3). The axis of the upwarping which has elevated these erosional surfaces is presumed to accord with the present main Scottish watershed, which runs north–south through the western Highlands and western Grampians. The durability of the granitic masses of the Cairngorm mountains

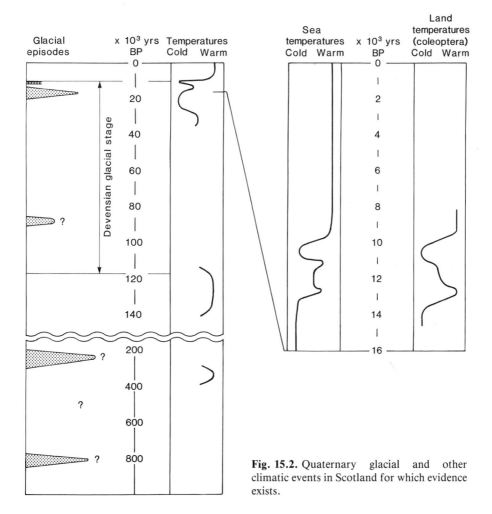

Fig. 15.2. Quaternary glacial and other climatic events in Scotland for which evidence exists.

has left these as a major upland mass lying to the east of the main watershed.

The most dramatic large scale erosional features are the deep U-shaped glens and sea lochs of the Highlands (Figs. 15.4, 15.5a). These have been produced, or at least deepened by glacial erosion. The deepest, the sea loch, Loch Morar, attains a depth of 300 m below sea level. Given typical rates of glacier erosion of the order of 1 mm/year, such a trough would require some 300,000 years of continuous glaciation for its excavation. Such troughs must clearly reflect repeated glacial occupancy if the length of glacial phases is no greater than those shown in Figure 15.1. We do not know however when the Scottish Highlands first began to suffer repeated widespread glaciation. It seems most likely to have been 0·75 million or 2·4 million years ago (Fig. 15.1).

Contrasts such as that between the highly indented fjord-coastline of the western Highlands and the smoother coastline of eastern Scotland are frequently seen in areas covered by the same ice sheet, between mountainous areas of recent uplift, which tend to be penetrated by fjords, and stable, topographically lower areas, where fjords have not been produced. It is a contrast seen between the western coast of Norway on the western side of the Scandinavian ice sheet, and the eastern coast of Sweden on its eastern side; or between the fjord-indented western coast and the smoother eastern coast of Canada. We assume that continuous uplift is likely to prevent development of a stable graded erosional profile and by maintaining strong erosion able to produce deep erosional trenches.

To a large degree, the orientations of the major over-deepened glacial troughs in the Highlands reflect structural or lithological contrasts in bedrock which define relatively easily eroded zones, such as the Great Glen Fault. However, there are other areas where glacier flow appears to have superimposed a large-scale erosional pattern on the landscape. One such area is in the

Fig. 15.3. The summit plateau and eastern corries of Braeriach. The Cairngorm mountain-top plateaux are thought to be remnants of Tertiary erosion surfaces. Deep corries in their flanks have been progressively excavated by small glaciers which have repeatedly occupied these hollows through the Quaternary. Photograph, Patricia Macdonald.

Fig. 15.4. The glacially eroded trough of the Lairig Ghru valley in the Cairngorms. Photograph, Patricia Macdonald.

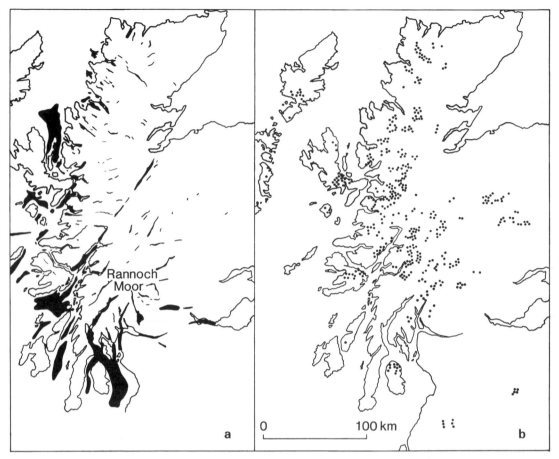

Fig. 15.5. a. Distribution of major glacially eroded troughs in Scotland. **b.** Distribution of corries in Scotland. (Re-drawn from Sissons 1967.)

Grampian Highlands (Fig. 15.6a) where a series of major troughs, Loch Leven, Glen Coe, Loch Etive, Loch Awe, Loch Fyne, Loch Long, Loch Lomond, Loch Katrine, Loch Voil, Loch Earn, Glen Dochert, Loch Tay, Glen Lyon, Loch Rannoch, Loch Ericht and Loch Treig, radiate from Rannoch Moor, suggesting that this area has been a persistent centre of ice sheet mass from which ice flow has radiated during the successive Quaternary glaciations and thereby imposed a dominant radial grain on the landscape.

Other features of glacial erosion which are too large to have been produced during the last glacial period alone, and must reflect repeated glaciation, are the corries of the high mountains (Figs. 15.3, 15.5b, 15.6). These were excavated by small glaciers at times when climate was not severe enough to develop an ice sheet or before the mountains were enveloped by growing ice sheets. They are more numerous and occur at lower altitudes in the

west Highlands than further east (Linton 1959), reflecting the importance of moisture-bearing westerly winds which nourished the corrie glaciers. The majority of corries face between north and east, an aspect which favours snow accumulation and glacier growth because of the protection of snow beds on north-easterly slopes from direct radiation, and blowing of snow from summit areas onto north-easterly facing lee-side slopes.

Remains of rock terraces cut by the sea and backed by fossil cliff lines occur on parts of the west coast (Fig. 15.7). These, the 'pre-glacial beaches' of the Geological Survey memoirs, are to be found at heights ranging from a few metres OD in Kintyre to 34 m on Islay and Jura and over 40 m on Colonsay. They are overlain by till in places and may have been glaciated more than once. It is believed that they were formed during periods of rapid combined marine and periglacial erosion. A more widespread member of this 'family' of landforms, the Main Late-

Fig. 15.6. A moraine formed at the terminus of a small glacier of the Loch Lomond Stadial in the north-eastern Corrie of Beinn Dearg Mor, Wester Ross. (Crown copyright.)

glacial Shoreline, is discussed below. The high cliffs of parts of the Moray Firth and North Sea coasts also seem to have been shaped in part prior to the last glaciation.

Sedimentary units on land

Sediments which date from the long Tertiary erosional period occur in NE Scotland as the Buchan Gravels Group. They are up to 25 m thick and deeply weathered. They contain clasts of flint and quartzite within a kaolinitic clayey silt, and appear to have derived some of their components from an intensely weathered rock which de-

veloped under a subtropical climate in Miocene times (Hall 1984).

Deeply weathered igneous, metamorphic and sedimentary rocks are to be found in several parts of the region, notably in the north-east. East of a line from Elgin to Strathmore, rock has been patchily but extensively decomposed to a gruss (granular sand), locally to depths of several tens of metres. Hall (1984) has suggested that this weathering took place under temperate conditions post-dating the Miocene, and as such is distinct from the clayey gruss associated with the Buchan Gravels Group. Similar deeply weathered rock occurs further west within

Fig. 15.7. Terrace and fossil sea-cliff cut by marine erosion during high relative sea levels. Lismore, Loch Linnhe. B.G.S. photograph.

the outcrop of the Foyers Granite and in the Gaick and Helmsdale areas.

There is evidence in Buchan (NE Scotland) for depositional events predating the main Late Devensian glaciation (Hall 1984). Tills and fossil soils were found at Kirkhill (Fig. 15.8) interbedded with sand and gravel and including solifluction deposits. They occupy basins and channels between tor-like prominences of partly decomposed felsite, of which many of the clasts in the gravels are formed. The occurrence of erratics of biotite-granite and red sandstone in the lowest beds is evidence for a glacial episode or episodes antedating the lower till. The tills themselves represent separate glaciations and the fossil soils interglacial and/or interstadial periods that are older than the limit of radiocarbon dating (about 40,000 BP). The upper till could have been deposited either during the last glaciation or during some earlier, possibly early Devensian, glacial phase. The presence of an erratic of rhomb porphyry in this till suggests that it post-dates the Saalian (penultimate glaciation of continental Europe –

see below). At Teindland (Fig. 15.8) a fossil soil that has yielded both interglacial and interstadial pollen is overlain by deposits which have been classified by different workers as solifluction deposits or till. At Crossbrae, near Turriff, a thin peat below soliflucted till has yielded radiocarbon ages of 29,000–26,000 BP. At Airdrie, two undated peats antedating Late Devensian till have yielded coleopteran evidence for severe, but possibly interstadial conditions (Coope 1962).

Shelly glacigenic deposits are widespread in parts of the north-east (Buchan, Caithness and Orkney) and the south-west (Ayrshire) and locally on low ground elsewhere. As many of the derived faunas are high-boreal to low-arctic with a few boreal-lusitanian species (Jamieson 1865, 1866) it is assumed that the sediments in which they occur have been reworked by glaciers. Shells of high-arctic species (see below) have been recorded in Islay and adjacent to a mass of marine clay at Clava (Peacock 1974, 1975) where they are considered to be ice-transported rafts of marine or glaciomarine sediment (Fig. 15.8 and

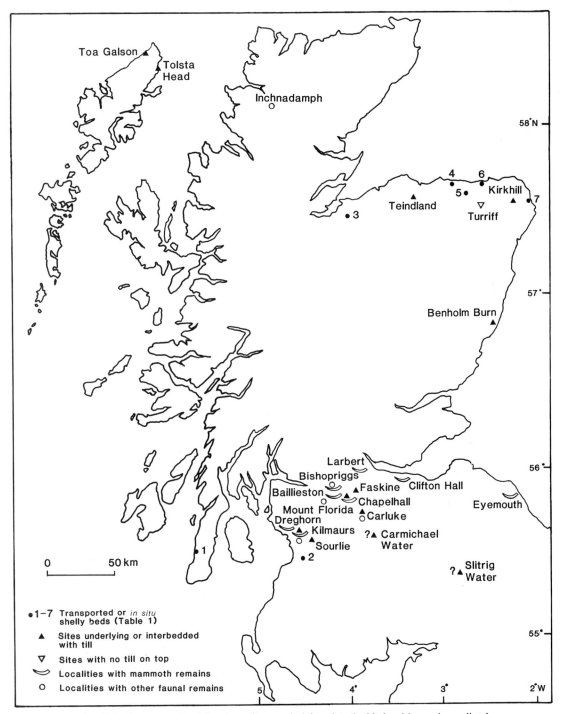

Fig. 15.8. Dates or fossil-bearing sites underlying interbedded with or immediately overlying the Main Devensian till in Scotland (see also Table 15.1).

Table 15.1. Notable sites for transported or *in situ* shelly beds (see Fig. 15.8)

Site	Fauna	Age [14C]	Age Amino-acid	Remarks
1. Cleongart	Chiefly derived			Glaciomarine
2. Afton Lodge	High-boreal/Low-arctic		Ipswichian?	Marine, possibly transported
3. Clava	High-boreal/Low-arctic		—	Marine, probably transported
4. Boyne Limestone Quarry	Probably arctic		—	Glaciomarine?, transported
5. King Edward	Boreal or arctic	>40 k	<80 k*	Marine or glaciomarine, 45·7 mOD. Said to be *in situ*, but base not seen
6. Castle Hill Gamrie	High-boreal/Low-arctic	>40 k	Devensian/ Ipswichian*	Marine, possibly *in situ*, but with some derived shells (e.g. *Arctica*)
7. Annachie	Possibly arctic		—	Marine/glaciomarine; widespread, thick deposits

1. Jessen (1905) with earlier references.
2. Eyles *et al.* (1949), D. Q. Bowen (personal communication 1984).
3. Horne *et al.* (1894), Peacock (1975), J. Merritt (personal communication 1988).
4. Peacock (1966, 1971 and unpublished).
5. Jamieson (1865, 1866), Miller *et al.* (1987), Sutherland (1984b).
6. Jamieson (1865), Peacock (1971), Sutherland (1984b).
7. Jamieson (1865), McMillan & Aitken (1981).

Table 15.2. Mammalian remains within or beneath glacial deposits (modified from Sutherland 1984)

Location	Site description	Remains
Bishopbriggs	In sands and gravels overlain by till	Left metacarpal, upper molar, left tibia and left humerus of woolly rhinoceros (*Coelodonta antiquitatis*)
Bishopbriggs	Uncertain. In till or sands and gravels under till	Molar tooth of mammoth (*Elephas primigenius*)
Baillieston	In bed of laminated sandy clay that laterally passes under till	Tooth of mammoth (*Elphas primigenius*)
Mount Florida	In sand and gravel overlain by till	Antler and bones of reindeer (*Rangifer tarandus*)
Chapellhall	In till	'Remains' of mammoth (*Elephas primigenius*)
Carluke	In till	Right antler of reindeer (*Rangifer tarandus*)
Larbert	In esker gravels	Tooth of mammoth (*Elephas primigenius*)
Clifton Hall	In till	Tusk of mammoth (*Elephas primigenius*)
Kilmaurs	In probable till at base of marine glacigenic deposits	Molar and 9 tusks of mammoth (*Elephas primigenius*) and antlers of reindeer (*Rangifer tarandus*)
Dreghorn	In sands and gravels under till	Tusk of mammoth (*Elephas primigenius*)
Eyemouth	Uncertain	Molar of mammoth (*Elephas primigenius*)
Inchnadamph	Resting on silty sand in inner chamber of cave	Antler and leg bone of reindeer (*Rangifer terandus*)
Sourlie	Organic silts in stratified sediments between tills	Antler fragments of reindeer; fragments of woolly rhinoceros

Table 15.1). 'Crag' fossils in the glacial deposits a few miles north of Aberdeen (Jamieson 1882) were presumably transported from seaward. Amino-acid analysis on *Arctica* shells from them suggests a middle to early Pleistocene age (Smith 1984).

Mammalian fossils have been recovered from 13 sites, chiefly from below till (Table 15.2 and Fig. 15.9). The best documented is at Sourlie, near Irvine where organic-rich silts within a 7 to 8 m thick stratified sequence, with lodgement till above and below, have yielded remains of reindeer (*Rangifer tarandus*) dated to about 30,000 BP, possibly together with woolly rhinoceros (*Coelodonta antiquitatis*) (Jardine *et al.* 1988). This find, together with those from Bishopbriggs (Rolfe 1966) and Inchnadamph (Lawson 1984) give clear support to the view that lowland Scotland was free of glacier ice in the 26,000–30,000 BP period. Sourlie appears to be the only site, apart from the Inchnadamph caves, where the deposits are *in situ*, though at others, such as Kilmaurs (Bishop & Coope 1977) the faunal remains may not have been transported far. The Inchnadamph caves have also yielded a Loch Lomond stadial fauna of reindeer, collared lemming (*Dicrostonyx torquatus*) and tundra vole (*Microtus gregalis*) (Lawson & Bonsall 1986).

Though the distribution of erratics can be largely attributed to the last glaciation in the central and western part of the area, the varied transport directions of those derived from major igneous bodies in NE Scotland is further evidence of a complex glacial history (Fig. 15.10). Boulders of Norwegian larvikite and rhomb porphyry have been found at several localities, notably on the coast south of Aberdeen where some occur in gravels below the main Late Devensian till. These may have been carried across the North Sea by the Scandinavian ice which reached the Durham coast during the Saalian glaciation, immediately prior to the Last (Ipswichian) Interglacial. The erratics of Jurassic and Cretaceous rocks, including the well-known Lower Greensand boulder at Moreseat, have probably been derived largely from the sea-floor of the Moray Firth during the last as well as earlier glaciations.

Sedimentary sequences in the offshore zone

The rhythmic growth of Quaternary ice sheets has been associated with global (*eustatic*) sea level lowering of up to 150 m below modern levels. In those areas where ice sheets have grown during glacial periods, such as Scotland, loading of the Earth's surface by up to 1–2 km of glacier ice has produced *isostatic* depression of the crust, which has often exceeded global eustatic lowering of sea level due to transfer of water from oceans to ice sheets. This has led to an increase of *relative* sea level (change of sea level in relation to a fixed point on the lithosphere) in the areas around the ice sheets. During the early parts of glacial periods, when ice may have been expanding else-

where on Earth, but before major ice sheet growth in Scotland, Scotland probably had relative sea levels lower than present. The growth of ice sheets on Scotland then produced crustal depression leading to high relative sea levels. The position of the shoreline has therefore fluctuated in a complex way. Shelf areas far from the ice sheet had low relative sea levels during glacial periods, whilst areas close to the ice sheet probably had low relative sea levels in early glacial times and high relative sea levels during the glacial maxima. These latter high levels led to marine deposition in areas currently far from the sea (cf. Fig. 15.23). Thus the marine Quaternary record depends on natural sections and boreholes on land and systematic seismic traverses controlled and supported by a network of shallow boreholes at sea. The stratigraphy is therefore a mixture of units based on seismic imaging, lithology and fossil faunal and floral evidence.

North Sea

The Quaternary stratigraphy for the central North Sea between 56°N and 58°N has recently been revised by Stoker *et al.* (1985) whose tentative conclusions are summarised in Figure 15.11. With the exception of the Holocene (see below), the majority of the North Sea fill is argillaceous, with or without isolated pebbles, thickening from a few metres near the coast to several hundred metres towards the east. Much of it was deposited in very cold water, partly under glaciomarine conditions, although undated warmer water microfaunas and floras at the base of the Ling Bank and Coalpit formations are assigned respectively to the Holsteinian (Hoxnian) and Eemian (Ipswichian) Interglacials. The Aberdeen Ground Formation, interpreted as delta front and prodelta/shallow water sediments (Stoker & Bent 1987), includes magnetic signatures correlated with the Brunhes – Matuyama boundary and the Jaramillo event (c. 930 ka) as well as micropalaeontological data suggesting that parts of it are of early Pleistocene age. A glacial episode in the Matayama reversed magnetic period, post-dating the Jaramillo event, has been reported, and a thick till at about the level of the Fisher Formation (otherwise thought to be glaciomarine; Stoker *et al.* 1985), has been attributed to the Saale of the continental sequence (Fig. 15.11) (Sejrup *et al.* 1987). Whilst this may reflect a time when the Scandinavian ice-sheet was coalescent with Scottish ice in the northern North Sea (Sejrup *et al.* 1987), it remains to be shown that the Fisher Formation includes till. It is not readily compatible with the apparent absence of widespread glaciation in England during the Saale/Wolstonian (cf. Rose 1988). The Ling Bank and Coalpit formations infill channels eroded into older deposits and a major erosion surface declining north-eastwards from about −60 m to −110 m is present at the base of the Marr Bank and Wee Bankie formation (Sutherland 1984a). The

Fig. 15.9. Bishopbriggs. Till from the last glaciation (3–4 m thick) resting on outwash gravels which elsewhere have yielded re-worked bones of woolly rhinoceros dated at 27,500 years before present. B.G.S. photograph.

Fig. 15.10. Glacial dispersion of erratics from distinctive bedrock sites in Scotland. Dispersal of Mesozoic sediments from the Moray Firth is also shown, as is the transport on-shore in northern Aberdeenshire of marine sediments derived from the western North Sea.

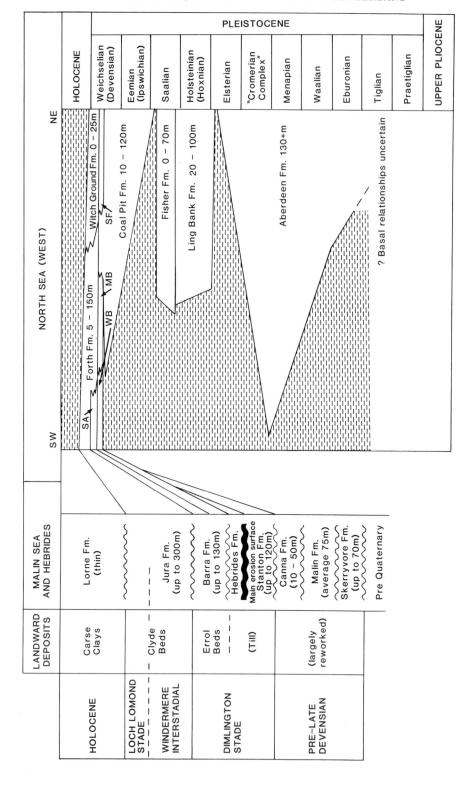

Fig. 15.11. Stratigraphy of Quaternary marine deposits in Scotland in the land area, in the Malin Sea and Sea of the Hebrides, and in the North Sea.

uppermost units are of Late Devensian and later age (Fig. 15.11).

Malin Sea and the Minches

Most of the Quaternary sediments, over 100 m thick in parts of the Malin Sea south of Skye, were deposited under glaciomarine, arctic conditions (Boulton *et al.* 1979, Davies *et al.* 1984). Four formations antedate the Late Devensian (Fig. 15.11). They rest on glaciated bedrock, but their age is uncertain. Davies *et al.* (1984) suggest that the two oldest are pre-Devensian, but that the Stanton Formation is of Early and/or Mid-Devensian age. The patchy distribution of Quaternary formations and the absence of the lower Pleistocene is probably the result of high bedrock relief (compare the low relief of the North Sea floor), repeated erosion in an area more heavily glaciated than further east and of tectonic uplift compared to the subsidence in the North Sea.

Hebrides shelf

Morainal bank complexes that occur near to the shelf edge south of St Kilda and off NW Scotland pass east-wards into glaciomarine strata (Selby 1989, Stoker 1988). The complex south of St Kilda is thought to mark the Late Devensian glacial limit (Selby 1989), but the more north-erly complex is attributed by Stoker (1988) to a period predating the last Scottish ice-sheet.

Late-Devensian glaciation

Landforms and sediments

Almost the whole of Scotland was overwhelmed by a great ice sheet during the Late Devensian. The landforms and sediments produced by it are ubiquitous on land and in the surrounding shallow seas. They occur within a larger scale erosional landscape that was moulded by at least 0·75 million years of repeated glaciation. Erosional landforms produced by this ice sheet are smoothed rock knobs (*roches moutonnées*, Fig. 15.12); striated and ice scoured surfaces (Fig. 15.12); and deep channels produced by meltwater draining from the ice sheet. Depositional features include streamlined, ice-parallel forms such as drumlins (Fig. 15.13) and flutes, and the drift "tails" lying on the lee side of many upstanding crags; moraines formed parallel to former ice margins (Fig. 15.14); kames and eskers formed of sediments deposited by meltwater flowing in contact with the ice and acquiring a characteristic form as the ice melted (Fig. 15.15); and the shorelines and deltas formed on the flanks of a number of Highland valleys from lakes damned-in by ice blocking the valley exit. Much of the low ground in the Central Lowlands and valley floors in hill country show considerable thicknesses of till, although some of the

Fig. 15.12. Glacier-scoured rock knobs of mica-schists (Eilde Flags) 4 miles south of Roy Bridge Station, Inverness-shire. They show striations cut by boulders embedded in the sliding sole of a glacier. B.G.S. photograph.

hard, resistant gneissic terrain of the north-west has produced little till. Most of the major valleys show exten-sive accumulations of sand and gravel (often associated with eskers and kames) laid down by outwash rivers from the advancing or decaying ice sheet (Fig. 15.15). Particu-larly impressive examples are to be found in the Teith Val-ley south of Callander, the upper Clyde Valley near Carstairs, the Spey Valley and the vicinity of Tyndrum.

Ice sheet build-up

Quaternary sea surface temperatures in the Rockall area inferred from foraminifer populations taken from a core on the Rockall Plateau (Ruddiman & Raymo 1988) prob-ably give the best indication of the progression of Scottish climate through the Devensian glacial stage. They show two major cold phases; one at about 70,000 years BP, and, after a warmer interstadial, a second more severe climatic phase at about 20,000 years BP (Fig. 15.1).

Scotland was probably glaciated during the early Devensian cold phase, but the extent of glaciation is

Fig. 15.13. Drumlins near New Galloway, Kirkcudbrightshire. Ice flow direction from top right to bottom left.

Fig. 15.14. Moraines east of Loch Tulla, Argyllshire. They formed successively at the margin of the Loch Lomond ice cap as it retreated, as an active glacier, from left to right. Photograph, Patricia Macdonald.

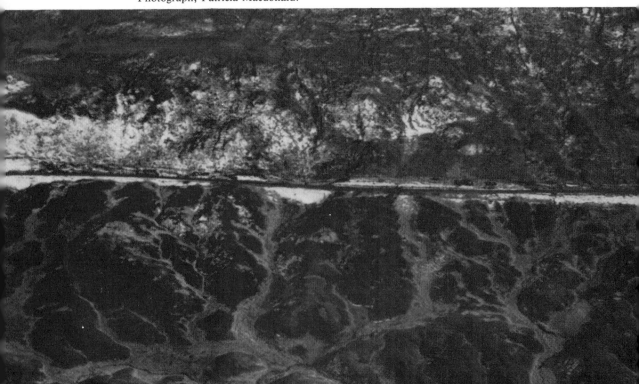

Fig. 15.15. An esker system at Gleneagles, Perthshire. Such ridges as these are often supposed to have formed from streams flowing beneath a glacier. However these examples probably formed from proglacial outwash beneath which dead glacier ice was buried. The major streams had incised through the ice, and when this finally melted, the major stream courses stood up as ridges. Photograph, Patricia Macdonald.

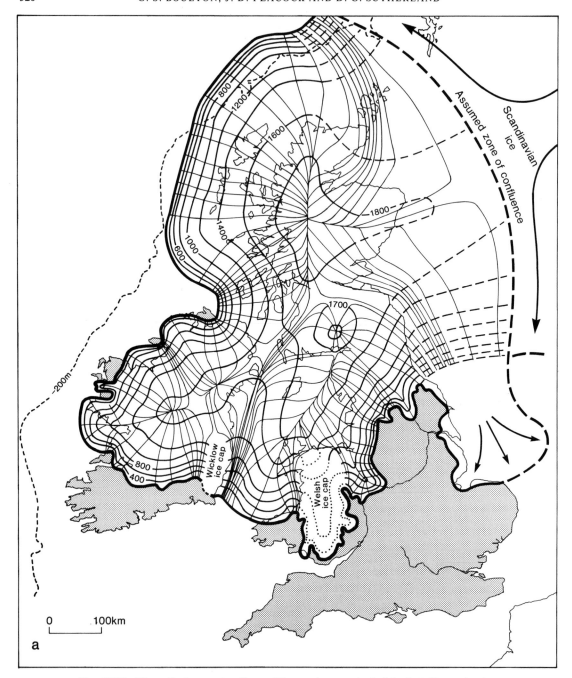

Fig. 15.16. Theoretical reconstructions of the maximum extent of the Late Devensian ice sheet over the British Isles. **a.** Model based on an ice sheet confluent with a Scandinavian ice sheet in the North Sea and using a 100 kPa basal shear stress. Contours (in metres) and flow lines are shown. No significant nunataks occur (Boulton *et al.* 1977).

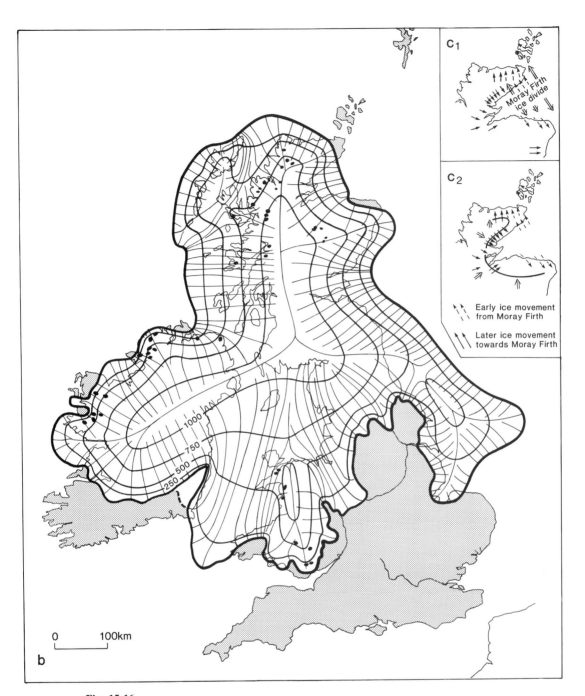

Within the figure:

C₁

Moray Firth ice divide

C₂

Early ice movement from Moray Firth

Later ice movement towards Moray Firth

1000

750

500

250

0   100km

b

Fig. 15.16.

b. Reconstruction based on limited ice sheet extent in the North Sea, a basal shear stress of 70 kPa on the land area and 30 kPa in the sea area. Principal nunatak areas are shown as dots.

c. Explanation of the evidence of last ice flow directions in the Moray Firth area. Evidence of landward flow in the outer Firth suggests an ice divide in the Firth (C_1), whilst seaward flow in the inner Firth (C_2) must have occurred after collapse of the ice dome. The change could reflect a Wolstonian/Late Devensian contrast (e.g. Sutherland 1984) or stages in decay of the Late Devensian ice sheet (e.g. Boulton *et al*. 1985).

Table 15.3. Radiometrically dated sediments closely predating the Late Devensian ice sheet

Site	Material	Lab. no.	Radiocarbon age (yr BP)
Sourlie	bone	SRR-3023	$29,900^{+430}_{-410}$
	plant debris	SRR-3146	$29,290 \pm 350$
	clay/silt	SRR-2147	$30,230 \pm 280$
	clay/silt	SRR-3148	$33,270 \pm 370$
Bishopbriggs	bone	GX-0597	$27,550^{+1,370}_{-1,680}$
Tolsta Head	peat	SRR-87	$27,333 \pm 240$
Garrabost	marine shell	SRR-2367	$26,300 \pm 320(o)$
			$23,000 \pm 230(i)$
Inchnadamph	bone	SRR-2103	$25,360^{+810}_{-740}$
	bone	SRR-2104	$24,590^{+790}_{-720}$
Assynt	speleothem	SUI-80A	$26,000 \pm 3,000$
Assynt	speleothem	AUZ-80	$30,000 \pm 4,000$
Assynt	speleothem	SUIZ-80B	$26,000 \pm 2,000$
Crossbrae Farm	peat	SRR-2401	$26,400 \pm 170$
			$22,380 \pm 250$

unknown. Sutherland (1981), Sissons (1982), Connell & Hall (1984) and Davies *et al.* (1984) have all presented data which can be interpreted in favour of an extensive early Devensian ice sheet, but unequivocal evidence is still lacking. A large part of Scotland was ice-free towards the end of the Middle Devensian, as is shown by sites where radiometrically dated organic materials underlie till (Table 15.3). The absence of dates younger than 25,000 BP on the mainland and 23,000 BP in the Outer Hebrides may indicate expansion of the ice sheet by these dates.

Along the northern margin of the Southern Uplands there are a number of sites where a lower till containing Highland erratics is overlain by an upper till with Southern Upland erratics (Kirby 1969, Sutherland 1984). They suggest that the growing ice sheet expanded initially most rapidly from Highland centres and that only later was there major growth of an ice centre on the Southern Uplands large enough to fend off Highland ice. The northern limit of Southern Upland erratics (Fig. 15.10) probably indicates the location of a zone of confluence with Highland ice at the maximum of glaciation.

The ice sheet had at least a third centre of outflow from domes located over the Outer Hebrides (Fig. 15.16). Striae and landforms parallel to former ice flow directions indicate a pattern of glacial radiation from Southern Lewis, suggesting that an ice dome was located over the area. Another dome probably lay over South Uist and Barra. These probably existed at the maximum of glaciation as evidence from the islands suggests that mainland ice may not have reached the Outer Hebrides during the Late Devensian.

The ice-sheet at its maximum

The time of maximum thickness and extent of the ice sheet over Scotland is not precisely known, but is presumed to have coincided with the coldest phase of the last glacial period at about 18,000–20,000 BP. Although the extent of the ice sheet further south in Britain and Ireland is reasonably well-documented, its extent around Scotland is poorly known (Fig. 15.17). It has been suggested that the Wee Bankie Formation, 60 km east of the mouths of the Forth and Tay estuaries, is a moraine laid down at the maximum Late Devensian extent of Scottish ice (Thomson 1978, Sutherland 1984, Stoker *et al.* 1985). This is consistent with the studies of Sejrup *et al.* (1987) who were not able to find evidence of coalescence of Late Devensian, British and Scandinavian ice in the Fladden Ground area of the central North Sea. A limited Devensian ice sheet extent in north-east Scotland is supported by the suggestions (e.g. Sutherland 1984, Hall 1984) that the southern shore of the Moray Firth was not covered by the Late Devensian ice sheet, that the shelly grey tills in eastern Caithness with Mesozoic erratics derived from the Moray Firth are of Wolstonian age, and that the line dividing these from tills with inland erratics to the west marked the easterly limit of the Late Devensian ice sheet (Sutherland 1984). However, Hall & Whittington (1989) have shown that the Moray Firth-derived tills in Caithness

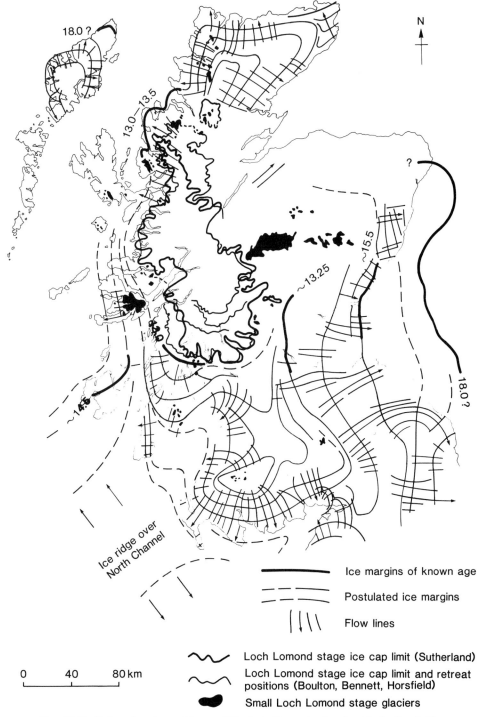

Fig. 15.17. Inferred pattern of decay of the Late Devensian ice sheet in Scotland. Heavy lines show ice margin position for which there is evidence of age in 1,000s of years. North-easterly trending features in N.E. Caithness may reflect rapid decay of an ice dome over Moray Firth, or may pre-date the Late Devensian ice sheet. Two versions of the extent and decay pattern of the Loch Lomond readvance ice sheet are given.

overlie the inland tills, and have argued that both are of Late Devensian age. If this is so, there must at least have been a major ice dome over the Moray Firth, and it is difficult to understand how its southern shore, and extensive areas of the Northern North Sea, could have remained unglaciated (Fig. 15.16).

Further to the north, the last direction of ice movement over the Orkney Islands was to the north-west (Mykura 1976), and although its age is unknown it may have been coeval with the north-westerly ice flow over eastern Caithness. In Shetland, the most recent glacier cover took the form of an ice dome on the main island (Mykura 1976). The presence of till overlying a presumed Ipswichian peat (Birks & Peglar 1979) suggest that the glacial phase was of Devensian, possibly Late Devensian age. Peacock (1984), Sutherland & Walker (1984) and von Weymarn (1979) have argued that the ice sheet in the Outer Hebrides did not cover the extreme north of Lewis, and Selby (1989) has suggested that the Outer Hebridean dome of the ice sheet extended near to the shelf edge south of St Kilda (see above).

The surface level of the ice sheet at its maximum extent may be definable in Easter and Wester Ross, where Ballantyne (in Ballantyne & Sutherland, 1987) has demonstrated the existence of a prominent contrast between the uppermost parts of some of the highest peaks which are mantled by thick frost-weathered *in situ* detritus, and lower slopes which show glacially scoured bedrock. The transition between the two is abrupt, occurring over a few tens of metres. Ballantyne suggests that this 'trim-line' marked the highest extent of the Late Devensian ice sheet surface. A glaciological model (Boulton *et al*. 1985) can be fitted to the available evidence of ice sheet extent, using a range of basal shear stresses (30–70 kilopascals) similar to those of modern glaciers in Iceland and Southern Greenland (Fig. 15.16). The modelled ice sheet at its maximum shows a prominent N–S ridge in the Highlands, a dome over the Western Isles, and a dome over the Southern Uplands which is connected by a ridge and saddle to Ireland. If the Moray Firth were an unglaciated embayment as some have claimed, the north-eastern side of the ice sheet would need to have been very cold and dry to ensure that a location near the centres of accumulation remained unglaciated whilst the ice sheet extended far to the south. An alternative model with a dome over Moray Firth and the Northern North Sea has also been constructed, which would be compatible with north-westward Late Devensian ice flow over Caithness and Orkney.

Deglaciation and the pattern of deglacial sea level change

The pattern of Late Devensian deglaciation of Scotland is poorly defined by the disposition of the few moraines which formed during halt or by readvances during the general ice sheet retreat. This evidence can be augmented by assuming that features such as drumlins and eskers, and most striae on bare rock surfaces, formed parallel to glacier flow immediately prior to retreat of the glacier from the area in which they are found, and thus that in areas of low relief, the trend of the retreating glacier margin was at right angles to the trend of these features. Using this approach (Boulton *et al*. 1985), it is possible to construct trend lines on successive glacier margins (Fig. 15.17). These trend lines can be translated into dated isochrons of deglaciation by considering the pattern and timing of Late-Glacial sea level change which plays such an important role in the Scottish Quaternary record.

Isostatic depression of the lithosphere by the ice-sheet load was sufficiently great to ensure that although global, eustatic sea levels were low during deglaciation, local relative sea levels around the Scottish ice-sheet were high. Figure 5.18 shows a series of relative sea level curves from points around the Scottish coast. Figure 15.19 shows the contemporary global eustatic sea level change (Fairbanks 1989) together with relative sea level change for the Clyde Sea area. The difference between the two (ignoring geoid changes) shows approximately the change in isostatic level of the crust compared to its present level. Using these relationships, and data from coastal sites, the pattern of relative sea level down the west and east coasts of Scotland has been plotted (Fig. 15.20) together with the rebound of the crust during and after deglaciation (Fig. 15.21). Features which formed at sea level have now been uplifted by postglacial isostatic rebound, and Figure 15.22 shows the current elevation of features which formed at sea level at the times shown (isobases). The sea invaded the areas progressively exposed by the retreating glacier. Shoreline features dating from the time of deglaciation were then elevated by isostatic uplift, so that the highest elevation of the sea (the marine limit) tends to date from the time of deglaciation. The marine limit is thus diachronous, intersects isobases, which are isochronous, and rises in the direction of glacier retreat (Fig. 15.22).

In eastern Scotland, the Errol Beds occur within the area of strong glacio-isostatic depression and later uplift (Peacock 1981) (Fig. 15.23). They are dominantly reddish-brown clays, silts and sands, sometimes laminated, with scattered ice-rafted pebbles and boulders (Paterson *et al*. 1981), and contain an arctic marine fauna. On land they are rarely more than a few metres thick, although several tens of metres have been recorded in deep channels and estuaries. The Errol beds are nearshore equivalents of the St Abbs Formation in the western North Sea and the Barra and Hebrides Formations in the western sea lochs and the sea of the Hebrides (Fig. 15.11). They probably formed between about 18,000 and 13,000 BP. Glaciomarine silts at St Fergus, near Peterhead, dated at 15,320±200 BP (Hall & Jarvis 1989), which contain an arctic molluscan fauna, are probably their lateral equivalent. They represent glaciomarine sediments

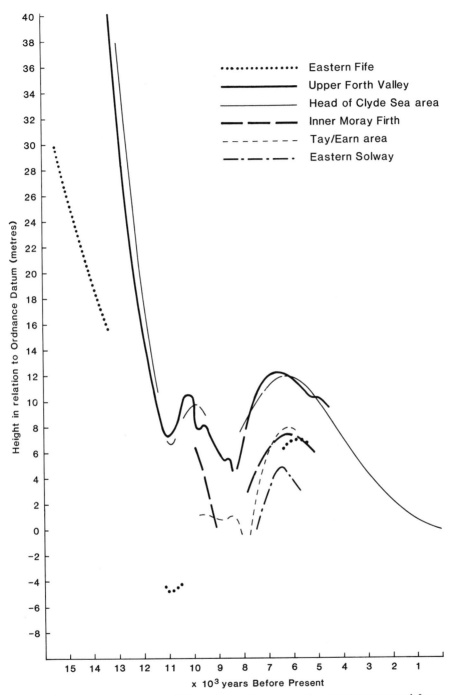

Fig. 15.18. Late Devensian and Flandrian relative sea level curves reconstructed from geological evidence from a number of areas in Scotland.

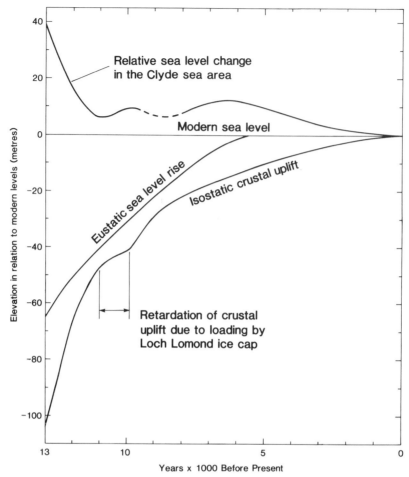

Fig. 15.19. Late Devensian and Flandrian relative sea levels and isostatic uplift in the Clyde Sea area. Relative sea level change, which is the local movement of sea level relative to a fixed point on the lithosphere surface, is a product of the interaction between global glacio-eustatic sea level and local glacio-isostatic crustal uplift. When the crustal uplift rate exceeds the rate of eustatic sea level rise, relative sea level falls. Retardation of crustal uplift by Loch Lomond ice cap expansion produces a sea level transgression just before 10,000 years BP.

formed in a sea area occupying a glacio-isostatic depression into which icebergs calved from the retreating ice mass (Figs. 15.21–15.23).

Rebound of the isostatically depressed area around the retreating ice sheet uplifted and warped beaches and other coastline features (Figs. 15.21, 15.22). The most detailed sequence of such shorelines occurs in SE Scotland (Sissons *et al.* 1966, Sissons 1983) and an approximate chronology for the ice retreat can be established on the basis of a simple model relating shoreline gradient to age (Andrews & Dugdale 1970; Sissons 1976, p. 120). Such a calculation suggests that the five shorelines identified by

Cullingford & Smith (1980) along the east cost of Central Scotland formed progressively between around 16,000 yr BP and about 14,000 yr BP. There was relatively little change in the position of the ice margin during a period of considerable change of shoreline gradient as the ice retreated through part of eastern Fife and along the Strathmore coast and hence the margin may be placed in this area at around 15,500 yr BP with some confidence (Fig. 15.17). These shorelines (gradients, 0·94 m/km to 0·54 m/km) and those formed up to the time of the Main Perth Shoreline (gradient, 0·43 m/km) (Sissons & Smith 1965, Cullingford 1977) were contemporaneous with the

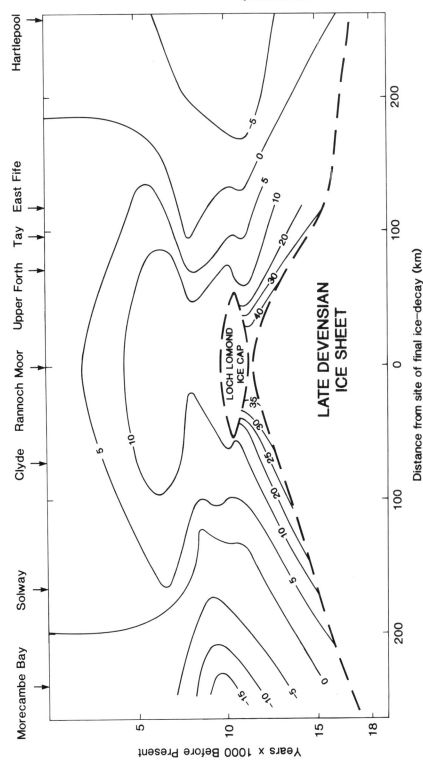

Fig. 15.20. Late Devensian and Flandrian relative sea level change along west and east coast transects. Numbered contours show sea level in relation to modern levels. The isostatic component of sea level change is more important nearer to the centres of ice growth. Falling sea level prior to 10 ka reflects the dominance of isostatic uplift during this period. Loch Lomond ice cap expansion and associated local uplift retardation produces a local transgression. The major transgression at about 7 ka reflects a more rapid rate of eustatic sea level rise (Fig. 15.19) and the subsequent regression in the area of former ice centres reflects residual slow crustal uplift after the cessation of glacio-eustatic sea level rise.

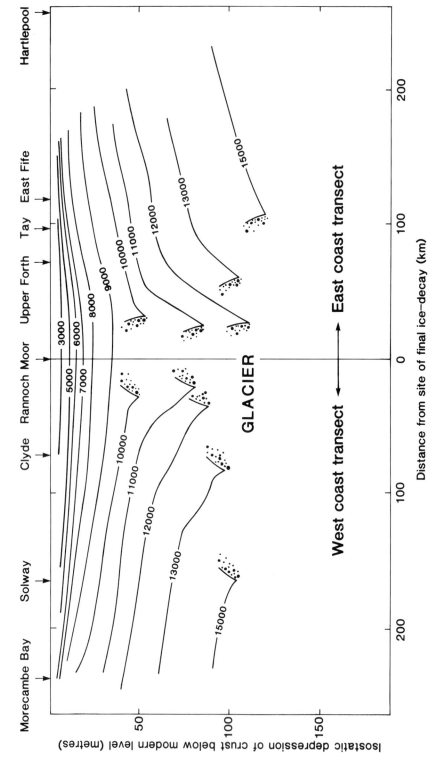

Fig. 15.21. Progressive crustal uplift along west and east coast transects as a consequence of Late Devensian ice sheet decay. This is reconstructed by substracting the eustatic sea level rise (Fig. 15.19) from the relative sea level record (Fig. 15.20).

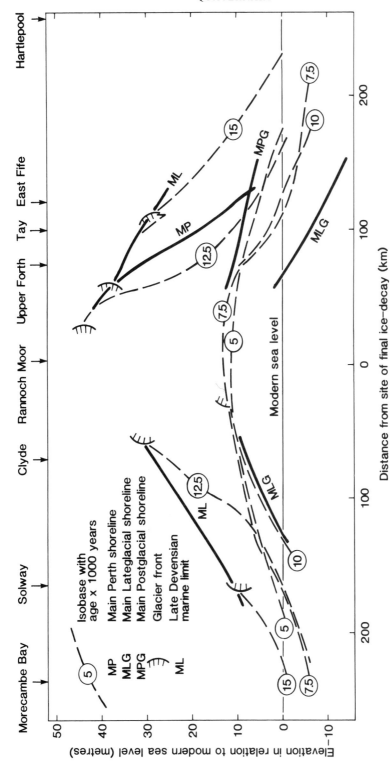

Fig. 15.22. Late Devensian and Flandrian isobases along west and east coast transects. They represent the present elevation of surfaces which lay at sea level at the times shown, and reflect the magnitude and pattern of crustal uplift since that time. The marine limit shows the highest point to which these features extend.

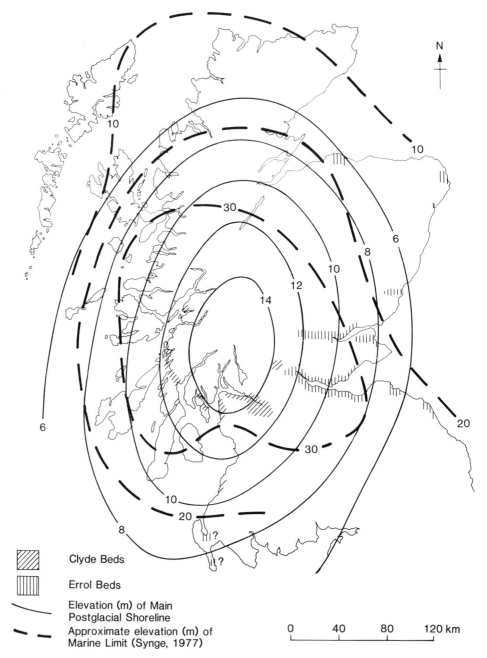

Fig. 15.23. Elevation of presumed isobases on the Main Postglacial Shoreline, and the elevation of the marine limit. Glaciomarine Clyde and Errol Beds, which probably formed more than 10–20 m below sea level, are now found above sea level where isostatic uplift has been greatest.

Errol Beds, and their offshore equivalent, the St Abbs Formation.

The formation of the Main Perth Shoreline and the end of the deposition of the Errol Beds occurred at around 13,000 yr BP. In places their inland limit can be related to outwash terraces or dead-ice terrain. In east central Scotland this inland limit approximately indicates the position of the ice at or slightly prior to 13,000 yr BP (Figs. 15.17, 15.23). In those areas where extensive searches have failed to find the Errol Beds (for example, much of the Clyde Sea area, the Cromarty Firth) it may be argued that there was ice-cover at the time of Errol Beds deposition. The shoreline sequences on the west coast are less clear than on the east but a tentative calculation of shoreline gradient against age can be made for the area of the South West Highlands and neighbouring islands. This suggests that the shoreline L1 of Dawson (1982) was formed at approximately 14,500 yr BP implying that parts of western Jura and NW Islay were deglaciated by this time.

For much of the deglacial phase, present evidence suggests continuous ice-sheet retreat around the coasts and through much of the lowland areas. Earlier postulated major readvances such as the Aberdeen–Lammermuir (Sissons 1967) and Perth (Simpson 1933; Sissons 1963, 1967) have been rejected (Paterson 1974, Sissons 1974) and there is no good morphological or stratigraphical evidence at present for a major readvance of the ice sheet in the lowland areas. In Wester Ross, however, there is an extensive end-moraine system (Robinson & Ballantyne 1979, Sissons & Dawson 1981). This end-moraine, though not dated directly, pre-dates the radiocarbon date of 12,800±155 yr BP at Loch Droma (Kirk & Godwin 1963) and may be tentatively dated to 13,000–13,500 yr BP. Possible correlative end moraines may also occur in Easter Ross (Sissons 1982, Sutherland 1984, Ballantyne et al. 1987).

The Clyde Sea area may have been covered by an ice dome centred over Arran at a late stage of the glaciation with ice flowing to the east across Ayrshire and to the west across the Kintyre peninsula. A minor local readvance of this ice is indicated by a distinct till unit in central Ayrshire. Deglaciation of the Clyde Sea area was followed by deposition of the Clyde Beds which are chiefly greyish brown clayey silts and sands with pebbles rafted by sea ice (Fig. 15.24). Like the Errol Beds these are thin onshore, but may be many metres thick in estuaries. Their deposition commenced by about 13,000–12,800 BP (Peacock & Harkness 1990, Sutherland 1986) as glaciers retreated to within the mouths of the sea lochs in response to the sharp climatic amelioration at this time. Rapid deglaciation of the Clyde Sea region appears to have taken place as a consequence of calving of icebergs as the ice front receded into the deeper waters of the inner sea area (Sutherland 1984). Elsewhere along the coast stillstands during ice-sheet retreat have been inferred from breaks in the marine limit near the mouths of sea lochs (Peacock 1970, Suther-

land 1981), and in the SW Highlands such a stillstand or minor readvance has been dated to approximately 13,000 yr BP (Sutherland 1981). It is possible that these breaks in the marine limit relate to topographic control on ice flow near the mouths of sea lochs (Sutherland 1984).

Windermere interstadial

At around 13,000 yr BP, conditions of arctic severity were rapidly replaced by a milder climate as the oceanic polar front in the North Atlantic migrated to the north of Scotland (Ruddiman & MacIntyre 1973) and North Atlantic Drift waters reached the Scottish shores (Peacock & Harkness 1990). Atmospheric temperatures, as inferred from fossil coleopteran assemblages, rose dramatically to close to present-day values (Bishop & Coope 1977, Atkinson et al. 1987). Vegetation responded slowly to the climatic change and the plant communities of this period were characterised by open-habitat taxa typical of pioneer vegetation on recently deglaciated terrain (Walker 1984, Pennington et al. 1972). As the interstadial progressed, plant succession led to the development of a closed vegetation cover throughout the lowlands and in the Highland valleys.

There was considerable diversity in the interstadial vegetational communities which reached their greatest complexity following a brief phase of climatic deterioration which may be correlated with the Older Dryas chronozone of the North West European sequence (Pennington 1975, Mangerud et al. 1974). The vegetation zones that developed during the main part of the interstadial occupied similar geographical areas to the principal vegetation zones that developed under the different climatic conditions of the middle Flandrian (Birks 1977) and which have been found in modern relict vegetation. Most of the country was covered by grasslands or dwarf-shrub heaths and tree development was limited to birch and pine in favoured localities (Gray & Lowe 1977). Tree birch had a northerly limit on the west coast in southern Skye (Birks 1973) but also occurred on the east coast in Aberdeenshire and possibly even Caithness (Peglar 1979) whilst pine has been recorded only locally in the eastern Central Lowlands and Aberdeenshire.

The vegetational succession during the early to middle insterstadial suggests that the climate became milder as the interstadial progressed. As with the opening of the interstadial, however, faunal evidence on land indicates otherwise and in the coleopteran record the main part of the interstadial has been found to be 2–3°C cooler than its opening phase at around 13,000 yr BP (Atkinson et al. 1987). Coope (1977) has suggested that temperatures fell off rapidly northwards across Scotland.

The marine fauna of the interstadial interval recorded in the Clyde Beds (the Windermere or Late-glacial Interstadial) is of high-boreal to low-arctic affinities and

12·9-13·4 m   13·4-13·9 m   13·9-14·4 m   14·4-14·9 m

Fig. 15.24. Core from the British Geological Survey borehole at Killearn, Strathblane, showing varved lacustrine sediments from between 12·92 and 14·9 m below surface, dating from a time when Loch Lomond Stadial glaciers dammed up a lake at the head of Strathblane. Scale–ruler 45·5 cm. They are underlain by glaciomarine sediments of the Clyde Beds at a depth of 46·3 to 47·75 m, which were deposited in high relative sea levels (due to glacio-isostatic depression of the crust) as the Late Weichselian ice sheet was retreating from the area. These are in turn underlain by till deposited by the ice sheet. B.G.S. photograph.

appears analogous to similar but less well documented assemblages which occur throughout the Quaternary. It is characterised on the one hand by several pan-arctic species such as *Yoldiella lenticula* and *Macoma calcarea* (mollusca), *Elofsonella concinna* and *Eucytheridea bradii* (ostracods) and *Elphidium clavatum* (forminiferida) and on the other by boreal, Atlantic molluscs (e.g. *Arctica islandica*, *Modiolus modiolus* and *Corbula gibba*) which extend to northern Norway today. The presence of such assemblages around the Scottish coast suggests colder water than at present (approximately 3°C during the Windermere Interstadial), a relatively weak North Atlantic Drift off Scotland (Peacock 1983) and normal marine rather than glaciomarine conditions (Peacock 1989). Faunas indicating slightly more temperate conditions have been recorded early in the Windermere Interstadial and also near 11,000 BP (Peacock & Harkness 1990).

The extent of deglaciation during the Lateglacial Interstadial has not been established because of the subsequent Loch Lomond Readvance. It can be argued that since there appears to have been a large ice mass present in the western Highlands at the time of the warmest climate at the opening of the interstadial, then some ice would have survived this brief period into the cooler late interstadial when conditions were suitable for glaciation of the western mountain areas. Thus complete deglaciation prior to the Loch Lomond Readvance may not have occurred (cf. Sissons 1976). Complete deglaciation of the western sea lochs is indicated by the distribution of fossiliferous Clyde Beds within the lochs as well as the environment of deposition of these sediments.

During the early interstadial, sea level around the coasts most affected by glacioisostasy was falling rapidly (e.g. Peacock *et al.* 1977, 1978) to levels below those subsequently attained by the sea during the Main Postglacial Transgression in the middle Flandrian (Figs. 15.18, 15.20). There is thus little direct evidence for the position of the shoreline during the middle to late interstadial but indirect fossil micro- and macrofaunal studies indicate that in the Highlands fringes sea level was relatively stable at altitudes within a few metres of its present level. In the outer islands and other areas peripheral to the centre of isostatic uplift sea level throughout the interstadial was below its present level.

Loch Lomond Stadial

At around 11,000 yr BP, the climatic decline that had become apparent on land in the latter part of the interstadial intensified and there was a return to conditions of arctic severity for approximately 1,000 years during the Loch Lomond Stadial. Mean summer temperatures during the stadial are believed to have been 7° to 9°C below those of the present (Sissons 1974, 1980; Sissons & Sutherland 1976; Bishop & Coope 1977). The oceanic polar front migrated to the south of the British Isles (Ruddiman & McIntyre 1973) and marine sediments with a restricted high arctic marine fauna were yet again deposited around the coasts (Peacock *et al.* 1978).

The Loch Lomond Stadial is a particularly interesting climatic phase in that although the tempo of Quaternary climatic change on a timescale of 10^4–10^5 years appears to be forced by insolation changes, the stadial coincides with a major insolation maximum in the northern hemisphere. It suggests a strongly non-linear environmental response to insolation forcing, and may be associated with changes in North Atlantic ocean circulation induced by melting of the North American ice sheet (Broecker *et al.* 1989).

The reality of a major Lateglacial phase of glacier growth, the Loch Lomond Readvance, has been established widely throughout Scotland by the mapping of moraine sequences (Sissons *et al.* 1974; Sissons 1979, 1983) and from stratigraphic evidence from the South West Highlands and Mull where sediments overlain by or incorporated into glacial deposits associated with the advance have been radiocarbon dated (Table 15.4), as have proglacial deposits characterised by marine faunas or terrestrial floras typical of severe climates (Peacock *et al.* 1989; Tipping 1986, 1989). These studies indicate that in the South West Highlands the readvance culminated after 10,500 yr BP. This chronology agrees with the age of the High Buried Shoreline of the Forth Valley, inferred from a shoreline gradient against age model, at approximately 10,200 yr BP. The High Buried Shoreline is contemporaneous with the Menteith Moraine formed at the maximum of the readvance (Sissons 1966, 1976).

In the Grampians, major terminal moraines were produced along many parts of the limit of the Readvance. Within the area occupied by the ice cap there are broad areas of 'hummocky moraine', often regarded as evidence of stagnation and rapid decay of inactive ice. However, Horsfield (1983) has demonstrated that the hummocky moraine frequently consists of linear elements analogous to modern moraines formed at actively retreating ice margins (Fig. 15.14). The moraines reflect retreat of an active ice margin at a rate not dissimilar to modern glacier retreat rates in southern Iceland. Many mountain areas which lay outside the Loch Lomond ice cap, had their corries occupied by small glaciers during the stadial (Sissons 1979) (Fig. 15.6).

The local re-loading of the lithosphere by the Loch Lomond ice cap appears to have retarded the rate of uplift (Fig. 15.21) so that it was exceeded by the rate of eustatic sea level rise, thereby producing a transgression near to the ice cap (Figs. 15.19, 15.20). Further from the ice cap (Fig. 15.20), where the loading effect was much less, the isostatic uplift rate remained greater than the rate of eustatic sea level rise, resulting in continuous lowering of relative sea level.

Table 15.4. Radiocarbon-dated materials overlain by or within glacial sediments of the Loch Lomond Readvance

Site	Ref.	Material Lab. no.	Age
Drymen	Sissons (1967b)	shells I-2235	$11,700 \pm 170$
Menteith	Sissons (1967b)	shells I-2234	$11,800 \pm 170$
Kinlochspelve	Gray & Brooks (1972)	shells I-5308	$11,330 \pm 170$
South Shian	Peacock (1971)	shells IGS-C14/16	$11,530 \pm 210$
South Shian	Peacock (1971)	shells IGS-C14/17	$11,805 \pm 180$
South Shian	Peacock (1971)	shells IGS-C14/18	$11,430 \pm 220$
Rhu	Rose (1980)	shells HAR-931	$11,110 \pm 250$
Loch Goil	Sutherland (1981)	shells T-1456	$12,260 \pm 150$
Balloch	Browne & Graham (1981)	shells SRR-1530	$10,920 \pm 140$
Helensburgh	Browne et al. (1983)	shells SRR-2006	$11,710 \pm 60$
Balure	Peacock et al. (1989)	shells SRR-3182	$10,105 \pm 100(o)$ $9,915 \pm 80(i)$
Balure	Peacock et al. (1989)	shells SRR-3204	$10,145 \pm 110$
Balure	Peacock et al. (1989)	shells OxA-1345	$10,555 \pm 130$
Croftaine	Rose et al. (1988)	organic Q-2673 detritus	$10,560 \pm 160$

The retreat and final disappearance of the glaciers of the Loch Lomond Readvance have been studied by examination of pollen sequences in sediments deposited immediately after deglaciation. Successively later pollen zones which reflect progressive deglaciation have been recorded at the base of such sequences in certain areas (Lowe & Walker 1981, Walker & Lowe 1981) but not at others (Tipping 1988) whilst the associated radiocarbon dating of the basal sediments or the relevant pollen horizons has been unsatisfactory (Walker & Lowe 1980, Sutherland 1980, Tipping 1987). Moreover, Amman et al. (1989) have presented evidence which suggests that the natural atmospheric CO_2 concentration decreased during the latter part of the Loch Lomond stadial and that a range of true ages prior to 10,000 BP will tend to show the same radiocarbon age of about 10,000 BP. For these reasons the deglaciation chronology is poorly known but it seems probable that complete deglaciation was achieved prior to the early Flandrian *Juniperus* maximum, which occurred at about 9,600 yr BP.

The vegetation that developed during the stadial was characterised by species typical of broken ground and open tundra although there appear to be no direct modern analogues for many of the plant communities (Walker 1984). Particularly notable have been variations in the abundance of pollen of species of *Artemisia* in stadial sediments. Geographically, very high values have been found in the eastern and central Grampians with declining values to the west and south (Birks & Mathewes 1978, Macpherson 1980, Tipping 1985). These variations have been interpreted as a direct reflection of the precipitation pattern during the stadial (Walker 1975, Birks & Matthewes 1978, Macpherson 1980) as certain species of

Artemisia are chionophobes and likely to flourish in areas of lower snowfall. The reduced levels of precipitation inferred from such data for the eastern and central Grampians correspond closely with the precipitation pattern deduced from glacier distribution during the stadial (Sissons 1980). Temporally, certain pollen profiles also show changes in *Artemisia* pollen values as the stadial progressed, with an early phase of low frequencies followed by a later drier period during the stadial (Macpherson 1980, Caseldine 1980, Tipping 1985).

During the stadial, slope processes were particularly active resulting in the general destruction of interstadial soil profiles as well as local burial by slumping or solifluction of interstadial peats and organic sediments (Donner 1957, Dickson et al. 1976, Clapperton & Sugden 1977). In lakes and enclosed basins the stadial is marked by a distinct horizon of minerogenic sediments frequently containing low concentrations of organic detritus derived from the interstadial soils (Pennington 1977). In the lowlands, river activity was enhanced and large alluvial fans deposited, as in the Lochwinnoch Gap, at the foot of the Ochil Hills and at Corstorphine in Edinburgh (Kemp 1971; Newey 1970; Sissons 1976, 1979).

Nearshore erosion during the stadial was severe. This is most clearly demonstrated in the Forth Valley where a distinct erosion surface has been identified over an area in excess of 28 km^2, cutting across Lateglacial marine sediments, till and bedrock, and directly overlain by early Flandrian sediments (Sissons 1969, 1976b). The inner margin of this surface defines the Main Lateglacial Shoreline (Fig. 15.22) which is isostatically tilted eastwards with a gradient of 0·17 k/km. Around the coast of the Southwest Highlands and neighbouring islands is another erosional shoreline isostatically tilted westwards, the Main Rock Platform, which has a similar gradient to the Main Lateglacial Shoreline (Gray 1974, 1978; Dawson 1980, 1988; Sutherland 1984). The similarities between the two features led Sissons (1974b) to conclude that they were both produced during the stadial, an hypothesis that received support from the evidence for rapid littoral erosion of bedrock around the Glen Roy ice-dammed lakes (Sissons 1978). Uranium-series dating of speleothem material overlying the Main Rock Platform (Gray & Ivanovich 1988) has suggested, however, that the platform may have been inherited in part from a pre-existing feature or features (Sutherland 1981, Browne & McMillan 1984).

Flandrian Interglacial

The climatic amelioration at the end of the Loch Lomond Stadial was abrupt and the rate of temperature rise may have been as great as 1°C per decade (Coope 1977, Atkinson et al. 1987). Offshore water temperatures may have risen from near zero to a little below modern values within a few decades (Peacock & Harkness 1990). The change in climate, dated to approximately 10,000 yr BP, coincided with the return of North Atlantic Drift waters to the Scottish coasts. The climatic ameliorations at the beginnings of both the Flandrian and the Windermere Interstadial were both of a similar magnitude and rapidity and resulted from the retreat of the oceanic polar front in the North Atlantic to the north of Scotland. At both times. the vegatation inherited from the previous cold periods was dominated by open-habitat taxa, these being succeeded by shrub and scrub vegetation. In contrast to the interstadial, however, when the initial mild phase only lasted a few hundred years, to be followed by a period in which temperatures were somewhat lower than at present, the mild climate of the early Flandrian was maintained, giving rise to a sequence of vegetational changes quite different from those of the interstadial.

The dwarf shrub and scrub phase saw successive dominance in much of the country of *Empetrum* and then *Juniperus* (this being particularly marked) with subsequent expansion of mixed deciduous woodland of birch and hazel. The arrival of the woodland species was earliest in the south of the country with tree birch being established there shortly after 10,000 yr BP and migrating northwards during the succeeding 1,000 years. Hazel arrived slightly later, spreading through the south and central lowlands by 9,300 yr BP and along the western seaboard and into the central Grampians prior to 8,500 yr BP. By this time the birch–hazel woodland in the southern lowlands and western Highland borders was being replaced by woodlands dominated by oak with some elm. In the Central and North-west Highlands, there was little development of mixed deciduous forest during the

early or middle Flandrian but rather of pine and birch woodland (Pennington et al. 1972, Birks 1977), as may also have occurred briefly in the hills of south-west Scotland (Birks 1989). The period of expansion of the pine forests was quite distinct in separate regions with an earlier development, at around 8,300 yr BP, in parts of the north-west, compared to the Central Highlands where pine expanded between 7,800 and 6,500 yr BP and southwest Scotland where it invaded from Ireland by 7,000 BP (Birks 1989) (Fig. 15.25). The reasons for this difference are not yet understood but as the pine in the north-west may be genetically distinct from that further south (Kinloch et al. 1986), it may have migrated from a different source area.

Throughout much of the country alder expanded into the forests during the middle Flandrian, appearing in the south at around 7,000 yr BP and achieving its maximum extent by around 6,500 yr BP. By then the Scottish forests had reached their greatest extent and diversity and the subsequent change of climate to a cooler, moister regime together with the impact of man, particularly around 5,000 yr BP onwards, resulted in forest decline and replacement by blanket peat, heaths and grasslands.

The milder and less stormy climate of the Flandrian compared to the preceding Loch Lomond Stadial, together with the corresponding development of vegetation and soil cover, resulted in a marked diminution of geomorphological activity compared to the earlier period. However, the glacial legacy of large volumes of unconsolidated debris, oversteepened slopes and disrupted drainage systems meant that the early Flandrian was a period of adjustment to the new conditions. There were a considerable number of rock slope failures (Holmes 1984) as well as extensive reworking by streams and flows of the available debris, producing river terraces and debris cones in many Highland valleys (Brazier et al. 1988). In contrast, the middle to late Flandrian was relatively quiescent. However, the last 500 years have witnessed a major change in the rates of geomorphological processes with increased fluvial as well as renewed mountain-top periglacial activity (Innes 1983, Brazier et al. 1988, Ballantyne 1986, Ballantyne & Whittington 1987). This recent resurgence may partly be due to the 'Little Ice Age' climatic deterioration but increased grazing pressure on the upland areas is an equally probable cause.

Around the coasts have been major changes in the position of the shoreline during the Flandrian due to the complex interplay of glacio-isostatic and eustatic sea-level movements (Figs. 15.19, 15.20). The differences in sea-level change experienced from place to place are primarily a function of distance from the centre of glacio-isostatic uplift in the western Highlands (Fig. 15.23). Thus in the Outer Isles where isostatic effects have been least, there has been a net sea-level rise during the Flandrian whilst around much of the mainland coasts there has been a net fall, although the pattern of change has, generally, been

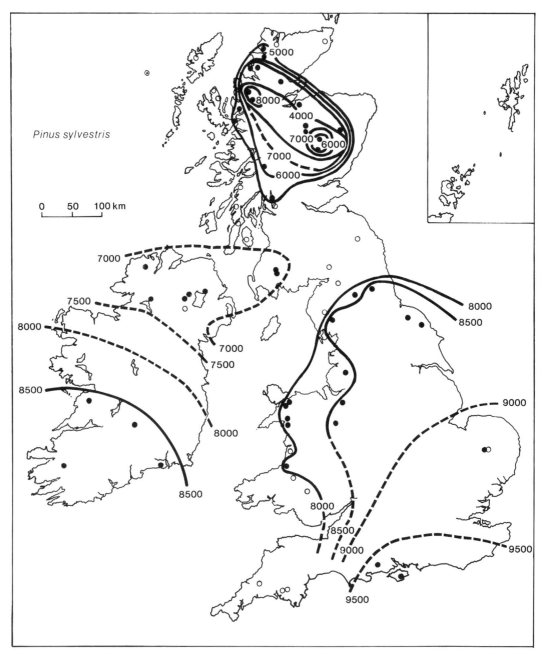

Fig. 15.25. Invasion of Scots Pine into Britain during current interglacial (Flandrian). Isochrons in years before present. Note the invasion of south-west Scotland from Ireland at about 7,000 BP and the spread from centres in the Highlands after 8,000 BP. (Redrawn from Birks 1989.)

complex with periods of sea-level fall during the early and late Flandrian being separated by a major transgression that reached its climax in the middle Flandrian (Fig. 15.19).

The major transgression, termed the Main Postglacial Transgression, was the result of the global sea-level rise consequent upon the final melting of the North American ice sheet. The maximum of the transgression was reached at different times in different parts of the country, being earliest in the western Forth Valley at around 6,800 yr BP (Sissons & Brooks 1971, Sissons 1983b) and generally later in areas further from the centre of isostatic uplift (Smith *et al.* 1983). A particularly prominent depositional shoreline, the Main Postglacial Shoreline, was formed at the maximum of the readvance. Isostatic tilting of that shoreline has given it a gradient of between 0·05 m/km and 0·08 m/km.

The littoral sedimentary sequences deposited in response to these movements of the shoreline are best developed in the estuaries of the east and south-west coasts. Here, silts and fine sands deposited during periods of marine influence are interbedded with terrestrial peats and fluvial sands and gravels (Sissons *et al.* 1966, Jardine 1980). Along the east coast a prominent fine-sand horizon that is present in both estuarine and near-shore terrestrial sediments was deposited at around 7,000 yr BP in response to either a major storm surge in the North Sea basin (Smith *et al.* 1985) or a tsunami resulting from a major submarine slide on the Norwegian continental slope (Dawson *et al.* 1988, Long *et al.* 1989).

REFERENCES

AMMAN, B. & LOTTER, A. F. — 1989 — Late glacial radiocarbon-dating and palynostratigraphy on the Swiss Plateau. *Boreas*, **18**, 109–126.

ANDREWS, J. T. & DUGDALE, R. E. — 1970 — Age prediction of glacio-isostatic strandlines based on their gradients. *Bull. Geol. Soc. Am.*, **81**, 3769–3771

ATKINSON, T. C., BRIFFA, K. R & COOPE, G. R. — 1987 — Seasonal temperatures in Britain during the past 22,000 years reconstructed using beetle remains. *Nature*, **325**, 587–592.

BALLANTYNE, C. K. — 1987 — The present-day periglaciation of upland Britain. *In* Boardman, J. (Ed.) *Periglacial landforms and processes in Great Britain and Ireland*. Cambridge University Press, 113–126.

BALLANTYNE, C. K. & SUTHERLAND, D. G. — 1987 — *Wester Ross Field Guide*. Quaternary Research Association, Cambridge.

BALLANTYNE, C. K. & WHITTINGTON, G. — 1987 — Niveo-aeolian sand deposits on An Teallach, Wester Ross, Scotland. *Trans. R. Soc. Edinb. Earth Sci.*, **78**, 51–64.

BIRKS, H. J. B. — 1989 — Holocene isochrone maps and patterns of tree-spreading in the British Isles. *J. Biogeogr.*, **16**, 503–540.

BOULTON, G. S., JONES, A. S., CLAYTON, K. M & KENNING, M. J. — 1977 — A British ice-sheet model and patterns of glacial erosion and deposition in Britain. *In* Shotton, F. W. (Ed.) *British Quaternary Studies*. Clarendon Press, Oxford.

BOULTON, G. S., CHROSTON, P. N. & JARVIS, J. — 1980 — A marine seismic study of late Quaternary sedimentations and inferred glacier fluctuations along the coast of western Inverness-shire, Scotland. *Boreas*, **10**, 39–51.

BOULTON, G. S., SMITH, G. D., JONES, A. S. & NEWSOME, J. — 1985 — Glacial geology and glaciology of the last mid-latitude ice sheets. *J. Geol. Soc. Lond.*, **142**, 447–474.

BRAZIER, V., WHITTINGTON, G. & BALLANTYNE, C. K. — 1988 — Holocene debris cone evolution in Glen Etive, Western Grampian Highlands, Scotland. *Earth Surf. Proc. Landforms*,

BISHOP, W. W. & COOPE, G. R. — 1977 — Stratigraphical and faunal evidence for Lateglacial and early Flandrian environments in south-west Scotland. *In* Gray, J. M. & Lowe, J. J. (Eds.) *Studies in the Scottish Lateglacial Environment*, 61–68. Pergamon Press, Oxford.

BIRKS, H. J. B.   1973   *The past and present vegetation of the Isle of Skye: a palaeo-ecological study*. Cambridge.

BIRKS, H. H. &   1978   Late Devensian and early Flandrian pollen and macrofossil
  MATHEWES, R. W.   stratigraphy at Abernethy Forest, Inverness-shire. *New Phytol.*, **80**, 455–484.

BIRKS, H. J. B. &   1979   Interglacial pollen spectra from Sel Ayre, Shetland. *New
  PEGLAR, S. M.   Phytol.*, **83**, 559–575.

BROWNE, M. A. E. &   1981   Glaciomarine deposits of the Loch Lomond Stade glacier in the
  GRAHAM, D. K.   Vale of Leven between Dumbarton and Balloch, west-central Scotland. *Quaternary Newsletter*, **34**, 1–7.

BROWNE, M. A. E. &   1984   Shoreline inheritance and coastal history in the Firth of Clyde.
  McMILLAN, A. A.   *Scott. J. Geol.*, **20**, 119–120.

BROWNE, M. A. E.,   1983   Blocks of marine clay in till near Helensburgh, Strathclyde.
  McMILLAN, A. A. &   *Scottish Journal of Geology*, **19**, 321–325.
  HALL, I. H. S.

CADELDINE, C. J   1980   A Lateglacial site at Stormont Loch, near Blairgowrie, eastern Scotland. *In* Lowe, J. J., Gray, J. M. & Robinson, J. E. (Eds.) *Studies in the Lateglacial of North-west Europe*. Pergamon, Oxford and New York, 69–88.

CLAPPERTON, C. M. &   1977   The Late Devensian glaciation of North-East Scotland. *In*
  SUGDEN, D. E.   Gray, J. M. & Lowe, J. J. (Eds.) *Studies in the Scottish lateglacial environment*, 1–13. Oxford.

CONNELL, E. R. &   1987   The periglacial history of Buchan, Scotland. *In* Boardman, J.
  HALL, A. M.   (Ed.) *Field Guide to Periglacial Landforms of Northern England*. Quaternary Research Association, Cambridge.

COOPE, G. R.   1962   A Pleistocene Coleopterous fauna with Arctic affinities from Fladbury, Worcestershire. *Q. J. Geol. Soc. London*, **118**, 103–123.

                               1962   *Coleoptera* from a peat interbedded between two boulder clays at Burnhead near Airdrie. *Trans. Geol. Soc. Glasg.*, **24**, 279–286.

                               1977   Fossil coleopteran assemblages of sensitive indicators of climatic change during the Devensian (last) cold stage. *Phil. Trans. R. Soc.*, **B280**, 313–337.

CULLINGFORD, R. A.   1977   Lateglacial raised shorelines and deglaciation in the Earn–Tay area. *In* Gray, J. M. & Lowe, J. E. (Eds.) *Studies in the Scottish lateglacial environment*, 15–32. Oxford.

DAVIES, H. C.,   1984   A revised seismic stratigraphy for Quaternary deposits on the
  DOBSON, M. R. &   inner continental shelf west of Scotland between 55°30′N and
  WHITTINGTON, R. J.   57°30′N. *Boreas*, **13**, 48–66.

DAWSON, A. G.   1980   Interglacial marine erosion in western Scotland. *Proc. Geol. Assoc., London*, **91**, 339–344.

                               1982   Late-glacial sea-level changes and ice limits in Islay, Jura and Scarba, Scottish Inner Hebrides. *Scott. J. of Geol.*, **18**, 253–266.

                               1988   The Main Rock Platform (Main Lateglacial Shoreline) in Ardnamurchan and Moidart, western Scotland. *Scott. J. Geol.*, **24**, 163–174.

DICKSON, J. H.,   1976   Three Late-Devensian sites in west-central Scotland. *Nature,*
  JARDINE, W. G. &   *London*, **262**, 43–44.
  PRICE, R. J.

DONNER, J. J.   1957   The geology and vegetation of late-glacial retreat stages in Scotland. *Trans Roy. Soc. Edinb.*, **63**, 221–264.

EYLES, V. A.,   1949   Geology of central Ayrshire, 2nd Edn. *Memoir of the*
  SIMPSON, J. B. &   *Geological Survey of Scotland*.
  MACGREGOR, M. C.

FAIRBANKS, R. G.   1989   A 17,000-year glacio-eustatic sea level record. *Nature*, **342**, 637–642.

GEIKIE, A.   1901   *The scenery of Scotland*. London. 1st ed. 1865, 2nd ed. 1887.

GEORGE, T. N.   1955   Drainage in the Southern Uplands: Clyde, Nith, Annan. *Trans. geol. Soc. Glasg.*, **22**, 1–34.

GODARD, A.   1965   *Recherches de géomorphologie en Ecosse du Nord-ouest.*

GRAY, J. M.   1974   Lateglacial and postglacial shorelines in western Scotland. *Boreas*, **3**, 129–138.

1978   Low-level shore platforms in the south-west Scottish Highlands: altitude, age and correlation. *Trans. Inst. Br. Geogr.*, **3**, 151–164.

GRAY, J. M. &   1972   The Loch Lomond Readvance moraines of Mull and Menteith.
BROOKS, C. L.   *Scottish Journal of Geology*, **8**, 95–103.

GRAY, J. M. &   1988   Age of the Main Rock Platform, western Scotland.
IVANOVICH, M.   *Palaeogeogr., Palaeoecol., Palaeoclimatol.*, **68**, 337–345.

GRAY, J. M. &   1977   The Scottish lateglacial environment: a synthesis. *In* Gray,
LOWE, J. J.   J. M. & Lowe, J. J. (Eds.). *Studies in the Scottish lateglacial environment*, 163–181. Oxford.

HALL, A. M.   1984   Introduction. *In* Hall, A. M. (Ed.) *Buchan Field Guide*. 1–26. Quaternary Research Association, Cambridge.

HALL, A. M. &   1989   A preliminary report on the Late Devensian glaciomarine
JARVIS, J.   deposits around St Fergus, Grampian Region. *Quaternary Newsletter*, **59**, 5–7.

HALL, A. M. &   1989   Late Devensian Glaciation of southern Caithness. *Scott. J. of*
WHITTINGTON, G.   *Geol.*, **25**, 307–324.

HOLLINGWORTH, S. E.   1938   The recognition and correlation of high-level erosion surfaces in Britain: a statistical study. *Quart. J. geol. Soc.*, **94**, 55–84.

HOLMES, G.   1984   *Rock slope failure in part of the Scottish Highlands*. Unpublished Ph.D. Thesis, University of Edinburgh.

HORNE, J.,   1893   The character of the high-level shell-bearing deposits at Clava,
ROBERTSON, D.,   Chapelhall and other localities. *Report of the British*
JAMIESON, T. F.,   *Association*, pp. 483–514.
FRASER, J.,
KENDALL, P. F. &
BELL, D.

HORSFIELD, W. B.   1983   *The deglaciation pattern of the western Grampians, Scotland.* Ph.D. Thesis, University of East Anglia.

IMBRIE, J., HAYS, J. D. &   1984   The orbital theory of Pleistocene climate: support from revised
MARTINSON, D. G. &   chronology of the marine ^{18}O record. *In* Berger *et al.* (Eds.)
SPECMAP GROUP   *Milankovic and Climate*. Reidel, Dordrecht, 269–306.

INNES, J.   1983   Lichenometric dating of debris flow deposits in the Scottish Highlands. *Earth Surf. Proc. Landforms*, **8**, 579–588.

JAMIESON, T. F.   1865   The history of the last geological changes in Scotland. *Q. Jl. Geol. Soc. Lond.*, **21**, 161–203.

1866   On the glacial phenomena of Caithness. *Q. Jl. Geol. Soc. Lond.*, **22**, 261–281.

1882   On the crag shells of Aberdeenshire and the gravel beds containing them. *Q. Jl. Geol. Soc. Lond.*, **38**, 145–159.

JARDINE, W.,   1988   A late Middle Devensian interstadial site at Sourlie, near
DICKSON, J. H.,   Irvine, Strathclyde. *Scott. J. Geol.*, **24**, 288–295.
HAUGHTON, P. D. W.,
HARKNESS, D. D.,
BOWEN, D. Q. &
SYKES, G. A.

JESSEN, A.   1905   On the shell-bearing clay in Kintyre. *Transactions of the Geological Society of Edinburgh*, **8**, 76–86.

KEMP, D. D.   1971   *The stratigraphy and sub-carse morphology of an area on the northern side of the River Forth, between the Lake of Menteith and Kincardine-on-Forth.* Univ. of Edinb. Ph.D. Thesis (unpubl.).

KIRBY, R. P.   1969   Till fabric analyses from the Lothians, central Scotland. *Geogr. Ann., Stockholm,* **51A**, 48–60.

KIRK, W. &   1963   A lateglacial site at Loch Droma, Ross and Cromarty. *Trans.*
GODWIN, H.   *R. Soc. Edinb.,* **65**, 225–249.

LAWSON, T. J.   1984   Reindeer in the Scottish Quaternary. *Quaternary Newsl.,* **42**, 1–7.

LAWSON, T. J. &   1986   Early settlement in Scotland: the evidence from Reindeer Cave,
BONSALL, C.   Assynt. *Quaternary Newsl.,* **49**, 1–7.

LINTON, D. L.   1959   Morphological contrasts between eastern and western Scotland. *In* Miller, R. & Watson, J. W. (Eds.) *Geographical essays in memory of Alan G. Ogilvie,* 16–45. Edinburgh.

LONG, D., SMITH, D. E. &   1989   A Holocene tsunami deposit in eastern Scotland. *J. of Quat.*
DAWSON, A. G.   *Sci.,* **4**, 61–66.

LOWE, J. J. &   1981   The early postglacial history of Scotland: evidence from a site
WALKER, M. J. C.   near Tyndrum, Perthshire. *Boreas,* **3**, 281–294.

McMILLAN, A. A. &   1981   The sand and gravel resources of the country west of
AITKEN, A. M.   Peterhead, Grampian Region. Description of 1:25,000 sheet NK04 and parts of NJ94, 95 and NK05, 14 and 15. *Mineral Assessment Report of the Institute of Geological Sciences,* No. 58.

MACPHERSON, J. B.   1980   Environmental change during the Loch Lomond Stadial: evidence from a site in the upper Spey valley, Scotland. *In* Lowe, J. J., Gray, J. M. & Robinson, J. E. (Eds.) *Studies in the Lateglacial of North-West Europe,* 89–102. Pergamon, Oxford.

MANGERUD, J.,   1974   Quaternary stratigraphy of Norden, a proposal for
ANDERSEN, S. TH.,   terminology and classification. *Boreas,* **3**, 109–128.
BERGLUND, B. E. &
DONNER, J. J.

MILLER, G. H.,   1983   Amino-acid ratios in Quaternary molluscs and foraminifera
SEJRUP, H. P.,   from western Norway. *Boreas,* **12**, 107–124.
MANGERUD, J. &
ANDERSON, B. G.

MYKURA, W.   1976   *British Regional Geology: Orkney and Shetland.* Stationery Office, Edinburgh.

NEWEY, W. W.   1970   Pollen analysis of Late-Weichselian deposits at Corstorphine, Edinburgh. *New Phytol.,* **69**, 1167–1177.

PATERSON, I. B.   1974   The supposed Perth Readvance in the Perth District. *Scott. J. Geol.,* **10**, 53–66.

PATERSON, I. B.,   1981   Quaternary estuarine deposits in the Tay–Earn area,
ARMSTRONG, M. &   Scotland. *Rep. Inst. Geol. Sci.,* **No. 81/7**.
BROWNE, M. A. E.

PEACH, B. N. & HORNE, J.   1930   *Chapters on the geology of Scotland.* Oxford.

PEACOCK, J. D.   1966   Note on the drift sequence near Portsoy, Banffshire. *Scottish Journal of Geology,* **2**, 35–37.

                                     1971   A re-interpretation of the Coastal Deposits of Banffshire and their place in the Late-glacial history of N.E. Scotland. *Geological Survey of Great Britain Bulletin,* **37**, 81–89.

                                     1975   Depositional environment of glacial deposits at Clava, northeast Scotland. *Bull. Geol. Surv. Gt. Br.,* **49**, 31–37.

                                     1981   Scottish Late-glacial marine deposits and their environmental significance. *In* Neale, J. & Flenley, J. (Eds.) *The Quaternary in Britain,* 222–236. Pergamon Press, Oxford.

PEACOCK, J. D. 1983 A model for Scottish interstadial marine palaeotemperature 13,000 to 11,000 BP. *Boreas*, **12**, 73–82.

1989 Marine molluscs and Late Quaternary environmental studies with particular reference to the Late-glacial period in north-west Europe: a review. *Quaternary Sci. Rev.*, **8**, 179–192.

PEACOCK, J. D. & GRAHAM, D. K. 1977 Evolution and chronology of lateglacial marine environments at Lochgilphead, Scotland. *In* Gray, J. M. & Lowe, J. J. (Eds.) *Studies in the Scottish lateglacial environment*, 89–100. Oxford.

PEACOCK, J. D., GRAHAM, D. K. & WILKINSON, I. P. 1978 Late-glacial and post-glacial marine environments at Ardyne, Scotland and their significance in the interpretation of the history of the Clyde Sea area. *Rep. Inst. Geol. Sci.*, **78/17**.

PEACOCK, J. D. & HARKNESS, D. D. 1990 Radiocarbon ages and the full-glacial to Holocene transition in seas adjacent to Scotland and Southern Scandinavia: a review. *Trans. R. Soc. Edinb. Earth Sciences* (in press).

PEACOCK, J. D., HARKNESS, D. D., HOUSLEY, R. A., LITTLE, J. & PAUL, M. A. 1989 Radiocarbon ages for a glaciomarine bed associated with the maximum of the Loch Lomond Readvance in west Benderloch, Argyll. *Scott. J. Geol.*, **25**, 69–79.

PEGLAR, S. M. 1979 A radiocarbon-dated pollen diagram from Loch of Winles, Caithness, north-east Scotland. *New Phytol.*, **82**, 245–263.

PENNINGTON, W. 1975 A chronostratigraphic comparison of Late-Weichselian and Late-Devensian subdivisions, illustrated by two radiocarbon-dated profiles from western Britain. *Boreas*, **4**, 157–171.

1977 Lake sediments and the lateglacial environment in northern Scotland. *In* Gray, J. M. & Lowe, J. J. (Eds.) *Studies in the Scottish lateglacial environment*, 119–141. Oxford.

PENNINGTON, W., HAWORTH, E. Y., BONNY, A. P. & LISHMAN, J. P. 1972 Lake sediments in northern Scotland. *Phil. Trans. R. Soc. London*, **264B**, 191–294.

ROBINSON, M. & BALLANTYNE, C. K. 1979 Evidence for a glacial readvance pre-dating the Loch Lomond Advance in Wester Ross. *Scott. J. Geol.*, **15**, 271–277.

ROLFE, W. D. I. 1966 Woolly rhinoceros from the Scottish Pleistocene. *Scott. J. Geol.*, **2**, 253–258.

ROSE, J. 1980 *In* Jardine, W. G. (Ed.) *Glasgow Region: Field Guide*. Quaternary Research Association.

1989 Stadial type sections in the British Quaternary. *In* Schluchter, C. (Ed.) *Quat. Type Sections*. Rotterdam, Balkema.

ROSE, J., LOWE, J. J. & SWITSUR, R. 1988 A radiocarbon date on plant detritus beneath till from the type area of the Loch Lomond readvance. *Scottish Journal of Geology*, **24**, 113–124.

RUDDIMAN, W. F. & McINTYRE, A. 1973 Time-transgressive deglacial retreat of polar waters from the North Atlantic. *Quat. Res.*, **3**, 117–130.

RUDDIMAN, W. F. & RAYMO, M. 1988 Northern hemisphere climate regimes during the last 3 Ma. *Phil. Trans. Roy. Soc. London*, **B318**, 411–430.

SELBY, I. 1989 *Quaternary Geology of the Hebridean Continental Margin*. Nottingham University Ph.D. Thesis (unpubl.).

SERJUP, H.-P., AARSETH, K. L., ELLINGSEN, E., REITHER, E., JANSEN, E., LOVLIE, R., BENT, A., BRIGHAM-GRETTE, J., LARSEN, E. & STOKER, M. 1987 Quaternary stratigraphy of the Fladen area, central North Sea: a multidisciplinary study. *J. Quaternary Sci.*, **2**, 35–58.

SIMPSON, J.   1933   The late-glacial readvance moraines of the Highland border
                                      west of the river Tay. *Trans. R. Soc. Edinb.*, **57**, 633–645.

SISSONS, J. B.   1963   The glacial drainage system around Carlops, Peeblesshire.
                                      Trans. Inst. Br. Geogr., **32**, 95–111.

                               1967   *The evolution of Scotland's scenery*. Edinburgh.

                               1969   Drift stratigraphy and buried morphological features in the
                                      Grangemouth–Falkirk–Airth area, central Scotland. *Trans.
                                      Inst. Br. Geogr.*, **48**, 19–50.

                               1974a A lateglacial ice-cap in the central Grampians, Scotland.
                                      Trans. Inst. Br. Geogr., **62**, 95–114.

                               1974b Lateglacial marine erosion in Scotland. *Boreas*, **3**, 41–48.

                               1976   Lateglacial marine erosion in South-East Scotland. *Scott.
                                      geogr. Mag.*, **92**, 17–29.

                               1978   The parallel roads of Glen Roy and adjacent glens. *Boreas*, **7**,
                                      229–244.

                               1979a The Loch Lomond Advance in the Cairngorm Mountains.
                                      Scott. geogr. Mag., **95**, 66–82.

                               1979b Palaeoclimatic inferences from former glaciers in Scotland and
                                      the Lake District. *Nature, London*, **278**, 518–521.

                               1980   The Loch Lomond Advance in the Lake District, northern
                                      England. *Trans. R. Soc. Edinb.: Earth Sciences*, **71**, 13–27.

                               1982   The so-called high 'interglacial' rock shoreline of western
                                      Scotland. *Trans. of the Inst. of Brit. Geogr.*, **NS 7**, 205–216.

                               1983   Quaternary. *In* Craig, G. Y. (Ed.) *The Geology of Scotland*.
                                      Scottish Academic Press, Edinburgh.

                               1983   The Quaternary geomorphology of the Inner Hebrides: a
                                      review and reassessment. *Proc. Geol. Assoc.*, **94**, 165–175.

SISSONS, J. B. &   1971   Dating of early postglacial land and sea-level changes in the
  BROOKS, C. L.   western Forth valley. *Nature (Phys. Sci.)*, **234**, 124–127.

SISSONS, J. B. &   1981   Former sea-levels and ice limits in part of Wester Ross, North-
  DAWSON, A. G.   West Scotland. *Proc. Geol. Assoc., London*, **92**, 115–124.

SISSONS, J. B. &   1965   Peat bogs in a postglacial sea and a buried raised beach in the
  SMITH, D. E.   western part of the Carse of Stirling. *Scott. J. Geol.*, **1**,
                                      247–255.

SISSONS, J. B.,   1966   Lateglacial and postglacial shorelines in South-East Scotland.
  SMITH, D. E. &   *Trans. Inst. Br. Geogr.*, **39**, 9–18.
  CULLINGFORD, R. A.

SISSONS, J. B. &   1976   Climatic inferences from former glaciers in the South-East
  SUTHERLAND, D. G.   Grampian Highlands, Scotland. *J. Glaciol. London*, **17**,
                                      325–346.

SISSONS, J. B. &   1974   Lateglacial site in the central Grampian Highlands. *Nature,
  WALKER, M. J. C.   London*, **249**, 822–824.

SMITH, D. E.   1984   Lower Ythan Valley. *In* Hall, A. M. (Ed.) *Buchan Field Guide*.
                                      Quaternary Research Association, 47–58.

STOKER, M. S.   1988   Pleistocene ice-proximal glaciomarine sediments in boreholes
                                      from the Hebrides Shelf and Wyville-Thomson Ridge, NW UK
                                      Continental Shelf. *Scott. J. Geol.*, **24**, 249–262.

STOKER, M. S. &   1987   Lower Pleistocene deltaic and marine sediments in boreholes
  BENT, A. J. A.   from the central North Sea. *J. Quaternary Sci.*, **2**, 87–96.

STOKER, M. S.,   1985   A revised Quaternary stratigraphy for the central North Sea.
  LONG, D. & FYFE, J. A.   *Rep. Br. Geol. Surv.*, **17(2)**.

SUTHERLAND, D. G.   1980   Problems of radiocarbon dating deposits from newly
                                      deglaciated terrain: examples from the Scottish Lateglacial. *In*
                                      Lowe, J. J., Gray, J. M. & Robinson, J. E. (Eds.) *Studies in
                                      the Lateglacial of North-West Europe*, 139–149. Pergamon,
                                      Oxford.

SUTHERLAND, D. G. 1981 The high-level marine shell beds of Scotland and the build-up of the last Scottish ice-sheet. *Boreas*, **10**, 247–254.

1984a The Quaternary deposits and landforms of Scotland and the neighbouring shelves: a review. *Quaternary Sci. Rev.*, **3**, 157–254.

1984b Modern glacier characteristics as a basis for inferring former climates with particular reference to the Loch Lomond Stadial. *Quat. Sci. Rev.*, **3**, 291–309.

SUTHERLAND, D. G. & 1984 A late Devensian ice-free area and possible interglacial site on
WALKER, M. J. C. the Isle of Lewis, Scotland. *Nature*, **309**, 701–703.

TIPPING, R. M. 1986 A late-Devensian pollen site in Cowal, south-west Scotland. *Scott. J. Geol.*, **22**, 27–40.

1986 Loch Lomond Stadial *Artemisia* pollen assemblages and Loch Lomond Readvance regional firn-line altitudes. *Quat. Newl.*, **46**, 1–11.

1987 The prospects for establishing synchroneity in the early postglacial pollen peak of *Juniperus* in the British Isles. *Boreas*, **16**, 155–163.

1989 Palynologial evidence for the extent of the Loch Lomond Readvance in the Awe valley and adjacent areas, S.E. Highlands. *Scott. J. of Geol.*, **25**, 325–338.

THOMSON, M. E. 1978 IGS studies of the geology of the Firth of Forth and its approaches. *Rep. Inst. Geol. Sci.*, **77/17**.

VON WEYMARN, J. 1979 A new concept of glaciation in Lewis and Harris. *Proc. R. Soc. Edinb.*, **77B**, 97–105.

WALKER, M. J. C. 1975 Two lateglacial pollen diagrams from the eastern Grampian Highlands of Scotland. *Pollen et Spores*, **17**, 67–92.

1984 Pollen analysis and Quaternary research in Scotland. *Quat. Sci. Rev.*, **3**, 369–404.

WALKER, M. J. C. & 1981 Postglacial environmental history of Rannoch Moor,
LOWE, J. J. Scotland. III. Early- and mid-Flandrian pollen stratigraphic data from sites on western Rannoch Moor and near Fort William. *J. of Biogeogr.*, **8**, 475–491.

WALKER, M. J. C. & 1977 Pollen analyses, radiocarbon dates and the deglaciation of
LOWE, J. J. Rannoch Moor, Scotland, following the Loch Lomond Advance. *In* Cullingford, R., Davidson, D. A. & Lewin, J. (Eds.). *Timescales in Geomorphology*, 247–259. Wiley, Chichester.

16

ECONOMIC GEOLOGY

R. Beveridge, S. Brown, M. J. Gallagher, J. W. Merritt

These four authoritative accounts document the remarkable changes that have taken place in the economic geology of Scotland since the first edition of this book was published 25 years ago. In 1962, coal was the dominant source of power, oil-shale mining had just stopped and the existence of oil and gas in the North Sea was unknown.

Now (1990) only two deep coalmines remain, coal production has shrunk from over 18 to some 5 million tonnes per annum, mostly from opencast production, whereas proven recoverable UK offshore oil reserves are at least 510 million tonnes and recoverable UK offshore gas reserves are at least 560 billion cubic metres. Further oil reserves of between 530 and 3,140 million tonnes and gas reserves of between 210 and 1,152 billion cubic metres may yet be found. A baryte mine is operating near Aberfeldy and an important gold deposit has been discovered at Tyndrum.

Scotland is a small but exceptionally beautiful country. Its politicians and planners have to cater for the diverse demands of its people, industry and tourism. They also need to be sensitive to environmental concerns. Beautiful buildings, towns and cities have been created by quarrying the countryside. New roads, bridges, buildings and industry continue to put pressure on the environment. All create jobs and all demand materials for construction which the extractive-minerals industry provides.

We use energy for power, heating, lighting and transportation. At the present rate of consumption the world is burning its reserves of fossil fuels gradually formed over the last 300,000,000 years or so in the brief period of 300–400 years. It has been called man's greatest geochemical experiment. Undoubtedly energy conservation and other sources of energy will play increasingly important roles in the 21st century.

It is clear from this chapter that applied geology has responded effectively to the diverse needs of people and industry in and furth of Scotland.

Editor

Coal

Coal has long been worked in Scotland (National Coal Board 1958). It is mentioned in documents of the reign of Malcolm IV (1153–65) and in the Charters of Holyrood, Newbattle, Dunfermline and Paisley Abbeys of the 12th and 13th centuries. Until the 16th century, working was local and primitive but from then on the industry gradually became more important. Pumping machinery was developed at the end of the 16th century and steam power for pumping and for raising coal appeared in the early 1700s. James Watt's engine of 1784 revolutionised the mining industry in Scotland as it made the working of deep, hitherto unreachable coals possible.

The remarkable expansion of the iron industry in the 19th century, the improvement of transport and communications and the general progress in the economic and social life of the country all helped to increase the demand for coal. From about 40,000 tons* per annum in 1550, 500,000 tons in 1700, the annual production of coal rose fairly steadily to reach a peak at the beginning of the First World War of about 43 million tons. The output of coal in Scotland over the past 136 years is shown in Fig. 16.1. In 1950 it was hoped that Scotland would produce about 30 million tons per year or about 13 per cent of the total British output (National Coal Board 1950). However, in the event the past 40 years has been a period of constant decline due to competition from the oil, natural gas and

* Figures of production that occurred before the introduction of S.I. units in 1977 are not converted (i.e. 1 ton = 1·01605 tonnes).

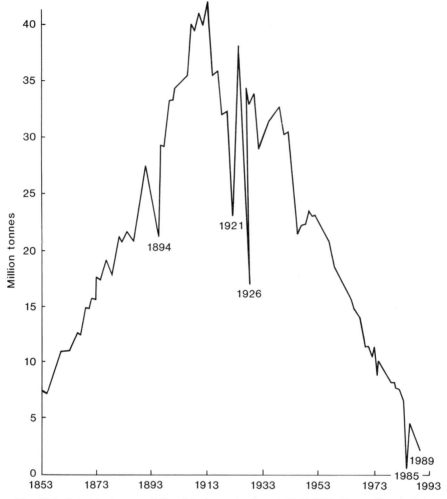

Fig. 16.1. Saleable output of Scottish deep-mined coal 1854–1989 (from Chamberlain (1961) and annual reports of the National Coal Board).

nuclear industries and in the financial year to March 1989 annual production had fallen to only 5·4 million tonnes of which 3·5 million tonnes was from opencast sources and only 1·9 million tonnes from deep mines. As financial pressures have grown so has low cost opencast output become of ever increasing importance. From 0·8 million tons (6 per cent of national output) in 1968/69 opencast output has steadily risen to its present level and now contributes 65 per cent of total national output. Originally conceived as a flexible means of meeting shortfalls in deep mines production and later as a means of cross-subsidising essentially uneconomic pits it has now become the dominant influence on the industry in Scotland. More than 200 deep mines have been closed since nationalisation in 1947. Currently only two remain

in operation and due to their relatively high costs of production and serious market constraints their future is doubtful. Opencast operations with their low production costs and assured markets are highly profitable and appear to have an attractive future.

Coals occur in all the Carboniferous groups (see Chapter 10) but with the exception of the Upper Hirst Seam of the Upper Limestone Group those in the Limestone Coal Group and the Productive Coal Measures are economically most important. Rocks of these groups are present over a considerable area of the Midland Valley (see Fig. 1.1). Smaller coalfields occur in Dumfriesshire and Argyllshire and a small coalfield of Middle Jurassic age was worked near Brora in Sutherland (Macgregor 1954 and p. 444 this volume).

Ayrshire

In Ayrshire there are only a few coals of importance in the Limestone Coal Group (Fig. 10.13) but up to 13 physically workable seams occur in the Coal Measures (Fig. 10.21). Most recently, underground mining was concentrated in the Coal Measures of the Ochiltree–Auchinleck area on the southern and eastern flanks of the Mauchline Basin particularly in the Lugar Main Seam. Production costs, however, were unacceptably high due to the extensive and intensive faulting characteristic of all the Ayrshire coalfields. Barony Colliery at Auchinleck was the last to operate and closed down in 1989.

Central Coalfield

The Central Coalfield in Lanarkshire, the historical heartland of industrial Scotland, had most of its Coal Measures seams (Fig. 10.20) worked to virtual exhaustion long ago; but seams of high quality coking coals in the Limestone Coal Group (Fig. 10.13) continued to be produced until fairly recently at Stepps, north-east of Glasgow (Cardowan Colliery) and at Whitburn (Polkemmet Colliery) in West Lothian. The struggle for survival proved unsuccessful in a climate of high production costs and diminishing demand from a declining steel industry. Polkemmet was the last colliery to close when it became seriously flooded during the miners' strike of 1984/85.

East Fife

In East Fife extensive undersea resources are present in the Coal Measures. The Dysart Main Seam reaches 9 metres thick in places while another eight seams in excess of 2 metres thick occur through the coal-bearing successions (Siddall 1979). Production ceased in 1988 with the closure of Seafield Colliery at Kirkcaldy due to outbreaks of spontaneous combustion which had affected the workings sporadically since the miners' strike of 1984/85. Access to the remaining undersea reserves could be achieved by the re-development of Frances Colliery (north of Kirkcaldy) but in the present financial circumstances of the industry the massive capital investment necessary seems unlikely to be forthcoming. It is of interest that in the late '50s prior to sinking the Seafield shafts the first offshore drilling in UK waters took place off the Fife coast as part of the initial exploration (Ewing & Francis 1960a, b). Further offshore boring took place in 1974/75 from a drillship (*Wimpey Sealab*).

Clackmannan-West Fife

The Clackmannan–West Fife area is the location of one of the most interesting developments in British coalmining since nationalisation in 1947. The Longannet Complex originally consisted of three separate mines (Solsgirth, Castlehill and Bogside) linked by a tunnel almost 9 km in length to Longannet Mine where the coal is brought to the surface alongside a 2,400 megavolt electricity generating station. The complex has access to large reserves in the Upper Hirst Seam of the Upper Limestone Group (Fig. 10.15) and with exploitation limited to a single seam the workings spread rapidly. A new shaft at Castlebridge near Clackmannan was opened in 1986 in order to mine additional areas in the west and Castlehill which had become over-extended was closed. Bogside mine was lost during the miners' strike of 1984/85. Currently the two Longannet units (Solsgirth and Castlebridge) are the only Scottish collieries remaining in operation. Although capable of producing some 2 million tonnes per annum the production costs are relatively high and they are extremely vulnerable to the commercial pressures being exerted by the South of Scotland Electricity Board, their only customer.

In the north of Fife at Westfield, the largest opencast mining project ever attempted in Britain commenced in 1956. The Passage Group in this area contains an abnormally thick development of coals (Chapter 10, p. 364). In 1959 the first Lurgi pressure gasification plant to be installed in Britain was opened on the site to convert the relatively low-grade coal into gas and by 1965 it was capable of producing one-fifth of Scotland's needs. The advent of natural gas from the southern part of the North Sea, however, changed the situation completely and by the '70s the Lurgi plant was used mainly as a pilot plant for testing coals from a variety of sources as potential gas producers. In the meantime Westfield continued to produce upwards of 1 million tonnes per annum for domestic and industrial use until 1987. During its lifetime it yielded a total of some 20 million tonnes.

Midlothian Coalfield

South across the Firth of Forth lies the Midlothian Coalfield syncline. The landward Coal Measures (Fig. 10.20) and the Limestone Coal Group (Fig. 10.13) along the steeply dipping limbs of the syncline have been extensively worked in the past. More recently reserves were worked in the deeper central part of the basin from Bilston Glen and Monktonhall Collieries but production ceased in 1989 with the closure of the former due to poor performance and unacceptably high costs. Monktonhall which has not produced since 1986 and was in process of linking to Bilston Glen could in theory be re-developed as an independent unit but its only potential market (Cockenzie Generating Station) is insecure and this development seems unlikely. Offshore drilling in 1975 proved an undersea extension of Coal Measures seams north of Musselburgh. Plans to provide access to these from Monktonhall were well advanced in the early '80s but foundered due to the excessive estimated development costs in an increasingly stringent financial climate.

Douglas, Sanquhar and Canonbie Coalfields

South of the main Central Coalfield lie the Douglas (Valley), Sanquhar and Canonbie fields. The first named contains 6–7 physically workable seams in the Limestone Coal Group at Coalburn and Bankhead while the same beds contain up to 12 seams at Ponfeigh. A large opencast site – the biggest in Europe – is presently being developed at Dalquhandy near Coalburn where it is planned to extract about 1 million tonnes per annum for the next 15 years. In the valley of the Douglas Water to the south-west Coal Measures coals have been important in the past (Fig. 10.20).

Sanquhar field lies south of the Southern Uplands Fault. Coal Measures lie unconformably on Lower Palaeozoic rocks, with in places, Passage Beds intervening. The Canonbie Coalfield lies on the Dumfries-shire–Cumbria border and is an important link between the successions in Scotland and the North of England. In the past coals were wrought from both Coal Measures and Limestone Coal Group. Poor exposures and faulting encouraged the British Geological Survey to drill the Archerbeck Borehole (p. 372) in 1960 to provide the first complete record of the coals formerly worked at Penton. Drilling by the N.C.B. in 1954/56 and in 1979/83 (Picken 1988) has revealed a field of Coal Measures extending for some 70 km^2 concealed below New Red Sandstone rocks. At least 5 seams are of physically workable thickness and reserves *in situ* exceed 400 million tonnes, sufficient to support two large collieries given a favourable economic climate.

Machrihanish and Brora Coalfields

At Machrihanish (Figs. 10.13 and 10.20) there is an outlier of Carboniferous beds ranging in age from the Limestone Coal Group to the Modiolaris zone of the Coal Measures. Mining was carried out there for several centuries, the principal seam being the 4·3 m 'Main Coal'.

The only workable seam of coal in Scotland not of Carboniferous age occurs at Brora in Sutherland. It is almost 1 m thick and occurs in Middle Jurassic strata (p. 444), but is rather high in pyrite and ash content, which combined with lack of nearby markets renders it unattractive for exploitation, though it has been worked intermittently since the end of the 16th century, including during the 1960s (Ministry of Power 1962) and the '70s.

Opencast versus deep mining

No contemporary account of the Scottish coalfields would be complete without more specific mention of opencast mining. Opencast projects in tonnage terms are usually fairly small and relatively short lived. Sites such as Westfield and Dalquhandy previously referred to are exceptional. Due to low production costs and high profitability they have become increasingly attractive in recent years. Currently 11 sites spread over all the coalfield areas are operating and produced 3·5 million tonnes in 1988/89. Planning permission for a further 45 million tonnes has either been granted or applied for. Criteria for the viability of opencast working obviously differ greatly from those applicable to deep mining. Notably the working ratio (i.e. the ratio of overburden and other barren strata to extractable mineral) is of much greater importance than the thickness of individual coal seams. In the opencast context it is often possible to extract seams too thin for underground exploitation.

One consequence of the drive for increased productivity in deep mining has been the concentration of output in fewer collieries, fewer working faces and fewer and thicker seams. Only the Upper Hirst seam commonly between 2·5 and 2·75 metres thick is being worked at present as progressively, thinner seams have been rendered uneconomic. Most of the coals worked now (including opencast) are used only for steam raising or for domestic purposes. The Upper Hirst output is even more restricted in use being suitable only for power stations.

In certain areas, particularly in the Central Coalfield between Denny and Stirling, and east of Kirkintilloch, coals approaching anthracite rank occur. Deep mining of these ceased in 1969 though a little has been produced by opencast methods since then. The anthracitisation was attributed to the proximity of igneous intrusions by Dinham & Haldane (1932) and Robertson & Haldane (1937). This view was challenged in Stirlingshire by Skipsey (1959) who suggested that the high rank coals were produced by increased temperature and pressure due to the higher level of regional metamorphism of the area, brought about by the 3,000 m downthrow of the Ochil Fault. It must be said, however, that in general heat due to high geothermal gradients rather than pressure (hydrostatic or tectonic) or proximity to local sources of heat is currently favoured as the prime cause of anthracitisation elsewhere (Stach *et al.* 1975, pp. 41–47).

R. Beveridge

Oil and gas

The commercial exploitation of oil and natural gas has a long and varied history in Scotland, dating back to 1851 when the chemist James Young (1811–1883) opened a plant at Bathgate to produce oil by the distillation of a sapropelic coal. This industry continued, using oil-shale as its principal raw material, until 1962. From 1919 to the present, there has been exploration for and, for a time, limited production of naturally occurring oil and gas from onshore sedimentary basins. The level of activity has always been low but more onshore acreage in Scotland is currently under licence to exploration companies than at any previous time.

In terms of its economic impact (and also its impact on the geological sciences in Scotland) the most important phase in this history began in 1969 with the discovery of the Montrose oil field by well 22/18–1, drilled *c.* 200 km east of Aberdeen (Fig. 16.2). Exploration and production in the North Sea basins east of Scotland continues apace today, and exploration efforts have also been directed at offshore basins to the west, from the Solway Firth to west of the Shetland Islands.

Oil from sapropelic coal and oil-shale

Employing a process known experimentally since the 17th century, James "Paraffin" Young opened a plant at Bathgate, West Lothian to produce oil by the destructive distillation of a sapropelic coal. He used coal from the Carboniferous Lower Coal Measures found at Boghead on the Torbane Hill estate, close to Bathgate (Fig. 16.3). This coal, of a variety termed a Boghead coal or Torbanite, after the locality, extended only over *c.* 1,000 hectares and was less than *c.* 50 cm thick; it was exhausted after 12 years, both through local use and through export of the coal to Germany and North America (MacGregor & Anderson 1923).

In 1858 the much more extensive oil-shales of the early Carboniferous Calciferous Sandstone Measures (Strathclyde Group of Paterson & Hall 1986) were discovered and recognised as an alternative source of oil. By 1865, when Young opened a plant to distil oil-shale at nearby West Calder, 120 plants were producing oil in east-central Scotland: an important local industry had grown up. By 1878. *c.* 500,000 barrels* of oil per year were being produced and, at the peak of the industry's output during the First World War, production of some 2·1 million barrels of oil per year was achieved (see Hallett *et al.* 1985).

Sapropelic coals are usually less widely distributed than the better known humic variety. Humic coals form largely from the *in situ* accumulation of woody tissue as a forest peat, whereas sapropelic coals, rich in spores, fragments of leaf cuticle and/or algal remains, accumulate initially as an organic-rich ooze in ponds, lakes or lagoons deficient in oxygen (Adams 1960, Tissot & Welte 1978). The sapropelic coals are typically dark brown to black, have a rather dull, greasy lustre, and a conchoidal fracture. They are characterised chemically by a high proportion of volatile hydrocarbon compounds. The variety of sapropelic coal known as a 'cannel' coal (from a Scots pronunciation of 'candle' and so named because it was once burnt to supply domestic lighting) has a large concentration of organic material from plant spores. The Boghead coal initially used by Young in his Bathgate plant resembles cannel coal in appearance and mode of formation but merits differentiation because of its richness in algal remains (dominantly from Botryococcus-related algae).

The principal raw material for this synthetic oil, however, was oil-shale. Tissot & Welte (1978) suggest that there is no precise lithological or chemical definition of an oil-shale, only an economic one: any rock yielding commercial amounts of oil on pyrolisis (heating to 500°C) can be termed an oil-shale. In contrast, Hutton (1982) defines an oil-shale as a rock containing 5 volume per cent or more of the spore- and algal-derived organic matter, liptinite. In central Scotland, the early Carboniferous oil-shales are argillaceous, dark brown or black rocks, exhibiting a brown streak and a resistance to weathering. They are very finely laminated (most readily seen in the spent shale) and sheets of shale may be flexible. Individual oil-shale bands can be up to 5 m thick. The main oil-shale development in the Lower Carboniferous stretches from just east of Linlithgow and Bathgate to the Pentland Hills (Fig. 16.3), with small workings also known from the Burntisland area in Fife (Carruthers *et al.* 1927).

The oil-shales produced 70 to 200 litres of oil per tonne of shale whereas the Boghead coal yielded 535 to 580 litres per tonne (Cameron & McAdam 1978). The distillation process also yielded ammonium sulphate (*c.* 13–26 kg per tonne of shale) which was used as a fertiliser. Extraction of the oil-shale was principally by the 'stoop and room' method (Carruthers *et al.* 1927) although opencast production also contributed from the 1940s onwards. In 1958 experiments were conducted to produce oil by *in situ* heating but these were unsuccessful and the work was discontinued (Duff 1983).

In 1962 HM Government gave notice of its intention to withdraw in 1964 the 1s 3d per gallon preferential tax concession to UK-produced oil. The Scottish oil-shale industry, which had been struggling for some time to compete with imports of naturally occurring oil, collapsed. Cumulative production of oil from oil-shale and coal in Scotland to 1962 is estimated at *c.* 75 million barrels (10 million tonnes: Hallett *et al.* 1985) which is equivalent to the recoverable reserves originally present in the relatively small, by North Sea standards, Buchan oil field (UK Block 21/1: Department of Energy 1988). Cameron & McAdam (1978) assessed the remaining reserves of oil-shale in the Lothians, the principal area of occurrence (Fig. 16.3): probable reserves were estimated at 120 million tonnes, of which *c.* 65 million tonnes were considered to be extractable. This would yield *c.* 37·5 million barrels (*c.* 5 million tonnes) of oil, or less than one third of the oil produced from the North Sea Brent field (UK Block 211/29) in 1987 alone. Additionally, 980 million tonnes of possible reserves and deep, unproven reserves of oil-shale were reported. There is little prospect, however, of commercial exploitation of the oil-

---

* 1 barrel is equivalent to 35 imperial gallons or 159 litres: 7·5 barrels is equivalent to 1 tonne of crude oil.

Fig. 16.2.

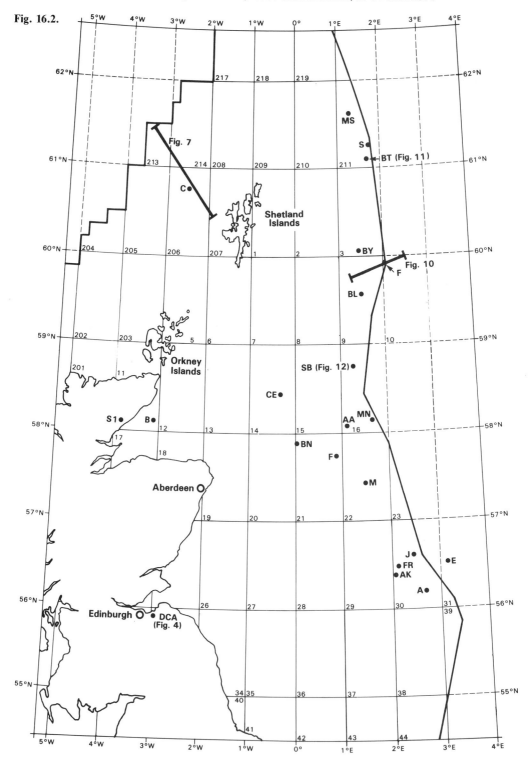

shales so long as North Sea production continues and secure supplies of naturally occurring oil remain plentiful worldwide.

Thin, presently uneconomic, oil-shales also occur in the mid-Jurassic Great Estuarine Group of Skye and Raasay. Yields of between 54 and 77 litres of oil per tonne of shale have been reported (see Lee 1920, Gibson 1922).

Onshore basins

Until the 1980s, onshore exploration for naturally occurring oil and gas in Scotland has been concentrated in the central and eastern part of the Midland Valley. The primary target was the early Carboniferous Calciferous Sandstone Measures, with sandstones the postulated reservoirs, the oil-shales as potential source of hydrocarbons, and the traps in anticlinal structures sealed by early Carboniferous shales. This set of favourable geological circumstances (or 'play') has been most successfully encountered in the area of the D'Arcy–Cousland anticline, Midlothian (Fig. 16.2), where commercial production, principally of natural gas, was achieved for a time. This Carboniferous play is still being explored but exploration interest now extends outside the Midland Valley and to other geological horizons.

Fig. 16.2. Location of oil and gas fields mentioned in the text.

A	Argyll oil field (Block 30/24)
AA	Alba oil field (Block 16/26)
AK	Auk oil field (Block 30/16)
B	Beatrice oil field (Block 11/30)
BL	Beryl oil field (Block 9/13)
BT	Brent oil field (Block 211/29)
BY	Bressay oil field (Block 3/28)
BN	Buchan oil field (Block 21/1)
C	Clair oil field (Block 206/8)
CE	Claymore oil field (Block 14/19)
E	Ekofisk oil field (Norwegian Block 2/4)
F	Frigg gas field (Block 10/1; extends into Norwegian territory)
FR	Fulmar oil field (Block 30/16)
FS	Forties oil field (Block 21/19)
J	Josephine discovery (Block 30/13)
M	Montrose oil field (Block 22/18)
MS	Magnus oil field (Block 211/12)
MN	Maureen oil field (Block 10/29)
S	Stratfjord oil field (Block 211/24; extends into Norwegian territory)
SB	South Brae oil field (Block 16/7)

Also shown is the location of the D'Arcy–Cousland Anticline, near Edinburgh, and the Sutherland No. 1 onshore exploration well in north-east Scotland.

Exploration began in 1919 (some 60 years after the world's first productive oil well was drilled in Pennsylvania, USA) when HM Government, concerned about the increasing level of oil imports, commissioned a programme of drilling which included 2 wells in Scotland. Both were drilled by S. Pearson and Sons, Ltd. for the Mineral Oil Production Department of the Ministry of Munitions. The West Calder No. 1 well (Fig. 16.4) in West Lothian reached a depth of 1,196 m (3,923 ft) but found few strata of reservoir quality in the Calciferous Sandstone Measures; minor traces of oil and gas were found. D'Arcy No. 1, in Midlothian, had a similar target and proved more successful. Gas was found at 221 m (724 ft) and flowed, on test, at c. 8,500 m³ per day. Oil was found at 552 m (1,810 ft) and about 50 barrels of waxy crude were recovered over a 2-month period in 1922 (see also Carruthers et al. 1927).

In 1918 Parliament passed the Petroleum Production Act establishing the requirement to obtain a licence to drill for petroleum in the UK. However, further exploration in Scotland did not occur until after a second Petroleum Production Act, in 1934, clarified matters of ownership of oil and rights of access to land which had earlier been the subject of dispute in England following an oil discovery on the Duke of Devonshire's estate (Hallett et al. 1985).

When exploration interest revived, the D'Arcy Exploration Company (exploration subsidiary of the Anglo–Iranian Oil Company, later to become British

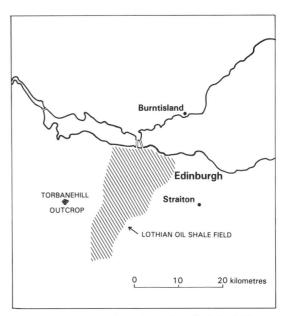

Fig. 16.3. Location of Lothian oil-shale field and the Torbane Hill outcrop.

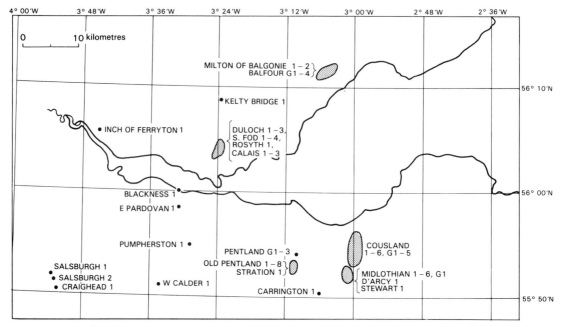

Fig. 16.4. Location of onshore hydrocarbon exploration wells in Scotland. All wells drilled to date lie in east-central Scotland except Sutherland No. 1 (not shown) located on the northern shore of the Moray Firth (see Fig. 16.2.).

Petroleum) and the Anglo–American Oil Company obtained licences to drill the NNE–SSW-trending D'Arcy–Cousland anticline. Cousland No. 1 was drilled by D'Arcy in 1937–38, *c.* 15 km SE of Edinburgh, and encountered gas in the Lower Carboniferous at 362–647 m (Lees & Taitt 1946). Rates of up to *c.* 170,000 m^3 of gas per day were obtained on test from sandstones with *c.* 15 per cent average porosity and average permeabilities of a few hundred millidarcies. Sandstones were impregnated with oil but very little flowed on test. Between 1938 and 1954, 8 appraisal wells were drilled and in 1957 commercial production from Cousland started, feeding gas by pipeline to the gas works at Musselburgh *c.* 8 km to the NW where it was mixed with town gas. Some 6·4 million m^3 of gas were produced in total before production ceased in 1965 (Holloway 1986). The area of the D'Arcy–Cousland anticline is currently under licence to Lasmo.

The initial Anglo-American Oil Company's well on the D'Arcy structure (Midlothian No. 1) was drilled in 1937–38. The structural closure tested at Lower Carboniferous level lies a few kilometres to the SSW of the Cousland structure, along the trend of the D'Arcy–Cousland anticline (Fig. 16.5). Oil was found and the discovery later appraised by 6 additional wells between 1938 and 1940. Cumulative production of *c.* 30,000 barrels of oil was achieved until shut-down in 1965. The Stewart No. 1 well

was also drilled on this anticline in 1981 but is reported to have been dry.

The D'Arcy company tested two other Carboniferous prospects in the Midland Valley before 1940, one within the Balfour anticline in Fife, by 4 wells in 1938–39, and the other at the Pentland Fault, by 3 wells in 1939 (see Lees & Taitt 1945, Tulloch & Walton 1958). The latter prospect was a potential structural trap beneath a sealing low-angle reverse fault but on drilling the fault proved to be too steep to form an effective trap.

Exploration resumed in 1944 with the drilling of D'Arcy's Salsburgh No. 1 near Airdrie. It encountered gas-bearing sandstones in the Calciferous Sandstone Measures at 835 m (2,739 ft) and flowed *c.* 9,300 m^3 per day on test. Wells were also drilled at Blackness and Easter Pardovan between Linlithgow and South Queensferry, in 1945–46 (Falcon & Kent 1960) but, with a gas flow of only 368 m^3 per day from Easter Pardovan No. 1, neither well proved commercial accumulations. A long gap in exploration drilling then lasted until D'Arcy's Pumpherston No. 1 well in 1962–63; this well discovered gas which was never produced. In 1966 the small American independent, Caprock Oil Limited, drilled 3 exploration wells in the area of the Rosyth anticline, Fife; Calais No. 1, near Dunfermline found oil shows, but both Duloch No. 1 and South Fod No. 1 were dry.

There was no commercial drilling for oil and gas in

Scotland during the 1970s but the 1980s have seen onshore exploration activity reach new peaks both in Scotland and elsewhere in the UK. Twenty-four wells have been drilled in Scotland in the period 1980 to 1987, including one appraisal well drilled in 1986 on a prospect in west Fife tested originally by Milton of Balgonie No. 1 in 1984–85. All onshore wells in Scotland have been aimed primarily at Carboniferous targets in the Midland Valley, with the exception of Premier Oil Company's Sutherland No. 1, drilled near Helmsdale, on the northern shore of the Moray Firth (see Fig. 16.6). The target here was presumably the Jurassic sandstones which, c. 30 km offshore to the east, provide the reservoir for oil in the Beatrice Field (UK Block 11/30).

The licensing regime relating to onshore UK changed in 1984 (see Hallett et al. 1985); the current system has 3 types of onshore licence (1) the exploration licence (EXL), valid for 6 years, which allows the licensee both to search and drill for petroleum, subject to planning agreement, (2) an appraisal licence (AL), valid for 5 years, necessary for the detailed assessment of a discovery, and (3) a development licence (DL), valid for 20 years, for the purpose of field development and production (Department of Energy 1984). Following the practice for offshore licence awards, applications for onshore exploration licences are now invited (1) within specified periods (or licensing rounds) and (2) for a block or blocks, each individual block to be 10 km × 10 km in area (Department of Energy 1984). To date, three rounds of onshore licensing [under the terms of the new Petroleum (Production) (Landward Areas) Regulations, 1984] have been held, in 1985–86, in 1987–88, and in 1989. As well as the licences covering areas in east and central parts of the Midland Valley, there is licensed acreage in the north between Inverness and Dornoch (Fig. 16.6). The target horizon in these northern areas is the Devonian, with hydrocarbon play concepts relying on intra-Devonian lacustrine mudstones as potential source rocks.

The Department of Energy includes parts of the inshore coastal waters around Scotland as 'onshore' areas for licensing purposes. Licences are currently held in the Minch east of Lewis, in north-west Skye and the

Fig. 16.5. Cross-section through the D'Arcy anticline (after Hallett et al. 1985). See Fig. 16.2 for location.

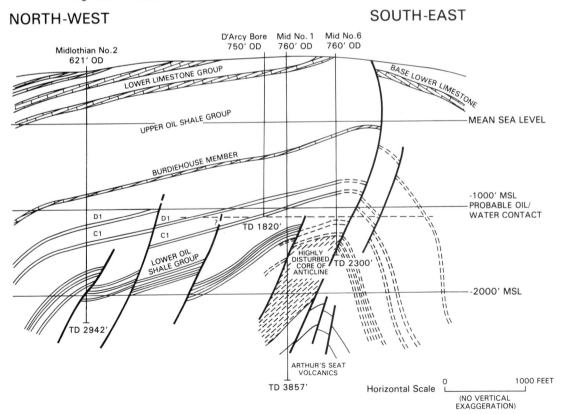

immediate offshore area, in the Sea of Hebrides south-east of Barra, and offshore Mull. Targets in these areas presumably lie within the Mesozoic succession. A licence is also held in the Firth of Clyde south of Arran where late Palaeozoic strata may be the target for exploration. No wells have been drilled in these areas to date but seismic surveys have been shot.

Finally, an area in the southern part of the Solway Firth

is licensed, part of a recent wider interest in the Solway basin and its eastern extension into the Northumberland Trough. Scott & Colter (1987) draw analogies between the Solway Basin, with its Carboniferous to Triassic fill, and the nearby Irish Sea Basin where Westphalian Coal Measures are the source of the gas in the Triassic sandstone reservoir of the Morecambe Bay Field. In the Solway Basin, Triassic sandstones have reservoir

Fig. 16.6. Areas of Scotland covered by hydrocarbon exploration or appraisal licences (April, 1988). The UK Department of Energy includes parts of inshore coastal waters around Scotland as 'onshore' areas for licensing purposes.

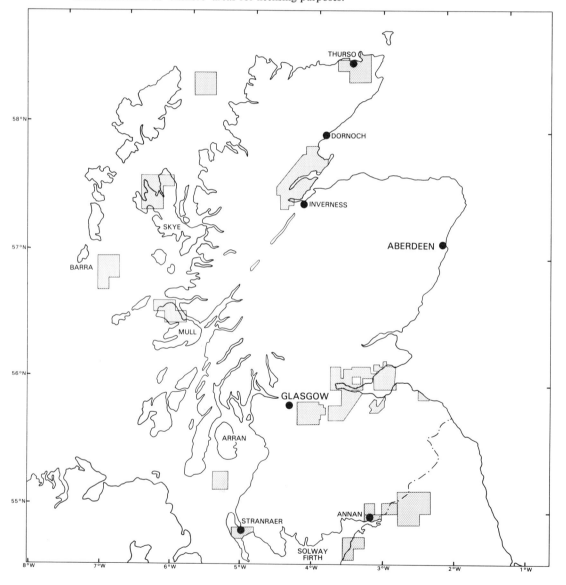

potential but the lateral impersistence of intra-formational marls limits the availability of caprocks to seal Triassic traps. A more favourable situation for hydrocarbon entrapment may be afforded by Permian shales overlying the Permian Penrith Sandstone (Scott & Colter 1987).

The onshore basins in Scotland remain to be fully explored for naturally occurring oil and gas. Only the more obvious anticlinal traps have been tested in the Carboniferous of the Midland Valley and all the skill and innovative thinking of the explorationist, together with a favourable economic climate, are now needed to promote further investigations.

Offshore basins west of Scotland

The UK designated continental shelf west of Scotland covers a vast area which has been little explored for hydrocarbons (Brooks 1986). Commercial drilling began in 1972 with well 206/12–1 (see Fig. 16.7), drilled c. 75 km west of Shetland, and to date 68 exploration wells have been drilled in this and adjacent areas to the north, which correspond structurally to the West Shetland Basin, the Rona Ridge and the Faeroe Basin (Figs. 16.8, 16.9).

The Rona Ridge separates the West Shetland Basin in the east from the Faeroe–Shetland Basin in the west. These features have a pronounced NE–SW structural grain and belong to a system of Mesozoic, and perhaps older, basins on the eastern margin of the North Atlantic that includes the Rockall Trough farther south and the Møre Basin off western Norway. The West Shetland Basin has a relatively thin Jurassic but thick Cretaceous sequence resting on Triassic and late Palaeozoic strata, and overlain in turn by a Tertiary succession which extends farther west, towards Shetland, to rest on

Fig. 16.7. Northern part of the UK Continental Shelf area (gridded). The grid has numbered quadrants, 1° latitude by 1° longitude in area. Each quadrant is subdivided in 30 blocks, each block 10′ latitude by 12′ longitude in area (inset). The grid is used to number wells, e.g. well 206/12–1 is the first well drilled in Block 12 within Quadrant 206.

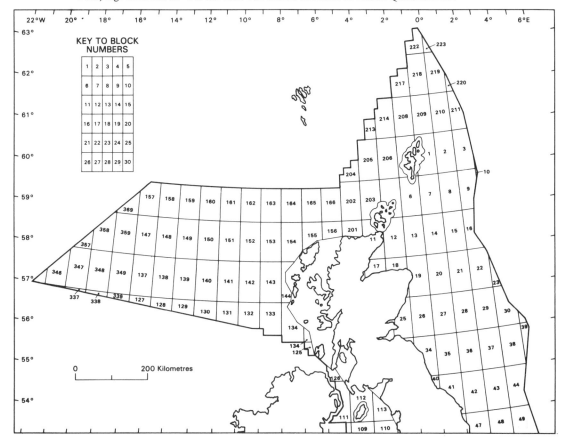

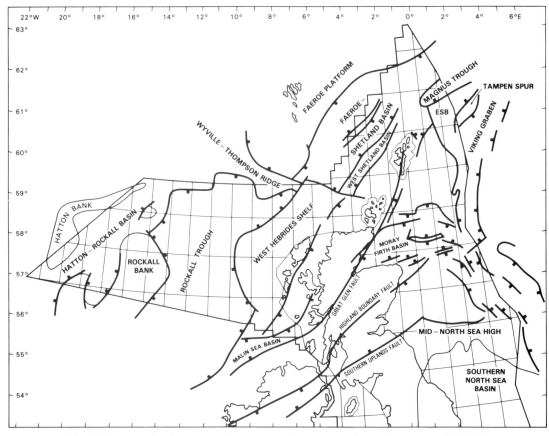

Fig. 16.8. Generalised map of major structural elements in the northern part of the UK Continental Shelf (gridded area). For explanation of the grid see Fig. 16.6.

Lewisian basement over the West Shetland Platform (see Hitchen & Ritchie 1987). To the west of the Rona Ridge, with its thinned late Cretaceous cover on Devono–Carboniferous, the Tertiary and Mesozoic section thickens; 8–10 km of Tertiary and Cretaceous strata are presently in the Faeroe Basin but the thickness of the

Jurassic and the nature of the pre-Jurassic section are yet to be determined (Mudge & Rashid 1987, Hitchen & Ritchie 1987).

The Tertiary succession includes thick (up to *c.* 6 km, Bott & Smith 1984) sequences of basaltic lavas and is intruded by sills; seismic resolution beneath the lavas and

Fig. 16.9. Cross-section through the West Shetland and Faeroe basins (after Hitchen & Ritchie 1987). The position of the Clair oilfield (UK Block 206/8) is shown, located over the Rona Ridge. Its Devono–Carboniferous (D–C) sandstone reservoir is overlain by Cretaceous shales. See Fig. 16.2 for location.

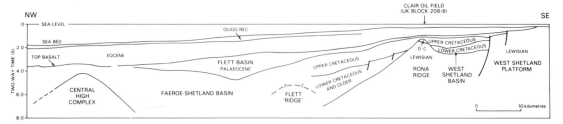

the distinction between sills and sedimentary features on seismic profiles cause problems during exploration. The lavas obscure the western margin of the Faeroe Basin and thick piles of lavas form the NW–SE-trending Wyville–Thomson Ridge (Roberts et al. 1983). The lavas extend westwards to the Faeroe Islands where c. 5 km are exposed.

Bailey et al. (1987) limit major hydrocarbon generation from Mesozoic strata to the Faeroe Basin where Kimmeridgian to Ryazanian marine shales have source potential. These authors suggest that shales present on the eastern edge of the Faeroe Basin may be mature for oil generation, whereas farther west, deeper burial may have taken the shales into the gas generating zone. Reservoir potential in the west Shetland area occurs at Devono–Carboniferous level (cf. Clair Field UK Block 206/8), in the Jurassic to Cretaceous where submarine fan or fan-delta systems derived from active fault scarps may produce suitable sand bodies (Meadows et al. 1987), and in the Tertiary, possibly also developed in submarine fan sequences (a summary of the stratigraphy is given in Fig. 16.10).

The first exploration success came in 1977 with the discovery of the Clair oil field by BP well 206/8-1A, drilled to test a prospect on the Rona Ridge, c. 80 km west of Shetland. The Clair structure is a NE–SW-trending fault block, with reservoir sandstones of Devono–Carboniferous age in red-bed fluviatile facies, and additional oil in subjacent fractured basement gneisses. The trap has a caprock of Cretaceous shales and the likely source of oil is the Upper Jurassic shales which occur in the flanking Faeroe Basin, lying to the west (Bailey et al. 1987). The large reserves, 3·5 to 4·0 billion (10^9) barrels of oil in place (467–533 million tonnes), make it the largest oil accumulation yet known on the UK continental shelf (Ridd 1981). There has, however, been no commercial production from the Clair accumulation to date. Ridd (1981) notes that a number of factors, the faulted nature of the reservoir, rapid lateral variation in reservoir quality, the 'heaviness' of the oil (22– 25° API) and the relatively shallow depth to the reservoir (c. 1,350 m), will make development difficult.

Other discoveries in this area classed as 'significant' by the Department of Energy (1990) include those made by British Gas well 214/30–1, located in the Faeroe Basin, and the Britoil well 206/1–2, located on the eastern edge of the Faeroe Basin; both are gas discoveries. Jackson (1986) notes that the 214/30–1 well was important in confirming the prospectivity of a Tertiary sand play in the west of Shetland area.

Despite these highlights, the results of exploration activity west of Shetland have to date been disappointing by the standards of the North Sea and when assessed against the level of economic return necessarily demanded by operating companies exploring in such a high risk, hostile environment. Advances in drilling technology and rig design continue to improve the industry's capability to explore in exposed, deep water areas; in 1986 the Chevron company drilled well 213/27–1 under 800 m of water in the Faeroe Basin, the greatest water depth yet for a commercial well in the UK sector.

The North Sea basins

In 1989, UK offshore oil production was 86·7 million tonnes, all from sedimentary basins lying broadly to the east and north-east of Scotland. With oil consumption in 1989 at c. 81·6 million tonnes, the UK remains a net exporter of oil (Department of Energy 1990).

Production in the UK North Sea east of Scotland started in 1975 from the Permian reservoir of the Argyll oil field (Block 30/24: Pennington 1975). The first exploration successes in the UK sector had occurred in 1965 when gas was discovered in Rotliegend (Lower Permian) aeolian sandstones in the Southern North Sea basin, east of the Humber estuary. The first exploration well in the basins off Scotland was 29/23-1, drilled in 1967 c. 250 km east of the Forth: the first exploration success came in 1969 with well 22/18-1 discovering the Montrose oil field, with its Palaeocene sandstone reservoir (Fowler 1975). At present 33 fields are in production in the area east and north-west of Scotland. During 1987, 46 exploration wells were spudded and 46 wells appraised previous discoveries in this area, evidence of substantial commercial interest persisting in the North Sea.

The production of gas from basins east of Scotland is relatively small. The only gas field in production is the Frigg field (UK block 10/1) which straddles the UK–Norwegian median line (Fig. 16.11). Gas has also been found along with oil in a number of other fields. This associated gas is currently being produced, together with oil, from late Jurassic sandstones in the Fulmar field (UK Block 30/16), from mid- and late Jurassic sandstone reservoirs in the East Shetland Basin, and from late Jurassic sandstones in fields in the eastern Moray Firth. Total gas production from the UK in 1989 was c. 45 billion (10^9) cubic metres, of which 3,363 million cubic metres came from the UK share of the Frigg field and c. 11,337 million cubic metres was associated gas obtained from oil fields, some of which was used on the offshore production platforms (Department of Energy 1990).

In the UK Southern North Sea, gas is derived from Carboniferous Coal Measures and mainly trapped in Rotliegend reservoirs by Zechstein (late Permian) salt. Some gas is also found in Triassic sandstones. In the basins east and north-east of Scotland, the hydrocarbon play types are different and more diverse: oil production is predominant. Reservoirs range in age from Devonian to Eocene (see Fig. 16.10) and occur in traps developed in both intra-basinal and basin margin locations, by one or a combination of faulting, truncation of strata by

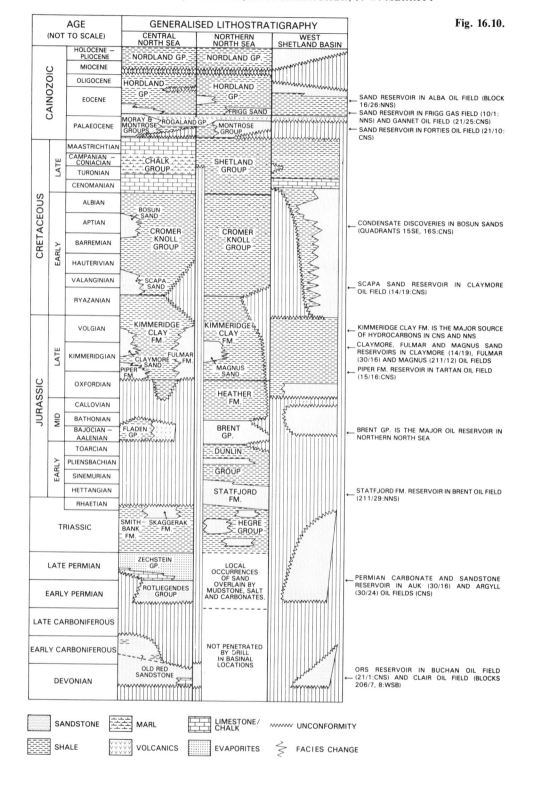

Fig. 16.10.

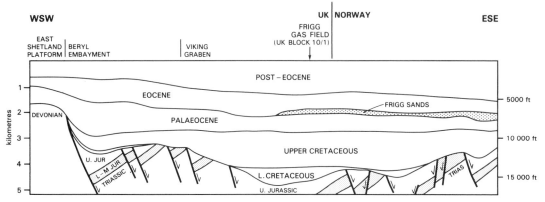

Fig. 16.11. Schematic cross-section through the Viking Graben, northern North Sea, showing the location of the Frigg gas field. See Fig. 16.2 for location.

erosion, movement of Zechstein salt, and local stratigraphic variation (see Parsley 1986). The key factors in the remarkable productivity of the North Sea oil province, and in the location of its hydrocarbon accumulations, are the nature, distribution and depth of burial of the dark, organic-rich marine shales of the late Jurassic to earliest Cretaceous Kimmeridge Clay Formation. The shales are widely and thickly developed at depths in excess of 3 km throughout much of the North Sea's Mesozoic rift system, consisting of the Viking and Central Grabens and the complex of fault-bounded sub-basins within the Moray Firth area (Fig. 16.8). These conditions are extremely favourable for the generation of large volumes of oil. The oil generating 'kitchen' in the North Sea grabens lies at depths of *c*. 3 to 4·2 km, oil-prone source rocks buried to even greater depths will yield gas (see Cornford 1986 for a fuller account). Deeply buried Kimmeridge Clay Formation is probably the source of much of the associated gas in the central and northern North Sea areas but Heritier *et al*. (1981) postulated a Lower Jurassic source for Frigg gas. An alternative source for the waxy oil in the early to mid-Jurassic sandstones of the Beatrice oil field (UK Block 11/30) has also been proposed; Devonian lacustrine shales and/or mid-Jurassic paralic shales are postulated. The Kimmeridge Clay Formation in this area has never been buried deeply enough to reach maturity for oil generation.

The oldest North Sea reservoir rocks in the basins east

Fig. 16.10. Stratigraphic summary for offshore areas, together with highlights of oil/gas occurrences. No formal or broadly accepted, informal lithostratigraphic terminology presently exists in the public domain for the succession in the West Shetland Basin.

of Scotland are sandstones of Devonian age and produce oil from fracture porosity in the Buchan field (UK Block 21/1). The Devonian reservoir, part of the sediment infill of the Orcadian Basin, occurs in a horst overlain by a Lower Cretaceous mudstone seal. Oil has migrated into the horst structure from the Kimmeridge Clay Formation present in the flanking down-faulted 'lows' (Butler *et al*. 1976, Richards 1985).

Carboniferous sandstones and Permian carbonates make a minor contribution to the oil accumulation in the Claymore oil field (Block 14/19), situated in the eastern part of the Moray Firth Basin; the main reservoir zones are in late Jurassic and early Cretaceous sandstones (Maher & Harker 1987). Permian sandstone and dolomite, the latter with vuggy porosity formed by leaching during phases of post-Permian erosion, form the reservoirs in the Auk (Block 30/16) and Argyll (Block 30/24) oil fields (Brennand & van Veen 1975, Pennington 1975). The fields are situated on the up-faulted flanks of the Central Graben and the Kimmeridge Clay Formation, found off-structure, was the source.

The Triassic has, to date, proved to be of limited economic importance within the oil province of the UK North Sea. In the central North Sea, an oil discovery has been made in Block 30/13 (Josephine structure) and oil has also been encountered in Blocks 22/16 and 22/9b (Fisher 1986). Farther north, Marcum *et al*. (1978) report some oil in Triassic beds below the Jurassic reservoirs in the Beryl oil field (Block 9/13) and, of most significance in the UK sector, the Brent and Stratfjord fields (Blocks 211/29 and 211/24 respectively) have substantial reserves of oil in the fluvial to shallow marine Statfjord Formation which spans the Triassic–Jurassic boundary. Condensate has been encountered in the Triassic in well 211/13-1 (Brooks 1977). Reservoir characteristics of Triassic sandstones in the central and northern North Sea are commonly poor, but Burley (1984) has reported secondary porosity generation in sandstones of the

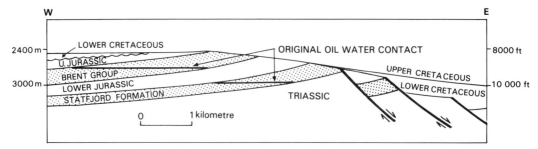

Fig. 16.12. Schematic cross-section through the Brent oil field (UK Block 211/29) in the northern North Sea (after Bowen 1975). It shows the tilted fault-block geometry of the trap, with up-dip truncation. Mid-Jurassic Brent Group and early Jurassic–Triassic, Stratfjord Formation sandstone reservoir units are truncated beneath late Jurassic–Cretaceous shales. For location see Fig. 16.2.

Skagerrak Formation in the central North Sea which may enhance Triassic prospectivity.

The most productive geological system in the North Sea as a whole is the Jurassic (Brown 1986) which accounts for *c.* 70 per cent of oil reserves in producing fields (Parsley 1986). In addition to the Statfjord Formation, reservoir sands occur in (1) the mid-Jurassic deltaic to shallow marine Brent Group (Brown *et al.* 1987) and its lateral equivalents, (2) in the shallow marine late Jurassic sands of the Piper Formation in the eastern Moray Firth and the Fulmar Sands of the Central Graben, and (3) in the late Jurassic submarine fan sands of, for example, the South Brae and Magnus oil fields (UK Blocks 16/7 and 211/12 respectively) in the Viking Graben (Fig. 16.11). The most common hydrocarbon trap is one formed within a fault-bounded block and sealed by late Jurassic to early Cretaceous shales (partially by late Cretaceous shales in the Magnus field, De'Ath & Schuyleman 1981).

The Brent oil field (UK Block 211/29) has sandstone reservoirs at two horizons, the mid-Jurassic Brent Group and the Triassic to early Jurassic Statfjord Formation. The trap is formed both by the structural tilt of a fault-bounded block (Fig. 16.12) and the truncation, over the crest of the structure, of the tilted strata which are in turn overlain unconformably by impermeable late Jurassic to Cretaceous shales. Bowen (1975) estimated the total displacement of the Brent fault-block, relative to the adjacent block to the east, as in excess of 1,800 m.

The South Brae oil field has a faulted anticlinal trap developed over the hanging wall of a basin margin fault (Fig. 16.13). The field has an oil column of *c.* 500 m, greater than the height of closure of the anticline; closure is partly achieved against Devonian strata, which have been faulted against the Upper Jurassic at the basin margin. The reservoir sandstones and conglomerates pass laterally to interbedded very fine sandstone, whose permeability is too low for it to be an effective reservoir, and dark marine shales. The dark shales belonging to the

Kimmeridge Clay Formation and are the likely source of the Brae oil. The Kimmeridge Clay Formation also acts as a top-seal. The Brae reservoir sediments were shed from the nearby platform area to the west of the field (Fladen Ground Spur) during the late Jurassic growth on the basin margin faults (see Turner *et al.* 1987).

Early Cretaceous sandstones in the UK central and northern North Sea are economically most important in the eastern Moray Firth where, as well as forming part of the pay zone in the Claymore oil field (see above), they also contain gas condensate in the SE part of Quadrant 15. The late Cretaceous to Danian Chalk in the Norwegian and Danish sectors of the Central North Sea has proved to be a very important oil reservoir (for example in the giant Ekofisk field in Norwegian waters), but no commercial accumulation of hydrocarbons has been found to date in the UK sector despite the widespread development, south of *c.* 59°N, of comparable Chalk facies. The key to this lack of success may be the poorer sealing characteristics of the overlying Palaeocene sediments; in the UK area strata above the Chalk consist of silt and sands, in contrast to the Palaeocene muds which occur in the Norwegian and Danish sectors. The muds were deposited at more distal locations in the lateral facies tract of a Palaeocene submarine fan system which had its main sediment source to the NW (Rochow 1981). The sealing capacity of the strata above potential chalk reservoirs is not only crucial to the entrapment of oil but is also important in sustaining an overpressuring of the chalk reservoir which helps to retain sufficient porosity and permeability between the grains of coccolith debris by inhibiting mechanical compaction during burial.

Palaeocene reservoirs are of major importance in the basins east of Scotland, most notably in the Forties oil field (Carmen & Young 1981; UK Block 21/10) which up to the end of 1989 has produced 265 million tonnes of oil, more than any other field in the UK North Sea

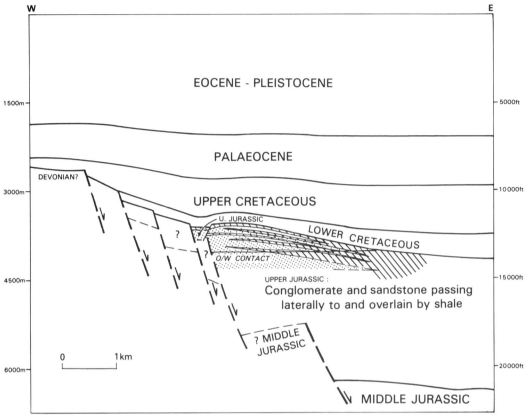

Fig. 16.13. Schematic cross-section through the South Brae oil field (UK Block 16/7), south Viking Graben (after Harms *et al.* 1981). The oil in late Jurassic sandstones and conglomerates is partly trapped against the bounding fault to the west. For location see Fig. 16.2.

(Department of Energy 1989). The Forties field has submarine-fan sandstones as its reservoir and a Palaeocene mudstone seal. The trap is formed by dip-closure which results from the Palaeocene draping over a deeper, intra-basinal horst. The oil is derived from the Kimmeridge Clay Formation. A second trap style at Palaeocene level is constructed by dip-closure over a Zechstein salt dome (e.g. in the Maureen oil field, UK Block 16/29). Palaeocene sands pinching out on the platform just west of the Viking Graben contain heavy oil (API gravity 20°) in the UK Block 3/28 (Bressay Field).

Until recently the economic significance of the Eocene sequence in the UK North Sea was restricted to the large Frigg gas field (UK Block 10/1), where the gas is present in a stratigraphic trap formed by differential compaction of submarine-fan channel and levee sands surrounded by siltstones and mudstones. Oil has now been discovered elsewhere in Eocene sands in UK Blocks 9/18 and 9/23 in

the central part of the Viking Graben, and in UK Block 16/26 (Alba discovery) in the south Viking Graben.

Throughout the North Sea, oil derived from the Kimmeridge Clay Formation is typically a medium gravity, naphththeno-paraffinic crude (*c.* 36° API on average), usually with a lower sulphur content (Cornford 1986). The heavy oils (20° API) which occur locally are probably formed by the bacterial degradation of 'normal' oil derived from the Kimmeridge Clay Formation.

Reserves

A survey of UK North Sea oil and gas discoveries by Brennand & Van Hoorn (1986) clearly shows a progressive reduction in size of discovery with time. Target prospects in exploration are becoming smaller, more subtle and/or deeper but despite this new discoveries continue to be made, including 16 in basins east of

Scotland during 1989 (Department of Energy 1990). The North Sea is therefore still capable of providing substantial rewards to the successful explorationist.

Reserve figures for the UK Continental Shelf as a whole, issued by the Department of Energy (1990), put the total remaining oil reserves in existing discoveries at between 510 (proven reserves) and 1,810 (maximum possible) million tonnes, the bulk of which are to be found in the North Sea to the east and north-east of Scotland. The current estimate of undiscovered recoverable oil reserves in basins offshore Scotland is around 510 to 2,620 million tonnes. For gas in the UKCS as a whole the remaining reserves in present discoveries are between 560 (proven) and 1,770 (maximum possible) billion cubic metres. The estimate of undiscovered recoverable gas reserves for the UKCS is between 210 and 1,152 billion cubic metres. Most gas production and remaining reserves lie to the south of Scottish waters, in the Southern North Sea and Irish Sea areas.

S. Brown

Construction and industrial minerals

In value terms, coal, gas and oil are dominant in the Scottish economy, but the construction and industrial minerals industries nevertheless make a very important contribution, providing direct employment in the extraction processes and indirect employment in downstream processing and manufacturing activities. In the UK as a whole, production of oil, gas and coal in 1987 was valued (BGS 1989) at £16·5 billion as against £1·7 billion for construction and industrial minerals and £0·2 billion for metals, but comparable figures for Scotland are not readily available.

Aggregates

Output of aggregates in Scotland (Fig. 16.14) reached an all-time peak in the early 1970s, fell dramatically in the mid 1970s and is now rising once more. The earlier peak relates to a period of considerable activity in the building and construction industry stimulated by the discovery of North Sea oil. Production of crushed rock aggregate is now clearly outstripping that of naturally occurring sand and gravel; 27·3 million tonnes of aggregate were produced in 1987 of which 62 per cent was crushed rock and 38 per cent was naturally occurring sand and gravel. The aggregates industry is very important to the Scottish economy, but the sand and gravel side, in particular, is placing increasing pressure on planning authorities to balance necessary mineral exploitation with competing uses of land and with conservation of the environment. Detailed resource appraisals are required in order that quarrying and extraction proposals can be examined objectively (Thurrell 1981). A comprehensive report covering all aspects of the aggregates industry in the UK has been published by the Geological Society of London (Collis & Fox 1985).

Crushed rock

Over 60 per cent of crushed rock aggregate produced at present in Britain is limestone, but in Scotland, 89 per cent comes from rocks classified as 'igneous and metamorphic', 9 per cent is sandstone (mainly greywacke) and only 2 per cent is limestone. Roadstone accounted for 56 per cent of Scottish output in 1987, of which 27 per cent was coated and 73 per cent uncoated; fill and 'various constructional uses' including rail ballast accounted for 37 per cent of output and the remaining 7 per cent was used in concrete.

The overriding property determining the suitability of a rock for making aggregate is its crushing strength, which is mostly governed by the nature of the rock, but can be affected quite significantly by the state of weathering and alteration. As long as a rock has a crushing strength equal to, or better than that of the 'average' Carboniferous limestone (of England), it will be suitable for applications accounting for 80 per cent or more of the total production of crushed rock aggregate. The mechanical properties, other than crushing strength, that are critical in more specific applications are resistance to polishing and wear, which governs the suitability of rock chippings for road surfacing, and drying shrinkage, which governs the versatility of aggregate for concreting purposes (Harris 1977).

Scotland is particularly well endowed with resources of rock suitable for producing aggregate (Merritt & Elliott 1984, Smith 1989a). In central Scotland, the most sought-after resources are dolerite sills of Carboniferous to Permian age, particularly the quartz-dolerites of the Midland Valley Sill Complex. Formerly quarried widely for making setts and curbs, these rocks are now very important sources of both general-purpose roadstone and premium-grade, road surfacing chippings. Sills of the kylite suite occurring in Ayrshire are exceptionally fresh, tough and versatile rocks. They are quarried near Troon for the complete range of crushed rock products, including low-shrinkage aggregate used in the manufacture of concrete oil platforms. Very low shrinkage, pale-coloured aggregate is produced from the friable Douglas Muir Quartz-conglomerate (Lower Carboniferous) of Milngavie and Strathblane, NE of Glasgow.

Felsite chippings were used extensively in road surfacing, especially in Lanarkshire and Fife, but they are rarely sufficiently skid-resistant to meet the more stringent specifications applying to heavily trafficked roads today. Felsites are often excellent sources of rail ballast and their commonly attractive red, pink or khaki colour can allow a market distribution much wider than ordinary roadstone or concreting aggregate. Other sources of coloured aggregate for harling and miscellaneous decorative uses include the Durness Limestone in Skye ('Skye marble') and Devonian sandstone in Shetland, which provide white and purple-coloured aggregates respectively.

Granites are potentially suitable for general road making and concreting applications and some of those occurring along the western seaboard have good export potential. Scotland's first coastal 'superquarry' exploits the Strontian Granite at Glensanda on the northern shore of Loch Linnhe; it is expected to produce over 7 million tonnes of aggregate per annum, all of which will be exported to southern Britain and overseas. Other candidates for coastal quarrying include the larger granite bodies of Shetland, Skye, Ballachullish and Etive together with the smaller intrusions by Lochs Fyne and Duich. The anorthosite within the metaplutonic complex of South Harris is expected to be capable of producing light-reflective, skid-resistant, road surfacing aggregate of a type that is particularly sought after on the continent. Other sources of road surfacing aggregate with export potential include the Old Red Sandstone of Shetland, the Torridonian Sandstone of NW Scotland and the greywackes of the Southern Uplands. Some limestones in NW Scotland and in Shetland might also be considered for coastal quarrying.

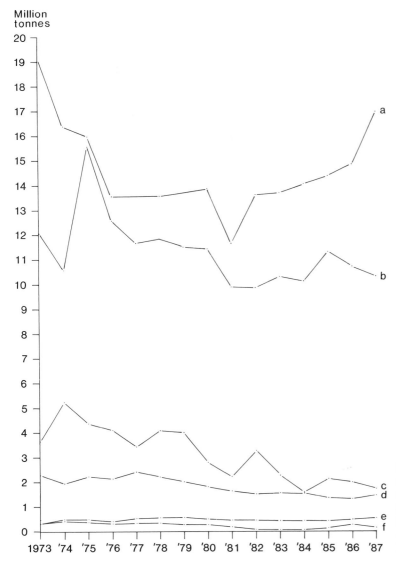

Fig. 16.14. Scottish production of (a) crushed rock aggregate, (b) common sand and gravel, (c) common clay and shale, (d) limestone, (e) silica sand, (f) fireclay. Source: British Geological Survey.

Sand and gravel

Costs of haulage make up a substantial proportion of the delivered price of sand and gravel; thus, despite rationalisation of the industry, the commodity is still exploited at localities scattered widely across the country. It is mainly derived from 'dry' workings in Quaternary fluvioglacial, alluvial, beach and aeolian deposits (in decreasing order of importance). Little aggregate is presently dredged offshore.

Production statistics (BGS 1989) indicate that 55 per cent of sand and gravel sold in Scotland was used in the manufacture of concrete and concrete products, for which it is traditionally preferred to crushed rock aggregate because of its lower cost and better workability. A further 19 per cent of production was used as 'fill', mainly in road sub-bases and embankments, 12 per cent was sold as 'mortaring' and 'sharp' sand to the building industry (including some used in the manufacture of calcium silicate bricks); fine-grained sand for mixing with bitumen to make asphalt accounted for another 8 per cent of production.

The quality of sand and gravel deposits varies significantly but modern methods of processing are capable of overcoming many problems, such as the removal of silt, clay and coal debris. Gravels containing weathered basic igneous rocks and certain types of greywacke derived from the Southern Uplands may, however, have adverse effects on the drying shrinkage and wetting expansion characteristics of concrete. Furthermore, cherts associated with the greywackes might cause alkali–silica reaction, certain phyllites may react with silicates, and dolomites might dedolomitise; these reactions can all cause potentially deleterious expansion in concrete.

The results of 17 detailed sand and gravel resource assessment surveys have been published by the BGS (Fig. 16.15) along with other, less detailed, reports covering the whole of Scotland.

Dimension stone

The demand for dimension stone for building virtually disappeared in the 1950s, but a general dissatisfaction with the aesthetics and lasting qualities of concrete, together with reawakening of architects to the visual appeal of natural stone, has led to an upsurge in the demand for granite and sandstone blocks. Until recently the main requirement has been for facing or cladding and for selected replacement of stone in restoration projects, but a market has now re-emerged for load-bearing cube stone.

In 1987, 68,000 tonnes of igneous rock and 5,000 tonnes of sandstone were quarried as dimension stone in Scotland, and in Britain as a whole there was an 83 per cent increase in production (BGS 1989). Granite is at present quarried on the Ross of Mull, several quarries

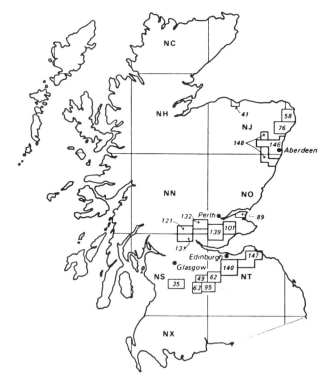

Fig. 16.15. Index to areas covered by detailed sand and gravel resource appraisals. Numbers refer to the sequential numbering of the British Geological Survey Mineral Assessment Report series.

work Permian sandstone near Elgin, Moray and around Dumfries, a sandstone in the Upper Limestone Group is worked near Airth, Stirling and another sandstone in the Upper Oil Shale Group is exploited at Burntisland, Fife (Bunyan *et al.* 1987). Flagstone is quarried near Thurso, Caithness.

Brickclay

The Scottish brick industry has undergone a major rationalisation in the past decade. In 1978 there were 54 brickworks, but by 1985 this number had been reduced to 18. Demand for bricks, particularly common bricks, has declined for several reasons, including fewer house starts and changing methods of building design and insulation. In Scotland, brickworks were traditionally associated with collieries and ironstone mines where carbonaceous mudstone spoil ('fiery blaes') was blended with specially dug mudstones ('virgin blaes'), and locally with boulder clay, to produce common, 'composition' bricks. Though quite strong, such bricks are typically black-cored and they effloresce or stain badly on exposure to the atmosphere. They are suitable for external walls only if they are rendered or harled. Today, the Scottish brick industry is accommodating an increasing demand for good quality and attractive-looking facing bricks.

Mudstones suitable for making facing bricks have been exploited towards the base of the Lower Carboniferous sequence around Carluke in Lanarkshire, and within the reddened Upper Coal Measures of the Central Coalfield (Elliot 1985). A large new facing brick plant in West Lothian is currently blending mudstone and fireclay dug from the basal parts of the Lower Coal Measures. Uniformity of raw materials, together with assured continuity of supply, affect the value of potential mudstone resources just as much as composition (Ridgeway 1982).

Plastic, stoneless clays from the Flandrian and late-Devensian marine and estuarine 'raised beaches' and 'carse clays' were formerly exploited extensively for the manufacture of pipes, pots, drainage tiles, roofing tiles and facing bricks, especially along the east coast of Scotland and in the lower Clyde valley. The last remaining plant exploits the late-Devensian Errol Beds at Errol, Tayside. The relatively high costs of firing superficial clays, together with the remoteness of many of the deposits, are the two factors mostly responsible for the decline of this sector of the industry.

Silica sand and silica rock

Silica sand (including disaggregated friable sandstone) is a relatively pure quartz-rich commodity that is used for purposes other than construction work, principally in glassmaking and for foundry moulding purposes. Silica rock (lump silica) is a hard, compact, even-grained rock consisting essentially of quartz from which silica bricks and other silica refractories are manufactured.

Demand for glassmaking sand has declined in the last decade because of the import of glass, the increased use of plastic substitutes and re-cycling. As with foundry sand, and with fireclay, the decline has affected the lower quality end of the market the most. At present, the only source of colourless glass sand in Britain suitable for high-grade domestic and decorative glassware is the Upper Cretaceous White Sandstone of Morvern, which is worked at Lochaline and transported by ship to the glassworks at Runcorn, Merseyside (Smith 1989b). It is a white, friable sandstone that contains up to 99·85 per cent silica and is mined on the room-and-pillar principle. Nationally important resources of silica sand also occur in the Passage Group of the Central Coalfield and Clackmannan Syncline (MacPherson 1986a), often in close association with the Lower Fireclays. Although used mainly in foundries, these white, friable sandstones (98·5–99·0 per cent SiO_2) can be acid-leached to produce a sand suitable for manufacturing colourless glass (Highley 1977).

Silica sand is used for moulds in the production of over 95 per cent of iron, steel and non-ferrous metal castings. High silica content is important, because impurities can affect refractoriness, but the size, shape and roundness of the sand grains are crucial properties that control the permeability and strength of the moulds. There is a trend towards the acceptance of finer grained sands that produce better finishes to the castings. However, demand for foundry moulding sand and silica rock have both decreased significantly over the past two decades, mainly as a result of the decline of heavy industry in Scotland.

Most sands that are exploited for glassmaking are now used also in foundries, because clean, washed sand is being increasingly preferred to naturally bonded (clayey) sands. The latter are now only obtained from the Upper Limestone Group in North Ayrshire and near Coatbridge. Blown dune sand is dug at Ardeer in Ayrshire for use in foundries and in coloured-glass manufacture. Silica sand is used in water filtration beds and some is exported from Lochaline to Scandinavia for the manufacture of silicon carbide ('carborundum'). Lower quality sands and tailings are sold for a variety of uses in the building and construction industry but particularly for making low-shrinkage concrete products. There is a potential market in the petroleum industry for well-graded, well-rounded siliceous sand in the hydraulic fracturing of oil-reservoir rocks.

Highly siliceous sand (<97·7 per cent SiO_2) forms the matrix to the Douglas Muir Quartz-conglomerate (see above). The pebbles in this deposit consist principally of quartz and they may be a potential source of lump silica, like the Dalradian Binnein Quartzite (<99·1 per cent SiO_2) of the Loch Leven and Glencoe areas and the Scaraben Quartzite (<98 per cent SiO_2) of the Moine near Helmsdale. The orthoquartzite at the base of the Cambro–Ordovician succession of the NW Highlands is an established source of high-grade silica rock (<97 per cent SiO_2).

Limestone and dolomite

Scotland does not possess large resources of thickly bedded, high-purity limestones like those of southern Britain, but limestones and dolomites are nevertheless quite widespread and some have good commercial potential (MacPherson 1986b, Grout & Smith 1989a). Limestone has a wide variety of uses (Harris 1982) that can be grouped under the headings chemical, construction and cement manufacture.

On the whole, Scottish limestones available for large-scale opencast extraction are insufficiently pure and homogeneous for use in glass manufacture or as metallurgical fluxes, though some dolomites have been used by the refractories industry and in the manufacture of fibrous insulating materials. Limestones were once widely burnt for making agricultural lime, but this practice ceased because the higher neutralising value obtained through burning is insufficient to offset the cost of fuel. At first ground limestone replaced lime in agriculture, but magnesian limestone (Ca, $MgCO_3$), which has better neutralising properties than pure calcium carbonate, is now generally preferred. Although there are vast resources of dolomitic limestone in the Cambrian Durness Group, which crop out between Skye and Loch Eriboll, and in the Dalradian assemblage on Shetland and between the Moray Firth and Kintyre, increasing amounts of ground magnesian limestone are being imported from NE England.

The treatment of acidic drinking water derived from peat-covered uplands is an established use of limestone in Scotland, and the material may play an important future role in tackling various environmental problems associated with acidic rain. There is an urgent requirement to reduce high levels of acidity in soils and surface waters in northern Britain, and limestone free from toxic impurities is the most suitable agent for this purpose. Should flue-gas desulphurisation (FGD) plants be fitted at Scottish coal-burning electricity generating stations there will be a further requirement which may help offset the current dwindling demand for indigenous limestone in Scotland.

Cement is manufactured by heating a slurry of limestone and an aluminosilicate-rich constituent such as clay, mudstone or shale in a ratio of about 3:1. The principal chemical constraint governing the suitability of a limestone is that it must contain less than about 3 per cent MgO, which rules out dolomitic limestones. An ideal limestone for cement-making should be thick (>3 m), pure (>85 per cent $CaCO_3$), homogeneous and gently dipping, but there are few meeting these preconditions in Scotland apart from the small outcrop of Carboniferous limestone being exploited at Dunbar. Here, at the site of

Scotland's only cement works, available reserves are being rapidly exhausted. Alternative resources capable of supporting modern, 'dry process' cement manufacturing plants are restricted to other Lower Limestone Group limestones, particularly those cropping on the eastern side of the Midlothian Coalfield and in North Ayrshire. Increasing quantities of cement are being imported into Scotland in the form of clinker from England and overseas.

The reduced demand for powdered limestone in agriculture and as dampening dust to suppress the risk of explosions in coal mines has largely been offset by an increasing requirement for limestone filler in the manufacture of asphalt. Brucite-bearing dolomitic marbles, produced by the thermal metamorphism of the Durness Limestone adjacent to the Loch Borralan igneous complex near Ledmore, have been evaluated as a potential source of high brightness carbonate for filler and possible paper coating applications.

Fireclay

Scotland has large resources of fireclay that include some of the best quality, high alumina (>40 per cent Al_2O_3, calcined) fireclays in Britain (Merritt 1985). Fireclays are poorly bedded mudstones consisting essentially of kaolinite; most are seatclays below coals, but many of the most sought-after aluminous beds are not associated with coal and do not contain roots.

Valuable fireclays are mostly restricted to the Passage Group and the lower part of the Lower Coal Measures. In the Ayrshire Coalfield, the most exploited seams of high-alumina fireclay occur at the base of the Passage Group SW of Dalry (Douglas Fireclays) and along the northern outcrop of the Ayrshire Bauxitic Clay Formation, which lies immediately above the Passage Group lavas. In the Central Coalfield and Clackmannan Syncline, valuable high-alumina fireclays are located towards the base of the Passage Group (Glenboig, Bents or Lower Fireclays). Less aluminous seams were exploited towards the top of the Passage Group (Bonnybridge, Drum or Upper Fireclays) but they attract little attention today.

Fireclay was once worked extensively underground, especially in the Central Coalfield, but it is now wholly derived from opencast workings. The high-alumina fireclays are worked separately, but others are extracted together with associated coal seams. Supply greatly exceeds demand at present.

Fireclay production has fallen dramatically since the mid 1950s, reaching an all-time low in the early 1980s. This is mainly a result of the decline of the Scottish iron and steel industry, but another factor is that modern high-temperature smelting processes require firebricks with better refractory and abrasion-resistance properties; these products are manufactured from more aluminous, imported raw materials and from alternative commodities

(Highley 1982). Nevertheless, firebricks are still produced from indigenous fireclays and there is an increasing demand for specialised refractory products and calcined, high-alumina fireclay aggregate (grog, chamottes), much of which is exported for blending purposes. Scottish fireclay is used as a binder in animal foodstuffs, cosmetics and fertilisers. It is also used as a component in the manufacture of engineering building bricks and buff-coloured facing bricks, for which there is increasing demand at present.

Talc

Talc (*sensu stricto*) is a hydrated magnesium silicate, $Mg_3Si_4O_{10}(OH)_2$, but it generally occurs in intimate association with magnesite ($MgCO_3$) and dolomite. The versatility of talc as an industrial mineral stems from the fact that it is soft, smooth, white, chemically inert and refractory, has a high surface-area and has good lubricative, absorptive and insulating properties. The largest application of talc is in pharmaceuticals and cosmetics, but is is also used in the manufacture of paint, paper, roofing felt, ceramics, animal feedstuffs, fertilisers, pencils, fire extinguishers, welding rods, insecticides and polymers as well as in the cable and cloth making industries (Highley 1974).

Talc occurs widely throughout the Scottish Highlands and Islands, but economic resources (industrial grade only) are mainly restricted to Unst in the Shetland Islands (Grout & Smith 1989b), where Britain's only active quarry is situated. Scottish occurrences of talc have formed as secondary replacement products in serpentinised ultrabasic igneous rocks that have been subjected to intense shearing and deformation. Such sources are unlikely to yield cosmetic grade talc, which is presently imported into the UK. New quarries are likely to be opened on Unst and in a talc–magnesite deposit at Cunningsburgh, on Mainland Shetland. The material is presently shipped away for milling.

Other minerals

There has long been interest in the potash-rich shales of the Lower Cambrian Fucoid Beds of the NW Highlands, which are about 8 m thick between Loch Eriboll and Loch Broom and contain on average 8·5 per cent potash (Bowie *et al*. 1966). The potash occurs in the form of fine-grained orthoclase feldspar (adularia) from which the pure chemical cannot presently be extracted commercially, but the shales are now being worked at Loch Awe, south of Inchnadamph, for direct use as a source of slow-release fertiliser. The nepheline-syenites of Lochs Loyal, Ailsh and Boralan may also have potential uses in agriculture as sources of phosphate.

J. W. Merritt

Metalliferous minerals

The contribution of metalliferous minerals to the economy of Scotland increased significantly with the opening of Foss baryte mine in 1984 and development of the Cononish gold deposit (Parker *et al*. 1989). Their value is small in the context of hydrocarbons and aggregates but large tracts of Scotland's 77,000 km^2 land area are presently being explored, particularly for gold. This work is aided by the availability of comprehensive geochemical cover (Plant & Slater 1986), recent metallogenic interpretations (Plant *et al*. 1983, Russell 1985, Hall in press, Rice in press) and improved techniques in the field (for example Smith *et al*. 1984) as well as in the laboratory (for example, Potts & Prichard 1986, Fisk 1986). A database of information collected by BGS, companies and universities is maintained for public use by BGS, in addition to its published Geochemical Atlases and Mineral Reconnaissance Programme reports.

Prior to the modern phase of exploration, total production from vein deposits had been 800,000–900,000 tonnes of baryte, 400,000 t of lead, 10,000 t of zinc, 200 t of antimony, 75 t of silver and from gravels probably a few tonnes of gold. Chromite and ores of copper, nickel and manganese have also been recovered. Gasswater mine was Britain's largest baryte producer until 1964; no further working of metalliferous and allied minerals took place until 1983 when a modern mill was sited on the Strontian Main Vein to exploit baryte, galena and sphalerite (Mason & Mason 1983). The following year, direct shipping-grade baryte ore was opencast at Foss Mine and in conjunction with underground mining more than 50,000 tonnes are now being extracted annually for drilling-muds offshore. Plentiful reserves of baryte have been established at Foss and on the adjacent Cluniemore property where lead, zinc and silver are potential by-products (Gallagher 1984). The longevity of ore-forming processes is evident from the age-range of the gold, base metal and baryte occurrences listed in Tables 16.1, 16.3 and 16.4, spanning the Lower Proterozoic to the Triassic.

Precious metals

In the last two decades, stratabound, vein and epithermal gold and ophiolite-hosted platinum group minerals have been located in Scotland with the aid of improved analytical techniques (Fig. 16.16, Table 16.1). Earlier, alluvial gold was known to be widespread in the Highlands and Southern Uplands, notably around Leadhills and near Helmsdale where past extraction is estimated as at least 0·8 t and 0·1 t of gold respectively.

The oldest known gold occurrence is at Gairloch in a highly metamorphosed supracrustal sequence, the 2·4–2·0 by Loch Maree Group, where stratiform massive sulphide of Besshi-type is developed. The youngest is at Rhynie where a plant-bearing sinter, a silicified fault zone

Table 16.1. Summary of main precious metal locations (see Fig. 16.16)

Location	Type and mineral association	Host rock lithology	Age of host	Reference
Unst, Shetland	Sperrylite, stibiopalladinite, hollingworthite, irarsite, laurite and other PGMs with native Cu and Au, copper sulphides and nickel sulphides and arsenides	Concentrated in chromitites enclosed in dunite lenses within harzburgite of Unst ophiolite complex	L. Palaeozoic	Gunn et al. (1985); Prichard et al. (1986); Lord & Prichard (1989)
Helmsdale area	Alluvial Au and rare grains in migmatite; the Au is alloyed with Ag	Granitic migmatite of Strath Halladale complex	(?) L. Palaeozoic	Plant & Coleman (1972)
Gairloch, west Sutherland	Au–Ag inclusions in chalcopyrite associated with sphalerite and iron sulphides	Quartz–carbonate schist in metavolcanics and metagreywackes	L. Proterozoic	Jones et al. (1987)
Rhynie, Aberdeenshire	Anomalous levels of Au, As and Sb in oxidised pyrite	Fossiliferous chert and silicified fault zone in Lower Old Red Sandstone	L. Devonian	Rice & Trewin (1988)
Glen Almond, Perthshire	Alluvial Au, Au anomalies in felsic breccia	Quartz–mica schists and metavolcanic rocks	U. Dalradian	R. S. Middleton (pers. comm.); Plant et al. (1989)
Comrie district, Perthshire a. Milton Burn b. Invergeldie	a. Au inclusions in chalcopyrite, disseminated and in veinlets in a shear zone with pyrite b. Au, chalcopyrite and bismuthinite inclusions in arsenopyrite	a. Diorite of Comrie mass b. Feldspathic rocks in contact with amphibolite sill at margin of Comrie mass	late Silurian U. Dalradian	R. E. Hazleton (pers. comm.); Plant et al. (1989)
Tyndrum, Perthshire a. Cononish	a. Au occasionally visible but mainly as 5–100 μm grains intergrown with fine-grained galena within micro-fractures in pyrite grains; sphalerite also present; Au:Ag ratio of ore is 1:4·5	Structure up to 4 m wide with quartz vein axial to hydrothermal alteration zone cutting psammite, pelite, impure limestone, amphibolite. Also graphitic schist	L. Dalradian	Parker et al. (1989)
b. Other auriferous structures	b. Electrum, hessite (Ag_2Te), sylvanite ($AuAgTe_4$) and petzite (Au_3AgTe_2) associated with galena, pyrite and K-feldspar	Quartz–carbonate veins, feldspathic alteration zones in Eilde Quartzite	M. Dalradian U. Proterozoic	Pattrick et al. (1988)

Location	Mineralization	Host	Age	References
Lagalochan, Kilmelford	Porphyry-style pyrite–chalcopyrite–molybdenite with electrum, native Ag and hessite; with peripheral Pb–Zn–As sulphides	Subvolcanic diatreme–porphyry–breccia with zoned alteration	late Caledonian	Harris et al. (1988)
Alva, near Stirling	Native Ag, argentite, erythrite [$Co_3(AsO_4)_2.8H_2O$] with Cu–Pb–Zn–Fe sulphides	Worked calcite–baryte–quartz veins in andesitic volcanics	L. Devonian	Wilson & Flett (1921); Francis et al. (1970); Hall et al. (1982)
Hilderston, near Bathgate	Native Ag, erythrite, annabergite [$Ni_3(AsO_4)_2.8H_2O$] with Ni–Cu–Fe sulphides near surface, Pb–Zn sulphides in depth	Worked baryte–calcite vein in sandstone and tuff intruded by quartz–dolerite sill	L. Carboniferous sediments, Permo–Carboniferous sill	Wilson & Flett (1921); Stephenson et al. (1983)
Leadhills area	Alluvial Au (worked); major lead–zinc veins in area (see Table 16.3)	Greywackes and shales with late quartz (–pyrite) veins	Ordovician–Silurian	Gillanders (1981)
Hare Hill, near Dalmellington	Au enrichment with arsenopyrite in veined greywacke. Stibnite–quartz vein (worked) in area	Greywackes intruded by granodiorite	Ordovician greywackes, late Silurian granodiorite	Dewey et al. (1920); Naden & Caulfield (1989)
Fore Burn, near Straiton	a. Native Au in As–Fe sulphides with native Bi, bismuthinite, chalcopyrite; native Ag in tennantite–tetrahedrite b. Free visible Au in arsenopyrite–chalcopyrite veinlets	Quartz–carbonate–tourmaline veins in rhyodacites, tourmaline breccias and porphyrites Porphyritic diorite, K-feldspar alteration around veinlets	L. Devonian	Allen et al. (1982); Charley et al. (1989)
Moorbrock Hill, near Carsphairn	Native Au in intergrown chalcopyrite and tetrahedrite, with sphalerite and silver sulphides	Quartz–pyrite–arsenopyrite veinlets in fault zones at granodiorite–greywacke contact	Ordovician greywackes, late Silurian granodiorite	Naden & Caulfield (1989)
Glendinning, near Langholm	Au in arsenopyrite of both stratabound (disseminated) and vein type, enriched in Sb; worked stibnite vein in area	Greywackes, intraformational breccias, breccia veins and quartz–dolomite veins	Silurian greywackes; (?) early Devonian veining	Dewey et al. (1920); Gallagher et al. (1983, 1989); Duller (1990)
Glenhead, near Loch Doon	a. Native Au in quartz veins with arsenopyrite b. Disseminated As–Fe sulphides containing gold	Quartz veins in greywackes near margin of Loch Doon granodiorite Disseminated sulphide in mainly monzonitic minor intrusions and adjacent greywackes	Ordovician greywackes, late Silurian igneous rocks	Leake et al. (1981); Stone & Gallagher (1984)

and strongly altered sediments and volcanics are interpreted as a Lower Devonian hot spring system. An igneous source for the precious metals is well documented by the extensive research of the serpentinised ultrabasic rocks of Unst (Fig. 16.17). Concentrations of up to 7 g/t Au and 12 g/t Ir have been recorded from analyses of chromite-rich rocks. The gold commonly forms an alloy with Cu or Pd as tiny inclusions which together with the PGM are often associated with silicates in chromite grains (see Table 16.1). The characteristics of the previous metal assemblage in Unst are more comparable with those of stratiform basic complexes such as the Bushveld than with

Fig. 16.16. Precious metal occurrences, base metal and baryte veins and other locations referred to in the text.

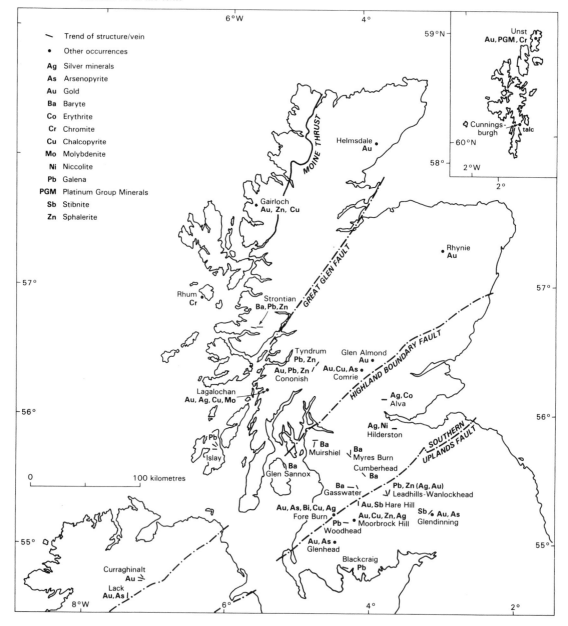

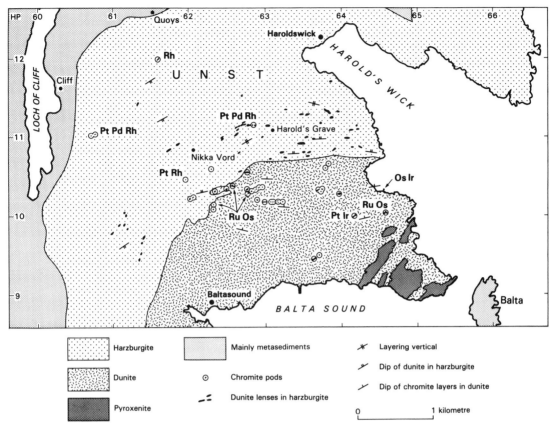

Fig. 16.17. Platinum-group mineral locations in the northern part of the Unst ophiolite.

ophiolitic podiform chromitites. Surface exploration (Gunn *et al.* 1985) led to the view that Pd–Pt-rich mineralisation is of hydrothermal origin, although associated with magmatic chromite segregations. Traces of gold in serpentinised harzburgite but no chromite or platinum group elements of economic significance were intersected in shallow drilling in 1985.

The most important gold deposit so far found in Scotland is the Cononish mine, near Tyndrum, where by end-1989, 1 mt of ore grading 12 g/t Au and 40 g/t Ag had been identified. The auriferous structure has been tested through 200 m of vertical interval and over 1 km of strike-length (Fig. 16.18). It post-dates metamorphism and deformation of the Dalradian host-rocks but is older than quartz veins containing baryte, carbonate and sulphides comparable with the main vein worked for lead at Tyndrum. The structure is best developed in psammitic schists of the Lower Dalradian but it also cuts the Middle Dalradian Ben Eagach Schist Formation. The Cononish deposit lies within a regional-scale As–Sb anomaly defined by geochemical stream sediment sampling (Plant

et al. 1989). Bi is also anomalous over the gold locations at Comrie and in Glen Almond, extending eastwards to the recently announced NNW-trending structures close to Aberfeldy. The Caledonian diatreme–porphyry–breccia complex at Lagalochan is the site of a zoned, shear-related Pb–Zn–Ag–Au–As–Sb-carbonate vein assemblage peripheral to an earlier hypogene concentration of Cu, Au and Mo. Alteration is widespread and in the main areas of mineralisation it is fracture controlled. Shearing in the Comrie diorite mass is also accompanied by gold. At one margin of this mass arsenopyrite mineralisation at the base of an epidiorite sill in the South Highlands Schists has a Cu–Au–Bi signature. Exploration of the Upper Dalradian, for example in the Glen Almond area, has been stimulated by successes in similar rocks of Northern Ireland (Fig. 16.16) and is expected to yield new discoveries.

There are several gold locations on either side of the Southern Uplands Fault, notably in the Fore Burn dioritic complex, where the alteration and mineralisation has sub-volcanic porphyry affinities. Tourmalinisation of tuff

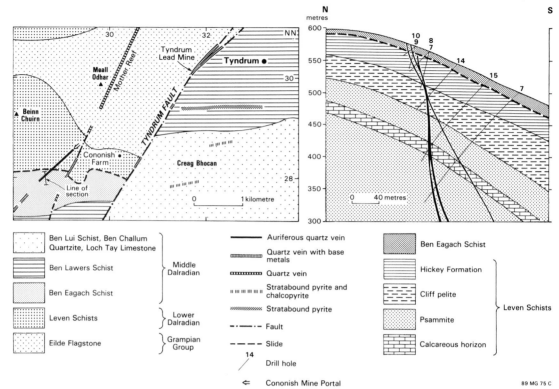

Fig. 16.18. Geological setting of the Cononish gold–silver deposit and cross-section.

and breccia is developed in the vicinity of auriferous sulphide veins (Fig. 16.19). Detrital gold is widespread in the Southern Uplands, as illustrated by the systematic surveys of Dawson *et al.* (1977, 1979) who suggested that provenance was chiefly local. This view is supported by recent discoveries of small gold-bearing quartz–pyrite arsenopyrite veins cutting Ordovician sediments close to the margins of the Caledonian granitic intrusions of Hare Hill (or the Knipe), Moorbrock Hill and Doon. The Glenhead gold veins at the southern end of the Doon complex post-date gold enrichment in minor intrusions and associated greywackes. Another style of gold enrichment is in disseminated and vein arsenopyrite in Silurian greywackes at Glendinning where some 200 t of antimony were produced in the past from a stibnite–quartz vein. Arsenopyrite containing up to 0·4 per cent Au follows the bedding of the host sediments and is also present in breccias. Arsenic levels are elevated in the greywackes of the mine area and in a strike-parallel dyke-swarm dominated by lamprophyres as well as around the Hare Hill and Moorbrook Hill prospects. Analyses of a small number of dykes revealed abnormal gold values (Rock *et al.* 1987), as at Glenhead, leading to the suggestion that the late Caledonian minor intrusions of the Southern

Uplands played an important role in As–Au–Sb metallogenesis.

At least 45 t of ore assaying 85 per cent Ag and substantial amounts of cobalt were produced from late Carboniferous veins in the Alva area in the past. The Ag–Co mineralisation occurs in fracture fillings of calcite and pyrite trending E–W, probably controlled by differential fracturing within a variegated Lower Devonian volcanic sequence. A younger Cu–Ba phase of mineralisation is recognisable. At Hilderston, silver enrichment in the near-surface part of a late Carboniferous hydrothermal vein gives way to galena and sphalerite with depth. Stratabound Pb–Zn mineralisation in Lower Limestone Group sediments nearby represents a potential source of the vein contents.

Metallogenesis

Evidence of origin has been referred to when summarising the Scottish gold locations, because of their diversity. The conclusion reached from the comprehensive study of the ancient stratiform Cu–Zn–Au deposit at Gairloch that it was exhalative and hosted by mafic volcanics is a major component in the metallogenic framework of Scotland as Russell (1985) has emphasised in his scheme of metal derivation (Fig. 16.20). Six locations are closely associated with late Silurian to Lower Devonian altered and veined volcanic and intrusive rocks of acid to intermediate composition (see Table 16.1). The Dalradian terrain of Scotland and Northern Ireland hosts

at least five deposits of significance (Fig. 16.16), suggesting that initial concentration of gold now associated with granitic intrusions or lineament systems (for example, Cononish) took place in Dalradian shale–turbidite sequences and tholeiitic volcanics (Plant *et al*. 1989). Problems remain, not least the derivation of alluvial gold in the Helmsdale area where the origin of the

Fig. 16.19. Auriferous sulphide veins in the Fore Burn igneous complex.

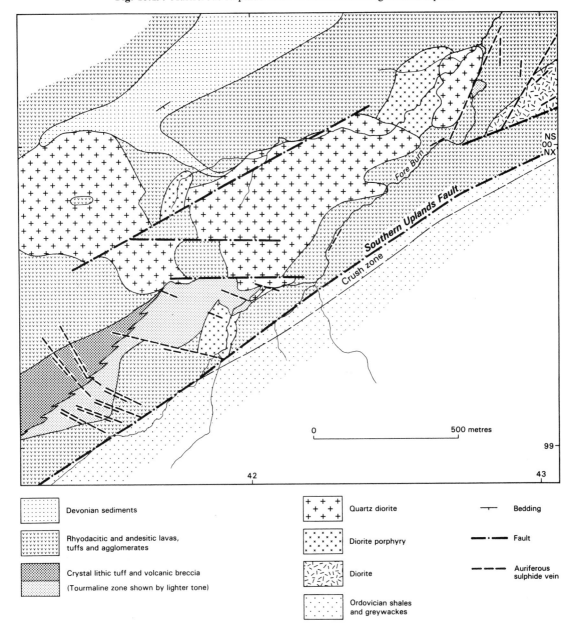

Legend:

- Devonian sediments
- Rhyodacitic and andesitic lavas, tuffs and agglomerates
- Crystal lithic tuff and volcanic breccia (Tourmaline zone shown by lighter tone)
- Quartz diorite
- Diorite porphyry
- Diorite
- Ordovician shales and greywackes
- Bedding
- Fault
- Auriferous sulphide vein

0   500 metres

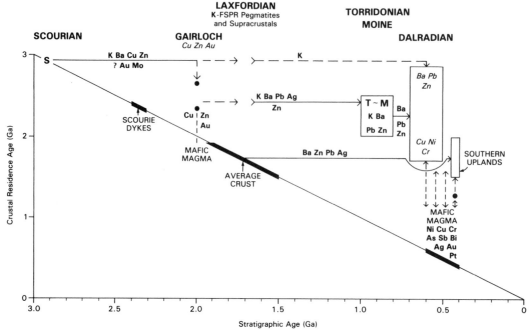

Fig. 16.20. Probable derivation of metals during succeeding mineralising events in Scotland (from Russell 1985, fig. 1).

most likely source rocks, a lit–par–lit granitic intrusion complex, is itself uncertain.

Stratabound baryte and base metals

More then 0·2 mt of direct shipping-grade baryte ore has been supplied to North Sea hydrocarbon operations since 1984 from Foss Mine, an underground and openpit working 7 km NW of Aberfeldy. Geologically inferred reserves are 2 mt and at least this quantity exists in the adjacent Cluniemore deposit, presently under development, where zinc and lead (with silver) approach economic grades in the wallrocks of the main baryte bed. The Middle Dalradian Ben Eagach Schist Formation also hosts bedded Ba–Zn–Pb mineralisation some 45 km along strike to the SW and NE of the Aberfeldy deposits, at Loch Lyon and Loch Kander. Stratabound sulphides are commonly developed in succeeding Middle Dalradian formations (Table 16.2).

Aberfeldy deposits

Stratabound Ba–Zn–Pb mineralisation extends at intervals over 7 km of the strike-length of the Ben Eagach Schist Formation in the mountains north of Aberfeldy (Figs. 16.21, 22). It was discovered by drainage geochemical sampling and subsequently delimited by mapping of sparse outcrops, geophysical definition of the conductive graphitic schist host-rocks beneath peat and glacial drift (Parker 1980) and shallow drilling (Coats, Pease & Gallagher 1984).

The dominant components of the mineralisation are in fact rare barium minerals, (a) barian muscovite containing up to 8 per cent BaO, widespread in graphitic muscovite–quartz schist and in other mica–schists, and (b) celsian (a hard, fine-grained quartz–celsian rocks (Fig. 16.23) of cherty appearance which characteristically form the walls of the sharply defined baryte beds, a factor significant in extraction. Sharp variations in the amount of dolomite and sulphides as well as in quartz and celsian account for the finely banded nature of these rocks.

Bedded, massive baryte is composed of anhedral to irregular baryte grains 0·1–2 mm across which commonly exceed 90 per cent by volume in the rock (Fig. 16.23). Fine banding is expressed by pyrite, minor amounts of zinc–lead sulphides and quartz. Carbonates and magnetite, occasionally associated with fuchsite, are also found in the baryte-rock. The Offshore Companies Materials Association (OCMA) specification is independent of mineralogical composition providing the density of the delivered product is 4·2 g.cm^{-3}. Associated are found carbonate rocks containing 12 per cent Zn + Pb over 4 m on Ben Eagach, breccias that are quite possibly intraformational and mucovite–schist. Graphitic

Table 16.2. Stratabound mineralisation in Middle Dalradian (see Fig. 16.21)

Formation	Lithology	Mineral location	Mineral assemblage Major	Minor	Reference
Tayvallich Subgroup					
Loch Tay Limestone	Adularia–baryte rock in limestone	Blacklunans, Glenshee (quarried for limestone or baryte)		Baryte, pyrrhotite, manganese oxide	Fortey (pers. comm.)
Crinan Subgroup					
Ben Lui Schist	Chromian schist in garnetiferous quartz–mica schist	Auchtertyre and Creag Bhocan, near Tyndrum, and Loch Lyon		Chromian mica, chromite, pyrrhotite, chalcopyrite, pentlandite	Fortey & Smith (1986)
Erins Quartzite	Quartz–mica schist, amphibolite	Meall Mor, Knapdale (worked at Abhainn Strathain)	Pyrite	Chalcopyrite, sphalerite, stibnite, trace of gold	Smith *et al.* (1978); Mohammed (1987)
Easdale Subgroup					
Ben Challum Quartzite	Quartzite, sometimes albitic, hornblende–schist	Ben Challum and Creag Bhocan (upper horizon); Auchtertyre (lower horizon). Near Tyndrum	Pyrite	Pyrrhotite, sphalerite, chalcopyrite; galena absent from lower horizon	Fisk (1986); Fortey & Smith (1986); Scott *et al.* (1988); Smith *et al.* (1988)
Ben Lawers Schist	Calcareous mica–schist, commonly hornblendic and garnetiferous	Tyndrum for 90 km NE to Glenshee	Pyrite	Cupriferous pyrite horizon in upper part of formation	Smith (1977)
Ardrishaig Phyllite	Mica-schist	Coillie Bhraghad and Craignure, Inveraray (worked)	Pyrite, pyrrhotite	Chalcopyrite, pentlandite	Wilson & Flett (1921)
		McPhun's Cairn, Loch Fyne (worked)	Pyrite, pyrrhotite	Galena, sphalerite	Willan & Hall (1980)
Ben Eagach Schist	Graphitic schist, quartzite, mica–schist, quartz–celsian rock (chert), thin limestones and metabasite sheets; carbonate rock and fragmental rock	Foss Baryte Mine, western sector of Aberfeldy mineralised zone; Cluniemore (eastern sector)	Barian muscovite, celsian (Ba[$Si_2Al_2O_8$]) baryte, pyrite. Sphalerite and galena relatively common in chert at Cluniemore	Sphalerite, galena, magnetite, hyalophane, cymrite, fuchsite, manganoan calcite	Coats *et al.* (1980, 1981); Moles (1982, 1985); Fortey & Beddoe-Stephens (1982); Willan & Coleman (1983); Gallagher (1984); Russell *et al.* (1984)
	Quartzite, graphitic quartz–schist, graphitic schist, banded calc–silicate schist	Coire Loch Kander and Allt an Loch, near Braemar	Baryte bed at Allt an Loch; sphalerite and galena in barian quartzite at Coire Loch Kander with highly pyritic schist	Hyalophane, armenite $BaCa_2Al_3$ ($Al_3Si_9O_{30}$). $2H_2O$ and other contact metamorphic minerals	Gallagher *et al.* (1989); Fortey *et al.* (in press)
	Graphitic schist, quartz–mica–garnet schist, limestone bands	Loch Lyon area – Beinn Heasgarnich and Allt Chall	Barian muscovite, manganoan calcite, baryte, celsian (locally)	Hyalophane, pyrite, sphalerite, galena, chalcopyrite	Coats *et al.* (1984)
	Quartzite, ptygmatic quartz veins, graphitic schist.	Lower Glen Lyon near Dericambus		Sphalerite, pyrrhotite, pyrite, galena	Coats & Pease (1984)
	Laminated quartzite, graphitic schist	Allt an Daimh, lower Glenshee		Sphalerite, galena in quartzite	Pease *et al.* (1986); Coats *et al.* (1987)

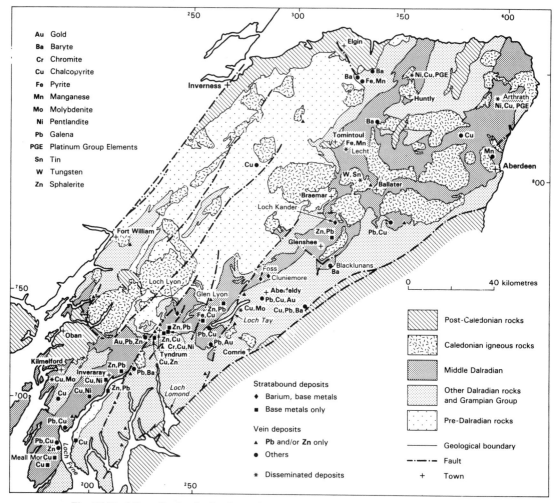

Fig. 16.21. Mineralisation in the Grampian Highlands, emphasising stratabound deposits in the Middle Dalradian.

muscovite–quartz schists with bands of detrital quartzite form the remainder of the host formation.

Foss Mine is presently operating in the western sector where mineralisation can be traced for 1·8 km in a highly folded, sub-vertical zone 60–110 m thick which has been tested by drilling through 250 m OD. The main baryte bed averages 4 m thick in the worked section of the deposit (Fig. 16.24) where proven reserves are 0·3 mt. Extraction has been successfully adapted to the pronounced folding of the baryte horizon underground but in the open pit, wallrock contamination is less easy to control.

On the Cluniemore property east of Foss, early exploration defined a zone of barium enrichment 100 m thick running for 0·6 km in the upper part of the Ben Eagach Schist which here structurally overlies the younger Ben Lawers Schist. Shallow drilling indicated that bedded baryte in the zone was impersistent in depth and along strike, in part at least due to faulting. Subsequently MI Great Britain Ltd. defined a second, stratigraphically lower zone of comparable thickness along 2 km of strike (Fig. 16.22). A high-quality baryte bed commonly 5–10 m thick (maximum 30 m) persists through this zone to depths of at least 200 m below surface. The wall-rocks of the baryte beds are characteristically quartz–celsian cherts in which sphalerite and galena appear to attain economic concentrations in the deeper drill intersections. The mineral potential of the Cluniemore property is therefore of world class; the overall barium content of the Aberfeldy mineralisation is probably the highest in the world.

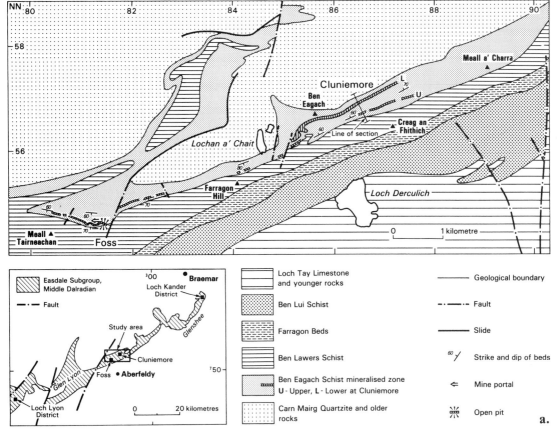

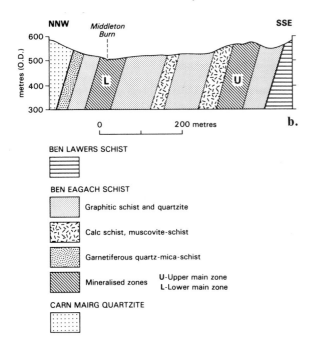

Fig. 16.22a. Setting of Foss baryte mine and the Cluniemore property in the Middle Dalradian near Aberfeldy.

Fig. 16.22b. Schematic section (line of section marked on map) through the Cluniemore baryte and base metal deposits.

a.

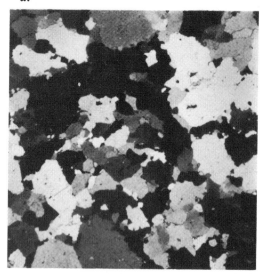

b.

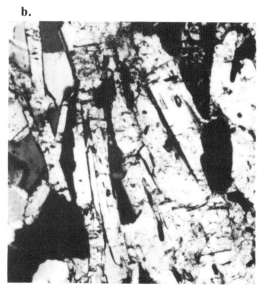

Fig. 16.23. a. Baryte rock, Foss Mine, Anhedral mosaic of baryte (shades of grey) and pyrite (opaque). Crossed nichols, field 6 mm. **b.** Celsian–quartz rock containing pyrite and sphalerite (opaque), Cluniemore. Aligned platey crystals of celsian up to 0·5 mm in length. Crossed nichols.

Other deposits

The stratabound Ba–Zn–Pb deposits at Loch Lyon and Loch Kander (Table 16.2) are similar in many respects to those at Aberfeldy. The mountain summit of

Beinn Heasgarnich on the south side of Loch Lyon (Figs. 16.21, 16.22) is encircled for 3 km by a bed of mineralised calcareous schist, 1–2 m thick, lying a few metres beneath the top of the Ben Eagach Schist. Barite and manganoan calcite are finely interbanded with silvery barian muscovite (up to 10·3 per cent BaO) and coexisting hyalophane in the bed which at its type section contains 4·5 per cent Ba and 0·2 per cent Zn. Only traces of galena and chalcopyrite accompany minor sphalerite. On the eastern side of the mountain a 1 m band of celsian–muscovite–pyrrhotite schist occurs in a faulted lens of Ben Eagach Schist. The barium-enriched schist outcrops on Beinn Heasgarnich lie on the inverted upper limb of a major fold – the Ben Lui fold. North-eastwards towards Loch Lyon, further outcrops are found over at least 1 km, and possibly up to 3 km, of strike length on the lower fold limb.

Beds of baryte and sulphide-rich quartzite found close to the top of a formation equivalent to the Ben Eagach Schist in the Loch Kander district, near Braemar (Fig. 16.25) lie 90 km NE along the Middle Dalradian strike from Loch Lyon. Baryte–quartz rock some 5 m thick is traceable for 0·7 km and barium enrichment in adjoining schist extends for double this strike length. Sphalerite and galena form mm-scale bands with iron sulphides in a highly folded cherty rock in cliffs SE of the loch. Galena has been remobilised into 3-cm-thick bands and among the associated minerals are armenite, new to the UK, hyalophane and a range of minerals attributable to contact metamorphism of a hydrothermally altered sediment or chemical exhalite. A small baryte–galena vein south of the loch (Fig. 16.25) has been thermally metamorphosed by a Devonian diorite thus making it the oldest such structure recorded in Scotland.

Stratabound sulphides without associated barium minerals occur in subeconomic quantities low in the Ben Eagach Schist, mainly in quartzite transitional to the underlying Carn Mairg Quartzite in Glenshee and Glen Lyon. They are traceable over 300 m of strike and for 200 m across strike in highly folded metasediments on the south side of Glen Lyon where remobilisation of stratabound sphalerite and galena into late metamorphic quartz segregations is much more widespread than has been observed elsewhere in the Middle Dalradian belt.

Higher in the lithostratigraphy around Loch Fyne, sulphides have been worked on small scale at McPhun's Cairn (lead), Coillie Bhraghad (nickel) and Meall Mor (copper) as part of extensive zones of stratabound pyrite development. Chalcopyrite is concentrated at the contacts of intensely epidotised metabasic sheets with quartzites and chlorite schists at an old copper mine near Meall Mor. The largest stratabound pyrite development is traceable over 90 km of regional strike in the upper part of the Ben Lawers Schist. Two pyritic horizons extend for 8 km or more in the succeeding Ben Challum Quartzite to the east and south of Tyndrum (see Table 16.2). A lower one

Fig. 16.24. Fold closure in finely bedded baryte ore with massive pyrite in core (right), Foss Mine, Aberfeldy.

(Auchtertyre) up to 80 m thick carries minor sphalerite and chalcopyrite. The upper one contains sphalerite and galena in the north-east on Ben Challum and chalcopyrite and sphalerite at Creag Bhocan farther south-west, attaining ore grade in thin units. Chromium minerals with traces of Cu–Ni sulphides are found close to the base of the younger Ben Lui Schist Formation in which serpentinised ultrabasic intrusions are known in the region.

Bedding-plane vein baryte at Blacklunans (Fig. 16.21), although post-metamorphic is hosted by an adularia–baryte assemblage of uncertain origin in the Loch Tay Limestone.

Metallogenesis

The Foss and Cluniemore deposits are believed to have formed by exhalation of warm saline brines containing Ba, Fe, Mn, Zn and Pb into small rifted basins floored by carbonaceous mud within a larger, second-order basin referred to as the Aberfeldy basin on Fig. 16.26. Stable isotope studies of Dalradian mineralisation (summarised by Hall in press) indicate that bedded baryte at Aberfeldy has a sulphur isotope value ($\delta^{34}S$ about $+33\%_{00}$) close to that of Dalradian seawater. Most of the Aberfeldy stratabound sulphides are also isotopically heavy (about $+24\%_{00}$) whereas pyrite in unmineralised country-rocks is relatively light (about $15\%_{00}$) indicating sulphur of bacteriogenic rather than hydrothermal origin. Earlier, sulphides were deposited in transitional quartzite facies of the Ben Eagach Schist Formation and lateral equivalents at Glen Lyon, also in the Aberfeldy basin, and at Allt an Daimh in the Glenshee basin. The scarcity of graphitic schist in the mineralised sequence near Loch Kander in the Glenshee basin suggests that stratigraphic availability of metal was as important as the presence of carbonaceous mud in the Glenshee basin. The barium-enriched band at Loch Lyon, lying within metres of the measurable lithostratigraphic position of the Loch Kander baryte bed, 90 km to the NE, is interpreted as having formed in

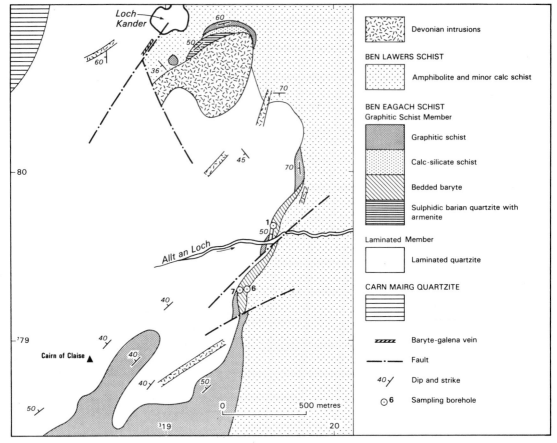

Fig. 16.25. Stratabound baryte and base metal deposits near Loch Kander, Braemar area.

a short-lived shelf area at the eastern margin of another second-order basin, deemed the Easdale basin (Fig. 16.26).

Base metal veins

Lead veins carrying smaller amounts of zinc and/or copper are widespread in Scotland but most abundant in the western parts of the Southern Uplands and of the Dalradian belt (Figs. 16.16, 16.21). The principal deposits (Table 16.3) and some of the veins noted in Tables 16.1, 16.4 were mined prior to Wilson and Flett's classical 1921 report although in the Leadhills–Wanlockhead district, the dominant source of lead in Scotland, exploitation continued into the '30s resuming briefly in 1958 (Dunham *et al.* 1979). Lead–zinc concentrates were recovered along with baryte at Strontian in the mid-1980s when an investigation of the lead veins of Islay (Fig. 16.16) was also carried out.

There is no discernible pattern in the recorded trends of the epigenetic base metal veins of Scotland. In the Leadhills–Wanlockhead mining field veins of highly variable trend (Table 16.3) cut greywackes adjacent to a major tectonic break, the Leadhills line (see Fig. 7.1 and Fig. 4 in Dunham *et al.* 1979). The Tyndrum fault is the locus of the principal lead veins in the area (Fig. 16.18). The fracture containing the Main Vein at Strontian corresponds approximately with the northern margin of the Strontian granitic pluton.

Galena is the principal sulphide in the banded fissure-filling veins which otherwise are quite variable in content. Baryte is more common, pyrite and chalcopyrite less so, in the Strontian veins where there is an indication of zinc enrichment relative to lead and baryte in depth.

Metallogenesis

Field relations and such isotopic data as are available suggest that the principal base metal veins of Scotland were emplaced in Carboniferous or Permian times. At Strontian and Blackcraig Permo–Carboniferous dykes pre-date the mineralisation while Tertiary dykes cut

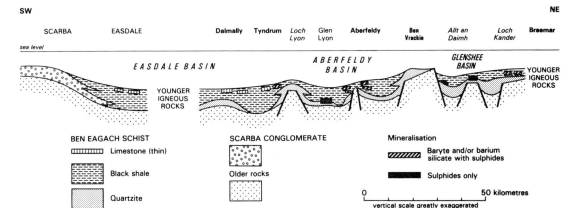

Fig. 16.26. Schematic section of the Dalradian basin at the time of Ben Eagach Schist deposition showing location of stratabound mineralisation.

the Main and Lurga veins at Strontian. Permo–Carboniferous basic intrusions, acting as heat sources, may have generated hydrothermal fluids which scavenged elements from a variety of country-rocks in rising along faults to form these deposits and the baryte veins listed in Table 16.3 (Russell 1985). For the Leadhills and Tyndrum deposits at least, Russell (op. cit.) envisages propagation of hydrothermal fluids along N–S zones of brittle fracture (geofractures) in the Lower Carboniferous. Fault-remobilisation of stratabound Fe–Zn–Pb–Cu sulphides in the Dalradian of the Tyndrum area (Fig. 16.18) could have been a source of metals for the veins (Fortey & Smith 1986). Ordovician black shales and andesitic volcanics are possible candidates for metals in the Leadhills–Wanlockhead vein swarm.

Baryte veins and disseminations

Of six principal vein deposits in Scotland (Fig. 16.16 and Table 16.3), those at Strontian, Myres Burn and Cumberhead were exploited in the 1980s but on a lesser scale than deposits worked prior to 1970 at Gasswater, Muirshiels and Glen Sannox which collectively produced 0·8–0·9 mt of baryte. Only the thicker and more easily accessible of several hundred baryte veins recorded in Scotland have been exploited and only at Strontian has there been significant recovery of lead and zinc (see previous section).

At Gasswater oreshoots up to 6 m thick occupied SE-trending faults in Devono–Carboniferous sediments over 1·2 km of strikelength and through 180 m of vertical interval. A similar trend is evident in the Glen Sannox veins on Arran. A NNE-trending vein at Muirshiel 6 m wide was traceable for 0·8 km then diverted into an E–W vein of up to 4 m worked for 0·3 km of strike. Both extended through 150 m of vertical interval in Lower

Carboniferous volcanic host-rocks and like the Gasswater veins contained only a little quartz and calcite in addition to baryte. Two vein sets are also present in similar country-rocks at Myres Burn, worked on a small scale for chemical-grade baryte in 1986–87. One of a group of veins in the River Nethan district was quarried at Cumberhead in 1984–85. A 3×1 km swarm runs NW–SE at right-angles to the regional strike of faulted Llandovery, Devonian and Lower Carboniferous sediments and Lower Devonian microgranodiorite sills.

Metallogenesis

The principal baryte veins of west-central Scotland, although cut by Tertiary dykes at Muirshiels, lie within the zone of Tertiary dyke intrusion which extends northwards to Mull, where a small baryte vein cuts Triassic sediments (Lowe 1965), and to the Strontian area where mineralisation occurred in the interval between Permo–Carboniferous and Eocene dyke intrusion (Table 16.3).

These veins were probably contemporaneous with faulting at more than one episode during this interval, resulting from mobilisation of sedimentary baryte concentrations or barium-rich intraformational waters present in Devono–Carboniferous sequences. Barium levels in the sediments may have been enhanced by supply from Middle Dalradian source rocks rich in barium silicates and baryte and mobilisation may have been assisted by high geothermal gradients associated with Permo–Carboniferous and early Tertiary igneous activity.

Baryte disseminated in Devonian sediments of the Orcadian and north-eastern Strathmore basins was sufficiently rich to be worked at Balfreish near Inverness (Gallagher 1984). Evaporitic concentration has been sug-

Table 16.3. Principal* vein deposits (see Fig. 16.16)

Location	Productive minerals	Production, 10³ t	Associated minerals	Trend	Geological age	Isotopic age	Host rocks	Reference
Strontian	Galena (historically) Sphalerite Baryte	3–5 30 tonnes	Calcite, quartz, minor strontianite and the barium zeolites harmotome and brewsterite	E–W	Pre Eocene dolerite dykes, post Permo–Carboniferous dolerite and camptonite dykes	U. Permian	Carboniferous–Permian dykes, late Caledonian Strontian granite, Moine gneisses	Wilson & Flett (1921); Gallagher (1964); Ineson & Mitchell (1974); Mason & Mason (1983)
Tyndrum	Galena Sphalerite	10 0·2	Quartz, baryte chalcopyrite, uraninite	NE–SW	Pre U. Carboniferous dolerite dyke; post U. Silurian felsite dyke	U. Carboniferous; U. Permian (uraninite)	Grampian Group gneisses; L. Dalradian schists	Ineson & Mitchell (1974); Pattrick (1985)
Muirshiel	Baryte	300	Minor hematite, quartz, mica, calcite, strontianite	NNE–SSW; E–W	Pre Eocene dolerite dykes, post U. Carboniferous dykes	Triassic	L. Carboniferous volcanics	MacGregor et al. (1944); Moore (1979); Stephenson & Coats (1983)
Glen Sannox	Baryte	52		NNW–SSE	Post L. Devonian	Triassic	L Devonian sandstones and conglomerates	MacGregor et al. (1944); Moore (1979)
Myres Burn	Baryte			NW–SE; NS to NNW–SSW	Post L. Carboniferous volcanics	—	L. Carboniferous trachytic tuffs and conglomerates	MacGregor et al. (1944)
Cumberhead	Baryte Galena (historically)		Hematite Minor quartz, pyrite, galena and sphalerite in associated veins	NW–SE	Post L. Carboniferous	Jurassic (palaeomagnetic)	L. Devonian felsite, Silurian sediments	Gallagher et al. (1982); Evans & El-Nikheli (1982)
Gasswater	Baryte	500	Calcite	NNW–SSE and WNW–ESE	Post L. Carboniferous	Late Carboniferous to Permian	L. Carboniferous sediments	Scott (1967)
Leadhills – Wanlockhead	Galena Sphalerite Silver	300 (metallic Pb) 10 25 tonnes	Quartz, calcite, dolomite, baryte, chalcopyrite, pyrite and some 50 minor primary and secondary minerals	More than 70 veins: NNW–SSE to NNE–SSW; also WNW–ESE to NW–SE	Post Ordovician	Mid-Carboniferous	Ordovician volcaniclastic greywackes, black shales and cherts	Wilson & Flett (1921); Temple (1956); Dunham et al. (1979); Gillanders (1981)
Woodhead	Galena	6·65 (metallic Pb)	Calcite, dolomite, quartz, sphalerite, chalcopyrite	WNW–ESE	Post Ordovician	—	Ordovician greywackes	Wilson & Flett (1921)
Blackcraig	Galena Sphalerite	5 1·2	Calcite, dolomite, baryte, chalcopyrite	WNW–ESE	Post Permo–Carboniferous	—	Silurian greywackes, Permo–Carboniferous dolerite dyke	Wilson & Flett (1921); Gallagher (1964)

* Other locations are included in Table 16.1.

gested by Parnell (1983), a process which may also have produced baryte cement in Triassic sandstones near Elgin (Peacock *et al.* 1968).

Mineralisation associated with 'Caledonian' intrusions

Igneous rocks ultrabasic to granitic in composition and early Lower Palaeozoic to early Devonian in age host not only many of the precious metal occurrences already described but also a wide variety of other minerals, dealt with here (Table 16.4).

Exploration Ventures Limited conducted a major investigation around 1970 for nickel (Wilks 1974) in the early Ordovician younger basic intrusions of Aberdeenshire (see Chapter 8). Apart from widespread disseminated Fe–Ni–Cu sulphides, massive sulphide was discovered in a structurally complex contact zone to the Knock mass north of Huntly and in the suboutcropping Arthrath mass north of Aberdeen. Sulphides are also concentrated in olivine-bearing cumulates and in graphitic, pyroxenitic pegmatites. Massive sulphide grading up to 3 per cent Ni, 6·5 per cent Cu consists of pyrrhotite and subordinate pentlandite and chalcopyrite and can reach a thickness of 20 m in the Littlemill ore zone at the SE margin of the Knock mass. Traces of platinum, palladium and gold have also been recorded (Fletcher & Rice 1989).

Complex mineral assemblages are found in and around late Caledonian granitic masses such as Grudie, Ratagain and Glen Gairn (see Table 16.4) but are only of academic interest. Porphyry-style mineralisation near Kilmelford exhibits some Mo–Cu levels of note. Small-scale mining for copper and nickel took place in the past in diorite bodies at Tomnadashan and Talnotry.

Metallogenesis

Spatial and most probably genetic relationships between many metalliferous mineral occurrences in Scotland and a wide variety of Caledonian igneous rocks are evident from Tables 16.1, 16.4. While these rocks are likely to continue to form exploration targets their known economic potential is small, apart from the Aberdeenshire basic complexes and the Unst ophiolite. For deposits such as the Cononish gold vein, occurring in rocks intruded by Caledonian granite, a significant role attributable to felsic magmatism may become apparent after fluid inclusion and isotopic work. Caledonian granitic plutons can also be regarded as potential sources of the lead, barium and zinc found in younger vein deposits such as Strontian and Woodhead.

Chromite and olivine

Marine deposits of chromite and olivine have recently been located in deltas 3 km^2 and 1 km^2 in area fed by streams draining the Tertiary ultrabasic and basic rocks of southern Rhum (Fig. 16.27). The surficial 1 m of sand represents about 9 million tonnes, containing some 70,000 tonnes of chrome spinel averaging 32 per cent Cr_2O_3 at a grade of nearly 1 per cent, together with 1·5–2 million tonnes of olivine averaging 47 per cent MgO at 25 per cent grade. Shell calcite forming about 20 per cent of the sand was removed prior to analysis (Gallagher *et al.* 1989). Accompanying minerals are ilmenite and vanadiferous magnetite and traces of platinum group elements have been detected. The sand size and freshness of the minerals are favourable for extraction. Comparable heavy mineral sands may underlie glacial sandy clays off SW Skye.

Previously worked chromite deposits on the island of Unst in northern Shetland have been intensively studied since Duff's (1983) summary, in part at least because of the associated precious metals (see Table 16.1). Podiform chromitites are near-vertical lenses parallel to the compositional layering in cumulate rocks or in dunite lenses within serpentised harzburgite where it is also a ubiquitous accessory mineral. They are commonest within dunite close to its contact with harzburgite in the northern part of the Unst ophiolite (Fig. 16.17) and as their form is in part tectonically controlled their subsurface continuity is unpredictable. A little chromite has been produced from a small ultrabasic intrusion at Corrycharmaig, near Killin where talc–magnesite rock is present (Harrison 1985). (Talc and magnesite are presently being extracted from altered ultrabasic rocks at Cunningsburgh, south of Lerwick and talc was worked until very recently at Queyhouse on the northern edge of the Unst ophiolite, see p. 567.)

Iron, manganese

The earliest traces of iron-making in Scotland are seen in the remains of 'bloomeries' where local bog iron-ore of recent origin was smelted. Several furnaces were set up in the 17th and 18th centuries in the central and western Highlands because of the availability of bog iron-ore and, more important, the plentiful supply of timber for charcoal manufacture. Even in these early days, however, imported ore, mainly from Cumberland, was used along with the local material. The year 1760 which saw the opening of the Carron Ironworks, near Falkirk, is perhaps the most significant one in the history of the Scottish iron industry as it marks the beginning of the extensive use of local coal to smelt local Carboniferous ore.

Carboniferous ironstone

Associated mainly with the shales that occur above coal seams are nodules and bands of clayband or blackband ironstone. Most of the clayband type consist of flattened spheres of ironstone occurring in laterally persistent

Table 16.4. Mineralisation associated with *Caledonian intrusions (see Chapter 8)

Location	Rock-type	Mineral assemblage	Reference
Helmsdale, Sutherland	Biotite–adamellite	Fracture-bound baryte, galena, kasolite (PbU_2SiO_4) and other uranium secondary minerals	Gallagher et al. (1971); Tweedie (1979)
Grudie and Shin, near Lairg, Sutherland	Adamellite and Moinian–Lewisian country-rocks	Disseminated fluorite, molybdenite; fracture-bound galena, fluorite, hematite, baryte, betechtinite [$Pb_2(Cu, Fe)_{21}S_{15}$], chalcopyrite, scheelite, members of bismuthinite–aikinite (Bi_2S_3; $PbS.Cu_2S.Bi_2S_3$) series	Gallagher et al. (1974)
Carn Chuinneag, near Bonar Bridge	Garnetiferous albite gneiss in granitic complex	Cassiterite–magnetite bands in gneiss	Gallagher et al. (1971)
Ratagain	Quartz monzonite	Quartz–fluorite veins with chalcopyrite, galena and sphalerite; traces of Au	Alderton (1989)
Huntly–Knock, Aberdeenshire	Basic and ultramafic intrusions–cumulates, granular and xenolithic gabbros	Submassive to massive pyrrhotite, pyrite, chalcopyrite and pentlandite, mainly in contact zone; traces of platinum group elements and Au	Fletcher et al. (1989); Fletcher & Rice (1989)
Arthrath, Aberdeenshire	Contaminated norite in basic intrusion	As for Huntly–Knock	Rice (1975)
Glen Gairn, Aberdeenshire	Lithium mica granite in diorite–granite complex	Hubnerite, [$(MnFe)WO_4$], cassiterite, scheelite, hydrotungstite ($WO_3.H_3O$)	Tindle & Webb (1989)
Ballater	Granite	Beryl, fluorite–quartz veins	
Beinn nan Chaorach, Kilmelford	Altered porphyritic dacite and intrusion breccia	Chalcopyrite, molybdenite	Ellis (1977)
Tomnadashan, Loch Tay	Diorite	Chalcopyrite, tetrahedrite, molybdenite	Pattrick (1984)
Talnotry, Newton Stewart	Diorite sill	Pyrrhotite, pentlandite, chalcopyrite, nickeline, gersdorffite, trace of Au	Stanley et al. (1987)
Black Stockarton Moor	Granodiorite sheets in subvolcanic complex	Chalcopyrite, molybdenite, trace of Au	Brown et al. (1979)

* Other locations are included in Table 16.1.

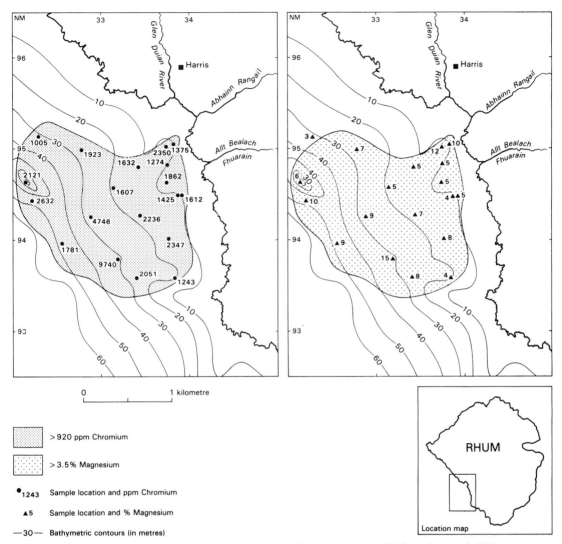

Fig. 16.27. Chromium (left) and magnesium distribution in surficial marine sand, SW Rhum.

layers, and are composed mainly of siderite, along with a variable amount of clay minerals, with an average iron content of 25–30 per cent. Nodules vary from 1 to 30 cm in diameter; sometimes they are closely packed to form an almost continuous layer of ironstone up to 30 cm thick, at other times they are sparsely distributed throughout a greater vertical thickness of shales. In some cases bands of darker ironstone occur. These blackband ironstones contain carbonaceous matter (which can make up to 20 per cent of the rock) in additon to siderite and clay.

In the early days of the industry clayband ironstone was worked at outcrop and was mined with coal whenever it

occurred within reasonable distance of the ironworks. In Fife for instance weathered-out nodules on the sea shore were collected and sold to the ironworks. The blackband ironstones became extremely important in the early part of the 19th century with the introduction of the hot blast technique, the presence of carbon in the ore making smelting of this hitherto refractory material relatively inexpensive. One of the earliest blackbands to be discovered was the Airdrie Blackband and by 1840 the Coatbridge district was the centre of the iron industry. In general the worked ironstone bands occur mainly in the Limestone Coal Group and the Coal Measures, the Upper

Limestone Group and the Calciferous Sandstone Measures containing few of economic interest. Under present economic conditions none of the Carboniferous ironstones is considered workable.

Curtis & Spears (1968) and Pearson (1979) have discussed at some length the possible origins of clayband ironstones. They suggest that the siderite is precipitated out of interstitial water within the compacting muds rather than at the water-sediment interface.

Liassic ironstone

A band of ironstone about 2·5 m thick (Lee 1920) is widely developed in Upper Liassic shales on Raasay (Chapter 13). Conditions of deposition were shallow marine as can be seen from the associated fossils. It is considered that the source of the iron was nearby well-vegetated land experiencing a warm humid climate, where preconcentration of iron by intensive leaching took place (Hallam 1975) and rivers carried iron along with other sediment to the sea. The iron is considered to have come out of solution and to form ooliths on slight topographic rises on the seafloor, while terrigenous sand and clay were deposited in surrounding depressions. The chamosite ore changes to siderite ore when traced at outcrop northwards. It seems from microscopic evidence that the siderite replaces the chamosite, but it is not known conclusively whether this is due to a change of environmental conditions shortly after precipitation of the chamosite or to later diagenitic changes. The ironstone was worked during the First World War ceasing in 1920.

It yielded 20–25 per cent Fe and reserves are estimated at about 10 mt (Macgregor 1940).

Haematite

Significant amounts of cryptomelane and other manganese minerals occur with hematite in breccias of the Lecht deposit near Tomintoul (Fig. 16.21), worked in the past for both iron and manganese (Nicholson 1989). Synsedimentary Mn–Pb–Zn enrichment in Middle Dalradian sediments has also been detected (Smith 1985). At Sand Lodge in Shetland copper and iron were worked from a vein in faulted Lower Devonian sediments. Oxidation of the vein led to gossan formation and limonite and brown haematite were mined to a depth of 30 m, below which mining continued for chalcopyrite.

Magnetite

In Skye magnetite occurs in altered Durness Limestone at its contact with the Beinn an Dubhaich granite, near Broadford, together with skarn minerals and minor pyrite and copper minerals. At Clothister Hill in Shetland, a steeply plunging lensoid magnetite body 53 m long and 6·7 m wide in quartzites and schists is surrounded by skarn rocks, indicating a pyrometasomatic origin. Infolded pods and lenticles of calcareous rocks may have been replaced by iron-rich fluids from an underlying magma and similar ore bodies may be present in the area. The ore is high grade (63 per cent Fe) and estimated reserves are 20,000 tonnes.

REFERENCES

References in this chapter are grouped under the four separate headings:

> COAL
> OIL AND GAS
> CONSTRUCTION AND INDUSTRIAL MINERALS
> METALLIFEROUS MINERALS

COAL

DINHAM, C. B. & HALDANE, D. 1932 The Economic Geology of the Stirling and Clackmannan Coalfield. *Mem. Geol. Surv. Scotland.*

EWING, C. J. C. & FRANCIS, E. H. 1960a Nos. 1 and 2 Off-Shore borings in the Firth of Forth (1955–56). *Bull. geol. Surv. G.B.,* **16**, 1–47.

1960b No. 3 Off-shore boring in the Firth of Forth (1956–57). *Bull. geol. Surv., G.B.,* **16**, 48–68.

MACGREGOR, M. The Coalfields of Scotland. *In* Trueman, A. (Ed.) *The Coafields of Great Britain*, 325–381. Edward Arnold, London.

NATIONAL COAL BOARD 1950 Plan for Coal.

NATIONAL COAL BOARD   1958   *A short history of the Scottish Coal-Mining Industry.* N.C.B. Scottish Division.

PICKEN, G. S.   1988   The concealed coalfield at Canonbie: an interpretation based on boreholes and seismic surveys. *Scott. J. Geol.,* **24**, 61–71.

SIDDALL, N.   1979   Centennial Address: The Mining Institute of Scotland. *Ming Eng. London,* **138**, 923–928.

SKIPSEY, E.   1959   Rank variations in the Coal Seams of North-East Stirlingshire. *Trans. Instn Ming Eng. Lond.,* **119**, 23–36.

STACH, E. *et al.*   1975   *Coal Petrology.* (Translated by Murchison, D. G., Taylor, G. H. and Zierke, F.) Gebruder Borntraeger, Berlin.

OIL AND GAS

ADAMS, P. J   1960   *The origin and evolution of coal.* H.M.S.O., London.

BAILEY, N. J. L., WALKO, P. & SAUER, M. J.   1987   Geochemistry and source rock potential of the west of Shetlands. *In* Brooks, J. & Glennie, K. W. (Eds.) *Petroleum geology of North West Europe.* Graham and Trotman, London, 711–721.

BOTT, M. H. A. & SMITH, P. J.   1984   Crustal structure of the Faeroe–Shetland Channel. *Geophys. J.R. Astron. Soc.,* **76**, 383–398.

BOWEN, J. M.   1975   The Brent oilfield. *In* Woodland, A. W. (Ed.) *Petroleum and the continental shelf of north-west Europe,* Vol. 1, 353–360. Applied Science Publishers, Barking, Essex.

BRENNAND, T. P. & VAN HOORN, B.   1986   Historical review of North Sea exploration. *In* Glennie, K. W. (Ed.) *Introduction to the petroleum geology of the North Sea,* 1–24. Blackwell, Oxford.

BRENNAND, T. P. & VAN VEEN, F. R.   1975   The Auk oil-field. In Woodland, A. W. (Ed.) *Petroleum and the continental shelf of north-west Europe,* vol. 1, 275–284. Applied Science Publishers, Barking, Essex.

BROOKS, J. R. V.   1977   Exploration status of the Mesozoic of the U.K. Northern North Sea. In *Mesozoic Northern North Sea symposium, Oslo,* 1/1–8. Norw. Pet. Soc.

    1986   Future potential for hydrocarbon exploration on the United Kingdom Continental Shelf. *In* Halbouty, M. T. (Ed.) *Future petroleum provinces of the world,* 677–698. Am. Assoc. Pet. Geol. Mem. 40.

BROWN, S.   1986   Jurassic. *In* Glennie, K. W. (Ed.) *Introduction to the petroleum geology of the North Sea,* 133–159. Blackwell, Oxford.

BROWN, S., RICHARDS, P. C. & THOMSON, A. R.   1987   Patterns in the deposition of the Brent Group (Middle Jurassic) UK North Sea. *In* Brooks, J. & Glennie, K. W. (Eds.) *Petroleum geology of North West Europe,* 899–914. Graham and Trotman, London.

BURLEY, S. D.   1984   Diagenetic modelling in the Triassic Sherwood Sandstone Group of England and its offshore equivalents, United Kingdom Continental Shelf. *Univ. Hull Ph.D. Thesis* (Unpubl.).

BUTLER, M., PHELAN, M. J. & WRIGHT, A. W. R.   1976   Buchan field: evaluation of a fractured sandstone reservoir. *Trans. fourth European Formation Evaluation Symp., Soc. Prof. Well. Log. Anal.,* 1–18.

CAMERON, I. B. & McADAM, A. D.   1978   The oil-shales of the Lothians, Scotland: present resources and former workings. *Rep. Inst. geol. Sci.,* 78/2.

CARMAN, G. J. & YOUNG, R.   1981   Reservoir geology of the Forties Oilfield. *In* Illing, L. V. & Hobson, G. D. (Eds.) *Petroleum geology and the continental shelf of north-west Europe,* 371–379. Heyden, London.

CARRUTHERS, R. G., CALDWELL, W., BAILEY, E. M. & CONACHER, H. R. J. 1927 The oil-shales of the Lothians. *Mem. geol. Surv. Scotland.*

CORNFORD, C. 1986 Source rocks and hydrocarbons of the North Sea. *In* Glennie, K. W. (Ed.) *Introduction to the petroleum geology of the North Sea*, 197–236, Blackwell, Oxford.

DE'ATH, N. G. & SCHUYLEMAN, S. F. 1981 The geology of the Magnus oilfield. *In* Illing, L. V. & Hobson, G. D. (Eds.) *Petroleum geology and the continental shelf of north-west Europe*, 342–351. Heyden, London.

DEPARTMENT OF ENERGY 1984 *Development of the oil and gas resources of the United Kingdom 1984.* H.M.S.O., London.

1990 *Development of the oil and gas resources of the United Kingdom 1989.* H.M.S.O., London.

DUFF, P. McL. D. 1983 Economic geology. *In* Craig, G. Y. (Ed.) *The geology of Scotland*, 425–454. Scottish Academic Press, Edinburgh.

DUINDAM, P. & VAN HOORN, B. 1987 Structural evolution of the West Shetland continental margin. *In* Brooks, J. & Glennie, K. W. (Eds.) *Petroleum geology of North West Europe*, 765–774. Graham and Trotman, London.

FALCON, N. L. & KENT, P. E. 1960 Geological results of petroleum exploration in Britain 1945–1957. *Mem. geol. Soc. London*, No. 2.

FISHER, M. J. 1986 Triassic. *In* Glennie, K. W. (Ed.) *Introduction to the petroleum geology of the North Sea*, 113–132. Blackwell, Oxford.

FOWLER, C. 1975 The geology of the Montrose Field. *In* Woodland, A. W. (Ed.) *Petroleum and the continental shelf of north-west Europe*, Vol. 1, 467–476. Applied Science Publishers, Barking, Essex.

GIBSON, W. 1922 Cannel coals, lignite and mineral oil in Scotland. *Spec. Rep. Miner. Resour. G.B.*, **24**, 56–63.

HALLETT, D., DURANT, G. P. & FARROW, G. E. 1985 Oil exploration and production in Scotland. *Scott. J. Geol.*, **21**, 547–570.

HARMS, J. C., TACKENBERG, P., PICKLES, E. & POLLOCK, R. E. 1981 The Brae oilfield area. *In* Illing, L. V. & Hobson, G. D. (Eds.) *Petroleum geology of the continental shelf of north-west Europe*, 352–357. Heyden, London.

HERITIER, F. E., LOSSEL, P. & WATHNE, E. 1981 The Frigg gas field. *In* Illing, L. V. & Hobson, G. D. (Eds.) *Petroleum geology of the continental shelf of north-west Europe*, 380–391. Heyden, London.

HITCHEN, K. & RITCHIE, J. D. 1987 Geological review of the West Shetland area. *In* Brooks, J. & Glennie, K. W. (Eds.) *Petroleum geology of North West Europe*, 737–749. Graham and Trotman, London.

HOLLOWAY, S. 1986 The natural gas occurrences of the United Kingdom landward area. *In* Schröder, L. & Schöneich, H. (Compilers) *International map of natural gas fields in Europe, explanatory notes*, 159–160. U.N. Economic Commission for Europe.

HUTTON, A. C. 1982 Petrographic classification of oil shales. *Int. J. Coal Geol.*, **8**, 203–231.

JACKSON, D. I. 1986 The natural gas occurrences of the United Kingdom offshore areas. *In* Schröder, L. & Schöneich, H. (Compilers) *International map of natural gas fields in Europe, explanatory notes*, 164–167. U.N. Economic Commission for Europe.

LEE, G. W. 1920 The Mesozoic rocks of Applecross, Raasay and north-east Skye. *Mem. geol. Surv. Scotland.*

LEES, G. M. & TAITT, A. H. 1946 The geological results of the search for oilfields in Great Britain. *Q. J. geol. Soc. London*, **101**, 255–317.

MAHER, C. E. & HARKER, S. D. 1987 Claymore oil field. *In* Brooks, J. & Glennie, K. W. (Ed.) *Petroleum geology of North West Europe*, 835–845. Graham and Trotman, London.

MACGREGOR, M. & ANDERSON, E. M. 1923 The economic geology of the Central Coalfield of Scotland, Area VI. *Mem. geol. Surv. Scotland.*

MARCUM, B. L., AL-HUSSAINY, R., ADAMS, G. E., CROFT, M. & BLACK, M. L. 1978 Development of the Beryl "A" Field. *Proc. European Offshore Petroleum Conference, London*, 319–321.

MEADOWS, N. S., MACCHI, L., CUBITT, J. M. & JOHNSON, B 1987 Sedimentology and reservoir potential in the west of Shetland, UK, exploration area. *In* Brooks, J. & Glennie, K. W. (Eds.) *Petroleum geology of North West Europe*, 723–736. Graham and Trotman, London.

MUDGE, D. C. & RASHID, B. 1987 The geology of the Faeroe Basin area. *In* Brooks, J. & Glennie, K. W. (Eds.) *Petroleum geology of North West Europe*, 751–763. Graham and Trotman, London.

PARSLEY, A. J. 1986 North Sea hydrocarbon plays. In Glennie, K. W. (Ed.) *Introduction to the petroleum geology of the North Sea*, 237–263. Blackwell, Oxford.

PATERSON, I. B. & HALL, I. H. S. 1986 Lithostratigraphy of the late Devonian and early Carboniferous rocks in the Midland Valley of Scotland. *Rep. Br. Geol. Surv.*, 18/3.

PENNINGTON, J. J. 1975 The geology of the Argyll Field. *In* Woodland, A. W. (Ed.) *Petroleum and the continental shelf of north-west Europe*, Vol. 1, 285–294. Applied Science Publishers, Barking, Essex.

RICHARDS, P. C. 1986 Upper Old Red Sandstone sedimentation in the Buchan oilfield, North Sea. *Scott. J. Geol.*, 226–237.

RIDD, M. F. 1981 Petroleum geology west of the Shetlands. *In* Illing, L. V. & Hobson, G. D. (Eds.) *Petroleum geology of the continental shelf of north-west Europe*, 415–425. Heyden, London.

ROBERTS, D. C., BOTT, M. H. P. & URUSKI, C. 1983 Structure and origin of the Wyville-Thomson Ridge. *In* Bott, M. H. P., Saxov, S., Talwani, M. & Thielde, J. (Eds.) *Structure and development of the Greenland–Scotland Ridge: new methods and concepts*, 133–185. Plenum Press, New York.

ROCHOW, K. A. 1981 Seismic stratigraphy of the North Sea 'Palaeocene' deposits. *In* Illing, L. V. & Hobson, G. D. (Eds.) *Petroleum geology of the continental shelf of north-west Europe*, 255–266. Heyden, London.

SCOTT, J. & COLTER, V. S. 1987 Geological aspects of current onshore Great Britain exploration plays. *In* Brooks, J. & Glennie, K. W. (Eds.) *Petroleum geology of North West Europe*, 95/107. Graham and Trotman, London.

TISSOT, B. P. & WELTE, D. H. 1978 *Petroleum formation and occurrence*. Springer-Verlag, Berlin.

TULLOCH, W. & WATSON, H. S. 1958 The geology of the Midlothian Coalfield. *Mem. geol. Surv. Scotland.*

TURNER, C. C., COHEN, J. M., CONNELL, E. R. & COOPER, D. M. 1987 A depositional model for the South Brae Oilfield. *In* Brooks, J. & Glennie, K. W. (Eds.) *Petroleum geology of North West Europe*, 853–864. Graham and Trotman, London.

CONSTRUCTION AND INDUSTRIAL MINERALS

BOWIE, S. H. U. 1966 Potassium-rich sediments in the Cambrian of Northwest Scotland. *Trans. Instn. Min. Metall.*, **75**, B125–145.

BRITISH GEOLOGICAL SURVEY   1989 *United Kingdom Minerals Yearbook 1988.* Keyworth, Nottingham: British Geological Survey.

BUNYAN, I. J.,   1987 *Building stones of Edinburgh.* Edinburgh Geological Society,
FAIRHURST, J. A.,   Edinburgh.
MACKIE, A. &
McMILLAN, A. A.

COLLIS, L. &   1985 Aggregates: sand, gravel and crushed rock aggregates for
FOX, R. A. (Eds.)   constructional purposes. The Geological Society of London, London.

ELLIOT, R. W.   1985 Central Scotland Mineral Portfolio: resources of clay and mudstone for brickmaking. Open-file Report, British Geological Survey, Edinburgh.

GROUT, A. &   1989a Scottish Highlands and Southern Uplands Mineral Portfolio: limestone and dolomite resources. Technical Report WF/89/5.
SMITH, C. G.   British Geological Survey, Edinburgh.

                 1989b Scottish Highlands and Southern Uplands Mineral Portfolio: talc resources. Technical Report WF/89/7. British Geological Survey, Edinburgh.

HARRIS, P. M.   1977 Igneous and metamorphic rock. Mineral dossier No. 19. Miner. Resour. Consult. Comm. H.M.S.O., London.

                 1982 Limestone and dolomite. Mineral dossier No. 23. Miner. Resour. Consult. Comm. H.M.S.O., London.

HIGHLEY, D. E.   1974 Talc. Mineral dossier No. 10. Miner. Resour. Consult. Comm. H.M.S.O., London.

                 1977 Silica. Mineral dossier No. 18. Miner. Resour. Consult. Comm. H.M.S.O., London.

                 1982 Fireclay. Mineral dossier No. 24. Miner. Resour. Consult. Comm. H.M.S.O., London.

MacPHERSON, K. A. T. 1986a Central Scotland Mineral Portfolio: special sand resources. Open-file Report, British Geological Survey, Edinburgh.

                 1986b Central Scotland Mineral Portfolio: limestone resources. Open-file Report, British Geological Survey, Edinburgh.

MERRITT, J. W. & 1984 Central Scotland Mineral Portfolio: hard rock aggregate
ELLIOTT, R. W.   resources. Open-file Report, British Geological Survey, Edinburgh.

MERRITT, J. W.   1985 Central Scotland Mineral Portfolio: fireclay resources. Open-file Report, British Geological Survey, Edinburgh.

RIDGEWAY, J. M.   1982 Common clay and shale. Mineral dossier No. 22. Miner. Resour. Consult. Comm. H.M.S.O., London.

SMITH, C. G.   1989a Scottish Highlands and Southern Uplands Mineral Portfolio: hard rock aggregate resources. Technical Report WF/89/4. British Geological Survey, Edinburgh.

                 1989b Scottish Highlands and Southern Uplands Mineral Portfolio: silica sand and silica rock resources. Technical Report WF/89/6. British Geological Survey, Edinburgh.

THURRELL, R. G.   1981 The identification of bulk mineral resources; the contribution of the Institute of Geological Sciences. *Quarry Management Products, London,* **8**, 181–193.

METALLIFEROUS MINERALS

(MRP denotes reports of the British Geological Survey (formerly Institute of Geological Sciences) in the Mineral Reconnaissance Programme series.)

ALDERTON, D. H. M.   1988 Ag–Au–Te, mineralization in the Ratagain complex northwest Scotland. *Trans. Instn Min. Metall. (Sect. B: Appl. earth sci.),* **97**, B171–180.